Magnetism

Magnetism

Principles and Applications

Derek Craik
University of Nottingham, Nottingham, UK

JOHN WILEY & SONS
Chichester • New York • Toronto • Brisbane • Singapore

Baffins Lane, Chichester,
West Sussex PO19 IUD, England

Telephone: National Chichester 01243 779777
International (+44) 1243 779777

Reprinted June 1997, June 1998

Other Wiley Editorial Offices

John Wiley & Sons, Inc., 605 Third Avenue,
New York, NY 10158-0012, USA

Jacaranda Wiley Ltd, 33 Park Road, Milton,
Queensland 4064, Australia

John Wiley & Sons (Canada) Ltd, 22 Worcester Road,
Rexdale, Ontario M9W 1L1, Canada

John Wiley & Sons (SEA) Pte Ltd, 37 Jalan Pemimpin #05-04,
Block B, Union Industrial Building, Singapore 2057

Library of Congress Cataloging-in-Publication Data

Craik, D. J. (Derek J.)
Magnetism : principles and applications / Derek Craik.
p. cm.
Includes bibliographical references and index.
ISBN 0 471 92959 X
ISBN 0 471 95417 9 (pbk.)
1. Magnetism I. Title.
QC753.2.C73 1995
538—dc20 93-38155
CIP

British Library Cataloguing in Publication Data

A catalogue record for this book is available from the British Library

ISBN 0 471 92959 X (cloth)
ISBN 0 471 95417 9 (paper)

Contents

Preface

It is hoped that this book will be of value to students, principally at undergraduate level, of physics, chemistry and engineering and also of interest to students of other subjects such as geology, biology and medicine. Also, it is noted that many postgraduate students have been found by the author to have a very poor appreciation of the subject beyond their particular research topics. Magnetism constitutes a fascinating study as such and has some role in virtually every branch of science and technology. There appears, for example, to be an increasing interest in the magnetic properties of organic materials, the influence of magnetic fields on chemical reaction rates and aspects of biochemistry and, naturally over the last decade, magnetic phenomena associated with superconductivity. While little specific attention can be paid to such topics in a brief work, the basic material surveyed should render them more comprehensible. Magnetic resonance imaging is dealt with at a little greater length, with the emphasis on simplicity of presentation and, in view of its outstanding value and potential for further development, constitutes an appropriate conclusion; the author is indebted to Sir Peter Mansfield for reading and commenting on the appropriate section.

The initial emphasis is on the clarification of basic principles such as the distinction between the solenoidal B field and the H fields which may in certain contexts be regarded as conservative, the artificiality but undoubted computational value of the pole concept and the equivalence of current and pole distributions. In this it seems preferable to err, possibly, on the side of being excessively painstaking. The magnetostatics is developed later to the extent of obtaining potential and field expressions of particular value when designing with modern magnets for which the approximation of uniform magnetization is good. Numerical methods, particularly the finite element approximation, are described to an extent that should permit the development of simple programs constituting a first step towards realistic design work using permeable materials together with magnets and coils, and encourage further study of the copious literature.

A simple treatment of relevant general quantum theory is followed by a more extensive account acknowledging the central role of spin in the statics and dynamics of the magnetization. Special mention should be made of *Density Matrix Theory and Applications* by Karl Blum (Plenum Press, New York and London, 1981) which the author found most useful here, particularly for the theory of relaxation. The way in which individual spins may couple to give an ordered ground state is described and an account of spin waves in a finite one-dimensional chain constitutes a simplification of a rather elaborate topic. The relations between spin wave excitations and excited states corresponding to direct solutions using the Heisenberg Hamiltonian are demonstrated with the objective of further clarification.

The significant development of magnetic materials spans more than half a century but continues intensely with the introduction, over the last decade or so, of rare earth–cobalt and Nd–Fe–B permanent magnets, amorphous and nanocrystalline high-permeability materials and multilayer exchange-coupled films for magneto-optical recording as examples. An attempt is made here to give some historical perspective as well as an account of more recent developments which may facilitate the study of the current literature.

Despite natural attempts to maintain a reasonably consistent level of presentation, it is evident that certain topics are inherently more difficult and just two sections, on crystal field theory and on relaxation, may be considered as somewhat compressed and call for recourse to the original sources cited for full comprehension.

The author is grateful for the permission to reproduce material from the following sources: *Journal of Applied Physics*; *Phys. Rev. Letters*; *Phil. Trans. Roy. Soc.*; *Phys. Rev.*; *Physics Letters*; *J. Am. Chem. Soc.*; *Proceedings Leeds Phil. Soc.*; Freeman and Watson, *Magnetism IIa*; *IEEE Trans. Magn*, *Physics World*, *Science*, *J. Phys. E*, Craik, *Magnetic Oxides* and Tebble and Craik, *Magnetic Materials*.

It is re-emphasized that magnetism constitutes an enormous topic which may only be described briefly in a somewhat intuitive manner. This book is dedicated with all sincerity to all those whose work, often of central importance, has not received due reference.

CHAPTER 1 General Survey

1 Preamble: Magnets, Poles and Dipoles

The most familiar observation relating to magnetism is that of the remarkably strong forces between (permanent) magnets or between a magnet and an iron body. The forces, for suitably prepared long bar magnets, are associated with regions at the ends which are colloquially called the poles; every magnet has poles at the two ends which are equivalent in that the force **F** on a selected test pole (i.e. end of a very long thin magnet) at a given distance is the same in magnitude for both, but **F** is reversed for the two, which can thus be considered to be of opposite sign. Taking long thin specimens of modern high-specification material it may readily be established that the forces between the poles correspond to the inverse square law familiar from electrostatics, suggesting the corresponding definition of a magnetic field H, such that $F \propto mH$ for a pole m, together with $\mathbf{H}(\mathbf{r}_2) = (1/4\pi)m\mathbf{r}/r^3$. Thus $\mathbf{r} = \mathbf{r}_2 - \mathbf{r}_1$ gives the direction, as in Figure 1.1(a), and the initial factor constitutes rationalization. This appears an obvious starting point for a systematic study of magnetism. It will be pursued as such, but only to the extent that it can be shown, in fact, that it is not suitable so that the present section is a preamble rather than an introduction.

The strength of the poles for a long bar of a particular material is found to be proportional to the cross-sectional area or the area of the end faces. Thus the poles appear to be surface poles, with a certain density per unit area. Free poles may exist, and are required by certain theories, but they do not exist in the present context. Any attempt to cut one pole away from a magnet results in the formation of new magnets, each with two poles. The most apparent way to account for these two observations is to postulate the presence of a vector quantity **M**, directed along the magnet axis, with magnitude M indicative of the magnet strength or magnetization for the particular material, so that the poles appear only to the extent that **M** intersects the surface. The pole strength is then $\sigma = \mathbf{M} \cdot \mathbf{n}$, with **n** the outward unit normal to the surface: $\sigma = 0$ along the sides and $\sigma = \pm M$ at the ends $\perp$ **M**.

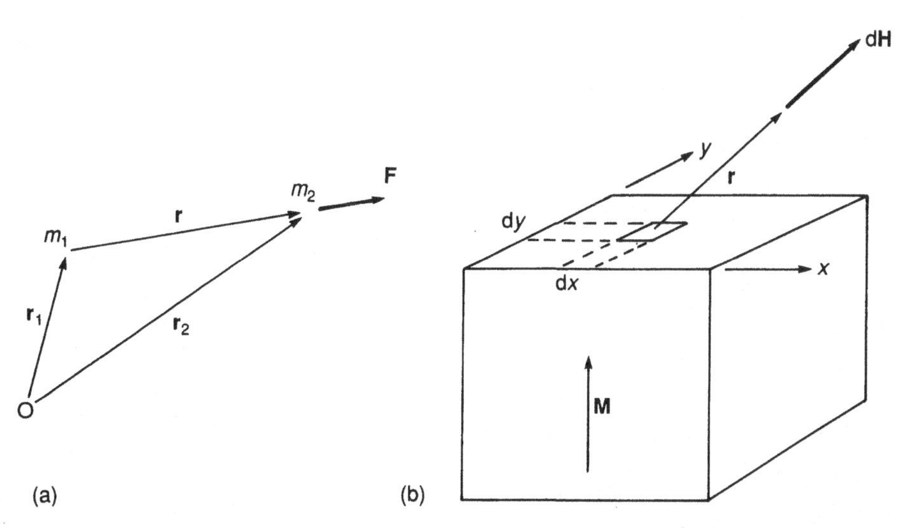

Figure 1.1 The forces between mythical point poles correspond to a magnetic equivalent to Coulomb's law (a). Applying this to the surface poles $\sigma\,dx\,dy$ and integrating, as in (b), gives the appropriate forces and fields indicating the utility of the surface (or volume) pole distributions as conventions

For a magnet of the shape indicated by Figure 1.1(b) the contribution to the field from the surface pole element $dm = \sigma\,dx\,dy$ is taken to be

$$d\mathbf{H}(\mathbf{r}_2) = \frac{1}{4\pi}\sigma\,dx\,dy\frac{\mathbf{r}}{\mathbf{r}^3}, \qquad \text{where } \mathbf{r} = \mathbf{r}_2 - \mathbf{r}_1$$

Integration over the surface, completed for the magnet by including the lower surface where $\sigma = -M$, gives $\mathbf{H}$ ($\mathbf{r}$). (It will be seen that this is very practicable for such surfaces.) The calculations could be made with, say, the upper surface slanted at θ to $\mathbf{M}$ with $\sigma = M\sin\theta$, or for any general shape. The field distribution occurring may be studied in principle by the forces on a test pole, but it will be accepted that such fields and means for their measurement do, in fact, exist. Taking a range of materials, from cobalt steels through Alnicos to more modern $SmCo_5$, say, the correlation would range from poor through moderate to near-exact. The approach is substantially justified and we also note that an approach seems to be made to what may be termed an ideal magnet: one with perfectly uniform $\mathbf{M}$, conforming exactly with the model.

Note that $\mathbf{H}$ as derived from sources in this way is conservative, is not solenoidal but is irrotational, as clarified later, and may thus be related to a scalar potential φ by $H = -\nabla\varphi = -\operatorname{grad}\varphi$. The above calculation could be made by integrating $d\varphi = (1/4\pi)dm/r$ and taking $\operatorname{grad}\varphi$.

In the absence of single or free poles, magnets are essentially macroscopic dipoles. The torque on an elementary dipole consisting of poles $\pm M\,d\mathcal{A}$ with separation l [Figure 1.2(a)] is

$$dT = \mu_0(M\,d\mathcal{A})lH\sin\theta, \qquad \text{where } \mu_0 = 4\pi\times10^{-7}\,\text{N A}^{-2} \qquad (M \sim H \sim \text{A m}^{-1})$$

The symbol $\sim$ is, throughout, to be read as 'corresponds to', here in relation to units, except when, as apparent by the context, it indicates very approximate or order-of-magnitude

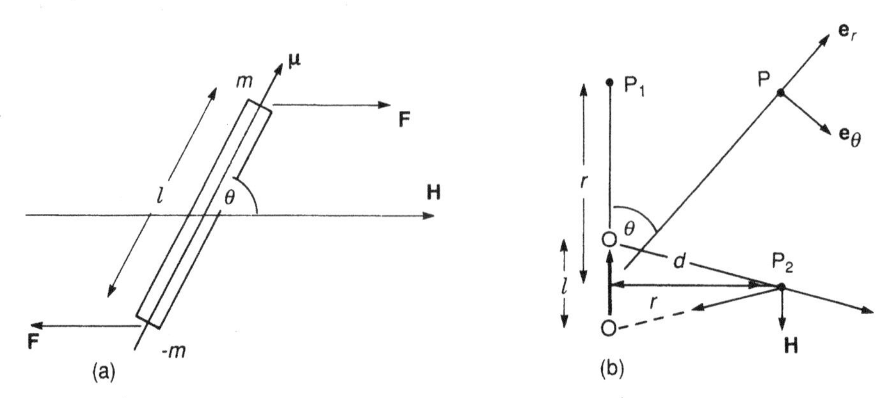

Figure 1.2 The torque on a magnet or macroscopic dipole also conforms with the pole convention (a). The fields from a point dipole considered as two point poles are readily calculated using the construction (b) and will be seen to correspond to those from a true magnetic dipole. (The scalar potential at P_1 is $\varphi = (m/4\pi)[1/(v - l/2) - 1/(v + l/2)]$. Divide the top and bottom by r and use $(1 + x)^{-1} \to 1 - x$; $x \to 0 : H = -\partial\varphi/\partial r$. At P_2, $\varphi = 0$ but H is readily calculated directly. H at general points follows immediately by resolving $\boldsymbol{\mu}$ into components $\parallel$ $\mathbf{e}_r$ and $\mathbf{e}_\theta$ and φ at general points may be inferred.)

equivalence. The origin of the units will become apparent later, as will the magnitude of the constant μ_0. It is apparent that a constant with these units is required, and also that $H \sim (M \cdot n)\mathrm{d}\mathcal{A}/r^2 \sim M$. Trivial integration of $\mathrm{d}T$ gives

$$T = \mu_0 M \mathcal{A} l H \sin\theta = \mu_0 \mu H \sin\theta, \qquad \text{where } \mu = M\mathcal{A}l = MV$$

or

$$\mathbf{T} = \mu_0 \boldsymbol{\mu} \times \mathbf{H}, \qquad \text{where } \boldsymbol{\mu} = \mathbf{M}V \tag{1}$$

as $\boldsymbol{\mu}$ is the total magnetic dipole moment or just the magnetic moment, $\mathbf{M}$ is the magnetic moment per unit volume. The magnet shape may be seen to be irrelevant.

If $\mathcal{A}$ becomes very small, $m = \sigma\mathcal{A}$ approximates a point pole. If $l \to 0$ but m increases so that $\mu = ml$ remains finite a point dipole results and for comparison with further developments this may be termed a magnetostatic dipole. Such literal two-pole magnetostatic dipoles arise only in this rather artificial way but, again for the sake of later comparisons, it is interesting to base a few calculations on them. Due to the symmetry, the scalar potential and field can be expressed in the radial and single angular coordinate as [see also equations (86) to (89), Section 2.9.2]

$$\varphi(\mathbf{r}) = \frac{\boldsymbol{\mu} \cdot \mathbf{r}}{4\pi r^3} = \frac{\mu \cos\theta}{4\pi r^2} \tag{2}$$

$$\mathbf{H}(\mathbf{r}) = \mathbf{e}_r \frac{2\mu\cos\theta}{4\pi r^3} + \mathbf{e}_\theta \frac{\mu \sin\theta}{4\pi r^3} \tag{3}$$

The derivation of these is left as an exercise which becomes very simple on realizing that it is only necessary to derive the cases for $\theta = 0$ and $\theta = \pi/2$ and to express $\boldsymbol{\mu}$ in two components [see Figure 1.2(b)]. The relation $\mathbf{H} = -\nabla\varphi$ here can be checked by recourse to Appendix 2. As further exercises it may be confirmed that a uniform field exerts no translational force on a dipole but in a field gradient there is a force

$$F_z = \mu_0 \mu_z \frac{dH}{dz} \tag{4}$$

with μ_z the component in the direction of the gradient, and, further, on integrating from a great distance, at which $\mathbf{H} = 0$ is assumed, the work term gives the energy in the field as

$$E = -\mu_0 \boldsymbol{\mu} \cdot \mathbf{H} \tag{5}$$

which could also be obtained by imagining a translation with $\boldsymbol{\mu} \perp \mathbf{H}$ followed by a rotation.

Despite their arbitrary basis, these conclusions are valid (save for certain important details such as the nature of the field when relaxation of the limits gives a quasi-point dipole, permitting consideration of the fields actually within the dipole), which remains to be demonstrated. A magnetized body generates fields of this nature as an approximation at substantial distances, so long as **M** is uniform or has axial symmetry. However, if it is accepted that a certain mass must be associated with μ, as described here, it is easy to see that the dynamic, undamped, response of a dipole to a field applied at angle θ to $\boldsymbol{\mu}$ is a sustained oscillation in a plane containing **H** and the original direction of $\boldsymbol{\mu}$, between $\pm\theta$, which is totally inappropriate.

The concentration of attention on dipoles rather than poles as such appears to constitute progress. **M** is pictured as an assembly of oriented dipoles (perfectly oriented for uniform **M**) with obvious cancellations of the poles except at surfaces. More meaningfully, the dipole fields for a regular array can be shown, by symmetry considerations, to average to zero except near to surfaces of the array, where they average to approximate the surface pole fields (avoiding the infinite contributions at the dipoles themselves). Eventually, however, attention must be drawn to the atoms that constitute the elementary dipoles, and if these are to be seen as pairs of poles the development must come to a halt.

A new starting point is suggested by a further observation. As indicated by the forces on a test magnet, the fields generated by a thin-walled coil, as illustrated in Figure 1.3, giving an approximation to a surface current sheet **I**, are, externally, identical to those of the magnet so long as $I = M$ in magnitude (hence the units of **M** and **H**). A systematic treatment which commences with the forces between, or fields from, currents, or charges in motion, encompasses the forces between magnets (whereas the converse does not apply) and is found to be appropriate generally.

Before proceeding to this some feeling for the magnitudes of magnetic fields and forces may be given, partly based on a further simple exercise. Taking the contribution to the potential from the element shown in Figure 1.4 as

$$d\varphi = \frac{\sigma}{4\pi} \frac{r\,dr\,d\theta}{R}, \qquad \text{where } R = (z^2 + r^2)^{1/2}$$

it follows by simple integration that the potential, and field, along the axis shown is

$$\varphi = \frac{\sigma}{2}[(z^2 + a^2)^{1/2} - z], \qquad \mathbf{H} = -\frac{\partial \varphi}{\partial z} = \mathbf{k}\frac{\sigma}{2}\left[1 - \frac{z}{(z^2 + a^2)^{1/2}}\right] \tag{6}$$

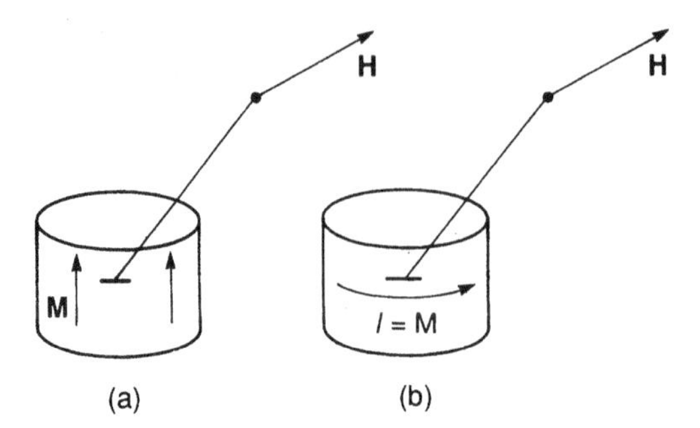

Figure 1.3 In appropriate limits the external fields from magnets (a) and from 'equivalent' coils (b) correspond at all points, suggesting the true origin of magnetic forces and fields

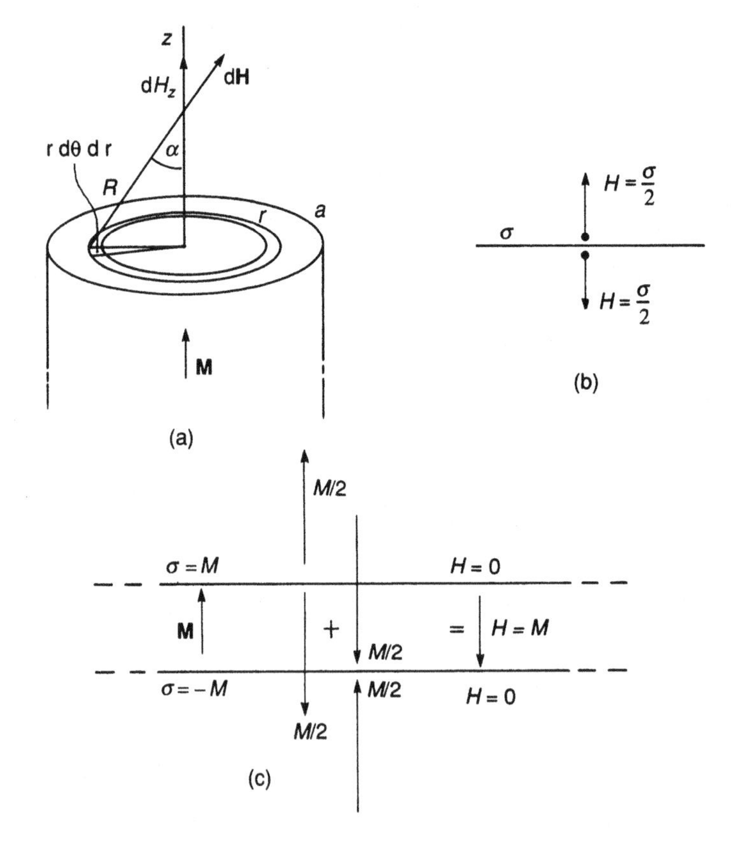

Figure 1.4 Construction for calculating the field on the axis from one end of a cylindrical magnet with $\sigma = M$ (a) and demonstrating the limiting results in (b). It follows that the fields external to a thin sheet as in (c) are zero and the internal field is as indicated

The field may be calculated directly at points along this axis where $\mathbf{H} = \mathbf{k}H$ by symmetry, using

$$\mathrm{d}H_z = \frac{\sigma}{4\pi}\frac{r\,\mathrm{d}r\,\mathrm{d}\theta}{R^2}, \qquad \cos\alpha = \frac{\sigma}{4\pi}\frac{zr\,\mathrm{d}r\,\mathrm{d}\theta}{R^3}$$

and integrating by parts with $\int(z^2+r^2)^{-3/2}\mathrm{d}r = r/[z^2(z^2+r^2)^{1/2}]$. As $z \to 0$ or $a \to \infty$, $H \to \sigma/2$ and it will be shown that this applies to the field component $\perp$ a single magnet surface (itself $\perp$ **M**) at all points adjacent to the surface.

$M = 1000\ \mathrm{kA\,m^{-1}}$ is a fair approximation for a permanent magnet material such as $SmCo_5$ (indicating the great magnitude of the equivalent, mythical, surface currents), so a pair of long magnets may produce such a field in a narrow gap. If two similar magnets are placed face to face and the distant end faces ignored, then the field from one acting on the pole density M of the other is $M/2$ and the force per unit surface area is $F = \frac{1}{2}\mu_0 M^2(\mathrm{N\,m^{-2}})$. This is clearly an upper limit for a pair of magnets, or for a magnet and its image in a permeable material (see Chapter 4, Section 1.3.1), but the magnitude is impressive: for the example, with sectional area of 1 $\mathrm{cm^2}$ only,

$$F = \pm\tfrac{1}{2}(4\pi \times 10^{-7}) \times 10^{12} \times 10^{-4} = 2\pi \times 10\ \mathrm{N} \sim 6\ \mathrm{kg}$$

Magnets handled carelessly may shatter on impact with steel structures and a modern specimen magnet placed on a steel cabinet, without spacers, may scarcely be removed without damage. Insecure electromagnet pole pieces are exceptionally dangerous and hand tools may become projectiles in the vicinity of the huge magnets used for magnetic resonance imaging, etc., unless these are carefully designed.

The following list gives some idea of field magnitudes in alternative units. Systems of units are discussed and compared in Appendix 1, the present text employing the SI, or Systeme Internationale, which is now familiar, although many of the articles to which reference will be made employ the c.g.s. and some publications, including books, perpetuate this system (or the e.m.u. and e.s.u. systems). For the present the tesla or unit of the induction field B is simply taken to be an alternative field unit and for the order of magnitude is equated to $10^6\ \mathrm{A\,m^{-1}}$, the correct factor being μ_0^{-1}. The oersted, Oe, is the c.g.s. unit and 1 Oe $= 10^{-4}$ T exactly. (Note that the rationalization factor 4π affects magnitudes but not units; its introduction in basic expressions has the effect of removing such factors from consequential expressions.)

10^8 T, $10^{14}\ \mathrm{A\,m^{-1}}$, 10^{12} Oe: existing at the surface of some neutron stars, as inferred from studies of hydrogen spectra. Such fields are indicated to induce grossly distorted orbitals and certainly do not constitute small perturbations.

10 T, $10^7\ \mathrm{A\,m^{-1}}$, 10^5 Oe: near the limit for steady laboratory fields, in superconducting solenoids. Larger fields may be produced, over small volumes, by pulsed coils.

1 T, $10^6\ \mathrm{A\,m^{-1}}$, 10^4 Oe: average electromagnet, or permanent magnet in a very favourable configuration.

10^{-4} T, $10^2\ \mathrm{A\,m^{-1}}$, 1 Oe: ambient at earth's surface; stray or accidental fields from machines, etc.

10^{-10} T, $10^{-4}\ \mathrm{A\,m^{-1}}$, 10^{-6} Oe: interstellar space; detectable by electromagnetic instruments (as by a flux gate which was placed on the surface of the moon).

10^{-14} T, $10^{-8}\ \mathrm{A\,m^{-1}}$, 10^{-10} Oe: around the limit for detection by superconducting quantum interference devices (SQUIDS); generated by the functioning of the human body (heart, brain, etc.).

Zero precisely: feasibly attained in an expanded superconducting enclosure or shield by sweeping out the last remaining quantum unit of flux (field times area) and specifically dependent on the quantization of the flux through a superconducting loop.

2 Introduction: B and H

A systematic account of magnetism may commence with the forces observed between current-carrying conductors as by Ampère, Oersted, Biot, Savart, etc., around the 1820s, or those forces between charges in motion which are distinct from the electrostatic forces. If two conducting rods are suspended side by side they attract each other when they carry currents in the same direction and the force is reversed if one current is reversed. When two charged particles pass close to each other the trajectories are distorted in a way that is only partially explained by electrostatic forces. The existence of these velocity-dependent forces can be understood by recourse to the principles of relativity, the invariance of charge with respect to its motion and Coulomb's law. Their directions and magnitudes are found to be given by

$$\mathbf{F}(\mathbf{r}_1) = \frac{\mu_0}{4\pi} q_1 \mathbf{v}_1 \times \left(q_2 \frac{\mathbf{v}_2 \times \mathbf{r}}{r^3} \right), \qquad \text{where } \mathbf{r} = \mathbf{r}_1 - \mathbf{r}_2 \tag{7}$$

assuming that the particles, or conductors, are situated in free space or, as a close approximation, in air. This is specifically the force on q_1 at $\mathbf{r}_1$. $\mathbf{r}$ is the vector extending *from* q_2 *to* q_1 since the vector sum of $\mathbf{r}$ and $\mathbf{r}_2$ gives $\mathbf{r}_1$. Figure 1.5 illustrates the cross-product and shows a couple of simple examples.

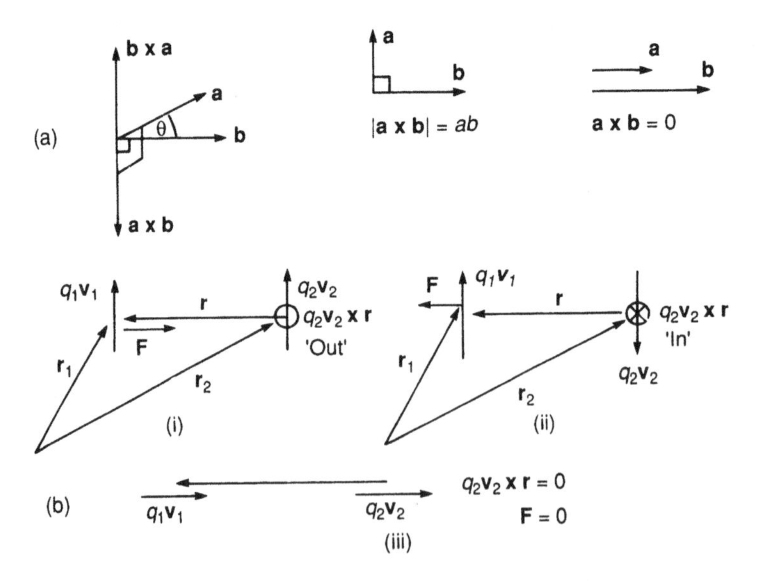

Figure 1.5 As a reminder of the cross-product (a), if the head of a corkscrew is rotated from **a** to **b** its direction of motion gives that of the vector $\mathbf{a}\times\mathbf{b}$. Note that, for the unit vectors, $\mathbf{i}\times\mathbf{j} = \mathbf{k}$, etc., while $\mathbf{j}\times\mathbf{i} = -\mathbf{k}$. Mentally transfer the vector $q_2\mathbf{v}_2\times\mathbf{r}$ in (b, i) to q_1 and apply rule (a) to see that the force is attractive. In (ii) or (iii) the force is repulsive or zero

The force can be alternatively expressed as

$$\mathbf{F}(\mathbf{r}_1) = q_1\mathbf{v}_1 \times \mathbf{B}(\mathbf{r}_1) = \mu_0 q_1 \mathbf{v}_1 \times \mathbf{H}(\mathbf{r}_1) \tag{8}$$

if

$$\mathbf{B}(\mathbf{r}_1) = \frac{\mu_0}{4\pi} q_2 \frac{\mathbf{v}_2 \times \mathbf{r}}{r^3}, \qquad \mathbf{H}(r_1) = \frac{q_2}{4\pi} \frac{\mathbf{v}_2 \times \mathbf{r}}{r^3} \tag{9}$$

defining the fields **H** and **B**, related in space by $\mathbf{B} = \mu_0 \mathbf{H}$.

Similarly, since an element of a circuit C_1 carrying a current of magnitude i_1 can be related to a moving charge by

$$i_1 \mathrm{d}\mathbf{r}_1 = q_1 \mathbf{v}_1$$

(because $\mathrm{d}\mathbf{r}/\mathrm{d}t = \mathbf{v}$ and $\mathrm{d}q/\mathrm{d}t = i$), we have

$$\mathrm{d}\mathbf{F}(\mathbf{r}_1) = \frac{\mu_0}{4\pi} i_1 \mathrm{d}\mathbf{r}_1 \times \left(i_2 \frac{\mathrm{d}\mathbf{r}_2 \times \mathbf{r}}{r^3} \right) = i_1 \mathrm{d}\mathbf{r}_1 \times \mathrm{d}\mathbf{B}(\mathbf{r}_1) \tag{10}$$

with

$$\mathrm{d}\mathbf{B}(\mathbf{r}_1) = \frac{\mu_0}{4\pi} i_2 \frac{\mathrm{d}\mathbf{r}_2 \times \mathbf{r}}{r^3}, \qquad \mathrm{d}\mathbf{H}(\mathbf{r}_1) = \frac{1}{4\pi} i_2 \frac{\mathrm{d}\mathbf{r}_2 \times \mathbf{r}}{r^3} \tag{11}$$

(Figure 1.6). Lacking access to current elements as such, this is meaningful to the extent that

$$\mathbf{B}(\mathbf{r}_1) = \frac{\mu_0 i_2}{4\pi} \int_{C_2} \frac{\mathrm{d}\mathbf{r}_2 \times \mathbf{r}}{r^3} \tag{12}$$

and in turn, integrating over the circuit C_1,

$$\mathbf{F}(\mathbf{r}_1) = i_1 \int_{C_1} \mathrm{d}\mathbf{r}_1 \times \mathbf{B}(\mathbf{r}_1) \tag{13}$$

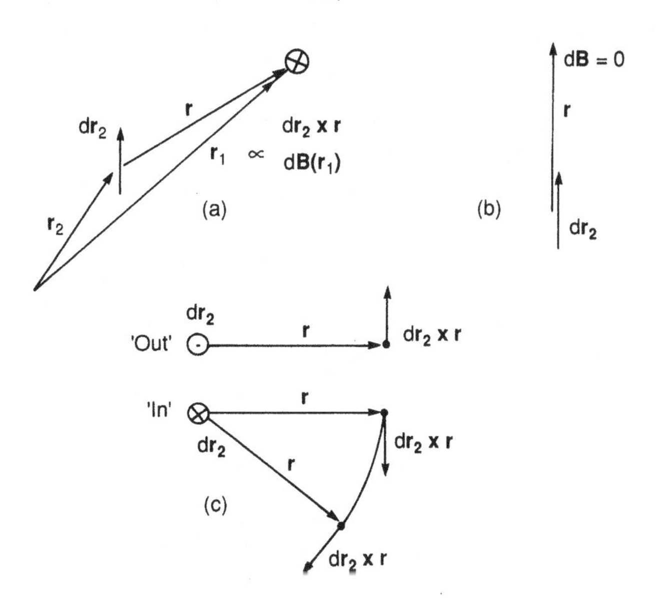

Figure 1.6 The direction of **dB** corresponding to equation (11); ⊗ indicates 'down' or into the diagram and ⊙ indicates 'up'. Note the axial symmetry w.r.t. **dr**

When the integrals are carried out, the force on the current in the circuit C_1 due to the current flowing in the circuit C_2 is that which would be measured in practice, and this may be taken to justify equations (10) and (11).

H is called the magnetic field and **B** the magnetic induction, or just the induction or induction field. Indeed, **B** itself is often referred to simply as the (magnetic) field. Both **H** and **B** are vector fields, being defined at all points in space, and both can be said to be of magnetic origin, so the confusion is understandable. So long as the concern is with a matter-free space (approximately with an air-filled space) containing the currents, **B** and **H** are linearly connected by $\mathbf{B} = \mu_0 \mathbf{H}$ with μ_0 constant, and can differ in no way other than in their magnitudes and units. For example, if it is shown that $\nabla \cdot \mathbf{B} = 0$ then $\nabla \cdot \mathbf{H} = (1/\mu_0)\nabla \cdot \mathbf{B} = 0$ also. Then the confusion is unimportant and it is reasonable to say, for example, that a certain solenoid gives a field of 6 T($B = 6.0$ T) or equally that it gives a field of (about) 5×10^6 A m^{-1} : the latter specification would, in fact, be unusual. Distinction may be made (apart from the units) by referring to 'B fields' or 'H fields'.

However, if the space of concern is occupied, wholly or partly, by matter that shows a certain response to the fields typified by a number μ_r which is called the relative permeability, a measure of the responsiveness discussed further in Section 2.4, then it is chosen to define **B** by

$$\mathrm{d}\mathbf{B}(\mathbf{r}_1) = \frac{\mu_r \mu_0}{4\pi} i_2 \frac{\mathrm{d}\mathbf{r}_2 \times \mathbf{r}}{r^3}$$

with d**H** as before [equation (11)] so that

$$\mathrm{d}\mathbf{B} = \mu_r \mu_0 \, \mathrm{d}\mathbf{H}, \qquad \text{and} \qquad \mathbf{B} = \mu_r \mu_0 \mathbf{H} = \mu \mathbf{H} \qquad \text{where } \mu = \mu_r \mu_0 \tag{14}$$

The immediate distinction between **H** and **B** is that **H** depends on current magnitudes and circuit geometries only whereas **B** depends also on the nature of the medium or the presence of magnetic (magnetically responsive) materials. Again, if μ_r is a constant throughout all space the distinction is trivial, but it will be seen that in more general situations (e.g. with some regions typified by $\mu_r = 1$, $\mu = \mu_0$ with a general μ_r or μ in other regions) **B** and **H** may differ basically in nature, when contributions to **H** other than those associated directly with current distributions are introduced.

Before pursuing this, one or two examples of field calculations may be illustrated and this is preceded by a note on field lines.

2.1 *Field or Flux Lines*

Field lines constitute a convention that is useful for envisaging fields or inductions and sometimes as an aid to the memory. Each line indicates the local direction of the field and the density of packing of the lines is proportional to the mean intensity of the field over the region. Thus a field line indicates the trajectory that would be followed by a mass-free pole, if one existed. (In fact, it is very easy to calculate trajectories by a prediction/correction method and this affords a useful numerical field line plotting method which the reader may choose to implement: take a small step in the local, initial, field direction, calculate the mean field magnitude and direction over the region of the step, return and take a corrected step and iterate.)

There are two rules having no foundation beyond the fact that they always correspond to observations or the results of meaningful calculations. Rule 1 is that the lines are akin

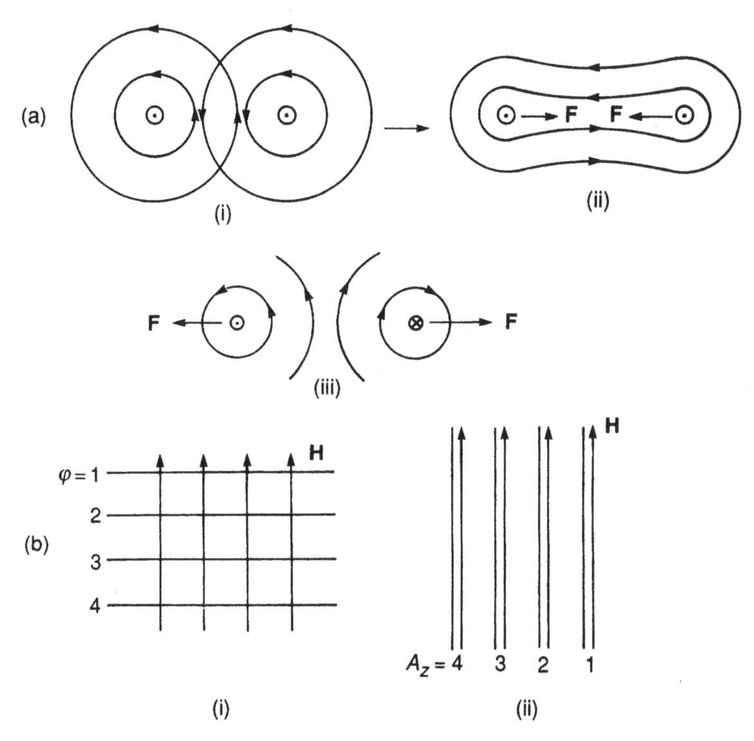

Figure 1.7 In (a) the field line convention is seen to give a useful reminder of the directions of the forces; (b) contrasts the relations between the field lines or directions and the equipotentials according to whether these are taken to be the scalar (i) or the vector potentials (ii). In (ii), **A** is taken to have a single component, normal to the diagram. If the spacing of the A equipotentials varies the field varies correspondingly and by choice of scale the field lines and equipotentials can be taken to coincide

to filamentary coiled springs tending always to contract in length, in correspondence with the attraction between unlike poles of magnets. Rule 2 is that the lines repel each other laterally, the more strongly the closer they are together. While these may seem naive they do permit the appreciation of the local magnitudes and directions of forces in field line plots. Consider the two currents or conductors shown in section in Figure 1.7(a). By symmetry alone the field lines for each separately must be circular. For the parallel currents the field lines annul between the wires and build up around the pair and 'pull the wires together'. For the antiparallel currents the fields build up in between and 'push the wires apart'. Despite their obvious artificiality such pictures do constitute a valid *aide-mémoire*. Field lines run normal to equipotentials which are planes of constant scalar potential φ, but if a vector potential (2.9) has a single component the equipotentials *are* the field lines [Figure 1.7(b)].

Distinction will be made in the following sections between conservative fields $\mathbf{H}_M$ having poles as sources as in Section 2.3.2 and 'true magnetic fields' **H** associated with currents (Section 2.3.1) which are solenoidal ($\nabla \cdot \mathbf{H} = 0$). Figure 1.8(a) shows a single sheet of surface pole density $\sigma = \rho t$ representing one end of a magnet and Figure 1.8(b) shows a current sheet of surface density $I = Jt$. The fields, as indicated, are similar to the extent that they are finite and defined at all points, passing through zero at the central planes of each sheet and only becoming discontinuous in the limit $t \to 0$ ($\rho \to \infty$, $J \to \infty$, with

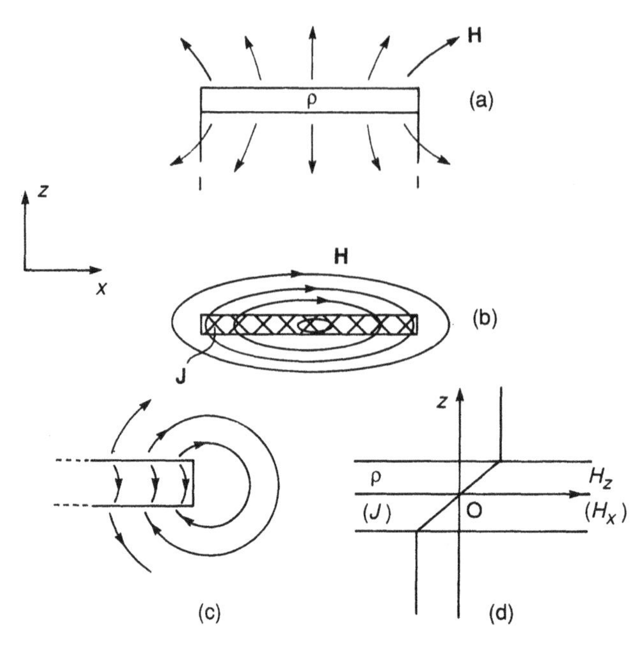

Figure 1.8 Conservative field lines have sources that may be taken to be the surface or volume pole densities at one end of a magnet as in (a). Solenoidal field lines run continuously around or throughout a current sheet as in (b). The abrupt changes in direction of the lines for the magnet represented by two pole sheets in (c) could not occur for a solenoidal field, while (d) compares the plots of the two field *magnitudes* on traversing the pole density or current density sheets and contrasts the directions

ρt, Jt constant). They are completely different in that the **H** lines form closed continuous loops while the $\mathbf{H}_M$ lines clearly do not. The pole sheets must obviously occur in pairs and then, as in Figure 1.8(c), the lines may be drawn as closed loops. However, these are not continuous field lines (which would be followed continuously by the friction-free pole) but rather consist of different sections with opposite directions, along the lines, applying to each section. The (solenoidal) **H** lines have components along the loops in the same direction throughout. Thus, though in a different sense both fields are continuous, the association may be made:

$$\text{Solenoidal} \sim \nabla \cdot \mathbf{H} = 0 \sim \text{continuous (lines)}$$

$$\text{Conservative} \sim \nabla \cdot \mathbf{H} \neq 0 \sim \text{discontinuous (lines)}$$

stressing that the concept applies to the field lines rather than to the fields as such.

2.2 Simple Field Calculations

The simplest integrated field expression of all will simply be quoted here since it is introduced as an example of the use of the vector potential in a later section. The field from a straight filamentary current of great length L is

$$\mathbf{H} = \mathbf{e}_\phi \frac{i}{2\pi r} \tag{15}$$

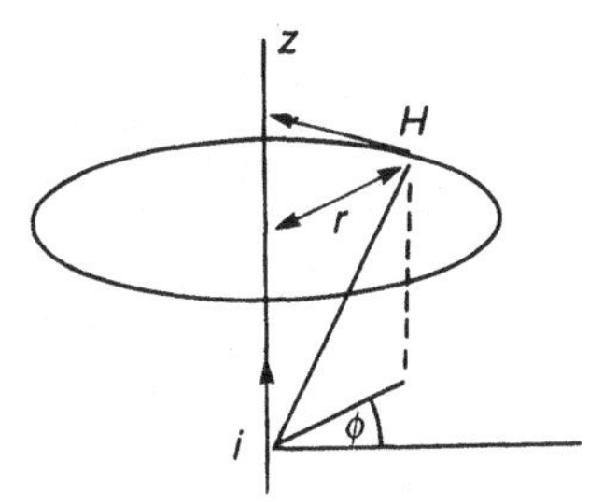

Figure 1.9 The field from a filamentary current in a long straight conductor

in a cylindrical system, as illustrated by Figure 1.9. In practice this would be the field at points close to a thin wire forming a circular circuit of very large radius. Also, $r = 0$ must be excluded because the current is taken to be filamentary with no finite current per unit cross-sectional area; different calculations must be made for a wire of finite section.

2.2.1 Circular current loop

Fields on the axis OZ normal to the plane of the ring of radius a (Figure 1.10) are considered, fields at general points being derived in Chapter 4. The element **dl** is paired with its reflection through the origin, **dl**$'$. From $i\,\mathbf{dl}$, $\mathrm{d}H = (1/4\pi)i\,\mathrm{d}l/r^2$ with the direction shown, in OX. Adding $\mathrm{d}H'$ from $i\,\mathrm{d}l'$, the remaining component is

$$\mathrm{d}H_z = \frac{1}{4\pi}\frac{2i\,\mathrm{d}l}{r^2}\sin\theta$$

and integrating trivially around the half-circle with contributions from the other half implied

$$H_z = \frac{i}{4\pi}\frac{2(\pi a)}{r^2}\sin\theta = \frac{1}{4\pi}2i\frac{\pi a^2}{r^3} = \frac{1}{4\pi}\frac{2i\mathcal{A}}{r^3} \tag{16}$$

where ($\sin\theta = a/r$, area $\mathcal{A} = \pi a^2$). The final form has a particular significance when a is small and $r \to z$, which should be apparent and will be discussed later [cf. equation (3), $\theta = 0$].

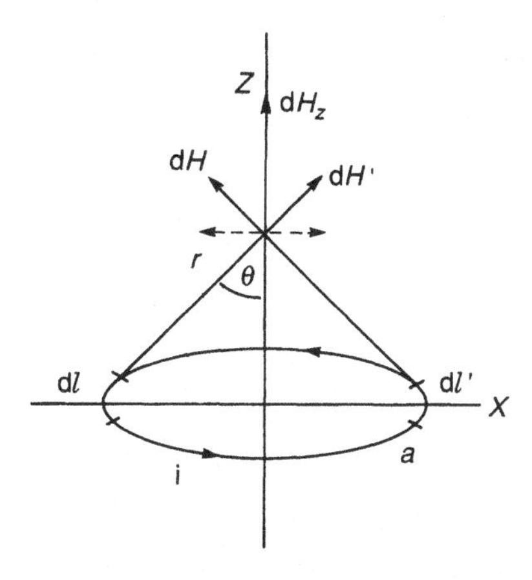

Figure 1.10 Construction for the fields on the axis of a circular current loop

2.2.2 Solenoids

Without qualification, solenoids are assumed to be circular-section cylinders with surface currents of density I per unit of length along the cylinder axis, flowing around the cylinder surface normal to the axis. They would be approximated by single-layer coils of fine wire: $I = ni$ with n turns m^{-1}.

For points on the axis, $H = H_z$ by symmetry. The element shown in Figure 1.11(a) constitutes a ring current:

$$i = \frac{I\,(r\,\mathrm{d}\theta)}{\sin\theta}$$

Using this in equation (16), the contribution from the surface element is

$$\mathrm{d}H_z = \frac{1}{4\pi}\frac{I r\,\mathrm{d}\theta}{\sin\theta}\frac{2\pi}{r}\left(\frac{a}{r}\right)\sin\theta = \tfrac{1}{2}I\sin\theta\,\mathrm{d}\theta$$

with $a/r = \sin\theta$ again. Integrating:

$$H_z = \tfrac{1}{2}I(\cos\theta_1 - \cos\theta_2) \tag{17}$$

At points distant from the ends of a long solenoid $\theta_1 \to 0$, $\theta_2 \to \pi$ and

$$H_z = \tfrac{1}{2}I[1-(-1)] = I \tag{18}$$

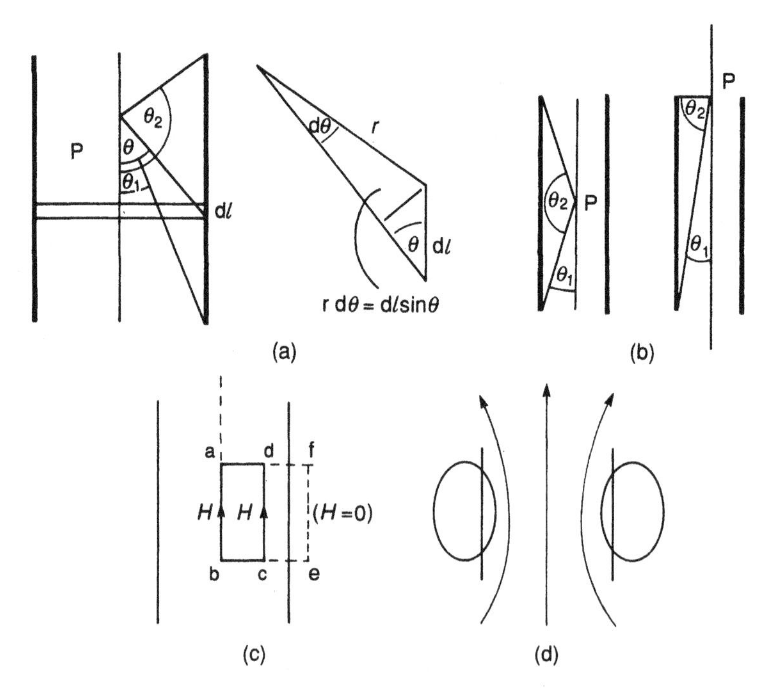

Figure 1.11 Construction for the fields on the axis of a solenoid (a). For a very long solenoid, at the centre, $\cos\theta_1 \to 1$, $\cos\theta_2 \to -1$ and at one end $\cos\theta_1 \to 1$ while $\cos\theta_2 = 0$ (b). In (c) the line integral along abcda must be zero and that along abefa is equal to ab $\times\, I$; (d) is an impression of the field lines for the short solenoid, as calculated later

whereas in line with one end of a semi-infinite solenoid (on the axis)

$$H_z = \tfrac{1}{2}I \tag{19}$$

[Figure 1.11(b)]. At (axial) points outside the semi-infinite solenoid,

$$H_z = \tfrac{1}{2}I(1 - \cos\theta_2) = \tfrac{1}{2}I\left(1 - \frac{z}{\sqrt{z^2 + a^2}}\right) \tag{20}$$

if the origin of z is at the end of the solenoid. Again, equation (20) shows similarities with an earlier expression [equation (6)].

Using the binomial expansion for small a/z:

$$\begin{aligned} H_z &= \tfrac{1}{2}I\left[1 - \left(1 + \frac{a^2}{z^2}\right)^{-1/2}\right] \\ &= \tfrac{1}{2}I\left[1 - \left(1 - \tfrac{1}{2}\frac{a^2}{z^2}\right)\right] = \frac{1}{4\pi}\frac{I(\pi a^2)}{z^2} \end{aligned} \tag{21}$$

which, rather surprisingly, is the inverse square law for the field from a pole of strength $I\mathcal{A}$, where $\mathcal{A} = \pi a^2 =$ the sectional area of the solenoid. Note that I is a current density so that the dimensions are consistent. This suggests why, in certain circumstances, fields may appear to be capable of calculation as if they arose from poles.

To depart from the axis, note first that if a line integral is taken around a circular path enclosing and centred on the current, in a plane normal to the current, in Figure 1.9 the result is

$$\oint \mathbf{H} \cdot \mathbf{dl} = \frac{i}{2\pi r}(2\pi r) = i \tag{22}$$

Any path may be taken since an arbitrary path may be divided into infinitesimal sections along each of which only one of the cylindrical coordinates varies and only those for which ϕ changes need be taken into account. The current need not be filamentary but may be a general distribution, and it then follows, by subdividing the distribution into elements or quasi-filaments and summing the effects, that the line integral is always equal to the total current encircled by the path. By reference to Figure 1.12 it is seen that equation (22) is consistent with the case that the current enclosed is zero. For the path abcd in Figure 1.11(c), with O the centre of the solenoid, the integral is zero, which is consistent with the field along cd being constant and equal in magnitude and direction to that along ab. This is the only interpretation that is consistent with the symmetry in the limit as the solenoid length increases indefinitely and the same conclusion can be reached from a general analysis commencing with the fields for a current ring at general points: around the centre of a long solenoid the field is constant and equal to that at the axis.

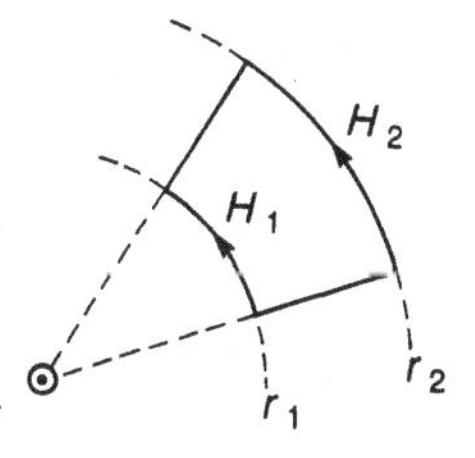

Figure 1.12 Since the lengths of the arcs are $\propto r$ and $H \propto 1/r$ the integral around the circuit enclosing no current is zero in the simple illustration

Taking the path abefa and again assuming that any field H_z' acting along ef must be constant, with $l \ll$ the solenoid length, then

$$lH_z + lH_z' = lI, \qquad \text{whence } H_z' = 0$$

and the field outside the solenoid is, in the limit, zero.

No such simple considerations apply to a short solenoid but it will be seen in Chapter 4 that the fields, at general points, may be readily calculated and have the form indicated in Figure 1.11(d).

2.3 Solenoidal and Conservative Fields

2.3.1 Solenoidal fields

A solenoidal field is, most directly, one that obeys, at any general point,

$$\nabla \cdot \mathbf{A} = 0 \tag{23a}$$

i.e. in rectangular coordinates,

$$\left(\mathbf{i}\frac{\partial}{\partial x} + \mathbf{j}\frac{\partial}{\partial y} + \mathbf{k}\frac{\partial}{\partial z}\right) \cdot (\mathbf{i}A_x + \mathbf{j}A_y + \mathbf{k}A_z) = \frac{\partial A_x}{\partial x} + \frac{\partial A_y}{\partial y} + \frac{\partial A_z}{\partial z} = 0 \tag{23b}$$

An equivalent specification is developed in relation to Figure 1.13. The field at P, at the centre of the parallelepiped, is **A** and, since the dimensions are taken to be small, at the faces normal to OX the x components can be taken to be $A_x \pm \frac{1}{2}\delta x \partial A_x/\partial x$ and similarly for A_y and A_z. The flux over the surface is the normal field component times the surface area (integrated in the general case) and the total flux from the closed surface shown is

$$\begin{aligned} F &= \left(A_x + \tfrac{1}{2}\delta x \frac{\partial A_x}{\partial x}\right)\delta y\,\delta z - \left(A_x - \tfrac{1}{2}\delta x\,\frac{\partial A_x}{\partial x}\right)\delta y\,\delta z + \cdots \\ &= \frac{\partial A}{\partial x}\delta x\,\partial y\,\partial z + \frac{\partial A_y}{\partial y}\delta y\,\delta x\,\delta z + \frac{\partial A_z}{\partial z}\delta z\,\delta x\,\delta y \end{aligned}$$

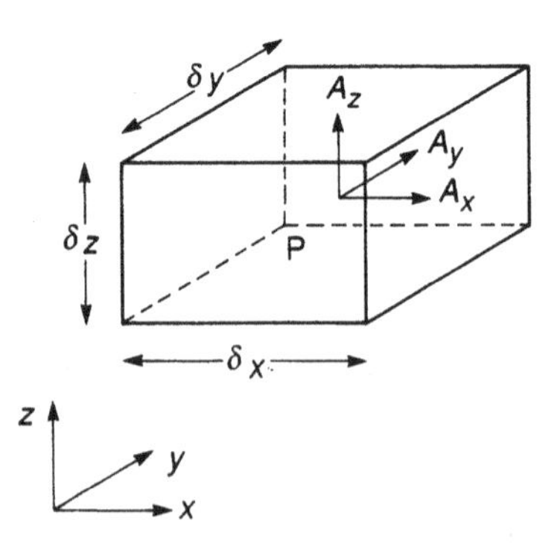

Figure 1.13 The vector field at the centre is **A** or (A_x, A_y, A_z) and the x component at the r.h. face is $A_x + \frac{1}{2}(\partial A_x/\partial x)\delta x$, etc

However, $(\delta x\,\delta y\,\delta z)$ is the volume and so the outward flux per unit volume, known as the divergence, or div( ), is

$$\text{div}\ \mathbf{A} = \frac{\partial A_x}{\partial x} + \frac{\partial A_y}{\partial y} + \frac{\partial A_z}{\partial z} = \nabla \cdot \mathbf{A} \tag{24}$$

In general the first equality may be regarded as an approximation in accordance with the assumptions made above; the equation applies strictly in the limit $\delta x,\ \delta y,\ \delta z \to 0$ and div $\mathbf{A} = \nabla \cdot \mathbf{A}$ then applies at the given point.

A sort of 'macroscopic divergence' is involved in Gauss's divergence theorem:

$$\int_S \mathbf{A} \cdot \mathbf{n}\, \mathrm{d}s = \int_V \nabla \cdot \mathbf{A}\, \mathrm{d}v \tag{25}$$

with $\mathbf{n}$ the positive or outwardly drawn normal to the surface S which encloses the volume V. If V shrinks to a size for which the field gradients can be assumed constant and $\int \nabla \cdot \mathbf{A}\, \mathrm{d}v \to \nabla \cdot \mathbf{A} \int \mathrm{d}v$, equation (24) is regained. In general V is unrestricted and the integral on the left may be described as the total normal outward flux. If $\mathbf{A}$ is solenoidal in that $\nabla \cdot \mathbf{A} = 0$ everywhere, then the flux must clearly be zero, but the converse does not necessarily follow. Nevertheless, if the flux can be seen to be zero over any conceivable surface it may be inferred that $\mathbf{A}$ is solenoidal.

From a pictorial point of view, if the calculated field lines are everywhere continuous and do not originate from sources or exhibit abrupt changes in direction then it is at least indicated that the field is solenoidal, and vice versa. The lines may be imagined to be surrounded by 'tubes of flux' and with continuity and freedom from sources every tube that enters any closed surface must also exit from some point on that surface. The **H** and **B** fields defined in Section 2 certainly appear, according to the illustrations given, to be solenoidal, the lines always forming closed circles or loops.

Secondly, if a surface with the shape of a closed cylinder is centred coaxially on the straight wire, for example, there is no flux over this surface because there are no components across it at any point. If a wedge is cut from this cylinder by planes passing through the axis then normal components arise over the plane surfaces, but it is only necessary to draw this new surface (as in Figure 1.12) and consider the symmetry to see that the total flux is zero. No surface to which this conclusion fails to apply can be devised.

Thirdly, and more specifically, the field from a current element at any general point can be abbreviated to

$$\mathrm{d}\mathbf{H}(\mathbf{r}) = b\frac{\mathbf{k} \times \mathbf{v}}{r^3} = \frac{b}{r^3}\mathbf{k} \times (\mathbf{i}x + \mathbf{j}y + \mathbf{k}z)$$

$$= -\mathbf{i}\frac{b}{r^3}y + \mathbf{j}\frac{b}{r^3}x + \mathbf{k}(0)$$

with Figure 1.5(a) to recall the **i**, **j**, **k** products. Thus

$$\nabla \cdot \mathrm{d}\mathbf{H} = -\frac{\partial}{\partial x}\frac{b}{r^3}y + \frac{\partial}{\partial y}\frac{b}{r^3}x$$

However,

$$\frac{\partial}{\partial x}r^{-3} = -3r^{-4}\frac{\partial r}{\partial x} = -3r^{-4}\tfrac{1}{2}(x^2 + y^2 + z^2)^{-1/2}(2x) = -\frac{3}{r^4}x$$

$$\frac{\partial}{\partial y} r^{-3} = -\frac{3}{r^4} y$$

and so

$$\nabla \cdot \mathbf{dH} = 0, \qquad \nabla \cdot \mathbf{dB} = 0 \tag{26}$$

[using $r = (x^2 + y^2 + z^2)^{1/2}$ and the chain law]. If $\mathbf{dH}$ is replaced by $(\delta H)_i$ and it is noted that $\nabla\cdot$ is linear, $\nabla \cdot [(\delta \mathbf{H})_1 + (\delta \mathbf{H})_2] = \nabla \cdot (\delta \mathbf{H})_1 + \nabla \cdot (\delta \mathbf{H})_2$, etc., then by considering the integral as the limit of the sum it is inferred that

$$\nabla \cdot \mathbf{H} = 0, \qquad \nabla \cdot \mathbf{B} = 0 \tag{27}$$

The reader may confirm that the field given by equation (15) obeys $\nabla \cdot \mathbf{H} = 0$, as above or taking advantage of the symmetry. Such demonstrations are only complete when the filamentary currents are replaced by current distributions (e.g. by cylinders with small but finite cross-section) and it is demonstrated that $\nabla \cdot \mathbf{H} = 0$ at points occupied by the distribution.

Finally, it will be seen that the expression for $\mathbf{dH}$ (and, on integrating, for $\mathbf{H}$) can be obtained by relating a vector potential $\mathbf{A}$ to the current distribution and in turn relating $\mathbf{H}$ to $\mathbf{A}$ by

$$\mathbf{H} = \left(\frac{1}{\mu_0}\right) \nabla \times \mathbf{A} \tag{28}$$

Generally, for *any* vector $\mathbf{A}$

$$\begin{aligned}
\nabla \cdot (\nabla \times \mathbf{A}) &= \left(\mathbf{i}\frac{\partial}{\partial x} + \mathbf{j}\frac{\partial}{\partial y} + \mathbf{k}\frac{\partial}{\partial z}\right) \cdot \left[\mathbf{i}\left(\frac{\partial A_z}{\partial y} - \frac{\partial A_y}{\partial z}\right)\right. \\
&\quad \left. + \mathbf{j}\left(\frac{\partial A_x}{\partial z} - \frac{\partial A_z}{\partial x}\right) + \mathbf{k}\left(\frac{\partial A_y}{\partial x} - \frac{\partial A_x}{\partial y}\right)\right] \\
&= \frac{\partial^2 A_x}{\partial x\, \partial y} - \frac{\partial^2 A_y}{\partial_x\, \partial_z} + \frac{\partial^2 A_x}{\partial y\, \partial z} - \frac{\partial^2 A_z}{\partial y\, \partial x} + \frac{\partial^2 A_y}{\partial z\, \partial x} - \frac{\partial^2 A_x}{\partial z\, \partial y} \\
&= 0
\end{aligned} \tag{29}$$

so long as the components have continuous derivatives so that $\partial^2 A_x/(\partial x\, \partial y) = \partial^2 A_x/(\partial y \partial x)$, etc. This condition can be accepted here due to the way in which A is related to the current distributions and so

$$\begin{aligned}
\nabla \cdot \mathbf{B} &= \nabla \cdot (\nabla \times \mathbf{A}) = 0 \\
\nabla \cdot \mathbf{H} &= \left(\frac{1}{\mu_0}\right) \nabla \cdot (\nabla \times \mathbf{A}) = 0
\end{aligned} \tag{30}$$

(These arguments may be reversed and if it is considered that $\nabla \cdot \mathbf{H} = 0$ has been established then it may be taken that $\mathbf{H}$ can be related to a further vector field, $\mathbf{A}$, in this way.)

It may be stressed that it is usually possible to 'demonstrate' that a conservative field (see the following section) is solenoidal by drawing some arbitrary surface or by studying $\nabla \cdot \mathbf{H}$ in a particular region and vice versa. The errors arise here because the demonstrations have lacked generality.

It has been noted, specifically for points along the axis, that the fields external to the end of a semi-infinite solenoid become identical to those from a pole, at the end, in the limit as the solenoid radius $\rightarrow 0$ (and the current density is correspondingly increased), and it may be demonstrated that this applies to general points. On centering a spherical surface at the end of the solenoid it might appear that the flux over this surface was readily calculable, as that from the pole, and was non-zero. This would necessarily be an error, the explanation of which of which is left to the reader.

2.3.2 Conservative fields

It appears from the foregoing that all true magnetic fields associated with currents or charges in motion are solenoidal whether these are designated **H** or **B**. This is essentially the case but it will be seen to be not only convenient but in some situations virtually necessary to introduce fields associated with the magnetization, i.e. with the pole distributions $\sigma = \mathbf{M}\cdot\mathbf{n}$ and $\rho = -\nabla\cdot\mathbf{M}$. These are the fields introduced in the preamble. They will be designated $\mathbf{H}_\mathrm{M}$ and initially regarded as being completely distinct from the fields **H** considered above. Clearly a major objective will be to rationalize these fields $\mathbf{H}_\mathrm{M}$ in relation to the true fields **H**.

A general expression for $\mathbf{H}_\mathrm{M}$ is

$$\mathbf{H}_\mathrm{M} = \frac{1}{4\pi}\int_S \mathbf{M}\cdot\mathbf{n}\frac{\mathbf{r}}{r^3}\mathrm{d}s + \frac{1}{4\pi}\int_V (-\nabla\cdot\mathbf{M})\frac{\mathbf{r}}{r^3}\mathrm{d}v \tag{31}$$

over all the surfaces and throughout the volume. The distribution $\rho = -\nabla\cdot\mathbf{M} \neq 0$ arises if $|\mathbf{M}|$ varies or if **M** varies in direction, as in the radially magnetized cylinder in Figure 1.14, for which (see the caption)

$$\rho = -\nabla\cdot\mathbf{M} = -\frac{M}{r}$$

The fields $\mathbf{H}_\mathrm{M}$ are not solenoidal. They have sources which are the distributions σ and ρ. The field lines do not consist of continuous loops and exhibit discontinuities. The total outward flux or divergence over a macroscopic surface is not necessarily zero but is equal to the pole strength enclosed by the surface. This is immediately seen to be so for a spherical surface enclosing a single pole at its centre:

$$\int_S \mathbf{H}_\mathrm{M}\cdot\mathrm{d}\mathbf{s} = \frac{1}{4\pi}\frac{p}{r^2}4\pi r^2 = p \tag{32}$$

and is readily seen to be general by summing over elements of pole density, noting that although placing the elements or quasi-point poles at the centre of a sphere facilitates the calculations, the particular disposition is immaterial. Thus if a region of uniform volume

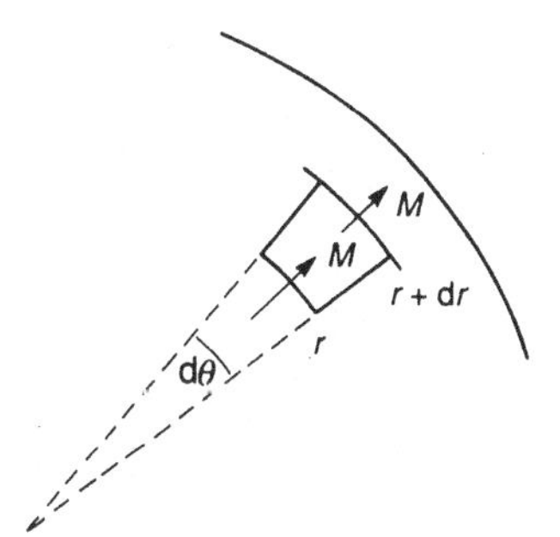

Figure 1.14 The outward flux for axial length dl is readily calculated and divided by the element volume to show that $\nabla\cdot\mathbf{M} = M/r$ in the limit

pole density is enclosed by the surface, or the volume enclosed becomes very small, the surface integral is equal to ρV, indicating that

$$\nabla \cdot \mathbf{H}_{\mathrm{M}}(\mathbf{r}) = \rho(\mathbf{r}) \tag{33}$$

where, in the limit, ρ is the pole density at the point $\mathbf{r}$. Note once more that if attention was restricted to points at which $\rho = 0$ the field would be fallaciously indicated to be solenoidal with $\nabla \cdot \mathbf{H}_{\mathrm{M}} = 0$; the reader may readily show that $\nabla \cdot \mathbf{e}_r(p/r^2) = 0$ for a monopole field, but only at points that are necessarily distant from the point pole. Again it is easy to see that the flux over a surface chosen not to include the pole is zero.

$\mathbf{H}_M$ is, however, irrotational, as defined by

$$\nabla \times \mathbf{H}_{\mathrm{M}} = \mathrm{rot}\ \mathbf{H}_{\mathrm{M}} = 0 \tag{34}$$

(rot $\equiv$ rotation). $\mathbf{H}$ is rotational and

$$\nabla \times \mathbf{H} = \mathbf{J} \tag{35}$$

at points occupied by a current density $\mathbf{J}$, although, in line with the above discussion, $\nabla \times \mathbf{e}_\phi(2I/v) = 0$ for example, as the reader may readily show by recourse to $\nabla \times$ in cylindrical coordinates (Appendix 2): points have implicitly been chosen at which the current density is zero. Stoke's theorem for any vector $\mathbf{A}$ is

$$\oint_C \mathbf{A} \cdot \mathrm{d}\mathbf{r} = \int_S \nabla \times \mathbf{A} \cdot \mathbf{n}\, \mathrm{d}s = \int_S \nabla \times \mathbf{A} \cdot \mathrm{d}\mathbf{s} \tag{36}$$

with the circuit to be taken in a positive sense and to bound the surface S. Clearly the line integral is zero if $\mathbf{A}$ is irrotational and it will be taken that the converse applies when the circuit and surface identify a region in space which is taken to be reduced to a point.

The particular line integral is the circulation and when the gradients can be taken to be constant over the infinitesimal region the circulation can be identified with the curl $\times\, \Delta S$. For the thin sheet in Figure 1.15 the internal field $\mathbf{H}_{\mathrm{M}}$ is uniform, as shown, and the external field is zero. It is thus obvious that the circulation about C_{a} is zero. The sheet may be indefinitely thin permitting $\Delta S \to 0$ and so $\nabla \times \mathbf{H} = 0$. Similar considerations apply to the circuit C_{b} crossing a single surface, since $\nabla \times \mathbf{H}'_{\mathrm{M}} = 0$ will always apply to contributions $\mathbf{H}'_{\mathrm{M}}$ from other surfaces and only the fields from the one surface need to be taken into account. A little consideration shows that the shapes and orientations of the circuits are immaterial.

As an example of the circulation along a macroscopic path, it will be shown that the internal, 'demagnetizing', fields within a uniformly magnetized sphere are uniform and

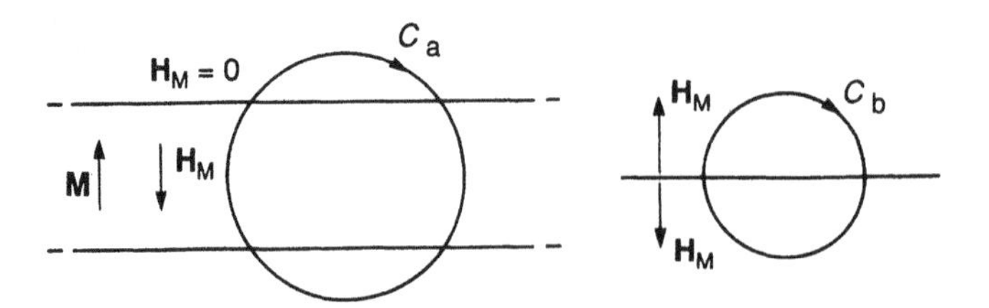

Figure 1.15 The circulation or line integral of $\mathbf{H}_{\mathrm{M}}$ resolved along C, taken around C, is clearly zero whether one (C_{b}) or more surfaces (C_{a}) are enclosed: $\mathbf{H}_{\mathrm{M}}$ is irrotational

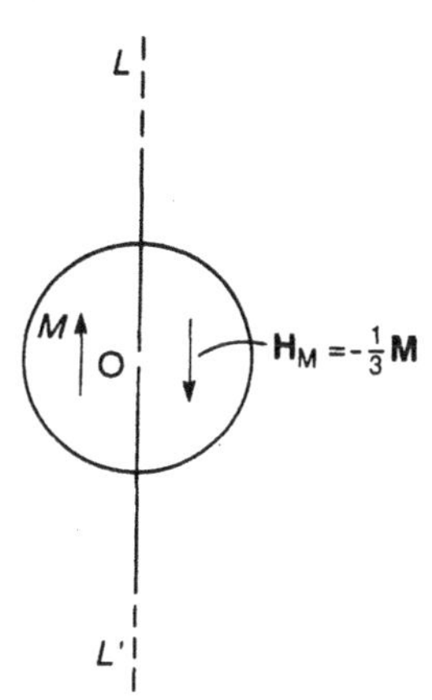

Figure 1.16 The line integrals of $\mathbf{H}_M$ along L, from 0 to ∞ or L' from $-\infty$ to 0, are zero

given by $\mathbf{H}_M^d = -(\frac{1}{3})\mathbf{M}$, while the external fields $\mathbf{H}_M^S$ are those of an effective dipole $\boldsymbol{\mu} = \mathbf{M} \times$ volume situated at the centre O. Taking L to be a line from 0 to infinity to parallel to $\mathbf{M}$ (Figure. 1.16),

$$\int_L \mathbf{H}_M \cdot \mathbf{r} = -\tfrac{1}{3}Ma + \int_a^{\infty} \frac{1}{4\pi} 2\left(\tfrac{4}{3}\pi a^3 M\right) \frac{1}{r^3}\,\mathrm{d}r$$
$$= -\tfrac{1}{3}Ma + \tfrac{2}{3}a^3 M\left(\frac{1}{2a^2}\right) = 0$$

The same applies to L' from $-\infty$ to 0 and C may be taken as $L + L'$ completed by lines at infinity ($\mathbf{H}_M = 0$), so that the circulation is zero, suggesting, though not here proving, that $\mathbf{H}_M$ is irrotational. (The fields referred to in this example are those calculated as effective pole fields arising from $\mathbf{M} \cdot \mathbf{n}$ at the sphere surface.)

As a diversion, in the present context, if the circulation is taken about a path enclosing a plane current sheet as in Figure 1.17 and either the path is very close to the current sheet or the sheet is unbounded in extent, the result is $2H\delta l$ with δl taken to be so small that $\mathbf{H}$ may be assumed uniform. Thus

$$2H\,\delta l = I\delta l, \qquad \text{whence } H = I/2 \tag{37}$$

giving the striking illustrations of Figure 1.17. The result for the two current sheets corresponds to that for a 'flattened solenoid' and the sectional shape of the solenoid is, in fact, immaterial (for long solenoids).

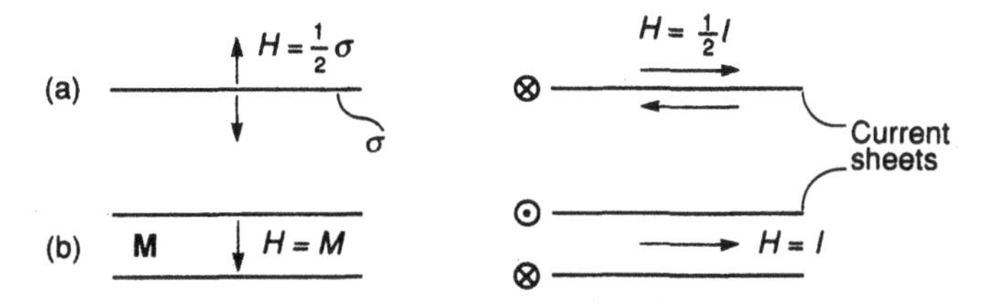

Figure 1.17 Contrasts between the fields from surface pole sheets and current sheet (a) and between the fields within pairs of such sheets (b)

Again a close parallel to the preceding section arises in that the reader may demonstrate similarly by expansion that

$$\nabla \times \nabla \varphi = 0 \tag{38}$$

where φ is a scalar field or scalar potential (a scalar quantity defined at all points in a given region of space having the continuous derivatives required). Thus setting

$$\mathbf{H}_{\mathrm{M}} = -\nabla \varphi \tag{39}$$

with the sign for convenience (so that **H** points 'downhill' in terms of φ) conforms with

$$\nabla \times \mathbf{H}_{\mathrm{M}} = -\nabla \times (\nabla \varphi) = 0 \tag{40}$$

[Conversely, since $\mathbf{H}_{\mathrm{M}}$ can be given either directly or equally by equation (39), with a suitable connection between the potential and the pole distributions, equation (40) may be taken as a general proof that $\mathbf{H}_{\mathrm{M}}$ is irrotational.] In the general case that a vector field is derived from a potential in this way, or equivalently that it is irrotational or yet again that it is associated with sources, it is referred to as a conservative field. For a conservative field it is apparent that $\mathbf{H}_{\mathrm{M}} \cdot \mathbf{dr}$ is an exact differential, since the line integral about a closed path is always zero and thus the general line integral is independent of the path taken between any two points. For solenoidal fields such as **H** or **B**, $\mathbf{H} \cdot \mathbf{dr}$ is not an exact differential since the line integrals around closed paths are not zero and in turn the general line integrals are path-dependent. The contrast is readily appreciated by taking suitably oriented semicircular paths, say, and rotating these about their diameters in the presence of either current distributions or pole distributions: the poles are scalars and the currents are vectors.

It follows from equations (33) and (39) that φ obeys the equations of Laplace or Poisson according to whether or not the pole density is zero at the point to which the equations apply: $\nabla \cdot \mathbf{H} = -\nabla \cdot \nabla \varphi = \rho$, i.e.

$$\nabla^2 \varphi = -\rho \qquad \text{(Poisson)} \tag{41}$$

$$\nabla^2 \varphi = 0 \qquad \text{(Laplace, with } \rho = 0\text{)} \tag{42}$$

For comparison with this, it will be seen later that the vector potential obeys

$$\nabla^2 \mathbf{A} = -\mu_0 \mathbf{J} \tag{43}$$

which can be read as a set of three component equations. The solution of these equations, with particular boundary conditions, constitutes the most general basis for magnetostatic calculations and is referred to in Chapter 4.

As an exercise on conservative fields, take a plate extending from $z = -2$ to $z = 2$ and otherwise unbounded, within which the pole density is ρ. Divide this into four plates of unit thickness and consider the density in each to be condensed on to sheets at the central planes with $\sigma = \rho$, each giving a field of $\pm 1/2\sigma$ at all points above and below the (infinite) sheets. Write down the values of H_z at $z = 0, \pm 1, \pm 2, \pm 3$ and deduce a general expression for $H_z(z)$; confirm this by a procedure of integration and confirm also that $\nabla \times \mathbf{H} = 0$, $\nabla \cdot \mathbf{H} = \rho$, $\nabla^2 \varphi = -\rho$. Show that the potentials and fields could, in fact, have been calculated (without reference to the result for the field from a sheet) by solving Poisson's equation with certain assumptions due to symmetry: the existence of $H_z(z)$ only, $\varphi = 0$ and $H = 0$ at $z = 0$.

The reader may also devise an analogous exercise to calculate H when ρ is replaced by a current distribution $\mathbf{J}$, showing that $\nabla\times\mathbf{H} = \mathbf{J}$, $\nabla\cdot\mathbf{H} = 0$. Note the conditions required for the existence of a field for which only one component varies.

It must seem strange that two magnetic fields which differ so basically can exist. The suspicion arises that one is the more fundamental and that the second is somehow to be related to this. Of course it is $\mathbf{H}$ that has been introduced in the more systematic manner while $\mathbf{H}_M$ was only introduced in relation to effective poles, so the choice is clear. Before pursuing this it is necessary to develop the concept of the magnetization which itself has only, thus far, been loosely described.

2.4 Magnetization

2.4.1 Magnetization and magnetic dipoles

The (uniform and permanent) magnetization vector $\mathbf{M}$ of an (ideal) magnet has, so far, been introduced in an arbitrary manner as that basic characteristic or property of the magnet required to obtain correlation between the fields calculated for magnets and for corresponding circuits or current distributions. No physical picture has been suggested. The association of $\mathbf{M}$ with surface poles, giving the appropriate fields, and the observation that these are not free poles akin to electric charges but must always exist in such a way that the total pole density is zero, immediately suggests that the magnet may be envisaged as an assembly of elementary or atomic magnetic dipoles aligned along the magnet axis. If, in turn, it was necessary to envisage these dipoles as pairs of elementary poles, $\pm m$ with spacing l and $\mu = ml$, (which would in fact give the correct results in appropriate limits), then an impasse would clearly be reached.

However, this is not, in fact, necessary because the equivalence illustrated by Figure 1.18 is readily demonstrated. The fields generated by a current loop are identical to those of the 'magnetostatic dipole' consisting of two formal poles in the limit $l \to 0$, as m increases correspondingly for $\mu = ml$ to remain finite and $\mathcal{A} \to 0$ and i increases similarly so long as

$$ml = \mu = i\mathcal{A} \qquad (\mathrm{A\,m^2}) \tag{44}$$

(It is, however, assumed that the limits preclude detailed comparisons of the fields in the immediate vicinity of the dipoles, in which case the appearance of the fields would be completely different for the two cases and, moreover, one field would be conservative and the other solenoidal.)

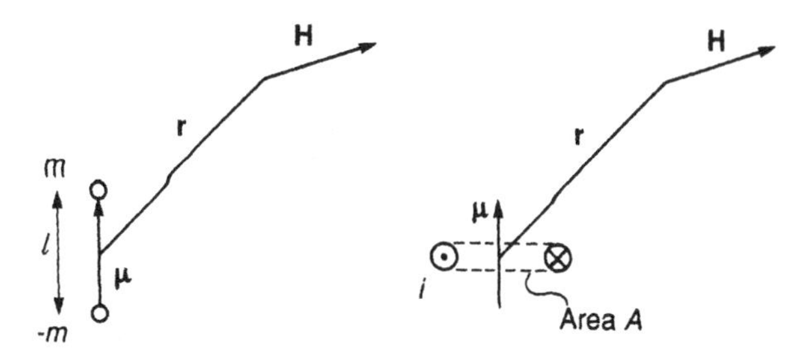

Figure 1.18 The fields $\mathbf{H}(\mathbf{r})$ from a 'magnetostatic' or literal two-pole dipole are identical to those from a true magnetic dipole or current loop in appropriate limits which preclude consideration of points in the immediate vicinity of the (point) dipoles

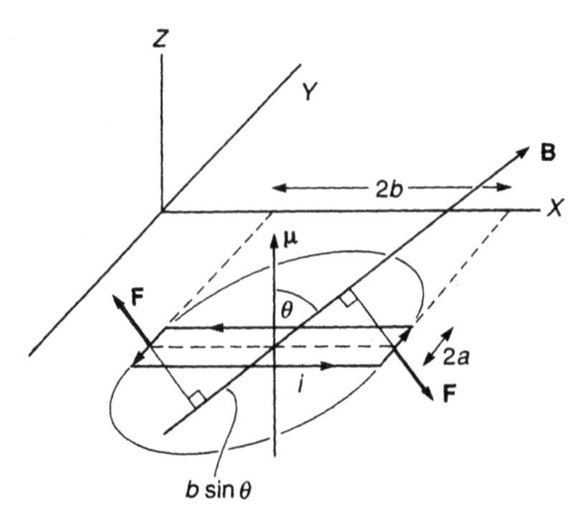

Figure 1.19 The axes are chosen to place **B** ∥ OXZ and the strip is taken ∥ OX. A general plane circuit in OXY it considered as an assembly of such strips with currents cancelling along the sides ∥ OX

For the fields at points restricted to lie on the axis of the loop on the line along the axis of the magnetostatic dipole, the correspondence follows by comparing the treatments of Sections 1 and 2.2.1. The vector $\boldsymbol{\mu}$ must be taken to be normal to the plane containing the circuit. The demonstration for general points is most readily achieved using the vector potential introduced later (see Section 2.9.2).

Taking a rectangular circuit as in Figure 1.19, the forces on the sides normal to **B** are

$$F = 2aiB \tag{45}$$

with the directions indicated, and the torque due to the field is

$$T = 2(2aiB)b\sin\theta = \mu B\sin\theta \qquad \text{where } \mu = 4abi = Ai$$

or, in view of the directions involved (see Figure 1.19),

$$\mathbf{T} = \boldsymbol{\mu}\times\mathbf{B} \tag{46}$$

[cp. equation (1)]. The forces on the other two sides are equal and opposite and coaxial and thus cancel completely. By suitably dividing a plane circuit of any shape into rectangles, which are taken to become increasingly narrow to approximate the true shape, and noting that the currents along lines common to neighbouring rectangles cancel, the reader may confirm the generality of equation (46) as a simple exercise and also, of course, recall the correspondence of the formula with that for a magnetostatic dipole.

In contrast to the case of the field calculations, no reference to the size of the circuit has been made. This is because **B** was implicitly uniform. Equation (46) will apply generally if $\mathcal{A}$ is sufficiently small for **B** to be assumed uniform over the circuit.

In view of these and other coincidences between the so-called magnetostatic dipole and the current loop the latter is said to constitute a true magnetic dipole (a point dipole in the limit). Thus the only significance to be attached to a pole as such is of an historical nature in leading to this particular designation of a current loop.

It is now much easier to picture a permanent magnet in terms of an ideally uniform array of aligned magnetic moments which are associated with the current loops corresponding to the orbiting electrons of the constituent atoms. This is most readily seen in terms of the Bohr or planetary model, but also applies in a more subtle way when general quantum theory is invoked and spin becomes of primary importance. The spontaneous mutual alignment of the atomic moments corresponding to ferromagnetism and the existence of a characteristic spontaneous magnetization of arbitrary orientation can be temporarily ascribed to the existence of a massive 'exchange field' existing in such materials. The alignment of $\mathbf{M}$ itself with respect to the magnet axis, which can be taken to be the common axis of the constituent aligned crystallites, can in turn be ascribed to the existence of 'anisotropy fields' which may be directed along these crystal axes, although both of these are effective and not real fields.

The magnitude M of the magnetization, or intensity of magnetization, is equal to the magnetic dipole moment per unit volume for the case of ideal alignment. If the dipoles $\boldsymbol{\mu}_i$ have variable orientations, each component M_j of the mean magnetization is given by

$$M_j = \sum_{i=1}^{N} \frac{\boldsymbol{\mu}_i \cdot \mathbf{e}_j}{V} \qquad \text{for } j = x, y, z \tag{47}$$

with $\mathbf{e}_j$ the appropriate unit vector and N dipoles in volume V. The units of M are clearly $\mathrm{A\,m^2/m^3} = \mathrm{A\,m^{-1}}$.

Suppose that the magnet is pictured as a general model of aligned current loops or dipole moments, which are not restricted to point dipoles or atomic dipoles, so that the detailed geometry of the loops is variable. They may be seen as elementary cylindrical surface currents (approximating point dipoles) which merge to form long cylinders extending throughout the specimen and then expand laterally and 'fuse' as in Figure 1.20(a) to (c). At each stage it is assumed that the sum over the currents and areas enclosed remains constant; i.e. the total dipole moment or magnetic moment remains constant. It follows that when the internal opposed currents are cancelled in going from (b) to (c) there remains only a surface current of magnitude M as shown, M being the original (point) dipole density or magnetization. In view of the additivity of the torques on, or fields from, assemblies of dipoles, (a) the torque on the equivalent surface current in a field must be the same as that on the original dipole assembly or magnet, and (b) the external fields generated, at a large distance at least, must be the same. To this extent a real surface current, i.e. a coil of fine wire or indeed any axial coil, may be said to have a magnetic moment. However, it could not be said to have an associated magnetization because, although the detailed nature of the

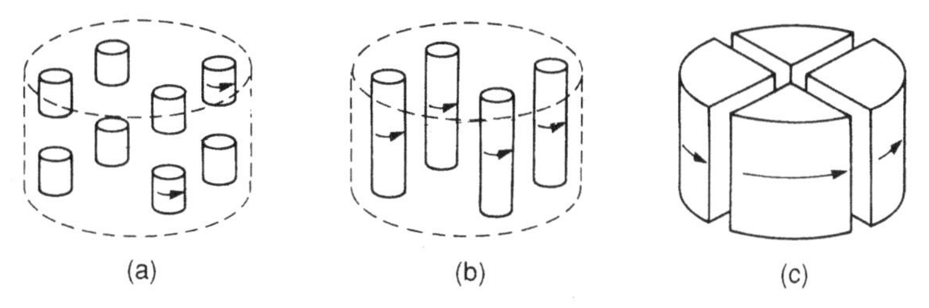

Figure 1.20 An array of dipoles is pictured as one of finite cylindrical current sheets (a) which may fuse longitudinally with implied adjustments giving the same dipole moment per unit volume (b). The cylinders then fuse laterally so that only the external surface currents remain (c)

dipoles giving rise to **M** need not be specified, they must be taken to approximate point dipoles to a reasonable extent in order for **M** to be treated as a continuous vector throughout the material. In considering M as the total dipole moment divided by the element of volume containing these dipoles, the element should be adequate to enclose a statistically significant number of dipoles.

The effective surface current M may be taken to be a 'representation' of the magnetization **M** just as the surface poles of density M give an alternative representation. Both are devices of value in magnetic calculations since any calculations based on Avogadro's number of individual dipoles would scarcely be practicable. Referring back to the discussion of Figure 1.3, it is now apparent why the surface current was assigned the same value as the magnetization.

2.4.2 *Induced magnetization: dipoles in applied fields*

So far only permanent spontaneous magnetization has been considered. The simplest account of induced magnetization is given by taking a material to consist of an array of permanent dipoles, devoid of any strong interactions leading to magnetic order, which may be termed a classical paramagnetic (or a true paramagnetic if the origin of the atomic magnetic dipoles is properly accounted for).

The energy of a dipole in an applied field is discussed in general terms in Chapter 4, but a simplified account can be given here. This calls for an applied field which is very simply described, but it is not possible to devise a solenoidal field having a variation in one direction only, since it is apparent that this would give a net flux (i.e. the integral of $\mathbf{B} \cdot \mathbf{n}$ with $\mathbf{n}$ the outwardly directed unit normal vector) over any closed surface and a corollary of div $\mathbf{B} = 0$ is that this flux is zero (see Section 2.3.1). The simplest model non-uniform field is one that varies with respect to a single radial coordinate r (in a cylindrical system) and has only the one component, $\mathbf{e}_r B$, as shown in Figure 1.21. The origin of such a field could be a plate-shaped magnet which has an indefinitely great length normal to **M** and to the plane of the figure and is also long parallel to **M** so as to produce a single effective

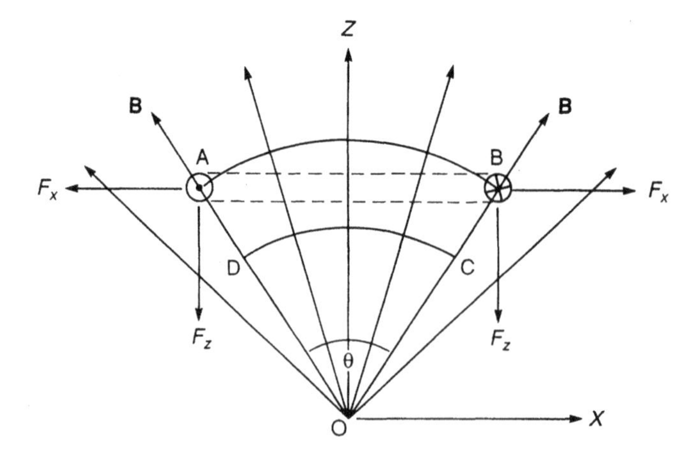

Figure 1.21 A model field $\mathbf{B}(r)$ depends only on a radial coordinate, origin O. The net force on the dipole, or current loop intersecting at A, B, arises from the existence of the x components of **B**

line of pole density, or an equivalent coil. Accepting that such a field can exist it must be given by

$$\mathbf{B} = \mathbf{e}_r \frac{C}{r} \tag{48}$$

with C a constant, which is immaterial for the present. This follows because, taking unit length normal to the diagram, the condition of zero net flux requires that

$$BL = Br(2\theta) = B_1 L_1 = B_1 r_1 (2\theta)$$

with L the length of arc AB and L_1 the length DC. Thus Br is constant and $B \propto 1/r$. Let two filamentary currents i flow normal to the diagram, 'out' at A and 'in' at B, these representing the section of a macroscopic dipole or square current loop $2a \times 2a$. The z components of $\mathbf{B}$ at A and B give equal and opposite forces F_x which cancel. Since, however, the components B_x are themselves opposite, these give (for the opposed currents) forces F_z in the directions indicated, and the total force on the dipole is

$$F_z = -2 \times 2a \times i B_x = -4ai B_x$$

$$= -4ai B \sin\theta = -4ai B \frac{a}{r} = -\frac{\mu B}{r}, \qquad \text{where } \mu = 4a^2 i$$

From equation (48) $\partial B/\partial r = -B/r$, and as a becomes small this can be identified with $\mathrm{d}B/\mathrm{d}z$ so that

$$F_z = \mu \frac{\mathrm{d}B}{\mathrm{d}z} \tag{49}$$

Note that according to the current directions in the diagram $\boldsymbol{\mu}$ is parallel to $\mathbf{B}$ and antiparallel to $\mathrm{d}\mathbf{B}/\mathrm{d}z$. If the currents were reversed, to reverse $\boldsymbol{\mu}$, then the force would be reversed. (It must, of course, be remembered that in taking $i\,\mathbf{dr} \times \mathbf{B}$, $\mathbf{dr}$ is to be taken as an element of length having the direction of the current. Current i may be taken as a positive quantity giving the magnitude of the current, and reversal of the current corresponds to changing the sign of $\mathbf{dr}$.)

This is another result which has been obtained, much more readily, for a magnetostatic dipole (with a similar result for an electric dipole in an electric field gradient). However, the correspondence is limited. Suppose a field everywhere parallel to OZ but yet having a finite $\mathrm{d}B/\mathrm{d}z$ were postulated. This would give a force on the magnetostatic dipole but not on the true (current loop) magnetic dipole. The latter force arises as a consequence of the components of $\mathbf{B}$ normal to $\mathrm{d}B/\mathrm{d}z$, the existence of which is in turn necessary for a true, solenoidal, field.

It is also left to the reader to demonstrate that if the dipole is rotated through 90° the force becomes zero (although a torque appears). It is the component of $\boldsymbol{\mu}$ in the direction of $\mathrm{d}\mathbf{B}/\mathrm{d}z$ that gives the force and more generally

$$F_z = \boldsymbol{\mu} \cdot \frac{\mathrm{d}\mathbf{B}}{\mathrm{d}z} \tag{50}$$

At great distances from the origin of the field, B, $\mathrm{d}B/\mathrm{d}z$ and F_z must be zero. The work done in taking a dipole from a distant point to the point z' at which the field has the value B, with $\boldsymbol{\mu} \parallel \mathbf{B}$, is

$$W = -\int_{\infty}^{z'} \mu \frac{\mathrm{d}B}{\mathrm{d}z}\,\mathrm{d}z = -\mu[B(z') - B(\infty)] = -\mu B$$

If, however, $\boldsymbol{\mu}$ were maintained at an angle to $\mathbf{B}$ and to $\mathrm{d}\mathbf{B}/\mathrm{d}z$, only its one component would be significant, and thus the energy of a dipole in an applied field becomes

$$E = -\boldsymbol{\mu} \cdot \mathbf{B} \tag{51}$$

The same result would be obtained by bringing the dipole into the field with an orientation such that the translational forces were zero and then rotating $\boldsymbol{\mu}$ and calculating the integral of $T\mathrm{d}\theta$.

2.4.3 *Magnetic susceptibility: paramagnetism*

If a field is applied to a randomly oriented assembly of magnetic dipoles which are assumed to have no mutual interactions and do not interact with any other system the dynamic response is complex (see Chapter 3, Section 2) but there is no change in the mean orientation along $\mathbf{B}$, i.e. no induced magnetization. The latter is defined as

$$M = \frac{1}{V}\frac{\sum \boldsymbol{\mu}_i \cdot \mathbf{B}}{B} = N\mu\langle\cos\theta\rangle \tag{52}$$

the latter form applying if all the $\boldsymbol{\mu}_i = \boldsymbol{\mu} \cdot \langle\cos\theta\rangle$ is the mean value with θ the angle between each $\boldsymbol{\mu}$ and $\mathbf{B}$ and N the number of dipoles per unit volume. However, a weakly interacting assembly may be taken to approach thermal equilibrium and equation (51) indicates that dipole orientations parallel to the field will be favoured. Since each value of $\cos\theta_j$ is associated with an energy term $E_j = -\mu B\cos\theta_j$, it follows that according to Boltzmann's statistics the mean of $\cos\theta$ is given by

$$\langle\cos\theta\rangle = \frac{\sum\limits_j \cos\theta_j \mathrm{e}^{-E_j/kT}}{\sum\limits_j \mathrm{e}^{-E_j/kT}} \rightarrow \frac{\int_{-1}^{1} x\mathrm{e}^{ax}}{\int_{-1}^{1} \mathrm{e}^{ax}\,\mathrm{d}x} \tag{53}$$

with the latter for a continuous distribution: $x = \cos\theta$ and $a = \mu B/kT$. The limits correspond to $\theta = \pi$ and 0, for all orientations. The integrals are standard and rearrangement gives

$$\langle\cos\theta\rangle = \coth a - \frac{1}{a} = \coth\frac{\mu B}{kT} - \frac{kT}{\mu B}, \qquad \text{where } \coth a = \frac{\mathrm{e}^a + \mathrm{e}^{-a}}{\mathrm{e}^a - \mathrm{e}^{-a}}$$

Expanding the exponentials, $\coth a \rightarrow 1/a$ as $a \rightarrow 0$ so that M is appropriately zero as $B \rightarrow 0$ or $T \rightarrow \infty$. Also, as $a \rightarrow \infty$, $\mathrm{e}^{-a} \rightarrow 0$ and $\langle\cos\theta\rangle \rightarrow 1$, $M \rightarrow N\mu$, with perfect orientation or saturation, as in the very similar plots in Figure 1.22. This clearly non-quantum or classical theory gives what might be termed classical or Langevin paramagnetism (since $\coth a - 1/a \equiv L(a)$ is the Langevin function). It applies to assemblies of particles with permanent freely rotating moments which are too large for quantum treatments to be significant (superparamagnetism) or equivalently as the limiting case for quantum treatments with very large quantum numbers.

Expanding the exponentials up to a^2,

$$\coth a - \frac{1}{a} = \frac{2 + a^2}{2a} - \frac{1}{a} = \frac{a}{2}$$

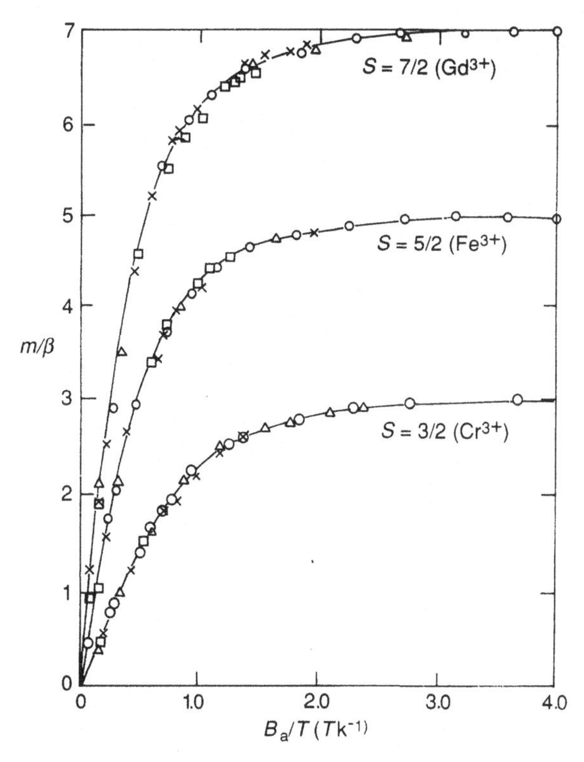

Figure 1.22 Experimental plots of induced magnetization for paramagnetics: I—potassium chromium alum, II—ferric ammonium alum and III—gadolinium sulphate octahydrate, having the same general features as the Langevin curve [Henry, W.E., Revs. Mod. Phys., **25**, 163, 1953]. (Reprinted by permission of John Wiley & Sons, Ltd, Magnetic Materials, R.S. Tebble & D.J. Craik, Wiley 1969)

and the plot of M versus B is linear:

$$M = n\mu\frac{a}{2} = \frac{N\mu^2 B}{2kT} \tag{54}$$

giving a 'zero-field' susceptibility, in this effectively linear region, as

$$\chi = \frac{M}{H} = \mu_0\frac{M}{B} = \frac{\mu_0 N\mu^2}{2kT} \tag{55}$$

To introduce a quantum mechanical treatment somewhat abruptly, the simplest case is that for which the moments associated with orbital motion of the electrons can be neglected (s-state atoms) and only the intrinsic spin magnetic moments, as discussed in Section 3.4, are taken into account. Effectively single-electron atoms such as sodium or, of course, hydrogen atoms are considered, so that the electrons are identifiable and not to be treated as an electron gas; i.e. Boltzmann statistics apply (contrast Section 5). Again assuming only weak interactions with other electrons or with other degrees of freedom of the system, attention is focused on one characteristic electron with the others and other degrees of freedom considered as a heat reservoir. The simplifying feature is the existence of only two permitted states, symbolically $|1\rangle$ and $|2\rangle$ or $|\beta\rangle$ and $|\alpha\rangle$, with each of which is associated an energy and magnetic moment component μ^z, in the field $\mathbf{B} = \mathbf{k}B$:

$$|\beta\rangle \equiv |1\rangle : \mu_1^z = \tfrac{1}{2}g\beta, \qquad E_1 = -\mu_1^z B = -\tfrac{1}{2}g\beta B$$

$$|\alpha\rangle \equiv |2\rangle : \mu_2^z = -\tfrac{1}{2}g\beta, \qquad E_2 = -\mu_2^z B = \tfrac{1}{2}g\beta B$$

(see Figure 1.23) with g a number $\doteqdot 2$ and β the quantum unit of magnetic moment (see Section 3). The probability that the electron should be in the state $|i\rangle$ with energy E_i is

$$P_i = \frac{e^{-E_i/kT}}{\sum e^{-E_i/kT}} = \frac{e^{-E_i/kT}}{Z} \tag{56}$$

with Z the partition functions. The mean magnetic moment component in the field direction is thus

$$\langle \mu_2 \rangle = P_1\mu_1^z + P_2\mu_2^z = \tfrac{1}{2}g\beta \frac{e^x - e^{-x}}{e^x + e^{-x}} = \tfrac{1}{2}g\beta \tanh x \tag{57}$$

with $x = \tfrac{1}{2}g\beta B/kT$. Thus the induced magnetization is

$$M = \tfrac{1}{2}Ng\beta \tanh\left(\frac{\tfrac{1}{2}g\beta B}{kT}\right) \tag{58}$$

For high values of the argument $\tanh x \to 1$, giving saturation in high fields or low temperatures. For low values of x, $\tanh x \to x$, a linear region occurs and the 'zero-field' values are

$$M = \frac{Ng^2\beta^2 B}{4kT}, \qquad \chi = \frac{\mu_0 N g^2 \beta^2}{4kT} = \frac{C}{T} \tag{59}$$

The zero-field susceptibility is defined as $(\partial M/\partial H)_{H \to 0}$, but can be taken approximately to be $\chi = M/H$ in the near-linear region. $\chi = C/T$ is known as the Curie law and C is the Curie constant.

Since the units of μ are $\mathrm{A\,m^2}$ (current times area) and $M \sim \mu/\text{volume} \sim \mathrm{A\,m^{-1}}$ and the division of M by (B/μ_0) is equivalent to dividing by $H(\sim \mathrm{A\,m^{-1}}$ also), then it is clear that χ is dimensionless, $\chi \sim 0$. The field $\mathbf{B}$ in this treatment is to be considered as an applied field related to $\mathbf{H}$ by $\mathbf{B} = \mu_0\mathbf{H}$ and is not itself influenced by the presence of the magnetizable material (see Section 2.6). The whole treatment could have been carried out by using an applied field $\mathbf{H}$ and expressing the energy as $-\mu_0\boldsymbol{\mu} \cdot \mathbf{H}$.

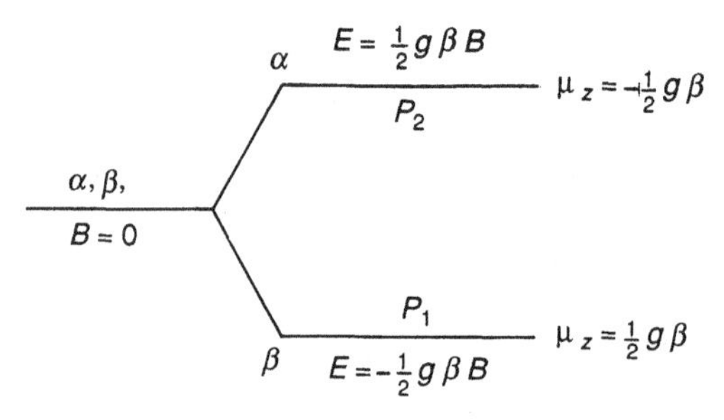

Figure 1.23 The energy level splitting for a single free electron spin in a field B; for a proton the level conventionally associated with α would be the lower

With

$$\frac{\frac{1}{2}g\beta}{k} \doteqdot \frac{\beta}{k} = \frac{9.274 \times 10^{-24}}{1.381 \times 10^{-23}} = 0.672$$

at $B/T = 0.1$, $x = 0.0672$ and $\tanh x = 0.0671 \doteqdot x$, while at $B/T = 4$, $\tanh x = 0.991 \doteqdot 1$. Thus the 'linear' region is reasonably extensive and conversely saturation is readily approached using a superconducting solenoid, giving $B \sim 10$ T with $T \sim 1.0$ K (see Figure 1.22).

This simple treatment is based on the acceptance of certain quantum mechanical results and the application of classical statistics. A more general account will be given in Chapter 3 and shown to justify the results obtained.

Choosing a very arbitrary 1 mole in 10^{-3} m^3 gives $\chi \approx 1.5 \times 10^{-5}$. For ferromagnetics, $\chi \to 0$ for good permanent magnets in which there is minimal response to small applied fields, but χ may range up to, say, 10^5 in 'soft' materials. Here the magnetization is induced at a macroscopic level, e.g. by rotation of the pre-existing and relatively huge magnetic moments of otherwise randomly magnetized particles, and it will be seen that this is very much easier than the induction at a microscopic level. Although χ is dimensionless, values in different systems may differ due to rationalization (Appendix 1).

2.5 *M, B, H* and H_M

The field H derived from current distributions is solenoidal, including H from dipoles or current loops. If a field, an 'applied field', is generated by a solenoid, say, which contains dipoles, which may respond in some way to this field, then clearly the total field remains solenoidal. The reason for introducing a conservative field or pole field is not immediately apparent.

The magnetization may be described as the resolved magnetic dipole moment per unit volume. The component of the magnetic moment along any axis is the product of the current and the area of the loop projected on a plane normal to that axis without any restriction on the size or shape (other than that it should be a single simple loop). However, the fields only have a simple form at distances $\gg$ a loop dimension and it is implied that a magnetic dipole is a loop of small radius, in relation to lengths of interest, which in the limit approaches a point. Without this implied restriction a macroscopic solenoid may be said to have a magnetic moment $\boldsymbol{\mu} = \mathbf{k}Il\mathcal{A}(I \perp OZ$, section area $\mathcal{A}$, length l) and the magnetic moment per unit volume would be $\mathbf{k}\mu/(l\mathcal{A}) = \mathbf{k}I$. This (magnitude) would be the turning moment per unit volume per unit normal applied field, but it would be unreasonable to associate this with a vector magnetization existing throughout the solenoid. Physically the elementary dipoles giving rise to $\mathbf{M}$ are associated with atoms or molecules distributed more or less uniformly throughout the material or space within which $\mathbf{M}$ exists and the corresponding loops must be taken to approach points, or at least be reasonably small, for $\mathbf{M}$ to be considered as a quasi-continuous vector field. Even then a 'point' in this space must be regarded as an infinitesimal region which is still large enough to contain a statistically significant number of dipoles.

An obvious difficulty arises if the elementary dipoles are considered as point dipoles because the fields within the loops cease to be finite. It is thus tempting to calculate the fields within and around an array of dipoles forming a simple cubic lattice, say, by averaging over small regions of space while necessarily avoiding the points occupied by the dipoles.

This is readily achieved, particularly if the dipoles are identical and oriented parallel to a common axis defining the direction of **M**, giving a field $\bar{H}_{\mathrm{d}}$, say.

The result is as illustrated by Figure 1.24. $\bar{H}_{\mathrm{d}}$ is just the field $\mathbf{H}_{\mathrm{M}}$ as calculated by taking $\mathbf{M} = N\boldsymbol{\mu}$ (N per unit volume), $\sigma = \mathbf{M} \cdot \mathbf{n}$ and $\mathrm{d}\mathbf{H} = (1/4\pi)(\mathrm{d}\sigma)\mathbf{r}/r^3$ and integrating; i.e. it is the pole field arising from the discontinuity in **M**. Thus $\bar{\mathbf{H}}_{\mathrm{d}}$ is conservative although it has been obtained by summing and averaging solenoidal fields.

The apparent contradiction arises because it was clearly not possible to include in the calculations the infinite contributions at the occupied points. Only a part of the total field was included and the nature of the result could not be presumed.

To circumvent such difficulties, suppose the dipoles are represented as small solenoids, say with length l equal to diameter $2r$ as in Figure 1.25, with N in unit volume, surface current densities I, 'lattice' spacing a. It is assumed that $r < a$ substantially and that $a \ll$ the dimensions of the array so that **M**, with $M = NI(\pi r^2)l$, approximates a vector field. Thus r is taken to be small but $r \rightarrow 0$ is relaxed.

It has been seen that the (axially directed) fields within a long solenoid approach uniformity with a magnitude equal to I and that the external fields become zero, but in line with the end of a semi-infinite solenoid on the axis, the field is reduced to $I/2$. This is what would be obtained by taking the sum of two contributions:

$$\mathbf{H} = \mathbf{H}_{\mathrm{s}} + \mathbf{H}_{\mathrm{c}} \tag{60}$$

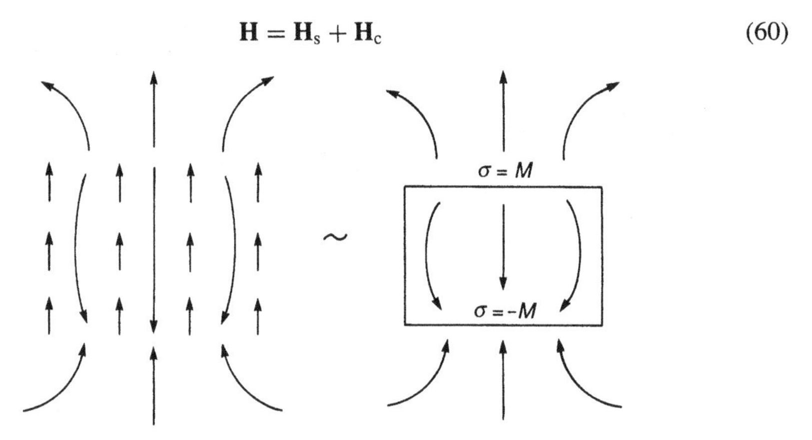

Figure 1.24 The averaged fields calculated from a regular array of dipoles approaches that calculated for a pair of sheets of pole density $\pm M$

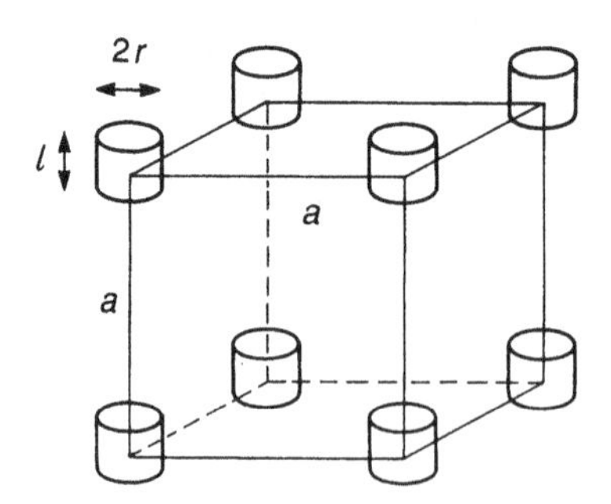

Figure 1.25 A regular cubic array of dipoles represented as small but finite cylindrical currents

where $\mathbf{H}_s$ is a field that is equal in magnitude to I within the solenoid and zero outside, as if the solenoid was still infinite, and $\mathbf{H}_c$ is a 'correction field' accounting for the discontinuity and is in fact the field that would be obtained by superimposing a surface pole density $\sigma = I$ on the end of the solenoid. This is seen, by comparing equations (6) and (17) to apply not only to points marginally within and without the solenoid but to all points on the axis. For a short solenoid equation (60) applies with $\mathbf{H}_c$ being that field which is obtained by superimposing the two imaginary discs of pole density as in Figure 1.26.

Since the fields are specified at all points along the axis of this cylindrical system and the fields at infinity must be zero, it may be stated that a solution of Laplace's equation with complete boundary conditions exists over the whole of space. By the uniqueness theorem, discussed in general works on Laplace's equation (but inevitably applying since otherwise the equation could never sensibly be solved), this is the only possible solution. Thus the division of the field from the solenoid into the two contributions in this way must apply to general points and not just to the axis. With recourse to explicit solutions for general points, as in Chapter 4, this can, of course, be confirmed.

To reiterate, the discs clearly have no physical reality and it is only stated that the fields from a short solenoid can equally be calculated directly from the currents or as if they were made up of these two contributions.

The advantage of this decomposition is that the fields within the solenoids or dipoles can now be taken into account. As the cylinders decrease in size the field $\mathbf{H}_c$ from the mythical discs becomes that from two poles, i.e. the dipole field (recalling that the field from the two-pole or magnetostatic dipole becomes identical to the true dipole field). Thus the calculations giving $\bar{\mathbf{H}}_d$ or $\mathbf{H}_M$ were in effect accounting for the $\mathbf{H}_c$ contributions. To obtain the true mean total field the contributions $\mathbf{H}_s$ must also be included and these give

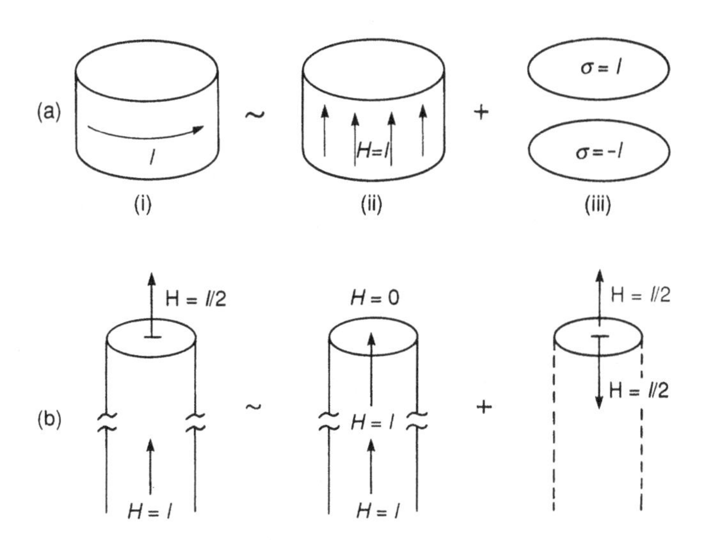

Figure 1.26 (a) The fields in and around a short solenoid (i) are those that would be given by assuming it to be a portion of an infinite solenoid (ii) with the addition of the fields from a pair of discs of pole density $\pm I$. A partial verification of this is made by noting that $H = I/2$ at the end of a long solenoid and $H = \pm I/2$ at points approaching the disc surface (b)

the average

$$NH_s\mathcal{A}l = NI\mathcal{A}l = N\mu = M$$

i.e. the dipole moment per unit volume, which is the magnetization. No matter how closely the solenoids approach point dipoles these same principles apply. Replacing $\bar{\mathbf{H}}_d$ by $\mathbf{H}_M$, the field derived from $\mathbf{M}\cdot\mathbf{n}$ at the surfaces, and denoting the total field $\mathbf{H}_T$,

$$\mathbf{H}_T = \mathbf{H}_M + \mathbf{M} \tag{61}$$

$\mathbf{H}_T$ must be solenoidal because it is the true averaged total dipole field, while $\mathbf{H}_M$ is conservative, which is permissible because it was derived by averaging an artificially selected part of the dipole field only. To emphasize the primary importance of the true total field $\mathbf{H}_T$ as compared with $\mathbf{H}_M$, which is contrived by this particular decomposition, and introducing a conversion of the units, $\mu_0\mathbf{H}_T$ is denoted the $\mathbf{B}$ field or induction field $\mathbf{B}$, i.e.

$$\mathbf{B} = \mu_0(\mathbf{H}_M + \mathbf{M}) \tag{62}$$

$\nabla\cdot\mathbf{B} = 0$ is consistent with $\nabla\cdot\mathbf{H}_M = \rho$ and $\nabla\cdot\mathbf{M} = -\rho$ at any point. An effective pole density is associated with a non-uniform $\mathbf{M}$ only to the extent that it is considered as a vector field. Statements, which may be encountered frequently, that $\nabla\cdot\mathbf{M} = 0$ necessarily because only dipoles exist are based on the implicit consideration of the dipoles as magnetostatic or 'two-pole' dipoles.

The model used has been implicitly that of a permanent magnet with a specified magnetization. The dipole array may be considered to be surrounded by a (real) solenoid, giving a (solenoidal) 'applied field' $\mathbf{H}_a$. This would simply add to $\mathbf{H}_T$ to give generally

$$\mathbf{B} = \mu_0(\mathbf{H}_a + \mathbf{H}_M + \mathbf{M}) = \mu_0(\mathbf{H} + \mathbf{M}), \qquad \text{where } \mathbf{H} = \mathbf{H}_a + \mathbf{H}_M \tag{63}$$

Note that this introduces an important change of convention, since in previous sections $\mathbf{H}$ alone was used for the true solenoidal field from current distributions, in distinction to the conservative $\mathbf{H}_M$. While $\mathbf{B}$ is necessarily solenoidal, $\mathbf{H}$ may or may not be so and in general

$$\nabla\cdot\mathbf{H} = \nabla\cdot\mathbf{H}_a + \nabla\cdot\mathbf{H}_M = \nabla\cdot\mathbf{H}_M \neq 0 \tag{64}$$

The elementary solenoids may also be taken to represent the dipoles, or dipole components, induced (diamagnetically) or aligned (paramagnetically) by the applied field according to $\mathbf{M} = \chi\mathbf{H} = \chi(\mathbf{H}_a + \mathbf{H}_M)$. In this case,

$$\mathbf{B} = \mu_0(\mathbf{H} + \mathbf{M}) = \mu_0(\mathbf{H} + \chi\mathbf{H}) = \mu_0\mu_r\mathbf{H} \tag{65}$$

defining the relative permeability

$$\mu_r = 1 + \chi \tag{66}$$

[compare $\mu_r = 1 + 4\pi\chi$, with $B = \mu_0(H + 4\pi M)$, without rationalization].

In space, $M = 0$, $\chi = 0$ and $\mu_r = 1$ so that $\mu = \mu_0$ is called the permeability of free space. It is *assigned* the value

$$\mu_0 = B/H \text{ (space)} = 4\pi\times 10^{-7}\ \mathrm{J\,A^{-2}\,m^{-1}(V\,A^{-1}\,s\,m^{-1})}$$

with units consistent with $B \sim \mathrm{J\,A^{-1}\,m^{-2}}$ = tesla, $H \sim \mathrm{A\,m^{-1}}$, themselves consistent with the definitions in equations (10), and (11).

To review, if a magnetic material could be modelled by a rather small number of relatively large current loops, placed in a larger solenoid for example, the fields throughout the whole system could be calculated, as could the response of the 'material', to some (non-statistical) extent. No conservative fields would arise or be required. However, the material modelled could not then be regarded as having a magnetization in the accepted sense. It is the decision to regard **M** as a vector field or near-continuum of quasi-point dipoles that precludes such a picture and leads to the introduction of conservative fields and thus of pole distributions $\sigma = \mathbf{M} \cdot \mathbf{n}$ and $\rho = -\nabla \cdot \mathbf{M}$. This in turn leads to the distinction between a total field **B** which must by definition always be solenoidal and a second field **H**, derived in part from σ and ρ and not generally solenoidal.

In particular, the treatment outlined gives a derivation of the vital relation (63), $\mathbf{B} = \mu_0(\mathbf{H} + \mathbf{M})$, obviously only meaningful with **M** a vector field, and this need not be introduced as a postulate or contrivance. The relation (65) $\mathbf{B} = \mu_0\mu_r\mathbf{H} = \mu\mathbf{H}$ indicates that **B** and **H** do have the same character if μ_r is constant in all space, but this is not the general situation. In the presence of a discrete magnetized specimen **M** is discontinuous and equation (65) indicates that, since **B** is continuous, **H** must then be discontinuous, or conservative. Conversely, the 'balance' of **H** and **M** ensures the continuity of **B**, as illustrated in Chapter 4.

Many phenomena can only be accounted for by recognizing that **B** must be treated as the total field, including **M**. If the magnetization is parallel to the surface of a thin sheet the field $\mathbf{H}_\mathrm{M} \to 0$ (considering two lines of negligible pole density). If electrons pass through a foil of cobalt, for example, their trajectories would not be affected if the Lorentz forces were dependent on **H**. In fact, in a situation where **M** periodically reverses in direction to form so-called magnetic domains, the electrons are deflected as indicated by Figure 1.27. In a suitably adjusted electron microscope alternate domain boundaries appear bright and dark and this gives rise to a method for the study of such structures. Clearly the Lorentz force must be taken to involve **B** rather than **H**.

It is scarcely necessary to point out that if the elementary moments were represented by magnetostatic 'two-pole' dipoles the fields in and around the arrays would again be the fields $\mathbf{H}_\mathrm{M}$, but the further contributions corresponding to **M** would be missing. It is such a viewpoint that leads to the *ad hoc* introduction of equation (65).

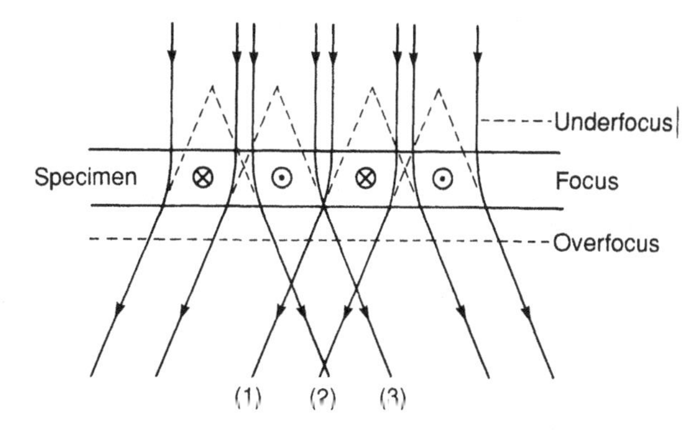

Figure 1.27 (a) Trajectories of electrons passing through a foil containing 180° domains, in line with the domain wall contrasts obtained when an electron microscope is defocused

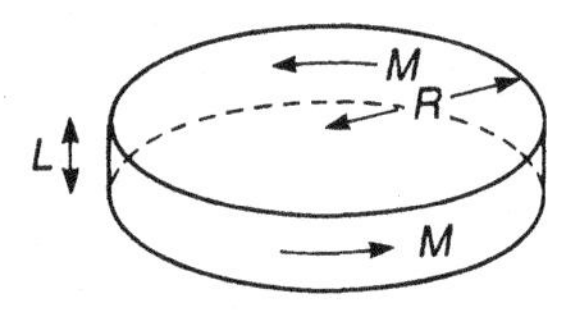

Figure 1.28 As $L \to 0$ and $R \to \infty$ the fields at the centre $\to 0$

As a final review, the geometry of the solenoids may be manipulated with corresponding adjustments of I. They may be extended to become of length a and thus fuse to form long narrow solenoids extending throughout the array, or magnet as in Figure 1.20(b). $\mathbf{H}_c$ for each is then derived from a disc of area $\mathcal{A}$ with $\sigma = \pm I$ and the average $\bar{\sigma} = \pm \mathcal{A}I/a^2$. Now that the length of each is a, $M = (1/a^3)\mathcal{A}Ia$ and so $\bar{\sigma} = \pm \mathbf{M} \cdot \mathbf{n}$. This is the requirement for $\bar{\mathbf{H}}_d$ to have the same form as $\mathbf{H}_M$, as assumed above.

If the elementary solenoids were now expanded laterally until they fused, all internal currents would cancel, leaving the (modified) surface current shown in Figure 1.20(c), the shape of the array being immaterial. Thus the induction in such a magnet is that calculated from a surface current $I = M$. If (see Figure 1.28) $L/R \to 0$ then $\mathbf{B} \to 0$ necessarily at the centre. $\mathbf{B}$ must be the same as that in a magnet having $M = I$. With $\mathbf{B} = \mu_0(\mathbf{H} + \mathbf{M})$, this indicates consistently that $\mathbf{H} = -\mathbf{M}$ within a transversely magnetized thin sheet.

2.6 Applied Fields

The basic simplicity of the discrimination between $\mathbf{B}$ and $\mathbf{H}$ indicated by equations (9) etc. has been lost. These still apply when only currents as such are involved, but when $\mathbf{M}$ is introduced and $\mathbf{H}$ is taken to include $\mathbf{H}_M$ then the field itself depends on μ, or at least on the way in which μ varies throughout space. $\mathbf{H}$ no longer has the attributes of an applied field and it is necessary to *define* an applied field as one from 'external' sources that is itself unaffected by the presence of magnetic material or the nature of the medium, as, for example, the field produced by an 'empty' solenoid specifically before the introduction of a magnetic specimen.

In this particular case there is no point in discriminating between $\mathbf{H}$ and $\mathbf{B}$ and the specification of an applied B field is as acceptable as specifying an applied (H) field as such. Designating the applied field as B_0, the relative permeability becomes $\mu_r = B/B_0$ and the susceptibility would be $\chi = \mu_0 M/B_0$.

Thus if it were not found necessary to introduce specifically conservative fields then it would not be necessary to refer to any field other than $\mathbf{B}$. It is the possibly conservative nature of H fields that makes them distinctive.

In the calculation of the paramagnetic susceptibility (Section 2.4.2) an applied field B was introduced, rather than H, so as to utilize the energy expression $E = -\boldsymbol{\mu} \cdot \mathbf{B}$ directly. To be precise, if the applied field is designated $\mathbf{B}_0$ then the energy involves the sum $\mathbf{B}_0 + \mathbf{B}_M$ with $\mathbf{B}_M = \mu_0 \mathbf{H}_M$ and $\mathbf{H}_M$ usually considered as the demagnetizing field: for a specimen that is very long in the direction of $\mathbf{M}$, $\mathbf{B}_M$ can be neglected. $\mathbf{B}_0$ is certainly not to be confused with $\mathbf{B} = \mu_0(\mathbf{H} + \mathbf{M})$. Suppose that originally $\mathbf{M} = 0$ and introduce the applied field $\mathbf{B}_0$. The calculation based essentially on the energy expression $E = -\boldsymbol{\mu} \cdot \mathbf{B}_0$ gives the equilibrium $\mathbf{M}$, $\mathbf{M}_0$ say, and the susceptibility χ. Now $\mathbf{B} = \mathbf{B}_0 + \mu_0 \mathbf{M}, > \mathbf{B}_0$, and if the energy of the dipoles was taken to be $E = -\boldsymbol{\mu} \cdot \mathbf{B}$ a second calculation would give an equilibrium $\mathbf{M} > \mathbf{M}_0$ and an increased $\chi = \mu_0 \mathbf{M}/\mathbf{B}_0$, and so forth. This would obviously be unacceptable. The field acting on each dipole must be taken to be the applied field B_0, as modified where appropriate by shape or demagnetizing effects but not by the presence of the magnetization.

2.7 Newton's Law, Scaling and Units

The reader may verify by expansion that $\mathbf{a}\times\mathbf{b}\times\mathbf{c} = \mathbf{b}(\mathbf{a}\cdot\mathbf{c}) - \mathbf{c}(\mathbf{a}\cdot\mathbf{b})$ so that equation (10) becomes

$$\mathrm{d}\mathbf{F}(\mathbf{r}_1) = \frac{\mu_0 i_1 i_2}{4\pi r^3}\left[\mathrm{d}\mathbf{r}_2(\mathrm{d}\mathbf{r}_1\cdot\mathbf{r}) - \mathbf{r}(\mathrm{d}\mathbf{r}_1\cdot\mathrm{d}\mathbf{r}_2)\right] \tag{67}$$

If $\mathrm{d}\mathbf{r}_1\cdot\mathbf{r} = 0$ and $\mathrm{d}\mathbf{r}_2\cdot\mathbf{r} = 0$ the interchange of $\mathbf{r}_1$ and $\mathbf{r}_2$ simply reverses $\mathbf{r}$ in the remaining term and $\mathrm{d}\mathbf{F}(\mathbf{r}_1) = -\mathrm{d}\mathbf{F}(\mathbf{r}_2)$: Newton's law applies in this and other particular cases but not generally. However, the law does apply, in a more meaningful context, to the forces obtained by integration around circuits. For charges in motion the apparent anomaly can be resolved when mutual accelerations and the consequent radiation of electromagnetic energy are taken into account.

The forces between two magnets may be calculated by integrating appropriate products $\mu_0\mathbf{H}(\mathbf{M}\cdot\mathrm{d}\mathbf{s})$ or $\mathbf{B}\times(\mathbf{M}\times\mathbf{n})\mathrm{d}s$ over the surfaces and, since each is equally the source of the fields, Newton's law is obeyed. A two-dimensional example is afforded by taking $\mathbf{H}$ from a thin-sheet magnet (Figure 1.29) as

$$\mathbf{H} = \mathbf{e}_r\frac{1}{4\pi}(2tM)\frac{1}{r} \tag{68}$$

The force on the upper magnet is easily calculated in terms of α_1 and α_2, as shown, so the field from this magnet, on the axis, may be inferred by Newton's law as an exercise.

Setting $\mathrm{d}\mathbf{r}_1 \to a\,\mathrm{d}\mathbf{r}_1$, $\mathrm{d}\mathbf{r}_2 \to a\,\mathrm{d}\mathbf{r}_2$, $\mathbf{r} \to a\mathbf{r}$, the force between two circuits is seen to be unchanged if all linear dimensions change by a constant factor (currents constant). The field at $k\mathbf{r}$ w.r.t. the centre of a magnet with characteristic dimension $k\mathbf{d}$ is that at $\mathbf{r}$ for dimension $\mathbf{d}$, as seen for a dipole approximation, since $H \propto r^{-3}$ and $\mu \propto r^3$. Thus the field from a thin-walled coil scales in this way if the currents are adjusted to maintain a constant surface current density.

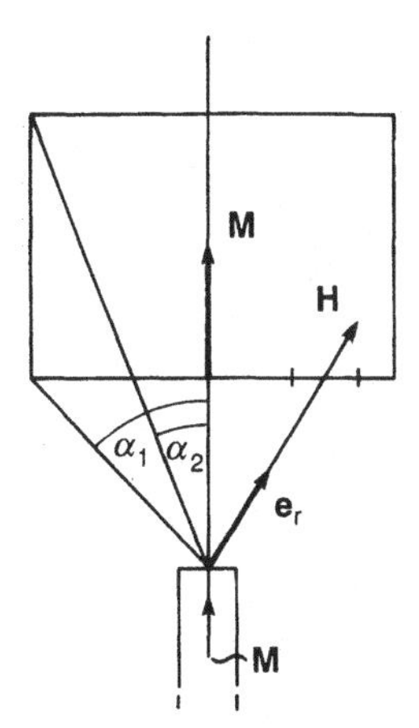

Figure 1.29 Newton's law always applies to the forces between complete circuits and thus to those between magnets. This affords a method for calculating the fields from the upper magnet, along its axis. Both are of infinite extent normal to the diagram

Since the lengths cancel, if μ_0 were omitted from equation (67) the dimensions of the r.h.s. would be current^2 and it is thus required that

$$\mu_0 \sim \frac{\text{energy}}{\text{current}^2 \times \text{length}} \sim \frac{\text{J}}{\text{A}^2\,\text{m}} \sim \frac{\text{henry}}{\text{m}}$$

and then

$$B \sim \mu_0 \times \frac{\text{current}}{\text{length}} \sim \frac{\text{J}}{\text{A}\,\text{m}^2} \sim \text{T (tesla)}$$

It will be seen (Chapter 4, Section 1.12) that the velocity of light *in vacuo* is given by $c_0 = (\varepsilon_0\mu_0)^{-1/2}$ and since ε_0 and μ_0 are non-physical it is arbitrarily chosen that $\mu_0 = 4\pi \times 10^{-7}$ with ε_0 consistent with this and the measured c_0.

2.8 Force and Susceptibility Measurements

If a specimen of spherical shape, say, and isotropic susceptibility χ_1 is placed in a field H then the induced magnetization is $M_1 = \chi_1 H_i$, where H_i is the field within the sphere. Since this field is given by $H_i = H - NM_1 = H - \frac{1}{3}M_1$ then, eliminating H_i,

$$M_1 = \frac{\chi_1 H}{1 + \frac{1}{3}\chi_1} \tag{69}$$

(In ellipsoids of revolution, $\mathbf{M}$, induced by a field $\parallel$ a principal axis, is uniform and the internal fields $\mathbf{H}_M$ are then uniform and given by $\mathbf{H}_M = -N\mathbf{M}$ with N a demagnetizing or shape factor. $N = \frac{1}{2}$ for a sphere; see Chapter 4, Section 1.1.)

Strictly for para- and diamagnetics, it may be assumed that $\frac{1}{3}\chi_1$ (or $N\chi_1$ generally, $N \leqslant 1$) can be neglected in relation to unity and $M = \chi H$ assumed. If the field has a gradient $\mathrm{d}H/\mathrm{d}z$ the force on the specimen will be

$$F = \mu_0 m \frac{\mathrm{d}H}{\mathrm{d}z} = \mu_0 v \chi_1 H \frac{\mathrm{d}H}{\mathrm{d}z} \tag{70}$$

with m the total moment and v the specimen volume. Measurement of F, as compared with that for a standard material of known χ, gives χ_1. In writing $M = \chi_1 H$ it is assumed that H may be taken as constant over the specimen volume.

Suppose now that the specimen is surrounded by a fluid of susceptibility χ_2 in which there is an induced magnetization $M_2 = \chi_2 H$ (Figure 1.30). The surface pole density becomes

$$\sigma = M_1 \cos\theta - \chi_2 H \cos\theta = H(\chi_1 - \chi_2)\cos\theta \tag{71}$$

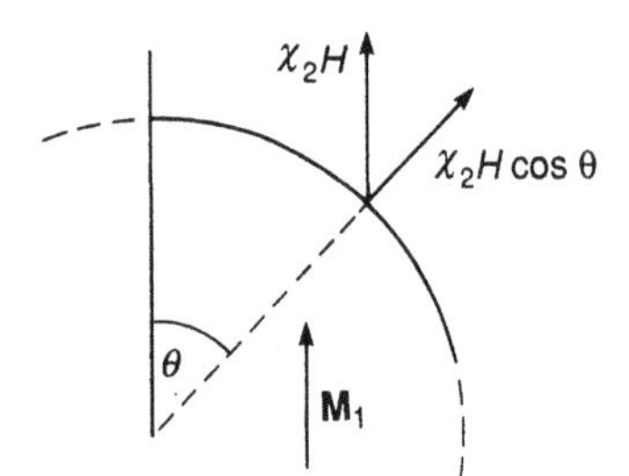

Figure 1.30 Illustrating the effect of the surrounding medium on the force on a spherical specimen in a field gradient

Since the force on a uniformly magnetized sphere ($M = \chi_1 H$) may be obtained by the integral of the product of σ and the local value of H over the surface, with σ for the sphere in space being given by $M_1 \cos\theta$, it follows that when σ is given by equation (71) the force becomes

$$F = \mu_0 v(\chi_1 - \chi_2) H \frac{\mathrm{d}H}{\mathrm{d}z} \tag{72}$$

The same conclusion can be reached by dividing the force into a contribution due to the specimen alone and one due to a spherical 'hole' in the fluid.

The approximation that $M_1 = \chi_1 H$ with H the *applied* field is equivalent to the neglect of the demagnetizing field in relation to the large applied field. Similarly, in the second stage of the development it is assumed that the stray fields from the specimen are negligible so that the magnetization of the surrounding medium is uniform and equal to $\chi_2\mathbf{H}$.

In practice it may be assumed that the magnetization in both components, of a colloid or emulsion say, will be approximately uniform and correspondingly the demagnetizing fields in the dispersed particles or droplets will be small. However, they are not zero and may be large enough to give measurable shifts in the nuclear magnetic resonance absorption frequencies (see Section 17.2) which, for any particular dispersed phase, will become dependent on the susceptibility of the continuous phase. Thus n.m.r. studies may serve as measurements of susceptibility, e.g. of χ_2 above if χ_1 is known or of differences in the χ_2 values, for different media.

2.9 Magnetic Vector Potential

If a potential, from which the induction or B field can be derived by some vector operation, is to be associated with a current element or moving charge then this must carry the necessary information on the direction of the current element or of the velocity of the charge. This potential must thus itself be a vector, the magnetic vector potential 'or just 'the vector potential' **A**. When reference is made simply to 'the magnetic potential' this is to be read as **A** rather than the scalar potential φ, reflecting the primary association of magnetic effects with currents rather than with poles.

Suppose we set

$$\mathbf{A}(\mathbf{r}_1) = \frac{\mu_0}{4\pi}\frac{q_2\mathbf{v}_2}{r}, \qquad \mathbf{r} = \mathbf{r}_1 - \mathbf{r}_2 \tag{73}$$

for the charge, and for a current element $i_2\,\mathrm{d}\mathbf{r}_2$,

$$\mathrm{d}\mathbf{A}(\mathbf{r}_1) = \frac{\mu_0}{4\pi} i_2 \frac{\mathrm{d}\mathbf{r}_2}{r}, \qquad \mathbf{A}(\mathbf{r}_1) = \frac{\mu_0 i_2}{4\pi}\int_{C_2}\frac{\mathrm{d}\mathbf{r}_2}{r} \tag{74}$$

This will then correspond to $\mathbf{B}(\mathbf{r}_1)$ being obtained according to

$$\mathbf{B} = \nabla \times \mathbf{A} \tag{75}$$

This equation is usually taken as the definition of **A** since **B** has been independently defined, and equation (74) may be considered to describe the calculation of **A**. The consistency of equations (75) and (11) is demonstrated by taking $\mathrm{d}\mathbf{r}_2$ to lie at the origin so that $\mathbf{r}_1 \equiv \mathbf{r} =$

$\mathbf{i}x + \mathbf{j}y + \mathbf{k}z$ and also $\mathrm{d}\mathbf{r}_2 = \mathbf{i}\,\mathrm{d}x_2 + \mathbf{j}\,\mathrm{d}y_2 + \mathbf{k}\,\mathrm{d}z_2$. Then

$$\nabla\times\frac{\mathrm{d}\mathbf{r}_2}{r} = \begin{vmatrix} \mathbf{i} & \mathbf{j} & \mathbf{k} \\ \dfrac{\partial}{\partial x} & \dfrac{\partial}{\partial y} & \dfrac{\partial}{\partial z} \\ \dfrac{\mathrm{d}x_2}{r} & \dfrac{\mathrm{d}y_2}{r} & \dfrac{\mathrm{d}z_2}{r} \end{vmatrix}$$

$$= \mathbf{i}\left[\mathrm{d}z_2\left(\frac{-y}{r^3}\right) - \mathrm{d}y_2\left(\frac{-z}{r^3}\right)\right] + \mathbf{j}\left[\mathrm{d}x_2\left(\frac{-z}{r^3}\right) - \mathrm{d}z_2\left(\frac{-x}{r^3}\right)\right]$$

$$+ \mathbf{k}\left[\mathrm{d}y_2\left(\frac{-x}{r^3}\right) - \mathrm{d}x_2\left(\frac{-y}{r^3}\right)\right]$$

using

$$\frac{\partial}{\partial x}\frac{1}{r} = \left(\frac{\partial}{\partial r}\frac{1}{r}\right)\frac{\partial r}{\partial x} = -\frac{1}{r^2}\frac{\partial}{\partial x}(x^2 + y^2 + z^2)^{1/2} = -\frac{x}{r^3}, \text{ etc.}$$

Also,

$$\frac{\mathrm{d}\mathbf{r}_2\times\mathbf{r}}{r^3} = \frac{1}{r^3}\begin{vmatrix} \mathbf{i} & \mathbf{j} & \mathbf{k} \\ \mathrm{d}x_2 & \mathrm{d}y_2 & \mathrm{d}z_2 \\ x & y & z \end{vmatrix} = \frac{1}{r^3}[\mathbf{i}(z\,\mathrm{d}y_2 - y\,\mathrm{d}z_2) + \mathbf{j}(x\,\mathrm{d}z_2 - z\,\mathrm{d}x_2) + k(y\,\mathrm{d}x_2 - x\,\mathrm{d}y_2)]$$

and it is clear that the direct calculation of d**B** by equation (11) is equivalent to the application of $\mathrm{d}\mathbf{B} = \nabla\times\mathrm{d}\mathbf{A}$ with d**A** given by equation (74). The same must apply for a complete circuit (considering the integration as the limit of summation). It is noted that the units of **A** are given by

$$\mathbf{A} \sim \mu_0 \times \text{current} \sim \mathrm{J\ A^{-1}\ m^{-1}}$$

consistent with $\mathbf{B} = \nabla\times\mathbf{A}$.

Despite its vector nature, **A**, or at least d**A**, is readily envisaged (see Figure 1.31) since, by equation (74), it is parallel to the current element. The magnitude varies as $1/r$ so the derivation of d**A** is rather like that of the scalar potential of a charge or pole with the additional assignment of the direction of the current element. The relation of **B** or **H** to **A** is somewhat more difficult to see than that of **H** to φ, for which $H = -\operatorname{grad}\varphi$ simply indicates that **H** is directly 'downhill'. A simple example should be helpful, using **A** as an intermediary for the calculation at **B**.

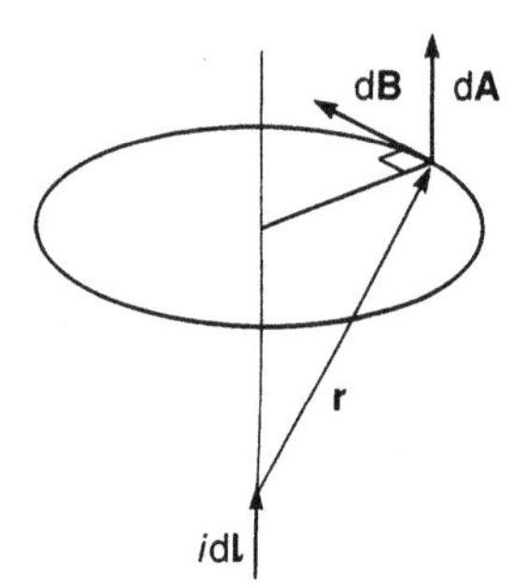

Figure 1.31 d**A** is $\parallel$ id**l** and thus $\mathrm{d}\mathbf{B} = \nabla\times\mathrm{d}\mathbf{A}$ is as shown

2.9.1 *A* and *B* for a straight line of current

In Figure 1.32 a current i flows along the Z axis from O to $z = a$ and the potential is to be calculated at $p = (r, 0)$. The contribution to $\mathbf{A}$ from the elementary current $i\mathbf{k}\,\mathrm{d}z$ is, by equation (74),

$$\mathrm{d}\mathbf{A} = \mathbf{k}\frac{\mu_0 i}{4\pi}\frac{\mathrm{d}z}{R}$$

and for the whole length

$$\mathbf{A} = \mathbf{k}\frac{\mu_0 i}{4\pi}\int_0^a \frac{\mathrm{d}z}{R} = \mathbf{k}\frac{\mu_0 i}{4\pi}\int_0^a \frac{\mathrm{d}z}{\sqrt{z^2 + r^2}} = \mathbf{k}\frac{\mu_0 i}{4\pi}\frac{1}{r}\int_0^a \frac{\mathrm{d}z}{\sqrt{z^2/r^2 + 1}}$$

Now

$$\frac{\mathrm{d}}{\mathrm{d}z}\ln\left(\frac{z}{r} + \sqrt{\frac{z^2}{r^2} + 1}\right) = \frac{\dfrac{1}{r} + \dfrac{z/r^2}{\sqrt{z^2/r^2 + 1}}}{\dfrac{z}{r} + \sqrt{\dfrac{z^2}{r^2} + 1}}$$

$$= \frac{\dfrac{(1/r)\sqrt{z^2/r^2 + 1} + z/r^2}{\sqrt{z^2/r^2 + 1}}}{\dfrac{z}{r} + \sqrt{\dfrac{z^2}{r^2} + 1}}$$

$$= \frac{1/r}{\sqrt{z^2/r^2 + 1}}$$

Thus

$$\mathbf{A} = \mathbf{k}\frac{\mu_0 i}{4\pi}\ln\left(\frac{a}{r} + \sqrt{\frac{a^2}{r^2} + 1}\right) \tag{76}$$

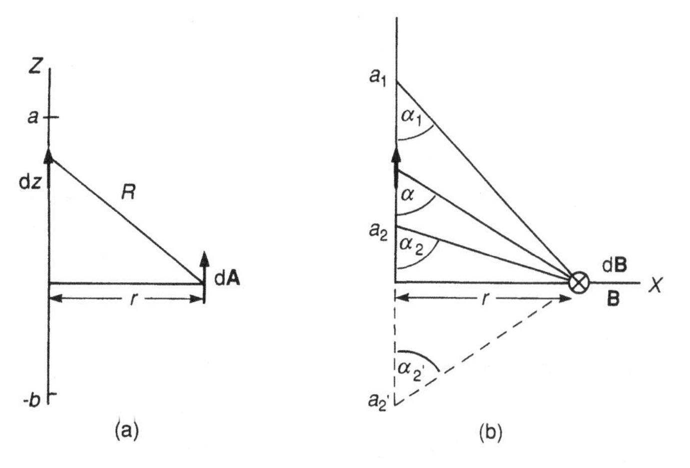

Figure 1.32 Construction for **A** and **B** from a straight linear filamentary current

To calculate **B** directly, from equation (11)

$$\mathrm{d}\mathbf{B} = \frac{\mu_0 i}{4\pi}\frac{(\mathbf{k}\,\mathrm{d}z)\times\mathbf{R}}{R^3}$$

$$= \mathbf{e}_\phi \frac{\mu_0 i}{4\pi}\frac{(\mathrm{d}z)R\sin\alpha}{R^3} = \mathbf{e}\frac{\mu_0 i}{4\pi}\frac{\sin\alpha\,\mathrm{d}z}{R^2}$$

$\mathbf{e}_\phi$ is a unit vector normal to the plane containing OZ and **R** and the direction is given by the convention for the cross-product. Alternatively, if a rectangular Cartesian system was superimposed on Figure 1.32 we would have $\mathbf{R} = \mathbf{i}r - \mathbf{k}z$ and $\mathrm{d}\mathbf{r}_2 = \mathbf{k}\,\mathrm{d}z$ in the notation of equation (11) and

$$\mathrm{d}\mathbf{r}_2\times\mathbf{R} = \begin{vmatrix} \mathbf{i} & \mathbf{j} & \mathbf{k} \\ 0 & 0 & \mathrm{d}z \\ r & 0 & z \end{vmatrix} = \mathbf{i}(0) + \mathbf{j}(r\,\mathrm{d}z) + \mathbf{k}(0) = \mathbf{j}R\sin\theta\mathrm{d}z$$

as in the above, save that rotation of the Cartesian system about OZ has no effect and $\mathbf{e}_\phi$ is preferable to **j**. Since $R^2 = z^2 + r^2$,

$$\mathrm{d}\mathbf{B} = \mathbf{e}_\phi \frac{\mu_0 i}{4\pi}\frac{r\,\mathrm{d}z}{(z^2+r^2)^{3/2}}$$

As can be checked by differentiation,

$$\int \frac{\mathrm{d}u}{(u^2+a^2)^{3/2}} = \frac{u}{a^2(u^2+a^2)^{1/2}}$$

and

$$\mathbf{B} = \mathbf{e}_\phi \frac{\mu_0 i}{4\pi}\left[\frac{z}{r(z^2+r^2)^{1/2}}\right]_0^a$$

$$= \mathbf{e}_\phi \frac{\mu_0 i}{4\pi}\frac{a}{r(a^2+r^2)^{1/2}} = \mathbf{e}_\phi \frac{\mu_0 i}{4\pi r}\cos\alpha_1 \tag{77}$$

Recourse to the standard integral may in fact be avoided by noting that $z = r/\tan\alpha$, $\mathrm{d}z = -(r/\sin^2\alpha)\mathrm{d}\alpha$ and

$$\frac{\mathrm{d}z\sin\alpha}{R^2} = -\frac{r}{\sin^2\alpha}\sin\alpha\frac{\sin^2\alpha}{r^2}\mathrm{d}\alpha = -\frac{1}{r}\sin\alpha\mathrm{d}\alpha$$

so that the integration is direct. Obviously if the integral was taken between $z = a_2$ and $z = a_2$ the result would be

$$\mathbf{B} = \mathbf{e}_\phi \frac{\mu_0 i}{4\pi r}(\cos\alpha_1 - \cos\alpha_2) \tag{78}$$

which would also correspond to superimposing a reverse current from a_2 to 0 on the original current from 0 to a [Figure 1.32(b)], in analogy with the principle of superposition of charges in electrostatics. However, if the current flows between a_2' and a this is equivalent to the sum of two separate currents, giving

$$\mathbf{B} = \mathbf{e}_\phi \frac{\mu_0 i}{4\pi r}(\cos\alpha_1 + \cos\alpha_2') \tag{79}$$

For an infinite line of current, $\alpha_1 = \alpha_2' \to 0$ and

$$\mathbf{B} = \mathbf{e}_\phi \frac{\mu_0 i}{2\pi r} \tag{80}$$

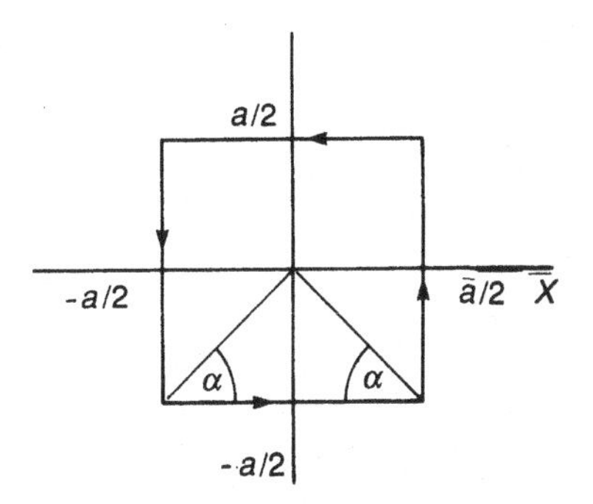

Figure 1.33 A simple circuit of four interconnected linear sections, giving a very simple expression for the field at the centre

A line of current that originates and terminates at points in space is in itself an unrealistic concept, for d.c. currents at least. However, equation (80) is meaningful in approximating **B** at points close to a current in, say, a circular circuit of large diameter. Also, equation (78) is meaningful when used in a sum for, for example, four appropriately arranged lines representing a thin wire circuit in the form of a square frame $a \times a$ (Figure 1.33). Putting $r = a/2$ and $\cos\alpha = 1/\sqrt{2}$, at the centre for example,

$$\mathbf{B} = 4 \times \mathbf{k}\frac{\mu_0 i}{4\pi}\frac{1}{a/2}\frac{2}{\sqrt{2}} = \mathbf{k}\frac{2\sqrt{2}\mu_0 i}{\pi a} \tag{81}$$

Note that $B \propto 1/a$ and that this applies at any point.

Since **B** can be found directly the relation $\mathbf{B} = \nabla\times\mathbf{A}$ can be used as a check of consistency. From equation (76), **A** has a single component A_z and, with constant a, is a function of r only. Thus from the expression for the curl in cylindrical coordinates in Appendix 2 it is only necessary to select terms containing A_z and $\partial/\partial r$:

$$\mathbf{B} = \nabla\times\mathbf{A} = \frac{1}{r}\left(r\frac{\partial A_z}{\partial r}\right)\mathbf{e}_\phi$$

and

$$\frac{\partial}{\partial r}\ln\left[\frac{a}{r} + \left(\frac{a^2}{r^2}+1\right)^{1/2}\right] = \frac{-\dfrac{a}{r^2} - \dfrac{a^2}{r^3}\left(\dfrac{a^2}{r^2}+1\right)^{-1/2}}{\dfrac{a}{r} + \left(\dfrac{a^2}{r^2}+1\right)^{1/2}}$$

$$= -\frac{\dfrac{a}{r^2}\left[\left(\dfrac{a^2}{r^2}+1\right)^{1/2} + \dfrac{a}{r}\right]\Big/\left(\dfrac{a^2}{r^2}+1\right)^{1/2}}{\dfrac{a}{r} + \left(\dfrac{a^2}{r^2}+1\right)^{1/2}}$$

$$= -\frac{a}{r(a^2+r^2)^{1/2}} = \frac{\cos\alpha}{r}$$

Thus **B** is obtained as by direct integration and the consistency of $\mathbf{B} = \nabla \times \mathbf{A}$ confirmed for the particular example.

There are cases for which **B** can only be derived by first calculating **A**, at least in a simple way; one of these arises in the following section.

2.9.2 The magnetic dipole

It may now be shown that the field or induction generated by a small closed current loop of vector area $\mathcal{A}$, current i, is everywhere $\propto \mathcal{A}i$ and has the same form as the field **H** calculated for the so-called magnetostatic dipole or elementary magnet. For this and other reasons the quantity $\boldsymbol{\mu} = \mathcal{A}i$ is called a magnetic dipole. It may be regarded as the true definition of the magnetic dipole since it is made in terms of base quantities and the magnetostatic dipole is relegated to the status of an illustration or convention. The units of μ are given by $\mu \sim$ current $\times$ area $\sim \mathrm{A\,m^2}$.

By 'small' we mean of linear dimension $\ll r$, the distance at which $\mathbf{B}(r)$ is calculated. Defining a point magnetic dipole by $\mathcal{A} \to 0$, $i \to \infty$ such that $\mathcal{A}i = \mu$ remains, the two fields referred to above correspond at all points. The vector $\mathcal{A}$ has obvious magnitude and a direction that follows the convention for angular momentum; i.e. replacing the current by a positive orbiting charge, $\mathcal{A}$ and $\boldsymbol{\mu}$ are $\parallel$ **L**. If the circulation is reversed $\boldsymbol{\mu}$ is reversed. The calculation affords a good demonstration of the value of the vector potential since the corresponding calculations of **B** directly are feasible but more complex.

The dimension b of the rectangular circuit in Figure 1.34(a) is to be considered small enough for $\mathbf{k}b$ to replace $d\mathbf{r}_2$ in equation (11). P is in OXZ and by symmetry the portions of the currents flowing $\parallel$ OX give equal and opposite contributions to **A** and are neglected [$i\, d\mathbf{r}\mathbf{i} \times \mathbf{r}$ cancels for pairs of elements at $(x, b/2, 0)$ and $(x, -b/2, 0)$].

Thus the potential at P is

$$\mathbf{A} = \mathbf{j}\frac{\mu_0 i b}{4\pi}\left(\frac{1}{r_1} - \frac{1}{r_2}\right) \tag{82}$$

If $r \gg a$, the vectors $\mathbf{r}$, $\mathbf{r}_1$ and $\mathbf{r}_2$ are closely parallel and approximately, by Figure 1.34(b), $r_1 = r - \delta$, $r_2 = r + \delta$ with $\delta = (a/2)\sin\theta$. Then

$$\mathbf{A} = \mathbf{j}\frac{\mu_0 i b}{4\pi}\left(\frac{1}{r-\delta} - \frac{1}{r+\delta}\right) = \mathbf{j}\frac{\mu_0 i}{4\pi}\frac{b}{r}\left(\frac{1}{1-y} - \frac{1}{1+y}\right)$$

with $y = \delta/r$. By the binomial series $(1+x)^k = 1 + kx + [k(k-1)/2!]x^2 + \cdots + \{k(k-1)\cdots[(k-n+1)/n!]\}x^n + \cdots$ and ignoring terms above x^2,

$$\mathbf{A} = \mathbf{j}\frac{\mu_0 i b}{4\pi r} \times 2y = \mathbf{j}\frac{\mu_0 i b}{4\pi r}\frac{a\sin\theta}{r} = \mathbf{j}\frac{\mu_0}{4\pi}\frac{iba\sin\theta}{r^2}$$

i.e. with iba = current $\times$ area enclosed $= \mu$:

$$\mathbf{A} = \mathbf{j}\frac{\mu_0}{4\pi}\frac{\mu\sin\theta}{r^2} \tag{83}$$

A current loop of general shape can be considered to be made up of a series of strips of width δy_i and length a_i with the currents cancelling along shared sections, so that the current

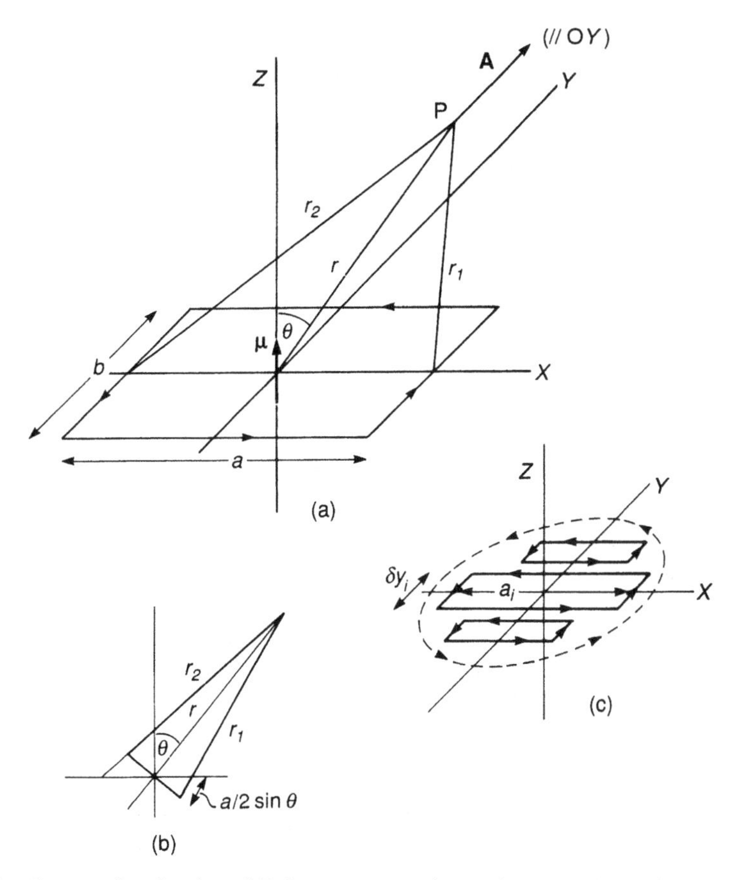

Figure 1.34 Construction for **A** and **B** from a current loop of general shape, in the limit of large r/a or r/b, by first considering the rectangular loop in (a)

effectively flows around the perimeter [Figure 1.34(c)]. As the $\delta y_i \to 0$, $\sum \delta a_i \delta y_i \to \mathcal{A}$, the total area enclosed by the current. Identifying each δy_i with b in the above, the potential for the whole loop is

$$\mathbf{A} = \sum \frac{\mu_0 i}{4\pi} \frac{b_i a_i \sin\theta}{r^2} \to \frac{\mu_0}{4\pi} \frac{i\mathcal{A}\sin\theta}{r^2} = \frac{\mu_0}{4\pi} \frac{\mu \sin\theta}{r^2} \tag{84}$$

though again with the obvious approximations justified only when r is great. This now applies to a loop of general shape and the restriction to points in OXZ is removed: for a general point **r** it is only necessary to consider the strips to be drawn ∥ the projection of **r** on the plane containing the loop. **A** is always normal to the plane containing $\boldsymbol{\mu}$ and to **r** [Figure 1.34(a)] and since $\mu r \sin\theta = |\boldsymbol{\mu}\times\mathbf{r}|$ equation (84) becomes

$$\mathbf{A} = \frac{\mu_0}{4\pi} \frac{\boldsymbol{\mu}\times\mathbf{r}}{r^3} \tag{85}$$

since $A = A(r, \theta)$ it is natural to use spherical coordinates (r, θ, ϕ), in which $\mathbf{A} = \mathbf{A}_\phi \mathbf{e}_\phi$ only. Restricting the general expression for $\nabla\times\mathbf{A}$ according to $A_r = A_\theta = 0$,

$$\begin{aligned}\mathbf{B} &= \nabla\times\mathbf{A}\\ &= \frac{\mu_0}{4\pi}\frac{1}{r^2\sin\theta}\left[\left(\frac{\partial}{\partial\theta}r\sin\theta A_\phi\right)\mathbf{e}_r + \left(-\frac{\partial}{\partial r}r\sin\theta A_\phi\right)r\mathbf{e}_\theta\right]\\ &= \frac{\mu_0}{4\pi}\frac{1}{r^2\sin\theta}\left[\left(\frac{\partial}{\partial\theta}\frac{\mu\sin^2\theta}{r}\right)\mathbf{e}_r - \left(\frac{\partial}{\partial r}\frac{\mu\sin^2\theta}{r}\right)r\mathbf{e}_\theta\right]\\ &= \frac{\mu_0}{4\pi}\frac{2\mu}{r^3}\cos\theta\mathbf{e}_r + \frac{\mu_0}{4\pi}\frac{\mu\sin\theta}{r^3}\mathbf{e}_\theta \end{aligned} \tag{86}$$

which corresponds to equation (3) for the *magnetostatic dipole* (save for the factor μ_0). Thus the alternative corresponding forms also hold:

$$\mathbf{B}(\mathbf{r}) = \frac{\mu_0}{4\pi}\left[\frac{3(\boldsymbol{\mu}\cdot\mathbf{r})\mathbf{r}}{r^5} - \frac{\boldsymbol{\mu}}{r^3}\right] \tag{87}$$

and

$$B_z = \frac{\mu_0}{4\pi}\mu\frac{3z^2 - r^2}{r^5}, \qquad B_x = \frac{\mu_0}{4\pi}\mu\frac{xz}{r^5}, \qquad B_y = \frac{\mu_0}{4\pi}\mu\frac{yz}{r^5} \tag{88}$$

to which may be added

$$\mathbf{B}(\mathbf{r}) = \frac{\mu_0}{4\pi}\nabla\left[\boldsymbol{\mu}\cdot\nabla\left(\frac{1}{r}\right)\right] = -\frac{\mu_0}{4\pi}\nabla\left[\frac{\boldsymbol{\mu}\cdot\mathbf{r}}{r^3}\right] \tag{89}$$

since this, when expanded, gives equation (86). (Use equation 8, Appendix 2)

One step in the above treatment calls for comment. This is the realization that, so long as the area of the general loop is considered to be appropriately subdivided, the potential is obtained at a general point (r, θ, ϕ) and the expressions are no longer restricted to points in a particular plane (i.e. *OXZ*). Suppose that we had taken

$$\mathbf{B} = \nabla\times\mathbf{A} = \nabla\times\left(\mathbf{j}\frac{\mu_0}{4\pi}\frac{\mu\sin\theta}{r^2}\right) = \frac{\mu_0\mu}{4\pi}\begin{vmatrix}\mathbf{i} & \mathbf{j} & \mathbf{k}\\ \dfrac{\partial}{\partial x} & \dfrac{\partial}{\partial y} & \dfrac{\partial}{\partial z}\\ 0 & \dfrac{\mu\sin\theta}{r^2} & 0\end{vmatrix}$$

This does not give a general result, because y does not appear, but it does not even give the correct restricted result. The expression implies that $\mathbf{A}$ has a y component only and carries no information on how $\mathbf{A}$ varies away from the specified plane. From a similar point of view it is fruitless to attempt to calculate $\mathbf{B} = \nabla\times\mathbf{A}$ on the axis OZ since it is easy to see by symmetry that on the axis $\mathbf{A} = 0$ (though of course the gradients are not zero). In fact, it is easy to calculate $\mathbf{B}$ directly, on the axis, for comparison with the general formula.

To clarify or formalize the direction of $\boldsymbol{\mu}$, it is conventional to define the vector area in such a way that

$$\boldsymbol{\mu} = i\int_S \mathrm{d}\mathbf{s} = \tfrac{1}{2}i\int_C \mathbf{r}\times\mathrm{d}\mathbf{r} \tag{90}$$

(which obviously holds for a circle with $r = a$ (constant) and the integral of $\mathrm{d}r = 2\pi a$ around the circuit). Taking i as the magnitude of the current, $i > 0$ always, and indicating the circulation by the direction of $\mathbf{dr}$, the direction of $\boldsymbol{\mu}$ follows (see Figure 1.35).

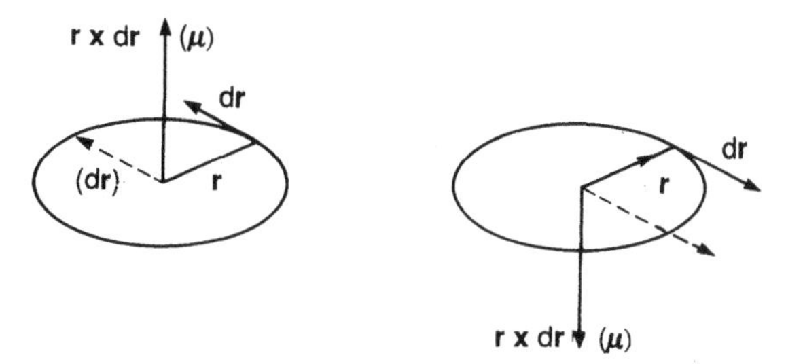

Figure 1.35 The direction of μ according to the expression given

3 Magnetic Moments

3.1 Bohr Magneton

The Bohr theory of atomic structure demands attention due to its legacy of important terms and concepts. In this planetary picture the electron in a hydrogen atom is seen strictly as a particle of mass m and charge $-e$ (with e the proton charge). The electrostatic force $e^2/(4\pi\varepsilon_0 r^2)$ is equated to the centripetal force mv^2/r to hold the electron in a circular orbit of radius r. This is inadequate since even if v is constant the direction of $\mathbf{v}$ varies continuously and electromagnetic energy is radiated by an accelerated charge. Bohr enforced stability in one of a number of discrete orbits by the postulate $mvr = nh/(2\pi) = n\hbar$ with $n = 1, 2, 3, \ldots$, where h is Planck's constant and $\hbar = h/(2\pi)$ is used for convenience. It is easy to eliminate v from these two equations to obtain r and further to find the total energy as $\frac{1}{2}mv^2 - e^2/(4\pi\varepsilon_0 r)$:

$$r = n^2 \frac{4\pi\varepsilon_0 \hbar^2}{me^2} = n^2 a_0, \qquad E_n = -\frac{1}{n^2}\frac{e^4 m}{2(4\pi\varepsilon_0)^2\hbar^2} \tag{91}$$

$a_0 (= 0.5292 \times 10^{-10}$ m) is the (first) Bohr radius and is frequently used as a unit of length in wave mechanics. The energies do coincide with the 'gross structure' values for hydrogen, though extension to other cases is impractical. Thus with the condition $hf = E_2 - E_1$ for the conservation of energy in a radiative transition, with frequency f, between states with energy E_1 and E_2 Bohr gave a reasonable account of the hydrogen spectra.

The presence of a charge corresponding to a current $i = -e/\tau$, where τ is the time for one orbit, associates a magnetic moment μ with the angular momentum L. Using $2\pi r/\tau = v$ to give τ and i and $\mu = \pi r^2 i$ together with the Bohr postulate, we obtain

$$\mu = n\frac{eh}{4\pi m} = n\frac{e\hbar}{2m} = n\beta \tag{92}$$

where $\beta = eh/(4\pi m)$ is the Bohr magneton.

Further, it is easy to show that the ratio

$$\frac{\text{magnetic moment}}{\text{angular momentum}} = \frac{\mu}{L} = \gamma = -\frac{e}{2m} \tag{93}$$

i.e. depends only on basic constants. This ratio, γ, is usually called the gyromagnetic ratio, although it may be argued that it should be called the magnetogyric ratio. It is re-introduced in Chapter 2.

3.2 Elementary Quantum Theory

Quantum theory relevant to magnetic moments is surveyed in Chapter 2. Here only the simplest results are mentioned to give an early introduction to atomic or ionic magnetic moments. Considering orbital motion for the example of a hydrogen atom, i.e. discounting spin, the observables for which definite values or eigenvalues of the appropriate operators can be calculated are the discrete energies E_n, the magnitude of the orbital angular momentum L and any one component of the angular momentum, chosen to be L_z; i.e. explicit operators $\hat{\mathcal{H}}$, $\hat{L}^2$ and $\hat{L}_z$ $(= -i\hbar\, \partial/\partial\phi)$ can be derived for these quantities and the values occur as the number such as a in relations that are typically $\hat{A}f(\mathbf{r}) = af(\mathbf{r})$, defining the eigenfunctions $f(\mathbf{r})$ by the particular form of the relation. The circumflex on $\hat{A}$ etc. denotes an operator, although this will be dropped later where the implications are clear. If two operators can be shown to commute, i.e. $\hat{A}\hat{B} = \hat{B}\hat{A}$, the values of the observables can be known simultaneously, and this determines the above list. Relations between operators are the same as those between the corresponding classical functions. The operator for the energy is called the Hamiltonian $\hat{\mathcal{H}}$ [$= \hat{p}^2/2m_1 + V(r)$ with $\hat{p}^2 = -\hbar^2\nabla^2$ for the momentum] and the eigenfunction is the wave function, usually denoted ψ or $\psi(\mathbf{r})$, and the above relation is then the wave equation:

$$\mathcal{H}\psi_{nlm}(r, \theta, \phi) = E_n\psi_{nlm}(r, \theta, \phi)$$

This constitutes a partial differential equation capable of solution to give the energies and functions for the case of the hydrogen atom and as approximations for other atoms. The subscripts n, l, and m are restricted to integral values by boundary conditions and the single index on E_n indicates that for hydrogen (gross structure) E depends only on n: in fact, the result is that given by the Bohr theory. The wave functions can be written as

$$\psi_{nlm}(r, \theta, \phi) = NR_{nl}(r)P_{lm}(\theta)\mathrm{e}^{im\phi} = NR_{nl}(r)Y_{lm}(\theta, \phi) \tag{94}$$

in which the $P_{lm}(\theta)$ are associated Legendre polynomials and the $Y_{lm}(\theta, \phi)$ are spherical harmonics exemplified in Table 1.1. ψ indicates the location of the electron to the extent that $P(\mathbf{r}) = \psi(\mathbf{r})\psi^*(\mathbf{r}) = |\psi(\mathbf{r})|^2$ is the probability density for it to exist at $\mathbf{r}$. Hence the factor N, contrived to ensure a total probability of 1 over all space: $\int \psi(r)\psi^*(r)\mathrm{d}v = 1 \times P(0)$, at the nucleus, is non-zero when $l = 0$; l may take integral values up to $n - 1$ and m ranges between $-l$ and l in integral steps. States having $l = 0, 1, 2, 3, \ldots$ are designated s, p, d, f-states so that the ground state for hydrogen is necessarily 1s where the leading number denotes n. The wave functions are necessarily also eigenfunctions of the operators $\hat{L}^2$ and L_z in this case and the eigenvalues are $\hbar^2 l(l + 1)$ and $m\hbar$ respectively.

Any particular set of values of the observables is known when n, l and m are known and it is thus reasonable to consider that the electron can exist in one of a number of states each of which is identified by a particular permissible set n, l, m. The symbol for the entity known as the state is $|\ \rangle$ or, in this case $|n\ l\ m\rangle$. Thus a 3d state is symbolically $|\,3\ 2\ m\,\rangle$ for any (permissible) m. These may also be regarded as (or associated with) vectors in a space of appropriate dimension, i.e. state vectors, and in this case there must exist an associated adjoint space containing vectors symbolized as $\langle\ |$ to permit taking scalar products as $\langle i|\ |j\rangle$ or just $\langle i|j\rangle$, where the labels are present to identify different vectors, or self-products $\langle i|i\rangle$ (see Appendix 2). Any state (vector) $|\ \rangle = c_1|1\rangle + c_2|2\rangle$ is a further state in the space containing $|1\ \rangle$ and $|2\rangle$, with coefficients c_i, and if any state in a particular space can be

Table 1.1 Spherical harmonics, with normalization according to

$$\int_0^{2\pi} d\phi \int_0^{\pi} Y_l^{m^*} Y_l^m \sin\theta d\theta = 1$$

$Y_0^0 = \sqrt{\frac{1}{4\pi}}$
$Y_1^0 = \sqrt{\frac{3}{4\pi}} \cos\theta$
$Y_1^{\pm 1} = \mp\sqrt{\frac{3}{8\pi}} \sin\theta e^{\pm i\phi}$
$Y_2^0 = \sqrt{\frac{5}{16\pi}} (3\cos^2\theta - 1)$
$Y_2^{\pm 1} = \mp\sqrt{\frac{15}{8\pi}} \sin\theta\cos\theta e^{\pm i\phi}$
$Y_2^{\pm 2} = \sqrt{\frac{15}{32\pi}} \sin^2\theta e^{\pm 2i\phi}$

expressed thus then the $|i\rangle$ are said to constitute a basis for that space. The $|i\rangle$ are taken to be linearly independent since otherwise one could be eliminated in terms of the others. The analogy with Cartesian space with unit vectors $\mathbf{i}, \mathbf{j}, \mathbf{k}$ is apparent.

Since any set n, l, m defines a wave function, a correspondence may be made as $|nlm\rangle \sim \psi_{nlm}(r)$ connecting but not equating, the state to the function. Sometimes $|\psi\rangle$, $|\varphi\rangle$, etc., are used to indicate the states with which the wave functions $\psi(r)$, $\varphi(r)$ may be associated and the Schrodinger wave equation is replaced by

$$\mathcal{H}|n\ l\ m\rangle = E_{nl}|n\ l\ m\rangle \tag{95}$$

or $\mathcal{H}|n\rangle = E_n|n\rangle$ (if E depends on n only). The $\langle i|j\rangle$, like the definite integrals, are numbers and the following equalities hold:

$$\langle\varphi_i|\varphi_j\rangle = \int \varphi_i^*(\mathbf{r})\varphi_j(\mathbf{r})dv$$

$$\langle\varphi_i|\hat{A}|\varphi_j\rangle = \int \varphi_i^*(\mathbf{r})\hat{A}\varphi_j(\mathbf{r})dv \tag{96}$$

$\langle i|i\rangle = 1$ indicates normalization and $\langle i|j\rangle = 0$ or $\int \varphi_i^*\varphi_j dv = 0$ indicates orthogonality. At the least the so-called 'Dirac notation' affords useful abbreviations.

Important demonstrations are that $\langle n|m\rangle = 0$ if the states are non-degenerate, i.e. give different eigenvalues and that the necessary condition that the eigenvalues of the operators, $\hat{A}$ say, for observables must be real is satisfied if $\langle i|\hat{A}|j\rangle^* = \langle j|\hat{A}|i\rangle$, $\hat{A}$ then being said to be Hermitian. Any linear combination of degenerate eigenvectors is easily seen to be a further eigenvector, since the operators are taken to be linear.

Available information is not limited to the occurence of eigenvalues. If a state $|\varphi\rangle$ is not an eigenstate of an operator $\hat{B}$, say, then the value of the quantity B may be evaluated as a mean or expectation value according to

$$\bar{B} = \langle \hat{B} \rangle_\varphi = \frac{\langle \varphi | \hat{B} | \varphi \rangle}{\langle \varphi | \varphi \rangle} \tag{97}$$

with $\langle \varphi | \varphi \rangle = 1$ for normalization.

3.3 Magnetic Moments

The facile approach, which is in fact acceptable as a first-order approximation, is to post-calculate magnetic moments from angular momentum results by applying the Bohr gyromagnetic ratio, so that

$$L = \hbar\sqrt{l(l+1)} \sim \mu = \frac{l}{2m_e}\hbar\sqrt{l(l+1)} = \beta\sqrt{l(l+1)}$$

$$L_z = m\hbar \sim \mu_z = -\beta m$$

This is equivalent to postulating operators for μ^2 and μ_z:

$$\hat{\mu}^2 = \gamma \hat{L}^2, \qquad \hat{\mu}_z = \gamma \hat{L}_z$$

Again to first order the energy shifts in an applied magnetic field are obtained by adding a term $-\gamma B \hat{L}_z$ to $\hat{\mathcal{H}}$ ($B \parallel OZ$ implied by use of L_z) to give $\mathcal{H} = \mathcal{H}_0 - \gamma B \hat{L}_z$ where $\mathcal{H}_0 = P^2/2m_e + V$ and, using $\hat{L}_z \psi = m\hbar\psi$, $E = E_0 - \gamma B m\hbar = E_0 + m\beta$. The resulting 'Zeeman splitting' expected for orbital motion alone does not correspond to observed spectra, even for isolated atoms, because other degrees of freedom exist that remain to be introduced, i.e. the electron spin.

In fact, for ions in crystals, i.e. in crystal fields, or in some molecules the orbital contributions to the magnetic moments may be negligible, or at first sight zero, with the moments depending almost wholly on the spins. The gross structure energies of hydrogen depend only on n. For multiple electron atoms the energies depend also on l but in the absence of a prescribed axis there can be no dependence on m. Thus any linear combination of the degenerate states, which differ only inasmuch as m differs, are also eigenstates of $\mathcal{H}$ and the real linear combinations described below are found to be appropriate in the presence of crystal fields.

For example, for $n = 2$, $l = 1$(2p):

$$|\varphi_{2p_x}\rangle = \frac{1}{\sqrt{2}}(|1\rangle + |-1\rangle) \tag{98}$$

(Only m need be specified here.) This is called the $2p_x$ orbital because, referring to Table 1.1, the reader may check that the function is (using $R_{21} = Nr\mathrm{e}^{-r/2a_0}$)

$$\varphi_{2p_x} = \frac{1}{4\sqrt{2\pi}}\mathrm{e}^{-r/2a_0}x \tag{99}$$

with $x = r\sin\theta\cos\phi$. This is obviously not an eigenstate of $\hat{L}_z$ and the mean value is

$$\bar{L}_z = \tfrac{1}{2}(\langle 1| - \langle -1|)\hat{L}_z(|1\rangle - |-1\rangle) = \tfrac{1}{2}(\langle 1| - \langle -1|)(\hbar|1\rangle + \hbar|-1\rangle) = 0 \tag{100}$$

assuming orthonormality. Similarly, d orbitals are (using $R_{32} = Nr^2 e^{-r/3a_0}$)

$$\varphi_{3\,d_{xy}} \sim \frac{1}{i\sqrt{2}}(|2\rangle - |-2\rangle) \sim NR_{32}\sin^2\theta\sin 2\phi = Nxye^{-r/3a_0}$$

$$\varphi_{3\,d_{xz}} \sim \frac{1}{\sqrt{2}}(|1\rangle + |-1\rangle) \sim NR_{32}\sin\theta\cos\theta\cos\phi = Nxze^{-r/3a_0}$$

$$\varphi_{3\,d_{yz}} \sim \frac{1}{i\sqrt{2}}(|1\rangle - |-1\rangle) \sim NR_{32}\sin\theta\cos\theta\sin\phi = Nyze^{-r/3a_0} \tag{101}$$

$$\varphi_{3\,d_{x^2-y^2}} \sim \frac{1}{\sqrt{2}}(|2\rangle + |-2\rangle) \sim NR_{32}\sin^2\theta\cos 2\phi = N(x^2 - y^2)e^{-r/3a_0}$$

$$\varphi_{3\,d_{z^2}} \sim |0\rangle \sim NR_{32}(3\cos^2\theta - 1) = Nz^2e^{-r/3a_0}$$

The shapes of these orbitals are shown in Figure 1.36. For different values of n the dependence of φ on r differs, but that on θ and ϕ, and thus the symmetry, is common and the orbitals may be designated p_x, p_y, d_{xy}, etc. For d_{xy}, for example,

$$\begin{aligned}\bar{L}_z &= \tfrac{1}{2}(\langle 2| - \langle -2|)\hat{L}_z(|2\rangle - |-2\rangle) = \frac{\hbar}{2}(\langle 2| - \langle -2|)(2|2\rangle + 2|-2\rangle) \\ &= \hbar(1 + 0 - 0 - 1) = 0\end{aligned} \tag{102}$$

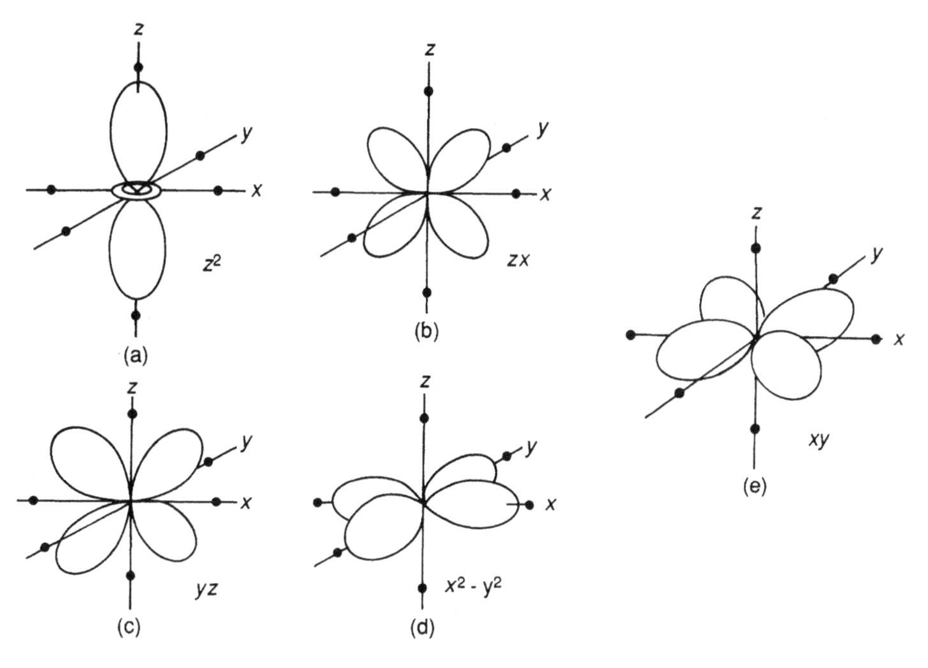

Figure 1.36 The shapes of the d orbitals. The s orbitals are spherical and p orbitals (e.g. P_z) similar to z^2. (Reprinted by permission of John Wiley & Sons, Ltd, Magnetic Oxides, ed. D.J. Craik, Wiley 1975)

In general, since the m values range symmetrically over positive and negative values, the result is zero for all the orbitals. This quenching of the orbital angular momentum, in crystals, appears to be universal and complete, but it will be seen that (a) it applies to 3d ions for which the electrons of importance are the outer electrons, the 4s electrons being lost on ionization, but not to 4f ions in which the 'magnetic electrons' are shielded from the crystal fields by the outer 5s electrons and (b) some orbital contributions can be restored by spin–orbit coupling.

In a different situation, a π electron in benzene is treated approximately as moving freely except that it is confined to a ring, taken to be $\perp OZ$. For this $\hat{\mathcal{H}} = \hat{p}^2/2m$ $(V = 0)$ is replaced by $\hat{\mathcal{H}} = \hat{L}_z^2/2I$, with I the moment of inertia. Obvious eigenfunctions correspond to

$$\mathcal{H}\psi = -\frac{\hbar^2}{2I}\frac{\partial^2}{\partial\phi^2}\cos m\phi = \frac{\hbar^2 m^2}{2I}\cos m\phi, \qquad E_m = m^2\frac{\hbar^2}{2I} \tag{103}$$

with m integral for $\cos m\phi = \cos m(\phi + 2\pi)$. The $\cos m\phi = \frac{1}{2}(e^{im\phi} + e^{-im\phi})$ are not eigenfunctions of $\hat{L}_z$ and it is left as a simple exercise to show that $\bar{L}_z = 0$. Pictorially, ψ is a superposition of states in which the electron circulates clockwise or anticlockwise with equal probability, and again no magnetic moment arises.

At this simple level it appears that ions in crystals, or in molecules corresponding to the simple model, should be magnetically inert with no response to an applied field. In fact benzene is strongly diamagnetic, i.e. in an applied magnetic field a magnetization is induced in a direction that is antiparallel to H so that $\chi = M/H$ is negative (and comparable in magnitude to paramagnetic χ values at room temperature, though diamagnetic effects are generally somewhat smaller and are independent of temperature). The implicit assumption has been made that the wave functions are not themselves affected by the magnetic fields, i.e. for a hydrogen-like atom, for example that they remain eigenfunctions of $\hat{L}_z$ and thus of $\hat{\mu}_z = \gamma\hat{L}_z$ and of $\hat{\mathcal{H}}$, when this includes a term for the field which is simply taken to be $-\gamma B\hat{L}_z$:

$$\hat{\mathcal{H}} = \hat{\mathcal{H}}_0 + \hat{\mathcal{H}}_B = -\frac{\hbar^2}{2m_e}\nabla^2 + V(r) - \gamma B\hat{L}_z$$

$$\hat{\mathcal{H}}|n\ l\ m\rangle = \hat{\mathcal{H}}_0|n\ l\ m\rangle - \gamma B\hat{L}_z|n\ l\ m\rangle = (E_0 - \gamma Bm\hbar)|n\ l\ m\rangle$$

The susceptibility follows on determining a statistical distribution over these levels and in this approximation no diamagnetism is predicted. $\hat{\mathcal{H}}_B$, above, should not be treated like $V(r)$ as a potential energy term that can simply be added to $\hat{\mathcal{H}}$ and it remains to be shown, in Chapter 2, Sections 3 and 4, why the Hamiltonian is valid, even as an approximation, and how diamagnetic effects arise.

Otherwise it can only be suggested that the orbiting electrons are pictured as ring currents which are expected to be modified, inductively, by the application of magnetic fields. The induced currents must be such as to give fields that oppose the applied field or the effect would be cumulative (Lenz's law). Thus the induced moments must be antiparallel to the applied field. This does give a crude explanation of the relatively large diamagnetic response of ring compounds, with current loops that are extensive compared to those in atoms.

3.4 Intrinsic or Spin Magnetic Moments

As indicated by the failure of an orbital theory alone to explain the details of atomic spectra and very directly by the Stern–Gerlach experiments discussed in Chapter 3, it must be assumed that an electron has a magnetic moment and associated angular momentum that is not dependent on its orbital motion and is thus intrinsic [1]. The above experiments, effectively on electrons having no orbital angular momentum (s-states) passing through gradient magnetic fields, indicate that the spin angular momentum may take just two components of equal magnitude along the axis prescribed by the field, and according to the general theory of angular momentum these must be $\pm\frac{1}{2}\hbar$. Thus, formally for the two states thus distinguished,

$$\begin{gathered} S_z|\tfrac{1}{2}\rangle = \tfrac{1}{2}\hbar|\tfrac{1}{2}\rangle, \qquad S_z|-\tfrac{1}{2}\rangle = -\tfrac{1}{2}\hbar|-\tfrac{1}{2}\rangle \\ (S_z|m_S\rangle = m_S\hbar|m_S\rangle, \text{ where } m_S = \pm\tfrac{1}{2}) \\ S_z|\alpha\rangle = \tfrac{1}{2}\hbar|\alpha\rangle, \qquad S_z|\beta\rangle = -\tfrac{1}{2}\hbar|\beta\rangle \end{gathered} \tag{104}$$

(the second using an alternative common convention) with S_z the operator for the component of the angular momentum. It may be taken that $\langle\alpha|\alpha\rangle = 1 = \langle\beta|\beta\rangle$ and that $\langle\alpha|\beta\rangle = 0$. It is further inferred that the magnitude of the spin angular momentum is given by

$$\begin{aligned} \hat{S}^2|\alpha\rangle &\equiv \hat{S}^2|\tfrac{1}{2}\rangle = \hbar^2 S(S+1)|\alpha\rangle \\ \hat{S}^2|\beta\rangle &\equiv \hat{S}^2|-\tfrac{1}{2}\rangle = \hbar^2 S(S+1)|\beta\rangle \end{aligned} \tag{105}$$

with $S = \frac{1}{2}$. Operators S_x and S_y for the other components must exist but the states $|\alpha\rangle$ and $|\beta\rangle$ are not eigenstates of these and in fact

$$\begin{aligned} S_x|\alpha\rangle &= \tfrac{1}{2}\hbar|\beta\rangle, \qquad S_x|\beta\rangle = \tfrac{1}{2}\hbar|\alpha\rangle \\ S_y|\alpha\rangle &= \tfrac{1}{2}i\hbar|\beta\rangle, \qquad S_y|\beta\rangle = -\tfrac{1}{2}i\hbar|\alpha\rangle \end{aligned} \tag{106}$$

The reader may show from these that

$$\langle S_x\rangle_\alpha = \langle S_x\rangle_\beta = \langle S_y\rangle_\alpha = \langle S_y\rangle_\beta = 0 \qquad (\langle S_x\rangle_\alpha = \langle\alpha|S_x|\alpha\rangle, \text{etc.})$$

and that eigenstates of S_x and of S_y exist as linear combinations of $|\alpha\rangle$ and $|\beta\rangle$:

$$\begin{gathered} |x,\tfrac{1}{2}\rangle = \frac{1}{\sqrt{2}}(|\alpha\rangle + |\beta\rangle), \qquad |x,-\tfrac{1}{2}\rangle = \frac{1}{\sqrt{2}}(|\alpha\rangle - |\beta\rangle) \\ |y,\tfrac{1}{2}\rangle = \frac{1}{\sqrt{2}}(|\alpha\rangle + i|\beta\rangle) \qquad |y,-\tfrac{1}{2}\rangle = \frac{1}{\sqrt{2}}(|\alpha\rangle - i|\beta\rangle) \\ (S_x|x,\tfrac{1}{2}\rangle = \tfrac{1}{2}\hbar|x,\tfrac{1}{2}\rangle, \text{etc.}) \end{gathered} \tag{107}$$

and further that

$$\begin{aligned} S^+|\alpha\rangle &= 0, \qquad S^+|\beta\rangle = \hbar|\alpha\rangle, \qquad \text{where } S^+ = S_x + iS_y \\ S^-|\alpha\rangle &= \hbar|\beta\rangle \qquad S^-|\beta\rangle = 0, \qquad \text{where } S^- = S_x - iS_y \end{aligned} \tag{108}$$

These latter spin raising and lowering operators are of considerable value, particularly due to the expansion $\mathbf{S}_1 \cdot \mathbf{S}_2 = S_1^z S_2^z + \frac{1}{2}(S_1^+ S_2^- + S_1^- S_2^+)$ with S_1 operating on 'spin 1', etc.

Another set of (Pauli) operators may be defined by $S_i = \frac{1}{2}\hbar\sigma_i$, for $i = x, y, z$. Obviously $\sigma_z|\alpha\rangle = |\alpha\rangle$, $\sigma_z|\beta\rangle = -1|\beta\rangle$ simply. Any spin state, $|\chi\rangle$ say, may be expressed as a linear combination or superposition of the arbitrarily chosen $|\alpha\rangle$ and $|\beta\rangle$ as

$$|\chi\rangle = c_1|\alpha\rangle + c_2|\beta\rangle, \qquad \langle\chi|\chi\rangle = 1 \tag{109}$$

The Pauli operators give the components of the spin polarization vector **P** according to

$$\begin{aligned} P_x &= \bar{\sigma}_x = \langle\chi|\sigma_x|\chi\rangle = c_1c_2^* + c_2c_1^* \\ P_y &= \bar{\sigma}_y = \langle\chi|\sigma_y|\chi\rangle = i(c_1c_2^* - c_2c_1^*) \\ P_z &= \bar{\sigma}_z = \langle\chi|\sigma_z|\chi\rangle = c_1c_1^* - c_2c_2^* \\ P^2 &= P_x^2 + P_y^2 + P_z^2 \end{aligned} \tag{110}$$

It can be seen that $P = 1$ for any spin state (for a single electron) while the direction of **P** differs for different states. For $|\alpha\rangle$, $P_x = 0 = P_y$ and $P_z = 1$ and for $|x, \frac{1}{2}\rangle$, $P_x = 1$ and $P_y = 0 = P_z$, etc. (Figure 1.37) If this brief account is not considered satisfactory for the present, recourse may be had to Chapter 3.

The gyromagnetic ratio must have the same dimensions as that for orbital motion and is found to be given by $\gamma = ge/2m_e$, with g determined experimentally as 2.002 319 304 386, approximated to 2.0 for many purposes. Thus the magnetic moments are given by, for example,

$$\begin{aligned} \hat{\mu}_x &= \gamma\hat{S}_x = \tfrac{1}{2}\hbar\gamma\hat{\sigma}_x = \tfrac{1}{2}g\beta\hat{\sigma}_x \\ \bar{\mu}_x &= \tfrac{1}{2}g\beta\bar{\sigma}_x = \tfrac{1}{2}g\beta P_x \doteqdot \beta P_x \end{aligned} \tag{111}$$

The operators $\hat{S}_x$, $\hat{S}_y$ and $\hat{S}_z$ do not commute and only one component (chosen as S_z) can be assigned a definite (eigen) value while the other two average to zero for the eigenstates $|\alpha\rangle$ and $|\beta\rangle$ of $\hat{S}_z$. Of course, the same applies to the $\hat{\sigma}_i$. The direction of the vector **S** can never be known. Thus the measured spin magnetic moments and their components must be related to the mean values of the $\hat{S}_i$ or $\hat{\sigma}_i$, i.e. the polarization components, since these can be determined for any spin state, as above (see Figure 1.37)

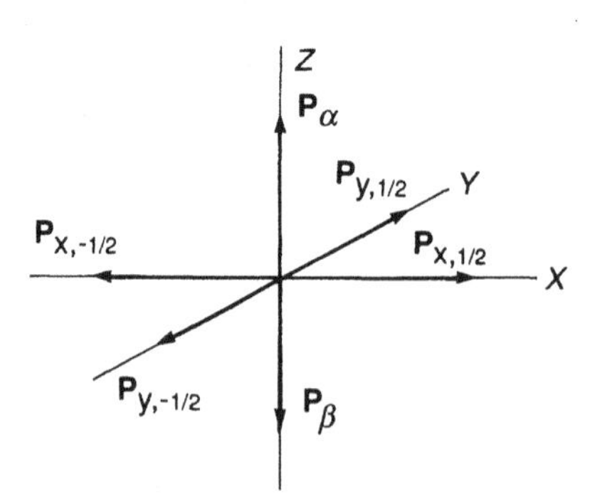

Figure 1.37 Polarization vectors for the states $|\alpha\rangle$, $|x, \frac{1}{2}\rangle$, etc., all of unit length. The vectors $\boldsymbol{\mu}$ are parallel or antiparallel to these

The clear distinction between the spin vector **S**, which can be envisaged as being in a state of continuous motion such that its direction can never be determined, and the polarization vector, which can be defined in direction and magnitude for any spin state, is vital to the understanding of magnetic phenomena.

A single electron is always characterized by the quantum number $S = \frac{1}{2}$. Greater values of S, integral or half-integral, arise when a number of individual spins are coupled, as discussed in Chapter 3. For two electrons in a 'ferromagnetic' type of coupling, $S = 1$, the magnitude of the angular momentum is $\hbar\sqrt{1(1+1)}$ and the determinate components are described as $M_S\hbar$ with $M_S = 1, 0, -1$. The associated magnetic moment components are $-g\beta M_S$.

Neutrons, protons and several other nucleii have spin $\frac{1}{2}$ and nucleons may couple to give higher spin quantum numbers. Conventionally $\hat{I}$ and $\hat{I}_z$ etc. replace $\hat{S}$ and $\hat{S}_z$ etc., but otherwise the same general principles apply. The values of γ, however, are very different (see Section 17).

3.5 Russell–Saunders Coupling: Hund's Rules

For all but the very heaviest atoms the total or atomic angular momenta are given by the application of Hund's rules. The way in which the individual electron contributions couple is known as Russell–Saunders coupling. The simple rules are verified by the appropriateness of the results obtained in relation to spectra, magnetic moments and susceptibilities, etc.

A set of electrons having the same quantum numbers n and l but identified by the permissible values of m_l and m_s constitute a shell—a closed or filled shell if all m_l and m_s values are involved. For 3d, $l = 2$, $m_l = 2, 1, 0, -1, -2$ and $m_s = \frac{1}{2}, -\frac{1}{2}$, the filled shell contains $n = 10$ identifiable electrons. For 4f, the maximum $n = 7 \times 2 = 14$. The first rule states that the system or atomic spin angular momentum is given by

$$|\mathbf{S}| = \hbar\sqrt{S(S+1)}, \qquad S_z = \hbar M_S, \qquad M_S = S, S-1, \ldots, -(S-1), -S \qquad (112)$$

where S is obtained as $\sum m_s$ over all the electrons with the individual m_s values assigned in such a way as to give the maximum value for S. Thus for $3d^n$, with n up to 5, it can be taken that $S = \frac{1}{2}n$, as in Figure 1.38, since the electrons, all with $m_S = \frac{1}{2}$, are identifiable by the five available m_l values. The rule may be compared with the results obtained for Heisenberg-coupled spins, for which it can be shown (Chapter 3) that, with n spins, $2n+1$ degenerate states result, characterized by the same value of S but by the different M_S values indicated above. For $n > 5$ it must be taken that the individual $m_s = -\frac{1}{2}$ for the sixth electron, etc., while the m_l values are repeated so that S falls to become zero for $n = 10$ as indicated (Figure 1.38). The number of M_S values is equal to $2S + 1$, which is called the spin multiplicity.

According to the second rule there is also a coupling between the individual orbital angular momenta leading to a total momentum characterized by L:

$$|\mathbf{L}| = \hbar\sqrt{L(L+1)}, \qquad L_z = \hbar M_L, \qquad M_L = L, L-1, \ldots, 0, \ldots, -(L-1), -L \quad (113)$$

where $L = \sum m_l$ again with the individual m_l values assigned so as to give the maximum sum permitted by the predominant rule 1. Thus for $n = 2$, $L = 2 + 2 = 4$ is proscribed by $m_s = \frac{1}{2}$ for both and $L = 2 + 1 = 3$. The remainder of the values as shown by Figure 1.38 follow for 3d and for 4f ($l = 3$, $m_l = 3, 2, 1, 0, -1, -2, -3$). Note particularly that for a

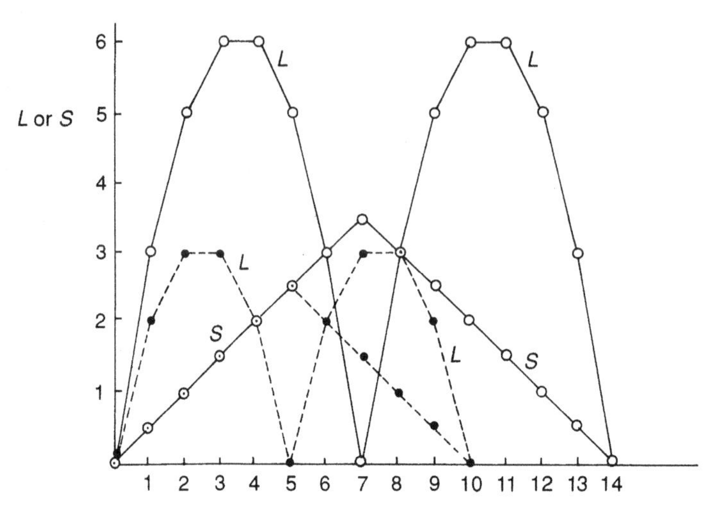

Figure 1.38 Hund's rules values of L and S for 3d (•) and 4f (○). (Reprinted by permission of John Wiley & Sons Ltd, Magnetic Oxides, ed. D.J. Craik, Wiley 1975)

half-filled shell $L = 0$ and the atom is said to be in an S-state (cf. 's-state' for one electron). For a filled shell $L = 0 = S$ and there is no permanent magnetic moment.

The atomic orbital and spin motions do not behave independently and there is a further spin–orbit coupling, discussed in Chapter 2, Section 5. The total atomic angular momentum is characterized by the quantum number J:

$$|\mathbf{J}| = \hbar\sqrt{J(J+1)}, \qquad J_z = \hbar M_J, \qquad M_J = J, (J-1), \ldots, -(J-1), -J \tag{114}$$

where the ground state J is given by

$$J = L + S \qquad \text{(shell more than half full)}$$

$$J = L - S \qquad \text{(shell less than half full)}$$

Hund's rule ground states are indicated symbolically by ${}^{2S+1}L_J$ with the explicit values of S and J and 'L' replaced by $S, P, D, F, G, \ldots$ for $L = 0, 1, 2, 3, 4, \ldots$.

If the atomic spin and orbital angular momenta could be considered separately the same values of g and of γ would apply to the atomic contributions and to the individual electrons. For strong spin–orbit coupling it is still possible to write

$$\mu = g_J \beta\sqrt{J(J+1)} \qquad (= p\beta, \text{ where } p = g_J\sqrt{J(J+1)}) \tag{115}$$

with

$$g_J = \frac{3J(J+1) + S(S+1) - L(L+1)}{2J(J+1)} \tag{116}$$

the derivation of g_J being given in Chapter 2, Section 5.3; p is the effective number of Bohr magnetons predicted, but conventionally p_e is usually taken to indicate the experimental value derived from susceptibility measurements, for example. Thus Table 1.2 shows immediately that p_e is far from the predicted values of p for the iron and palladium group

Table 1.2 The configurations, atomic quantum numbers, ground terms and predicted ionic magnetic moments for isolated atoms or ions of (a) the iron group transition metals, (b) the palladium group and (c) the rare earths, and the ionic moments in units of Bohr magnetons p_e as derived from measurements on ionic crystals (Reprinted by permission of John Wiley & Sons, Ltd, Magnetic Oxides, ed. D.J. Craik, Wiley 1975)

(a) Iron group, 3d; doubly ionized and neutral atoms

Ion (atom)	Configuration	S	L	J	Ground term	$g\sqrt{J(J+1)}$	p_e
Ca	—	0	0	0	^1S		
Ca^{2+}	$4s^2$	0	0	0	$^1S_0$	0	(dia)
Sc	$3d4s^2$	$\frac{1}{2}$	2	$\frac{3}{2}$	^2D		
Sc^{2+}	3d	$\frac{1}{2}$	2	$\frac{3}{2}$	$^2D_{3/2}$	1.55	1.7
Ti	$3d^24s^2$	1	3	2	^3F		
Ti^{2+} (V^{3+})	$3d^2$	1	3	2	$^3F_2$	1.63	2.8
V	$3d^34s^2$	$\frac{3}{2}$	3	$\frac{3}{2}$	^4F		
V^{2+} (Mn^{4+})	$3d^3$	$\frac{3}{2}$	3	$\frac{3}{2}$	$^4F_{3/2}$	0.77	3.8
Cr	$3d^54s$	2	0		^5S		
Cr^{2+} (Mn^{3+})	$3d^4$	2	2	0	$^5D_0$	0.0	4.9
Mn	$3d^54s^2$	$\frac{5}{2}$	0	$\frac{5}{2}$	^6S		
Mn^{2+} (Fe^{3+})	$3d^5$	$\frac{5}{2}$	0	$\frac{5}{2}$	$^6S_{3/2}$	5.92	5.9
Fe	$3d^64s^2$	2	2	4	^5D		
Fe^{2+} (Co^{3+})	$3d^6$	2	2	4	$^5D_4$	6.70	5.4
Co	$3d^74s^2$	$\frac{3}{2}$	3	$\frac{9}{2}$	^4F		
Co^{2+} (Ni^{3+})	$3d^7$	$\frac{3}{2}$	3	$\frac{9}{2}$	$^4F_{3/2}$	6.64	4.8
Ni	$3d^94s$	1	2		^3D		
Ni^{2+}	$3d^8$	1	3	4	$^3F_4$	5.59	3.2
Cu	$3d^{10}4s$	$\frac{1}{2}$	0		^2S		
Cu^{2+}	$3d^9$	$\frac{1}{2}$	2	$\frac{5}{2}$	$^2D_{5/2}$	3.55	1.9
Zn	$3d^{10}4s^2$	0	0		^1S		
Zn^{2+} (Cu^+)	$3d^{10}$	0	0		$^1S_0$	0.0	(dia)

(b) Palladium group, 4d; triply ionized

Ion (atom)	Configuration	S	L	J	Ground term	$g\sqrt{J(J+1)}$	p_e
Y^{3+}	—	0	0	0	$^1S_0$	0	
Zr^{3+}	4d	$\frac{1}{2}$	2	$\frac{3}{2}$	$^2D_{3/2}$	1.55	
Nb^{3+}	$4d^2$	1	3	2	$^3F_2$	1.63	0.7
Mo^{3+}	$4d^3$	$\frac{3}{2}$	3	$\frac{3}{2}$	$^4F_{3/2}$	0.77	3.6
Tc	—						
Ru^{3+}	$4d^5$	$\frac{5}{2}$	0	$\frac{5}{2}$	$^6S_{5/2}$	5.92	2.1
Rh^{3+}	$4d^6$	2	2	4	$^5D_4$	6.70	0.06
Pd^{3+}	$4d^7$	$\frac{3}{2}$	3	$\frac{9}{2}$	$^4F_{9/2}$	6.64	0.1
Ag^{3+}	$4d^8$	1	3	4	$^3F_4$	5.59	
Cd^{3+}	$4d^9$	$\frac{1}{2}$	2	$\frac{5}{2}$	$^2D_{5/2}$	3.55	

(c) Rare earth series, 4f; triply ionized

Ion (atom)	Configuration	S	L	J	Ground term	$g\sqrt{J(J+1)}$	p_e
La^{3+}	$5s^25p^6$	0	0	0	$^1S_0$	0.00	(dia)
Ce^{3+} (Pr^{4+})	$4f^15s^25p^6$	$\frac{1}{2}$	3	$\frac{5}{2}$	$^2F_{5/2}$	2.54	2.4

continued overleaf

Table 1.2 (*continued*)

Ion (atom)	Configuration	S	L	J	Ground term	$g\sqrt{J(J+1)}$	p_e
Pr^{3+}	$4f^2$–	1	5	4	$^3H_4$	3.58	3.6
Nd^{3+}	$4f^3$–	$\frac{3}{2}$	6	$\frac{9}{2}$	$^4I_{9/2}$	3.62	3.8
Sm^{3+}	$4f^5$–	$\frac{5}{2}$	5	$\frac{5}{2}$	$^6H_{5/2}$	0.84	1.5[a]
$Eu^{3+}(Sm^{2+})$	$4f^6$–	3	3	0	$^7F_0$	0.00	3.6[a]
$Gd^{3+}(Eu^{2+})$	$4f^7$–	$\frac{7}{2}$	0	$\frac{7}{2}$	$^8S_{7/2}$	7.94	7.9
Tb^{3+}	$4f^8$–	3	3	6	$^7F_6$	9.72	9.6
Dy^{3+}	$4f^9$–	$\frac{5}{2}$	5	$\frac{15}{2}$	$^6H_{15/2}$	10.63	10.6
Ho^{3+}	$4f^{10}$–	2	6	8	$^5I_8$	10.60	10.4
Er^{3+}	$4f^{11}$–	$\frac{3}{2}$	6	$\frac{15}{2}$	$^4I_{15/2}$	9.59	9.4
Tm^{3+}	$4f^{12}$–	1	5	6	$^3H_6$	7.57	7.3
Yb^{3+}	$4f^{13}$–	$\frac{1}{2}$	3	$\frac{7}{2}$	$^2F_{7/2}$	4.54	4.5
$Lu^{3+}(Yb^{2+})$	$4f^{14}$–	0	0	0	$^1S_0$	0.00	(dia)

[a]Room temperature values only.

ions *in crystals*, i.e. subject to crystal fields, and a better general correlation is obtained for $p_e = 2\sqrt{S(S+1)}$ for 3d, but reasonably $p_e \approx p$ for the 4f series.

This does not necessarily imply that the Russell–Saunders coupling scheme is inappropriate. At a descriptive level the 'magnetic' 4f electrons are shown by neutron diffraction studies to occupy a relatively small fraction of the space occupied by the ions and to be so shielded against the environment by the outer 5s and 5p electrons as to behave like isolated ions. Noting the loss of the 4s electrons on ionization, the 3d electrons are fully exposed to the crystal fields, i.e. the effects of neighbouring anions and, as shown in Chapter 2, Section 7 and already suggested, the orbital angular momentum may be 'quenched'. This quenching is not universal for the 3d series, and for Co^{2+} in particular the orbital motion is taken to be only partially quenched. However, $p_e \approx p$ does not give very good correlation and this appears to be a case where the crystal field effects overcome the spin–orbit coupling so that the moments are given by vector addition as $\mu^2 = \mu_S^2 + \mu_L^2$ and

$$p_e = [L(L+1) + 4S(S+1)]^{1/2} = 5.20 \tag{117}$$

whereas $p = 6.64$, $2\sqrt{S(S+1)} = 3.87$ and $p_e = 4.8$ to 5.2 by experiment.

Despite such complications the 'spin-only' rule constitutes a reasonable first approximation for 3d ions. In ferromagnetically ordered materials as $T \to 0$ or approximately at temperatures well below the Curie points where the spins are spontaneously aligned to the greatest possible extent, corresponding to the maximum value of $|M_S| = S$, this means that the magnetic moment component per ion is simply

$$\mu_z = -g_e \frac{e}{2m}\hbar M_S \doteqdot 2\beta S = n\beta \tag{118}$$

where n is the number of unpaired electrons. (Measurements of the spontaneous magnetization may be taken to give n_e, the effective number of unpaired electrons, such that $\mu_z = n_e\beta$ in practice (taking $g_e = 2$) and the spin-only picture applies with complete quenching to the extent that $n_e = n$.) The same principles apply with other types of magnetic order. The ionic moment (component) of Co^{2+} in CoO is 5.1β and in cobalt ferrite $\mu_z = 3.94\beta$ compared with the spin-only 3β, so it is clear once more that cobalt is exceptional and that the moments may depend on the detailed environment.

3.6 Paramagnetic Susceptibility and R–S Coupling

In an approximate treatment as discussed above the permitted values of the magnetic energy for the states $|JM_J\rangle$ are

$$E = -\mu_z B_0 = +g\beta M_J B_0 \tag{119}$$

($g \equiv g_J$). The probability P_M of finding an atom in the state M_J is thus

$$P_M = C\mathrm{e}^{-g\beta M_J B_0/kT} \tag{120}$$

The magnetic moment component in the state M_J is $\mu_z = -g\beta M_J$ and thus the mean μ_z of the representative atom is

$$\bar{\mu}_z = \frac{\displaystyle\sum_{M_J=-J}^{J} \mathrm{e}^{-g\beta M_J B_0/kT} \times (-g\beta M_J)}{\displaystyle\sum_{M_J=-J}^{J} \mathrm{e}^{-g\beta M_J B_0/kT}} \tag{121}$$

The denominator is the partition function Z for one atom. Differentiating this with respect to B_0 will give a sum similar to that in the numerator. By the chain law,

$$\frac{\partial Z}{\partial B_0} = \sum_{M_J=-J}^{J} \mathrm{e}^{-g\beta M_J B_0/kT} \times \left(\frac{-g\beta M_J}{kT}\right)$$

and the numerator is $(kT)\partial Z/\partial B_0$. Thus the mean moment is

$$\bar{\mu}_z = kT\frac{1}{Z}\frac{\partial Z}{\partial B_0} = kT\frac{\partial}{\partial B_0}\ln Z \tag{122}$$

Again, set $x = g\beta B_0/kT$, the ratio of magnetic to thermal energy. Then

$$\begin{aligned}
Z &= \sum_{M_J=-J}^{J} \mathrm{e}^{-xM_J} = \sum_{M_J=J}^{-J} \mathrm{e}^{-xM_J} \\
&= \mathrm{e}^{-xJ} + \mathrm{e}^{-x(J-1)} + \mathrm{e}^{-x(J-2)} + \cdots + \mathrm{e}^{-x(1-J)} + \mathrm{e}^{-x(-J)} \\
&= \mathrm{e}^{-xJ} + \mathrm{e}^{-xJ}\mathrm{e}^{x} + \mathrm{e}^{-xJ}\mathrm{e}^{2x} + \cdots + \mathrm{e}^{-xJ}\mathrm{e}^{(2J-1)x} + \mathrm{e}^{-xJ}\mathrm{e}^{2Jx} \\
&= \mathrm{e}^{-xJ}(1 + y + y^2 + \cdots + y^{2J-1} + y^{2J}), \qquad \text{where } y = \mathrm{e}^{x}
\end{aligned}$$

The sum of the geometric series is standard, i.e. $(1 - y^{2J+1})/(1 - y)$ and so

$$Z = \mathrm{e}^{-xJ}\frac{1 - \mathrm{e}^{x(2J+1)}}{1 - \mathrm{e}^{x}} = \frac{\mathrm{e}^{-xJ} - \mathrm{e}^{x(J+1)}}{1 - \mathrm{e}^{x}} = \frac{\mathrm{e}^{-x(J+\frac{1}{2})} - \mathrm{e}^{x(J+\frac{1}{2})}}{\mathrm{e}^{-x/2} - \mathrm{e}^{x/2}}$$

the final form being obtained by multiplying above and below by $\mathrm{e}^{-x/2}$. Using $\sinh y \equiv \frac{1}{2}(\mathrm{e}^{y} - \mathrm{e}^{-y})$ and later $\cosh y \equiv \frac{1}{2}(\mathrm{e}^{y} + \mathrm{e}^{-y})$ so that $\mathrm{d}/\mathrm{d}y \sinh ay = a\cosh ay$, $\coth y = \cosh y/\sinh y$, then

$$\ln Z = \ln\sinh(J + \tfrac{1}{2})x - \ln\sinh\tfrac{1}{2}x$$

and by equation (122),

$$\bar{\mu}_z = kT\frac{\partial}{\partial B_0}\ln Z = kT\left[\frac{(J+\frac{1}{2})\cosh(J+\frac{1}{2})x}{\sinh(J+\frac{1}{2})x} - \frac{\frac{1}{2}\cosh\frac{1}{2}x}{\sinh\frac{1}{2}x}\right]\frac{g\beta}{kT} \tag{123}$$

This is usually written as

$$\bar{\mu}_z = g\beta J B_J(x) \tag{124}$$

where the Brillouin function of x for a specified J is

$$B_J(x) \equiv \frac{1}{J}\left[(J+\tfrac{1}{2})\coth(J+\tfrac{1}{2})x - \tfrac{1}{2}\coth\tfrac{1}{2}x\right] \tag{125}$$

The limiting behaviour may again be studied. When y is large, e^{-y} is negligible compared with e^y and $\coth y \to 1$. In this limit $B_J(x) \to (1/J)(J+\frac{1}{2}-\frac{1}{2}) = 1$ and $\mu_z \to g\beta J$. Since J is the largest possible value of M_J this is as expected.

Conversely, as y becomes small the exponentials may be truncated to give $\cosh y = 1 + y + y^2/2 + y^3/6 + (1 - y + y^2/2 - y^3/6) = 2(1 + y^2/2)$ and $\sinh y = 2(y + y^3/6)$. Using the binomial expansion up to terms quadratic in y,

$$\coth y = \frac{1+(y^2/2)}{y}\left(1+\frac{y^2}{6}\right)^{-1} = \frac{1}{y}\left(1+\frac{y^2}{2}\right)\left(1-\frac{y^2}{6}\right)$$

$$= \frac{1}{y}[1 + y^2(\tfrac{1}{2} - \tfrac{1}{6})] = \frac{1}{y} + \frac{y}{3} \tag{126}$$

Note that the correct expression is not obtained if the exponentials are approximated to lower terms since this excessively restricts the approximations in the further development. Replacing the coth functions in equation (125), a little rearrangement gives

$$B_J(x) = \frac{J+1}{3}x, \qquad \text{where } x = g\beta B_0/kT \ll 1 \tag{127}$$

In the 'zero-field' limit and with a density of atoms N m^{-3}, the magnetization is

$$M = N\bar{\mu}_z = Ng\beta J B_J(x) = Ng\beta J\frac{J+1}{3}\frac{g\beta B_0}{kT} = \frac{Ng^2\beta^2 J(J+1)B_0}{3kT} \tag{128}$$

and the zero-field susceptibility is $M/H_0 = M/(B_0/\mu_0)$, i.e.

$$\chi = \frac{\mu_0 N g^2\beta^2 J(J+1)}{3kT} \tag{129}$$

which is the more general form of the Curie law, $\chi = C/T$. Note that if we have one electron per atom in an s-state with $L = l = 0$ and thus $J = S = s = \frac{1}{2}$, then $\chi = [\mu_0 N g^2\beta^2(\frac{3}{4})]/3kT = \mu_0 N g^2\beta^2/(4kT)$, as obtained previously. The levels and magnetic moments for $J = 1$ are shown in Figure 1.39. The reader may derive χ directly for this case, using $e^y = 1 + y$, and compare the result with equation (129).

To derive the general zero-field susceptibility directly, in the approximation $e^y = 1 + y$ the partition function is $1 - M_J x$ summed over all the $2J + 1$ levels and by symmetry

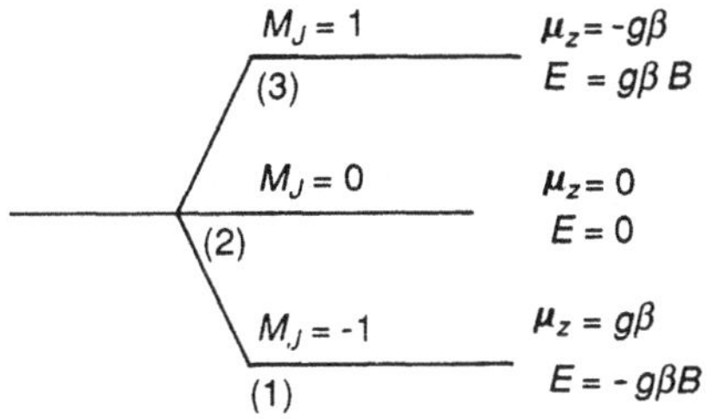

Figure 1.39 Energy level splitting in a magnetic field for coupled states, $J = 1$

$\sum M_J = 0$ so that $Z = 2J + 1$ simply. We then have

$$\bar{\mu}_z = \sum_{M_J=-J}^{J} P_{M_J}(-M_J g\beta) = \sum_{M_J=-J}^{J} \frac{1 - M_J x}{2J + 1}(-M_J g\beta) = \sum_{M_J=-J}^{J} M_J^2 x g\beta$$

again using $\sum M_J = 0$. Of course $\sum M_J^2$ does not vanish and it can be shown (or verified by simple examples) that

$$\sum_{M_J=-J}^{J} M_J^2 = \frac{J(J+1)(2J+1)}{3}$$

Substituting for $\sum M_J^2$ and for x gives $\bar{\mu}_z$ and χ, as above.

3.7 General Hamiltonian for Motion in a Field: Diamagnetism

Returning to orbital motion in a magnetic field, it has been noted that the arbitrary addition of $-\gamma B L_z$ to $\mathcal{H}_0$ corresponds to an approximation. The correct modification is

$$\mathcal{H}_0 = \frac{p^2}{2m} + V \rightarrow \mathcal{H} = \frac{1}{2m}(\mathbf{p} - e\mathbf{A})^2 + V \tag{130}$$

for each electron, to be summed overall for a system of electrons. As discussed in Chapter 2, the field may be introduced explicitly by confirming that $\mathbf{B} = \mathbf{k}B$ corresponds to a vector potential

$$\mathbf{A} = \mathbf{i}(-\tfrac{1}{2}yB) + \mathbf{j}(\tfrac{1}{2}xB)$$

using $\mathbf{B} = \nabla \times \mathbf{A}$, and the Hamiltonian is then obtained in the form

$$\begin{aligned} \mathcal{H} &= \frac{p^2}{2m} - \gamma B L_z + B^2 \frac{e^2}{8m}(x^2 + y^2) + V \\ &= \mathcal{H}_0 - \gamma B L_z + B^2 \frac{e^2}{8m}(x^2 + y^2) \end{aligned} \tag{131}$$

and the nature of the foregoing approximation is apparent.

It cannot now be assumed that the eigenstates are those of the zero-field $\mathcal{H}_0$. Perturbation theory can be applied to obtain the energies and thus the mean values of the magnetic moments follow according to

$$\mu = -\frac{\partial E}{\partial B}$$

When this is carried out three terms are obtained. The first gives the simple paramagnetic moment as obtained previously. The next two are proportional to B, the first of these being positive and shown by Van Vleck to give an additional contribution to the paramagnetic χ, which is, however, independent of temperature — 'temperature independent' or 'Van Vleck' paramagnetism. Although of particular interest in some cases it is a small effect and is frequently omitted.

The last term (to the order of approximation usually accepted) is

$$-B\langle j|\sum_i \frac{e_i^2}{4m_i}(x_i^2 + y_i^2)|j\rangle$$

summed over the electrons. Since the coordinate operators are simply the classical coordinates, this gives

$$\mu_{\rm d} = -B\frac{e^2}{4m}(\bar{x}_i^2 + \bar{y}_i^2) \qquad (132)$$

(all $e_i = e$ and $m_i = m$ for the electrons). Taking $x^2 + y^2 + z^2 = r^2$ and $\bar{x}^2 + \bar{y}^2 = \frac{2}{3}\bar{r}^2$, the susceptibility contribution is

$$\chi_{\rm d} = \frac{N\mu_{\rm d}}{H} = -\frac{N\mu_0 e^2}{6m}\bar{r}_i^2 \qquad (133)$$

(N atoms in unit volume). This is the diamagnetic (negative) susceptibility which sums directly with the paramagnetic contribution:

$$\chi = \chi_{\rm para} + \chi_{\rm d}$$

$\chi_{\rm d}$ is about an order smaller in magnitude for atoms or small molecules for which $\mu \approx \beta$ and is often regarded as a small empirically estimated correction, but closed-shell atoms or ions (L, S and $J = 0$) and most molecules with paired spins and quenched orbital angular momentum (e.g. benzene, as previously noted) are purely diamagnetic. The classical picture consists of the induction of circulating currents, which must have a sense that produces fields through the loops that oppose the applied field, giving induced moments antiparallel to the applied field.

If the electrons are taken to be localized then the molar $\chi_{\rm d}$ can be taken to be the sum of contributions for the constituent atoms in a molecule. However, if some of the electrons are considered free to move around, e.g. conjugated π-electron rings as in benzene, the sum above should be taken using the radius of the ring giving relatively large effects. The x and y coordinates in equation (132) are specifically those in a plane normal to the applied field and the effects are clearly anisotropic with the largest effects if the plane of the ring is taken to be normal to the field, as implied above. For generality, χ should be interpreted as a tensor with

$$\mathbf{M} = \boldsymbol{\chi}\mathbf{H} \sim \begin{pmatrix} M_x \\ M_y \\ M_z \end{pmatrix} = \begin{pmatrix} \chi_{xx} & \chi_{xy} & \chi_{xz} \\ \chi_{yx} & \chi_{yy} & \chi_{yz} \\ \chi_{zx} & \chi_{zy} & \chi_{zz} \end{pmatrix} \begin{pmatrix} H_x \\ H_y \\ H_z \end{pmatrix} \qquad (134)$$

If $\mathbf{H}$ is applied along a principal axis χ is diagonal and if all the elements χ_{ii} are equal then

$$\boldsymbol{\chi} = \mathbf{I}\chi, \qquad M_z = \chi H_z, \qquad \text{etc.}$$

and the isotropic case is resumed. For the anisotropic case an average χ is sometimes taken to be

$$\chi = \tfrac{1}{3}(\chi_{xx} + \chi_{yy} + \chi_{zz}) \tag{135}$$

Taking the isotropic case with

$$M = \chi H, \qquad \chi = \left(\frac{\mathrm{d}M}{\mathrm{d}H}\right)_{H=0}$$

the work done to increase the magnetization by $\mathrm{d}M$ is

$$\mathrm{d}w = -\mu_0 H\, \mathrm{d}M$$

and the energy built up on increasing the field from zero to a value H is, with $\mathrm{d}M = \chi \mathrm{d}H$,

$$E = -\mu_0 \int_0^H H\chi \mathrm{d}H = -\tfrac{1}{2}\mu_0 \chi H^2 = -\tfrac{1}{2}\mu_0 H M \tag{136}$$

From this

$$\frac{\mathrm{d}E}{\mathrm{d}H} = -\mu_0 \chi H = -\mu_0 M, \qquad \frac{\mathrm{d}^2 E}{\mathrm{d}H^2} = -\mu_0 \chi \tag{137}$$

and, by expansion, for example,

$$\chi_{xx} = -\frac{1}{\mu_0}\left(\frac{\partial^2 E}{\partial H_x^2}\right)_{H_x=0}$$

and

$$M_i = -\frac{1}{\mu_0}\left(\frac{\partial E}{\partial H_i}\right)_0, \qquad \mu_i = -\frac{1}{\mu_0}\left(\frac{\partial E}{\partial H_i}\right)_0 \qquad \text{for } i = x, y, z \tag{138}$$

taking E as the energy per unit volume or per particle respectively, with moment per particle $\boldsymbol{\mu}$. The energy of a particle with permanent ($\boldsymbol{\mu}$) and induced moments can be written as

$$\begin{aligned} E &= E_0 - \mu_0 \boldsymbol{\mu} \cdot \mathbf{H} - \tfrac{1}{2}\mathbf{H} \cdot \boldsymbol{\chi} \cdot \mathbf{H} \\ &= E_0 + \sum_i \left(\frac{\partial E}{\partial H_i}\right) H_i + \tfrac{1}{2} \sum_{i,j} \left(\frac{\partial^2 E}{\partial H_i \partial H_j}\right)_0 H_i H_j \end{aligned}$$

The anisotropy of the ring compounds may be indicated by the difference

$$\Delta = \chi_{zz} - \tfrac{1}{2}(\chi_{xx} + \chi_{yy}) \tag{139}$$

with χ_{zz} measured normal to the plane of the molecules. χ_{xx} and χ_{yy} should be the contributions due to the localized electrons only. As χ_{zz} depends on both these and the ring or mobile electrons, Δ should correlate with the number of mobile electrons in the ring, as in Table 1.3 [2]. Table 1.4 indicates the anisotropy of graphite, as expected, and gives some values to illustrate magnitudes of χ_d for metals.

Some organic materials, such as charge-transfer complexes of TCNQ (tetracyanoquinodimethane), have substantial concentrations of free spins due to the charge transfer, up to one per molecule, giving a paramagnetic χ_p. However, this χ_p may be reduced by band

Table 1.3 Indication of magnetic anisotropy of ring compounds [2]

Molecule	$\chi_{zz} - \frac{1}{2}(\chi_{xx} + \chi_{yy})$ $(10^{-9}\ \mathrm{m^3\ mol^{-1}})$	Number of free electrons
(benzene ring)	−0.750	6
(S ring)	−0.630	5
(triangle)	−0.214	3

Table 1.4 Diamagnetic susceptibility of graphite, with fields or parallel or perpendicular to the hexagonal c axis and of some metals ($\chi \times 10^6$)

Graphite (‖)	−260	Graphite (⊥)	−3.8		
Zinc	−1.8	Mercury	−2.1	Copper	−1.1
Silver	−2.4	Gold	−1.8	Germanium	−1.3

formation or by interactions favouring antiparallel alignment of neighbouring spins, and due to the presence of ring structures χ_d is large in magnitude. Thus at certain temperatures it is found that $\chi(T) = \chi_p(T) + \chi_d = 0$, with no response to applied fields in the (paramagnetic) linear region. However, χ_d is temperature-independent, or has a very small dependence due to the population of excited states as T increases, and even for materials that are net diamagnetics throughout the behaviour of $\chi_p(T)$ can be determined.

4 Larmor and Cyclotron Frequencies

For a single orbiting electron the energy associated with the induced diamagnetic moment is

$$E_d = \tfrac{1}{2}B^2 \frac{e^2}{4m}(x^2 + y^2)$$

corresponding to $\mu = -\partial E/\partial B$, $B = \mu_0 H$. With $x^2 + y^2 = r^2$ as the radius of the orbit in OXY and $I_z = mr^2$ for the moment of inertia,

$$E_d = \tfrac{1}{2}\left(\frac{eB}{2m}\right)^2 I_z \qquad (\text{cf. } E = \tfrac{1}{2}I_z\omega^2)$$

the latter being the classical expression with ω the angular velocity. Thus the classical angular velocity is

$$\omega = \frac{eB}{2m}, \qquad f = \frac{\omega}{2\pi} = \frac{eB}{4\pi m} \tag{140}$$

and f is called the Larmor frequency.

This is very similar to the cyclotron frequency. From Figure 1.40(a) the magnetic force on the particle with mass m and charge e is

$$\mathbf{F} = e\mathbf{v}\times\mathbf{B}, \qquad F = evB$$

with $\mathbf{v} \perp \mathbf{B}$. Equating this to the centripetal force. $F = mv^2/r = mr\omega^2$ $(v = \omega r)$ gives

$$\omega = \frac{eB}{m} \tag{141}$$

which is the cyclotron frequency.

Since there is no force component in the direction of $\mathbf{v}$ there is no charge in v so that the force balance is maintained, giving a circular orbit about an origin specified by the initial $\mathbf{v}$ and $\mathbf{B}$. For a general trajectory $\mathbf{v}$ may be resolved into $\mathbf{v}_\perp$ and $\mathbf{v}_\parallel$ in relation to $\mathbf{B}$ with $\mathbf{v}_\parallel \times \mathbf{B} = 0$, giving a superposition of the unchanged $\mathbf{v}_\parallel$ and the above circular orbiting motion with $\mathbf{v}$ replaced by $\mathbf{v}_\perp$, i.e. a spiral about $\mathbf{B}$ as in Figure 1.40(b). In practice energy is radiated due to the continuous changes in the vector velocity, even with v constant.

The angular velocity ω is independent of v but the radius of the trajectory, as viewed along $\mathbf{B}$, depends upon v. Using e and m for an electron gives the frequency as

$$\frac{f}{B} = \frac{\omega}{2\pi B} = 30 \text{ GHz } T^{-1}$$

e.g. for a free electron in the ionosphere with $B \sim 10^{-4}$ T, $f \sim 3$ MHz.

Cyclotron resonance can be observed with radiation of appropriate frequencies for free electrons and electrons in bands, giving valuable information on the band structure of metals. Ions absorb at their (much lower) cyclotron frequencies, giving ion cyclotron resonance.

When very high frequencies are involved allowance should be made for the relativistic correction of the rest mass m_0 according to

$$m = m_0 \left(1 - \frac{v^2}{c^2}\right)^{-1/2}$$

giving the cyclotron frequency

$$\omega = \frac{eB}{m_0}\left(1 - \frac{v^2}{c^2}\right)^{1/2} \tag{142}$$

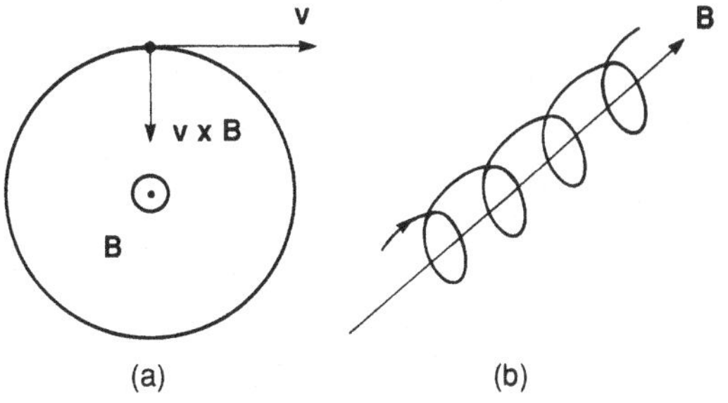

Figure 1.40 The cyclotron effect, B normal to the diagram (a), leading to the spiral motion in (b) when the initial velocity is neither $\parallel$ nor normal to B

and this is often termed the synchrotron frequency. Even if v is considered constant, $\mathbf{v}$ varies continuously and the energy radiated from ionized particles travelling at relativistic frequencies constitutes a major source of the radiofrequency energy reaching the earth.

Electron beams can be focused by appropriate field gradients and the magnetic control of electron trajectories is a vital feature of many instruments from electron microscopes to microwave tubes, in which latter the fields may be unidirectional or periodically reversed along the path of the beam. A quantum treatment, as in Chapter 2, will be seen to give the same cyclotron frequency.

5 Paramagnetism of Metals

Metals such as sodium may first be seen as an assembly of closed-shell diamagnetic ion cores in a sea of free electrons, the valence electrons, which may respond to a magnetic field. With $V = 0$ and $\mathcal{H} = -(\hbar^2/2m)\nabla^2$ it is apparent that $\mathcal{H}\psi = E\psi$ is satisfied by $\psi = Ne^{i(k_x x + k_y y + k_z z)} = Ne^{i\mathbf{k}\cdot\mathbf{r}}$ and that $E = (\hbar^2/2m)k^2 = (\hbar^2/2m)(k_x^2 + k_y^2 + k_z^2)$, where $\mathbf{k}$ is the wave vector. For a cubic crystal $L \times L \times L$ the boundary conditions indicate that $E = (\hbar^2\pi^2/2mL^2)(n_x^2 + n_y^2 + n_z^2)$ with n_x, n_y, n_z integral. For any substantial L the levels form a near-continuous band. Set $r^2 = n_x^2 + n_y^2 + n_z^2$ so that in the space defined by the n's as components, unit volume corresponds to a single state $|n_x n_y n_z\rangle$ and each value of r to a particular energy. The number of states with energy up to $E(r)$ is $N = 2 \times (\frac{1}{8})(\frac{4}{3}\pi r^3)$ with the factor of 2 for spin and $(\frac{1}{8})$ since one octant of this space is considered. The density of states is $D(E) = \mathrm{d}N/\mathrm{d}E = (\mathrm{d}N/\mathrm{d}r)(\mathrm{d}r/\mathrm{d}E)$ and the reader may show very simply that this gives

$$D(E)\mathrm{d}E = \frac{4\pi V(2m)^{3/2}}{h^3} E^{1/2}\,\mathrm{d}E = KE^{1/2}\,\mathrm{d}E \tag{143}$$

$(V = L^3)$ so that the density of states curve is parabolic as in Figure 1.41. The electrons are to be treated collectively so Fermi–Dirac statistics apply with the probability of occupation of states with energy E_i:

$$F(E_\mathrm{i}) = \frac{1}{1 + e^{(E_\mathrm{i} - E_\mathrm{f})/kT}} \tag{144}$$

E_f, the Fermi level, is seen by examination to be the highest occupied level at $T \to 0$ and $F(E_f) = \frac{1}{2}$. The electrons may have spin components parallel or antiparallel to a field and may be considered to occupy two 'half-bands' α and β shifted by the field by

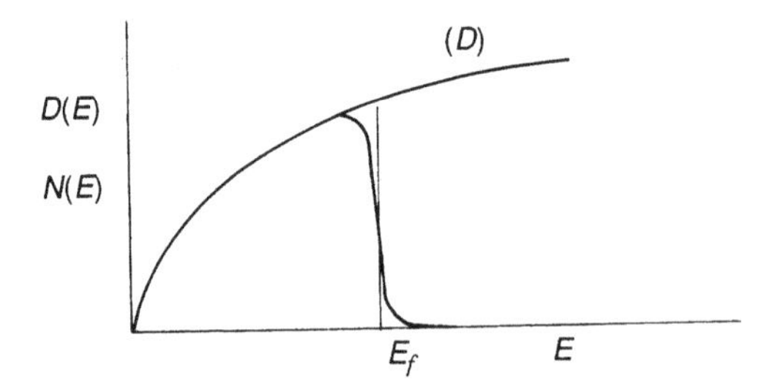

Figure 1.41 Density of states $D(E)$ or of occupied states, $N(E) = F(E) \times D(E)$

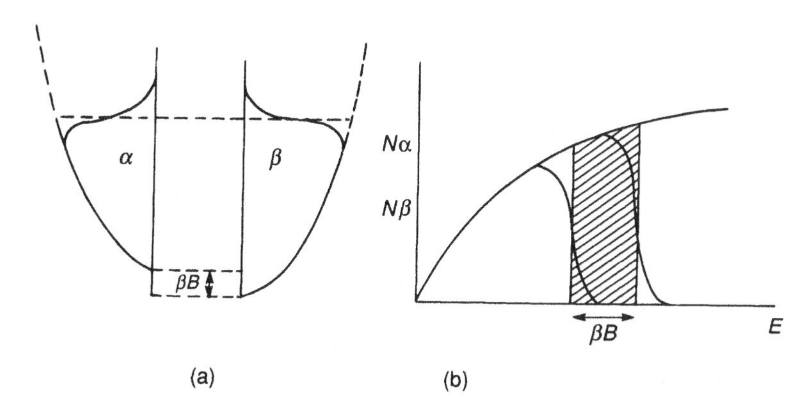

Figure 1.42 (a) The field-induced shift in the 'half-bands' is indicated for the α and β electrons and (b) shows the two superimposed with a common origin

$\pm\beta B$, as in Figure 1.42, in which $F(E)$ is superimposed on the parabolic bands to give $N(E) = F(E)D(E)$. The difference in populations $n_\beta - n_\alpha$ is approximated by the shaded area and M and χ are

$$M = (n_\beta - n_\alpha)\beta = D(E_f)\beta^2 B, \qquad \chi = \chi_p = \mu_0 D(E_f)\beta^2 = \frac{\mu_0 N \times 3\beta^2}{2kT_f} \tag{145}$$

the second since $dN/dE|_{E_f} = KE_f^{1/2}$ so that $D(E_f) = \frac{3}{2}N/E_f$. This is Pauli paramagnetism, χ_p being smaller than the value that would be given by applying Boltzmann statistics to identifiable particles and, in this approximation, independent of temperature [3]. A correction by Wilson [4] introduces a diamagnetic contribution with one-third of the above magnitude, giving $\chi_p = \mu_0 N\beta^2/(KT_f)$. Accounting for the diamagnetism of the cores, Cu, Ag and Au are net diamagnetics with $\chi_m = -0.086, -0.19, -0.136 \times 10^{-6}$, while Li and Na are weakly paramagnetic.

A surface of fixed energy in k space, such as that at E_f, i.e the Fermi surface, is spherical for free electrons. (This is a reciprocal space since $\mathbf{k} \cdot \mathbf{r}$ must be dimensionless.) When the periodic potential from the cores is taken into account the simple travelling waves become Bloch functions $\psi(\mathbf{r}) = Au_k(\mathbf{r})e^{i\mathbf{k}\cdot r}$, $u_k(\mathbf{r})$, having the periodicity of the lattice. Fermi surfaces are distorted from the spherical. Energy gaps arise at $k = n\pi/a$, separating a number of bands, typically, for metallic behaviour, a filled valency band and a half-filled ($T \to 0$) conduction band. An empty conduction band, separated by a large gap from the valence band, corresponds to absence of conductivity or paramagnetic response since, by the Pauli principle, neither the $\mathbf{k}$ values (and momenta) nor the spin components can change in the filled band. If the gap is small ($\sim kT$) a Fermi distribution indicates some population of the upper band with resulting semiconductivity and a (calculable) weak paramagnetic response. For metals $E \propto k^2$ may still apply approximately with distortions towards the band edges, with $D(E)$ that for free electrons, with a correction effected by writing $E = \hbar k^2/2m^*$, with $m^* = \hbar^2/(d^2E/dk^2)$ for the particular band structure. For Li and Na, $m^*/m = 1.5$ and 0.99, giving calculated $\chi_p = 1.5$ and 0.8×10^{-5} compared with 2.6 and 1.2×10^{-5} by e.s.r., the discrepancies being ascribed to enhancement by exchange coupling by Pines [5]. The effective mass m^* varies over the Fermi surface and in the anisotropic case must be

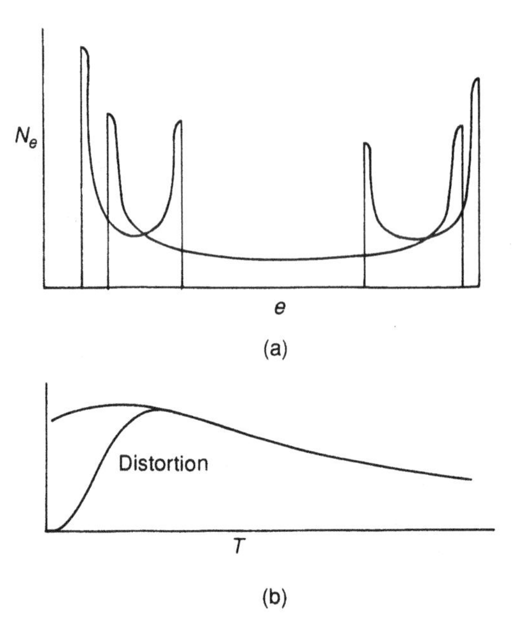

Figure 1.43 (a) Density of states curve for a narrow band for electrons in a one-dimensional periodic potential together with the splitting into two bands with a central gap consequent on the introduction of a Peierls distortion or symmetry reduction. (b) $\chi(T)$ for the original (half-filled) band and the reduction at low temperatures caused by the introduction of the gap and the metal–semiconductor transition (very similar to the reduction due to antiferromagnetic coupling)

described by $(m_{ij}^*)^{-1} = (1/\hbar^2)\partial^2 E_k/\partial k_i \partial k_j$ for $i, j, k = x, y, z$. The components may be measured and the Fermi surface studied by cyclotron resonance.

The bands may also be seen as being derived from the d and s levels of the separate atoms, with narrow d bands corresponding to substantial localization. They may overlap in energy, and high values of $N(E_f)$ in the d band have been invoked to explain the relatively high, and weakly temperature-dependent, χ values of non-ordered transition metals.

In one dimension, if alternate atoms approach in pairs the period becomes $2a$ and new gaps may open at half the original k values. If such a gap corresponded to the Fermi level then some of the electron energies would be pulled down so that such a Peierls distortion [6] might occur spontaneously, to give a metal–semiconductor transition with a drop in χ (Figure 1.43). Many observations of discontinuities in $\chi(T)$ in quasi-one-dimensional charge-transfer crystals containing separate stacks of electron-donor and of electron-acceptor molecules have been explained in this way.

6 Magnetoconductivity, Hall, de Haas–van Alphen Effects

The force on an electron in the presence of electric and magnetic fields is

$$\mathbf{F} = -e(\mathbf{E} + \mathbf{v}\times\mathbf{B}) = \frac{d\mathbf{p}}{dt} = \hbar\frac{d\mathbf{k}}{dt} \tag{146}$$

using $\mathbf{p} = m\mathbf{v} = \hbar\mathbf{k}$. When $\mathbf{B} = 0$,

$$\mathbf{k}(t) - \mathbf{k}(c) = -e\frac{\mathbf{E}t}{\hbar}$$

and the Fermi surface is moved by

$$\delta\mathbf{k} = -\frac{e\mathbf{E}t}{\hbar} \tag{147}$$

along the k_x axis in k space, if $\mathbf{E} = \mathbf{i}E$, as in Figure 1.44. Since $\mathbf{v} = (\hbar/m)\mathbf{k}$ this indicates current flow which in practice is limited by scattering by imperfections or phonons, giving a steady state static displacement. If the collision time is τ the consequent change in velocity is

$$\delta\mathbf{v} = \frac{\hbar}{m}\delta\mathbf{k} = -\frac{e\tau}{m}\mathbf{E} \tag{148}$$

and with n electrons per unit volume and current density $\mathbf{J}$, Ohm's law is obtained:

$$\mathbf{J} = n(-e)\delta\mathbf{v} = \frac{ne^2\tau}{m}\mathbf{E} \tag{149}$$

For the case $\mathbf{E} = 0$ and $\tau \to \infty$ (zero resistance) the component equations of motion with $\mathbf{B} = \mathbf{k}B$, $\delta\mathbf{v} = (\hbar/m)\delta\mathbf{k}$ and $F = m\,\mathrm{d}\mathbf{v}/\mathrm{d}t$ are

$$m\frac{\mathrm{d}}{\mathrm{d}t}\delta v_x = -eB\delta v_y, \qquad m\frac{\mathrm{d}}{\mathrm{d}t}\delta v_y = eB\delta v_x \tag{150}$$

Differentiating one of these and substituting in the other gives 'SHM equations' in the separate variables with solutions

$$\delta v_x = v_0 \cos\omega_c t, \qquad \delta v_y = v_0 \sin\omega_c t, \qquad \text{where } \omega_c = \frac{eB}{m} \tag{151}$$

Resonant absorption can occur at the cyclotron frequency ω_c. If τ is finite but large and $\tau^{-1} = \lambda$ is small, the equations contain a damping term $m\lambda\,\mathrm{d}(\delta v_x)/\mathrm{d}t$ which broadens the

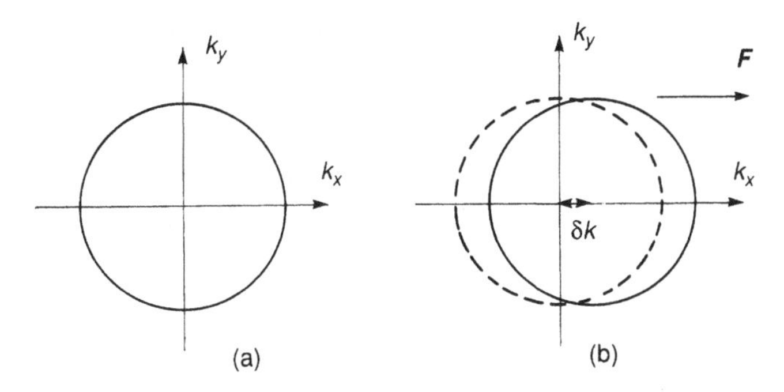

Figure 1.44 (a) Fermi surface in the absence of a field or at $t = 0$ when the field is switched on. For each $\mathbf{k} = \mathbf{k}'$, $\mathbf{k} = -\mathbf{k}'$ exists and there is no net momentum or velocity. (b) At time t with a field $\mathbf{E}$ and force $\mathbf{F}$ the sphere is displaced by $\delta\mathbf{k} = \mathbf{F}t/\hbar$. The progressive displacement is arrested by scattering processes

resonance. For $B = 1.0$ T, $f_c = \omega_c/2\pi \approx 20$ GHz (~ 2 mm waves). Low temperatures are required for τ to be sufficiently long ($\tau = 10^{-9}$ s at 4 K in copper) for the resonance to be observed.

With simultaneous fields $\mathbf{B}(=\mathbf{k}B)$ and $\mathbf{E}$ (general) the equations of motion are

$$m\left(\frac{d}{dt}+\frac{1}{\tau}\right)\delta v_x = -e\left(E_x + B\delta v_y\right)$$

$$m\left(\frac{d}{dt}+\frac{1}{\tau}\right)\delta v_y = -e\left(E_y - B\delta v_x\right)$$

$$m\left(\frac{d}{dt}+\frac{1}{\tau}\right)\delta v_z = -eE_x \tag{152}$$

When the steady state is achieved the relevant equations are the above, with the omission of d/dt. The solutions are then readily obtained as

$$\delta v_x = -\frac{e\tau/m}{1+(\omega_c\tau)^2}(E_x - \omega_c\tau E_y)$$

$$\delta v_y = -\frac{e\tau/m}{1+(\omega_c\tau)^2}(E_y + \omega_c\tau E_x) \tag{153}$$

with implicit dependence on B, from ω_c. The current density is

$$\mathbf{J} = n(-e)\delta\mathbf{v} \tag{154}$$

Setting $\sigma_0 = ne^2\tau/m$ and $\mathbf{J} = \boldsymbol{\sigma}\mathbf{E}$ with a conductivity tensor $\boldsymbol{\sigma}$, the components are

$$J_x = \sigma_{xx}E_x + \sigma_{xy}E_y = \frac{\sigma_0}{1+(\omega_c\tau)^2}(E_x - \omega_c\tau E_y)$$

$$J_y = \sigma_{yx}E_x + \sigma_{yy}E_y = \frac{\sigma_0}{1+(\omega_c\tau)^2}(\omega_c\tau E_x + E_y)$$

$$J_z = \sigma_{zz}E_z = \sigma_0 E_z$$

Noting that J_z depends on E_z only, these can be written as

$$\begin{pmatrix} J_x \\ J_y \\ J_z \end{pmatrix} = \frac{\sigma_0}{1+(\omega_c\tau)^2}\begin{pmatrix} 1 & -\omega_c\tau & 0 \\ \omega_c\tau & 1 & 0 \\ 0 & 0 & 1+(\omega_c\tau)^2 \end{pmatrix}\begin{pmatrix} E_x \\ E_y \\ E_z \end{pmatrix} \tag{155}$$

For the geometry of Figure 1.45, $J_y = 0$ and a field component E_y must arise, given by

$$E_y = -\omega_c\tau E_x = -\frac{eB\tau}{m}E_x \tag{156}$$

E_y is the Hall field and since it is proportional to the component of the magnetic field across the strip or crystal, measurement of E_y constitutes a valuable and much-used vector field measurement.

The Hall constant is defined as

$$R_H = \frac{E_y}{J_x B} = -\frac{eB\tau E_x/m}{ne^2\tau E_x B/m} = -\frac{1}{ne} \tag{157}$$

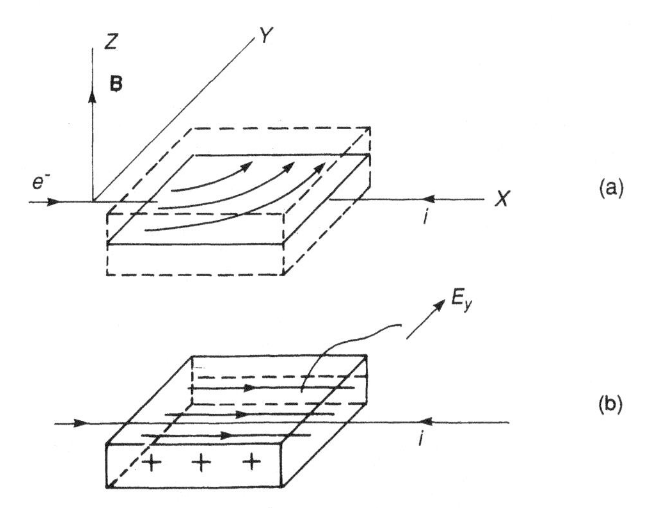

Figure 1.45 The initial motion of electrons in a magnetic field (a) and the steady state motion (b) in the presence of the Hall field

using equations (149) and (156). With e the proton charge, R_H is negative for simple metals and its experimental value gives the concentration of conduction electrons, n, as simply one per atom for Li, Na and K. In the semiconductor $R_H < 0$ for n-type and $R_H > 0$ for p-type materials in which the carriers are electrons or holes respectively.

In semi-metals such as Sb and Bi (with relatively low conductivity due to a small population of an overlapping upper band) and in semiconductors, the magnitude of R_H may be many orders higher than those for Na, etc., due to the smaller n. With $E_y = R_H J_x B$ the sensitivity of such crystals in field measuring devices is correspondingly high.

The effect may be envisaged as the build-up of charge on the crystal edges as in Figure 1.45(a) as the current commences. In the steady state the consequent E fields balance the deflecting effects of B and the carriers move $\parallel OX$ [Figure 1.45(b)]. If the current, and thus **v**, is reversed, **F** is reversed and the charges are reversed. By applying suitable electrodes and using high-impedance circuits the Hall effect may be advantageously observed as a high-frequency voltage proportional to the relevant component of **B**.

Combining the expressions for R_H and σ,

$$R_H\sigma = \frac{1}{ne}\frac{ne^2\tau}{m^*} = \frac{e\tau}{m^*} = \mu \tag{158}$$

with μ the carrier mobility in terms of which $\sigma = ne\mu$. Thus measured values of R_H and σ give mobilities.

As shown in Chapter 2, Section 4.2, an electron confined to move in the OXY plane but otherwise free, with a field $\mathbf{B} \parallel OZ$, moves in orbits with quantized energies:

$$E_l = \frac{e\hbar B}{m}(l + \tfrac{1}{2}) \qquad \text{for } l = 1, 2, 3, \ldots$$

and assuming the transition rule $\Delta l = \pm 1$ this also corresponds to cyclotron resonance at

$$f = \frac{1}{h}\frac{e\hbar B}{m} = \frac{\omega_c}{2\pi} = f_c \tag{159}$$

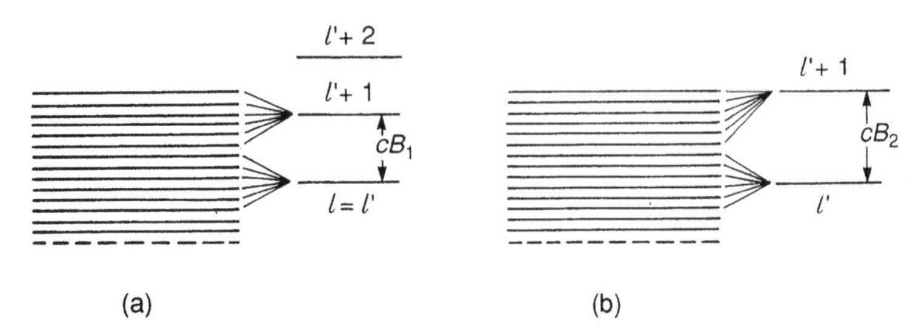

Figure 1.46 Assuming that all the levels are occupied, the condensation of all the levels on the left on to the highly degenerate levels on the right in (a) leaves the energy unchanged, but in (b) leads to an increase in energy

The spacing between the levels is proportional to B and as B is varied the correlation of these Landau levels with the Fermi energy (for free electrons in the absence of a field) will change as shown in Figure 1.46. In Figure 1.46(a), E_f is taken to lie midway between two levels. The degeneracy of the levels is such that all the electrons in the band ($B = 0$) can be accommodated in the Landau levels ranging up to E_f. In the presence of degeneracy Fermi–Dirac statistics can be applied simply by treating the g-fold degenerate levels as g separate levels with minimal spacing. The population of the Landau levels can be associated with the band levels as indicated in the diagram. The raising and lowering of the individual levels cancels out and there is no net effect of **B** on the total energy. For the situation shown in Figure 1.46(b), where the increase in B, and thus the level spacing, has caused one level to coincide with E_f, there is a net increase in energy due to the field, which is optimum for this coincidence. $\partial E/\partial B$ is positive on going from a to b, but then $\partial E/\partial B$ becomes negative as B is further increased. $\partial^2 E/\partial B^2$, and thus the susceptibility, fluctuates as B changes. This effect can be observed at high fields and low temperatures and is known as the de Haas–van Alphen effect [7]. Full analyses show that the association of the periodicity of the effect with the crossing of the Fermi surface by the Landau levels gives valuable information on the shapes of Fermi surfaces.

A particular 'giant magnetoresistive effect' is described in Chapter 6, Section 4.

7 Thermodynamics

7.1 General Principles

For changes at constant volume the conventional, mechanical, work term $P\,\mathrm{d}V$ is zero and supposing that magnetic work may arise the first law of thermodynamics becomes

$$đQ = \mathrm{d}U + đ\omega = \mathrm{d}U - \mu_0 H\,\mathrm{d}M \tag{160}$$

The stroke indicates that the symbol đ does not denote a true or exact differential but rather an infinitesimal change, for which the line integral between two points may depend on the path chosen. The line integral around a closed path is not necessarily zero.

In the usual way it is assumed that while any one thermodynamic variable depends to an extent upon all others, an adequate number of relations exists to permit any one to be expressed as a function of two others (for a fixed quantity of material), e.g.

$$U = U(T, M), \qquad \mathrm{d}U = \left(\frac{\partial U}{\partial T}\right)_M \mathrm{d}T + \left(\frac{\partial U}{\partial M}\right)_T \mathrm{d}M \tag{161}$$

and from equation (160),

$$đQ = \left(\frac{\partial U}{\partial T}\right)_M \mathrm{d}T + \left[\left(\frac{\partial U}{\partial M}\right)_T - H\right] \mathrm{d}M \tag{162}$$

so that the heat capacity at constant M is

$$C_M = \left(\frac{\partial Q}{\partial T}\right)_M = \left(\frac{\partial U}{\partial T}\right)_M = T\left(\frac{\partial S}{\partial T}\right)_M \qquad \left(\text{cf. } C_V = \left(\frac{\partial U}{\partial T}\right)_V\right. \tag{163}$$

(where $\mathrm{d}S = đQ/T$ for reversible changes with S the entropy).

The enthalpy $\mathcal{H} = U + PV$ is replaced by (V and P constant)

$$\mathcal{H} = U - \mu_0 HM \tag{164}$$

Differentiating and using equation (160),

$$\mathrm{d}\mathcal{H} = \mathrm{d}U - \mu_0 H\,\mathrm{d}M - \mu_0 M\,\mathrm{d}H = đQ - \mu_0 M\,\mathrm{d}H \tag{165}$$

and the heat capacity at constant field is

$$C_H = \left(\frac{\partial Q}{\partial T}\right)_H = \left(\frac{\partial \mathcal{H}}{\partial T}\right)_H = T\left(\frac{\partial S}{\partial T}\right)_H \qquad \left(\text{cf. } C_P = \left(\frac{\partial \mathcal{H}}{\partial T}\right)_P\right. \tag{166}$$

At a magnetic transition the state of order changes rapidly and C_H exhibits a peak.

The Helmholtz free energy F is

$$F = U - TS, \quad \mathrm{d}F = \mathrm{d}U - T\,\mathrm{d}S - S\,\mathrm{d}T = \mu_0 H\,\mathrm{d}M - S\,\mathrm{d}T = \mu_0 H\,\mathrm{d}M \quad (T \text{ constant}) \tag{167}$$

The Gibbs free energy is

$$G = \mathcal{H} - TS, \qquad \mathrm{d}G = T\,\mathrm{d}S - \mu_0 M\,\mathrm{d}H - T\,\mathrm{d}S - S\,\mathrm{d}T = -\mu_0 M\,\mathrm{d}H - S\,\mathrm{d}T \tag{168}$$

Because G is a function of state and $\mathrm{d}G$ an exact differential

$$\mu_0 \left(\frac{\partial M}{\partial T}\right)_H = \left(\frac{\partial S}{\partial H}\right)_T \tag{169}$$

because of the correspondence

$$\mathrm{d}z = P\,\mathrm{d}x + Q\,\mathrm{d}y \sim \mathrm{d}z = \left(\frac{\partial z}{\partial x}\right)_y \mathrm{d}x + \left(\frac{\partial z}{\partial y}\right)_x \mathrm{d}y$$

and the rule that for continuous functions

$$\frac{\partial^2 f(x, y)}{\partial x\,\partial y} = \frac{\partial^2 f(x, y)}{\partial y\,\partial x}$$

Expressing S as $S(T, H)$,

$$T\,\mathrm{d}S = T\left(\frac{\partial S}{\partial T}\right)_H \mathrm{d}T + T\left(\frac{\partial S}{\partial H}\right)_T \mathrm{d}H$$
$$= C_H \mathrm{d}T + T\left(\frac{\partial M}{\partial T}\right)_H \mathrm{d}H \tag{170}$$

using equations (166) and (169). Similarly, with $S = S(T, M)$,

$$\mathrm{d}F = \mu_0 H\,\mathrm{d}M - S\,\mathrm{d}T, \qquad T\,\mathrm{d}S = C_M \mathrm{d}T - T\left(\frac{\partial H}{\partial T}\right)_M \mathrm{d}M \tag{171}$$

Using equation (170), for an adiabatic change, $đQ = T\,\mathrm{d}S = 0$,

$$\mathrm{d}T = -\frac{T}{C_H}\left(\frac{\partial M}{\partial T}\right)_H \mathrm{d}H, \qquad \Delta T = -\frac{T}{C_H}\left(\frac{\partial M}{\partial T}\right)_H \Delta H \tag{172}$$

with implicit assumptions.

M usually decreases as T rises, at constant H, and certainly does so for a Curie law paramagnetic with $M = HC/T$. Thus an increase in H produces a rise in temperature and vice versa: the magneto caloric effect. If a paramagnetic is equilibrated in pumped liquid helium at $\sim 1K$, in a strong field, and the field is then removed, the temperature may fall to $\sim$ 1 mK in an insulated enclosure [8]. Polarized nucleii may be even more effective due to the weaker interactions and give $T \sim 1\mu$K. Nuclear susceptibilities are very small due to the relative values of the Bohr and nuclear magnetons (a factor of 10^3 giving 10^6 in χ), but a high level of polarization can be achieved by making use of the interactions between the electron and nuclear spins [9]. Since the Curie law holds down extremely low temperatures for certain salts, χ measurements afford a method for measuring such temperatures.

7.2 Statistical Thermodynamics

If all the particles in a system or ensemble are distributed over the accessible states according to the canonical distribution

$$p_i = \frac{\mathrm{e}^{-E_i/kT}}{\sum \mathrm{e}^{-E_i/kT}} = \frac{\mathrm{e}^{-bE_i}}{Z} \qquad (b \equiv 1/kT) \tag{173}$$

then all the thermodynamic functions can be related to the partition function Z. Since the mean energy per particle is

$$\bar{E} = \sum_i p_i E_i = \frac{\sum E_i \mathrm{e}^{-bE_i}}{Z}$$

and noting

$$\sum E_i \mathrm{e}^{-bE_i} = -\sum \frac{\partial}{\partial b}\mathrm{e}^{-bE_i} = -\frac{\partial Z}{\partial b}$$

it follows that

$$\bar{E} = -\frac{1}{Z}\frac{\partial Z}{\partial b} = -\frac{\partial \ln Z}{\partial b} \tag{174}$$

From this it can be shown also that

$$S = k(\ln Z + b\overline{E}) = k \ln Z + \frac{\bar{E}}{T} \tag{175}$$

For N particles and setting $U = N\bar{E}$, the Helmholtz free energy becomes

$$F = U - TS = -NkT \ln Z \tag{176}$$

If the magnetic moment component for the state i is μ_i then the mean moment in a field B is

$$\bar{\mu} = \sum p_i \mu_i = \frac{\sum e^{\mu_i B b} \mu_i}{Z} \tag{177}$$

and by the same reasoning that gave $\bar{E}_i$ it is seen that

$$\bar{\mu} = kT \frac{\partial \ln Z}{\partial B}, \qquad M = NkT \frac{\partial \ln Z}{\partial B} \tag{178}$$

if N is the number of moments per unit volume. Comparing this with equation (176),

$$M = -\frac{\partial F}{\partial B}, \qquad \chi = \mu_0 \frac{\partial M}{\partial B} = -\mu_0 \frac{\partial^2 F}{\partial B^2} \tag{179}$$

Quantities such as the magnetization and free energy, which are first derivatives, are continuous at Curie or Néel points. Quantities such as χ and the specific heat, which constitute second derivatives, are discontinuous. Thus the magnetic order–disorder transitions are known as second-order transitions or phase changes.

Z would be obtained by taking the trace (sum of diagonal elements) of a matrix for which $A_{ii} = \exp(-bE_i)$, $b = 1/kT$. Such a matrix would be obtained as the representation, in orthonormal eigenstates of a Hamiltonian $\mathcal{H}$, of an operator called the thermal average density matrix operator, or just the density matrix

$$\rho_0 = e^{-b\mathcal{H}} \tag{180}$$

as is most easily seen in the high-temperature approximation:

$$\langle i | e^{-b\mathcal{H}} | i \rangle \doteqdot \langle i | 1 - b\mathcal{H} | i \rangle = \langle i | 1 - bE_i | i \rangle = 1 - bE_i \doteqdot e^{-bE_i}$$

Thus

$$Z = \sum e^{-bE_i} = \mathrm{Tr}\,\{e^{-b\mathcal{H}}\} = \mathrm{Tr}\,\{\rho\} \tag{181}$$

(The curly brackets indicate the matrix and are often implicit.) If ρ_0 is redefined as

$$\rho_0 = \frac{e^{-b\mathcal{H}}}{Z} \tag{182}$$

it may then be considered as normalized inasmuch as $\mathrm{Tr}\,\rho_0 = 1$ necessarily, Z being a number. As shown in Chapter 2, Section 2.2, the thermal average of an observable for which the operator is A is given by

$$\langle A \rangle_T = \mathrm{Tr}\, A_\rho = \mathrm{Tr}\, \rho A \tag{183}$$

The spin density matrix is that for which $\mathcal{H}$ is the spin Hamiltonian (see Chapter 3). In the high-temperature approximation

$$\rho_0 = \frac{1 - b\mathcal{H}}{Z}, \quad \text{where } Z = (2I + 1)^{N_I}(2S + 1)^{N_S} \tag{184}$$

for a system of N_I nuclear spins I and N_S electron spins S. Z is just the trace of a unit matrix and its value is thus the dimension required to represent the whole system: e.g. $Z = 2$ for a single spin $\frac{1}{2}$.

It is possible to make the following *definitions* in terms of the density matrix:

$$\begin{aligned} &\text{Magnetic moment components} : \mu_i = \operatorname{Tr} \rho \hat{\mu}_i = \operatorname{Tr} \rho \gamma \hat{S}_i \qquad \text{for } i = x, y, z \\ &\text{Energy} : E = \operatorname{Tr} \rho \mathcal{H} = -b \operatorname{Tr} \mathcal{H}^2 \\ &\text{Entropy} : S = -k \operatorname{Tr} \rho \ln \rho \end{aligned} \tag{185}$$

For example, expressing $\hat{\rho}$ and $\hat{\mu}_z = -\gamma \hat{S}_z$ in the eigenstates $|\alpha\rangle$ and $|\beta\rangle$ of S_z and with $\mathcal{H} = -\gamma B S_z$ ($\mathbf{B} = \mathbf{k}B$),

$$\rho = \tfrac{1}{2}\begin{pmatrix} 1 + b\gamma B(\frac{1}{2}\hbar) & 0 \\ 0 & 1 + b\gamma B(-\frac{1}{2}\hbar) \end{pmatrix} \{\mu_z\} = -\gamma(\tfrac{1}{2}\hbar)\begin{pmatrix} 1 & 0 \\ 0 & -1 \end{pmatrix} \tag{186}$$

in the high-temperature approximation. The reader may readily complete the treatment to show that

$$\mu_z = \frac{\gamma^2\hbar^2 B}{4kT} = \frac{g^2\beta^2 B}{4kT}, \qquad \chi = \frac{\mu_0 N g^2 \beta^2}{4kT} \tag{187}$$

as obtained previously. The reader may similarly show that the more general expressions for μ_z and χ may be readily obtained, and illustrate the other definitions similarly (avoiding the approximation in the case of the entropy). The later accounts should clarify the role of the density matrix.

8 Exchange Interactions and Paramagnetic Susceptibilities

The Heisenberg exchange interaction, effectively though not directly between two spins, is

$$\mathcal{H} = -\frac{\mathcal{J}}{\hbar^2}\mathbf{S}_1 \cdot \mathbf{S}_2 \tag{188}$$

If this were read as a classical expression for the energy of two spins it would indicate

$$E = -\frac{\mathcal{J}}{\hbar^2}S_1 S_2, \qquad \mathbf{S}_1 \parallel \mathbf{S}_2$$

$$E = +\frac{\mathcal{J}}{\hbar^2}S_1 S_2, \qquad \mathbf{S}_1 \text{ antiparallel to } \mathbf{S}_2$$

indicating that a positive $\mathcal{J}$ favours parallel alignment (a tendency towards ferromagnetic order) and a negative $\mathcal{J}$ favours antiparallel alignment (tending to antiferromagnetic order). (The ideas of order are relevant only if $\mathcal{H}$ is supposed to apply similarly to all neighbours in a large array.)

The actual solutions for two spins $\frac{1}{2}$ are readily obtained (see Chapter 3) and the energy levels, for $\mathcal{J} > 0$, are shown, together with the splittings due to an applied field, in Figure 1.47(a). The magnetic moment components associated with the states are also indicated. The partition function is

$$Z = \mathrm{e}^{+g\beta B/(kT)} + \mathrm{e}^{0} + \mathrm{e}^{-g\beta B/(kT)} + \mathrm{e}^{-\mathcal{J}/(kT)} = 3 + \mathrm{e}^{-\mathcal{J}/(kT)}$$

in the Curie approximation [assuming $g\beta B \ll kT$ but that $\mathcal{J}/(kT)$ may be substantial]. The induced molar magnetization is thus

$$M = \frac{L}{2} g\beta(n_1 - n_3) \doteqdot \frac{Lg\beta}{2} \frac{2g\beta B/(kT)}{3 + \mathrm{e}^{-\mathcal{J}/(kT)}} \tag{189}$$

and the zero-field susceptibility is

$$\chi = \frac{\mu_0 L g^2 \beta^2}{kT(3 + \mathrm{e}^{-\mathcal{J}/(kT)})} \tag{190}$$

n_1 and n_3 are the occupation probabilities of the states, as indicated, and the number of systems or dimers is $L/2$ to give the susceptibility per mole of electrons.

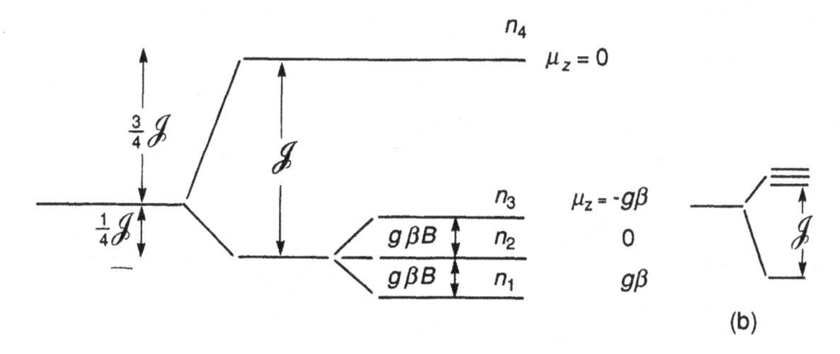

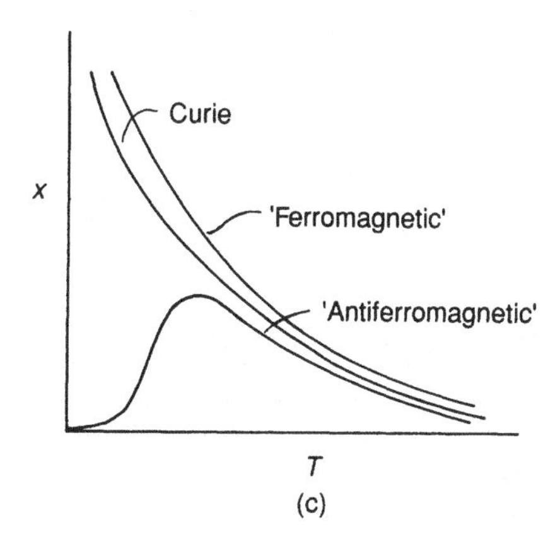

Figure 1.47 Energy levels for two Heisenburg coupled electrons (left) with the additional splittings due to an applied magnetic field arbitrarily superimposed on the right. A change in the sign of $\mathcal{J}$ interchanges the singlet and triplet. If the triplet is lowest the exchange interaction increases χ above the Curie value and for case (b) the effect can be a dramatic reduction in χ, towards zero at low $T/\mathcal{J}$ (c)

The reader may confirm that the same result is obtained for the case when the levels are inverted (with respect to the shifts due to the exchange) due to a negative value of $\mathcal{J}$. The implications are completely different, however. When $\mathcal{J} > 0$, $e^{-\mathcal{J}/(kT)} \to 0$ as $T \to 0$, giving

$$\chi = \frac{\mu_0 L g^2 \beta^2}{3kT} = \tfrac{4}{3}\frac{C}{T}, \qquad \text{where } C = \frac{\mu_0 L g^2 \beta^2}{4k} \tag{191}$$

Thus the 'ferromagnetic' interaction increases the susceptibilities that would apply in the absence of the interaction. This applies at all temperatures since $e^{-\mathcal{J}/(kT)} < 1$ necessarily, but the effect is greatest at low temperatures (strictly at high values of $\mathcal{J}/T$), and at high temperatures (low $\mathcal{J}/T$), when $e^{-\mathcal{J}/(kT)} \doteqdot 1$, the simple Curie law is followed. The approach to the Curie law for low $|\mathcal{J}|/T$ clearly also applies when $\mathcal{J}$ is negative, but it is seen that in this case χ must fall to zero as $T \to 0$ for any finite $\mathcal{J}$. Thus as T is reduced there must be a peak in χ and the plot is in fact as shown in Figure 1.47(c). From equation (190) it is seen that $\chi\mathcal{J}$ is a function of $\mathcal{J}/T$, i.e. a plot of $\chi\mathcal{J}$ versus $\mathcal{J}/T$ is universal and $T_m \propto \mathcal{J}$ where T_m is the temperature at which χ is maximum.

There is a less obvious major contrast between the cases of $\mathcal{J} > 0$ and $\mathcal{J} < 0$. The solutions for the Heisenberg Hamiltonian, extended to cover a substantial number n of spins $\frac{1}{2}$, with coupling between nearest neighbours and with the geometry of the system indicated by the number of 'bonds' chosen, is straightforward and limited in scope only by available computing facilities: $n = 12$ is very reasonable. It is found that when $\mathcal{J} < 0$ and n is even the plots are scarcely affected by n. However the 'ferromagnetic enhancement' for $\mathcal{J} > 0$ will be shown to be strongly dependent on n at low temperatures or high $\mathcal{J}/T$.

The difference between the two cases arises because in the range of $\mathcal{J}/T$ values for which the enhancement is strongly effective the susceptibility is close to zero for $\mathcal{J} < 0$ and though the relative differences in χ, for different values of n, are substantial the absolute differences are necessarily small.

It should be noted that the convention $\mathcal{H} = -2\mathcal{J}\mathbf{S}_i \cdot \mathbf{S}_j$ is sometimes used, with obvious implications.

9 Magnetic Materials

9.1 Magnetic Order

An exchange interaction with positive $\mathcal{J}$ is said to be ferromagnetic in nature and a negative $\mathcal{J}$ may be said to be antiferromagnetic.

What is meant by ferromagnetism is that in a low-temperature range the spins have a spontaneous alignment in zero field and in a higher range the alignment is lost and the exchange simply modifies the paramagnetic susceptibility that would arise with no interactions. The transition occurs at a Curie temperature which, like a melting point, may be considered a characteristic for any material. The alignment only approaches perfection as $T \to 0$ with a corresponding spontaneous magnetization $M_s(0) = N\mu$ with N moments μ per unit volume and $\mu = n\beta$ for a hypothetical ionic material with n unpaired spins per ion and completely quenched angular momentum ($g = 2.0$ implied). For metals it is conventional to write $\mu = n_e\beta$ and the effective number of spins may, due to band formation, be non-integral. As suggested later by Figure 1.54, for example, $M_s(T)$ falls slowly initially and if T_c is a few hundred K above room temperature then the room temperature M_s is close to $M_s(0)$: for iron T_c = 1043 K, $M_s(0)$ = 1752 and $M_s(293)$ = 1710 × 10 A m^{-1}; Ni ~ T_c = 631 K, $M_s(0)$ = 510, $M_s(293)$ = 485 × 10^3 A m^{-1}; Co ~ T_c = 1394 K, $M_s(0)$ = 1446, $M_s(293)$ = 1431 × 10^{3}A m^{-1}; CrO_2 ~ T_c = 393 K, $\sigma_s(0)$ = 315, $\sigma_s(293)$ = 200 A m^2 kg^{-1}. Room temperature is usually implied in quoting M_s or σ_s (per kg).

In the simpler (colinear) antiferromagnetics [Figure 1.48(b)] the ordered low-temperature state or phase is one in which the neighbouring moments are spontaneously aligned in an antiparallel manner to give zero M_s (with all moments the same). Above the critical (Néel) temperature T_N paramagnetism is exhibited. Antiferromagnetism is very common, in transition metal oxides for example, and the list of known materials is immense. For MnO, FeO, CoO and NiO, which are approximately colinear, T_N = 122, 198, 291 and 523 K.

In ferrimagnetics such as magnetite, Fe_3O_4, the exchange interactions are primarily antiferromagnetic but the crystals contain ions with differing moments ordered in the lattice in such a way that all the 'A' moments, say, become parallel to one direction, with all the 'B' moments antiparallel to this. Two, or more, magnetic sublattices are formed, each alone having a ferromagnetic character, and M_s arises as the difference between the sublattice magnetizations [Figure 1.48(c)]. The number of ions on the different sites need not be equal. The variation of $M_s(T)$ may be extremely complex and $M_s(T)$ may rise with T in certain ranges. The most typical ferrimagnetics are the spinel ferrites (e.g. ferrous ferrite, or Fe_3O_4), the magnetic garnets and certain rare earth–transition metal alloys.

A further basically antiferromagnetic order, giving a small M_s, is that in which the alternate moments are nearly antiparallel but a small canting occurs to give $\mathbf{M}_s$ in a direction normal to the approximate spin axis, as indicated in Figure 1.48(d). This applies to haematite, α-Fe_2O_3, and yttrium and rare-earth orthoferrites.

Usually the order itself is not influenced substantially by applied fields, the exchange coupling being equivalent to 'exchange fields', which are many orders greater than conventional real fields. However, in *meta*-magnetics, transitions between antiferromagnetism and ferromagnetism may occur in applied fields (e.g. [10]).

In the general case there are particular crystal axes, 'easy axes', along which M_s tends to lie, or with respect to which the spins are ordered in antiferromagnetics, secondary

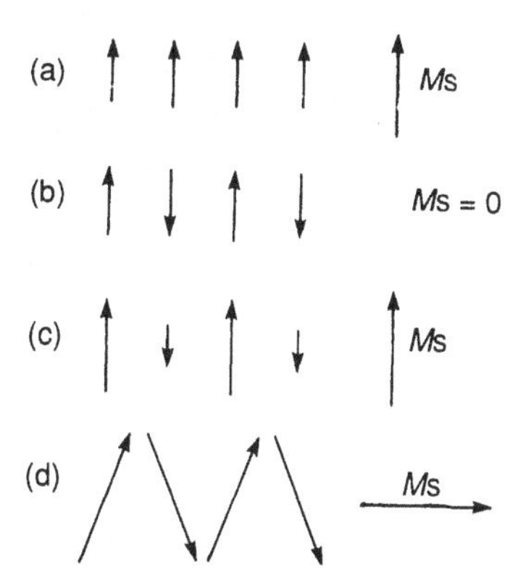

Figure 1.48 Formal one-dimensional representations of the atomic magnetic moments in (a) ferromagnetics, (b) antiferromagnetics, (c) ferrimagnetics and (d) canted spin materials with $M_s \perp$ the ordering axis. The simple colinear cases are assumed and real configurations may be much more complex (See also Figure 5.15: ferrimagnetism)

interactions giving a so-called magnetocrystalline anisotropy (see Section 11 and later). The easy axes in iron and nickel are [100] and [111] respectively and in hexagonal cobalt M_s tends to lie along the c axis. The effect can, to a certain extent, be represented by effective anisotropy fields which are comparable to available applied fields so that rotations away from the easy axis may be affected.

The existence and nature of the ordering and the type and magnitude of the anisotropy can, to a large extent, be inferred from measurements of magnetization, including the effects

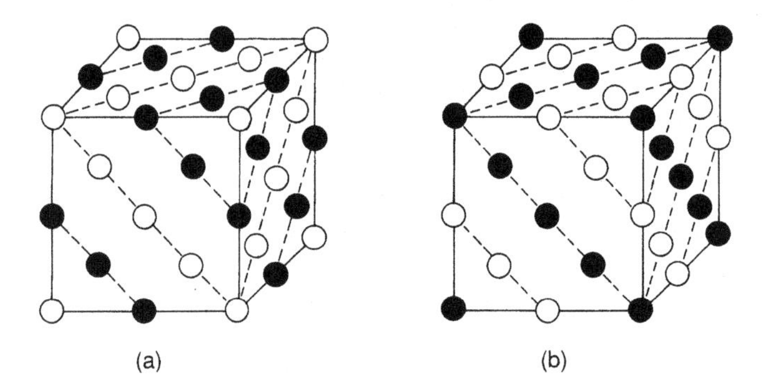

Figure 1.49 Alternative magnetic structures applying to MnO and other Tm oxides. The full and open circles indicate ions with opposite spin orientations and the oxygens lie in (111) layers between the cations [11]. (Reprinted by permission of John Wiley & Sons, Ltd, Magnetic Materials, R.S. Tebble and D.J. Craik, Wiley 1969)

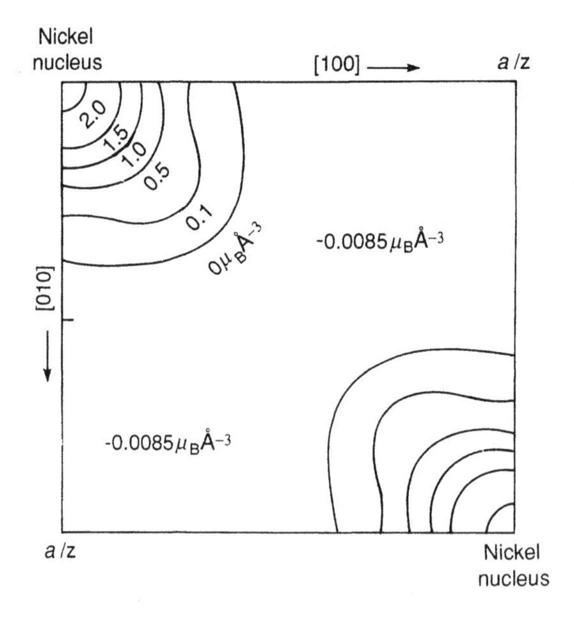

Figure 1.50 The distribution of the magnetic moment (spin density) in the (100) plane of a nickel crystal according to the diffraction of polarized neutrons [12]. The atomic moment was indicated to be derived as 0.65β for the 3d spin, 0.055β for the (quenched) orbit and -0.105β for negatively exchange polarized 4s electrons (see Chapter 3)

of the orientation of the applied fields w.r.t. the (single-crystal) axes. In detail, and particularly for antiferromagnetics with $M_s = 0$, Mossbauer spectroscopy and nuclear magnetic resonance can give valuable information (Sections 17.2.10 and 18) and neutron diffraction is of special importance. Neutron beams, originally available from atomic piles, have associated wavelengths comparable to X-rays. Neutrons are scattered strongly by nucleii whereas X-rays are scattered by the atomic electrons so that neutron diffraction can supplement X-ray studies. However, neutrons, like electrons, have intrinsic spin and magnetic moment and also interact with the moments in a solid; if these are ordered extra lines appear in the diffraction patterns. The lines of magnetic origin disappear as, for example, MnO is taken through T_N [11]. The structure indicated for MnO is that shown in Figure 1.49. Just as X-ray diffraction can give electron density distributions, neutrons can give spin densities [12], indicating the localization and orientation of the moments (Figure 1.50).

Such methods indicate that some of the rare earth metals order in an extremely complex way with ranges of temperature over which they may be ferromagnetic, antiferromagnetic and, of course, paramagnetic. Gadolinium is ferromagnetic below $T_c = 293$ K, virtually room temperature, and $M_s(T)$, as in Figure 1.51(a) [13], is interpreted in terms of simple axial alignment of the moments (7.5 β) along the c axis of the hexagonal structure on cooling through T_c, followed at lower temperatures by a structure in which the moments lie on the surface of a cone whose axis is the c axis. Figure 1.51(b) shows a plot of specific heat versus temperature for gadolinium, to draw attention to the general occurrence of such discontinuities at the Curie point. Terbium is ferromagnetic at low temperatures [$M_s(0) \sim 9.34\beta$] but becomes antiferromagnetic at 221 K and has a Néel temperature of 229 K. In the latter state the spins have a helical arrangement, the vectors rotating about the c axis with the twist angle between layers varying from 18.5° at 221 K to 20.5° at 229 K [14].

Such complex arrangements and behaviour are not very common. In the great majority of materials only one type of order occurs and in most materials of technical interest only simple colinear ferro- or ferrimagnetism is involved.

Above the Curie or Néel points the paramagnetic susceptibility of ferro- or antiferromagnetics does not correspond to the Curie C/T law, since the exchange coupling itself does not disappear. In Figure 1.52(a) line I represents $1/\chi \propto T$ for an arbitrary value of C. As indicated previously, a ferromagnetic interaction increases χ above the Curie value, i.e. reduces $1/\chi$, giving line II indicating $\chi \to \infty$ at T_c approximately. Line III applies to antiferromagnetics, with an opposite shift corresponding to reduced values of χ. The intercept θ, the Weiss temperature, cannot now correspond to the ordering temperature T_N. Below T_N magnetization can only be induced when the field overcomes the exchange torques tending to maintain $M_s = 0$, and χ falls; $1/\chi$ rises as shown in Figure 1.52(b). The sketch applies to a polycrystalline specimen or powder. For single crystals χ measured normal to the axis of ordering remains constant and χ measured parallel to the axis falls very rapidly.

Far above T_c a ferrimagnetic is expected to behave like an antiferromagnetic, but this must combine with an approach to $1/\chi = 0$ as T_c is approached, as for a ferromagnetic. Thus experimentally observed plots such as that shown as IV are to be expected and it can be shown that these should in fact be parabolic.

Plot III is represented by

$$\chi = \frac{C}{T - \theta} \qquad (\theta < 0) \tag{192}$$

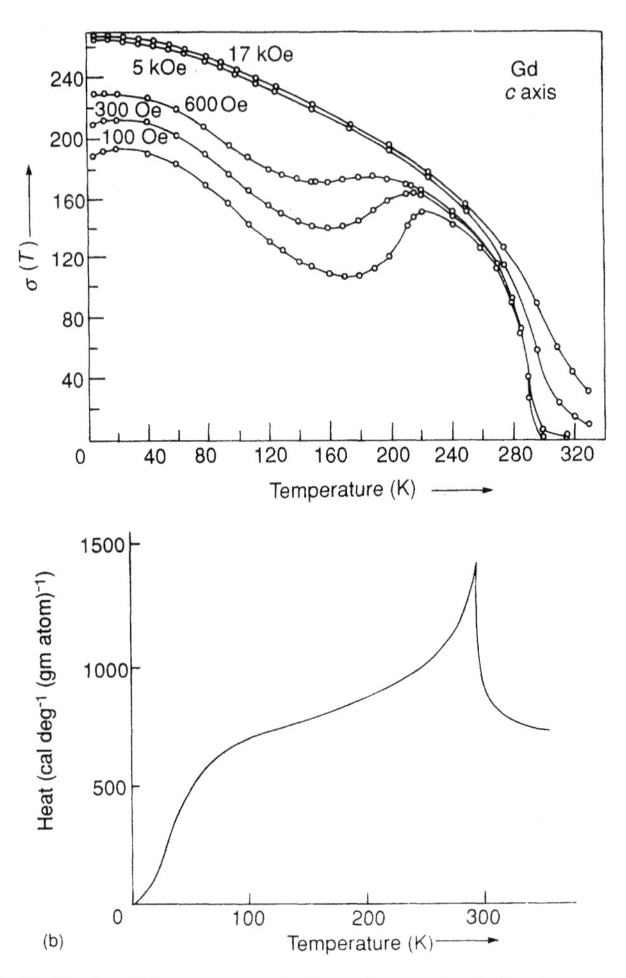

Figure 1.51 (a) $M_S(T)$ for Gd as measured along the c axis, indicating complex spin ordering changes in the different magnetic fields (corrected for demagnetizing effects) as noted; (b) shows the specific heat with a peak at T_c [13]. (Reprinted by permission of John Wiley & Sons, Ltd, Magnetic Materials, R.S. Tebble and D.J. Craik, Wiley 1969)

the Curie–Weiss law, and plot II may be represented in the same way if T_c replaces θ. However, the intercept of plot II is not always exactly at T_c and it may be taken that equation (192) applies to ferromagnetics with $\theta > 0$ and $\theta \doteqdot T_c$.

It should be stressed that the application of the Heisenberg Hamiltonian indicates an approach to an antiferromagnetic ground state as $T \rightarrow 0$ if $\mathcal{J} < 0$, and there is also, for $\mathcal{J} > 0$, an approach to extremely high values of χ, but in neither case is a distinct, second-order, phase change indicated. The problems involved in accounting for phase changes of any kind are formidable and for the very complex basic theories reference should be made to works such as those by Mattis [15] White [16] and White and Geballe [17]. The only well-established and straightforward approach is the Weiss molecular field theory, as follows.

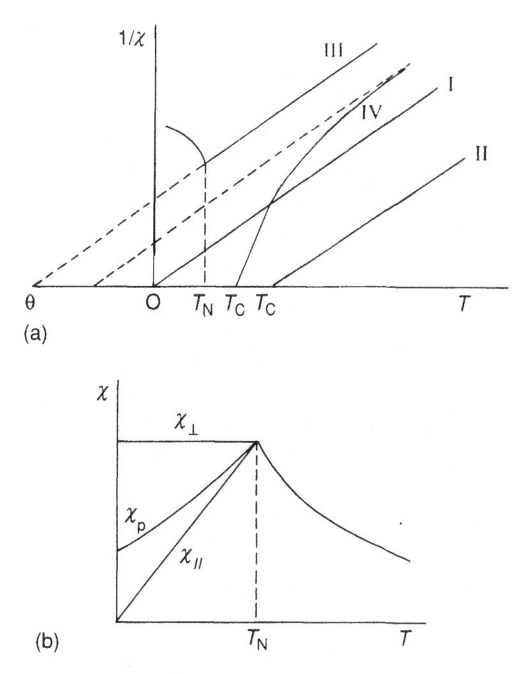

Figure 1.52 (a) Plots of χ^{-1} versus T as discussed. (b) The expected plots of $\chi(T)$ measured $\|$ and $\perp$ to the ordering direction in a colinear antiferromagnetic, with χ_p for a powder

Finally, it may be noted that small lattice distortions accompany the onset of ordering. For example, the monoxides of Mn, Fe, Ni and Co are cubic above T_N but below T_N the first three become rhombohedral while CoO becomes tetragonal, $C/a < 1$.

9.2 Ferromagnetic Order: Spin Waves and Molecular Fields

It will be seen in Chapter 3 that the matrix of the Heisenberg Hamiltonian

$$\mathcal{H} = -\mathcal{J}\sum_{i=1}^{n} \mathbf{S}_i \cdot \mathbf{S}_{i+1} = -\mathcal{J}\sum[S_i^z S_{i+1}^z + \tfrac{1}{2}(S_i^+ S_{i+1}^- + S_i^- S_{i+1}^+)] = -\mathcal{J}(\mathcal{H}_1 + \mathcal{H}_2)$$

can be partitioned into $n+1$ submatrices, for example for a ring of four spins $\frac{1}{2}$. Choosing the first basis state as $|1\rangle = |\alpha\ \alpha\ \alpha\ \alpha\rangle$, the first 'matrix' is just the single element and $|1\rangle$ can be seen directly to be an eigenstate since it is an eigenstate of $\mathcal{H}_1$ and the effect of $\mathcal{H}_2$ is to give zero: the eigenvalue is clearly $-\mathcal{J}n(\frac{1}{4}\hbar^2)$ or $-\mathcal{J}(\hbar^2)$ here. The total z component is $4 \times \frac{1}{2}\hbar = 2\hbar$. The second submatrix is 4×4 over basis states with one reversal, i.e. $|\alpha\ \alpha\ \alpha\ \beta\rangle$, etc.: total $S_z = 3 \times \frac{1}{2}\hbar - \frac{1}{2}\hbar = \hbar$. These are all eigenstates of $\mathcal{H}_1$ with eigenvalue $\frac{1}{4}\hbar^2 + \frac{1}{4}\hbar^2 - \frac{1}{4}\hbar^2 - \frac{1}{4}\hbar^2 = 0$. Each gives two different states ($\times\hbar^2$) within the subset when operated on by $\mathcal{H}_2$, corresponding to the interchange of neighbouring α and β labels. On writing these out, it is seen that the state $\sum_{i=2}^{5} c_i|i\rangle$ with all $c_i = 1$, the 'uniform' state, is an eigenstate of $\mathcal{H}$ with eigenvalue $-\mathcal{J}\hbar^2$ again. Similarly, for the third matrix over the six basis states $|\alpha\ \alpha\ \beta\ \beta\rangle$, etc.: two reversals, $S_z = 0$, the simple sum

$\sum_{i=6}^{11} |i\rangle$ is seen, at a little greater length, to be an eigenstate of $\mathcal{H}$ with the same energy as the above. Proceeding, there are $n+1$ uniform states, one for each submatrix, all giving the energy $-n \times \frac{1}{4}\hbar^2 \mathcal{J}$, and this can be shown to be the lowest of the energies. The states are distinguished inasmuch as $S_z = 2\hbar, \hbar, 0, -\hbar, -2\hbar$, but all can be shown to be characterized by the same value of the total spin $S = 2$ or $S = n \times \frac{1}{2}$ generally.

The principle is general. For n spins $\frac{1}{2}$ with Heisenberg coupling and $\mathcal{J} > 0$, the ground state is $(n+1)$-fold degenerate and the individual states all give $S = \frac{1}{2}n$ but differ in that $S_z = M_S\hbar$ with $M_S = S, S-1, \ldots, \ldots, (S-1), -S$. (The ground-state energy depends upon the number of 'bonds' taken into account, as well as n, of course.) So long as only the ground states are occupied the system behaves as a giant spin and the moments are perfectly ferromagnetically ordered. The ground state is certainly not uniquely $|\alpha\ \alpha \cdots \alpha\rangle$ or $|\beta\ \beta \cdots \beta\rangle$. The reversal of one (or more) spins does not change the total spin so long as this is equally associated with all the spins (i.e. $k = 0$).

The first meaningful matrix in the example is 4×4 and this gives three further eigenvalues, of higher energy than the uniform state, which are associated with spin waves for which $k = 1, 2$ and 3, i.e. the coefficients of the subset of four basis states indicate that the spin reversal is delocalized in a wave-like manner: for consistency the ground state with equal coefficients may be denoted $|k = 0>$. As n increases, more and more levels are packed into a similar energy range, or conversely the spacings between the spin wave energies decrease. Similar considerations apply to the other submatrices.

The state of perfect order with, say, $M = N \times \frac{1}{2}\, g\beta = M_0$ can only exist at $T \to 0$ K. As T increases M falls due to the generation of spin waves or, equally, the progressive occupation of the excited states in the direct solution, because these states are not characterized by $S = n \times \frac{1}{2}$ but by progressively lower values of S. It will be seen that M falls as $M = M_0(1 - bT^{3/2})$, at rather low temperatures.

The state $|\alpha\ \beta\ \alpha\ \beta \cdots \alpha\ \beta\rangle$ is itself an eigenstate of $\mathcal{H}$ (As readily seen for a chain, it is an eigenstate of $\mathcal{H}_1$, and $\mathcal{H}_2$ simply has the effect of moving the chain through one period, i.e. it has no effect on the state save for the factor produced.) In fact it gives ($\mathcal{J} > 0$) the highest energy, which becomes the lowest for antiferromagnetic coupling, but in this case the ground state is naturally a singlet with $S = 0 = S_z$.

Returning to $\mathcal{J} > 0$, the simple spin wave treatments do not apply in practice, or perhaps in principle, when the system becomes chaotic (colloquially and perhaps specifically) at higher temperatures for which the ordered state is breaking up and M is falling rapidly. The only simple account of the onset of order at a critical, Curie, temperature is given by molecular field treatments. Imagine that the atoms are moved apart, $\mathcal{J} \to 0$, and a small field is applied to give $M = \chi_p H_a$, with $\chi_p = C/T$, with C the Curie constant. Resuming the normal spacing, a characteristic spin tends to be aligned with the mean net moment of the others, i.e. with M. H_a can be taken to become $H_a + H_e$, with $H_e = \gamma M$ the exchange field, which is equivalent in effect to the exchange coupling. γ is the Weiss or molecular field coefficient. The induced magnetization becomes $M = \chi_p(H_a + \gamma M)$, i.e. $M = \chi_p H_a/(1 - \gamma\chi_p)$. It is implied that T is high so that $M \propto H$ still. The measured χ is now

$$\chi = \frac{M}{H_a} = \frac{\chi_p}{1 - \gamma\chi_p} = \frac{C/T}{1 - \gamma(C/T)} = \frac{C}{T - \gamma C} \qquad (T > \gamma C)$$

i.e.

$$\chi = \frac{C}{T - T_c}, \qquad \text{where } T_c = \gamma^C \tag{193}$$

which is the Curie–Weiss law for the paramagnetic region.

Relaxing the limitation to the linear region and introducing the Brillouin function ($S = \frac{1}{2}$ for simplicity), and taking $H_a = 0$, $H = H_e$ only,

$$M = \tfrac{1}{2} N g\beta \tanh\left(\frac{\frac{1}{2} g\beta\mu_0 H_e}{kT}\right) = \tfrac{1}{2} N g\beta \tanh\left(\frac{\frac{1}{2} g\beta\mu_0 \gamma M}{kT}\right) \tag{194}$$

Noting M on the right, this is a transcendental equation with no apparent solution other than $M = 0$, since $\tanh 0 = 0$. Using $M_0 = \frac{1}{2} N g\beta$ as the maximum for perfect alignment and $\gamma C = T_c$, the argument of tanh is

$$\frac{\frac{1}{2} g\beta\mu_0 \gamma M}{kT} = \mu_0 \gamma \frac{N g^2 \beta^2}{4kT} \frac{1}{\frac{1}{2} N g\beta} M = \frac{M}{M_0} \frac{T_c}{T} = \frac{M'}{T'}$$

Thus

$$M' = \tanh \frac{M'}{T'} \tag{195}$$

in terms of the dimensionless $M' = M/M_0$, $T' = T/T_c$. Plots of the straight line L_1 of unit slope, M' versus M' and of $\tanh M'/T'$, L_2, are superimposed in Figure 1.53. Any intersection of the two lines other than at zero indicates the existence of a non-zero solution for M and thus of a spontaneous magnetization (in zero field). Recalling $\tanh x = (e^x - e^{-x})/(e^x + e^{-x})$ and $d/dx \tanh x = 1 - \tanh x \to 1$ as $x \to 0$, it follows simply that $d/dM' \tanh M'/T' = 1/T'$ when $M' \to 0$. L_2 thus has the same (unit) slope at the origin as L_1 if $1/T' = 1 : T = T_c$. If $T > T_c$ the initial slopes are such that there is no intersection and T_c is thus the temperature below which a spontaneous magnetization exists. As $T < T_c$ increasingly, the point of intersection moves out along L_1 to the point 1, 1($\tanh x \to 1$ as $x \to \infty$), i.e. as $T \to 0$, $M' \to 1$ and $M \to M_0$. Thus the shape of experimental plots as in Figure 1.54 can be understood generally and there is consistency over temperatures above and below T_c [18, 19]. The points in Figure 1.54(a) are relevant to Section 17.2.10.

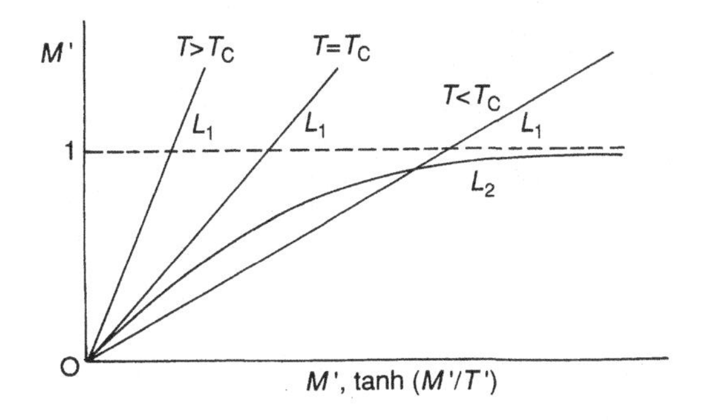

Figure 1.53 Construction for the molecular field calculations with an implied change of scale for each of the lines L_1

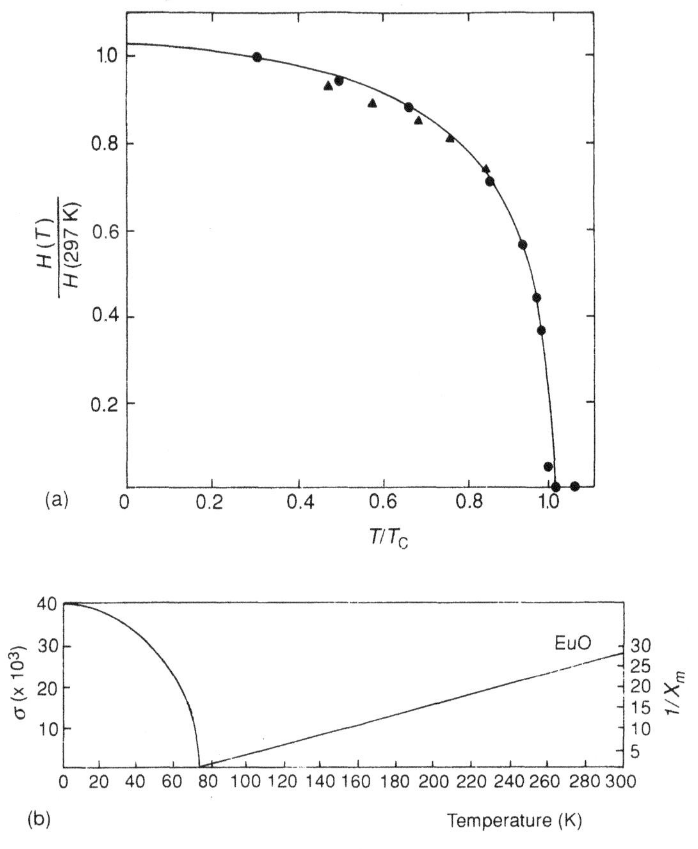

Figure 1.54 (a) $M_S(T)$ as measured directly together with points indicating the effective internal or hyperfine fields obtained by n.m.r., as discussed later [18]. (b) $M_S(T)$ and χ^{-1} for EuO, the latter corresponding to 7β per ion [19]. (Reprinted by permission of John Wiley & Sons, Ltd, Magnetic Oxides, ed. D.J. Craik, Wiley 1975)

It may be noted that the free energy is $F = -kT \ln Z$ and the partition function Z, e.g. for $S = \frac{1}{2}$, is $2[\cosh \frac{1}{2} g\beta B_m/(kT)]$. When $M = 0$, and $B_m = 0, \cosh(\ ) = 1$ and $Z = Z_0 = 2$ but otherwise $\cosh(\ ) > 1$ and $Z > Z_0$ with $F < F_0$ and the zero solution noted above may be discarded on these grounds.

An alternative approach is to consider $M = \chi_p H_a$ with $H_e = 0$, in a small field H_a, introduce $H_e = \gamma M$ so that $M \rightarrow \chi_p(H_a + \gamma \chi_p H_a)$ and to proceed iteratively to give

$$M = \chi_p H_a(1 + \gamma\chi_p + \gamma^2\chi_p^2 + \gamma^3\chi_p^3 + \cdots)$$

The series ceases to converge at the limiting value $\gamma\chi_p = \gamma C/T = 1$, giving $T = T_c = \gamma C$. For $T < T_c$ the series diverges, indicating that M can be finite even for $H_a = 0$. A simple method of computation is to set $M^{(1)} = 0$ and assign

$$M^{(i)} = \tanh \frac{\mu_0 g\beta}{2kT}(H_a + \gamma M^{(i-1)})$$

with n adequate for $M^{(i)} - M^{(i-1)} \to 0$. H_a may be set to a negligibly small value and the result used as an approximation for numerical solution of the transcendental equation, with $H_a = 0$. It may be demonstrated to any chosen accuracy that $\partial M/\partial T$ is discontinuous at $T = T_c$, and this is associated with the observation that L_2 approaches linearity near zero. As $H_a \to 0$ the Curie–Weiss plot remains linear down to arbitrarily small values of χ^{-1} and its intersection and that of $M(T)$ can be brought into coincidence. However, while the effect of moderate fields on the Curie–Weiss plot is to introduce a small curvature only, that on M close to T can be very substantial, as shown in Figure 1.55.

In addition to the contribution from lattice vibrations etc., there are contributions to the internal energy U and specific heat C that can be treated as the energies of the spins in the molecular field. The specific heat and internal energy are here taken to relate purely to these 'magnetic contributions'. The internal energy for n spins $\frac{1}{2}$ is

$$U = -\tfrac{1}{2}n(\tfrac{1}{2}g\beta B_m)\tanh\left(\frac{\frac{1}{2}g\beta B_m}{kT}\right) \tag{196}$$

Here B_m is the molecular field according to

$$B_m = \lambda \tanh\left(\frac{\frac{1}{2}g\beta B_m}{kT}\right)$$

(since the tanh function is proportional to the magnetization), in relation to which

$$T_c = \frac{\lambda g\beta}{2k}$$

as is readily shown by setting

$$\frac{\partial}{\partial B_m}\lambda \tanh\left(\frac{\frac{1}{2}g\beta B_m}{kT}\right) = \frac{\lambda}{\cosh^2 \frac{1}{2}g\beta B_m/(kT)}\frac{\frac{1}{2}g\beta}{kT} = 1$$

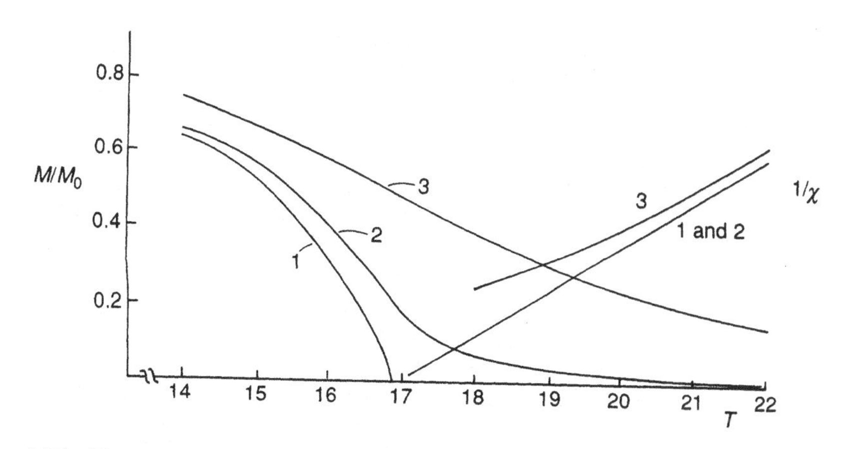

Figure 1.55 Plots 1 correspond to effectively zero field and 2 and 3 show the effects of increasing fields predicted by the molecular field theory

as $B_m \to 0$ and $\cosh^2(0) = 1$. Substituting for the tanh function in equation (196),

$$U = -\frac{Ng\beta B_m}{4}\frac{B_m}{\lambda}$$

$$= -\tfrac{1}{2}NkT_c\frac{B_m^2}{\lambda^2} = -\tfrac{1}{2}NkT_c r^2 \qquad (197)$$

with $r = B_m/\lambda$. Since the molecular field B_m is proportional to the magnetization and in the zero applied field the magnetization is zero above T_c, it is clear that

$$c = \frac{1}{N}\frac{\partial U}{\partial T} = -\tfrac{1}{2}kT_c\frac{\partial r^2}{\partial T} \qquad (198)$$

is discontinuous at T_c. (where c is the specific heat per spin). Having solved the transcendental equation to find the magnetization and B_m, c follows. The change in the specific heat at T_c can be shown to have the magnitude

$$\Delta c = 5k\frac{S(S+1)}{S^2 + (S+1)^2} \qquad (199)$$

The molar heat capacity follows by multiplying by Avogadro's number. For $S = \frac{1}{2}$, $\Delta c = 3k/2$ and in the classical limit of $S \to \infty$, $(S + 1 \to S)$, $\Delta c = 5k/2$.

Although c is discontinuous, as is the zero-field susceptibility, at $T = T_c$, the internal energy falls continuously to zero. Also, c and χ have been seen to be second-order differentials of the partition function, and phase changes involving such discontinuities are said to be of second order. The spin order–disorder transition is thermodynamically second order.

10 The Exchange Integral and Exchange Constant

The classical expression for the exchange energy, valid by analogy with the quantum results, for an atom i coupled with nearest neighbour atoms j, is the sum over the neighbours of

$$W_{ij} = -2\mathcal{J}_{ij}\mathbf{S}_i \cdot \mathbf{S}_j \qquad (200)$$

the factor of 2 being arbitrary but usually included. $\mathcal{J}$ is known as the exchange parameter. Assuming that the exchange is isotropic and that all the $\mathcal{J}_{ij} = \mathcal{J}$,

$$W_{ij} = -2\mathcal{J}S_iS_j\cos\varphi_{ij} \doteqdot \mathcal{J}S^2\varphi_{ij}^2 \qquad (201)$$

with all $S_i = S$ (for identical atoms), using $\cos 2\varphi = 2\cos^2\varphi - 1$, assuming small angles φ_{ij} between neighbouring spins and dropping constant terms. A more general expression for the energy for a unit cell of a simple cubic, f.c.c. or b.c.c. structure is obtained on carrying out the summations as

$$W = 2\mathcal{J}Sa^2\left[(\nabla\alpha_1)^2 + (\nabla\alpha_2)^2 + (\nabla\alpha_3)^2\right] \qquad (202)$$

in terms of the direction cosines α_1, α_2, α_3 relating the spin directions to rectangular coordinate axes.

An important situation is that in which the spins remain parallel within each of a set of neighbouring planes in the crystals with the common orientation changing from plane to plane, as illustrated for complete rotation in four equal steps in Figure 1.56: usually a great

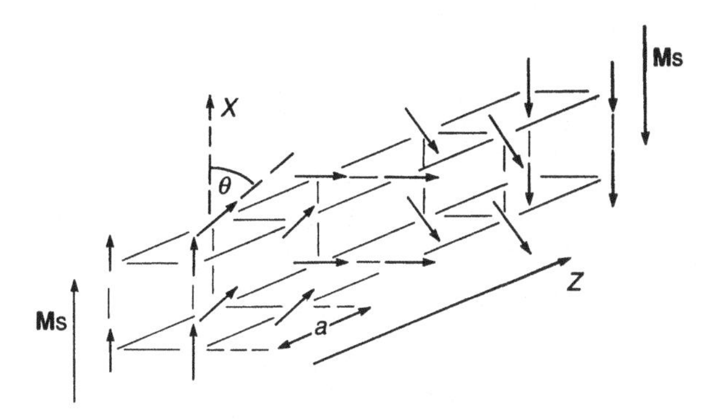

Figure 1.56 All the spins in planes ∥ OXY are assumed parallel. In the general case n (= 4 here) is very large, θ can be taken to be continuous and $\partial\theta/\partial z$ is not constant

number, n, of steps is appropriate. For a simple cubic lattice $a \times a \times a$ there are $1/a^2$ atoms per unit area of plane and the exchange energy per unit area is

$$E = n \times \frac{1}{a^2} \times \mathcal{J}S^2\frac{\pi^2}{n^2} \propto \frac{1}{n} \tag{203}$$

summing equal terms over the n planes involved with all $\varphi_{ij} = \pi/n$. The more gradual the rotation, i.e. the larger n becomes, the lower is the energy.

More generally, if θ is the angle between the common spin direction in the plane positioned along OZ as in Figure 1.56 and a reference axis OX, the change in θ between planes is $a(\partial\theta/\partial z)$ and the energy per spin pair is

$$W = \mathcal{J}S^2a^2\left(\frac{\partial\theta}{\partial z}\right)^2$$

The energy per unit area (for the simple cubic lattice) is thus

$$E = \int_{-\infty}^{\infty}\frac{\mathcal{J}S^2}{a}\left(\frac{\partial\theta}{\partial z}\right)^2 \mathrm{d}z = A\int_{-\infty}^{\infty}\left(\frac{\partial\theta}{\partial z}\right)^2 \mathrm{d}z, \qquad \text{where } A = \frac{\mathcal{J}S^2}{a} \tag{204}$$

The same expression is seen to apply for other cubic lattices with

$$A = \frac{2\mathcal{J}S^2}{a} \qquad \text{(b.c.c.)} \tag{205}$$

$$A = \frac{4\mathcal{J}S^2}{a} \qquad \text{(f.c.c)} \tag{206}$$

A is the exchange constant and is important in characterizing domain walls as in Section 12.2.

11 Magnetocrystalline Anisotropy and Magnetostriction

The tendency for M_s to lie along an easy axis can be described using the magnetocrystalline anisotropy constants K_i. For a purely uniaxial crystal, with no anisotropy normal to the single easy axis, an approximate expression for the anisotropy energy density is

$$E_K = K \sin^2 \theta \tag{207}$$

θ being the angle between $\mathbf{M}_s$ and a coordinate axis parallel to the easy axis. E_K is zero when $\mathbf{M}_s$ is parallel or antiparallel to the arbitrarily directed coordinate axis and maximum for $\theta = \pi/2$, i.e. has appropriate symmetry. This would apply with other even terms and

$$E_K = K_1 \sin^2 \theta + K_2 \sin^4 \theta + \cdots \tag{208}$$

is more general, the number of terms chosen depending on the accuracy with which measurements can be made. For cobalt at room temperature (the K_i vary rapidly with temperature),

$$K_1 = 4.1 \times 10^5 \mathrm{J\,m^{-3}}, \qquad K_2 = 1.0 \times 10^5 \mathrm{J\,m^{-3}}$$

and it is clear that the simple equation (207) can be used reasonably, particularly for small deviations of $\mathbf{M}_s$ from the easy axis for this and in fact other materials. Exact measurements of the torque on $\mathbf{M}_s$ as a crystal is rotated in large saturating fields reveal a small anisotropy in the basal plane accounted for by

$$E_K = K_1^u \sin^2 \theta + K_2^u \sin^4 \theta + K_3^u \sin^6 \theta + K_3 \sin^6 \theta \cos^4 \phi \tag{209}$$

with θ and ϕ the polar angle (as above) and azimuthal angle w.r.t. the c axis. The superscript distinguishes the uniaxial terms. This E_K is minimum, for hexagonal crystals, with $M_s \parallel$ [0001] for

$$K_1 + K_2 > 0 \qquad \text{and} \qquad K_1 > 0$$

However, for

$$0 \leqslant K_1 < -K_2 \qquad \text{or} \qquad -K_1 > 2K_2 \qquad \text{and} \qquad K_1 < 0$$

E_K is lowest when $\mathbf{M}_s$ lies in the (0001) basal plane, giving an easy-plane or planar material.

For cubic crystals an expression with appropriate symmetry is (truncated)

$$E_K = K_1(\alpha_1^2\alpha_2^2 + \alpha_2^2\alpha_3^2 + \alpha_3^2\alpha_1^2) + K_2(\alpha_1^2\alpha_2^2\alpha_3^2) \tag{210}$$

In the case that

$$K_1 > 0 \qquad \text{and} \qquad K_1 > -\tfrac{1}{9}K_2$$

the [100] axes are the easy axes and if

$$K_1 < 0 \qquad \text{and} \qquad K_1 < -\tfrac{4}{9}K_2 \qquad \text{or} \qquad 0 < K_1 < -\tfrac{1}{9}K_2$$

the [111]'s are easy axes. However, it is usually adequate to accept that, formally,

$$K_1 > 0 \sim [100], \qquad K_1 < 0 \sim [111] \text{ easy axes}$$

K_1 is positive for iron and negative for nickel.

Anisotropy fields may be regarded as those fields which, when applied along an easy axis, give the same torques as those corresponding to the anisotropy and thus may be taken to represent the anisotropy. The correspondence only applies when the magnetization deviates from an easy direction only by a small angle $\Delta\theta$ for which $\sin\Delta\theta = \Delta\theta$ is acceptable and the torque is

$$T_{H_K} = -\frac{\partial}{\partial\theta}E_{H_K} = -\frac{\partial}{\partial\theta}\mu_0 H_K M_s \cos\theta = \mu_0 H_K M_s \Delta\theta$$

The torque due to the anisotropy can be expanded and truncated as

$$T_K = \frac{\partial E_K}{\partial\theta} = \left(\frac{\partial E_K}{\partial\theta}\right)_{\theta=0} + \left(\frac{\partial^2 E_K}{\partial\theta^2}\right)_{\theta=0}\Delta\theta = \left(\frac{\partial^2 E_K}{\partial\theta^2}\right)_{\theta=0}\Delta\theta$$

since $\partial E_K/\partial\theta = 0$ when M_s lies along an easy direction. Equating the two,

$$H_K = \frac{1}{\mu_0 M_s}\left(\frac{\partial^2 E_K}{\partial\theta^2}\right)_{\theta=0} \tag{211}$$

(H_A is often used to denote a general anisotropy field with H_K used to indicate that magnetocrystalline anisotropy is specifically considered.) Using equation (208),

$$\begin{aligned} H_K &= \frac{1}{\mu_0 M_s}\left[2K_1(-\sin^2\theta + \cos^2\theta) + \frac{1}{\mu_0 M_s}4K_2(-\sin^4\theta + 3\sin^2\theta\cos^2\theta)\right]_{\theta=0} \\ &= \frac{2K_1}{\mu_0 M_s} \end{aligned} \tag{212}$$

Similarly, it can be shown that

$$H_K = \frac{2K_1}{\mu_0 M_s} \qquad [100] \text{ easy direction} \tag{213}$$

$$H_K = -\frac{(\frac{4}{3}K_1 + \frac{4}{9}K_2)}{\mu_0 M_s} \qquad [111] \text{ easy direction} \tag{214}$$

There is no lower limit to H_K since, for certain materials (e.g. certain Ni–Fe alloys), $K_1 \to 0$. The wide range of values is indicated by

Iron: 45, Cobalt: 4200, $SmCo_5$: 20 000 kA m^{-1} and other values are given in Chapter 5.

Magnetostrictive strains are anisotropic or there would be no saturation magnetostrictive length change δl observed as the $\mathbf{M}_s$ vectors rotate to give saturation, from an initial ideal demagnetized state with equal volumes magnetized along each of the easy directions. This effect is described by the saturation magnetostriction coefficient $\lambda_s = \delta l/l$. Typically, $\lambda_s = -10$ to -100×10^{-6} but the anisotropic coefficients may be positive or negative. The measured deformation $\lambda - \lambda_d$, with λ_d for demagnetization, depends upon the direction cosines $\alpha_1, \alpha_2, \alpha_3$ specifying the direction along which $\mathbf{M}_s$ is eventually aligned and the β_i specifying the direction along which the measurement is made. Usually the two coincide and $\lambda - \lambda_d$ is designated $\lambda_{100}(\lambda_{111})$ when the direction is a cube edge (diagonal), etc. The λ_d themselves can be expressed (e.g. [20]) as

$$\lambda_d = B_0 + \tfrac{1}{3}B_1 + \tfrac{1}{3}B_4$$

for [100] easy directions, $K_1 > 0$ as for iron, or

$$\lambda_d = B_0 + \tfrac{1}{3}B_1 + \tfrac{1}{3}B_3 + \tfrac{1}{6}B_4$$

for [111] easy directions, $K_1 < 0$ as for nickel, and further

$$\lambda_{100} = \tfrac{2}{3}B_1 + \tfrac{2}{3}B_4$$
$$\lambda_{111} = \tfrac{1}{3}B_2 + \tfrac{1}{3}(B_3 - \tfrac{2}{3}B_4) + \tfrac{1}{9}B_5$$

for iron, etc., and

$$\lambda_{100} = \tfrac{2}{3}B_1 - \tfrac{1}{3}B_3 + \tfrac{8}{9}B_4$$
$$\lambda_{111} = \tfrac{1}{3}B_2 + \tfrac{1}{9}B_5$$

for nickel, etc. The B_i are magnetostriction constants related to S_{ij} and L_i, which are respectively the elastic compliance moduli and the magnetoelastic coupling constants. The deformation of any cubic crystal can be written in an approximation (for powers of α_i up to the second) as

$$\frac{\delta l}{l} = \tfrac{2}{3}\lambda_{100}(\alpha_1^2\beta_1^2 + \alpha_2^2\beta_2^2 + \alpha_3^2\beta_3^2 - \tfrac{1}{3}) + 3\lambda_{111}(\alpha_1\alpha_2\beta_1\beta_2 + \alpha_2\alpha_3\beta_2\beta_3 + \alpha_1\alpha_3\beta_1\beta_3)$$

For a random polycrystal the longitudinal magnetostriction is obtained by averaging (with all $\alpha_i = \beta_i$) as

$$\lambda = \tfrac{2}{5}\lambda_{100} + \tfrac{3}{5}\lambda_{111}$$

For uniaxial crystals Bozorth [21] used

$$\frac{\partial l}{l} = k_1(\alpha_1^2 - 1)\beta_1^2 + k_2(\alpha_2^2\beta_2^2 + \alpha_3^2\beta_3^2) + k_3(\alpha_3^2\beta_2^2 + \alpha_2^2\beta_3^2) + 2(k_2 - k_3)\alpha_2\alpha_3\beta_2\beta_3$$
$$+ 2k_4\alpha_1\beta_1(\alpha_3\beta_3 + \alpha_2\beta_2)$$

with $k_1 = -110$, $k_2 = -45$, $k_3 = -95$, $k_4 = -235$, (all $\times\ 10^{-6}$) for cobalt.

12 Magenetostatic Energy and Domain Walls

12.1 Magnetostatic Energy and Domains

The state in which a ferromagnetic is usually found is one of zero magnetization. In a polycrystal this could be because each crystallite is uniformly magnetized, but the directions of $\mathbf{M}_s$ are distributed randomly. More generally, and necessarily in single crystals, it may be inferred that the crystal, or each crystallite, is divided into a number of regions or domains of volume v_i each with uniform magnetization of magnitude $M_i = M_s$ such that in the demagnetized state $\sum v_i M_j^i = 0$, for $j = x, y, z$. Obvious minimal examples are shown in Figure 1.57. The domain boundaries or domain walls are sheets of thickness usually much less than the domain width, in which the magnetization rotates (usually) gradually between the directions in the two domains, the width and surface energy density of the walls being calculable in terms of A, K and the wall orientations.

The tendency towards demagnetization by the formation of domains accords with the consequent reduction of the total energy, initially considered as the sum of the magnetostatic self-energy and the domain wall energy (for uniaxial crystals with $\mathbf{M}$ parallel to easy directions forming 180° domains). The magnetostatic energy E_m may be considered as that

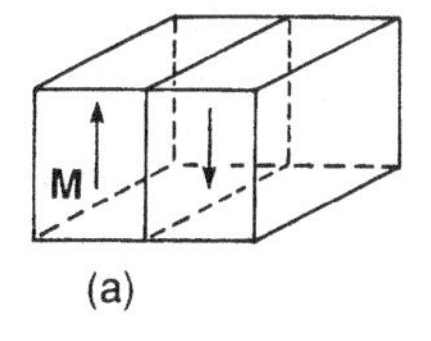

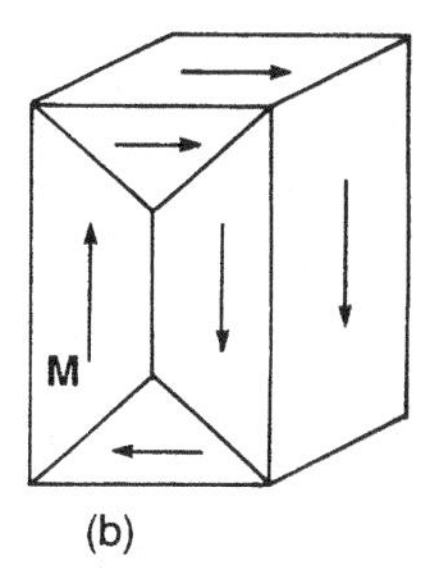

Figure 1.57 Rudimentary 180° domain structures in a uniaxial crystal (a) and in a cubic (iron) crystal (b) with 90° domains affording flux closure and minimal magnetostatic energy

required to bring up the constituent dipoles from infinity, i.e. as the energy of formation; as discussed in Chapter 4, it is calculated by evaluating the integral of the product of pole density and potential over the specimen, the potential being that generated by the magnetization. Alternatively and equivalently,

$$E_{\mathrm{m}} = \tfrac{1}{2} \int \mu_0 \mathbf{H}_{\mathrm{d}} \cdot \mathbf{M}\, \mathrm{d}v \tag{215}$$

over the specimen volume, where $\mathbf{H}_{\mathrm{d}}$ is the internal demagnetizing field.

If the specimen has the shape of an ellipsoid of revolution, $\mathbf{H}_{\mathrm{d}}$ is uniform if $\mathbf{M}$ is uniform. For oblate or prolate spheroids with $\mathbf{M}$ parallel to a principal direction [22],

$$\mathbf{H}_{\mathrm{d}} = -N\mathbf{M}$$

with N a (positive) dimensionless demagnetizing factor. For ellipsoids of revolution (Figure 1.58 with $b = c$) with $\mathbf{M} \parallel$ the long axis a (prolate spheroids)

$$N_a = \frac{1}{q^2 - 1}\left[\frac{q}{\sqrt{q^2 - 1}} \ln(q + \sqrt{q^2 - 1} - 1)\right] \rightarrow \frac{1}{q^2}(\ln 2q - 1) \tag{216}$$

with $q = a/b$ and the second expression for $q \gg 1$. For flat ellipsoids with two long axes a and b nearly equal and much greater than the short axis c,

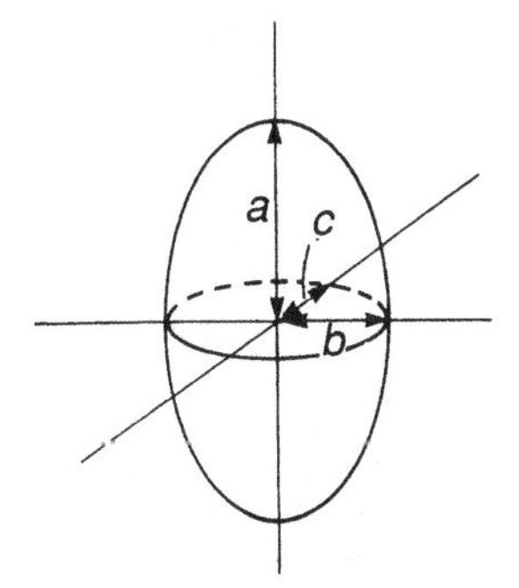

Figure 1.58 A general ellipsoid with Cartesian axes taken to coincide with the principal axes

$$N_a = \frac{\pi}{4}\frac{c}{a}\left[1 - \tfrac{1}{4}\frac{a-b}{a} - \tfrac{3}{16}\left(\frac{a-b}{a}\right)^2\right]$$
$$N_b = \frac{\pi}{4}\frac{c}{a}\left[1 + \tfrac{5}{4}\frac{a-b}{a} + \tfrac{21}{16}\left(\frac{a-b}{a}\right)^2\right] \tag{217}$$

For oblate spheroids with **M** parallel to the circular plane,

$$N = \tfrac{1}{2}\left[\frac{q^2}{(q^2-1)^{3/2}}\sin^{-1}\frac{\sqrt{q^2-1}}{q} - \frac{1}{q^2-1}\right] \tag{218}$$

with q = diameter/thickness.

The relation

$$N_a + N_b + N_c = 1 \tag{219}$$

applies to the three factors appropriate to the three axes of an ellipsoid. Thus, by symmetry, for a sphere

$$N_a = N_b = N_c = \tfrac{1}{3} \tag{220}$$

For a long cylinder with **M** $\parallel$ the long axis a, $N_a = 0$ because the 'end poles' are distant and thus

$$N_b = N_c = \tfrac{1}{2} \tag{221}$$

which applies when **M** is normal to the long axis. For a thin plate, considered as an ellipsoid with $k = 0$, two factors are zero and $N_c = 1$, as inferred previously.

If the principal axes of an ellipsoid coincide with the Cartesian axes as in Figure 1.58, the surface poles due to the x component of a generally directed **M** can only give a field $\parallel OX$, etc., and thus

$$\begin{pmatrix} H_x \\ H_y \\ H_z \end{pmatrix} = \begin{pmatrix} N_b & 0 & 0 \\ 0 & N_c & 0 \\ 0 & 0 & N_a \end{pmatrix}\begin{pmatrix} M_x \\ M_y \\ M_z \end{pmatrix} = \begin{pmatrix} N_{11} & 0 & 0 \\ 0 & N_{22} & 0 \\ 0 & 0 & N_{33} \end{pmatrix}\begin{pmatrix} M_x \\ M_y \\ M_z \end{pmatrix} \tag{222}$$

If the axes do not coincide it is necessary to take the components of **M** along the ellipsoid axes, find the field components along these axes and then resolve these along OX, OY, OZ so that, formally, $\mathbf{H}_\mathrm{d} = \mathbf{NM}$, where **N** is the demagnetizing tensor which is diagonal only in the above special case.

Magnetostatic or demagnetizing energies of short cylinders or rectangular blocks with uniform magnetization are readily calculated numerically and effective demagnetizing factors can be defined in relation to these, or by

$$\bar{N} = -\bar{H}_\mathrm{d}/M$$

where $\bar{H}_\mathrm{d}$ is the mean demagnetizing field calculated. These can only be used in approximate calculations.

If a cube (or sphere) is divided equally into two domains as in Figure 1.57(a), E_m is halved and if it were divided into a great number of domains, it is clear that H_d and E_m would tend towards zero. However, the wall energy density per unit volume E_w would become very great and an equilibrium domain width is expected to correspond to a minimum of $(E_m + E_w)$. However, since E_m is the energy per unit volume it is easily seen that for particles below a certain critical size it may not be energetically favourable to introduce even a single domain wall and very small particles or crystallites are expected to exhibit 'single-domain behaviour' as noted in Chapter 5, Section 2.1, the single- and two-domain energies are equal at $r = r_c = 9\gamma/(\mu_0 M_s^2)$ and for iron $r_c \approx 1$ nm. γ is the wall energy per unit area.

If the particle size is reduced even further the peak value of W_k as $\mathbf{M}_s$ rotates, e.g. of $K \sin^2\theta \times$ particle volume, may become less than kT at room temperature, say. The energy barriers to rotation become relatively ineffective and the distribution of the magnetic moments of a particle assembly may be given by applying Boltzmann statistics giving superparamagnetic behaviour, discussed in Chapter 5, Section 2.2.

12.2 Domain Walls

The more gradual the rotation of $\mathbf{M}$ the lower is the exchange energy. If $\mathbf{M}_s$ were arbitrarily pinned at oppositely magnetized faces of a crystal it might be expected that the rotation should occupy the whole specimen. However, this would give a high value of E_K (particularly apparent for a uniaxial crystal) and it is thus expected that a wall with a particular (asymptotic) width $\delta\,(A, K)$ and surface energy $\gamma\,(A, K)$ should be formed. Usually (unless $K \to 0$), $\delta \ll l$, the domain width, and $\delta = 0$ is assumed in calculating E_m.

Taking the rotation to occur along OZ and to be indicated by the single angle $\theta(z)$ as in Figure 1.56 (but with a great number of steps so that θ can be considered as a continuous variable), the anisotropy energy for each layer per unit surface area of wall may be designated $f(\theta)$ and (per unit area)

$$\gamma = \gamma_e + \gamma_K = \int_{-\infty}^{\infty} \left[A \left(\frac{\partial \theta}{\partial z} \right)^2 + f(\theta) \right] \mathrm{d}z \tag{223}$$

Assuming that $\theta(z)$ is in fact that function which gives the minimum energy and introducing a small change $\delta\theta$, the change in γ is

$$\delta\gamma = \int_{-\infty}^{\infty} \left[2A \frac{\partial \theta}{\partial z} \frac{\partial}{\partial z} \delta\theta + \frac{\partial f(\theta)}{\partial z} \delta\theta \right] \mathrm{d}z \tag{224}$$

In the usual way it is assumed that the minimum energy (stable configuration) corresponds to $\delta\gamma = 0$. Integrating the first term by parts

$$\int_{-\infty}^{\infty} 2A \frac{\partial \theta}{\partial z} \frac{\partial}{\partial z} \delta\theta \,\mathrm{d}z = \left[2A \frac{\partial \theta}{\partial z} \delta\theta \right]_{-\infty}^{\infty} - \int_{-\infty}^{\infty} 2A \frac{\partial^2 \theta}{\partial z^2} \delta\theta \,\mathrm{d}z$$

and assuming $\mathbf{M}$ to be uniform within the domains, where the limits effectively apply, the first term is zero. The condition $\delta\gamma = 0$ is

$$\int_{-\infty}^{\infty} \left[\frac{\partial f(\theta)}{\partial \theta} - 2A \frac{\partial^2 \theta}{\partial z^2} \right] \delta\theta \,\mathrm{d}z = 0$$

and since $\delta\theta$ is arbitrary, this means that

$$\frac{\partial f(\theta)}{\partial \theta} - 2A \frac{\partial^2 \theta}{\partial z^2} = 0 \tag{225}$$

(which could have been written directly as the Euler equation of the one-dimensional variational problem). Multiplying by $\partial\theta/\partial z$ and integrating,

$$f(\theta) = A\left(\frac{\partial\theta}{\partial z}\right)^2, \qquad \mathrm{d}z = \sqrt{A}\frac{\mathrm{d}\theta}{\sqrt{f(\theta)}} \tag{226}$$

[since $\theta = \theta(z)$ only]. Integrating this formally,

$$z = \sqrt{A}\int_0^\theta \frac{\mathrm{d}\theta}{\sqrt{f(\theta)}}$$

Noting from the above that the exchange and anisotropy terms are equal,

$$\gamma_{180} = 2\sqrt{A}\int_0^\pi \sqrt{f(\theta)}\,\mathrm{d}\theta \tag{227}$$

e.g.

$$\gamma_{180} = 2\sqrt{A}\int_0^\pi \sqrt{K}\sin\theta\,\mathrm{d}\theta = 4\sqrt{AK} \qquad \text{(simple uniaxial case)} \tag{228}$$

A typical value of A may be obtained from the Bloch $T^{3/2}$ law as $10^{-11}\,\mathrm{J\,m^{-1}}$, giving ($K = 1.3 \times 10^7\ \mathrm{J\,m^{-3}}$)

$$\gamma_{180}(\mathrm{SmCo_5}) = 46 \times 10^{-3}\ \mathrm{J\,m^{-2}}$$

which is practically an upper limit, with no lower limit since $K \to 0$ may be envisaged.

Taking both z and θ to be zero at the wall centre, $f(\theta) = K\cos^2\theta$ and

$$z = \sqrt{\frac{A}{K}}\int_0^\theta \frac{\mathrm{d}\theta}{\cos\theta} = \sqrt{\frac{A}{K}}\ln\tan\left(\frac{\theta}{2} + \frac{\pi}{4}\right) \tag{229}$$

using a standard integral. $\theta(z)$ throughout the 180° wall is as shown in Figure 1.59 with asymptotic approaches to **M** in the bulk of the domains. Truncation may be achieved by extrapolation using $\partial\theta/\partial z$ at the wall centre as shown, giving

$$\delta = \pi\frac{\partial\theta}{\partial z}\bigg|_{z=0} = \pi\sqrt{\frac{A}{K}} \tag{230}$$

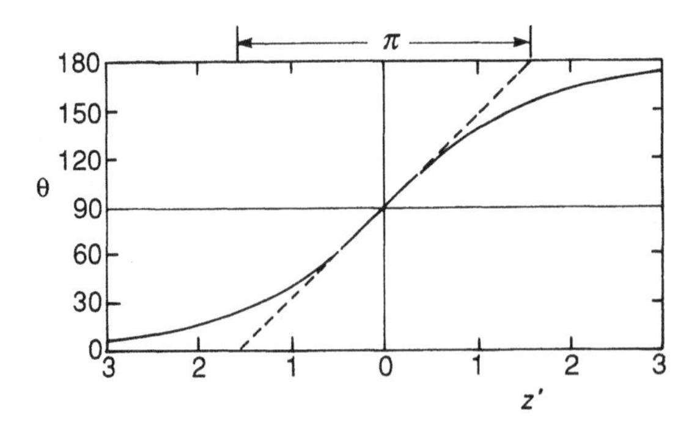

Figure 1.59 The variation of the moment orientation throughout a 180° domain wall

Taking A as before and $K = 4 \times 10^4 \text{ J m}^{-3}$ for iron,

$$\delta_{180}(\text{Fe}) = 50 \text{ nm}$$

$$\delta_{180}(\text{SmCo}_5) = 2.8 \text{ nm}$$

For iron the continuum model is justified but for $SmCo_5$ most of the rotation is predicted to occur over a few lattice planes only, and then a small threshold field (intrinsic coercive field) is required to move the wall, even in a perfect crystal [23]. Again, if $K \to 0$ there is no upper limit, but even when efforts are made to achieve $K \approx 0$ in practice (to optimize χ as shown later) a variety of observations indicate that walls with finite though large widths occur.

Even in cubic materials, 180° walls, e.g. between [100] and [$\bar{1}$00] domains, are common but in iron 90° walls separate [100] and [010] domains, for example, and in nickel etc. ($K_1 < 0$) 71° and 109° walls separate domains magnetized along $\langle 111 \rangle$ as in Figure 6.3. For a 180° wall in iron with the spins rotating in a (100) plane,

$$\gamma_{180} = 2\sqrt{AK} \tag{231}$$

while for a 90° wall

$$\gamma_{90} = \sqrt{AK} \tag{232}$$

The spins at the centre of a 180° wall can follow an easy direction. Further results are given by Lilley [24], following Landau and Lifshitz [25, 26], by Néel [27] and by Kaczer [28].

Where walls intersect surfaces $\parallel$ **M** narrow strips of pole density may form which can attract the fine particles (~ 10 nm diameter) of a colloid of magnetite, such colloids or

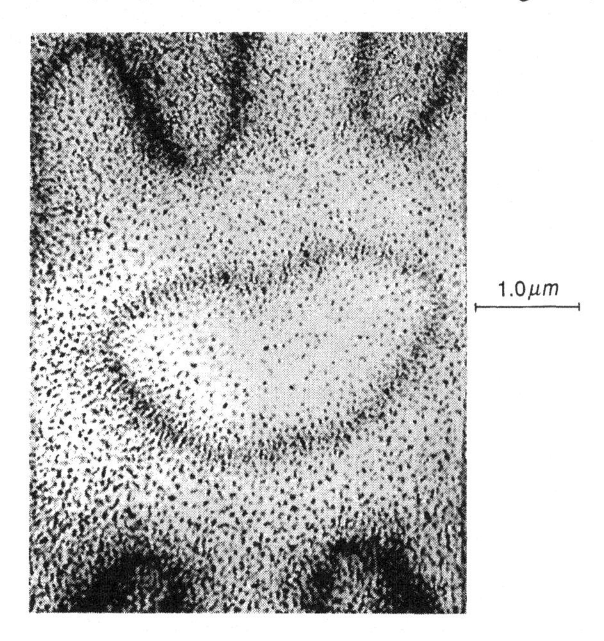

Figure 1.60 The formation of chains of minute magnetite particles in the strong stray fields above the intersection of a narrow 180° domain wall with a crystal surface $\perp$ $\mathbf{M}_s$ of barium ferrite (giving rise to interesting optical effects in a polarizing microscope)

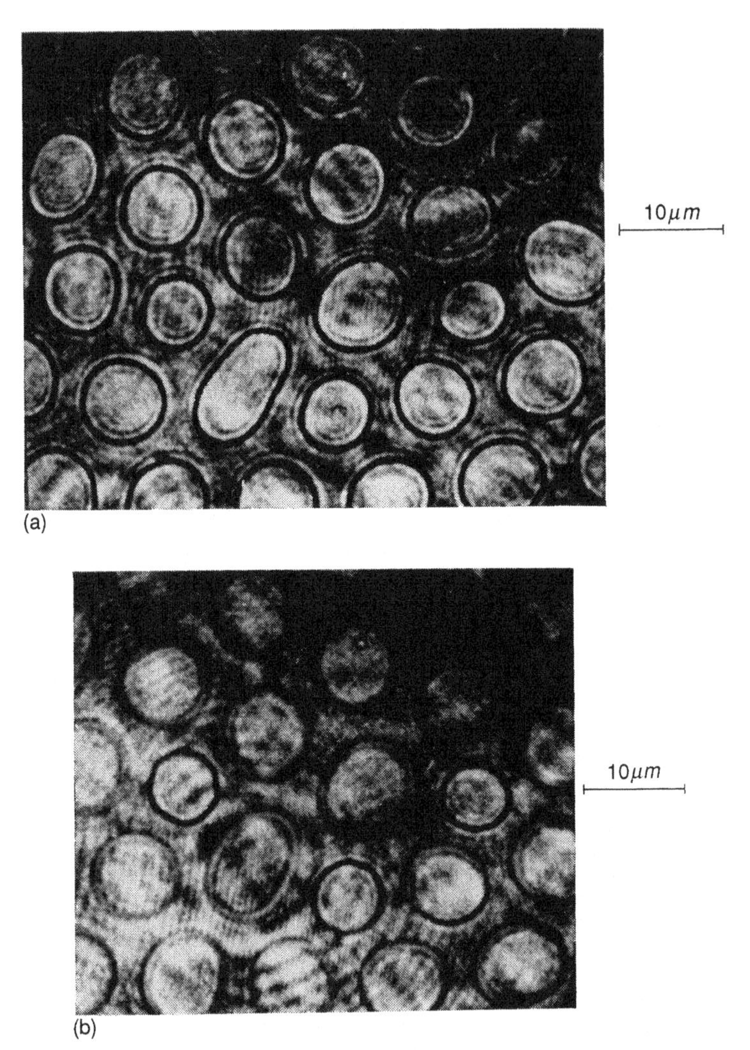

Figure 1.61 (a) An array of 180° cylindrical, bubble, domains with M normal to the surface of a thin single-crystal film of $Tb_{0.3}\ Eu_{0.3}\ Y_{2.4}\ Ga_{1.1}\ Fe_{3.9}\ O_{12}$ revealed by the Faraday effect using a laser source (see Chapter 6). (b) In a 50 Hz, 15 Oe field the walls of the larger domains are seen to oscillate freely but those of the smaller domains, assumed to have a complex and high energy (Bloch line) structure, remain static [29]

'ferrofluids' being readily prepared. Strong field gradients arise above the intersections of walls with surfaces $\perp$ **M** and lines of particles may be formed to visualize these walls as in Figure 1.60. Figure 1.61 [29] shows 180° walls in a uniaxial garnet crystal, revealed by the Faraday effect (Chapter 5); the caption indicates the complexities that may arise. Reference may be made to specific works on domains [30, 31] and to Chapter 4, Section 2 and Chapter 6, Section 1.

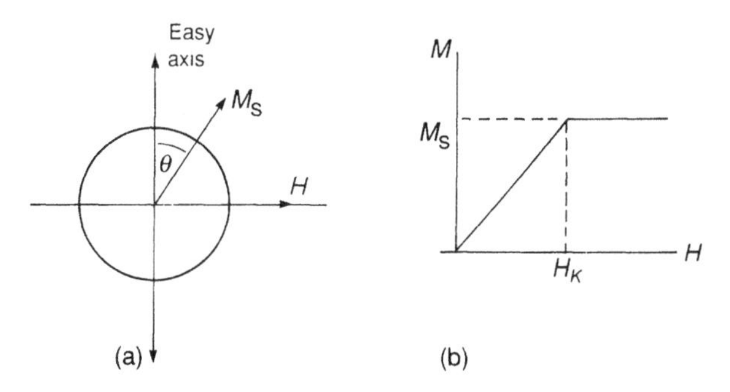

Figure 1.62 For a uniaxial single-domain sphere the equilibrium θ depends in a simple way on the balance of the torques due to the field and the magnetocrystalline anisotropy (a), giving the result in (b) where M is the component of $\mathbf{M}_s$ along $\mathbf{H}$

13 Magnetization Processes

13.1 Rotation

For the uniaxial sphere in Figure 1.62 the θ-dependent energy density is $K\sin^2\theta - \mu_0 H M_s \sin\theta$ and equilibrium corresponds to

$$2K\sin\theta\cos\theta - \mu_0 H M_s \cos\theta = 0, \qquad \text{i.e. } \sin\theta = \frac{\mu_0 H M_s}{2K}$$

The magnetization induced in the field direction is

$$M = M_s \sin\theta = \mu_0 \frac{H M_s^2}{2K}, \qquad \chi = \frac{M}{H} = \frac{\mu_0 M_s^2}{2K} \tag{233}$$

χ is constant up to saturation at $H = H_K = 2K/(\mu_0 M_s)$ as in Figure 1.62(b). Using two constants it is easy to show that the equation for the equilibrium M is

$$\mu_0 H M_s = 2\frac{M}{M_s}\left[K_1 + 2K_2\left(\frac{M}{M_s}\right)^2\right] \tag{234}$$

from which M and χ for any H can be computed. Curve-fitting $M(H)$ can give estimates of K_1 and K_2. For initial susceptibilities, extrapolated to zero H or measured in very small fields $(M/M_s)^2$ can be ignored and χ is as before.

With a field at angle α to an easy direction the initial susceptibility is

$$\chi_i = \mu_0 \frac{M_s^2 \sin^2\alpha}{2K} \tag{235}$$

for a uniaxial or cubic, $K_1 > 0$, material. For $K_1 < 0$ the anisotropy energy for small deflections is $E_K = \frac{2}{3}|K_1|\theta^2$, giving

$$\chi_i = \frac{3\mu_0 M_s^2 \sin^2\alpha}{4|K_1|} \tag{236}$$

For a random assembly of particles, using $\langle \sin^2 \alpha \rangle = \frac{2}{3}$,

$$\chi_i = \frac{\mu_0 M_s^2}{3K_1} \quad \text{(uniaxial, cubic } K_1 > 0) \tag{237}$$

$$\chi_i = \frac{\mu_0 M_s^2}{2K_1} \quad \text{(cubic, } K_1 < 0) \tag{238}$$

13.2 Switching: Coherent Reversal

With **H** precisely parallel to an easy axis and antiparallel to $\mathbf{M}_s$, $\mathbf{H}\times\mathbf{M}_s = 0$. If $\mathbf{M}_s$ is displaced by a small angle $\delta\theta$ from the easy axis the anisotropy field concept applies and if $H > 2K/(\mu_0 \mathbf{M}_s)$ the net field is in the direction of **H** and adequate to overcome energy barriers during rotation, i.e. even with $\delta\theta = 0$ the original direction is metastable. Thus it may be inferred that abrupt switching occurs in a reverse field if

$$H = H_c = H_K = \frac{2K}{\mu_0 M_s} \tag{239}$$

This can be confirmed by plotting the energies and torques as H is varied.

The magnetization loop is thus as shown in Figure 1.63. The remanent magnetization remaining after removing a saturating field is $M_r = M_s$. The *coercivity* or *coercive field* H_c, required to give zero magnetization, is in this case the switching field since $M = 0$ is an unstable state.

If **H** is applied at 45° to the easy axis the reader may readily confirm that $\mathbf{M}_s$ rotates reversibly until a discontinuity occurs at a field given by

$$\mu_0 H M_s \sin\frac{\pi}{2} = 2K \sin\frac{\pi}{4}\cos\frac{\pi}{4}, \qquad \text{i.e. } H_c = \frac{K}{\mu_0 M_s}$$

since beyond the corresponding angle the torque due to the field increases with θ and that due to the anisotropy decreases. The loop is as shown in Figure 1.64, after Stoner and Wohlfarth [32].

The dependence of the magnetostatic energy of an ellipsoid of revolution on the angle between **M** and the a axis is shown in Chapter 4 to be the same as that of E_K on the angle between **M** and the (single) easy axis, and with $K_s = \frac{1}{2}\mu_0 M_s^2(N_b - N_a)$ for the shape

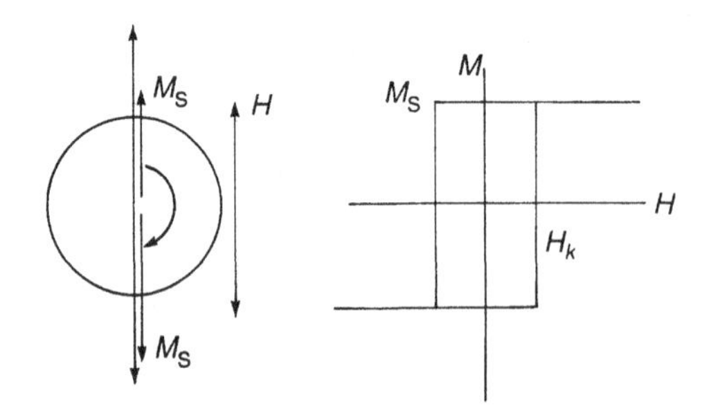

Figure 1.63 **H** ∥ the easy axis gives no torque but discontinuities are expected to occur to give the loop shown

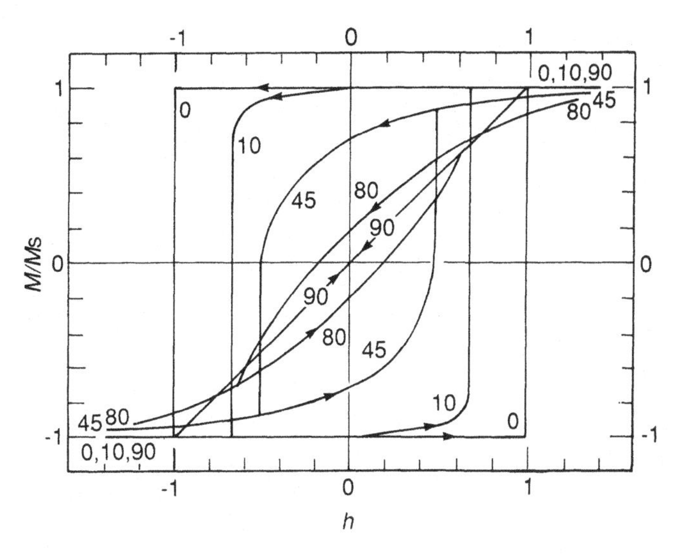

Figure 1.64 A set of 'Stoner–Wohlfarth' loops for the field orientations indicated, given in terms of $h = H/H_K$ or $h = H/H_s$ for either magnetocrystalline or shape anisotropies [32]

anisotropy the simple results applying to crystal anisotropy can be applied to shape effects as, for example,

$$\chi = \frac{\mu_0 M_s^2}{2K_s} = \frac{1}{N_b - N_a}, \qquad H_c = \frac{2K_s}{\mu_0 M_s} = M_s(N_b - N_a) \tag{240}$$

Further, a shape anisotropy field may be derived as before as

$$H_s = M_s(N_b - N_a) \tag{241}$$

and the set of magnetization curves in Figure 1.64 can be given in terms of $h = H/H_A$ with $H_A = H_K$ or $H_A = H_s$, applying equally to either anisotropy. A virtual upper limit to $H_c = H_s$, for very long iron particles, is

$$H_c = (N_b - N_a)M_s = (\tfrac{1}{2} - 0)M_s = 850 \text{ kA m}^{-1}$$

If both types coexist, as in an anisotropic ellipsoid, the sum of the three appropriate energy terms can be minimized to predict the behaviour. If a single easy axis coincides with the *a* axis of the ellipsoids the two anisotropy fields may be summed. Shape effects can increase or decrease values of H_c due to crystal anisotropy alone.

It has been observed that lattices generally deform when ordering occurs and these deformations are anisotropic and depend on the direction of the magnetization. If an initially demagnetized specimen of length l, with randomly oriented M_s vectors, is magnetized there is a small change of length δl in the direction of magnetization and the magnetostriction coefficient is defined as

$$\lambda = \frac{\delta l}{l} \tag{242}$$

with λ_s for saturation. If the magnetization lies at an angle θ to the direction of measurement of δl it is found that

$$\frac{\delta l}{l} = \tfrac{3}{2}\lambda_s\left(\cos^2\theta - \tfrac{1}{3}\right) \tag{243}$$

If the magnetostrictive deformation occurs in the presence of an applied tensile stress σ, then considering the work done the variable part of the corresponding energy is

$$E_\sigma = -\tfrac{3}{2}\lambda_s\sigma\cos^2\theta = -\tfrac{3}{2}\lambda_s\sigma(1 - \sin^2\theta) \tag{244}$$

When $\lambda_s > 0$, E_σ is minimum for $\theta = 0$ and $\theta = 2\pi$ and the magnetization tends to be aligned along the (tensile) stress axis, giving a uniaxial anisotropy analogous to a crystal anisotropy with K replaced by

$$K_\sigma = \tfrac{3}{2}\lambda_s\sigma \tag{245}$$

with an anisotropy field

$$H_\sigma = \frac{3\lambda_s\sigma}{\mu_0 M_s} \tag{246}$$

Most of the principles developed here are relevant to single-domain particles with no domain wall processes involved, and are developed later as appropriate.

13.3 Remanence

For aligned single-domain particles it has been seen that M_r/M_s can be unity. For a randomly oriented assembly of s–d crystallites, if it is assumed that, on removal of a saturating field, $\mathbf{M}_s$ in each crystallite relaxes back to the nearest easy direction, then averaging gives (as shown by Gans [33])

Anisotropy	Uniaxial	Cubic, $K_1 > 0$	Cubic, $K_1 < 0$
M_r/M_s	0.5	0.832	0.866

13.4 Induced Magnetization and Domain Wall Motion

Coeroivities predicted for coherent rotation for iron, nickel and cobalt range from 2.5 to 850 kA m^{-1} but measured values, for presumed multidomain specimens, are usually orders of magnitude smaller than these. Calculated rotational susceptibilities are of the order of 10 to 100 compared to measured values up to 10^6 for single crystals of iron and measured remanences may be virtually zero. Domain wall processes are much 'easier' than rotation.

A field applied parallel to a 180° wall may be considered to exert a pressure on the wall

$$P_H = 2\mu_0 H M_s \tag{247}$$

in view of the change in energy as the wall moves. For a structure such as that shown in Figure 1.57(a) this may be balanced by a restoring pressure due to demagnetizing fields, but if the specimen becomes indefinitely elongated the latter becomes negligible and χ increases indefinitely. (This neglects the intrinsic wall coercivities which may exist even in perfect crystals if the walls are exceptionally narrow [23] and the continuum model is not fully applicable [34].) In general, if it is assumed that the wall remains plane it is easy to calculate and minimize the sum of E_m and $E_H = -\mu_0\mathbf{H}\cdot\mathbf{M}$ to give the equilibrium wall positions and magnetization. This gives the anhysteretic magnetization curve since the crystal is implicitly perfect with no variation in the domain wall energy with its position. $M_r = 0 = H_c$ and there is no hysteresis loss, defined by

$$W_H = \mu_0 \oint H \, dM \tag{248}$$

around the 'loop'; this is the net work expended by the field in changing the magnetization which is equal to the area of the M/H loop (in this case zero).

If a specimen contains a large number of walls the magnetization may be taken to be homogeneous as an approximation. For an ellipsoidal specimen with low crystal anisotropy the induced magnetization and susceptibility are, as previously noted,

$$M = \frac{\chi H_a}{1 + N\chi}, \qquad \chi_e = \frac{M}{H_a} = \frac{\chi}{1 + N\chi} \to \frac{1}{N} \qquad (N\chi \gg 1)$$

Thus if $\chi = M/H$, with H the internal field, is high then the effective susceptibility is controlled by demagnetizing effects unless $N \to 0$.

In technical applications overall demagnetizing effects are avoided by assembling strips of material into rectangular frames, winding ribbons or tapes into toroids or producing toroids of sintered materials such as ferrites. The properties measured for such specimens can be taken to be characteristics of the material (and its particular microstructure), although some internal demagnetizing effects which arise where $\mathbf{M}_s$ changes in direction at grain boundaries may still be involved.

In such specimens, however, χ and $\mu_r = 1 + \chi$ may be controlled by impediments to free domain wall motion caused by any imperfections or irregularities that render the wall energy position-dependent. These include any variations in A or K or, more drastically, the presence of pores (or non-magnetic inclusions) or inclusions of foreign magnetic phases. A wall tends to remain in a 'trough' where its energy density is anomalously low or to be 'repelled' by regions where γ increases. A pore or inclusion reduces the local wall energy density by effectively removing a small area of the wall. Moreover, if a relatively large spherical inclusion is bisected by a narrow wall the associated magnetostatic energy is halved, since it is effectively an isolated sphere with M_s opposite to that of the surroundings, and this clearly impedes the wall.

The wall motion thus proceeds by a series of Barkhausen jumps between energy minima, and if the specimen loads a coil the effect will be the production of noise in the voltage generated. In a randomly oriented polycrystal only a certain fraction of saturation can be achieved by wall motion (corresponding to the remanence) and this must be followed by rotation. If a core is biased by a static field so that an a.c. field gives small excursions of the magnetization at a high level, it is found that the noise decreases, i.e. the signal-to-noise improves, although the susceptibility and signal per unit a.c. field amplitude falls.

Many materials may be expected to contain uniformly distributed inclusions, e.g. of iron carbide in iron, with narrow size ranges, and thus the field required to initiate wall motion will move the walls past successive inclusions. When a near-saturating field is removed and reversed this same characteristic 'wall-impedance field' magnitude will be required to cause demagnetization and thus constitutes the coercivity. This gives, for a grain-oriented specimen, a virgin curve from, say, an a.c. demagnetized state (achieved by gradually reducing the amplitude of an initially large a.c. field), and an M/H loop, somewhat as

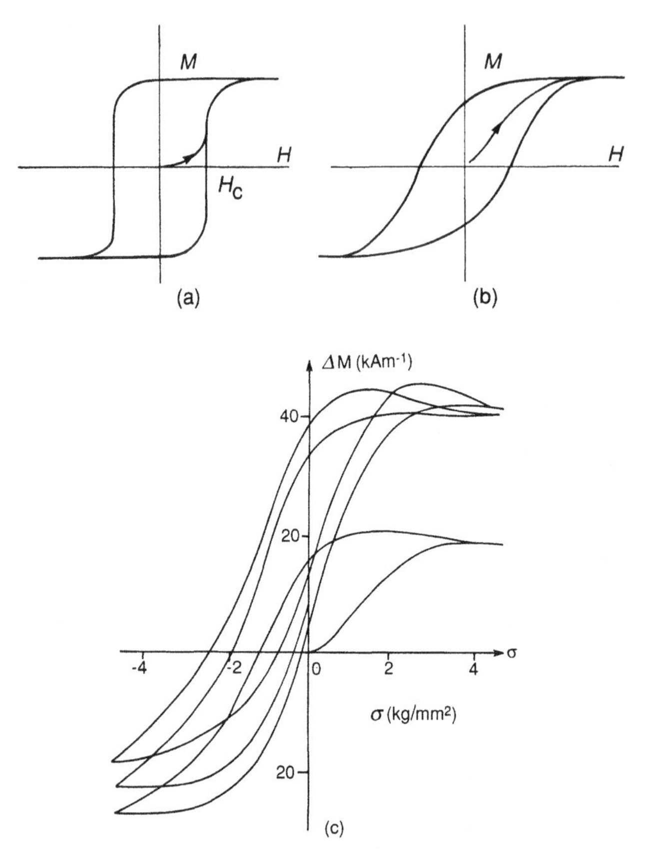

Figure 1.65 (a) The expectation for a grain-oriented material with a characteristic wall impedance field H_c and (b) a sigmoid loop for a randomly oriented polycrystal with rotational processes taking over at the higher fields; (c) indicates how applied tensile stresses may induce a magnetization, and remanence, in small fields [30]. Repeated stress cycles for random SiFe (3 per cent)

shown in Figure 1.65(a). The curvatures are due to (a) some variation in the wall impedance fields and (b) demagnetizing effects, both internal in association with grain boundaries and associated with small components of $\mathbf{M}_s$ across the external surfaces, since the orientation is never perfect. In such cases the following approximations obviously apply:

$$
\begin{aligned}
W_h &\doteqdot 4\mu_0 H_c M_s \\
\chi_m &\doteqdot \frac{M_s}{H_c} \\
\mu_r^m &= 1 + \frac{M_s}{H_c}
\end{aligned}
\tag{249}
$$

with W_h the hysteresis loss per cycle and χ_m the maximum susceptibility, i.e. the tangent of the greatest angle of slope of the line drawn from 0 to points on the loop.

For randomly oriented polycrystals the loops are more as shown in Figure 1.65(b). The pronounced curvatures giving a typically sigmoid virgin curve are due, at low fields, to the existence of a range of wall pressures due to varying angles between the field and the walls and, at higher fields, demagnetizing effects and the merging of wall motion and rotational processes.

In the development of soft magnetic materials, with high permeabilities and low hysteresis, the main objectives are the elimination of features impeding wall motion or the reduction of their effectiveness. The first is achieved by purification and homogenization, as well as stress relief since local stresses cause variations in the total anisotropy. The second is achieved largely by minimizing the anisotropy, (and magnetostriction) as in Ni–Fe alloys or permalloys, to give minimal γ and very wide walls. The inclusions etc. may then be much smaller than the wall width and their effectiveness is reduced [35, 36]. In extreme cases the walls may occupy a substantial proportion of the specimen and the merging of the wall motion and rotation occurs at low levels of M, giving initial susceptibilities χ_i comparable to maximum susceptibilities. [For the loops in Figure 1.65(a) applicable to grain-oriented silicon iron, $\chi_i < \chi_{max}$.]

A wall impeded at a particular point will tend to be plucked like a bow string, but the curvature is opposed both by the increase in total wall energy and the introduction of magnetostatic energy due to the components of M_s across the wall. Thus the walls in high M_s, high γ materials such as SiFe tend to remain planar while those in very soft ferrites and alloys bow more readily.

If the crystallites are relatively free of imperfections but the walls are restricted at grain boundaries, either naturally due to magnetostatic effects or due to intergranular porosity etc., as occurs in ferrites, then the initial permeability may be largely associated with wall bowing. The greater the span the easier is the motion in relation both to the surface tension effect and the magnetostatic energy involved. It is frequently found that μ rises approximately linearly with grain diameter (see 2.3, Chapter 5).

One way to minimize the anisotropy is to reduce the grain size so extremely that, over a distance for which $\mathbf{M}_s$ is constrained to be nearly uniform by the exchange, the orientation of the crystal axes changes so that the anisotropy averages out, i.e. to produce nanocrystalline materials (see Chapter 5, Section 2.3.1, final part) or, ultimately, amorphous materials (Chapter 5, Section 2.3.1). There appears currently to be rather a contest between the development of these newer materials and the improvement of conventional materials.

The application of a tensile stress to a specimen in a field which itself gives little induced magnetization, e.g. the earth's field, may, by affecting the total anisotropies, favour the growth of certain domains and induce a substantial magnetization remaining, partially, as a remanence when the stress is removed (see, for example, [37, 38] and the theory of Brown [39]), as in Figure 1.65(c). Motor vehicles, ships and submarines acquire magnetic signatures which may be removed by elaborate 'de-Gaussing', and the historic production of compass needles by hammering in the earth's field doubtless depended on stress-induced magnetization.

13.5 Domain Nucleation

If complete saturation is attained a new problem arises since to achieve demagnetization new domains must nucleate and grow, though in many cases residual domains may persist in association with pores, defects on surfaces even when $M \to M_s$ closely, and the problem is then only apparent.

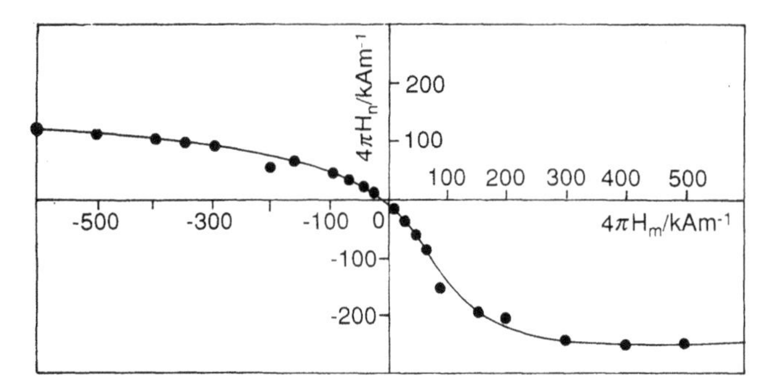

Figure 1.66 The dependence of the nucleation fields H_n in crystals of $EuFeO_3$ with damaged surfaces, on the magnitude of the saturating field and its direction, suggesting exchange coupling with foreign phases, exchange anisotropy (by C. Tanasoiu and the author)

Since the magnetization must reverse locally to form the nucleus of a 180° domain, for example, reverse fields of the magnitude of H_K would appear to be required. However, these would cause complete reversal by rotation and it is known that nucleation fields H_n are usually much lower than H_K; indeed, in many soft materials domains appear to nucleate spontaneously in zero reverse applied fields. A major factor appears to be the reinforcement of reverse applied fields by local stray or demagnetizing fields, including the effects of field components transverse to the applied fields and to M_s (see, for example, [40]).

In hard magnetic materials, permanent magnets, the coercivities may be controlled by extremely strong domain wall pinning due to deliberately contrived multiphase structures or fine precipitates, but in high-anisotropy single-phase materials H_c appears to be controlled by nucleation, as discussed in Chapter 5.

In a famous demonstration by Sixtus and Tonks [41] a wire of permalloy was stretched to give a stress-induced easy axis along its length ($\lambda_s > 0$). After saturation in a 'positive' direction the specimen remained saturated, $M_r = M_s$, even in small negative fields. However, when a negative pulse field was applied along a small length of the wire a domain was nucleated and a single wall swept along the wire to give a giant Barkhausen jump encompassing complete reversal, as inferred from the voltage pulses produced at different times in two pickup coils around the wire. Thus information was obtained both on nucleation and on domain wall velocities.

In certain cases the nucleation fields depend on the previously applied saturation fields, as originally demonstrated in orthoferrite crystals by the author and McIntyre [42] (Figure 1.66). This is associated with the presence of foreign phases formed during specimen preparation and identifiable by Faraday microscopy, exchange-coupled to the bulk. It is only as this material becomes saturated that the nucleation fields approach high values (see Chapter 6, Section 1). It has been suggested that related effects are important in a number of permanent magnet materials.

14 Classical Magnetization Dynamics

The principles governing the approach of **M** to its equilibrium magnitude and direction in an applied field, a state assumed in the foregoing treatments, are discussed in Chapter 3.

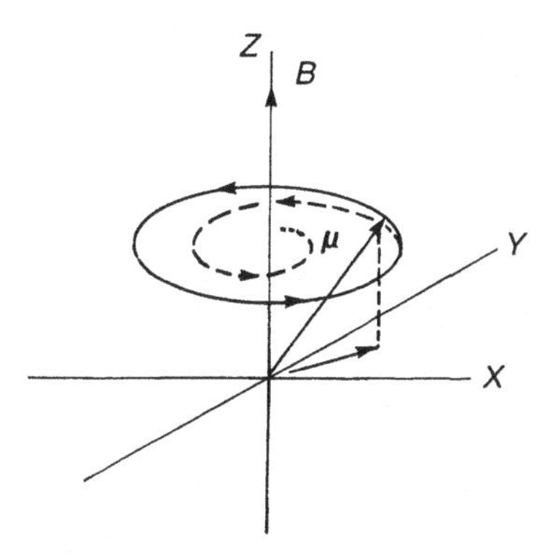

Figure 1.67 The (classical) precessional motion of an undamped spin or magnetic moment in an applied field with longitudinal and transverse components of constant magnitude. The broken line indicates the effect of damping

For a moment $\boldsymbol{\mu}$, e.g. $\mathbf{M} \times$ volume for a single-domain particle, the undamped equation of motion is $\mathrm{d}\boldsymbol{\mu}/\mathrm{d}t = \gamma \boldsymbol{\mu} \times \mathbf{B}$ with $\mathbf{B}$ a field applied at an angle θ to $\boldsymbol{\mu}$. Solutions (for $\mathbf{B} \parallel OZ$) are shown to be $\mu_x = \mu \sin\theta \cos\omega_0 t$, $\mu_y = \mu \sin\theta \cos\omega_0 t$, $\mu_z = \mu \cos\theta =$ constant, where $\omega_0 = \gamma B$. The moment $\boldsymbol{\mu}$ moves on a cone or precesses, at ω_0, and no magnetization is induced in the field direction. This is natural since energy changes consequent on changes in $\boldsymbol{\mu} \cdot \mathbf{B}$ clearly require the introduction of damping or relaxations transferring energy to a reservoir or 'lattice'. When damping is introduced the precession is modified and $\boldsymbol{\mu}$ spirals towards $\mathbf{B}$ (Figure 1.67). The gyromagnetic response arises from the inevitable association of angular momentum with $\boldsymbol{\mu}$ or $\mathbf{M}$ and is thus a very basic feature of magnetic behaviour.

For single-domain particles or saturated specimens $\boldsymbol{\mu}$ may be considered the total magnetic moment. Kikuchi [43] showed that the minimum switching time for reversal of the magnetization was $\tau_{\min} = 2/(\gamma B_0)$ and that this was achieved by setting $\lambda = \gamma M_s$ with λ the damping parameter introduced by Landau and Lifshifz [25].

If a second smaller field is applied in the *OXY* plane and rotates in phase with $\boldsymbol{\mu}$ at $\omega = \omega_0$ then it is conceivable that $\boldsymbol{\mu}$ may be caused to approach the direction of B_0 with an energy change, giving a crude picture of ferromagnetic resonance as developed in Chapter 3.

For the dynamics of domain walls a familiar damped equation of motion may be used:

$$m\ddot{x} + \beta\dot{x} + \alpha x = 2M_s B(t) \tag{250}$$

with m the effective wall mass, β a damping coefficient, α a stiffness constant and the right-hand side the impressed periodic force. As a wall moves the spins rotate in a manner akin to precession in an effective field; the integral of the square of this field was used by Doring [44] and Becker [45] to calculate an excess wall energy proportional to $\dot{x}^2$ and an effective mass by equating this energy to $\frac{1}{2}m\dot{x}^2$. It can be shown that the damping (retarding) force is directly proportional to velocity if eddy currents control the damping. Finally, if $\ddot{x}$ and $\dot{x} \to 0$, $\alpha x = 2M_s B/A$, with A the total wall area, and since the induced magnetization is $M = 2M_s A x$, α can be related to the low-frequency susceptibility by

$$\alpha = 4\frac{\mu_0 M_s}{\chi} \tag{251}$$

The solution of the widely applicable equation (250) is standard. Clearly if $\beta = 0$ and the r.h.s = 0 the frequency of free oscillation is

$$\omega_0 = \sqrt{\frac{\alpha}{m}} \tag{252}$$

When $\beta \neq 0$ but $\beta^2 < 4\alpha m$ (and the r.h.s. is included), a resonant response occurs with peak absorption at

$$\omega = \left(\frac{\alpha}{m} - \frac{\beta^2}{4m^2}\right)^{1/2} \tag{253}$$

As β increases from a minimal value the resonance peak is first broadened and then disappears to be replaced by a gradual fall in the response, which is termed a relaxation.

When pulse fields of constant peak value are applied the acceleration period is usually negligible and the first term in equation (250) can thus be neglected. Once the walls have broken free from the impeding sites at $H = H_c$ the third term is irrelevant and the wall can be considered to move in the excess field according to

$$\beta \dot{x} = 2\mu_0 M_s(H - H_c), \qquad v = R(H - H_c) \tag{254}$$

with R the wall mobility. The switching speed depends on the number of walls but is inversely proportional to $(H - H_c)$: $\tau^{-1} = S(H - H_c)$.

15 High-Frequency Susceptibilities and Losses

The energy stored in a coil or inductor can be expressed in terms of the inductance L and current flowing i, or the fields, according to (Chapter 4, 1.8 and 1.11):

$$E_t = \tfrac{1}{2}\int \mathbf{B}\cdot\mathbf{H}\,\mathrm{d}v = \tfrac{1}{2}Li^2 \tag{255}$$

For the energy density E, assuming $B \parallel H$,

$$\mathrm{d}E = H\,\mathrm{d}B = \mu_0(H\,\mathrm{d}H + H\,\mathrm{d}M) \tag{256}$$

If $B = \mu H$ and μ is constant, $\mathrm{d}E = \mu H\,\mathrm{d}H$ and integrating as the field is built up from zero $E = \frac{1}{2}\mu H^2 = \frac{1}{2}BH$ with this restriction. The expression for $\mathrm{d}E$ takes account of the energy of the magnetization in the field and also shows that even in the absence of magnetizable material, i.e. in space or effectively in an air-filled coil, work must be done to create the field. However, this 'field term' is reversible and there is no associated net energy flow from a source to an inductor which presents an impedance but no effective resistance. The second term, $\mu_0 H\,\mathrm{d}M$, may also lead to reversible energy changes throughout a cycle with no net energy flow or losses if the magnetization is induced reversibly, but more generally the hysteresis loss per cycle [equation. (248)] is the area of the M/H loop ($\times\mu_0$) and in the general case is non-zero due to finite coercivity. Usually the term 'hysteresis loss' is confined to the loss for the quasi-static loop. As the cycle frequency increases the loops may be seen to expand, with additional high-frequency losses. Low-amplitude loops can frequently be approximated as

$$M = \chi_0 H \pm \tfrac{1}{2}\eta H^2 \tag{257}$$

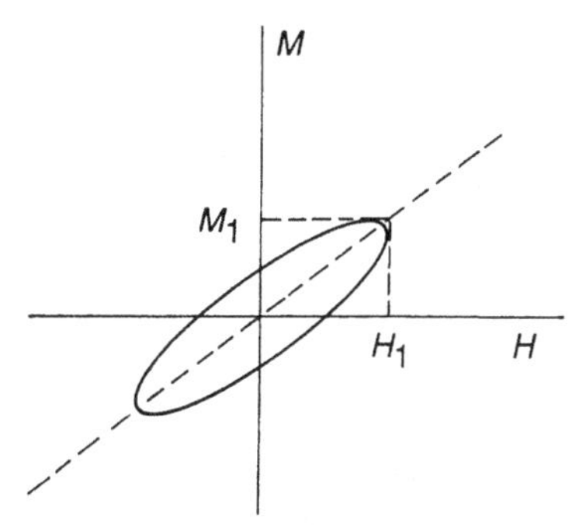

Figure 1.68 Approximated low-amplitude loop:$M = \chi_0 H \pm \frac{1}{2}\eta H^2$ with χ_0 the slope of the broken line and η an empirical Rayleigh constant. The loss is $\propto \eta H_1^3$

(see Figure 1.68) with χ_0 the anhysteretic initial susceptibility and η the Rayleigh constant. The hysteresis loss per cycle is then

$$\omega = \tfrac{4}{3}\mu_0^{\mathrm{r}}\eta H_1^3 \tag{258}$$

with H_1 the field amplitude and μ_0^r the initial relative permeability.

It is clear by simple geometry that as loops resembling that in Figure 1.68 are traversed dynamically there must be a phase difference or lag δ between M and H as indicated by

$$\begin{aligned} H(t) &= H_0 \cos\omega t \\ M(t) &= M_0 \cos(\omega t - \delta) \\ &= M_0 \cos\omega t \cos\delta + M_0 \sin\omega t \sin\delta \\ &= \frac{M_0 \cos\delta}{H_0} H_0 \cos\omega t + \frac{M_0 \sin\delta}{H_0} H_0 \sin\omega t \\ &= \chi' H_0 \cos\omega t + \chi'' H_0 \sin\omega t \end{aligned} \tag{259}$$

defining

$$\chi' = \frac{M_0 \cos\delta}{H_0}, \qquad \chi'' = \frac{M_0 \sin\delta}{H_0} \tag{260}$$

In complex notation with $H(t) = H_0 \mathrm{e}^{i\omega t}$, assuming that

$$\begin{aligned} M(t) &= \chi H(t) = (\chi' - i\chi'')H_0\,\mathrm{e}^{i\omega t} \\ &= (\chi' - i\chi'')H_0(\cos\omega t + i\sin\omega t) \\ &= \chi' H_0 \cos\omega t + \chi'' H_0 \sin\omega t + i(\chi' H_0 \sin\omega t - \chi'' H_0 \cos\omega t) \end{aligned}$$

then the real part of M corresponds to equation (259). Thus χ' and χ'' can be interpreted as the real and imaginary parts of a complex susceptibility:

$$\chi = \chi' - i\chi'' = |\chi|\mathrm{e}^{i\delta} = |\chi|\cos\delta - i|\chi|\sin\delta \tag{261}$$

with $|\chi|^2 = \chi'^2 + \chi''^2$.

The power loss per cycle per unit volume is

$$W_c = \int_0^T \mu_0 H \, dM = \int_0^T \mu_0 H \frac{dM}{dt} dt$$

$$= -\int_0^T \mu_0 \chi' H_0^2 \omega \cos \omega t \sin \omega t \, dt + \int_0^T \mu_0 \chi'' H_0^2 \omega \cos^2 \omega t \, dt$$

$$= \pi \chi'' \mu_0 H_0^2 \tag{262}$$

over the cycle time $T = 2\pi/\omega$, the first term giving zero and the second evaluated as $(1/\omega)(\frac{1}{2}\omega t + \frac{1}{4}\sin 2\omega t)$. Thus the power loss per second per unit volume is

$$P = \tfrac{1}{2}\omega \mu_0 \chi'' H_0^2 \tag{263}$$

This is clearly the a.c. power loss since $P \to 0$ as $\omega \to 0$, explicitly and because $\chi'' \to 0$ also, as $\delta \to 0$, and it is assumed that M follows H exactly at its equilibrium value. Thus $\chi_0 = M_0/H_0$ is the anhysteretic initial susceptibility. Such treatments are usually taken to apply specifically to initial susceptibilities or the corresponding initial permeabilities.

If a zero-resistance air-filled coil has inductance L_0 then $L = \mu L_0$ when it is filled by a core of constant permeability μ. When the permeability is complex the complex impedance is

$$Z = R + iX = i\omega L_0 \mu = i\omega L_0 \mu' + \omega L_0 \mu''$$

$$R = \omega L_0 \mu''$$

$$X = \omega L_0 \mu'$$

with R the resistive and X the reactive component of the impedance.

Plots of μ'' and μ' against frequency are termed permeability spectra and examples are given in Figures 1.71 and 5.18. They may be indicative of so-called relaxations with gradual fall of μ' and rising μ'' or of resonances with a peak in μ'' as μ' falls rapidly (although relaxation is also, of course, involved in the latter). The resonances require the existence of restoring torques or forces as afforded by anisotropy. The relaxations may be associated with the diffusion of carbon atoms in iron or the diffusion or hopping of electrons between ferrous and ferric ions in ferrites, and these may also lead to magnetic after-effects, an appreciably slow approach to the equilibrium magnetization, and disaccommodation of the permeability, slow decreases in the real part being due to the progressive stabilization of certain states of the magnetization by the diffusion. In very highly insulating oxide materials with minimal Fe^{2+} content the formation of spin-waves coupling directly to the lattice vibrations (phonons) is invoked.

16 Eddy Currents

In metals and relatively highly conductive oxide materials, eddy current losses are expected to predominate. Empirically the major distinction is that eddy current losses depend strongly on the specimen dimension: strip or sheet thickness or powder particle size. Eddy current effects are not unique to, but are enhanced in, magnetic materials and form the basis of induction heating, techniques from the surface tempering of steels to delicate medical treatments.

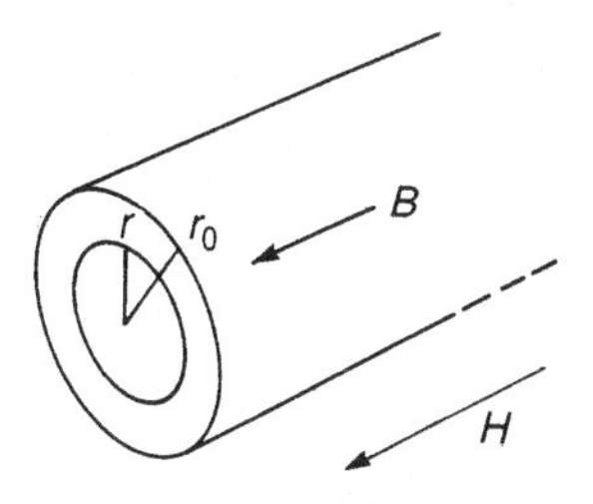

Figure 1.69 Construction for eddy currents in a cylinder

Applying the relation for the electric field E, which derives from Maxwell's laws:

$$\oint E_l \, \mathrm{d}l = -\int\int \frac{\mathrm{d}B_\mathrm{n}}{\mathrm{d}t} \, \mathrm{d}S \tag{264}$$

to the infinitely long cylinder of Figure 1.69 gives

$$2\pi r E(r) = -\pi r^2 \frac{\mathrm{d}B}{\mathrm{d}t}, \qquad \text{i.e. } E(r) = -\frac{r}{2}\frac{\mathrm{d}B}{\mathrm{d}t} \tag{265}$$

The surface integral applies to the area enclosed by the radius r and it is assumed as a simplifying approximation that $\mathrm{d}B/\mathrm{d}t$ is uniform. The induced current is

$$i(r) = \frac{E(r)}{\rho} = -\frac{r}{2\rho}\frac{\mathrm{d}B}{\mathrm{d}t} \tag{266}$$

with ρ the resistivity. The power loss in an annulus between r and $r + \mathrm{d}r$ is $Ei(2\pi r \, \mathrm{d}r)$ per unit length of the cylinder. Integrating, the instantaneous loss per unit volume is

$$P = \frac{1}{\pi R^2}\int_0^R \left(-\frac{r}{2}\frac{\mathrm{d}B}{\mathrm{d}t}\right)\left(-\frac{r}{2\rho}\frac{\mathrm{d}B}{\mathrm{d}t}\right) 2\pi r \, \mathrm{d}r$$

$$= \frac{1}{2\rho R^2}\left(\frac{\mathrm{d}B}{\mathrm{d}t}\right)^2 \int_0^R r^3 \, \mathrm{d}r = \frac{R^2}{8\rho}\left(\frac{\mathrm{d}B}{\mathrm{d}t}\right)^2 = \frac{R^2}{8\rho}\mu^2\left(\frac{\mathrm{d}H}{\mathrm{d}t}\right)^2$$

Integrating over a cycle with $H(t) = H_0 \cos \omega t = H_0 \cos 2\pi f t$, the mean power loss becomes

$$\bar{P} = \frac{\pi^2 R^2 \mu^2 f^2 H_0^2}{4\rho} \tag{267}$$

This is quoted by Smit and Wijn [46] for example who also gave

$$\bar{P} = \frac{2\pi^2 R^2 \mu^2 f^2 H_0^2}{3\rho} \qquad \text{(plate, thickness } R\text{)} \tag{268}$$

$$\bar{P} = \frac{\pi^2 R^2 \mu^2 f^2 H_0^2}{5\rho} \qquad \text{(sphere, radius } R\text{)} \tag{269}$$

(The expressions quoted by Smit and Wijn contained a factor of 10^{-16} to give the quantity in W cm^{-3} when H is taken to be in Oe and ρ in Ω cm, but such factors are unnecessary in the SI and the losses are automatically given in W m^{-3}, remembering that $\mu = \mu_\mathrm{r}\mu_0$.) The losses are now dependent on ω^2, rather than ω, and the dependence on the square of the characteristic dimension is introduced.

The losses as calculated above are usually found to be too low by a factor of about 3 since homogeneous magnetization was assumed and in practice intense losses are associated with moving domain walls, except at extremely high frequencies, at which the walls are effectively demobilized by these effects and (homogeneous) rotation takes over. The magnitude of the anomaly depends on the widths and spacings of the walls and these are usually unknown.

It may be repeated that in the foregoing dB/dt, and by implication B, was taken to be independent of r, i.e. uniform throughout the cylinder. This will never be wholly true. In a thin annular skin on the outside of the cylinder H can be identified with the applied field due to the continuity of the tangential component. The currents induced in this annulus create a reverse field within (and only within) this annulus which detracts from the applied field and thus the penetration of the field is limited. The equation governing the field is

$$\nabla^2 \mathbf{H} = \frac{\mu}{\rho}\frac{\mathrm{d}\mathbf{H}}{\mathrm{d}t} \tag{270}$$

which may readily be solved in the one-dimensional case (Figure 1.70) assuming the usual separation $H = f_1(t) \times f_2(y)$ with boundary conditions corresponding to $H = H_0 \mathrm{e}^{i\omega t}$ at the surface, $y = 0$:

$$H = H_x(y, t) = H_0 \mathrm{e}^{-\alpha y} \mathrm{e}^{i(\omega t - \alpha y)}, \qquad \text{where } \alpha = \sqrt{\frac{\mu_r \omega}{2\rho}} \tag{271}$$

The frequency is unchanged but a phase shift proportional to y is introduced. The magnitude decreases exponentially within the material and H is virtually excluded for y substantially greater than α. A skin depth δ is defined as that at which the amplitude is reduced by $1/e$:

$$\alpha\delta = 1, \qquad \text{i.e. } \delta = \sqrt{\frac{2}{\mu_r \omega}} \tag{272}$$

e.g. at 50 Hz for silicon iron, $\delta \approx 1$ mm, which is much less than the value for copper (≈ 10 mm) with a lower value of ρ ($2 \times 10^{-10} \Omega$ m) but $\mu_r = 1$. The currents may be shown to behave in a similar manner:

$$i = i_z = i_0 \mathrm{e}^{-\alpha y} \mathrm{e}^{i(\omega t - \alpha y)}$$

and the power loss per unit surface area may be obtained by integration as

$$P = \tfrac{1}{2}\sqrt{\frac{\mu\omega\rho}{2}}\, \bar{H}_0^2 \tag{273}$$

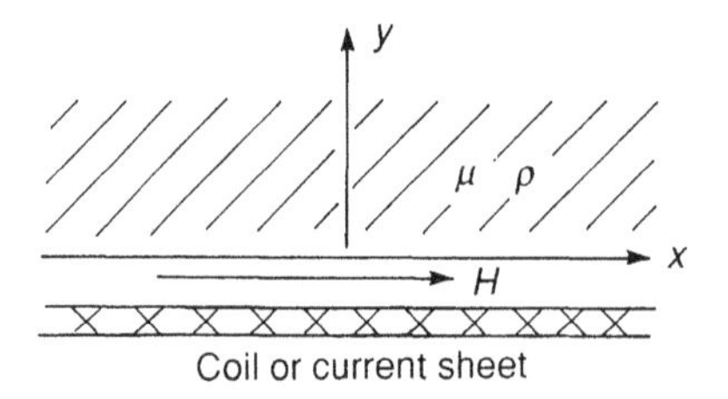

Figure 1.70 The tangential field H must be the same just inside the material but then falls off exponentially

with $\bar{\mathbf{H}}_0$ the r.m.s. value. It may also be seen that the dissipation is very non-uniform and that, for example, selective surface heating of steels may be achieved in an important technical application. Full analyses can also be made for plates of finite thickness and for cylinders: with cylindrical symmetry the diffusion equation becomes Bessell's differential equation and the results are obtained as Bessell functions.

It is now seen that the simple results given in equations (267) to (269) apply for $R < \delta$ substantially, so that uniform penetration of the field may be assumed.

The governing equations have the form of the diffusion equation. Since the functional for the diffusion equation is known this may be solved numerically by the finite element method (see Section 4.3, Chapter 5) and a very substantial literature on this may readily be located in the obvious journals. (The time-dependent Schrodinger equation has the form of the diffusion rather than the wave equation and the author has found the solution of this to be straightforward.)

In the usual case the objective is to reduce eddy current losses by the use of thin strips (in the past by introducing wires or metal powders) or, of course, by attention to the resistivity values of metals or the substitution of metals by the relatively insulating oxides or ferrites. (For ferrites ρ ranges from 10^5 for $MgFe_2O_4$ through $\sim$ 10 for NiZn ferrites to $10^{-5}\Omega$ m for Fe_3O_4.) However, eddy currents may be viewed more positively with applications for induction heating ranging from metallurgy to medicine.

The treatments of relaxation/resonance losses and eddy current losses are distinct. The former depend on the material and the latter depend also on the geometry. The use of complex permeabilities is not appropriate for eddy currents since the phase lag varies spatially. The short-range motion of electrons in the hopping mechanism must be distinguished from the long-range motion giving the extensive currents assumed in the eddy current analyses, but if the activation energy for hopping is progressively reduced the two

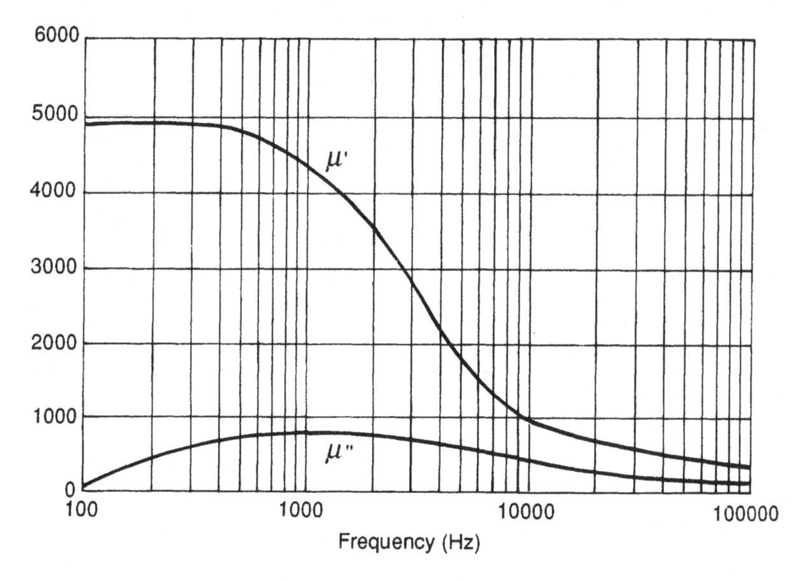

Figure 1.71 The real and imaginary parts of the permeability for a magnetite single crystal; typical relaxation effects as opposed to resonant absorption [47] (cp Figure 5.18)

approaches may be imagined to merge. Rather remarkably Galt [47] interpreted the losses in Fe_3O_4 crystals, as indicated by Figure 1.71, in terms of relaxation and calculated that the eddy current effects were virtually negligible. However, eddy current losses are considered to be appreciable in commercial MnZn ferrites, which contain ferrous ions to reduce the anisotropy and increase the low-frequency permeability at the expense of the useful frequency range, with $\rho = 10^{-3}$ to 10^{-4} Ω m. Due to this, reoxidation is used to give layers of Fe_2O_3 at the grain boundaries, or impurities such as calcium are introduced, which segregate at the grain boundaries and form insulating layers.

17 Survey of Spin Resonance Techniques

Ferromagnetic resonance and domain wall resonance have been introduced briefly and are considered further in Chapter 3. The important techniques of nuclear magnetic resonance (n.m.r.) and electron spin resonance (e.s.r.) can only be treated quantum mechanically and while general features are considered in Chapter 3 a simplified outline is considered appropriate here.

At the simplest level, e.s.r. for a single electron spin $\frac{1}{2}$ is described with reference to Figure 1.72 by the spectroscopic condition:

$$\Delta E = \gamma \hbar B_0 = g\beta B_0 = hf = \hbar\omega, \qquad \text{where } \omega = \omega_0 = \gamma B_0 \tag{274}$$

as suggested by the classical treatment. If the electron also has orbital angular momentum the simple two-level system would not be directly applicable and paramagnetic resonance would be a more appropriate description. However, for an ion in crystal fields the ground state may be a doublet well separated from higher states and an effective spin $\frac{1}{2}$ may be defined by setting $2S + 1$ equal to the degeneracy and using an appropriate value of g with $\gamma = ge/(2m)$ (Chapter 2, Sections 8 and 9.1). A magnetic field of angular frequency

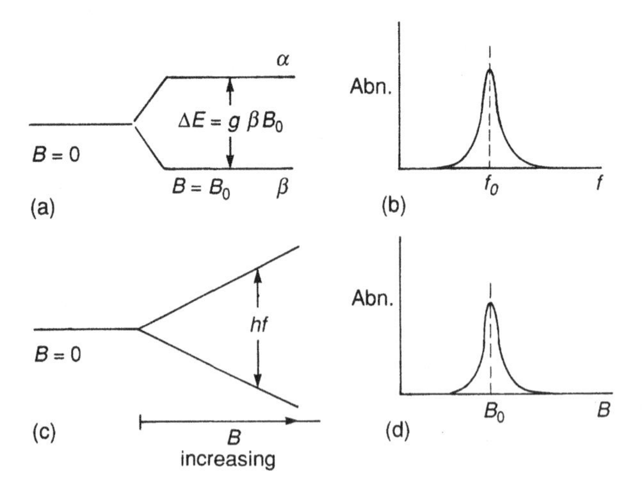

Figure 1.72 At a simple level the splitting for an electron spin (or proton, inverted) in an applied field (a) leads to resonant absorption at $f = f_0 = g\beta B/h$ or, if B is varied and f is constant, (c) to resonance at $B = B_0 = hf/g\beta$ (d)

ω_0 may induce $\beta \rightarrow \alpha$ transitions with absorption of energy, as indicated by a coil or cavity becoming lossy at this frequency (a change in Q). The $\alpha \rightarrow \beta$ transitions are equally induced and there is a net flow of energy only when $n_\beta > n_\alpha$ is maintained, the system remaining at or near equilibrium in the field B_0 with transfer of energy to the lattice.

The assumption that underlies simple descriptions, such as the present, is that the large static field B_0 and the small field B_1, oscillating or rotating with angular frequency ω, can be treated separately: the Hamiltonian $-\gamma B_0 S_z$ or $-\gamma B_0 I_z$ gives the energy levels as indicated and $B_1(t)$ is taken to induce transitions between these in a rather arbitrary manner. Since the spin is equally exposed to both fields, both should appear in the Hamiltonian (as $-\gamma B_1 S_x \cos \omega t$, etc.) and this time-dependent $H(t)$ controls the behaviour of the system in a manner that is naturally more difficult to describe (see Chapter 3). It is the acceptance of the simple picture (relying on the great difference in the magnitudes of B_0 and B_1) that permits a brief description of a wide range of phenomena or techniques.

For a free electron spin, $g = 2$, with $B_0 = 1.0$ T,

$$\omega_0 = \gamma = \frac{2\beta}{\hbar} \doteqdot 2 \times 10^{11}, \qquad f_0 \doteqdot 2 \times 10^{10} \text{ Hz}, \qquad \lambda = 10^{-2} \text{ m} = 1 \text{ cm}$$

and microwaves are involved. (Reduction of B_0 for greater convenience is usually precluded since sensitivity would be lost: but see Section 4, Chapter 6.) It is easier to sweep a field than to sweep a microwave frequency and Figure 1.72(c) is the more appropriate. The first e.s.r. studies were made by Zavoiskey [48].

The simple description of n.m.r. corresponds to the above, if attention is confined to the hydrogen nucleus or proton or other $S = \frac{1}{2}$ nucleii, except that (a) α and β are interchanged because γ_N is positive ($-e \rightarrow e$) and (b) since

$$\gamma_N = g_N \frac{e}{2m_p} = g_N \frac{\beta_N}{\hbar}, \qquad \beta_N = \frac{e\hbar}{2m_p} \tag{275}$$

with $g \sim 1$ and m_p the proton mass, ω_0 is about 1000 times lower than for e.s.r. and the h.f. radiation, in the radiofrequency, may be afforded by a coil. Thus the plot of absorption versus frequency is more appropriate here.

The e.s.r. and n.m.r. transitions are magnetic dipole transitions involving the **B** component of the h.f. field, $\mathbf{B}_1(t) = \mathbf{B}_1 e^{i\omega t} \cdot \mathbf{B}_1$ must be in a plane normal to $\mathbf{B}_0$. The linearly oscillating field $\mathbf{B}_1(t)$ may be decomposed into the circularly contra-rotating vectors of magnitude $\frac{1}{2}B_1$ (see Figure 1.73) and in a semi-classical picture one rotating component $B_1^{(1)}$ (at ω_0) remains in phase with the moment that is precessing in $\mathbf{B}_0$ while the other $B_1^{(2)}$ changes phase rapidly. To an observer following the precession (i.e. in a frame F_{ω_0} rotating at ω_0) the in-phase component is static and moreover B_0 is effectively zero because to this observer the precession appears to have ceased. Thus it is at least conceivable that, in F_{ω_0}, precession around $B_1^{(1)}$ may occur in such a way as to change the component of $\boldsymbol{\mu}$ along the z axis and thus the energy, while $B_1^{(2)}$ can only cause a 'wobble'. (Arguments that are sometimes based on the 'direction of the torque' due to $B_1^{(1)}$ are not valid, confusing the direction of the vector representing the torque with the direction of its effect: the torque tends to turn the magnetic moment towards the field direction although this is not the gyromagnetic response which results.) The effect of $B_1^{(1)}$ is by no means simple and is only clarified by lengthy analyses (Chapter 3). Relaxation effects continuously re-establish the equilibrium magnetization to give a steady state absorption.

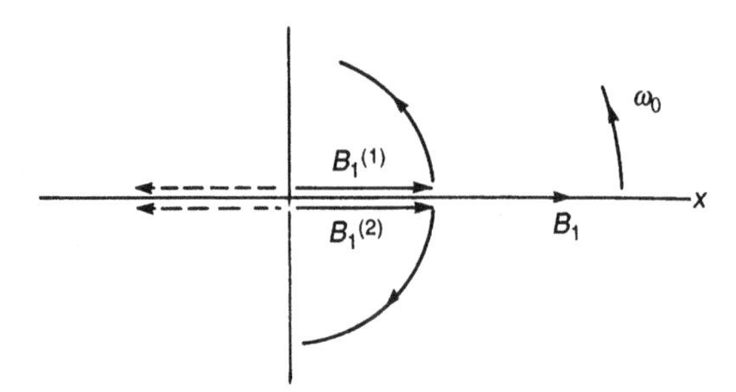

Figure 1.73 The equivalence of two rotating vectors of magnitude $B_1^{(1)} = B_1^{(2)} = \frac{1}{2}B_1$ to a linearly oscillating field (frequency ω_0) of amplitude B_1. To an observer carried by the frame rotating at ω_0, $B_1^{(1)}$ is stationary and $B_1^{(2)}$ rotates at $2\omega_0$

The lines are shown in Figure 1.72 with a finite width ΔB or Δf, at the half-height, due partly to imperfections in the uniformity of $\mathbf{B}_0$, each spin in an assembly seeing a slightly different field. Also, the spins themselves generate fields that will be seen to destroy the coherence of a precessing spin system and lead to line broadening ('T_2 effects'). In extreme cases the line effectively disappears. Interactions between electron spins are much stronger than those between nuclear spins and e.s.r. is only practicable when either the unpaired electrons are well separated as in organic free radicals or when transition metal ions are artificially diluted by doping them into diamagnetic host crystals, such as alumina, as in ruby or in other special cases, with no such restrictions applying to n.m.r.

17.1 Electron Spin Resonance

For electrons in S-states, $g = 2$ is still expected to apply, as for an isolated electron ($g = 2$ implies $g = 2.0023\ldots$ since significant shifts in the g value are usually orders greater than 0.0023). Otherwise, if the orbital quenching is incomplete or some orbital angular momentum is restored by applied fields (see Chapter 2), it is still found possible to write

$$\mathcal{H} = -\gamma \mathbf{B} \cdot \mathbf{S} = -g\frac{\beta}{\hbar}\mathbf{B} \cdot \mathbf{S} \tag{276}$$

with a value of g which accounts for the orbital effects. Such spin Hamiltonians are contrived in this way to apply to systems that are not wholly spin systems. The shifts from $g = 2$ depend on the energy levels in the crystal fields (where appropriate) and their measurements have a vital connection with basic crystal field theory as outlined by Stevens and Bates for example [49] (see Chapter 2, Section 9). In the general anisotropic case the scalar g is to be replaced by a dyadic (Chapter 3).

Structure may arise in the e.s.r. spectra for ions in crystal fields (distinct from hyperfine effects as below). If the crystal fields are strong the spin–orbit coupling can initially be ignored. For $Cr^{3+}(3d^3)$ by Hund's rule $L = 2 + 1 + 0 = 3$ and $S = 3 \times \frac{1}{2} = \frac{3}{2}$. In cubic fields an orbital singlet is lowest and the spin degeneracy is 4, but if an axial component is superimposed the spin quartet splits into a ground state with $M_s = \pm\frac{3}{2}$ and a second spin doublet with $M_s = \pm\frac{1}{2}$, only a little higher. This zero (magnetic) field splitting is designated 2D in Figure 1.74 and the field splitting is indicated by the lines with slopes of $\pm 3\beta$ and $\pm\beta$. The transition rule $\Delta M_s = \pm 1$, implied for the single electron, is general so the transitions

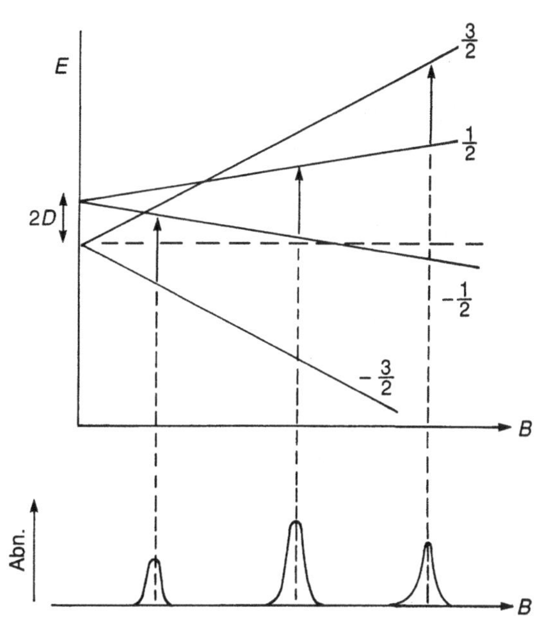

Figure 1.74 Energy level splitting due to crystal fields giving fine structure in the e.s.r. spectra: $\Delta M_s = \pm 1$

are as shown and it is left as a simple geometric exercise to show that the splittings are as indicated. This is an example of the 'fine structure' as opposed to the hyperfine structure due to nuclear interactions which may be observed in hydrogen atoms and in organic radicals for which the orbital angular momentum can be considered completely quenched.

17.1.1 Electron-Proton Interaction

The electron–proton interaction may be expected to correspond to the Heisenberg form $\mathbf{S} \cdot \mathbf{I}$ but it is found that $(B_0 \parallel OZ)$

$$\mathbf{S} \cdot \mathbf{I} = S_x I_x + S_y I_y + S_z I_z \rightarrow S_z I_z \tag{277}$$

in the so-called high-field approximation as discussed later. A basis for the joint space of the two spins consists of the product states (i.e. products of eigenstates of S_z and I_z, as discussed in Chapter 3)

$$|\alpha\rangle|\alpha\rangle \equiv |\alpha\alpha\rangle, |\alpha\beta\rangle, |\beta\alpha\rangle, |\beta\beta\rangle$$

with the convention that the first symbol gives the state of the electron ($|\alpha\beta\rangle \sim$ 'e in $|\alpha\rangle$ and p in $|\beta\rangle$') and the rule that S_z applies to the first factor and I_z to the second. Equation (277) is now seen to represent a great simplification because $\mathcal{H}$ becomes, with the coupling constant A,

$$H = -\gamma B S_z - \gamma B I_z + \frac{A}{\hbar^2} S_z I_z = \frac{g\beta B}{\hbar S_z} - g_N \beta_N B I_z + \frac{A}{\hbar^2} S_z I_z \tag{278}$$

and it is apparent that the basis states are eigenstates with energies

$$E_{\alpha\alpha} = \tfrac{1}{2} g\beta B - \tfrac{1}{2} g_N \beta_N B + \tfrac{1}{4} A$$

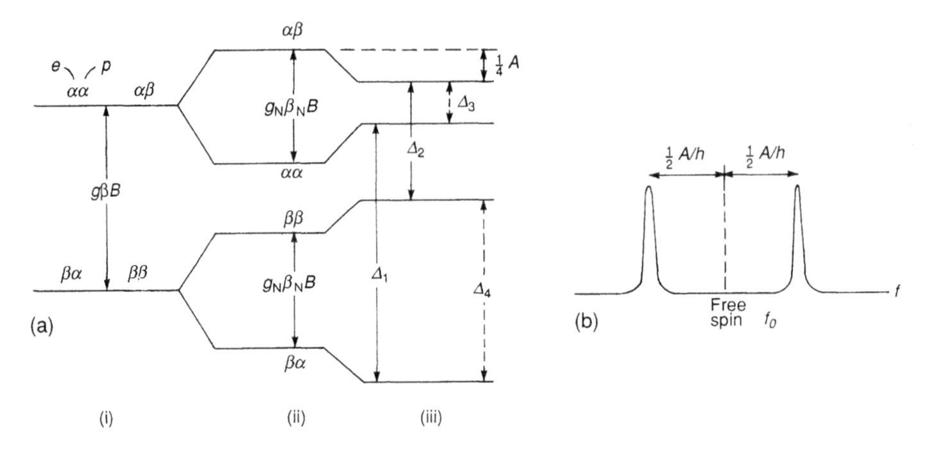

Figure 1.75 Simplified energy levels for an electron and a proton (i) as if only the electron existed, (ii) adding the energy of a non-interacting proton and (iii) adding the hyperfine interaction. For H as such the lines cross. Δ_1 and Δ_2 correspond to an electron spin reversal only and Δ_3 and Δ_4 to a proton spin reversal. The e.s.r. spectrum is as shown in (b). With $\beta \sim 1000 \times \beta_N$ such diagrams are far from being to scale

$$E_{\alpha\beta} = \tfrac{1}{2}g\beta B + \tfrac{1}{2}g_N\beta_N B - \tfrac{1}{4}A$$

$$E_{\beta\alpha} = -\tfrac{1}{2}g\beta B - \tfrac{1}{2}g_N\beta_N B - \tfrac{1}{4}A$$

$$E_{\beta\beta} = -\tfrac{1}{2}g\beta B + \tfrac{1}{2}g_N\beta_N B + \tfrac{1}{4}A$$

The first two terms in each case would apply for distant non-interacting spins and the energy level diagram can be built up as in Figure 1.75. This is clearly not to scale (cf. β, β_N) and with large values of A the lines may cross. For the e.s.r. transitions indicated $\Delta_1 = g\beta B - \frac{1}{2}A$, $\Delta_2 = g\beta B + \frac{1}{2}A$ and with Δ = hf the lines occur as shown at

$$f_1 = \frac{g\beta B}{h} - \frac{1}{2}\frac{A}{h}, \qquad f_2 = \frac{g\beta B}{h} + \frac{1}{2}\frac{A}{h} \tag{279}$$

with a symmetrical splitting about the free electron spin frequency giving directly the hyperfine coupling constant $A' = A/h$. In terms of a swept field the hyperfine splitting constant a is defined as that in B, i.e.

$$a = \frac{A}{g\beta}, \qquad A' = a\frac{g\beta}{h}$$

17.1.2 Generalization

For an electron coupled equivalently to two protons the eigenstates are such as

$$|\alpha\rangle|\beta\rangle|\alpha\rangle \equiv |\alpha\ \beta\ \alpha\rangle \sim \text{‘}e \text{ in } \alpha,\ \text{proton 1 in } \beta,\ \text{proton 2 in } \alpha\text{’}$$

The zero-coupling levels can be assigned intuitively and the coupling shifts added according to, for example,

$$\frac{A}{\hbar^2}(S_z I_{z_1} + S_z I_{z_2})|\alpha\ \alpha\ \alpha\rangle = \frac{A}{\hbar^2}(\tfrac{1}{2}\hbar \times \tfrac{1}{2}\hbar + \tfrac{1}{2}\hbar \times \tfrac{1}{2}\hbar)|\alpha\ \alpha\ \alpha\rangle = \frac{A}{2}|\alpha\ \alpha\ \alpha\rangle$$

with the rule that I_{z_1} acts on the first proton and I_{z_2} on the second. The other shifts are clearly

$$E^A_{\alpha\alpha\beta} = 0 = E^A_{\alpha\beta\alpha}, \qquad E^A_{\alpha\beta\beta} = -\frac{A}{2}$$

with equal and opposite shifts if the electron is in β, giving the levels, transition and spectra in Figure 1.76. The satellites have half the height of that at the free electron frequency due to the different number of transitions involved.

For n protons the coupling term

$$\frac{A}{\hbar^2}\sum_{i=1}^{n} S_z I_{z_i} = \frac{A}{\hbar^2} S_z \sum_{i=1}^{m} I_{z_i} \tag{280}$$

depends on the total spin component of the protons which is obtained by setting $I = n \times \frac{1}{2}$ with components $-I$ to I in integral steps, e.g.

$$n = 3, \qquad I = \tfrac{3}{2}, \qquad M_I = -\tfrac{3}{2}, -\tfrac{1}{2}, \tfrac{1}{2}, \tfrac{3}{2} \qquad (n+1 \text{ values})$$

The shifts are easily seen to be

$$\Delta E_A = \pm\tfrac{1}{2} M_I A \tag{281}$$

with the signs according to the electron in $|\alpha\rangle$ or in $|\beta\rangle$ so that the line shifts from the free electron spin value for $\beta \to \alpha$ transitions are

$$\Delta f = \frac{1}{h}(\tfrac{1}{2} + \tfrac{1}{2}) M_I A = \frac{1}{h} M_I A \tag{282}$$

For $n = 3$ this gives Figure 1.77, with the number of lines equal to $n + 1 = 4$. The relative intensities follow because only the one nuclear state $|\alpha\ \alpha\ \alpha\rangle \sim M_I = \frac{3}{2}$ but $M_I = \frac{1}{2}$ is associated with $|\alpha\ \alpha\ \beta\rangle$, $|\alpha\ \beta\ \alpha\rangle$ and $|\beta\ \alpha\ \alpha\rangle$, etc. Generally, the degeneracy G of a level, with $A = 0$, is

$$G = \frac{n!}{(n-r)!r!}$$

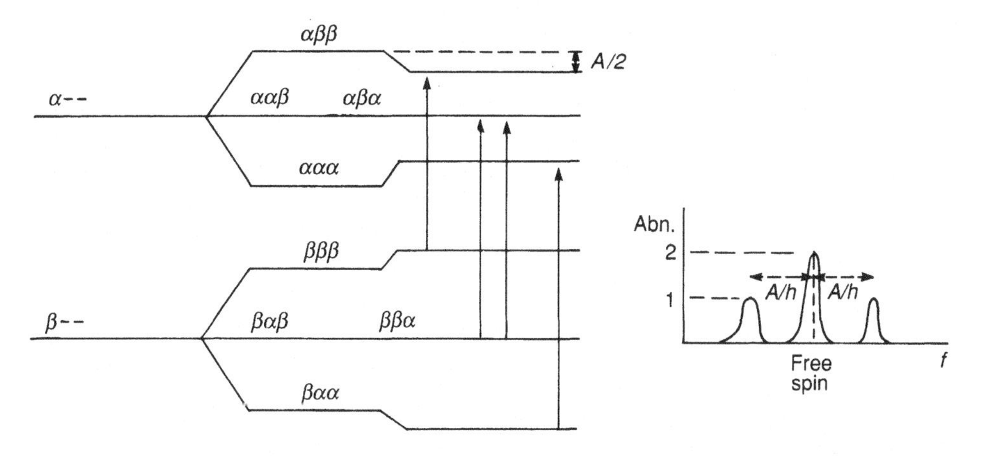

Figure 1.76 Splitting diagram and e.s.r. spectrum for an electron and two proton spins

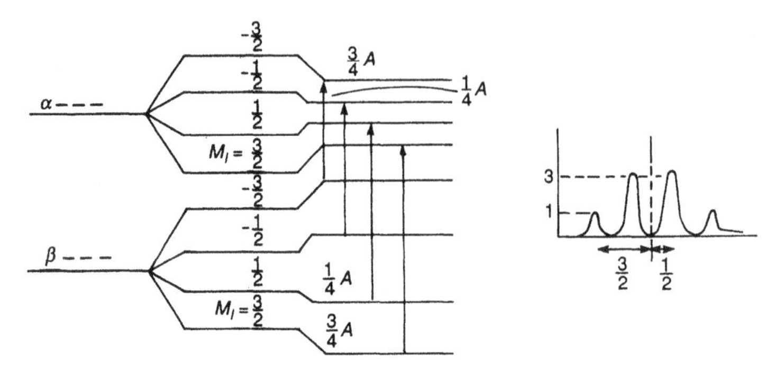

Figure 1.77 Splitting diagram and e.s.r. spectrum for an electron coupled equivalently to three proton spins

where r counts the number of α or β values, whichever is smaller in number, representing the proton states. By following these empirical rules the spectra involving any n equivalent protons can be drawn up automatically.

If the electron interacts with, say, two groups of protons with interactions with those of group 1 and of group 2 characterized by A_1 and A_2, as might be supposed for the free radical in Figure 1.78(a), then the spectrum is as shown in Figure 1.78(b). No new principles are involved and it is left to the reader to (a) derive the splitting diagram and spectrum explicitly, and (b) derive a simple set of rules, corresponding to the above, for the shifts and intensities in terms of n_1 and n_2.

Free radicals with a net free spin may be formed by breaking a bond, in which the spins are otherwise paired with no e.s.r. signal, and e.s.r. is thus relevant to chemical reaction studies. Free spins also arise in charge-transfer complexes involving an electron acceptor A and an electron donor D: $A + D \rightarrow A^{-}D^{+}$; the extent of the transfer and the characteristics of the electrons may be studied by e.s.r. measurements. The area under the absorption curve is proportional to the spin susceptibility and additional measurements of χ by force methods thus allow the diamagnetic χ contributions to be calculated.

17.2 Nuclear Magnetic Resonance

Protons and neutrons have spin $\frac{1}{2}$ and generally nucleons couple to give nuclear spins which are integral or half-integral or zero (but often, and significantly for ^{1}H itself, of course, non-zero). The interactions are weaker than for electron spins and obviously there is no equivalent to the spin pairing in bonds so that n.m.r. is of great importance and of wide application. For an odd mass number the spin is half-integral. For an even mass number I is integral with z odd and $I = 0$ with z even (^{12}C, ^{26}O, ^{32}S). Table 1.5 gives examples.

The nuclear magneton is

$$\beta_N = \frac{eh}{4\pi m_p} = \frac{e\hbar}{2m_p} = 5.050\,787 \times 10^{-27}\ \mathrm{A\,m^2}(\sim \mathrm{J\,T^{-1}}) \tag{283}$$

and comparing

$$\boldsymbol{\mu} = \gamma \mathbf{I}, \qquad \text{where } \boldsymbol{\mu} = g_N \frac{e}{2m_p}\mathbf{I} = g_N \frac{\beta_N}{\hbar}\mathbf{I}$$

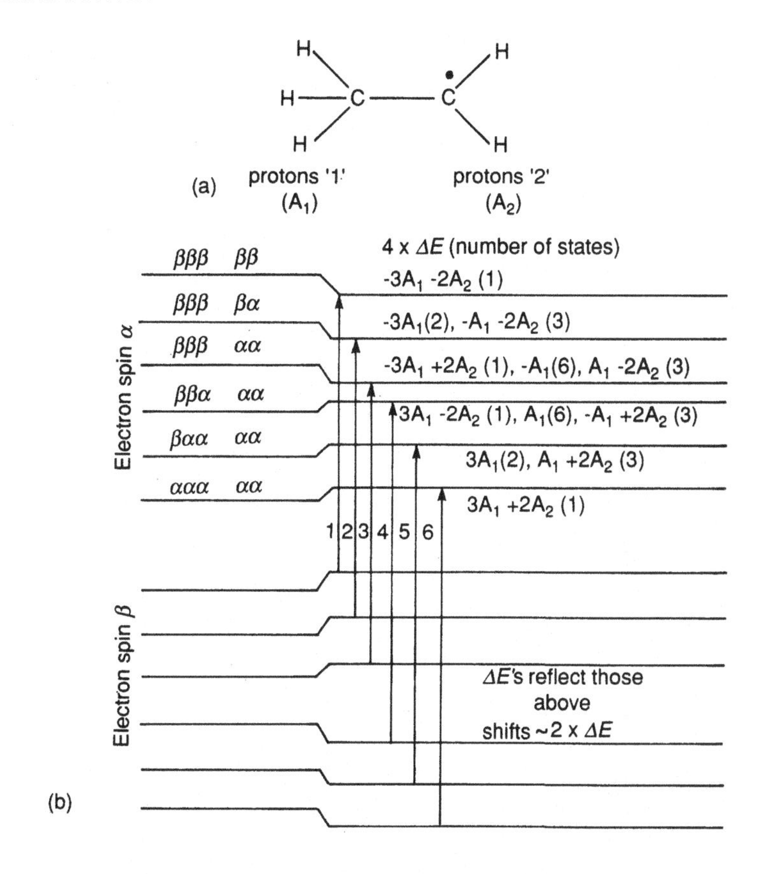

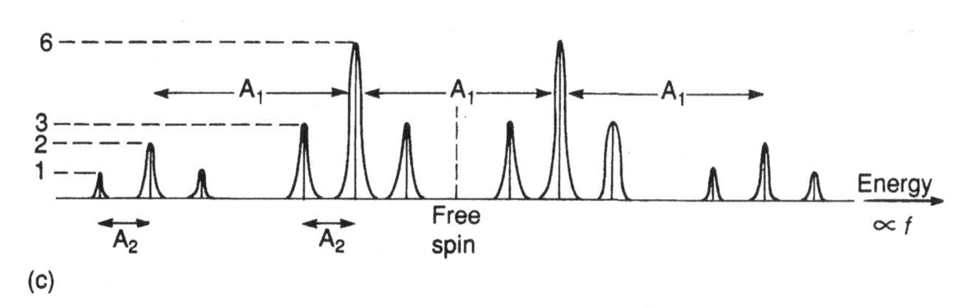

Figure 1.78 A radical in which the free electron is coupled with different strengths to each of two groups of protons, the splitting diagram and e.s.r. spectrum taking account of the degeneracies involved. $A_1 \gg A_2$ is assumed here

gives

$$g_N = \frac{\hbar}{\beta_N}\gamma \quad (= 5.5865 \text{ for } {}^1H) \tag{284}$$

The n.m.r. measurements give approximations to γ or g for bare nucleii but it is the shifts from these expected frequencies and the splittings or structure that are of widest interest.

Table 1.5 Nuclear spins with the neutron (n) and electron (e) for comparison: $f_0 \sim 100$ MHz for ^1H and $f_0 \sim 10^5$ MHz for the electron in $B_0 \approx 2.0$ T. r is the relative sensitivity, in relation to ^1H, at a field of 1.0 T

	Z	I	g_N	$\gamma/10^7$	r
^1H	1	$\frac{1}{2}$	4.83724	26.7519	1.0
^2H	1	1	1.2126	4.1066	0.02
^7Li	3	$\frac{3}{2}$	4.2039	10.3975	
^{11}B	5	$\frac{3}{2}$	3.4708	8.5843	
^{13}C	6	$\frac{1}{2}$	1.2166	6.7283	0.03
^{14}N	7	1	0.5710	1.9338	
^{17}O	8	$\frac{5}{2}$	−2.2407	−3.6279	
^{19}F	9	$\frac{1}{2}$	4.5532	25.1810	0.9
^{23}Na	11	$\frac{3}{2}$	2.8627	7.0801	
^{27}Al	13	$\frac{5}{2}$	4.3084	6.9760	
^{31}P	15	$\frac{1}{2}$	1.9602	10.841	0.1
^{59}Co	27	$\frac{7}{2}$	5.234	6.317	
n		$\frac{1}{2}$	−3.3136	−18.3257	
e		$(\frac{1}{2})$		−17608.4	

The n.m.r. linewidths may be extremely small, due to the effects of rapid molecular motion, in non-viscous liquids, on the relaxation (see Chapter 3, Section 1.12), permitting the clear resolution and measurement of these effects even when they are very small. Nuclear magnetic resonance was first observed by Purcell *et al.* [50] and Bloch *et al.* [51] in 1946.

The foregoing hyperfine interaction affords an introduction. This will not always be isotropic and

$$\begin{aligned} \mathcal{H} &= \frac{\beta}{\hbar}\mathbf{B}\cdot\mathbf{g}\cdot\mathbf{S} - \gamma_N\mathbf{B}\cdot\mathbf{I} + \mathbf{I}\cdot\mathbf{A}\cdot\mathbf{S} \\ &= \frac{\beta}{\hbar}\mathbf{B}\cdot\mathbf{g}\cdot\mathbf{S} - \frac{\beta_N}{\hbar}g_N\mathbf{B}\cdot\mathbf{I} + \mathbf{I}\cdot\mathbf{A}\cdot\mathbf{S} \end{aligned} \tag{285}$$

with

$$\mathbf{A} = \mathbf{i}A_x\mathbf{i} + \mathbf{j}A_y\mathbf{j} + \mathbf{k}A_z\mathbf{k} \tag{286}$$

reducing to the form used above, $A \equiv A_z$, in the high field approximation. The difference between the two Zeeman terms is noted, with a scalar or single-component g_N ascribed to the non-existence of orbital angular momentum for the nucleus.

The nuclear transitions according with $\Delta m_I = \pm 1$ and $\Delta m_s = 0$ are appended in Figure 1.75, giving the n.m.r. spectrum of Figure 1.79(a) centred on $\omega = \gamma_N B_0$. The implicit assumption is that the nuclear Zeeman splitting is greater than the hyperfine splitting (with the electron Zeeman splitting thus the greatest). The alternative case may be illustrated as in Figure 1.79(b), giving the spectrum in Figure 1.79(c). The results indicate, in either case, the hyperfine splitting and the nucleons involved (γ_N) and elucidate the nature of electron trapping sites in crystals containing colour centres, for example.

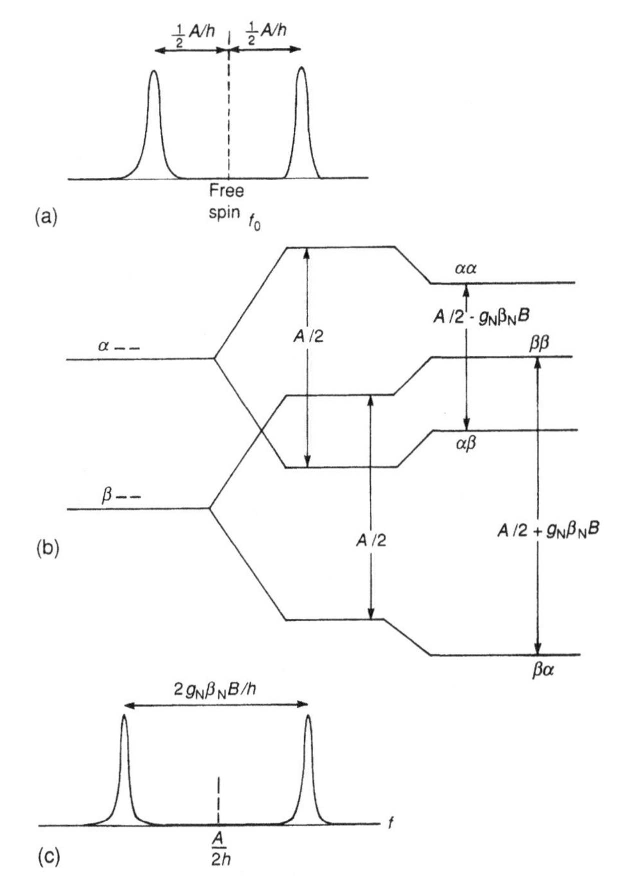

Figure 1.79 (a) The n.m.r. spectrum corresponding to the nuclear transitions shown in Figure 1.75. When the 'A terms' predominate and are reasonably displayed first as in (b), the resulting spectrum is as in (c)

17.2.1 Chemical shift and shielding

Even in the simple case of diamagnetic atoms the field acting on the nucleus will not be the applied field B_0 since this will be shielded, or affected by the fields arising from the circulation of the surrounding electrons, itself induced by B_0 (the diamagnetic effect). Classically, the field at the centre of an induced current loop ($\boldsymbol{\mu}$ antiparallel to $\mathbf{B}$) opposes the applied field. $\mathbf{B}$, at the nucleus, may be related to B_0 by

$$B = B_0(1 - \sigma) \tag{287}$$

a positive σ indicating reduction of B_0. The assumption of linearity corresponds to the classical calculation of the diamagnetic induction.

If the wave functions are known, as for hydrogen, or accepted as Hartree–Fock atomic functions for example, σ may be calculated as

$$\sigma = \frac{e^2}{3m}\langle 0|\sum_i \frac{1}{r_i}|0\rangle \tag{288}$$

The values usually given are $\sigma/10^6$ and may be quoted colloquially as, for example, $\sigma = 17$ ppm or 'parts per million'. For molecules, calculations become increasingly difficult beyond H_2 ($\sigma = 32 \times 10^{-6}$) with no simple correlation between σ and χ. Values of σ giving the field at O in CO, at N in N_2 and at F in F_2 are negative. The shielding varies for different orientations of **B** with respect to the axes of symmetry of the molecules, but the tensor $\boldsymbol{\sigma}$ is usually taken to reduce to an average scalar value due to molecular motion.

Writing

$$\mathcal{H} = -\gamma B_0(1-\sigma)I_z, \qquad \omega_0 = \gamma B_0(1-\sigma) \tag{289}$$

the σ values may in principle be determined by n.m.r. measurements (ω_0 at two fields giving γ as well as σ), but they are very small in relation to the accuracy with which the fields can be independently measured, so that it is customary to proceed as follows.

In constant applied fields the shifts in the resonant frequency due to screening or shielding differ for different protons, say, in different molecular environments. (Most of the examples will be given for proton n.m.r.) The differences may be only $\sim$ 60 Hz at about 60 MHz, i.e. 1 in 10^6, but these can be measured due to the high resolution possible and the inherent accuracy of frequency measurement. For protons in two or more different environments, two 'types' of protons, the frequencies are

$$f_i = \left(\frac{\gamma}{2\pi}\right) B_0(1-\sigma_i) \tag{290}$$

an example being given in Figure 1.80 for $(CH_3)_2C(OCH_3)_2$, with two in equivalent sites. The signal heights are found to be

$$S_i \propto N_i \gamma^4 [B_0(1-\sigma_i)]^2 B_1 \frac{g(f)}{T} \tag{291}$$

where $g(f)$ characterizes the line shape and the N_i are the numbers or molar concentrations of nucleii in the sites i. (This refers to the signal measured as the rate of change of the induced magnetization component along a receiver coil with the axis conventionally $\parallel OY$ with the a.c. field of amplitude B_1 along OX and $B_0 \parallel OZ$. It indicates why B_0 should be high, in addition to the implication of equation (287), and why B_1 should be high with qualifications relating to saturation, to be discussed, and also why ^1H is favoured by the high value of γ, giving a relative sensitivity of 33 compared to ^{13}C.) If Figure 1.80 applied to an unknown compound the indications that the protons equally occupied two different sites would be of obvious value.

For the frequency shifts to be of general value they must be considered as shifts away from f_0 for protons in a chosen compound, giving a single line reasonably separated in frequency

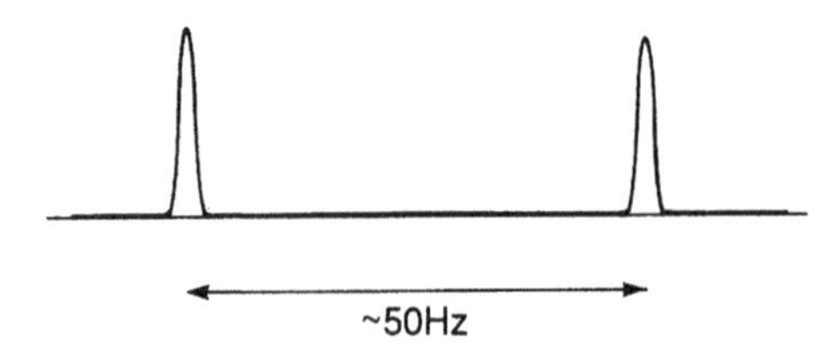

Figure 1.80 An impression of practical proton n.m.r. linewidths in relation to chemical shifts (motional narrowing implied)

from lines to be studied, a standard which is also chemically inert, non-polar and symmetrical (to avoid solvation effects). A favoured standard for proton n.m.r. is tetramethylsilane (TMS) with all the protons effectively equivalent due to rapid rotation around the Si—C bonds.

Only the shifts per unit applied field can be characteristic. For protons '1' and '2' the resonant frequencies are

$$\omega_1 = \gamma B_0(1-\sigma_1), \qquad \omega_2 = \gamma B_0(1-\sigma_2), \qquad \text{with } \Delta f = \frac{\gamma B_0}{2\pi}(\sigma_2 - \sigma_1) \tag{292}$$

[The reader may show, making use of the binomial approximation $(1-\sigma_1)^{-1} \doteqdot 1+\sigma_1$, that if the field was swept at a constant frequency the corresponding field shift would be $\Delta B = B_0(\sigma_1 - \sigma_2)$.] The frequency shift per unit field is $\Delta f/B_0$ and, to about one part in 10^6, B_0 may be identified with either $\omega_1/(2\pi\gamma)$ or $\omega_2/(2\pi\gamma)$. Taking proton 2 as the reference, $f_2 \equiv f_r$, the *chemical shift* may conveniently be defined as

$$\delta = 10^6 \frac{f_s - f_r}{f_r} \doteqdot 10^6(\sigma_r - \sigma_s) \tag{293}$$

since δ is dimensionless and a small number. Accepting the definition, δ should be quoted as a number but it is sometimes irregularly given as 'x ppm'. The value of the site-dependent chemical shifts in analysis and structure determination is apparent and is extended by the further details observed at high resolution ($\sim$ 1 Hz) as below. This requires field homogeneity over the sample, about 1 in 10^8, and the effective homogeneity may be improved, by averaging, by spinning the sample at low angular velocities (cf. spinning at high velocities to suppress particular line broadening effects; see Chapter 3, Section 1.12).

17.2.2 Spin–spin coupling and splittings

The empirical nuclear spin–spin interaction is found to correspond to a Heisenberg term, giving the Hamiltonian for two identical spins, e.g. protons 'a' and 'b', as

$$\mathcal{H} = -\gamma B_a I_a^z - \gamma B_b I_b^z + \left(\frac{J}{\hbar^2}\right) \mathbf{I}_a \cdot \mathbf{I}_b \tag{294}$$

with $B_a = B_0(1-\sigma_a)$, $B_b = B_0(1-\sigma_b)$. The solution is simple for the case of equivalent protons, $B_a = B_b$, as already indicated but, as illustrated by Figure 1.81(a), this gives splitting of the energy levels according to the value of J but no spectral splitting, due to the suppression of certain transitions. However, solution is also simple in the linear case corresponding to $\hbar^2(\omega_b - \omega_a)^2 \gg J^2$ since it is shown that $\mathcal{H}$ simplifies to

$$\mathcal{H} = -\omega_a I_a^z - \omega_b I_b^z + \left(\frac{J}{\hbar^2}\right) I_a^z I_b^z \tag{295}$$

(writing $\omega_a = \gamma B_a$, $\omega_b = \gamma B_b$). The simple spin products $|\alpha\alpha\rangle$, $|\alpha\beta\rangle$, $|\beta\alpha\rangle$ and $|\beta\beta\rangle$ are now eigenstates of the interaction term as well as of the Zeeman terms, i.e. of $\mathcal{H}$ itself. The energies are clearly:

$$E_{\beta\beta} = \tfrac{1}{2}\hbar(\omega_a + \omega_b) + \tfrac{1}{4}J$$

$$E_{\alpha\beta} = \tfrac{1}{2}\hbar(\omega_b - \omega_a) - \tfrac{1}{4}J$$

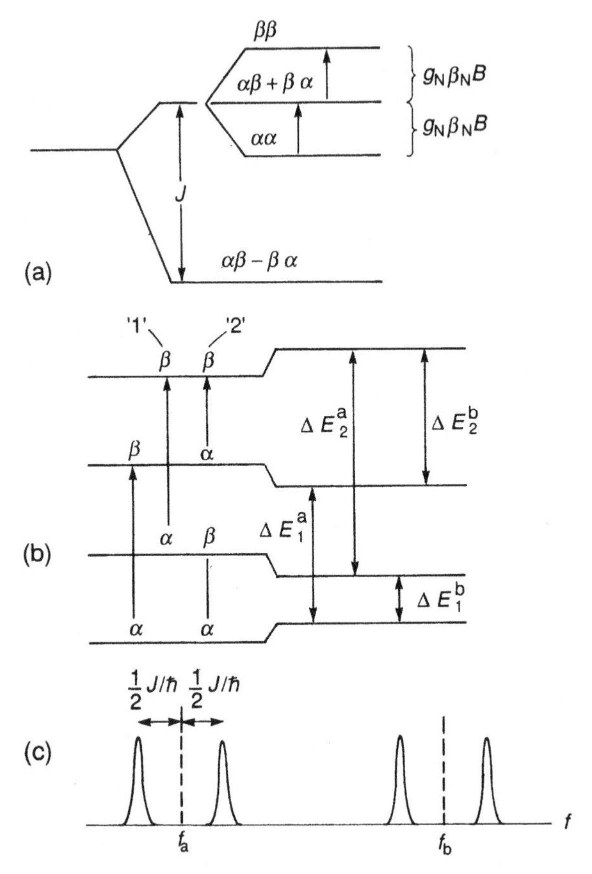

Figure 1.81 (a) The level splitting for two equivalent protons gives no spectral splitting since the allowed transitions are equivalent. (b) Splitting for two protons in different effective fields, in the approximation discussed, with an indication of the transitions and the spectrum (c)

$$E_{\beta\alpha} = \tfrac{1}{2}\hbar(\omega_a - \omega_b) - \tfrac{1}{4}J$$
$$E_{\alpha\alpha} = -\tfrac{1}{2}\hbar(\omega_a + \omega_b) + \tfrac{1}{4}J$$

with the convention that $|\beta\alpha\rangle \sim$ 'a' in β and 'b' in α, etc., as illustrated, with the spectrum, in Figure 1.81(b) and (c).

Extension is very simple in this case. Consider $n = 3$ protons with the $2^n = 8$ eigenstates being all the combinations such as $|\alpha\beta\alpha\rangle$. The coupling strengths may be different and so

$$\mathcal{H} = -\omega_1 I_1^z - \omega_2 I_2^z - \omega_3 I_3^z + \left(\frac{J_{12}}{\hbar^2}\right) I_1^z I_2^z + \left(\frac{J_{13}}{\hbar^2}\right) I_1^z I_3^z + \left(\frac{J_{23}}{\hbar^2}\right) I_2^z I_3^z$$

The reader may readily confirm that the energies are

$$\left.\begin{matrix} |\beta\ \beta\ \beta\rangle \cdots + \frac{1}{4}J_{12} + \frac{1}{4}J_{13} + \frac{1}{4}J_{23} \\ |\alpha\ \beta\ \beta\rangle \cdots - \frac{1}{4}J_{12} - \frac{1}{4}J_{13} + \frac{1}{4}J_{23} \end{matrix}\right\} \Delta E = \hbar\omega_1 + \tfrac{1}{2}J_{12} + \tfrac{1}{2}J_{13}$$

$$\left.\begin{array}{l}|\beta\ \beta\ \alpha\rangle \cdots + \frac{1}{4}J_{12} - \frac{1}{4}J_{13} - \frac{1}{4}J_{23} \\ |\alpha\ \beta\ \alpha\rangle \cdots - \frac{1}{4}J_{12} + \frac{1}{4}J_{13} - \frac{1}{4}J_{23}\end{array}\right\} \Delta E = \hbar\omega_1 + \frac{1}{2}J_{12} - \frac{1}{2}J_{13}$$

$$\left.\begin{array}{l}|\beta\ \alpha\ \beta\rangle \cdots - \frac{1}{4}J_{12} + \frac{1}{4}J_{13} - \frac{1}{4}J_{23} \\ |\alpha\ \alpha\ \beta\rangle \cdots + \frac{1}{4}J_{12} - \frac{1}{4}J_{13} - \frac{1}{4}J_{23}\end{array}\right\} \Delta E = \hbar\omega_1 - \frac{1}{2}J_{12} + \frac{1}{2}J_{13}$$

$$\left.\begin{array}{l}|\beta\ \alpha\ \alpha\rangle \cdots - \frac{1}{4}J_{12} - \frac{1}{4}J_{13} + \frac{1}{4}J_{23} \\ |\alpha\ \alpha\ \alpha\rangle \cdots + \frac{1}{4}J_{12} + \frac{1}{4}J_{13} + \frac{1}{4}J_{23}\end{array}\right\} \Delta E = \hbar\omega_1 - \frac{1}{2}J_{12} - \frac{1}{2}J_{13}$$

The first three terms have been omitted since these are the same in each pair save that the first term in the first of each is $\frac{1}{2}\hbar\omega_1$ and that in the second of each pair is $-\frac{1}{2}\hbar\omega_1$, due to the order chosen; only the first spin in each pair of products differs. In writing the ΔE values as shown it is implied that these are associated with allowed transitions and this is the case since for the basis states the transition rule becomes $\Delta m_t = \pm 1$ or, colloquially, only one spin reverses. Furthermore, the intensities are all seen to be equal.

The ΔE's listed correspond to four lines centred on $f_1 = \omega_1/2\pi$ and the remainder follow by similarly choosing transitions in which the other spins switch. This is, of course, schematic and the groups may overlap. The spectra may clearly be regarded as n groups of 2^{n-1} lines.

Extension to any n is simple for the linear case and spectra can be predicted numerically in the general case, for a reasonable value of n, since the matrices are readily drawn up systematically. Thus high-resolution n.m.r. spectra give information on the nature and magnitude of the exchange couplings which are carried through the electrons in the bonds in addition to the chemical shifts and make an enormous contribution to chemical structure analysis, as described in the many specialized works on the subject.

17.2.3 Relaxation and damping

A certain time is required to establish the equilibrium magnetization $M_0 = \chi H_0$ of an electron or nuclear spin system in a field. More precisely, since it is found that there is an exponential approach to M_0 a certain rate constant is always involved which is related reciprocally to a characteristic time T_1, which can be regarded as the time required to give a substantial magnetization. For times $t \gg T_1$ it can be assumed that $M = M_0$ has been achieved. T_1 is known as the longitudinal relaxation time. It has been seen, classically at least, that, without damping, individual spins precess but do not become aligned with the field. The damping corresponds to transfer of energy to the surroundings, generally designated the lattice as in the case of a crystal, and thus T_1 is also called the spin–lattice relaxation time. It will be shown explicitly in Chapter 3 that when equilibrium is achieved there are no transverse magnetization components, and this can be viewed as a lack of any coherence in the precession, i.e. of any common phase for the precession of individual spins.

Suppose that M_0 has been established in $B_0 \parallel OZ$. A rotating field of magnitude B_1 and frequency ω is switched on, instantaneously defining the axis OX. In a frame of axes $OX'Y'Z$ rotating around OZ at ω, B_1 is stationary and, according to observations in this frame, the apparent precession is modified so that if $\omega = \omega_0 = \gamma B_0$ it appears that B_0 does not exist. Thus M_0 commences to precess around B_1 alone as in Figure 1.82(a) and if B_1 is applied as pulses of appropriate duration M_0 may be rotated to OY' or to $-OZ$ ($\pi/2$ and π pulses). There is a simultaneous tendency for the equilibrium $M_0 \parallel OZ$ to be restored

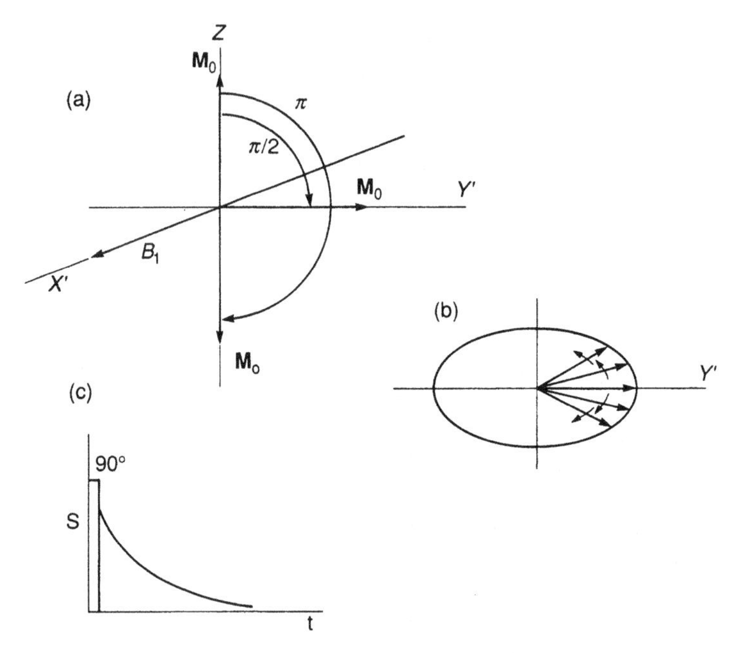

Figure 1.82 In a suitably rotating frame (a) $\mathbf{B}_1$ appears to be static along OX', say, while B_0 appears to be zero, and according to the magnitude of B_1 (i.e. of $\omega_1 = \gamma B_1$) and the pulse duration, M_0 may precess through $\pi/2$, π, etc. M_0 decays due to 'T_2 effects' (b), giving a free induction decay signal in a coil coaxial with OX or OY as in (c)

and thus the pulse duration ($\propto B_1^{-1}$) must be much less than T_1. Assuming this, following a $\pi/2$ pulse M_0 remains initially $\parallel OY'$ and thus rotates at ω_0 in the laboratory frame and generates a signal in an appropriate pickup coil.

Eventually, in a time $t \approx T_1$, this would decay as the equilibrium $M_0 \parallel OZ$ is restored. However, the transverse magnetization generally disappears more rapidly than this, at a rate characterized by a transverse relaxation time T_2, because even if the individual spin components constituting M_0 remain in OXY ($T_1 \rightarrow \infty$) the component which is initially static along OY' must decay if the spins precess at different rates about B_0, in the laboratory frame, and thus fan out in the rotating frame [Figure 1.82(b)]. This loss of coherence arises if each spin sees a different field due to the inhomogeneity of B_0, or, in a random walk manner, when the spins are exposed to varying fields from neighbouring spins. Thus T_2 is also called the spin–spin relaxation time.

The decay of the signal following the r.f. pulse [Figure 1.82(c)] is called the free induction decay (f.i.d.) since it occurs in the zero r.f. field. If all the nucleii are equivalent and the radio frequency matches the single resonance frequency the f.i.d. approximates $\exp(-T_2/t)$, but the analyses of f.i.d.'s may be extremely complex. T_1 can also be determined by pulse methods, e.g. by π pulses followed by carefully timed 90° pulses which 'read' the redevelopment of M_0.

As well as controlling the f.i.d. it can be seen that spin–spin relaxation has a controlling influence on the line widths (widths of the absorption lines at the half-heights) observed by continuous-wave (c.w.) n.m.r.:

$$\Delta f = \frac{1}{\pi T_2} \tag{296}$$

(the spectra and f.i.d.'s being directly related by Fourier transformations).

Since the rotation giving the initial transverse components constitutes precession in B_1 and $B_1 \ll B_0$, a typical pulse duration of $\sim$ 10 μs contrasts with the precession period $1/f_0 \sim 10^{-11}$ s (for 100 MHz n.m.r.). The pulses are well defined. A d.c. pulse with short rise and fall times can be synthesized by summing a (Fourier) series of sinusoidal oscillations, and the squarer the pulse the wider the frequency range required. Conversely, a square pulse can be mathematically or practically analysed to give a series of frequencies. An r.f. pulse of frequency f_0 is equivalent to a distribution of frequencies spreading symmetrically about f_0 and indicated by a function of frequency which is the Fourier transform of the function of time which represents the pulse. With a pulse width τ the frequency ranges approximately over $\pm 1/\tau$. Thus $\tau = 10$ μs gives a frequency range of $\sim 10^5$ Hz covering the resonant frequencies for all protons in a typical sample with a range of shifts and splittings. The f.i.d. is thus a superposition of the f.i.d.'s for all the protons which respond and the Fourier transform gives the spectrum itself as indicated by Ernst and Anderson in 1966 [52]. Multiple pulses can be used to enhance the sensitivity while there are strict limitations to the scan speed for c.w. operation. Many ($\sim$ 1000) f.i.d.'s may be summed before transformation by minicomputers. The pulse spacing should be more than T_1 to allow the attainment of equilibrium at each step, while the duration $\tau \sim T_2$, which is necessarily less than T_1.

The field at one nucleus (e.g. proton) due to a second distant $r = 0.1$ nm is typically

$$\Delta B = \frac{\mu_0}{4\pi}\frac{\mu}{r^3} = \frac{\mu_0}{4\pi}\frac{\frac{1}{2}g_N\beta_N}{r^3} = 2.7 \times 10^{-3}\ \mathrm{T}$$

with corresponding $\Delta f = \gamma \Delta B/2\pi = 114$ kHz. This is a very large effect compared to chemical shifts ($\Delta B = 0.01 \times 10^{-3}$ T for $B = 1.0$ T, $\sigma = 10 \times 10^{-6}$) and spin–spin splittings, and is thus expected to give extreme linewidths. It is only due to the averaging out of such effects, due to rapid molecular motion in suitable liquids, that high-resolution n.m.r. becomes feasible and the features of interest can be resolved. This 'motional narrowing' is described in Chapter 3.

17.2.4 Nuclear quadrupoles

Due to the composition of the nucleus the charge may not be symmetrically distributed and many nucleii (not ^{1}H of course) have associated electric quadrupoles, as well as higher moments which are usually discounted. While an electric dipole is seen to be fully specified by

$$\boldsymbol{\mu} = \sum q_i \mathbf{r}_i \tag{297}$$

with the $\mathbf{r}_i$ the position vectors of the q_i with origin at the centre of charge [Figure 1.83(a)], a quadrupole is specified by

$$Q = \sum q_i \mathbf{r}_i^2 \tag{298}$$

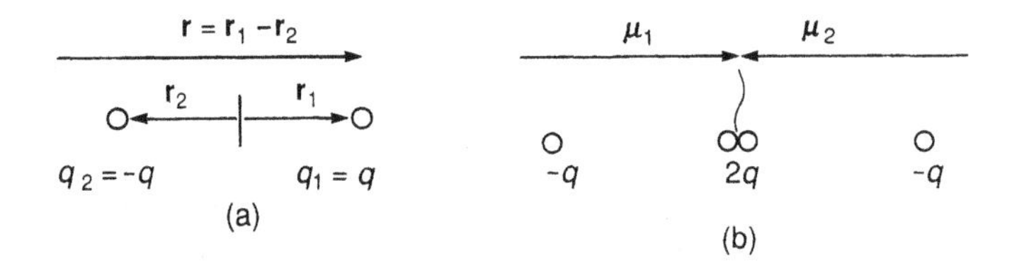

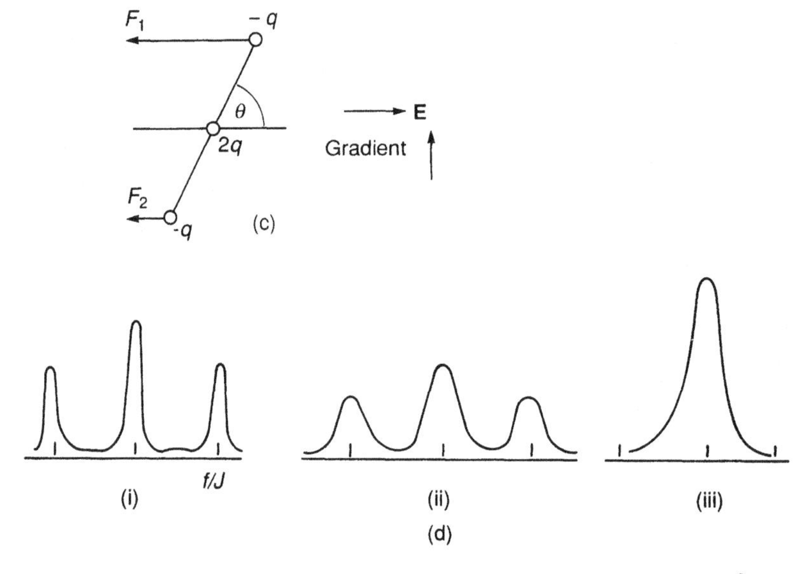

Figure 1.83 In (a), $\boldsymbol{\mu} = q_1\mathbf{r}_1 + q_2\mathbf{r}_2 = q\mathbf{r}$ and $Q = 0$. In (b), $\mu = 0$ and $Q = 2qr^2$. In an electric field gradient with $E_1 \neq E_2$, $F_1 \neq F_2$, an electric quadrupole is subject to a torque (c). (d) indicates splitting and broadening of a proton line due to coupling to a nucleus with a quadrupole moment

and is most readily envisaged as in Figure 1.83(b). Since the signs disappear on taking the $\mathbf{r}_i^2$, $Q = 0$ for Figure 1.83(a) but

$$Q = 2qr^2 \tag{299}$$

for Figure 1.83(b) (with $\mu = 0$). The quadrupole is often taken as that per unit proton charge:

$$Q = \frac{2qr^2}{e} \qquad \text{for quadrupole moment} = eQ \tag{300}$$

With a nuclear $r \sim 10^{-14}$ m, $Q \sim 10^{-28}$ m^2 is expected. The axis of the distortion, from spherical symmetry, of the charge is that corresponding to the spin angular momentum.

There is no torque on a quadrupole in a uniform electric field but in a field gradient [Figure 1.83(c)] there is a torque and an energy term varying with the orientation of the axis of the charges and of **I**. This can also be seen if the quadrupole is regarded as two oppositely directed point dipoles separated along a common axis with the energy of each given by $-\mathbf{E} \cdot \boldsymbol{\mu}$. With the appropriate substitutions the Hamiltonian for the quadrupole interaction with a field gradient is

$$\mathcal{H}_Q = \chi h \frac{3\hat{I}_z^2 - \hat{I}^2}{4\hbar^2 I(2I-1)} \tag{301}$$

with quantum number I. χ is the coupling constant expressed as a frequency:

$$\chi = \frac{e^2 Q}{h} V_{zz}, \qquad \text{where } V_{zz} = -\frac{1}{e}\frac{\mathrm{d}E}{\mathrm{d}z} = \frac{1}{e}\frac{\mathrm{d}^2 V}{\mathrm{d}z^2} \tag{302}$$

More generally,

$$\mathcal{H}_Q = \hat{\mathbf{I}} \cdot \mathbf{Q} \cdot \hat{\mathbf{I}}, \qquad \text{where } \mathbf{Q} = \frac{eQ}{2I(2I-1)\hbar}\mathbf{V} \tag{303}$$

with $\mathbf{V}$ represented by a 3×3 matrix containing all the components of the field gradients. The diagonal elements of $\mathbf{V}$ are related by Laplace's equation since

$$V_{xx} + V_{yy} + V_{zz} = \frac{\partial^2 V}{\partial x^2} + \frac{\partial^2 V}{\partial y^2} + \frac{\partial^2 V}{\partial z^2} = \nabla^2 V = 0$$

i.e. the trace of $\mathbf{V}$ is zero and only two elements are independent.

With $\hat{I}_z|Im_I\rangle = m_I\hbar|Im_I\rangle$ and $\hat{I}^2|Im_I\rangle = \hbar^2 I(I+1)|Im_I\rangle$, the eigenvalues are

$$E = h\chi \frac{3m_I^2 - I(I+1)}{4I(2I+1)} \tag{304}$$

The axis of quantization is that of the direction of the field gradient. For ^{14}N in NH_3, $\chi = 3$ MHz and nuclear quadrupole resonance as such may be observed.

In a magnetic field the quadrupole strongly influences the *magnetic* resonance. To first order the energies for a single spin are

$$E = -\omega_0\hbar m_I + \frac{\chi h[3m_I^2 - I(I+1)]}{8I(2I-1)}(3\cos^2\theta - 1)$$

with θ the angle between the electric field gradient from, for example, non-cubic crystal fields and $\mathbf{B}_0$. The selection rule is still $\Delta m_I = \pm 1$ and with $m^2 - (m-1)^2 = 2m - 1$ the resonance frequencies are

$$f = f_0 - \frac{3\chi(2m_I - 1)(3\cos^2\theta - 1)}{8I(2I-1)} \tag{305}$$

giving $2I + 1$ lines equally separated by

$$\Delta f = \frac{3\chi}{4I(2I-1)}(3\cos^2\theta - 1) \tag{306}$$

This splitting, for ^{14}N, for example, may be very large.

Despite the contrasting origins there is a remarkable similarity between the quadrupole and dipole–dipole splittings, with the same trigonometrical factor. For $I = 1$, for example, the quadrupole effects are as given by the dipole formulae on replacing R by $(\frac{3}{4})\chi$. For a solid powder, or an assembly of molecules with randomly oriented field gradients, a single broad line may similarly result. Small site distortions from cubic symmetry may give substantial field gradients and quadrupole broadening. Although the effects are partially averaged out by molecular tumbling in liquids the linewidths may still range up to 10 kHz, being given by π/T_2^Q with

$$\frac{1}{T_2^Q} = \frac{1}{T_1^Q} = \frac{3}{10}\pi^2 \frac{2I+3}{I^2(2I-1)}\chi^2\tau_c \tag{307}$$

with τ_c a correlation time characterizing the rate of tumbling.

Quadrupolar effects couple the nucleii to the lattice via the crystal field gradients and thus give effective spin–lattice (T_1) relaxation. Such effects are expected to make the primary contributions to spin–lattice relaxation whenever $Q \neq 0$. When T_1 is short it is to be expected that $T_2 \doteqdot T_1$, as above, because the spin–lattice relaxation also destroys the coherence and transverse components.

Protons have no quadrupole moments, but if a proton is coupled to, say, a ^{14}N nucleus with $I = 1$, the three sharp lines that might be expected for the proton resonance may be substantially broadened or even replaced by a single very broad line, according to the environment and Q of the ^{14}N nucleus, as in Figure 1.83(d). The proton resonances may be considered to switch between the three possibilities at a rate corresponding to the relaxation rate of the ^{14}N, i.e. to $1/T_1^{(N)}$. By the uncertainty principle, $\Delta t \Delta E \approx \hbar$, the proton levels involved in each transition should no longer be considered as sharp but to be broadened to an extent depending on Δt or $T_1^{(N)}$. The lower this T_1 value the greater the consequent spread of frequencies or broadening of the proton peaks. Thus spin–spin coupling can effectively transfer quadrupolar effects.

17.2.5 Chemical or site exchange

Chemical shifts depend on the local environments and spin–spin splittings depend on the way in which nucleii are 'connected' or coupled. Both may vary in time due to internal rotation in molecules, the breaking of bonds during reactions, the interchange of bound and free protons and so forth.

The result of a slow rate of exchange of a nucleus between two sites is the superposition of the relevant spectra. As the rate increases the lines may broaden in a way that may again be accounted for by invoking the uncertainty principle. If the lifetime in site A is T_A, the uncertainty in a particular level is $\Delta E = \hbar/T_A$, corresponding to

$$\Delta f = \frac{\Delta E}{h} = \frac{1}{\pi T_A} \tag{308}$$

In more detail, it can be shown that the absorption line maintains its characteristic (Lorentzian) shape and that this Δf is the width at the half height, so that comparing $\Delta f = 1/(\pi T_2)$ the effect will be appreciable if $T_A < T_2$ where T_2 is associated with other effects such as field inhomogeneities.

As T_A decreases, i.e. the rate of exchange k rises, the lines corresponding to two inequivalent protons may broaden until they merge and the single line may become Lorentzian with a width proportional to the square of the chemical shift difference and k^{-1}. As k rises further this single line may be seen to narrow, and this can also be accounted for by a fuller analysis. Such effects are clearly of great importance in the study of chemical and molecular kinetics and are discussed at length in many specialized works.

17.2.6 Spin echoes

Following the application of a 90° pulse and the free induction decay as the individual moments 'fan out' in the rotating frame, the application of a suitable 180° pulse can rapidly rotate the moments to new positions in $OX'Y'$ from which their continued rotation constitutes an approach to a common axis. As discussed in Chapter 3, the fanning-out is replaced by coalescence with restoration of the transverse component and thus, remarkably, of the signal. If the 180° pulse is applied at a time τ after the 90° pulse this so-called echo occurs after a further time τ and decays, as does the original signal. A sequence of n pulses may be applied, schematically:

$$90^\circ_x[\tau - 180^\circ_y - \tau - \text{echo}]_n$$

(the Carr–Purcell–Mulboom–Gill (CPMG) sequence [53]) to give a series of echoes with an exponentially decaying envelope from which a 'natural' or intrinsic T_2 can be derived, one dependent on the specimen and not on the equipment (i.e. accidental variations in the applied field).

17.2.7 Knight shifts

Large shifts of frequency from the nuclear free resonance value, some 10 times most chemical shifts, are observed in metals and designated as the Knight shifts. ΔB is proportional to the applied field and nearly independent of temperature and is larger the larger the atomic number.

According to Knight [54], the shift can be explained in terms of an effective field at the nucleus resulting from the polarization of s-state conduction electrons, i.e. the paramagnetic induced magnetization, the wave function for s electrons being non-zero at the nucleus (see Section 3). The effective field is the equivalent of the hyperfine interaction and the shift is given by

$$\Delta B = \frac{\mu_0 \chi H_0 \overline{|\psi(0)|^2}}{N} \tag{309}$$

with χ the (temperature-independent) paramagnetic spin susceptibility, N the number of atoms per unit volume and $\overline{|\psi(0)|^2}$ the mean value of the probability density at the nucleus. Values of χ thus obtained, with calculated values of $\psi(0)$ for Na or Li, for example, compare well with those predicted as in Section 6.

Experimental n.m.r. studies of metals, first achieved in copper by Pound [55], are restricted by the limited penetration of h.f. fields, and most studies are made on fine powders with particle size comparable to the skin depth.

Knight shifts are also referred to in Section 17.2.10.

17.2.8 Nuclear magnetic resonance in solids

The Hamiltonian is

$$\mathcal{H} = -\omega_0(I_1^z + I_2^z) + \frac{A(\theta)}{\hbar^2}[-I_1^z I_z^z + \tfrac{1}{4}(I_1^+ I_2^- + I_1^- I_2^+)] \tag{310}$$

for two identical and equivalent nucleii, neglecting spin–spin coupling. The essential resemblance of this to $\mathcal{H}$ for simple exchange-coupled spins lies in the raising and lowering operators:

$$\begin{aligned}
(I_1^+ I_2^- + I_1^- I_2^+)(|\alpha\beta\rangle + |\beta\alpha\rangle) &= \hbar^2(|\beta\alpha\rangle + |\alpha\beta\rangle) \\
(I_1^+ I_2^- + I_1^- I_2^+)(|\alpha\beta\rangle - |\beta\alpha\rangle) &= \hbar^2(|\beta\alpha\rangle - |\alpha\beta\rangle) \\
&= -\hbar^2(|\alpha\beta\rangle - |\beta\alpha\rangle)
\end{aligned} \tag{311}$$

and it may be immediately inferred that eigenstates are

$$\begin{aligned}
|\beta\beta\rangle = |1\rangle, &\qquad E_1 = \omega_0\hbar - \left(\frac{A}{\hbar^2}\right)(\tfrac{1}{4}\hbar^2) = \omega_0\hbar - \tfrac{1}{4}A \\
|\alpha\beta\rangle + |\beta\alpha\rangle = |2\rangle, &\qquad E_2 = 0 - \frac{A}{\hbar^2}(-\tfrac{1}{4}\hbar^2 - \tfrac{1}{4}\hbar^2) = \tfrac{1}{2}A \\
|\alpha\beta\rangle - |\beta\alpha\rangle = |3\rangle, &\qquad E_3 = 0 - \frac{A}{\hbar^2}(-\tfrac{1}{4}\hbar^2 + \tfrac{1}{4}\hbar^2) = 0 \\
|\alpha\alpha\rangle = |4\rangle, &\qquad E_4 = -\omega_0\hbar - \frac{A}{\hbar^2}(\tfrac{1}{4}\hbar^2) = -\omega_0\hbar - \tfrac{1}{4}A
\end{aligned} \tag{312}$$

The energy levels and transitions are shown in Figure 1.84(a). Levels 1 and 3 are not shifted equally as for spin–spin coupling and thus a splitting does result in this case. Replacing A, the resonances occur at

$$f = f_0 \pm \tfrac{3}{4}R(3\cos^2\theta - 1), \qquad \text{where } R = \frac{\mu_0}{4\pi}\gamma^2\frac{\hbar}{2\pi}\frac{1}{r^3} \tag{313}$$

As $r \to \infty$ the lines coalesce at $f_0 = \gamma B_0/2\pi$ and the splitting, for a given θ, is symmetrical about this. The lines also coalesce when

$$3\cos^2\theta - 1 = 0 \text{ for } \theta = 54.73\ldots^\circ$$

On rotating a crystal in which the two protons occur in H_2O, the water of crystallization, for example, the variation of the splittings affords a means of determining the disposition of these molecules in the lattice. For gypsum, $CaSO_4{\cdot}2H_2O$, two different orientations of the H—H axis in the lattice are indicated by the occurrence of two doublets [56].

The displacement of either line can be positive or negative between the limits (for the positive sign)

$$\begin{aligned}
\theta = \frac{\pi}{2}, &\qquad \Delta f = -\tfrac{3}{4}R \\
\theta = 0, &\qquad \Delta f = \tfrac{3}{2}R
\end{aligned} \tag{314}$$

Considering the line between the nucleii to lie with equal probability along any one of a large number of lines from the origin to all the equal surface elements into which a spherical surface around the origin is divided, it is seen that the number of these lines is greatest for $\theta = \pi/2$ and least for $\theta = 0$. Adding individual contributions for all these discrete orientations and drawing an envelope corresponding to their density, as for a density of states curve, the total absorption is obtained as in Figure 1.84(c). The same considerations apply to the second member of the doublet, so for an assembly of randomly oriented pairs, as in a powder, the result is as shown in Figure 1.84(d).

While spin–spin coupling occurs only between nucleii in the same molecule, because it is carried by the intervening electron orbitals, dipolar effects apply between spins in different

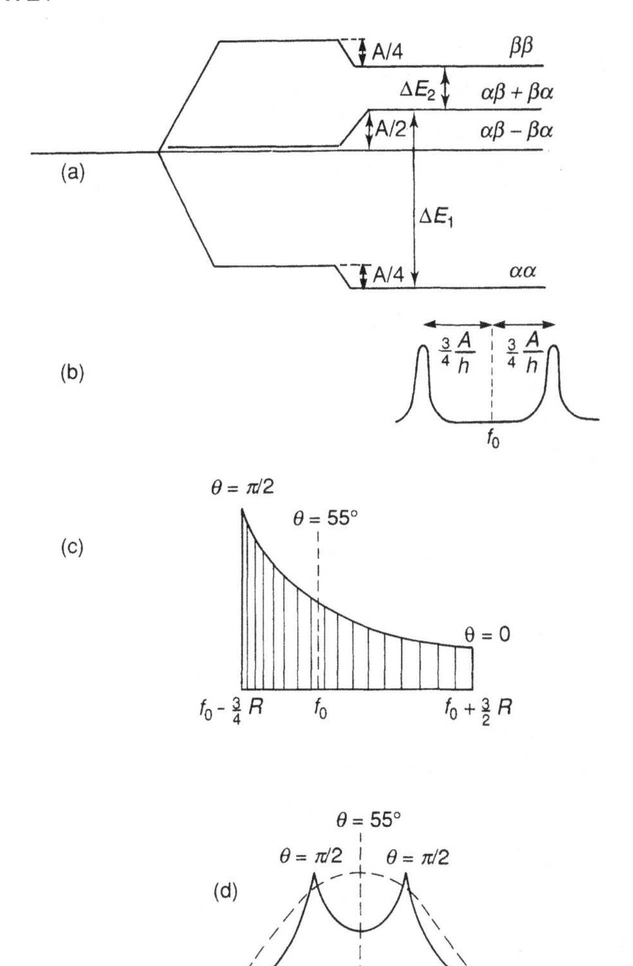

Figure 1.84 $\Delta E_1 = hf_0 + \frac{3}{4}A$ and $\Delta E_2 = hf_0 - \frac{3}{4}A$: the level for $|\alpha\ \beta\rangle + |\beta\ \alpha\rangle$ is not midway between $|\alpha\ \alpha\rangle$ and $|\beta\ \beta\rangle$ as for spin–spin coupling [cf. Figure 1. 81(a)] so a splitting occurs as in (b). Considering the varying orientations in a stepwise manner the spectra shown in (c) and (d) can be built up (the intensity corresponding to the density of the arbitrarily separated lines)

molecules, though the $1/r^3$ dependence is to be noted. If these additional interactions are taken into account a single line may be produced for a powder with width $\approx 2R$, as shown by the broken line in Figure 1.84(d). It is recalled that R may be very large.

17.2.9 High-resolution spectra for solids

The factor $(3\cos^2\theta - 1)$ occurs in terms associated with both dipole–dipole and quadrupole broadening. Andrew [57] showed that if the specimen was spun about an axis inclined at

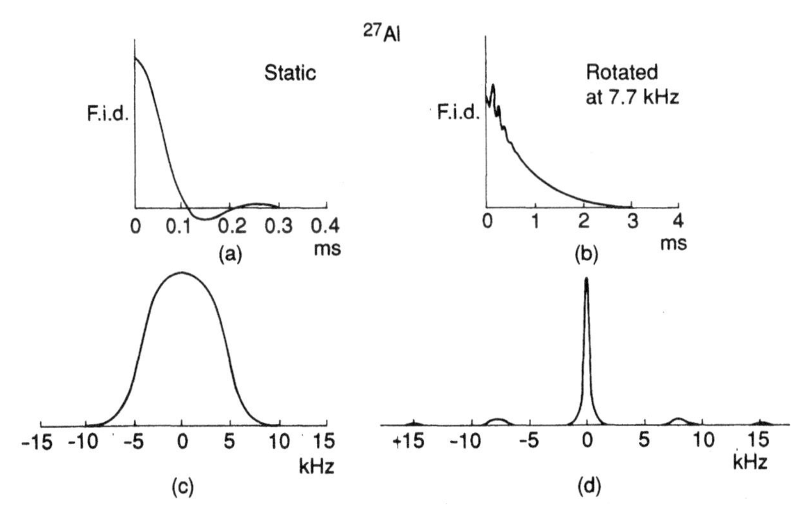

Figure 1.85 The free induction decay signal and its Fourier transform (i.e.the absorption line) for a powder of Al, with and without magic angle spinning [58]. Note the difference in the f.i.d. time scales and the way in which this is reflected in the linewidths, as well as the appearance of sidebands

a particular angle β to $\mathbf{B}_0$ an effect analogous to motional narrowing in liquids could be produced, the interactions averaging out. It is found (see Chapter 3, Section 2.12) that the mean value of θ (the angle between lines joining spin pairs and $\mathbf{B}_0$) contains $(3\cos^2\beta - 1)$ as a factor and so with $\cos\beta = 1/\sqrt{3}$, $\beta = 54.73°$, the interactions should average to zero. The spinning angular velocity ω_r must be larger than the static linewidth, presenting severe technical problems. Sidebands arise, spaced at ω_r and $2\omega_r$ from the centreline as shown for ^{27}Al in the metal powder form in Figure 1.85 [58]. Knight shifts remain and can be determined, as, for example, 1640 p.p.m. for Al, by recording the spectrum of the powder in an $AlCl_3$ solution.

A great advance was made in 1966 when Mansfield and Ware [59] and Ostroff and Waugh [60] developed multiple-pulse methods (cf. the two-pulse spin echo) to achieve high resolution without the limitations that are found to apply to the above methods. The effects of dipolar interactions can be reduced by a factor of two by two pulses and further reduced by building up an 'MW-2 sequence', after Mansfield and Ware [61] and Waugh *et al.* [62]. The quadrupolar as well as dipolar effects are greatly diminished in the multiple-pulse methods whereas the shifts and scalar splittings are preserved.

17.2.10 Nuclear magnetic resonance in magnetically ordered materials

The hyperfine interaction of the nucleus with electrons for which the wave function is non-zero at the nucleus, i.e. s-state electrons with $|\psi_s(0)|^2 \neq 0$, can be denoted

$$\mathcal{H}_{\text{hf}} = \mathbf{S}\cdot\mathbf{A}\cdot\mathbf{I} \rightarrow A\mathbf{S}\cdot\mathbf{I} \rightarrow AS_zI_z \tag{315}$$

for the general, isotropic or high-field (approximation) cases. For an assembly of electrons with a certain mean value $\langle S_z\rangle$, the operator S_z may be eliminated from this term, leading to a nuclear spin Hamiltonian, for $\mathbf{B} = \mathbf{k}B$:

$$\mathcal{H} = -\gamma_N B I_z + A\langle S_z\rangle I_z \tag{316}$$

The mean $\langle S_z\rangle$ is supposed to correspond to the partial polarization of the electron spins by a magnetic field at a certain temperature and is thus simply related to the magnetization and susceptibility:

$$\begin{aligned} M(T) &= N\mu_z = N\gamma\langle S_z\rangle = \frac{Ng\beta\langle S_z\rangle}{\hbar} \\ \chi &= \frac{\mu_0 Ng\beta\langle S_z\rangle}{\hbar B} \end{aligned} \tag{317}$$

where B can be taken as the static resonance field. $\mathcal{H}$ becomes

$$\mathcal{H} = -\gamma_N B I_z + \frac{A\chi B\hbar}{\mu_0 Ng\beta} I_z = -\gamma_N\left(B - \frac{a\chi}{\mu_0\hbar\gamma_N Ng\beta}\right) I_z \tag{318}$$

with $A = a/\hbar^2$. The second term in the parentheses represents the Knight shift for metals.

The s-electrons may contribute to a spontaneous magnetization if they are conduction electrons polarized by localized ordered spins according to the RKKY mechanism (see Chapter 2, 6.1) or mixed into a d-band. As $T \to 0$ K their magnetization becomes

$$M_s(0) = \tfrac{1}{2}Ng\beta \qquad (\langle S_z\rangle = \tfrac{1}{2}\hbar) \tag{319}$$

with N the volume concentration of s-electrons (so that this M_s is not that of the crystal but can be taken to be proportional to it). Generally,

$$\langle S_z\rangle = \frac{\hbar M_s(T)}{Ng\beta} = \tfrac{1}{2}\hbar\frac{M_s(T)}{M_s(0)} \tag{320}$$

using equation (319), and so

$$\mathcal{H} = -\gamma_N\left(B - \tfrac{1}{2}\hbar\frac{M_s(T)}{M_s(0)}\frac{A}{\gamma_N}\right) I_z = -\gamma_N(B - \Delta B)I_z \tag{321}$$

The shift is, vitally, not now dependent on the applied field. The last term in equation (321) may similarly be expressed as

$$-\frac{\frac{1}{2}\hbar A}{\mu_0\gamma_N}\frac{\chi B}{M_s(0)} = -\frac{\frac{1}{2}\hbar A(\chi H)}{\gamma_N M_s(0)}$$

so the magnitudes of the two effects are compared by the ratio

$$r = \frac{M_s(T)}{\chi H} = \frac{M_s(T)}{M_p} \doteqdot \frac{M_s(0)}{M_p} \tag{322}$$

if T well below T_c is assumed, with M_p the Pauli paramagnetism.

Taking an order of magnitude Fermi temperature $T_f = 10^4$ K and $B = 1.0$ T,

$$M_p \doteqdot \frac{N\beta^2 B}{kT_f} = 10^{19}N\beta^2, \qquad r \doteqdot \frac{N\beta}{10^{19}N\beta^3} \approx 10^4$$

For several metals the Knight shift can be expressed as ~ 0.1 per cent, i.e. in $B = 1.0$ T the shift is equivalent to a field of 1.0 mT. It follows that ΔB due to the ordered spins is of the order of 10 T.

Rather than being a small change, this is similar to, or greater than, the fields usually used to observed n.m.r., and indeed it is found that n.m.r. may be observed in ordered materials even in the absence of applied fields [63] with

$$\mathcal{H} = \gamma_N B_n I_z, \qquad \text{where } B_n = \tfrac{1}{2}\hbar \frac{M_s(T)}{M_s(0)} \frac{A}{\gamma_N} \tag{323}$$

B_n or $H_n = B_N/\mu_0$ is the nuclear field and is, of course, a conventional or effective field. With resonance at $\omega = \gamma B_n$ a plot of $\omega(T)/\omega(0)$ gives the temperature dependence $M_s(T)$ *in the zero applied field*, as in the earlier Figure 1.54.

It is assumed that M_s for the conduction electrons is proportional to M_s for the material. However, attention should not be focused only on conduction electrons. The closed shells of 3d ions, for example, contain 1s, 2s and 3s electrons. The 1s electrons with very large $|\psi(0)|$ potentially give very large nuclear fields, though at first sight these should cancel out since the interactions are with oppositely directed spins. A general rule, inferred from a range of calculations, is that electrons with antiparallel spins tend to occupy the same region of space, i.e. 'attract', while parallel spin electrons tend to occupy distant regions, i.e. 'repel'. For H_2, for example, the bonding orbital is that for which $|\psi|^2$ is large between the nuclei, in a rather small region of space, and this is the singlet symbolically shown as '$\uparrow\downarrow$'. For the antibonding orbital $|\psi|^2$ spreads over larger regions of space and this is the triplet '$\uparrow\uparrow$'. (Despite the obvious comparison with side-by-side dipoles or magnets, this is not a magnetostatic effect.) This is sometimes called the Pauli effect since it is connected with the Pauli exclusion principle and is associated with spin-pairing in localized bonds generally. Consider a polarized, localized, 3d electron in an ordered material (spin '$\uparrow$') moving in an orbit external to the 1s electrons (Figure 1.86). As indicated, it tends to move more closely to the '$\downarrow$' 1s electron than to the '$\uparrow$' electron. Thus, according to the Coulomb forces exerted the '$\downarrow$' electron is pushed towards the nucleus and vice versa:

$$|\psi_s \downarrow (0)|^2 > |\psi_s \uparrow (0)|^2$$

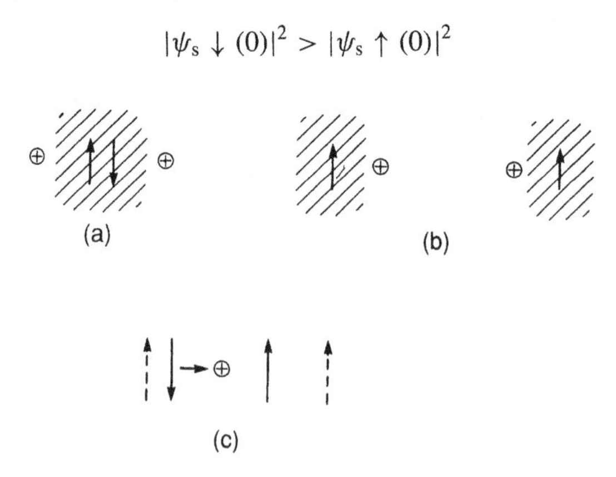

Figure 1.86 In a crude illustration of d–s polarization the d spin (broken arrow $\uparrow$) moves more closely to the '$\downarrow$' s spin and 'pushes it towards the nucleus' due to the Coulomb forces (c). The effective attraction of antiparallel spins, and vice versa, is not magnetostatic but may be inferred by recourse to the bonding (a) and antibonding (b) configurations for the H_2 molecule

Table 1.6 Effective fields (tesla) at the nuclei due to exchange polarization [63]

	$Mn^{2+}(3d^5)$		$Fe^{3+}(3d^5)$	$Fe^{2+}(3d^6)$
$1s\uparrow$ $1s\downarrow$	250 284 −250 287	= −3	−5	−3
$2s\uparrow$ $2s\downarrow$	22 667 −22 808	= −141	−179	−131
$3s\uparrow$ $3s\downarrow$	3 121 −3 047	= +74	+1215	+79
Total		−70	−63	−55

This d–s polarization is shown by realistic calculations to be the predominant effect, Freeman and Watson [63] giving the calculated values in Table 1.6. The magnitudes are similar to the above estimates for conduction electrons.

Assuming that other effects, e.g. dipolar, are small or weakly temperature-dependent and that $A \neq A(T)$, the n.m.r. $\omega(T)$ again gives $M_s(T)$ by direct proportionality. However, this is now seen to apply to any particular set of 3d electrons and $M_s(T)$ for individual *sublattices* can be measured. Such curves can be obtained, as in Figure 1.87 [64], even for antiferromagnetics with zero overall M_s, and the full value of the method becomes apparent. Many references to early work are given by Feldmann *et al.* [65].

In ferromagnetics in zero applied static fields very intense absorptions can be observed, as if the effective h.f. B_1 were much greater than that applied. This enhancement disappears or is reduced in static fields approaching saturating values, suggesting an association with oscillating domain walls. Figure 1.88 shows two spins in 180° domains and one that is at the centre of the wall at $t = 0$. At the high frequencies involved the wall will usually be

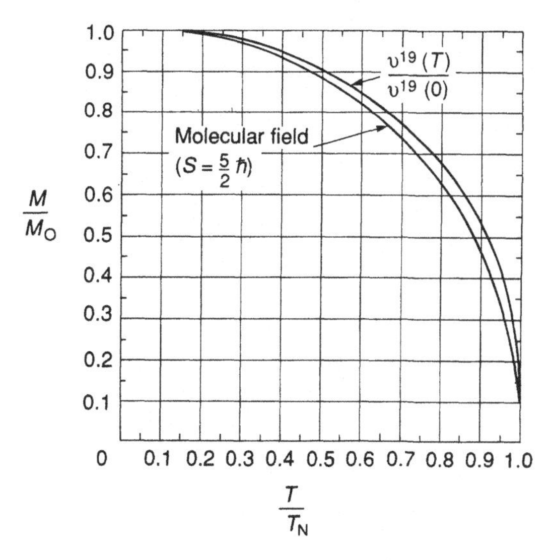

Figure 1.87 The temperature dependence of the sublattice magnetization of the antiferromagnetic MnF_2 as indicated by the ^{19}F resonance frequency [64]. The second curve corresponds to the molecular field theory as applied to antiferromagnetics

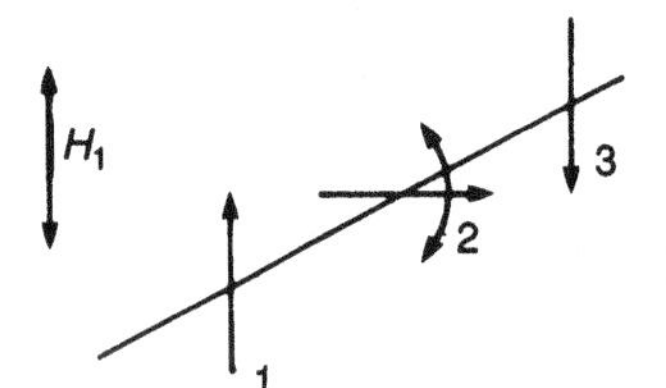

Figure 1.88 Spin 2 is at the centre of a 180° domain wall and as the wall oscillates slightly the nucleus to which it is coupled experiences a large oscillating exchange field

heavily damped and the central spin (or plane of spins) oscillates as indicated as the wall oscillates in the h.f. field H_1. This is equivalent to modulating the local nuclear field not by H_1 but by $H_1^{(n)}$, an oscillating *nuclear* field that is, of course, very much higher. Note that the direction of H_n varies throughout the wall. For cobalt the enhancement may be as high as $H_1^{(n)}/H_1 \sim 1000$. A quantitative account would require information on the wall densities and amplitudes of oscillation.

Some enhancement remains in the absence of domain walls due to the rotation of the magnetization induced by H_1. For small rotations ($\sin\theta = \theta$) a field H_1 applied normal to an easy axis gives $\theta = H_1/H_k$ as the angle between M_s and the easy axis and

$$H_1^{(n)} = H_n \sin\theta \doteqdot \frac{H_1 H_n}{H_k} = H_1 \frac{\chi}{M_s} H_n \tag{324}$$

18 Mossbauer Spectroscopy and Ordered Materials

Nuclei other than protons may have spin $\frac{1}{2}$ in the ground state but have excited states with different values of the nuclear spin, since they contain protons and neutrons with $I = \frac{1}{2}$ which may couple in different ways. For example, ^{57}Fe, very fortuitously, has an $I = \frac{1}{2}$ ground state and an $I = \frac{3}{2}$ excited state at 14.4 keV (Figure 1.89). This very large gap corresponds to the emission of a γ-ray for excited to ground-state transitions. For an isolated atom, a single nucleus, the energy of the γ-ray, will correspond to $E_\gamma \doteqdot \Delta E = 14.4$ keV approximately, only because the momentum carried is significant. Momentum must be conserved and the nucleus recoils; the recoil energy of the nucleus $E_r^{(n)} = p^2/(2m_n)$ with m_n the nuclear mass constitutes a 'correction' for E_γ. The natural linewidth of the γ-ray is

$$W = \frac{\hbar}{\tau} = 4.7 \times 10^{-9} \text{ eV} \qquad (\tau = 0.14\ \mu_s \text{ for } ^{57}\text{Fe})$$

Since this is so small, resonant absorption by a second identical nucleus will not occur (i.e, the required overlap does not exist, particularly since recoil must be considered once more for the absorption). For a nucleus in a solid, however, the recoil is shared by the whole crystal and although the momentum involved is the same the energy shift is much smaller:

$$E_\gamma = \Delta E - \frac{p^2}{2m_c} \tag{325}$$

with m_c the mass of the crystal: $m_c \gg m_n$. This is effectively recoil-less emission and resonant absorption by a second nucleus becomes possible.

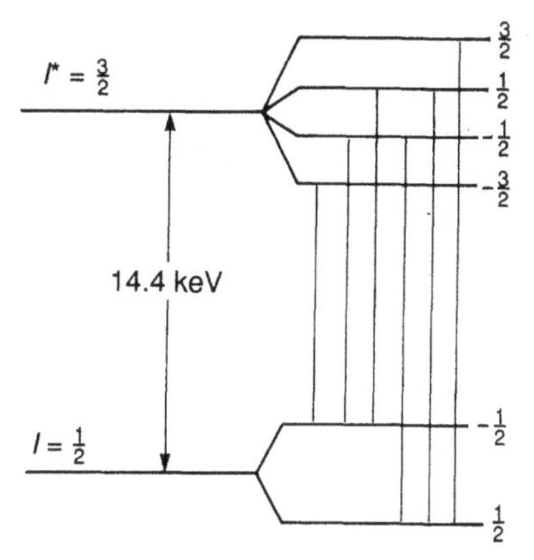

Figure 1.89 Zeeman splitting of the nuclear energy levels for an $I = \frac{1}{2}$ ground state and an $I = \frac{3}{2}$ excited state, with the allowed transitions

Suppose now that both levels are split by a magnetic field (or by quadrupole splitting). The transitions shown in Figure 1.89 give six frequencies very slightly displaced from $\Delta E/h$. A crystal of the same material but with no applied field and high crystallographic symmetry (no splittings) absorbs at $\Delta E/h$. However, if the absorber, or source, is moved with velocity v towards or away from the source (absorber) the Doppler shift in energy Ev/c may be contrived to compensate for each of the shifts away from ΔE, in turn, so that absorptions will occur. The v values are expected to be small. This is the basis of Mossbauer spectroscopy [66]: source and absorber are interchangeable.

The first requirement is a radioactive source containing nuclei, which are those to be studied but with no state splitting so that a single frequency is produced, equipped with a precise velocity drive. The single-line γ-rays pass through the specimen and the quanta are counted. The plot of count rate or transmitted intensity constitutes the spectrum. Since $v \propto E$ the connection between the spectrum and the energy levels is simple. With the abbreviations

$$g_{\mathrm{N}}^{(3/2)}\beta_{\mathrm{N}}B_{\mathrm{n}} = a, \qquad g_{\mathrm{N}}^{(1/2)}\beta_{\mathrm{N}}B_{\mathrm{n}} = b$$

for the upper and lower splitting magnitudes the energies and shifts for the transitions allowed by $\Delta m_I = 0, \pm 1$ are

$$1.\ \Delta m_I = -1, \qquad \Delta E = \Delta - \tfrac{3}{2}a - \frac{b}{2}, \qquad \delta E' = -7$$

$$2.\ \Delta m_I = 0, \qquad \Delta E = \Delta - \tfrac{1}{2}a - \frac{b}{2}, \qquad \delta E' = -3$$

$$3.\ \Delta m_I = 1, \qquad \Delta E = \Delta + \tfrac{1}{2}a - \frac{b}{2}, \qquad \delta E' = -1$$

$$4.\ \Delta m_I = -1, \qquad \Delta E = \Delta - \tfrac{1}{2}a + \frac{b}{2}, \qquad \delta E' = +1$$

$$5.\ \Delta m_I = 0, \qquad \Delta E = \Delta + \tfrac{1}{2}a + \frac{b}{2}, \qquad \delta E' = +3$$

$$6.\ \Delta m_I = 1, \qquad \Delta E = \Delta + \tfrac{3}{2}a + \frac{b}{2}, \qquad \delta E' = +7$$

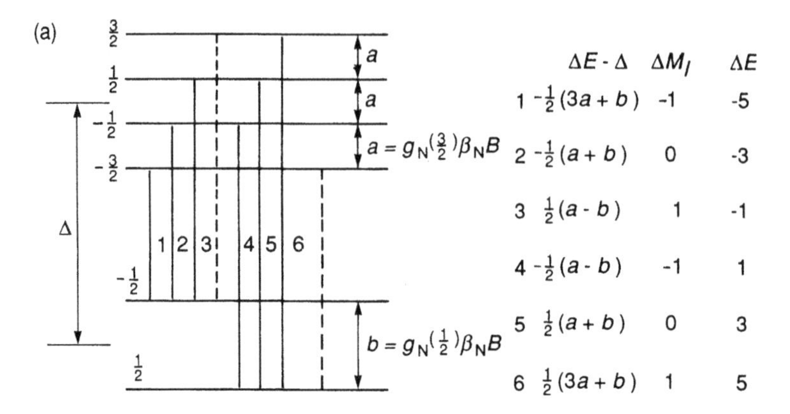

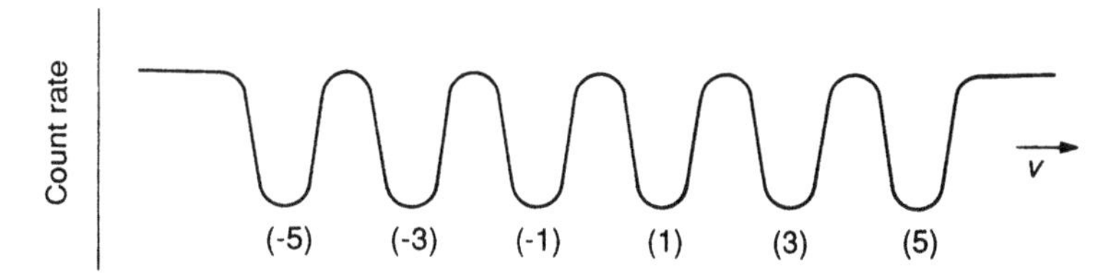

Figure 1.90 A model Mossbauer spectrum based on the structure of Figure 1.89 setting $a = 2$ and $b = 4$ purely for illustration

Here $\delta E'$ is a number representing the shift from $\Delta E = \Delta$ obtained by arbitrarily setting $b = 4$ and $a = 1$ to obtain the illustration of Figure 1.90.

^{57}Co decays to give ^{57}Fe which is radioactive and in the second excited state ($I = \frac{5}{2}$), and this decays ($\tau \sim 10^{-7}$ s) largely to the first excited state, that of interest, which gives a reasonable γ-ray intensity and appropriate linewidth. The parent lifetime (270 days) is appropriate. Other isotopes suitable for Mossbauer effect studies, and for n.m.r. are listed by Feldman *et al.* [65, Figure 7].

While n.m.r. is intrinsically difficult in solids, Mossbauer spectroscopy (MS) is specifically appropriate to solids. MS is also particularly appropriate for ordered materials in which the effective nuclear fields are larger than most conventional laboratory fields.

If there are no splittings associated with the specimen (or source) the Mossbauer line does not necessarily occur at zero v. This is only the case when the nuclei are not only identical but in identical environments. This is due to the isomer shift δ, a consequence of the difference in the radii of the nuclei in the ground state (r) and excited state (r^*) together with the dependence of $\psi(0)$ on the environment:

$$\delta = E_s - E_a = \frac{2}{3}\pi Z e^2 |\langle r^* \rangle^2 - \langle r \rangle^2 | [|\psi_a(0)|^2 - |\psi_s(0)|^2] \tag{326}$$

For ^{57}Fe, $(r - r^*)/r \approx 10^{-13}$. The involvement of r^* associates this effect specifically with γ-ray spectroscopy. Since it is ΔE for the unsplit levels which is affected, the net effect when the absorber levels are split is a uniform shift in the spectrum when the absorber nuclei are in environments that differ from those of the emitter. The 'environment' encompasses both the chemical or crystallographic environment and the nature of the ion containing the nuclei (e.g. Fe^{2+} or Fe^{3+}), since both features can affect the value of $|\psi(0)|^2$ at the nucleus. It follows that if the source nucleii are, as usual, all identical but the absorber contains, say, two types of ion or one type of ion in two different lattice sites then the resulting spectrum consists of the superposition of the two separate spectra, in effect to two different nuclear fields.

For example, below a so-called 'Vervey temperature' of 119 K the Mossbauer spectrum of magnetite contains separate lines for the octahedral site Fe^{2+} and Fe^{3+} ions. Above 119 K the MS consists of two superimposed six-line spectra, one due to Fe^{3+} in tetrahedral sites (which also appears in the above) and one that can be ascribed to octahedral Fe^{3+} and Fe^{2+}, giving an average associated with very rapid electron exchange between the two (see Chapter 5, Section 2.3). The temperature dependence of the line-broadening of the averaged spectrum gives an activation energy for the electron hopping process of 0.065 eV [67]. The temperature dependence of the nuclear fields at the two sublattices correlates with that of the sublattice magnetizations according to neutron diffraction [68].

The collapse of the Mossbauer spectrum in extremely fine grain specimens is indicative of the rapid fluctuations in time of the magnetization vectors due to superparamagnetic behaviour, with a relaxation time of 10^{-8} s [69]. Exchange fields in ordered organic materials may be detected and the environments of cobalt ions on the surfaces of oxide particles, used for magnetic recording studied by Mossbauer spectroscopy. These few examples should be adequate to indicate the power of the Mossbauer effect in relation to detailed static and dynamic magnetic studies.

References

1. Dirac, P.A.M., *Proc. Roy. Soc.*, **A132**, 60 (1936); *Phys. Rev.*, **74**, 817 (1948); see also Fleischer, R.L., Hart, H.R., Jr, Jacobs, I.S., Price, P.B., Schwartz, W.M. and Woods, R.T., *J. Appl. Phys.*, **41**, 958 (1970).
2. Sutter, D.H. and Flygare, W.H., *J. Am. Chem. Soc.*, **61**, 6859 (1969).
3. Pauli, W., *Z. Physik*, **41**, 81 (1927).
4. Wilson, A.H., *The Theory of Metals*, Cambridge University Press, 1953.
5. Pines, D., *Phys. Rev.*, **95**, 1090 (1954).
6. Peierls, R.E., *Quantum Theory of Solids*, Clarendon, Press Oxford, 1965.
7. de Hass, W.J. and van Alphen, P.M., *Commun. Kammerlingh Onnes Lab., Univ. Leiden*, 208d (1930) and 220d (1933).
8. Debye, P., *Ann. Phys. Leipzig*, **81**, 1154 (1926); Giauque, W.F., *J. Am. Chem. Soc.*, **49**, 1864 (1927).
9. Kurti, N., Robinson, F.N.H., Simon, F.E. and Spohr, D.A., *Nature*, **178**, 450 (1956).
10. Kaczer, J. and Shalnikova, T., *Proceeding International Conference on Magnetism*, Nottingham, p. 589, 1964.
11. Shull, C.G., Strauser, W.A. and Wollan, E.O., *Phys. Rev.*, **83**, 333 (1951).
12. Mook, H.A. and Shull, C.G., *J. Appl. Phys.*, **37**, 1034 (1966).
13. Nigh, H.E., Legvold, S. and Spedding, F.H., *Phys. Rev.*, **132**, 1092 (1963).
14. Koehler, W.C., *J. Appl. Phys.*, **36**, 1078 (1965).
15. Mattis, D.C., *The Theory of Magnetism*, Springer-Verlag, Berlin, 1981.
16. White, R.M., *Quantum Theory of Magnetism*, McGraw-Hill, New York, 1970.
17. White, R.M. and Geballe, T.H., *Long Range Order in Solids*, Academic Press, New York, 1979.
18. Nagle, D., Frauenfelder, H., Taylor, R.D., Cochran, D.R.F. and Mathias, B.T., *Phys. Rev. Lett.*, **5**, 364 (1960).
19. Mathias, B.T., Bozorth, R.M. and Van Vleck, J.H., *Phys. Rev. Lett.*, **7**, 160 (1961).

20. Birss, R.R. and Isaac, E.D., *Magnetic Oxides*, Ed. D.J. Craik, John Wiley and Sons, London and New York, p. 289, 1975.
21. Bozorth, R.M., *Phys. Rev.*, **96**, 311 (1954).
22. Osborne, J.A., *Phys. Rev.*, **67**, 351 (1954); Stoner, E.C., *Phil. Mag.*, **36**, 803 (1954).
23. van den Broek, J.J. and Zijlstra, H., *IEEE Trans. Magn.*, **MAG-7**, 226 (1971).
24. Lilley, B.A., *Phil. Mag.*, **41**, 792 (1950).
25. Landau, L. and Lifshitz, E., *Physik Z. Sowjet.*, **8**, 153 (1935).
26. Lifshitz, E., *J. Phys. USSR*, **8**, 337 (1944).
27. Néel, L., *Cahiers Phys.*, **25**, 1 (1944).
28. Kaczer, J., *Czech. J. Phys.*, **7**, 557 (1957).
29. Craik, D.J. and Myers, G., *Phil. Mag.*, **31**, 489 (1975).
30. Craik, D.J. and Tebble, R.F., *Ferromagnetism and Ferromagnetic Domains*, North Holland, Amsterdam, 1965.
31. Carey, R. and Isaac, E.D., *Magnetic Domains*, English Universities Press, London, 1966.
32. Stoner, E.C. and Wohlfarth, E.P., *Phil. Trans. Roy. Soc.*, **A240**, 599 (1948).
33. Gans, R., *Ann. Physik.*, **15**, 28 (1932).
34. Craik, D.J., Hill, E.W. and Harrison, A.J., *IEEE Trans. Magn.*, **MAG-11**, 1379 (1975).
35. Néel, L., *Cahiers Phys.*, **25**, 19 (1944); *Ann. Univ. Grenoble*, **22**, 299 (1946).
36. Kersten, M., *Z. Physik*, **124**, 714 (1947); *Z. Angew. Phys.*, **7**, 397 (1955).
37. Craik, D.J. and Fairholme, R.J., *J. de Physique*, **32** (suppl. 2–3), 681 (1971).
38. Craik, D.J. and Wood, M.J., *J. Phys. D*, **3**, 1009 (1970); See also Carey, R., *Proc. Phys. Soc.*, **80**, 934 (1962).
39. Brown, W.F., *Phys. Rev.*, **75**, 147 (1949).
40. Shtrikman, S. and Treves, P., *J. Appl. Phys.*, **31**, 72S (1960).
41. Sixtus, K.J. and Tonks, L., *Phys. Rev.*, **37**, 930 (1931) and **43**, 70 (1933).
42. Craik, D.J. and McIntyre, D.A., *Phys. Lett.*, **21**, 288 (1966).
43. Kikuchi, R., *J. Appl. Phys.*, **27**, 1352 (1956).
44. Doring, W., *Naturforsch*, **3**, 373 (1948).
45. Becker, R., *J. Phys. Rad.*, **12**, 332 (1951).
46. Smit, J. and Wijn, H.P.J., *Adv. in Electr. and Electr. Phys.*, **6**, 69 (1954); *Ferrites*, Philips Technical Library, Eindhoven, 1959.
47. Galt, J.K., *Phys. Rev.*, **85**, 664 (1952).
48. Zavoiskey, E., *J. Phys. USSR*, **9**, 211 (1945); Cummerow, R.L. and Halliday, D., *Phys. Rev.*, **92**, 212 (1953).
49. Stevens, K.W.H. and Bates, C.A., *Magnetic Oxides*, Ed. D.J. Craik, John Wiley and Sons, London and New York, p. 141, 1975.
50. Purcell, E.M., Torrey, H.C. and Pound, R.V., *Phys. Rev.*, **69**, 37 (1946).
51. Bloch, F., Hansen, W.W. and Packard, E.M., *Phys. Rev.*, **69**, 127 (1946).
52. Ernst, R.R. and Anderson, W.A., *Rev. Sci. Instr.*, **37**, 93 (1966).
53. Carr, H.Y. and Purcell, E.M., *Phys. Rev.*, **94**, 630 (1954); Mulboom, S. and Gill, D., *Rev. Sci. Instr.*, **29**, 6881 (1958).
54. Knight, W.D., *Phys. Rev.*, **76**, 1259 (1949).
55. Pound, R.V., *Phys. Rev.*, **73**, 112 (1948).
56. Pake, G.E., *J. Chem. Phys.*, **16**, 327 (1948).
57. Andrew, E.R., *Prog. NMR Spectrosc.*, **8**, 1 (1971); *Int. Rev. Phys. Chem.*, **1**, 195 (1981).
58. Andrew, E.R., Hinshaw, W.S. and Riffen, R.S., *Phys. Lett.*, **46A**, 57 (1973).
59. Mansfield, P. and Ware, D., *Phys. Lett.*, **22**, 133 (1966).
60. Ostroff, E.D. and Waugh, J.S., *Phys. Rev. Lett.*, **16**, 1097 (1966).
61. Mansfield, P. and Ware, D., *Phys. Rev.*, **168**, 318 (1968); Mansfield, P., Richard, K.H.B. and Ware, D., *Phys. Rev.*, **B1**, 2048 (1970).
62. Waugh, J.S., Hubert, L.M. and Haeberlen, U., *Phys. Rev. Lett.*, **20**, 180 (1968).
63. Freeman, A.J. and Watson, R.E., *Magnetism*, Vol. II A, Eds. G.T. Rado and H. Suhl, Academic Press, New York, p. 168, 1965.
64. Heller, P. and Benedeck, G.B., *Phy. Rev. Lett.*, **8**, 428 (1962); see also Jaccarino, V., *Magnetism*, Vol. II A Eds G.T. Rado and H. Suhl, Academic Press, New York and London, p. 307, 1965.

65. Feldman, D., Kirchmayr, H., Schmoltz, A. and Velicescu, M., *IEEE Trans. Magn.*, **MAG-7**, 61 (1971).
66. Mossbauer, R.L., *Z. Physik*, **151**, 124 (1958).
67. Sawatzki, G.A., Coey, J.M.D. and Morrish, A.H., *J. Appl. Phys.*, **40**, 1402 (1969).
68. Riste, T. and Tenzwe, L., *J. Phys. Chem. Solids*, **19**, 117 (1961).
69. McNab, T.K., Fox, R.A. and Boyle, A.J.F., *J. Appl. Phys.*, **39**, 5703 (1968).

CHAPTER

2 Quantum Theory and Magnetism

1 Introduction

This chapter surveys a few of the quantum mechanical treatments relevant to the study of magnetism, with obvious restrictions. The central role of spin in magnetic effects is discussed separately in Chapter 3. Spin must obviously be involved in the present treatment which in some respects serves as an introduction to the following chapter, as in the brief preliminary general reviews.

2 Review of Wave Mechanics

Since no observation can be made without some interaction with the system observed, and for small systems this may constitute a substantial disturbance, a measurement of the quantity A followed by one of B, formally 'AB', may give different results to 'BA' (B measured first). This suggests the association of operators $\hat{A}$, $\hat{B}$, ... with observables A, B ..., which operators may or may not commute. They may initially be expected to comprise differential operators and functions of coordinates, since the order in which they were applied would then clearly be of importance. Such operators cannot return numbers on assigning values to arguments, as can functions, but numbers are associated with operators in the form of eigenvalues, typified by a in a characteristic equation: $\hat{A}f(\mathbf{r}) = af(\mathbf{r})$, in which $f(\mathbf{r})$, the eigenfunction, is specifically unchanged save for the appearance of a.

Entities a and b commute if $ab = ba$, $ab - ba = 0$ or $[a, b] = ab - ba = 0$, defining the commutator of a and b. Commutation is important and the reader may confirm by expansion that

$$[b, a] = -[a, b]$$

$$[a + b, c] = [a, c] + [b, c]$$

$$[ab, c] = a[b, c] + [a, c]b$$

An association is made between commutation and the Heisenberg uncertainty principle, i.e. that if $[\hat{A}, \hat{B}] \neq 0$, A and B are conjugate variables and the values A and B cannot be known simultaneously and precisely.

It may be inferred from the proposed de Broglie wave function for a freely moving electron that for components of the linear momentum the operators should be $\hat{p}_x = -i\hbar\,\partial/\partial x$, etc. By application to a general $f(x)$ and finally cancelling the function to leave the operator equation the reader may show that $[x, \hat{p}_x] = i\hbar$. Since this may be taken to be a general statement of the uncertainty principle the operator $\hat{x}$ (etc.) is identified with the coordinate itself:

$$\hat{p}_x \rightarrow -i\hbar\frac{\partial}{\partial x}, \qquad \hat{x} \rightarrow x$$

For quantum and classical results to merge as the size of the system increases, the relations between the operators should be the same as those between the corresponding classical functions—the correspondence principle. This indicates that $\hat{K} = \hat{p}^2/2m_e$ for the kinetic energy and, since $\hat{x} = x$, $\hat{V}(\mathbf{r}) = V(\mathbf{r})$, the classical potential energy, the Hamiltonian for the total energy of an electron is

$$\mathcal{H} = \frac{\hat{p}^2}{2m_e} + V(\mathbf{r}) = -\frac{\hbar^2\nabla^2}{2m_e} + V(\mathbf{r})$$

($p^2 = p_x^2 + p_y^2 + p_z^2$ and $p_x = -i\hbar\,\partial/\partial x$, etc.). The operators are assumed linear in general, obeying the distribution law as $\hat{A}(c_1 f_1 + c_2 f_2) = c_1\hat{A}f_1 + c_2\hat{A}f_2$, etc., and associative, $\hat{A}\hat{B}f(\mathbf{r}) = \hat{A}[\hat{B}f(\mathbf{r})]$, and to obey the index law to the extent that $\hat{A}^2 f(\mathbf{r}) = \hat{A}[\hat{A}f(\mathbf{r})]$ indicates repeated application. Thus a function of an operator expandable as a polynomial presents no difficulty and if $\hat{A}f(\mathbf{r}) = af(\mathbf{r})$ the reader may show by expansion that

$$e^{\hat{A}}f(\mathbf{r}) = \cdots = e^{a}f(\mathbf{r})$$

Assuming linearity, if $\hat{A}f_1 = a_1 f_1$ and $\hat{A}f_2 = a_2 f_2$, then $c_1 f_1 + c_2 f_2$ is not generally an eigenfunction of $\hat{A}$ but it is such if $a_1 = a_2 = a$ with $\hat{A}(c_1 f_1 + c_2 f_2) = a(c_1 f_1 + c_2 f_2)$, and any linear combination of eigenfunctions that are degenerate (give the same eigenvalue) is also an eigenfunction. A further basic demonstration is that if f_1 and f_2 are non-degenerate and have real eigenvalues then $\int f_1^*(\mathbf{r}) f_2(\mathbf{r})\, \mathrm{d}\tau = 0$ and f_1 and f_2 are said to be orthogonal.

If f is an eigenfunction common to both $\hat{A}$ and $\hat{B}$ with $\hat{A}f = af$ and $\hat{B}f = bf$, then $\hat{A}\hat{B}f = \hat{A}(\hat{B}f) = \hat{A}bf = b(\hat{A}f) = baf$ and similarly $\hat{B}\hat{A}f = baf$ since $[a, b] = 0$. Conversely, it is assumed that if $[\hat{A}, \hat{B}] = 0$ then the eigenvalues of $\hat{A}$ and of $\hat{B}$ can be determined simultaneously and precisely. For example, using $\mathbf{L} = \mathbf{r} \times \mathbf{p}$, the $\hat{p}_i$ and the correspondence principle, the operators $\hat{L}_i$ and $\hat{L}^2 = \sum \hat{L}_i^2 (i = x, y, z)$ can be derived, e.g. $\hat{L}_z = -i\hbar\, \partial/\partial\phi$ simply due to the choice in spherical coordinate systems, with ϕ only being involved for rotations around OZ. It can then be demonstrated that

$$[\hat{L}^2, \hat{L}_z] = 0 = [\hat{\mathcal{H}}, \hat{L}^2], \qquad [L_z, L_x] \neq 0, \qquad [L_z, L_y] \neq 0 \tag{1}$$

specifically for a central potential, $V(\mathbf{r}) \to V(r)$, as for the hydrogen atom, so that $\hat{\mathcal{H}}$, $\hat{L}^2$ and $\hat{L}_z$ represent the list of observables determined as eigenvalues, the constants of the motion. (This term arises because wave functions have a time-dependence but the energy, etc., remains constant.) Beyond this list, information can still be obtained in the form of mean values.

2.1 General Formulation

It is initially inferred from the Stern–Gerlach experiments (see Section 1.1, Chapter 3) that an electron has a spin angular momentum that can have just two components, $\frac{1}{2}\hbar$ or $-\frac{1}{2}\hbar$ or $m_s\hbar$, $m_s = \pm\frac{1}{2}$, along a prescribed axis (OZ, say). Dealing with s-state electrons ($L = 0$) in their ground states, the electrons may differ in no other way and so may reasonably be said to be in states defined by these values and identified symbolically by state symbols $|\frac{1}{2}\rangle$ (or $|\alpha\rangle$) and $|-\frac{1}{2}\rangle$ (or $|\beta\rangle$). It is further inferred that other states may, after all, exist, e.g. those with definite components w.r.t a different axis, OX say, symbolically $|x, \frac{1}{2}\rangle$ or $|x, -\frac{1}{2}\rangle$ and also that these behave as if the electron had, at the same time, an equal probability of being in the states $|\frac{1}{2}\rangle$ and $|-\frac{1}{2}\rangle$. The only way to account for this seems to be to assume that the $|\pm\frac{1}{2}\rangle$ (implicitly the $|z, \pm\frac{1}{2}\rangle$) are to be regarded arbitrarily as the 'primary' states and that the $|x, \pm\frac{1}{2}\rangle$ or any other states are somehow 'made up of' the primary states. In order to formulate this concept in mathematical terms it is suggested that the states are associated with vectors of a general nature so that $|\ \rangle$ now stands for a vector or state vector and $|\frac{1}{2}\rangle$ and $|-\frac{1}{2}\rangle$ are the basis states in terms of which any single spin state can be written as a linear combination or superposition:

$$|\chi\rangle = c_1|\tfrac{1}{2}\rangle + c_2|-\tfrac{1}{2}\rangle \tag{2}$$

This particular vector space is thus a linear space of dimension 2.

It is recalled (see Appendix 2) that the important feature of vector space is not any intrinsic attribute of the entities called vectors, or what physical quantities may be associated with them, but the rules relating the vectors to each other, in particular the way in which each is linearly related to the others or to a minimal (linearly independent) number of vectors called the basis vectors. (Linear independence ensures that one cannot be eliminated in terms of the others: the minimal number required to give all vectors in the space, i.e. to span the space, is the dimension of the space.) It is apparent that in the present case the choice of basis was arbitrary and simply corresponded to the initial choice of the axis.

The otherwise vague state symbols are now to be regarded as, or associated with, vectors, and 'state' and 'state vector' may be used equivalently once the meaning is clear. The quantum mechanical operators or transforms are defined within the space of state vectors, inasmuch as they give different state vectors in that space, $\hat{A}|i\rangle = |j\rangle$, and in particular if $\hat{A}|i\rangle = a_i|i\rangle$ or $\hat{A}|a_i\rangle = a_i|a_i\rangle$ then $|a_i\rangle$ is an eigenvector of $\hat{A}$ and a_i one of its eigenvalues.

As discussed in Appendix 2, a space of column matrices $\boldsymbol{\chi}$ has a one-to-one association with the space of general vectors $|\chi\rangle$ which are now referred to as state vectors (or just states since the symbol implies the meaning). Adjoints $\boldsymbol{\chi}^\dagger$ (row matrices) are obtained from the $\boldsymbol{\chi}$ and any matrix $\mathbf{A}$ which operates on the $\boldsymbol{\chi}$ has adjoint $\mathbf{A}^\dagger$ which operates equivalently on the $\boldsymbol{\chi}^\dagger$ (as $\boldsymbol{\chi}^\dagger\mathbf{A}^\dagger$) so that, in turn, adjoint states $\langle\chi|$ may be associated with the $\boldsymbol{\chi}^\dagger$ (and thus in a one-to-one manner with the $|\chi\rangle$). In this indirect manner operators $\hat{A}^\dagger$ on the $\langle\chi|$ (as $\langle\chi|\hat{A}^\dagger$) may be considered adjoints of the operators $\hat{A}$ on the $|\chi\rangle$.

The concepts of normalization and orthogonality of states also become more meaningful in relation to the corresponding rules for matrices, with orthonormality indicated by $\langle i|j\rangle = \delta_{ij}$. Normalization may generally be considered to be contrived while orthogonality arises in connection with the demonstration (Appendix 2) that $\langle a_i|a_j\rangle = 0$ if $\hat{A}|a_i\rangle = a_i|a_i\rangle$, $\hat{A}|a_j\rangle = a_j|a_j\rangle$ and $a_i \neq a_j$. If, for example, $|\chi\rangle = a_1|1\rangle + a_2|2\rangle$ and $|\varphi\rangle = b_1|1\rangle + b_2|2\rangle$ then multiplication on the left by $\langle 1|$ and then by $\langle 2|$ of the equation $A|\chi\rangle = |\varphi\rangle$ gives equations that combine as $\mathbf{A}\boldsymbol{\chi} = \boldsymbol{\varphi}$, in which $A_{ij} = \langle i|\hat{A}|j\rangle$, $\boldsymbol{\chi} = (a_1a_2)^{\mathrm{T}}$ and $\boldsymbol{\varphi} = (b_1b_2)^{\mathrm{T}}$. $\mathbf{A}$ is called the matrix representation, or just the matrix, of $\hat{A}$. All the rules that apply to the $\hat{A}$, $\hat{A}^\dagger$, $|\ \rangle$, $\langle\ |$ are those that apply to the $\mathbf{A}$, $\mathbf{A}^\dagger$, $\boldsymbol{\chi}$, $\boldsymbol{\chi}^\dagger$. $\hat{A}$ gives real eigenvalues if $\mathbf{A}^\dagger = \mathbf{A}$ which applies if $\hat{A}^\dagger = \hat{A}$. The reader may demonstrate that if $[\hat{A}, \hat{B}] = 0$ then $[\mathbf{A}, \mathbf{B}] = 0$.

In the study of spins as such (with quenched or non-existent orbital angular momentum, e.g. s-state electrons, protons) the state vectors receive primary consideration and the operators are first considered in matrix form (see Section 1.3, Chapter 2). Operators such as the $\hat{S}_i$ in the vector space are defined as entities which operate in correspondence with the matrices. Alternatively, in wave mechanics explicit operators are derived and eigenfunctions found by solving differential equations. Wave functions expanded as, for example, $\psi(\mathbf{r}) = c_1\varphi_1(\mathbf{r}) + c_2\varphi_2(\mathbf{r})$ (or $\sum_{i=1}^{n} c_i\varphi_i$, where n may not be finite) show clear similarities to the state vectors and, with certain restrictions which often apply naturally when the φ_i are eigenfunctions, can be considered to constitute a vector space. Orthonormality of the basis is now defined by $\int \varphi_i^*(\mathbf{r})\varphi_j(\mathbf{r})\mathrm{d}\tau = \delta_{ij}$ and column matrices $\boldsymbol{\psi} = (c_1c_2\ldots)^{\mathrm{T}}$ can again be associated with the $\psi(\mathbf{r})$ according to the prescription

$$c_i = \int \varphi_i^*(\mathbf{r})\psi(\mathbf{r})\mathrm{d}\tau \qquad (\text{cp. } c_i = \langle i|\chi\rangle)$$

and similarly matrices of operators are expressed as

$$A_{ij} = \int \varphi_i^*(\mathbf{r})\hat{A}\varphi_j(\mathbf{r})\mathrm{d}\tau \qquad (\text{cp. } \langle i|\hat{A}|j\rangle)$$

multiplication by $\langle i|$ being replaced, in the foregoing, by multiplication by φ_i^* together with integration over all space. Since the $\psi(\mathbf{r})$ are associated with the $\boldsymbol{\psi}$ which may then be associated with the states $|\psi\rangle$ the association $\psi(r) \sim |\psi\rangle$ also arises. In the wave mechanical approach the numbers which are the matrix elements may be calculated from the explicit operators and functions, whereas in the approach more appropriate to spins the matrices are derived or inferred initially and the effects of the operators follows, with no necessity to assign explicit expressions to these entities.

2.2 The Density Operator and Density Matrix

As noted, any entity of the form $|\ \rangle\langle\ |$ constitutes an operator. The density operator, sometimes itself called the density matrix, is defined for any state $|\chi\rangle$ as

$$\rho = |\chi\rangle\langle\chi| \tag{3}$$

and is of particular interest and value. It is Hermitian because

$$\rho^\dagger = (|\chi\rangle\langle\chi|)^\dagger = ((\langle\chi|^\dagger)(|\chi\rangle^\dagger) = |\chi\rangle\langle\chi|$$

but this is not of great interest since its eigenvector is obviously $|\chi\rangle$ with eigenvalue unity. Maintaining the $n = 2$ illustration, the matrix of ρ is

$$\rho = \begin{pmatrix} \langle 1|\chi\rangle\langle\chi|1\rangle & \langle 1|\chi\rangle\langle\chi|2\rangle \\ \langle 2|\chi\rangle\langle\chi|1\rangle & \langle 2|\chi\rangle\langle\chi|2\rangle \end{pmatrix} = \begin{pmatrix} c_1c_1^* & c_1c_2^* \\ c_2c_1^* & c_2c_2^* \end{pmatrix} \tag{4}$$

since $\langle 1|$ selects c_1 from $|\chi\rangle = c_1|1\rangle + c_2|2\rangle$, etc. Generalization is obvious; ρ is usually referred to as the density matrix with the distinction between the operator and its matrix remaining implicit.

The way in which mean values may be related to ρ may be written down and verified:

$$\langle A\rangle_\chi = \mathrm{Tr}\,(\rho A) = \mathrm{Tr}\,(A\rho), \qquad \text{where } \rho = |\chi\rangle\langle\chi| \tag{5}$$

The trace (Tr) of a matrix is the sum of the diagonal elements, while the trace of an operator is the sum of those elements of the representation for which the index is repeated. The reader should demonstrate by simple examples that for any two matrices $A = \{a_{ij}\}$ and $B = \{b_{ij}\}$, the traces of the two products $P_1 = AB$ and $P_2 = BA$ simply contain the same terms in a different order, i.e. $\mathrm{Tr}\,(AB) = \mathrm{Tr}\,(BA)$ regardless of whether $AB = BA$. Expanding the mean value expression:

$$\begin{aligned}\langle A\rangle_\chi &= (c_1^*\langle 1| + c_2^*\langle 2|)A(c_1|1\rangle + c_2|2\rangle) \\ &= c_1^*c_1\langle 1|A|1\rangle + c_1c_2^*\langle 2|A|1\rangle + c_2c_1^*\langle 1|A|2\rangle + c_2c_2^*\langle 2|A|2\rangle\end{aligned}$$

(the basis not being restricted to eigenvectors of A). This is seen to be the result obtained on writing out the matrix $\langle i|A|j\rangle$, taking the product with ρ as in equation (4) and taking the trace. Thus the postulated equation 5 is justified.

Although a mixed state clearly cannot be written as $|\chi\rangle = p_1|\chi_1\rangle + p_2|\chi_2\rangle$ it may be suggested that ρ for the mixed state can be written as

$$\rho = p_1\rho_1 + p_2\rho_2 + \cdots \tag{6}$$

where $\rho_1 = |\chi_1\rangle\langle\chi_1|$, $\rho_2 = |\chi_2\rangle\langle\chi_1|, \ldots$ with probability p_1 of the system being found in $|\chi_1\rangle$, etc. The mean value of A for the mixed state would then be

$$\langle A\rangle_{\chi_1,\chi_2} = \mathrm{Tr}\,(\rho A)$$

$$= \mathrm{Tr}\,[(p_1\rho_1 + p_2\rho_2 + \cdots)A]$$
$$= p_1\,\mathrm{Tr}\,(\rho_1 A) + p_2\,\mathrm{Tr}\,(\rho_2 A) + \cdots$$
$$= p_1\langle A\rangle_{\chi_1} + p_2\langle A\rangle_{\chi_2} + \cdots \tag{7}$$

which is clearly appropriate. The density matrix is thus of central importance in the treatment of mixed states and therefore in quantum statistics and, as seen later, in the evolution in time of such states.

Suppose that a system (electron, atom, etc.) can be expected to be found in just one of two states, because these are eigenstates of $\mathcal{H}$, $|E_1\rangle$ and $|E_2\rangle(\mathcal{H}|E_i\rangle = E_i|E_i\rangle)$ with much lower energies than any other states. Taking these two as a basis the Hamiltonian matrix is simply

$$\mathcal{H} = \begin{pmatrix} E_1 & 0 \\ 0 & E_2 \end{pmatrix} \tag{8}$$

The density matrix for any state $|\chi\rangle = c_1|E_1\rangle + c_2|E_2\rangle$ is as given in equation (4). If the state is specifically $|E_1\rangle$ then $c_1 = 1$ and $c_2 = 0$ and if the state is $|E_2\rangle$, $c_2 = 1, c_1 = 0$. With a probability p_i for the system to be in $|E_i\rangle$ the density matrix for this particular mixture is

$$\rho = p_1\rho_1 + p_2\rho_2 = p_1\begin{pmatrix} 1 & 0 \\ 0 & 0 \end{pmatrix} + p_2\begin{pmatrix} 0 & 0 \\ 0 & 1 \end{pmatrix}$$

As an obvious example, for thermal equilibrium it may be expected that

$$p_1 = \frac{\mathrm{e}^{-E_1/(kT)}}{Z}, \qquad p_2 = \frac{\mathrm{e}^{-E_2/(kT)}}{Z}, \qquad Z = \sum_{i=1}^{2} \mathrm{e}^{-E_i/(kT)}$$

so that

$$\rho = \rho^0 = \frac{1}{Z}\begin{pmatrix} \mathrm{e}^{-E_1/(kT)} & 0 \\ 0 & \mathrm{e}^{-E_2/(kT)} \end{pmatrix} \tag{9}$$

which is the statistical or equilibrium density matrix. It is normalized to the extent that

$$\mathrm{Tr}\,(\rho) = \frac{\sum \mathrm{e}^{-E_i/(kT)}}{Z} = 1$$

The mean value of the energy would be $\bar{E} = \langle\mathcal{H}\rangle = \mathrm{Tr}\,(\rho\mathcal{H}) = p_1E_1 + p_2E_2$ as expected, using equations (8) and (9). The matrix as such in equation (9) (ignoring Z) is the matrix in the basis $|E_i\rangle$ of the operator

$$\rho^0 = \mathrm{e}^{-\mathcal{H}/(kT)} \tag{10}$$

assuming of course that $\mathcal{H}^2|\chi\rangle = \mathcal{H}(\mathcal{H}|\chi\rangle)$, etc. The matrix elements of ρ^0 as given in equation (10) are thus

$$\langle E_i|\mathrm{e}^{-\mathcal{H}/(kT)}|E_0\rangle = \mathrm{e}^{-E_j/(kT)}\langle E_i|E_j\rangle = \delta_{ij}\mathrm{e}^{-E_j/(kT)}$$

as required [equation (9)]. The normalized operator can be written as

$$\rho_N^0 = \frac{\rho^0}{\mathrm{Tr}\,(\rho^0)}, \qquad \text{where } \rho^0 = \mathrm{e}^{-\mathcal{H}/(kT)} \tag{11}$$

In the present context the density matrix is applied mainly to spins, as in the following chapter.

2.3 Angular Momentum

In the absence of charge, angular momentum is not associated with a magnetic moment, but magnetic moments are necessarily associated with angular momentum according to $\boldsymbol{\mu} = \gamma \mathbf{L}$ and the topic is thus of central importance here. The moment of the momentum about a particular axis has the magnitude $L = pr = mvr$ with r the distance between the trajectory and the chosen axis. Since the magnitude of the torque T is force times normal distance from the axis of the torque,

$$T = \frac{\mathrm{d}p}{\mathrm{d}t}r = \frac{\mathrm{d}L}{\mathrm{d}t}$$

for motion in a circle of radius r.

Angular velocity ω is the rate of change of angle : $\omega = \mathrm{d}\theta/\mathrm{d}t$. Using $r\mathrm{d}\theta = \mathrm{d}x$, $\omega r = \mathrm{d}x/\mathrm{d}t = v$ and

$$L = mr^2\omega = I\omega, \qquad \text{where } I = mr^2$$

with I the moment of inertia. Vectorially [Figure 2.1(a)]

$$\mathbf{L} = \mathbf{r} \times \mathbf{p} = m\mathbf{r} \times \mathbf{v}$$

In Figure 2.1(b) the only component of $\mathbf{L}$ is clearly $L_z = xp_y - yp_x$ and similar sketches confirm that

$$L = \begin{vmatrix} \mathbf{i} & \mathbf{j} & \mathbf{k} \\ x & y & z \\ p_x & p_y & p_z \end{vmatrix} = \mathbf{r} \times \mathbf{p}$$

As noted in Section 2, the components of the linear momentum are given by

$$p_q = -i\hbar\frac{\partial}{\partial_q}, \qquad [q, p_q] = i\hbar, \qquad q = x, y, z \tag{12}$$

Noting that $[q, p_s] = 0$ if $q \neq s$, because then the coordinate behaves as a constant w.r.t the partial differentiation, and similarly $[p_q, p_s] = 0$, it is simple though tedious to take the

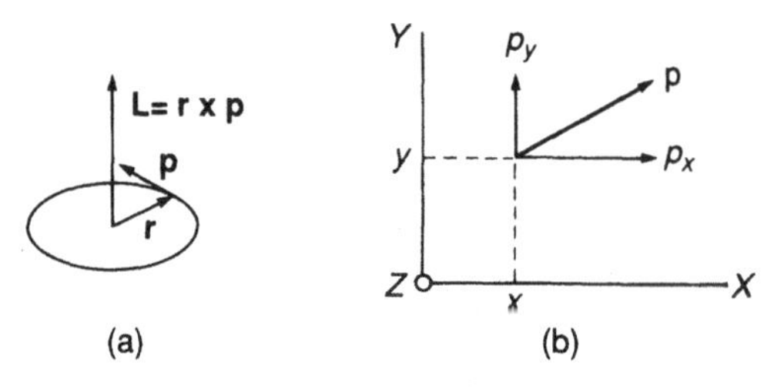

Figure 2.1 The direction of $\mathbf{L}$ in relation to $\mathbf{p}$. In (b) the only $\mathbf{L}$ component is L_z

components from equations (12) and expand the commutators to show that

$$\begin{aligned}[L_x, L_y] &= (yp_z - zp_y)(zp_x - xp_z) - \cdots \\ &= i\hbar(xp_y - yp_x) = i\hbar\ L_z \\ [L_y, L_z] &= i\hbar\ L_x \\ [L_z, L_x] &= i\hbar\ L_y\end{aligned} \tag{13}$$

The cyclic nature of these may be expressed symbolically by '$x,\ y \to z;\ y,\ z \to x;\ z,\ x \to y$'. These can be seen, by expanding the product, to correspond to

$$\mathbf{L}\times\mathbf{L} = i\mathbf{L}. \tag{14}$$

Due to the generality of the derivation these commutation relations may be taken to define the quantal angular momentum whatever its nature. In particular they apply to both orbital and spin motion. The relations of equation (13) can be used to indicate that

$$[L^2, L_q] = 0 \qquad \text{for } q = x, y, z \tag{15}$$

Further demonstrations which the reader may complete, at a little length, are that

$$[\nabla^2, L_q] = 0, \qquad [\nabla^2, L^2] = 0 \qquad \text{for } q = x, y, z \tag{16}$$

so that the first term in $\mathcal{H}$ for a single electron atom commutes with these operators (taking $\mathcal{H} = -(\hbar^2/2m)\nabla^2 + V(\mathbf{r})$ simply). The second term, $V(\mathbf{r})$ or $V(r, \theta, \phi)$, will not generally comply with this. However, on transferring to polar coordinates

$$L_x = i\hbar\left(\sin\phi\frac{\partial}{\partial\theta} + \cot\theta\cos\phi\frac{\partial}{\partial\phi}\right) \tag{17a}$$

$$L_y = i\hbar\left(-\cos\phi\frac{\partial}{\partial\theta} + \cot\theta\sin\phi\frac{\partial}{\partial\phi}\right) \tag{17b}$$

$$L_z = -i\hbar\frac{\partial}{\partial\phi} \tag{17c}$$

and since these do not contain r, if $V(\mathbf{r}) \to V(r)$, then

$$[V(r), L_q] = 0, \qquad [V(r), L^2] = 0 \tag{18}$$

The condition $V = V(r)$ corresponds to spherical symmetry or a central potential as for an electron in a hydrogen atom: $V(r) = -\mathrm{e}^2/(4\pi\varepsilon_0 r)$, with r the relative coordinate of the electron in relation to the nucleus if m in $-\hbar^2/(2m)$ is taken to be the reduced mass of the electron and proton. For two (or more) electrons the radial symmetry will be broken by an interaction term $V(\mathbf{r}_1, \mathbf{r}_2)$. However, the interaction may be approximately accounted for by assuming that, for electron (1), the presence of electron (2) may be represented by a charge distribution which is approximately spherically symmetric and $V(\mathbf{r}_1)$ can be taken as $-Z'e^2/(4\pi\varepsilon_0 r)$ with $Z'e$ the shielded nuclear change (e.g. $Z' = \frac{5}{4}$ for He).

In those cases where a central potential can be assumed then, it is accepted that $\mathcal{H}$ commutes with L^2, L_x, L_y and L_z, but it has been noted that the L_q do not commute among themselves. Thus the constants of the orbital motion are E, L^2 and L_z, the latter being chosen because the conventional choice of spherical coordinates leads to the remarkably simple expression in equation (17c). (This could almost have been guessed by comparison with $p_x = -i\hbar\partial/\partial x$ since it is the only variation of the coordinate ϕ that gives a component of angular momentum along OZ and the replacement of x by ϕ at least seems natural.)

If m is an eigenvalue of L_z and $|m\rangle$ the corresponding eigenket, we have $L_z|m\rangle = m|m\rangle$ or equivalently $L_z\Phi_m = -i\hbar\, \partial\Phi_m/\partial\phi = m\Phi_m$, where Φ_m is the eigenfunction. By separating the variables, or just by inspection, the solution is

$$\Phi_m(\phi) = A\mathrm{e}^{im\phi} \tag{19}$$

If it is assumed that Φ is a single-valued function (which is in fact true though not apparent since the only general rule is that the probability density $\Phi\Phi^*$ must be continuous) then a rotation of 2π must leave Φ unchanged and

$$\mathrm{e}^{im\phi} = \mathrm{e}^{im(\phi+2\pi)} = \mathrm{e}^{im\phi}\mathrm{e}^{i2\pi m}, \qquad \text{where } \mathrm{e}^{i2\pi m} = 1$$

Thus $\cos 2\pi m + i\sin 2\pi m = 1$, enforcing

$$m = 0, \pm 1, \pm 2, \ldots$$

The normalization constant is given by $\int_0^{2\pi} \Phi_m^*(\phi)\Phi_m(\phi)\mathrm{d}\phi = 1$ as $A = \sqrt{1/2\pi}$ for any m and thus

$$\Phi_m(\phi) = \sqrt{\frac{1}{2\pi}}\mathrm{e}^{im\phi} \tag{20}$$

L^2 can be written in spherical coordinates using equations (17) as

$$\begin{aligned} L^2 &= -\hbar^2\left[\frac{1}{\sin\theta}\frac{\partial}{\partial\theta}\left(\sin\theta\frac{\partial}{\partial\theta}\right) - \frac{1}{\sin^2\theta}\frac{\partial^2}{\partial\phi^2}\right] \\ &= -\hbar^2\left[\frac{1}{\sin\theta}\frac{\partial}{\partial\theta}\left(\sin\theta\frac{\partial}{\partial\theta}\right) + \frac{1}{\sin^2\theta}\frac{L_z^2}{\hbar^2}\right] \end{aligned} \tag{21}$$

This itself indicates the commutation with L_z and also that the eigenfunctions corresponding to $|l\ m_l\rangle$ must be functions of θ and ϕ but not of r. Further, since ϕ does not occur other than in L_z^2 the functions must have the form

$$Y_{lm}(\theta, \phi) = P_{lm}(\theta)\Phi_m(\phi) = P_{lm}(\theta)\mathrm{e}^{im\phi}$$

which are explicitly eigenfunctions of L_z. The functions satisfying $L^2Y_{lm}(\theta,\phi) = \hbar^2 l(l+1)Y_{lm}(\theta,\phi)$ and $L_zY_{lm}(\theta,\phi) = \hbar mY_{lm}(\theta,\phi)$ are normalized spherical harmonics and examples have been given in Chapter 1, Section 3. While no attempt is made to indicate the method of solution they may be confirmed by substitution. The θ dependence is given by associated Legendre functions, expressed as

$$P_{lm}(\cos\theta) = \frac{(-1)^l}{2^l l!}\sin^2\theta\frac{\mathrm{d}^{l+m}\sin^{2l}\theta}{\mathrm{d}(\cos\theta)^{l+m}} \qquad (0 \leqslant m \leqslant l)$$

and the normalized functions are

$$Y_{lm}(\theta,\phi) = \frac{1}{\sqrt{2\pi}}\sqrt{\frac{2l+1}{2}\frac{(l-m)!}{(l+m)!}}P_{lm}(\cos\theta)\mathrm{e}^{im\phi} \tag{22}$$

Using the expression for ∇^2 in spherical coordinates (Appendix 2) the Hamiltonian for one electron in a central potential is

$$\mathcal{H} = \frac{-\hbar^2}{2m_e}\left[\frac{1}{r^2}\frac{\partial}{\partial r}\left(r^2\frac{\partial}{\partial r}\right) + \frac{1}{r^2 \sin\theta}\frac{\partial}{\partial\theta}\left(\sin\theta\frac{\partial}{\partial\theta}\right) + \frac{1}{r^2\sin^2\theta}\frac{\partial^2}{\partial\phi^2}\right] + V(r)$$

and with reference to equation (21) this can be written as

$$\mathcal{H} = \frac{-\hbar^2}{2m_e}\frac{1}{r^2}\frac{\partial}{\partial r}\left(r^2\frac{\partial}{\partial r}\right) + \frac{L^2}{2m_e r^2} + V(r) \tag{23}$$

Since the eigenfunctions of $\mathcal{H}$ are also eigenfunctions of L^2 and it is seen that $\mathcal{H}$ does not depend on θ or ϕ other than in L^2, the wave functions must have the form

$$\psi_{nlm}(r, \theta, \phi) = R_{nl}(r)Y_{lm}(\theta, \phi) \tag{24}$$

Since the effect of L^2 [in equation (23)] on Y_{lm} and thus on ψ is known, $\mathcal{H}\psi = E\psi$ becomes an ordinary differential equation in r only with solutions $R(r)$. The energies are found to be

$$E_n = -\tfrac{1}{2}\frac{Z^2e^2}{a_0}\frac{1}{n^2}$$

where n is an integer and $n > l$ and $a_0 = \hbar^2/(m_e e^2)$ is the Bohr radius: the (gross structure) energies for the hydrogen atom are thus those found by the early Bohr theory. The radial functions may be expressed in terms of $\rho = \alpha_n r$, with $\alpha_n^2 = 8m_e|E_n|/\hbar^2$, as

$$R_{nl}(r) = N_{nl}e^{-\rho/2}L_{n+1}^{2l+1}(\rho) \tag{25}$$

in which the $L(\ \ )$ are associated Laguerre polynomials.

The states may now be designated $|n\ l\ m_l\rangle \sim \psi_{nlm_l}$. States with $n = 1, 2, 3, \ldots$ may be designated $K, L, M, \ldots$ states, and the states with specific n and l (and any m_l) are indicated by ns, np, nd, nf, etc., in which n is replaced by the relevant number and s, p, d, $f, \ldots$ correspond to $l = 0, 1, 2, 3, \ldots$. It is important in a magnetic context (e.g. in n.m.r.) to note that for s states ($l = 0$), ψ is finite at the origin (i.e. at the nucleus). It is seen that the probability $|\psi|^2$ at the origin, $r = 0$, is proportional to $(1/a_0)^3$ for the s states.

For each value of n there are n permitted values of $l = 0, 1, 2, \ldots, n-1$ and for each l there are $2l + 1$ permitted values of $m_l = 0, \pm 1, \ldots, \pm l$. The total number of distinct states for a given n, i.e. a given energy, is thus

$$c_1 = \sum_{l=0}^{n-1}(2l+1) = 1 + 3 + 5 + \cdots + (2n-1) = n^2$$

i.e. the energy levels have this degeneracy. (There is an essential or geometric degeneracy for all central potential situations of $2l + 1$ associated with the arbitrary assignment of an axis in space and here, specifically with the Coulomb potential, the degeneracy is higher than this.)

Together with the selection rules, which state that l must change by ± 1 simultaneously with a change in n and are thus trivial in this case, the foregoing accounts for the gross structure of the hydrogen spectrum and gives a crude approximation to that of 'effective one-electron atoms' such as Na.

2.3.1 Angular momentum raising and lowering operators

The more general treatment of any quantal angular momentum satisfying $\mathbf{J}\times\mathbf{J} = i\mathbf{J}$ introduces the new operators (which are not Hermitian)

$$J^+ = J_x + iJ_y, \qquad J^- = J_x - iJ_y \tag{26}$$

These satisfy, for example,

$$\begin{aligned}[J_z, J^+] &= J_zJ_x + iJ_zJ_y - J_xJ_z - iJ_yJ_z \\ &= [J_z, J_x] + i[J_z, J_y] \\ &= J_x + iJ_y\end{aligned}$$

i.e.

$$[J_z, J^+] = J^+, \qquad [J_z, J^-] = -J^-, \qquad [J^+, J^-] = 2J_z \tag{27}$$

the latter being readily demonstrated. Here $\hbar$ is implied; i.e. only the numbers giving the angular momentum are considered.

The first of equations (27) gives $J_zJ^+ = J^+J_z + J^+$ and applying this to the states $|n\ m\rangle$, such that $J^2|n\ m\rangle = n|n\ m\rangle$ and $J_z|n\ m\rangle = m|n\ m\rangle$, gives $J_z(J^+|n\ m\rangle) = J^+m|n\ m\rangle + J^+|n\ m\rangle = (m+1)(J^+|n\ m\rangle)$. Thus $J^+|n\ m\rangle$ is an eigenvector of J_z which gives the eigenvalue $m+1$ if J_z is assumed to give the eigenvalue m. Note that in this treatment the latter assumption does not imply any initial restrictions on m. Using the second of the equations (27) it is readily seen that $J^-|n\ m\rangle$ is similarly an eigenvector of J_z with eigenvalue $m-1$:

$$\begin{aligned}J^+|n\ m\rangle &= \alpha_+|n\ (m+1)\rangle \\ J^-|n\ m\rangle &= \alpha_-|n\ (m-1)\rangle\end{aligned}$$

where the $\alpha_{+/-}$ are to be determined. Thus the eigenvalues of J_z differ by unity (the actual quantities by $\hbar$). Because J^+ generates the state with eigenvalue $m+1$ from that with m it is called a raising operator while J^- is a lowering operator. They are very widely useful, as are the analogous spin operators S^+ and S^-.

As m is raised by successive integers a maximum value, say m', must be reached for which a further increment would lead to a value of J_z^2 which exceeded J^2, which is not, of course, permissible. However, the equation $J_z(J^+|n\ m\rangle) = (m+1)(J^+|n\ m\rangle)$ is satisfied for the particular case $J^+|n\ m\rangle = 0$ and so it is inferred that this applies when $m = m'$, $J^+|n\ m'\rangle = 0$. Equally for some minimum value m'', $J^-|n\ m''\rangle = 0$.

Now

$$\begin{aligned}J^-J^+ &= (J_x - iJ_y)(J_x + iJ_y) \\ &= J_x^2 + J_y^2 + i(J_xJ_y - J_yJ_x) \\ &= J_x^2 + J_y^2 - J_z \\ &= J^2 - J_z^2 - J_z\end{aligned}$$

so that

$$J^- J^+ |n\ m'\rangle = (J^2 - J_z^2 - J_z)|n\ m'\rangle = [n - m'(m'+1)]|n\ m'\rangle = 0$$

$$J^+ J^- |n\ m''\rangle = [n - m''(m''-1)]|n\ m''\rangle = 0$$

so that

$$n - m'(m'+1) = 0 \tag{28a}$$

$$n - m''(m''-1) = 0 \tag{28b}$$

(the second following similarly). Subtracting these and rearranging,

$$(m' + m'')(m'' - m' - 1) = 0$$

and since $m' > m''$ by implication, the first bracket must be zero, i.e.

$$m'' = -m'$$

The values of m are thus symmetrically spaced in pairs centred on zero and since they differ by unity must form series such as

$$\begin{array}{ccccccccc} -\frac{5}{2} & -\frac{3}{2} & -\frac{1}{2} & & \frac{1}{2} & \frac{3}{2} & \frac{5}{2} & (m' = \frac{5}{2}) \\ -3 & -2 & -1 & 0 & 1 & 2 & 3 & (m' = 3) \end{array}$$

Set $m' = j$, $m'' = -j$. $m' - m'' = 2j$ must be an integer, i.e. j can only be integral or half-integral, as suggested above:

$$j = 0, \tfrac{1}{2}, 1, \tfrac{3}{2}, \ldots$$

Equation (28a) gives the eigenvalue of J^2, i.e. gives n, corresponding to the maximum $m = m' = j$ as

$$n = j(j+1)$$

The states may now be labelled $|j\ m\rangle$ with

$$J^2|j\ m\rangle = j(j+1)|j\ m\rangle, \qquad J_z|j\ m\rangle = m|j\ m\rangle$$

In this general treatment j may be assigned any integral or half-integral value and for each j there are $2j+1$ values of m:

$$\begin{array}{ll} m = \pm j, \pm(j-1), \ldots, 0 & (j \text{ integral}) \\ m = \pm j, \pm(j-1), \ldots, \pm\frac{1}{2} & (j \text{ half-integral}) \end{array}$$

Only $\mathbf{J}\times\mathbf{J} = i\mathbf{J}$, and its consequences, has been used. However, some connection must be made with other aspects of the theory of orbital motion or with spectroscopic or other observations. It is then inferred that, for orbital motion ($J \to L$), e.g. in the hydrogen atom, the states occur in groups designated $|n\ l\ m\rangle$, with n now the principle quantum number, in which L is integral and restricted by the specified value of the principle quantum number n.

Conversely, for a freely moving electron, there is clear evidence (see Chapter 3) of an angular momentum that is not associated with orbital or spatial motion and, moreover, can only take two components w.r.t. a specified 'z' axis. This requires that $j = \frac{1}{2}$ only with $m = \pm\frac{1}{2}$. To stress the distinction, this alternative mode of motion is called electron spin and the symbol j is replaced by S, with $S = \frac{1}{2}$.

Returning to J^+ and J^- generally, particularly to the factors α_+ and α_-, it is considered adequate to quote these, giving specifically

$$J^+|j\ m\rangle = [(j-m)(j+m+1)]^{1/2}|j\ (m+1)\rangle \tag{29a}$$

$$J^-|j\ m\rangle = [(j+m)(j-m+1)]^{1/2}|j\ (m-1)\rangle \tag{29b}$$

and noting the automatic satisfaction of $J^+|j\ m\rangle = 0$ for $m = j$ and of $J^-|j\ m\rangle = 0$ for $m = -j$.

Matrices for L^+ and L^- are easily derived noting that L^+ connects only states such as $\langle L\ m|$ with $|L\ (m-1)\rangle$ and only the $\langle L\ m|L^-|L\ (m+1)\rangle$ are non-zero. For $L = 1$ all the numerical factors are $1/\sqrt{2}$ (or zero):

$$L^+ = \frac{\hbar}{\sqrt{2}}\begin{pmatrix} 0 & 1 & 0 \\ 0 & 0 & 1 \\ 0 & 0 & 0 \end{pmatrix}, \qquad L^- = \frac{\hbar}{\sqrt{2}}\begin{pmatrix} 0 & 0 & 0 \\ 1 & 0 & 0 \\ 0 & 1 & 0 \end{pmatrix} \tag{30}$$

ordering the basis states as $|1\ 1\rangle, |1\ 0\rangle, |1\ -1\rangle$, and reintroducing $\hbar$ at this stage. These permit the derivations of $L_x = \frac{1}{2}(L^+ + L^-)$ and $L_y = 1/(2i)(L^+ - L^-)$ (while L_z can be written directly):

$$L_x = \frac{\hbar}{\sqrt{2}}\begin{pmatrix} 0 & 1 & 0 \\ 1 & 0 & 1 \\ 0 & 1 & 0 \end{pmatrix}, \qquad L_y = \frac{i\hbar}{\sqrt{2}}\begin{pmatrix} 0 & -1 & 0 \\ 1 & 0 & -1 \\ 0 & 1 & 0 \end{pmatrix}, \qquad L_z = \begin{pmatrix} 1 & 0 & 0 \\ 0 & 0 & 0 \\ 0 & 0 & -1 \end{pmatrix} \tag{31}$$

for the example of $L = 1$. The corresponding state matrices are

$$|1\ 1\rangle = \begin{pmatrix} 1 \\ 0 \\ 0 \end{pmatrix}, \qquad |1\ 0\rangle = \begin{pmatrix} 0 \\ 1 \\ 0 \end{pmatrix}, \qquad |1\ -1\rangle = \begin{pmatrix} 0 \\ 0 \\ 1 \end{pmatrix} \tag{32}$$

It is a simple exercise to derive the matrix for J^2, to check that the matrices themselves conform to the commutation relations and to show that $\langle L_x\rangle = \langle L_y\rangle = 0$ for all three states. This is left to the reader. L^+ and L^- may be replaced by S^+ and S^- (e.g. for $S = 1$) and the matrices S_x, S_y and S_z follow as in equation 31.

3 The Hamiltonian for a Charged Particle in a Magnetic Field

According to the correspondence principle the Hamiltonian for a particle moving in a field-free space is related to the operator for linear momentum by $\mathcal{H} = p^2/(2m)$, the correspondence being with $p = mv$ and (kinetic) energy $= \frac{1}{2}mv^2$. If the particle has charge q and moves in a magnetic field $\mathbf{B}$, i.e. is in the presence of a magnetic potential $\mathbf{A} : \mathbf{B} = \nabla\times\mathbf{A}$, then

$$\mathcal{H} = \left(\frac{1}{2m}\right)(\mathbf{p} - q\mathbf{A})^2 \tag{33}$$

which in view of its central importance calls for some elucidation.

In effect the classical momentum (for $A = 0$), i.e. $p_x = m\dot{x}$ is replaced, in single-component form, by

$$p_x = m\dot{x} - qA_x \tag{34}$$

Taking the force to be the time-derivative of the momentum,

$$F_x = m\ddot{x} - \frac{\mathrm{d}}{\mathrm{d}t}qA_x = F_x^{\mathrm{N}} - F_x^{\mathrm{B}} \tag{35}$$

As suggested, the first term can be considered the Newtonian force (mass times acceleration) and the second the magnetic force. Considering equation (33) as a postulate, it must be shown that this is reasonable.

Neither q, nor $\mathbf{A}$ itself, are functions of time, the field being implicitly static, but the charge is moving in a space in which $\mathbf{A}$ varies. $\mathbf{A}$ may be taken to vary linearly, so as to give a uniform field, so that the replacements may be made:

$$\frac{\mathrm{d}}{\mathrm{d}t}qA_x = \frac{\delta(qA_x)}{\delta t} = q\frac{\delta A_x}{\delta t} \tag{36}$$

where the δ's represent small changes: if these are very small and $\mathbf{A}$ varies smoothly the restriction to uniform $\mathbf{B}$ can be dropped. For simplicity suppose that A_x is the only component existing and that it varies only with z, $A_x \equiv A_x(z)$:

$$\mathbf{B} = \begin{vmatrix} \mathbf{i} & \mathbf{j} & \mathbf{k} \\ — & — & \dfrac{\partial}{\partial z} \\ A_x & 0 & 0 \end{vmatrix} = \mathbf{j}\frac{\partial A_x}{\partial z}$$

and $\mathbf{B}$ is directed along OY (Figure 2.2). Suppose that q moves from O to p in time δt, in the plane $OXZ \perp \mathbf{B}$ and that δA_x is the consequent change in A_x:

$$q\frac{\delta A_x}{\delta t} = q\frac{\delta A_x}{\delta z}\frac{\delta z}{\delta t} \rightarrow qv_z\frac{\partial A_x}{\partial z} \tag{37}$$

returning to differentials as $\delta z, \delta t \rightarrow 0$. The general force is $\mathbf{F}^B = q\mathbf{v}\times\mathbf{B}$, in this case

$$\mathbf{F}^B = -\mathbf{i}v_z B_y = -\mathbf{i}v_z\frac{\partial A_x}{\partial z} \tag{38}$$

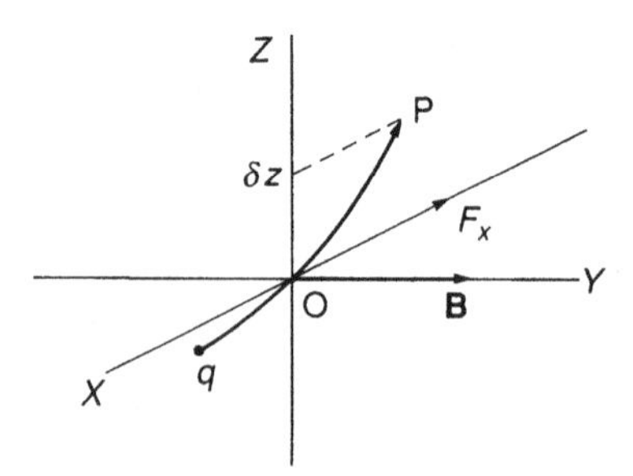

Figure 2.2 The trajectory is in OXZ with motion $O \rightarrow P$ in time δt

Displacements δy are irrelevant and displacements δx do not affect F_x^B. Comparing equations (35), (37) and (38), it is seen that the postulated 'effective momentum' [equation (34)] and thus the magnetic field Hamiltonian are verified.

The reader may choose to generalize the foregoing for any field direction and also to demonstrate that the great difference between this case and that of motion in an electric scalar potential φ (which simply calls for the addition of a potential energy term $V = q\varphi$ to the kinetic energy term as given) derives from the different ways in which the terms depend on the *motion* of the particle through the fields.

3.1 Motion of Free Electrons in a Magnetic Field

The Hamiltonian ($V = 0$) is

$$\mathcal{H} = \frac{(\mathbf{p} + e\mathbf{A})^2}{2m} \tag{39}$$

If $\mathbf{B}$ is taken to be $\mathbf{k}\ B$ then $\mathbf{A}$, corresponding to $\mathbf{B} = \nabla\times\mathbf{A}$, must lie in OXY with no z component so that

$$\begin{aligned}\mathcal{H} &= \frac{(p_x + eA_x)^2}{2m} + \frac{(p_y + eA_y)^2}{2m} + \frac{p_z^2}{2m}\\ &= \mathcal{H}_{xy} + \mathcal{H}_z\end{aligned} \tag{40}$$

with $\mathcal{H}_z = p_z^2/2m$. This can be abbreviated to

$$\mathcal{H} = \frac{P_x^2}{2m} + \frac{P_y^2}{2m} + \frac{p_z^2}{2m} \tag{41}$$

with

$$\begin{aligned}P_x &= p_x + eA_x\\ P_y &= p_y + eA_y\end{aligned} \tag{42}$$

The commutator $[P_x, P_y]$ is of interest:

$$[P_x, P_y] = (p_x + eA_x)(p_y + eA_y) - (p_y + eA_y)(p_x + eA_x)$$

Noting that

$$[p_x, p_y] = 0, \qquad [A_x, A_y] = 0$$

the latter being inferred because of the way in which $\mathbf{A}$ may be related to a particle velocity, i.e. to $\mathbf{p}/m$, rearrangement gives

$$\begin{aligned}[P_x, P_y] &= -i\hbar e\left(-\frac{\partial A_y}{\partial x} + \frac{\partial A_x}{\partial y}\right)\\ &= -i\hbar e(\nabla\times\mathbf{A})_z = -i\hbar eB\end{aligned} \tag{43}$$

using $p_x = -i\hbar\, \partial/\partial x$, etc. It follows that

$$\left[(eB)^{-1/2}P_x, -(eB)^{-1/2}P_y\right] = i\hbar, \qquad \text{i.e. } [X, P] = i\hbar$$

so that

$$X = (eB)^{-1/2} P_x, \qquad P = -(eB)^{-1/2} P_y \tag{44}$$

behave as conjugate variables with $X \sim x$ and $P \sim p_x$, recalling that $[x, p_x] = i\hbar$, as noted by, for example, Alonso and Valk [1, p. 168].

The Hamiltonian may now be written as

$$\mathcal{H} = \mathcal{H}_{xy} + \mathcal{H}_z = eB\left(\frac{P^2}{2m} + \frac{X^2}{2m}\right) + \frac{p_z^2}{2m} \tag{45}$$

The Hamiltonian for a simple harmonic oscillation (in one dimension) is

$$\mathcal{H} = \frac{p^2}{2m} + V(x) = \frac{p^2}{2m} + \tfrac{1}{2}Kx^2 \tag{46}$$

with K the stiffness or force constant in terms of which the restoring force is proportional to displacement, as $F = -Kx$, and the potential energy is thus $\frac{1}{2}Kx^2$, x being the displacement from the classical equilibrium position. The Schrodinger equation now constitutes one of the notable examples that are capable of direct solution to give the eigenfunctions $\psi_n(x)$ and eigenvalues:

$$E_n = (n + \tfrac{1}{2})\hbar\omega, \qquad \text{where } \omega = \left(\frac{K}{m}\right)^{1/2} \quad \text{and } n = 0, 1, 2, 3, \ldots$$

the value of $E_0 = \frac{1}{2}\hbar\omega$ being the zero point energy corresponding to the minimal oscillation.

It is seen that $\mathcal{H}_{xy}/eB$ has the same form as $\mathcal{H}$ in equation (46) with the replacement $K \sim 1/m$, and thus the corresponding energies (for $\mathcal{H}_{xy}$ itself) must be

$$E_n = eB(n + \tfrac{1}{2})\hbar\omega, \qquad \text{where } \omega = \frac{1}{m} \text{ and } n = 0, 1, 2, 3, \ldots$$

i.e.

$$E_n = \frac{eB}{m}(n + \tfrac{1}{2})\hbar = (n + \tfrac{1}{2})\hbar\omega_c, \qquad \omega_c = \frac{eB}{m} \tag{47}$$

It may also be inferred that the transition rule is that for the harmonic oscillator, i.e. that $\Delta n = \pm 1$, giving rise to a single resonant absorption at frequency

$$f = \left(\frac{1}{h}\right)\hbar\omega_c = \frac{\omega_c}{2\pi} = f_c \tag{48}$$

i.e. at the cyclotron frequency. Thus the quantum treatment corresponds to the classical picture in which the angular frequency of the radiation must match the angular frequency of the orbiting particle.

It is noted that the solution of one part of the Hamiltonian only, i.e. of $\mathcal{H}_{xy}$, has been obtained. However, $\mathcal{H}_z$ commutes with $\mathcal{H}_{xy}$ and it is inferred that the eigenfunctions of H should be eigenfunctions of both $\mathcal{H}_{xy}$ and $\mathcal{H}_z$. This applies if

$$\psi(x, y, z) = \phi(x, y)e^{ik_z z}$$

corresponding to free motion along the z axis (i.e. the field axis). The eigenvalues of H are then

$$E = (n + \tfrac{1}{2})\hbar\omega_c + \frac{\hbar^2 k_z^2}{2m} = (n + \tfrac{1}{2})\hbar\omega_c + \frac{p_z^2}{2m} \tag{49}$$

and any components of the motion along OZ do not affect the resonance.

4 Motion of Electrons in a Magnetic Field and a Central Potential

The Hamiltonian with $\mathbf{B} = \nabla\times\mathbf{A}$ and electrostatic potential energy $V(r)$ is

$$\mathcal{H} = \frac{1}{2m}(\mathbf{p} + e\mathbf{A})^2 + V(r) \tag{50}$$

Considering the first term:

$$(\mathbf{p} + e\mathbf{A})^2 = (\mathbf{p} + e\mathbf{A})(\mathbf{p} + e\mathbf{A}) = p^2 + e^2A^2 + e\mathbf{p}\cdot\mathbf{A} + e\mathbf{A}\cdot\mathbf{p}$$

noting that $\mathbf{A}$ must be treated as an operator and $[\mathbf{p}, \mathbf{A}] = 0$ cannot be presumed. The vector operator $\mathbf{p}$ is

$$\mathbf{p} = \mathbf{i}p_x + \mathbf{j}p_y + \mathbf{k}p_z = -i\hbar\left(\mathbf{i}\frac{\partial}{\partial x} + \mathbf{j}\frac{\partial}{\partial y} + \mathbf{k}\frac{\partial}{\partial z}\right) = -i\hbar\nabla$$

so that $\mathbf{p}\cdot\mathbf{A} = -i\hbar\nabla\cdot\mathbf{A}$. Consider the effect of $\nabla\cdot\mathbf{A}$ on a scalar function $f \equiv f(x, y, z)$:

$$\begin{aligned}
\nabla\cdot\mathbf{A}f &= \left(\mathbf{i}\frac{\partial}{\partial x} + \mathbf{j}\frac{\partial}{\partial y} + \mathbf{k}\frac{\partial}{\partial z}\right)\cdot(\mathbf{i}A_x f + \mathbf{j}A_y f + \mathbf{k}A_z f) \\
&= A_x\frac{\partial f}{\partial x} + f\frac{\partial A_x}{\partial x} + A_y\frac{\partial f}{\partial y} + f\frac{\partial A_y}{\partial y} + A_z\frac{\partial f}{\partial z} + f\frac{\partial A_z}{\partial z} \\
&= \left(\frac{\partial A_x}{\partial x} + \frac{\partial A_y}{\partial y} + \frac{\partial A_z}{\partial z}\right)f + \left(A_x\frac{\partial}{\partial x} + A_y\frac{\partial}{\partial y} + A_z\frac{\partial}{\partial z}\right)f \\
&= (\nabla\cdot\mathbf{A})f + (\mathbf{A}\cdot\nabla)f
\end{aligned}$$

Since $\nabla\cdot\mathbf{A}$ is to be evaluated first, the condition $\nabla\cdot\mathbf{A} = 0$ can now be applied. Thus

$$e\mathbf{p}\cdot\mathbf{A} + e\mathbf{A}\cdot\mathbf{p} = 2e\mathbf{A}\cdot\mathbf{p}$$

and the Hamiltonian becomes

$$\mathcal{H} = \frac{p^2}{2m} + \frac{e}{m}\mathbf{A}\cdot\mathbf{p} + \frac{e^2}{2m}A^2 + V(r) \tag{51}$$

The earlier demonstration that $[\mathcal{H}, L^2] = 0$ and $[\mathcal{H}, L_z] = 0$ depended on the presence of a central potential only and it cannot now be assumed that L^2 and L_z are constants of the motion.

Fields from external magnets are assumed to be uniform over the orbits. Expanding $\mathbf{B} = \nabla\times\mathbf{A}$:

$$\mathbf{B} = \mathbf{i}\left(\frac{\partial A_z}{\partial y} - \frac{\partial A_y}{\partial z}\right) + \mathbf{j}\left(\frac{\partial A_x}{\partial z} - \frac{\partial A_z}{\partial x}\right) + \mathbf{k}\left(\frac{\partial A_y}{\partial x} - \frac{\partial A_x}{\partial y}\right)$$

If

$$A_y = cx, \quad A_x = -cy, \quad A_z = 0$$

then

$$\mathbf{B} = \mathbf{i}(0) + \mathbf{j}(0) + \mathbf{k}(2c)$$

and so

$$\mathbf{A} = \mathbf{i}\left(-\frac{B}{2}\right)y + \mathbf{j}\left(\frac{B}{2}\right)x \sim \mathbf{B} = \mathbf{k}B$$

a field of magnitude B along OZ. The reader may show by expansion and taking the curl that the generalization is that $A = \frac{1}{2}\mathbf{B}\times\mathbf{r}$ is the vector potential corresponding to the values assigned to the components of $\mathbf{B}$. Thus, for the uniform field,

$$\frac{e}{m}\mathbf{A}\cdot\mathbf{p} = \frac{e}{2m}\mathbf{B}\times\mathbf{r}\cdot\mathbf{p} = \frac{e}{2m}\mathbf{B}\cdot\mathbf{r}\times\mathbf{p} = \frac{e}{2m}\mathbf{B}\cdot\mathbf{L}$$

using $\mathbf{r}\times\mathbf{p} = \mathbf{L}$ and making use of $(\mathbf{a}\times\mathbf{b})\cdot\mathbf{c} = \mathbf{a}\cdot(\mathbf{b}\times\mathbf{c})$ which is a general vector relation that may be readily confirmed by expansion. The Hamiltonian becomes

$$\begin{aligned}\mathcal{H} &= \frac{p^2}{2m} + \frac{e}{2m}\mathbf{B}\cdot\mathbf{L} + \frac{e^2}{2m}A^2 + V(r) \\ &= \left[\frac{p^2}{2m} + V(r)\right] - \gamma\mathbf{B}\cdot\mathbf{L} + \frac{e^2}{2m}A^2 \\ &= \mathcal{H}_0 + \mathcal{H}_1 + \mathcal{H}_2 \end{aligned} \tag{52}$$

with $\gamma = -e/(2m)$. If $\mathbf{B} = \mathbf{k}B$ then $\mathbf{B}\cdot\mathbf{L} = BL_z$.

The effect of $\mathcal{H}_2$ is less than that of $\mathcal{H}_1$ in most laboratory fields. The neglect of $\mathcal{H}_2$ constitutes the weak or low field approximation. $\mathcal{H} = \mathcal{H}_0$ would correspond to $\mathbf{B} = 0$ and $\mathcal{H} = \mathcal{H}_0 + \mathcal{H}_1$ and $\mathcal{H} = \mathcal{H}_0 + \mathcal{H}_2$ could be thought of as the 'paramagnetic' and the 'diamagnetic Hamiltonian' respectively.

4.1 Weak Field Case: $\mathcal{H}_2 = 0$

For hydrogen-like atoms the ψ_{nlm} are eigenfunctions of $\mathcal{H}_0$ and of L_z and so

$$-\gamma BL_z\psi_{nlm} = -\gamma Bm\hbar\psi_{nlm} = \beta Bm\psi_{nlm} \tag{53}$$

using $-\gamma\hbar = -(-e/2m)\hbar = e\hbar/(2m) = eh/(4\pi m) = \beta$, the Bohr magneton. Thus

$$\mathcal{H}\psi_{nlm} = (\mathcal{H}_0 - \gamma BL_z)\psi_{nlm} = (E_{0_{nl}} + m\beta B)\psi_{nlm} \tag{54}$$

E_0 generally depends upon l as well as n but cannot depend upon m in a spherically symmetrical environment. The magnetic field defines an axis giving meaning to the components of the angular momentum and gives an energy contribution that is different for each value of m, i.e. lifts the degeneracy in m—the Zeeman effect. The energy term $m\beta B$ corresponds to a magnetization component of $\mu_z = -m\beta$.

Each E_0 splits to give $2l + 1$ levels and the spacing is always $\Delta E = \beta B$. The term 'Zeeman splitting' applies conventionally to the splitting of the observed spectral lines. By the selection rules for the transition $n, l, m \to n', l', m'$, $\Delta m = 0$ or ± 1, each line in the zero field should be split into three ($l = 1$) by the field with symmetrical shifts of $\Delta f = \pm(1/h)\beta B$ for two of the lines. Of course, such a simple splitting is not observed in practice because only the orbital motion has been considered here and spin has to be taken into account.

The operator giving the magnetic moments directly would have the form

$$\boldsymbol{\mu} = \gamma \mathbf{L}, \qquad \mu^2 = \gamma^2 L^2 \tag{55}$$

giving the eigenvalues of magnitude

$$\mu = \gamma\hbar\sqrt{l(l+1)} = \beta\sqrt{l(l+1)} \tag{56}$$

This general treatment should be contrasted with the *ad hoc* treatment used in the introduction, in which it was assumed that since certain magnitudes and components of angular momentum could be calculated, magnetic moments followed by applying a classical gyromagnetic ratio γ and energy terms from the classical $-\boldsymbol{\mu} \cdot \mathbf{B}$. While the *ad hoc* approach is validated to an extent, it is now seen to constitute an approximation. Note also that the quantities γ and β arise naturally and can be taken to be *defined* by the present treatment.

The simple modification of $\mathcal{H}_0$ by adding a term to represent the magnetic energy, analogous to $V(r)$, is never basically justified because such a term does not represent a contribution to the potential energy of the electron.

4.2 The 'Diamagnetic Term' $\mathcal{H}_2$

For the uniform field $\mathbf{k}B$,

$$\mathcal{H}_2 = \frac{e^2}{2m_e}A^2 = \frac{e^2}{2m_e}\frac{B^2}{4}(x^2 + y^2) = \frac{e^2}{2m_e}\frac{B^2}{4}r^2 \tag{57}$$

with r the distance from the origin in *OXY*. The simplest illustration is made by taking r to be constant, as in a model for the motion of a π electron in a benzene ring with potential energy V taken to be constant around the ring but to rise to large values otherwise so as to confine the electron. The plane of the ring is *OXY* and $\mathbf{L} = \mathbf{k}L_z$ only. For motion around the ring the kinetic energy can be expressed as $L^2/(2I)$, with I the moment of inertia [cf. $p^2/(2m)$]. Using $L_z = -i\hbar\, \partial/\partial\phi \to -i\hbar \mathrm{d}/\mathrm{d}\phi$,

$$\mathcal{H} = \mathcal{H}_0 + \mathcal{H}_1 + \mathcal{H}_2 = -\frac{\hbar^2}{2I}\frac{\mathrm{d}^2}{\mathrm{d}\phi^2} + V + i\beta B\frac{\mathrm{d}}{\mathrm{d}\phi} + \frac{\beta^2 I B^2}{2\hbar^2} \tag{58}$$

with $I = m_e r^2$, $\beta = e\hbar/(2m_e)$ for $\mathcal{H}_2$.

For this particularly constrained motion $\mathcal{H}_2$ is a constant (w.r.t. the coordinates) which cannot affect the eigenfunctions and simply shifts the energies according to the value of B.

The obvious solutions for $\mathcal{H}_0$ are

$$\psi_m = \left(\frac{1}{2\pi}\right)^{1/2} \mathrm{e}^{im\phi}, \qquad E_{0_m} = \frac{m^2\hbar^2}{2I}, \qquad \text{for } m = 0, \pm1, \pm2, \ldots \tag{59}$$

as verified by substitution. The discrete values of m correspond to $\mathrm{e}^{im(\phi+2\pi)} = \mathrm{e}^{im\phi}$, $\cos 2\pi m + i \sin 2\pi m = 1$ or to the general discussion of angular momentum. The

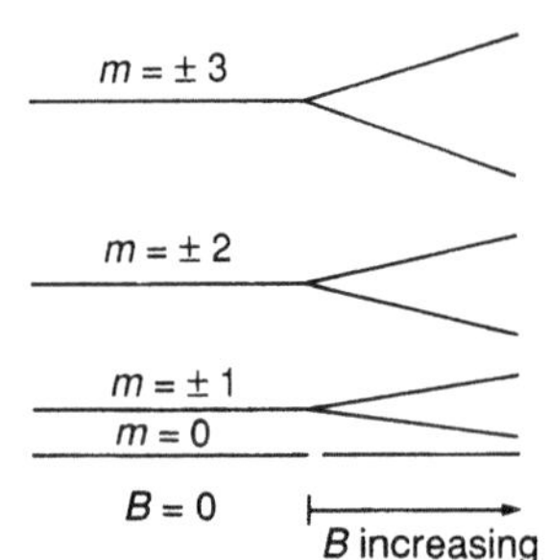

Figure 2.3 Zeeman splitting for an electron moving in a ring. Note the different slopes

ψ_m are also seen to be eigenfunctions of $\mathcal{H}_1$, giving, for $\mathcal{H}_1$ alone, $E_1 = m\beta B$. Since $E_0 \propto m^2$ there is a double degeneracy for $m \neq 0$ which is lifted by the field (Figure 2.3).

Assuming occupation of the ground state only, $m = 0$ and $E_1 = 0$ and the energy of the singlet level is $\beta^2 I B^2/(2\hbar)^2$. The susceptibility is thus

$$\chi = -\mu_0 \frac{\partial^2 E}{\partial B^2} = -\mu_0 \frac{\beta^2 I}{\hbar^2} \tag{60}$$

and the induced magnetic moment is

$$\mu = -\frac{\partial E}{\partial B} = -\frac{\beta^2 I B}{\hbar^2} \tag{61}$$

The induced moment is antiparallel to the applied field and χ is negative, i.e. $\mathcal{H}_2$ corresponds to the diamagnetic response. χ is independent of the field magnitude and of temperature.

Considering just one electron in this system, concentration on $m = 0$ would be realistic because $\Delta E_0(m = 0 \rightarrow m = \pm 1)$ is $\hbar^2/(2I) \sim 10^{-18}$ J using $r = 1.4 \times 10^{-10}$ m for the benzene ring and $kT \sim 10^{-23}$ J at room temperature. The arbitrary single-electron system would exhibit simple diamagnetism. More realistically, the full complement of electrons are to be fed into the levels (or states) in pairs on account of the spin with the spins assumed to be antiparallel and to make no contribution to the susceptibility, and the situation is more complex.

Purely as a model, consider one electron in a doublet level, $m = \pm 1$, split by the field as shown in Figure 2.4. The partition function may be taken as

$$Z = \mathrm{e}^{aB} + \mathrm{e}^{-aB}, \qquad \text{where } a = \frac{\beta}{(kT)}$$

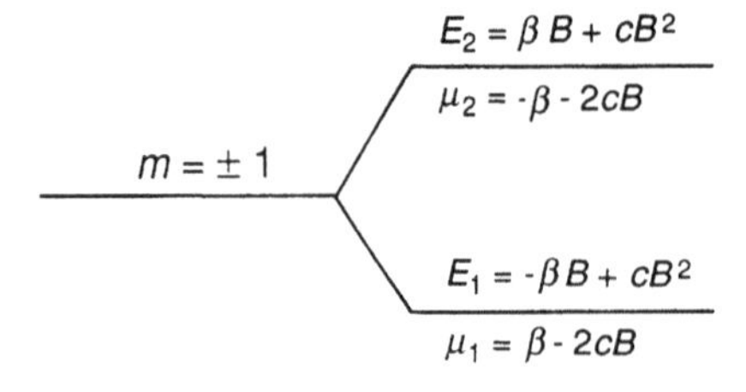

Figure 2.4 Energies and magnetic moments for $m = 1$

ignoring a constant due to the equal shifts cB^2. The energy per electron at equilibrium is

$$E = \frac{e^{aB}}{Z}(-\beta B + cB^2) + \frac{e^{-aB}}{Z}(\beta B + cB^2) = -\beta B \frac{e^{aB} - e^{-aB}}{Z} + cB^2 \left(\frac{Z}{Z}\right)$$

Differentiating by parts gives

$$\frac{\partial E}{\partial B} = -akTB \left[a - a\frac{(e^{aB} - e^{-aB})^2}{(e^{aB} + e^{-aB})^2} \right] - akT \frac{e^{aB} - e^{-aB}}{e^{aB} + e^{-aB}} + 2cB$$

Approximating $e^{aB} = 1 + aB$ gives

$$\frac{\partial E}{\partial B} = -akTB(a - a^3B^2 + aB) + 2cB$$

Approximating again since aB is small except at high fields and low temperatures, the induced moment per electron is

$$\mu = -\frac{\partial E}{\partial B} = \frac{\beta^2 B}{kT} - 2cB \qquad (62)$$

In the present approximation the molar susceptibility is $\chi = L\mu/H = L\mu_0\mu/B$, and substituting for c (with L being Avogadro's number),

$$\chi = \frac{\mu_0 L\beta^2}{kT} - \frac{\mu_0 L\beta^2 I}{\hbar^2} = \chi_{\text{para}} + \chi_{\text{dia}} \qquad (63)$$

The first term would be the expression for the paramagnetic susceptibility obtained in the weak field limit and the second is that for the diamagnetic susceptibility obtained when $m = 0$. Thus it is seen that the two contributions are simply additive and, in general, may be calculated separately and summed to give the net response.

Using the numbers given above, it is clear that the first term in equation (63) is much greater than the second, except at very high temperatures, and on a single-electron approximation the system might be expected to be paramagnetic. However, even for a single electron in the second level, $m = \pm 1$, since the states $|1\rangle$ and $|-1\rangle$ are completely equivalent it is to be expected that linear combinations such as $|+\rangle = 1/\sqrt{2}(|1\rangle + |-1\rangle)$ and $|-\rangle = 1/\sqrt{2}(|1\rangle - |-1\rangle)$ are preferred. These are acceptable because of the degeneracy of $|1\rangle$ and $|-1\rangle$. No angular momentum components are associated with these, i.e. the angular momentum is quenched. Alternatively, it is noted that the expectation value of the induced magnetic moment is, for example,

$$(\langle 1| + \langle -1|)(-\gamma BLz)(|1\rangle + |-1\rangle) = -\gamma B\hbar[1 + 0 + 0 + (-1)] = 0$$

However, the diamagnetic term remains unaffected and even from this simple point of view (ignoring coupling) benzene and other 'ring compounds' are clearly diamagnetic. Molar susceptibilities of the order of 10^{-10} mol^{-1} are predicted (and observed) and these are much greater in magnitude than atomic diamagnetic susceptibilities due to the dependence on $I = m\langle r^2 \rangle$.

5 Dirac Spin and Spin–Orbit Coupling

Quantum theory in which the Hamiltonian is based on the non-relativistic classical Hamiltonian clearly does not directly introduce spin. The general theory of angular momentum taken together with the results of the Stern–Gerlach experiment indicate the existence of a form of electronic angular momentum that is not associated with the orbital motion. Dirac [2] succeeded in formulating a relativistically correct quantum theory which naturally introduced a quantity designated **S**, which could be shown to be associated with a (non-orbital) angular momentum in that it obeyed $\mathbf{S}\times\mathbf{S} = i\mathbf{S}$. **S** is thus the intrinsic or spin angular momentum, although the concept of a literally spinning charge is not necessarily invoked. (The magnetic moment can be calculated by dividing a spinning charge into current loops and it has been estimated that $\mu \approx \beta$ implies velocities on the surface exceeding that of light.)

Dirac's Hamiltonian for an electron moving in a potential V and magnetic field **B** is

$$\mathcal{H} = \frac{1}{2m}(\mathbf{p} + e\mathbf{A})^2 + V - \frac{p^4}{8m^3c^2} + \frac{e}{m}\mathbf{S}\cdot\mathbf{B} + \frac{1}{2m^2c^2r}\frac{\mathrm{d}V}{\mathrm{d}r}\mathbf{S}\cdot\mathbf{L} - \frac{1}{4m^2c^2}\frac{\mathrm{d}V}{\mathrm{d}r}\frac{\partial}{\partial r} \tag{64}$$

The third term is an approximate correction to the kinetic energy. The last term, called the Darwin term, with no classical counterpart applies only to the case $l = 0$ and gives a shift in the s levels.

The fourth term is the direct counterpart of $e/(2m)\mathbf{L}\cdot\mathbf{B}$ (which is obtained if the first term is expanded). It justifies the assumption that the total magnetic moment of a particle with spin may be obtained from

$$\boldsymbol{\mu} = \gamma_{\mathrm{L}}\mathbf{L} + \gamma_{\mathrm{s}}\mathbf{S} = \left(\frac{\beta}{\hbar}\right)(g_{\mathrm{L}}\mathbf{L} + g_{\mathrm{s}}\mathbf{S})$$

with $g_{\mathrm{L}} = 1$ precisely and $g_{\mathrm{s}} \doteqdot 2$ if the particle is an electron, and also the otherwise arbitrary assumption that, for an s-state electron, for example, the classical energy term for the spin in a field $-\boldsymbol{\mu}_{\mathrm{s}}\cdot\mathbf{B}$ may be incorporated directly in the Hamiltonian.

The spin and orbital magnetic moments are seen to differ fundamentally in that for spin there is a permanent intrinsic moment only and no induced moment comparable to that which gives the diamagnetic contribution for orbital motion. The identification of the 'Dirac spin' with the existence of general angular momenta of a non-orbital nature with components of, for example, $\frac{1}{2}\hbar$ and magnitudes $\hbar\sqrt{(1/2)((1/2)+1)}$ is apparent. There must be states (e.g. single-electron s states) for which

$$S_Z|n\ l\ m_l\ S\ m_{\mathrm{s}}\rangle = m_{\mathrm{s}}\hbar|n\ l\ m_l\ S\ m_{\mathrm{s}}\rangle, \qquad \text{where } m_{\mathrm{s}} = +\tfrac{1}{2}, -\tfrac{1}{2}$$

and since the spin is independent of the coordinates these states can be written as

$$|n\ l\ m_l\ S\ m_{\mathrm{s}}\rangle = |n\ l\ m_l\rangle|S\ m_{\mathrm{s}}\rangle = |n\ l\ m_l\rangle|m_{\mathrm{s}}\rangle$$

with $S = \frac{1}{2}$ implied in the latter. We may then write

$$S_z|\tfrac{1}{2}\rangle = \tfrac{1}{2}\hbar|\tfrac{1}{2}\rangle, \qquad S_z|-\tfrac{1}{2}\rangle = -\tfrac{1}{2}\hbar|-\tfrac{1}{2}\rangle$$

since there is no dependence of the spin components on the orbital motion.

However, the fourth term indicates that, in the general case, the orbital and spin motions are not independent, i.e. a spin–orbit or LS coupling exists. This is usually written

$$\mathcal{H}_{\rm LS} \equiv \mathcal{H}_{\rm SO} = \lambda \mathbf{L} \cdot \mathbf{S} = \lambda \mathbf{S} \cdot \mathbf{L}$$

Some such coupling is expected semi-classically. In a frame in which the electron is stationary the charged nucleus moves according to the orbital occupied by the electron, generating magnetic fields that influence the energy due to the intrinsic spin magnetic moment. LS coupling is a magnetic effect.

5.1 Spin–Orbit Coupling

If the electron at $\mathbf{r}$ w.r.t. the nucleus at O moves with velocity $\mathbf{v}$, the field at $\mathbf{r}$ is that which would act on a stationary electron with the nucleus moving at $-\mathbf{v}$ (Figure 2.5). With charge Ze,

$$B(\mathbf{r}) = -\frac{\mu_0}{4\pi} Ze \frac{\mathbf{v}\times\mathbf{r}}{r^3} \tag{65}$$

Setting $\mathbf{v}\times\mathbf{r} = vr$, $Z = 1, r = 10^{-10}$ m and, $v = 10^6$ m s^{-1}, B may be estimated as $\sim$ 1.6 T, i.e. a typical 'laboratory' or electromagnet field — a substantial effect.

Since

$$\mathbf{v}\times\mathbf{r} = -\mathbf{r}\times\mathbf{v} \qquad \text{and} \qquad \mathbf{r}\times m\mathbf{v} = \mathbf{r}\times\mathbf{p} = \mathbf{L}$$

we have, from equation (65),

$$\mathbf{B} = \frac{\mu_0}{4\pi}\frac{Ze}{r^3}\frac{\mathbf{r}\times\mathbf{p}}{m} = \frac{\mu_0 Ze}{4\pi m r^3}\mathbf{L} \tag{66}$$

Assuming that $\mathcal{H}_{\rm LS} = -\mathbf{B}\cdot\boldsymbol{\mu}_{\rm s}$ and $\boldsymbol{\mu}_{\rm s} = -g_{\rm e}[{\rm e}/(2m)]\mathbf{S}$ should give the required result. Unfortunately, this simple treatment is in error and a relativistic correction gives a negative term which has just half the magnitude of that calculated and

$$\mathcal{H}_{\rm LS} = \frac{\mu_0}{4\pi}\frac{g_{\rm e}Ze^2}{2m^2r^3}\mathbf{L}\cdot\mathbf{S} \rightarrow \frac{\mu_0}{4\pi}\frac{g_{\rm e}Ze^2}{4m^2r^3}\mathbf{L}\cdot\mathbf{S} = \lambda\mathbf{L}\cdot\mathbf{S} \tag{67}$$

(The correction should strictly be applied to the field calculation.)

For a central potential the electric field is

$$\mathbf{E} = \frac{Ze\mathbf{r}}{4\pi\varepsilon_0 r^3}, \qquad \frac{Ze\mathbf{r}}{r^3} = 4\pi\varepsilon_0\mathbf{E}$$

so that

$$\mathbf{B} = -\frac{\mu_0}{4\pi}4\pi\varepsilon_0\mathbf{v}\times\mathbf{E} = -(\mu_0\varepsilon_0)\mathbf{v}\times\mathbf{E} = -\frac{1}{c^2}\mathbf{v}\times\mathbf{E}$$

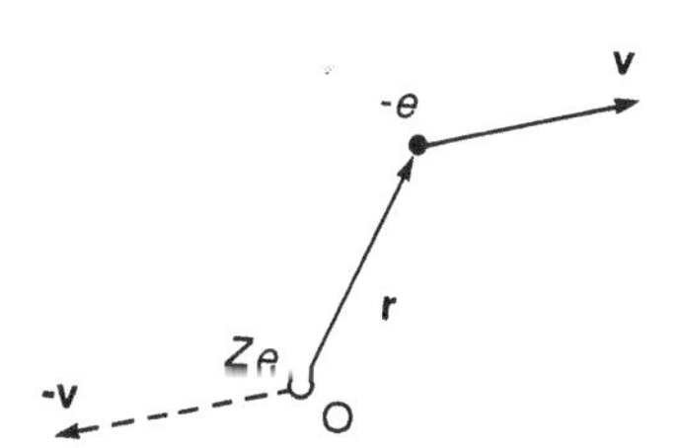

Figure 2.5 Relative velocities for electron and nucleus

making use of $\mu_0\varepsilon_0 = 1/c^2$ with c the velocity of light. (Note that the factor $1/c^2$ arises naturally in the SI; the c.g.s. expression involves $1/c$.) Since the electric field is given by [for $V = V(r)$]

$$\mathbf{E} = \frac{1}{e}\frac{\mathbf{r}}{\bar{\mathbf{r}}}\frac{\mathrm{d}V}{\mathrm{d}r}$$

(remembering that V is the potential energy = charge times electric potential and that the charge is $-e$):

$$\mathbf{B} = -\frac{1}{c^2 er}\frac{\mathrm{d}V}{\mathrm{d}r}\mathbf{v}\times\mathbf{r} = \frac{1}{emrc^2}\frac{\mathrm{d}V}{\mathrm{d}r}\mathbf{L} \tag{68}$$

and

$$\mathcal{H}_{\mathrm{LS}} = g_{\mathrm{e}}\frac{\mathrm{e}}{2m}\mathbf{B}\cdot\mathbf{S} = \frac{g_{\mathrm{e}}}{2m^2rc^2}\frac{\mathrm{d}V}{\mathrm{d}r}\mathbf{S}\cdot\mathbf{L} \rightarrow \frac{g_{\mathrm{e}}}{4m^2rc^2}\frac{\mathrm{d}V}{\mathrm{d}r}\mathbf{S}\cdot\mathbf{L} = \lambda\mathbf{S}\cdot\mathbf{L} \tag{69}$$

again with the relativistic correction. It is easy to show (using $\mu_0\varepsilon_0 = 1/c^2$ once more) that with $V = -Ze^2/(4\pi\varepsilon_0 r)$,

$$\mathcal{H}_{\mathrm{LS}} = \frac{g_{\mathrm{e}}Ze^2}{(4\pi\varepsilon_0)4m^2r^3c^2}\mathbf{S}\cdot\mathbf{L} \tag{70}$$

5.2 The High-Field Limit

Expressing ∇^2 and L^2 in spherical coordinates, a simplified Hamiltonian in the presence of a magnetic field $\mathbf{k}B$ [see equations (23) and (70)] is

$$\begin{aligned}\mathcal{H} &= \left\{-\frac{\hbar^2}{2m_{\mathrm{e}}}\left[\frac{1}{r^2}\frac{\partial}{\partial r}\left(r^2\frac{\partial}{\partial r}\right) - \frac{1}{r^2\hbar^2}L^2\right] - \frac{Ze^2}{4\pi\varepsilon_0 r}\right\} + \lambda\mathbf{S}\cdot\mathbf{L} + \frac{\beta}{\hbar}(g_{\mathrm{e}}S_z + L_z)B \\ &= \mathcal{H}_0 + \mathcal{H}_{\mathrm{LS}} + \mathcal{H}_{\mathrm{B}}\end{aligned} \tag{71}$$

For $\lambda = 0 = B$ and $V \equiv V(r)$, L^2, S^2, L_z and S_z are constants of the motion and this clearly still applies for $B \neq 0$ so long as $\lambda = 0$. The appropriate uncoupled states are $|n\ l\ m\rangle|S\ m_{\mathrm{s}}\rangle$ which are eigenstates of $\mathcal{H}^1 = \mathcal{H}_0 + \mathcal{H}_{\mathrm{B}}$:

$$(\mathcal{H}_0 + \mathcal{H}_{\mathrm{B}})|n\ l\ m\rangle|S\ m_{\mathrm{s}}\rangle = [E_0 + \beta B(g_{\mathrm{e}}m_{\mathrm{s}} + m)]|n\ l\ m\rangle|S\ m_{\mathrm{s}}\rangle \tag{72}$$

The Zeeman effect in this uncoupled limit is thus obtained simply by adding the magnetic energy contributions as shown on the right in Figure (2.6). Since $\mathcal{H}_{\mathrm{B}}$ includes B and $\mathcal{H}_{\mathrm{LS}}$ does not, as B is increased $\mathcal{H}_{\mathrm{B}}$ must predominate and the neglect of $\mathcal{H}_{\mathrm{LS}}$ is the more justified; this is termed the high-field limit.

A small spin–orbit coupling treated as a perturbation gives, to first order, using the uncoupled states:

$$E_{\mathrm{LS}}^{(1)} = \frac{g_{\mathrm{e}}Ze^2}{4c^2m_{\mathrm{e}}^2(4\pi\varepsilon_0)}\mathcal{G}\langle l\ m\ S\ m_{\mathrm{s}}|\mathbf{S}\cdot\mathbf{L}|l\ m\ S\ m_{\mathrm{s}}\rangle \tag{73}$$

The integral $\mathcal{G}$ is the mean value $\langle r^{-3}\rangle$ for hydrogen-like atoms evaluated as

$$\mathcal{G} = a_{nl} = \frac{1}{a_0^3}\frac{Z^3}{n^3 l(l+1)(l+\frac{1}{2})} \qquad (l \neq 0) \tag{74}$$

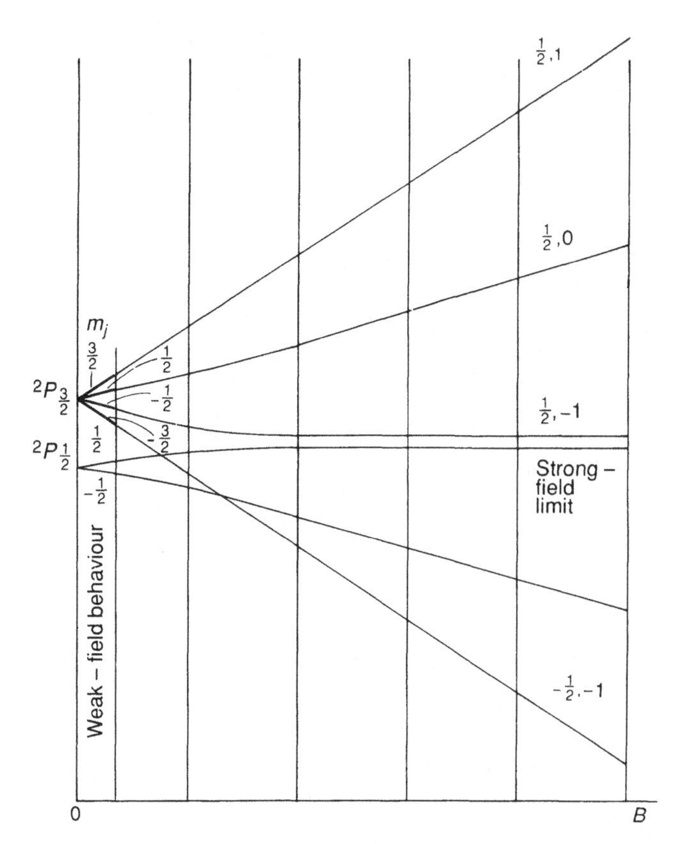

Figure 2.6 The transition from the weak field (strong coupling) to the strong field (decoupled) Zeeman splitting. Note the symmetrical slopes at low fields

with a_0 the first Bohr radius. Expanding $\mathbf{S} \cdot \mathbf{L} = S_x L_x + S_y L_y + S_z L_z$ and noting that only the S_z and L_z do not produce new states (consider L_x as $\frac{1}{2}(L^+ + L^-)$, etc.), the last factor in equation (73) reduces to

$$\langle l\ m | \langle S\ m_s | S_z L_z | l\ m \rangle | S\ m_s \rangle = \hbar^2 m m_s$$

Thus the effect of a small spin–orbit interaction to first order using the uncoupled states is to give the energy as

$$E = E_0 + \beta B(g_e m_s + m) + \frac{g_e Z^4 e^2 \hbar^2 m m_s}{4c^2 m_e^2 a_0^3 n^3 l(l+1)(l+\frac{1}{2})} \tag{75}$$

It is assumed, of course, that both $\mathcal{H}_B$ and $\mathcal{H}_{LS}$ are small compared to $\mathcal{H}_0$, even when B is moderately high, so that $\mathcal{H}_B \gg \mathcal{H}_{LS}$ and the perturbation in the uncoupled states is justified.

5.3 Coupled Basis, J and g_J

In the zero magnetic field or a small field such that $\mathcal{H} \doteqdot \mathcal{H}_0 + \mathcal{H}_{LS}$, the constants of the motion differ from the above. L^2 and S^2 (see Section 2, 3) commute with each of their components and thus with $\mathbf{L}\cdot\mathbf{S} = L_xS_x + L_yS_y + L_zS_z$. However,

$$\begin{aligned}[L_z, \mathbf{S}\cdot\mathbf{L}] &= [L_z, S_xL_x] + [L_z, S_yL_y] + [L_z, S_zL_z] \\ &= S_x[L_z, L_x] + S_y[L_z, L_y] + 0 \\ &= i\hbar(S_xL_y - S_yL_x)\end{aligned}$$

and

$$[S_z, \mathbf{S}\cdot\mathbf{L}] = \cdots = i\hbar(S_yL_x - S_xL_y) \tag{76}$$

so that L_z and S_z are not constants of the motion. Now

$$(\mathbf{L}+\mathbf{S})^2 = L^2 + S^2 + 2\mathbf{S}\cdot\mathbf{L}$$

each term commuting with $\mathbf{S}\cdot\mathbf{L}$. Thus there is a new constant of the motion

$$J^2 = (\mathbf{L}+\mathbf{S})^2, \qquad \mathbf{J} = \mathbf{L}+\mathbf{S} \tag{77}$$

the total angular momentum:

i.e.
$$[J^2, \mathcal{H}_0 + \mathcal{H}_{\mathrm{LS}}] = 0 \tag{78}$$

Since the components add simply, $J_z = S_z + L_z$:

$$[J_z, \mathbf{S}\cdot\mathbf{L}] = [L_z, \mathbf{S}\cdot\mathbf{L}] + [S_z, \mathbf{S}\cdot\mathbf{L}] = 0$$

so that J_z is also a constant of the motion and appropriate states are now $|n\ l\ s\ j\ m_j\rangle$ or ψ_{nlsjm}, with

$$J^2\psi = \hbar^2 j(j+1)\psi, \qquad J_z\psi = \hbar m_j\psi \tag{79}$$

It is easily seen that J_x and J_y are not constants of the motion.

ψ separates as a radial part which depends on n, l and j, i.e. $R_{nlj}(r)$ and a part χ_{lsjm_j}, which may in turn be expressed as functions of θ and ϕ multiplying spin states. The R_{nlj} are not strictly the familiar R_{nl} from the hydrogen atom account. However, if $\mathcal{H}_{\mathrm{LS}}$ is replaced by a value using the approximation that $\mathbf{S}\cdot\mathbf{L}$ gives eigenvalues of $\sim\hbar^2$ this may be compared with V:

$$\frac{\lambda\hbar^2}{V} = \frac{\mu_0}{4\pi}\frac{g_e Ze^2\hbar^2/(4m^2r^3)}{Ze^2/(4\pi\varepsilon_0 r)} = \frac{g_e\hbar^2}{4m^2r^2c^2} \approx 10^{-2}$$

again with $\mu_0\varepsilon_0 = 1/c^2$ and $r = 10^{-10}$ m. Thus the R_{nlj} would be obtained by solving a differential equation using a value of V which is only slightly modified by the interaction and it is acceptable to replace them by the original R_{nl}. The first-order correction to the energy, due to the spin–orbit coupling, is then

$$\Delta E = \int_0^\infty R_{nl}^*(r)\frac{1}{r^3}R_{nl}(r)r^2\,\mathrm{d}r\langle l\ s\ j\ m_j|\lambda'\mathbf{S}\cdot\mathbf{L}|l\ s\ j\ m_j\rangle \tag{80}$$

where $\lambda' = r^3\lambda$. Since

$$J^2 = (\mathbf{L}+\mathbf{S})\cdot(\mathbf{L}+\mathbf{S}) = L^2 + S^2 + 2\mathbf{S}\cdot\mathbf{L}$$

we have

$$(\mathbf{S}\cdot\mathbf{L})|l\ s\ j\ m_j\rangle = \tfrac{1}{2}(J^2 - L^2 - S^2)|l\ s\ j\ m_j\rangle$$
$$= \tfrac{1}{2}\hbar^2[j(j+1) - l(l+1) - s(s+1)]|l\ s\ j\ m_j\rangle$$

The integral is as given in equation (74) and so

$$E = E_\mathrm{n} + \Delta E = E_\mathrm{n} + \frac{g_\mathrm{e} Z^4 e^2 \hbar^2}{(4\pi\varepsilon_0)\times 4c^2 m_\mathrm{e}^2 a_0^3 n^3}\,\frac{j(j+1) - l(l+1) - s(s+1)}{l(l+1)(l+\frac{1}{2})}$$
$$= E_\mathrm{n} + a_{nl}\frac{j(j+1) - l(l+1) - s(s+1)}{l(l+1)(l+\frac{1}{2})} \tag{81}$$

depending on n, l, s and j but not on m_j. With $S = \frac{1}{2}$ the possible values of j are $j = l + \frac{1}{2}$ and $j = l - \frac{1}{2}$. For attractive potentials as in atoms it is found that $a_{nl} > 0$ and the level with $j = l - \frac{1}{2}$ is the lower. For example, for a p state, $l = 1$, substitution in equation (81) shows that the original level is split by the interaction into levels raised by $a_{nl}/3$, for $j = l + \frac{1}{2}$, and lowered by $\frac{2}{3}a_{nl}(j = l - \frac{1}{2})$, as shown on the left of Figure 2.6. This is the low-field limit, due to the assumptions made in this section. The states are designated by the symbols $^{2S+1}(L)_j$ so in this case the levels are designated $^2\mathrm{P}_{\frac{3}{2}}$ and $^2\mathrm{P}_{\frac{1}{2}}$.

When the LS coupling is sufficiently strong for the coupled states to be appropriate, the first-order energy shift due to an applied magnetic field cannot be calculated directly as

$$\Delta E_\mathrm{B} = \langle l\ s\ j\ m_j|\frac{\beta}{\hbar}(g_\mathrm{e}S_z + L_z)|l\ s\ j\ m_j\rangle$$

because the states are not now eigenstates of S_z and L_z. Progress may be made by recourse to the 'vector model' which is justified by the projection theorem or Wigner–Eckart theorem. In Figure 2.7 **L** and **J** are treated as vector quantities. They are not static but drawn at an arbitrary time. **J** is precessing relatively slowly around OZ (i.e. **B**) to correspond with the existence of a definite z component only with $\omega \propto \beta B$ and **L** (and **S**) precess rapidly about **J** with $\omega \propto \lambda\mathbf{L}\cdot\mathbf{S}(\gg \beta B)$. The mean L_z can only be obtained by forming the vector $\mathbf{L}_J$, say, with the direction of **J**, the component of **L** along **J**, and taking $\mathbf{L}_J\cdot\mathbf{k}$. The magnitude of $\mathbf{L}_J$ is

$$L_J = \frac{\mathbf{L}\cdot\mathbf{J}}{J}$$

and

$$\mathbf{L}_J = \frac{\mathbf{L}\cdot\mathbf{J}}{J}\frac{\mathbf{J}}{J} = \frac{(\mathbf{L}\cdot\mathbf{J})\mathbf{J}}{J^2}, \qquad L_z = \mathbf{L}_J\cdot\mathbf{k} = \frac{(\mathbf{L}\cdot\mathbf{J})J_z}{J^2}$$

with $\mathbf{J}\cdot\mathbf{k} = J_z$. Using $\mathbf{L}\cdot\mathbf{J}$ as given by

$$\mathbf{J} = \mathbf{L} + \mathbf{S}, \qquad \mathbf{S}\cdot\mathbf{S} = (\mathbf{J}-\mathbf{L})\cdot(\mathbf{J}-\mathbf{L}) = J^2 + L^2 - 2\mathbf{L}\cdot\mathbf{J}$$

L_z becomes

$$L_z = \frac{(J^2 + L^2 - S^2)J_z}{2J^2}$$

and similarly

$$S_z = \frac{(J^2 - L^2 + S^2)J_z}{2J^2}$$

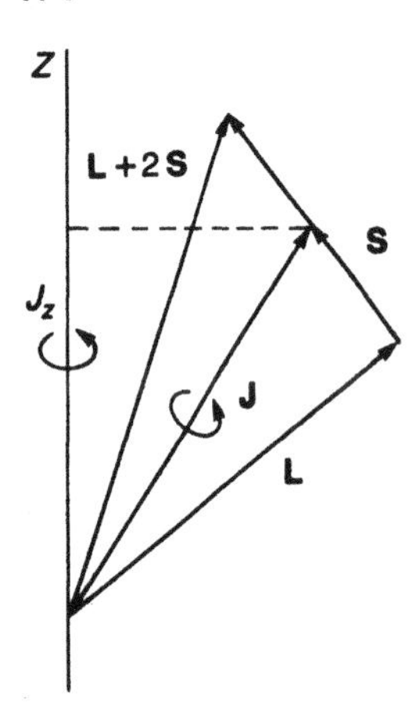

Figure 2.7 The vector model for *SO* coupling

The vectors are now replaced by the operators, except that in the denominator J^2 is replaced by $\hbar^2 j(j+1)$ because it is there used as a scaling factor. L_z and S_z have been successfully replaced by constants of the motion and the energy shift due to the field can now be calculated:

$$\begin{aligned}\Delta E &= \langle l\ s\ j\ m_j|\frac{\beta B}{\hbar}\tfrac{1}{2}[(\mathbf{J}^2+\mathbf{L}^2-\mathbf{S}^2)+g_e(\mathbf{J}^2-\mathbf{L}^2+\mathbf{S}^2)]\frac{J_z}{\hbar^2 j(j+1)}|l\ s\ j\ m_j\rangle \\ &= \beta B m_j\left\{\frac{j(j+1)+l(l+1)-s(s+1)}{2j(j+1)}+g_e\frac{[j(j+1)-l(l+1)+s(s+1)]}{2j(j+1)}\right\} \\ &= \beta B m_j F(l,s,j) \end{aligned} \tag{82}$$

A total magnetic moment operator may be defined by

$$\boldsymbol{\mu} = -\frac{\beta}{\hbar}(\mathbf{L}+g_e\mathbf{S}), \qquad \mathcal{H}_B = -\boldsymbol{\mu}\cdot\mathbf{B} \tag{83}$$

Clearly $\boldsymbol{\mu}$ is parallel to neither **L** nor **S**. We may also write, for the coupled states,

$$\boldsymbol{\mu}_J = \gamma_J\mathbf{J} = -g_J\frac{\beta}{\hbar}\mathbf{J} \tag{84}$$

where $\boldsymbol{\mu}_J$ is the projection of $\boldsymbol{\mu}$, as given by equation (83), along **J**, i.e.

$$\boldsymbol{\mu}_J = (\boldsymbol{\mu}\cdot\mathbf{J})\frac{\mathbf{J}}{J} = \frac{\beta}{\hbar}[(\mathbf{L}+g_e\mathbf{S})\cdot\mathbf{J}]\frac{\mathbf{J}}{\hbar^2 j(j+1)}$$

Thus

$$g_J = \langle l\ s\ j\ m_j|\frac{(\mathbf{L}+g_e\mathbf{S})\cdot\mathbf{J}}{\hbar^2 j(j+1)}|l\ s\ j\ m_j\rangle$$

We already have

$$\Delta E = \langle l\ s\ j\ m_j|\frac{\beta B}{\hbar}\left[\frac{(\mathbf{L}+g_e\mathbf{S})\cdot\mathbf{J}}{\hbar^2 j(j+1)}\right]J_z|l\ s\ j\ m_j\rangle = \beta B m_j F(l,s,j)$$

and since

$$\left(\frac{\beta B}{\hbar}\right) J_z|l\ s\ j\ m_j\rangle = \beta B m_j|l\ s\ j\ m_j\rangle$$

then by inspection

$$g_J = F(l, s, j) = \frac{j(j+1) + l(l+1) - s(s+1)}{2j(j+1)} + g_e\frac{j(j+1) - l(l+1) + s(s+1)}{2j(j+1)} \tag{85}$$

With the approximation $g = 2$,

$$g_J = \frac{3j(j+1) - l(l+1) + s(s+1)}{2j(j+1)} \tag{86}$$

It is clear that if $\boldsymbol{\mu} = -g_J(\beta/\hbar)\mathbf{J}$ then

$$\mu_z = -g_J\left(\frac{\beta}{\hbar}\right) J_z$$

Thus in the strongly coupled or weak-field case, for the coupled states, the magnetic moments and components are given by the eigenvalues

$$\mu = g_J\beta\sqrt{j(j+1)}, \qquad \mu_z = -g_J\beta m_j \tag{87}$$

with g_J as given (g_J is sometimes called the Landé splitting factor).

This treatment can be greatly abbreviated if the projection theorem as such is accepted as the starting point. An elaborate demonstration making use of the Wigner–Eckart theorem (see, for example [1], p. 248) shows that for any vector operator (tensor operator of order unity) such as **J**, **L**, **S** and thus $\boldsymbol{\mu} = -\beta(\mathbf{L} + g_e\mathbf{S})$, the mean value for the state $|j\ m\rangle$ is

$$\langle j\ m|\boldsymbol{\mu}|j\ m\rangle = \frac{\langle j|\boldsymbol{\mu}\cdot\mathbf{J}|j\rangle\langle j\ m|\mathbf{J}|j\ m\rangle}{j(j+1)} \tag{88}$$

Since $\boldsymbol{\mu}\cdot\mathbf{J}$ is a scalar, this indicates that the mean value of $\boldsymbol{\mu}$, $\langle\boldsymbol{\mu}\rangle$, is parallel to $\langle\mathbf{J}\rangle$, although $\boldsymbol{\mu}$ is not parallel to **J**. In general the theorem relates the matrix elements of $\boldsymbol{\mu}$ to those of the projection of $\boldsymbol{\mu}$ along **J**, i.e. of $(\boldsymbol{\mu}\cdot\mathbf{J})\mathbf{J}/J^2$. Thus it can be taken that

$$g_J = \frac{\langle j|\boldsymbol{\mu}\cdot\mathbf{J}|j\rangle}{j(j+1)}$$

and the completion is left to the reader, with some reference to the foregoing.

The existence of the sodium D-line doublet constitutes an early indication of the inadequacy of a theory that neglects spin. It corresponds to transitions from the spin–orbit split $^2\mathrm{P}_{\frac{1}{2}}$ and $^2\mathrm{P}_{\frac{3}{2}}$ levels to the ground state $^2\mathrm{S}_{\frac{1}{2}}$ level. While sodium can be regarded as an effective one-electron atom the coupling is much stronger than that in hydrogen and the doublet is resolved as such by eye at low resolution. The Zeeman splittings are very much finer at $B \sim 1.0$ T. The fine structure or doublet width is quoted as $\lambda^{-1} \approx 1000\ \mathrm{m}^{-1}$: $\Delta E_{\mathrm{LS}} = \mathrm{hf} = h(c/\lambda) \approx 6 \times 10^{-34} \times 3 \times 10^8 \times 10^3$ J. Using $\Delta E_{\mathrm{B}} \approx \beta B$, the value of B corresponding to $\Delta E_{\mathrm{B}} = \Delta E_{\mathrm{LS}}$ would be $\Delta E_{\mathrm{LS}}/(10 \times 10^{-24}) \approx 20.0$ T. Thus the weak field condition is satisfied by fields that still give a resolvable Zeeman splitting, say 0.1 T. For $^2\mathrm{S}_{\frac{1}{2}}$, $^2\mathrm{P}_{\frac{1}{2}}$ and $^2\mathrm{P}_{\frac{3}{2}}$, $g_J = 2$, $\frac{2}{3}$ and $\frac{4}{3}$ respectively. Thus the energy levels and associated Zeeman spectra are as shown in Figure 2.8 ($\Delta j = \pm 1, 0$; $\Delta m = \pm 1, 0$; $\Delta l = \pm 1$).

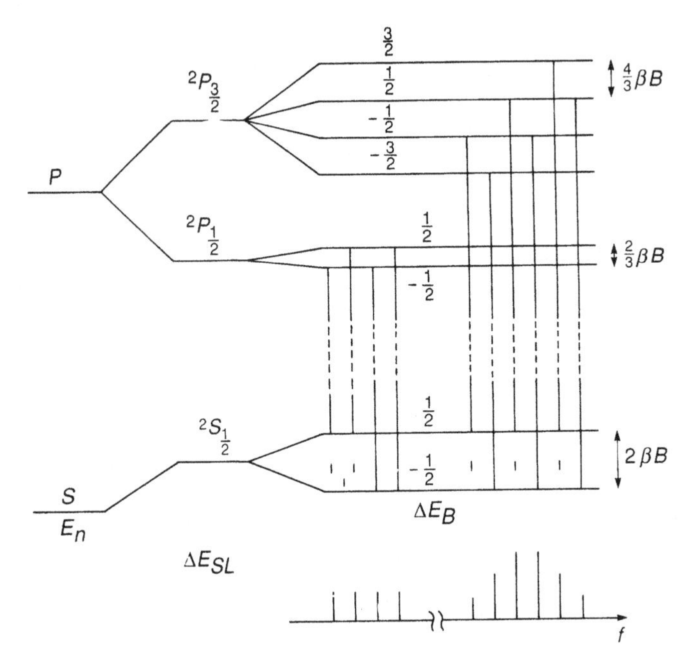

Figure 2.8 Zeeman splitting for S and P states and the associated spectrum, M_J as indicated

With spin–orbit coupling of any strength the coupled states should always be appropriate. The postulation of a high-field limit is an acknowledgement that this rule can be effectively overcome and decoupling achieved by an adequate field—the Paschen–Back effect. Continuity between the two limits may be achieved by perturbation theory.

6 Permutations and Symmetry

When two or more particles are to be considered the interactions are usually represented as a sum of pair interactions and the Hamiltonian is

$$\mathcal{H} = \sum \mathcal{H}(i) + \sum{}' \mathcal{H}(i, j) \tag{89}$$

$\sum$ indicating the sum over the N particles and $\sum'$ the sum over all pairs avoiding duplication, equivalent to $i > j$ or $j > i$. $\mathcal{H}(i) \equiv \mathcal{H}(\mathbf{r}_i)$ depends on the coordinates of the ith particle:

$$\mathcal{H}(i) = \frac{p_i^2}{2m} + V(\mathbf{r}_i)$$

while $\mathcal{H}(i, j)$ or $\mathcal{H}(\mathbf{r}_i, \mathbf{r}_j)$ depends on the coordinates of a pair of particles.

For $N = 2$, say, the state vector must be that labelled by the quantum numbers for both particles as

$$|n_1\ l_1\ m_1\ n_2\ l_2\ m_2\rangle$$

indicating conventionally that particle '1' is specified as being in the state indicated by the particular values of n_1 l_1 and m_1 and '2' is in the state $n_2 l_2 m_2$. This can be contracted to

$|n_1\ n_2\rangle$ if the n_i are taken to be *collections* of quantum numbers, or if only one varies in a group and the remainder are implied; it can be simplified even further to $|n\ n\rangle$ if it is implied that the first symbol corresponds to particle 1 and the second to particle 2. If only the changes in the spin component are of interest and $S = \frac{1}{2}$ is implied, the state $|\frac{1}{2}\ -\frac{1}{2}\rangle$ or $|\alpha\ \beta\rangle$ would indicate '1' in α and '2' in β. $|\beta\ \alpha\rangle$ would be a distinct state, because it indicates 1 in β and 2 in α, even if the particles are identical (e.g. both electrons).

Define a permutation operator P_{12} that transforms one state to the other:

$$P_{12}|\alpha\ \beta\rangle = |\beta\ \alpha\rangle, \qquad P_{12}|\beta\ \alpha\rangle = |\alpha\ \beta\rangle$$

or P_{ij} that interchanges the symbols in positions i and j for any number of particles. These interchanges are equivalent to interchanging the implied order for the same list of symbols. The state could be given as $|1\ 2\ 3\ \ldots N\rangle$ or $\varphi(1\ 2\ 3\ \ldots N)$, meaning that the first of any list of quantum numbers applies to particle 1, the second to 2, etc., while the state $P_{12}\varphi$ would be that in which the role of particles 1 and 2 were interchanged: $P_{12}\varphi(12) = \varphi(21)$.

Hamiltonians of the above type are invariant to permutations since this only affects the order in which the terms are written:

$$P\mathcal{H} = \mathcal{H} \tag{90}$$

Since the same numbers of particles are in the same numbers of states before and after permutation the energy must be unchanged if φ is replaced by $P\varphi$ and moreover permuting the whole equation is just a formality, so that both

$$\mathcal{H}(P\varphi) = E(P\varphi) \qquad \text{and} \qquad P(\mathcal{H}\varphi) = P(E\varphi) = E(P\varphi) \tag{91}$$

and thus

$$\mathcal{H}P = P\mathcal{H}, \qquad [\mathcal{H}, P] = 0 \tag{92}$$

A repeated permutation restores the original functions so that, applying P once more to the first equations (91) and using the commutation (92),

$$P\mathcal{H}P\varphi = E\varphi \qquad \text{or} \qquad \mathcal{H}P^2\varphi = \mathcal{H}\varphi$$

which shows that

$$P^2\varphi = \varphi \tag{93}$$

Thus either

$$P_{ij}\varphi = \varphi \qquad \text{(symmetric)} \tag{94}$$

or

$$P_{ij}\varphi = -\varphi \qquad \text{(antisymmetric)} \tag{95}$$

A state such as $|\alpha\ \beta\rangle$ is neither symmetric nor antisymmetric, but if $|\chi\rangle = |\alpha\ \beta\rangle - |\beta\ \alpha\rangle$,

$$P_{12}|\chi\rangle = |\beta\ \alpha\rangle - |\alpha\ \beta\rangle = -|\chi\rangle$$

then $|\chi\rangle$ is antisymmetric and obviously $|\chi\rangle = |\alpha\ \beta\rangle + |\beta\ \alpha\rangle$ is symmetric. Note that $P_{ij}\varphi = \pm\varphi$ ensures that

$$|P_{ij}\varphi|^2 = |\varphi|^2$$

as is obviously required.

Consider now a system with two (identical) particles i and j in the same state. P_{ij} must leave φ unchanged, $P_{ij}\varphi \equiv \varphi$. For φ to be antisymmetric (φ_a), $P_{ij}\varphi_a = -\varphi_a = \varphi_a$, indicating that $\varphi_a = 0$ so for φ_a to exist no two particles must occupy the same state. Conversely, it may be concluded that the states or wave functions for electrons (Fermions) subject to the Pauli exclusion principle must properly be antisymmetric. [For particles such as photons, with spin 0 and obeying Bose–Einstein statistics (i.e. Bosons), the system wave functions are symmetric.]

6.1 Symmetry and Exchange

For a two-electron system with

$$\mathcal{H} = \mathcal{H}_1 + \mathcal{H}_2 + \mathcal{H}_{12} = \mathcal{H}_0 + \mathcal{H}_{12} \tag{96}$$

we may hope to approximate system states by products of the single-electron eigenstates (according to either $\mathcal{H}_1|n\rangle = E_n|n\rangle$ or $\mathcal{H}_2|n\rangle = E_n|n\rangle$), to be written as either $|n\rangle|m\rangle \equiv |n\ m\rangle$ or $|m\rangle|n\rangle \equiv |m\ n\rangle$. The products satisfy $\mathcal{H}_1|n\ m\rangle = E_n|n\ m\rangle$, $\mathcal{H}_2|n\ m\rangle = E_m|n\ m\rangle$, etc.; i.e. $\mathcal{H}_1$ is taken to depend on the coordinates of electron 1 and the quantum number written first applies specifically to electron 1, etc. If the states are replaced by functions of $\mathbf{r}_1$ and of $\mathbf{r}_2$ and $\mathcal{H}_1 = [-\hbar^2/(2m)]\nabla_1^2 + Ze^2/(4\pi\varepsilon_0 r_1)$, etc., it is easy to see how this applies. The adjoint states $\langle n\ m|$ will be written with the same ordering, i.e. $\langle n\ m| = |n\ m\rangle^\dagger$. If the single-electron states are taken to be orthonormal then $\langle n\ m|n\ m\rangle = \langle n|n\rangle\langle m|m\rangle = 1$ and $\langle n\ m|m\ n\rangle = \langle n|m\rangle\langle m|n\rangle = 0$ if both electrons are centred on the same nucleus.

These states are eigenstates of $\mathcal{H}_0 = \mathcal{H}_1 + \mathcal{H}_2$ with energies $E_n + E_m$, but not of $\mathcal{H} = \mathcal{H}_0 + \mathcal{H}_{12}$. The matrix equation for the energies in this approximation becomes

$$\begin{pmatrix} \langle n\ m|\mathcal{H}_0|n\ m\rangle + \langle n\ m|\mathcal{H}_{12}|n\ m\rangle & \langle n\ m|\mathcal{H}_0|m\ n\rangle + \langle n\ m|\mathcal{H}_{12}|m\ n\rangle \\ \langle m\ n|\mathcal{H}_0|n\ m\rangle + \langle m\ n|\mathcal{H}_{12}|n\ m\rangle & \langle m\ n|\mathcal{H}_0|m\ n\rangle + \langle m\ n|\mathcal{H}_{12}|m\ n\rangle \end{pmatrix} \begin{pmatrix} c_1 \\ c_2 \end{pmatrix} = E \begin{pmatrix} c_1 \\ c_2 \end{pmatrix} \tag{97}$$

giving the secular equation

$$\begin{vmatrix} E_0 + Q - E & J \\ J & E_0 + Q - E \end{vmatrix} = 0 \tag{98}$$

with $E_0 = E_n + E_m$, the energy of the non-interacting pair and

$$Q = \langle n\ m|\mathcal{H}_{12}|n\ m\rangle = \langle m\ n|\mathcal{H}_{12}|m\ n\rangle \tag{99}$$

called the Coulomb integral since by replacing the states by functions the integral becomes that of the total charge times electron probability density for electron 1 times the potential given by electron 2. The other abbreviation is

$$J = \langle n\ m|\mathcal{H}_{12}|m\ n\rangle = \langle m\ n|\mathcal{H}_{12}|n\ m\rangle \tag{100}$$

known as the exchange integral due to the involvement of states for which the roles of the two electrons are interchanged. The final steps in equations (99) and (100) are apparent on inspection of the integrals.

The solution to equation (98) is

$$(E_0 + Q - E)^2 - J^2 = 0, \qquad \text{i.e. } E = E_0 + Q \pm J \tag{101}$$

so that Q gives an energy shift and J a splitting. When $E = E_0 + Q + J$ the pair of equations

$$(E_0 + Q - E)c_1 + Jc_2 = 0$$
$$Jc_1 + (E_0 + Q - E)c_2 = 0$$

is satisfied by $c_1 = c_2$. When $E = E_0 + Q + J$, $Jc_1 + Jc_2 = 0$ and $c_1 = -c_2$. Thus the resulting states are properly symmetric or antisymmetric.

The assumption of orthogonality is acceptable when both electrons move in the same central potential and the above treatment gives results appropriate to the He atom (as a rather crude approximation, since it can be seen to be equivalent to first-order perturbation theory and $\mathcal{H}_{12}$ is not really small). For the two electrons of the hydrogen molecule orthogonality cannot be assumed. Abbreviate the general combination to

$$|\varphi\rangle = c_1|n\ m\rangle + c_2|m\ n\rangle = c_1|1\rangle + c_2|2\rangle$$

Multiply $\mathcal{H}|\varphi\rangle = E|\varphi\rangle$ first by $\langle 1|$ from the left and then by $\langle 2|$ and write the two resultant equations in matrix form:

$$\begin{pmatrix} \langle 1|\mathcal{H}|1\rangle & \langle 1|\mathcal{H}|2\rangle \\ \langle 2|\mathcal{H}|1\rangle & \langle 2|\mathcal{H}|2\rangle \end{pmatrix} \begin{pmatrix} c_1 \\ c_2 \end{pmatrix} = E \begin{pmatrix} \langle 1|1\rangle & \langle 1|2\rangle \\ \langle 2|1\rangle & \langle 2|2\rangle \end{pmatrix} \begin{pmatrix} c_1 \\ c_2 \end{pmatrix} \tag{102}$$

which is the general eigenvalue problem. Setting $\langle 1|1\rangle = 1$ and $\langle 1|2\rangle = \alpha^2$ and making other substitutions as before gives

$$\begin{pmatrix} Q + E_0 & J + \alpha^2 E_0 \\ J + \alpha^2 E_0 & Q + E_0 \end{pmatrix} \begin{pmatrix} c_1 \\ c_2 \end{pmatrix} = E \begin{pmatrix} 1 & \alpha^2 \\ \alpha^2 & 1 \end{pmatrix} \begin{pmatrix} c_1 \\ c_2 \end{pmatrix} \tag{103}$$

which would, of course, become equation (97) with $\alpha^2 = 0$. Thus α^2 is the overlap integral for the product states:

$$\alpha^2 = \langle 1|2\rangle = \langle n\ m|m\ n\rangle = \langle n|m\rangle\langle m|n\rangle = \langle n|m\rangle^2$$
$$\alpha = \langle n|m\rangle \tag{104}$$

and absence of overlap of the wave functions is equivalent to orthogonality. The major distinction between He and H_2 should now be apparent.

(As an exercise the reader may show that this same equation [103] is obtained by the variational method using $|\varphi\rangle$, as given, as the trial wave function. In the absence of normalization the expectation value of the energy is given by $\langle E\rangle_\varphi = \langle\varphi|\mathcal{H}|\varphi\rangle/\langle\varphi|\varphi\rangle$ and it is assumed that the 'best' function is that which minimizes $\langle E\rangle$. Further, it is assumed that since the best approximation to $\langle E\rangle$ is the minimum value, minimum may be read as stationary. Thus multiply through by $\langle\varphi|\varphi\rangle$, expand $\langle\varphi|$ and $|\varphi\rangle$ and obtain two equations in the variables c_1 and c_2 according to

$$\frac{\partial}{\partial c_i}(E\langle\varphi|\varphi\rangle) = E\frac{\partial}{\partial c_i}\langle\varphi|\varphi\rangle + \langle\varphi|\varphi\rangle\frac{\partial E}{\partial c_i}$$
$$= \frac{\partial}{\partial c_i}\langle\varphi|\mathcal{H}|\varphi\rangle, \qquad \text{with } \frac{\partial E}{\partial c_i} = 0 \qquad \text{for } i = 1, 2$$

Although equation (102) is the matrix representation of $\mathcal{H}|\varphi\rangle = E|\varphi\rangle$ it is not exact because the basis is restricted.)

While the equation is now more difficult to solve it may be suggested by the foregoing example that solutions might correspond to $c_1 = c_2 = 1$ and to $c_1 = 1$, $c_2 = -1$. By the first hypothesis,

$$Q + E_0 + J + \alpha E_0 = E(1 + \alpha^2)$$

$$J + \alpha E_0 + Q + E_0 = E(\alpha^2 + 1)$$

Since these are the same, the first suggestion is verified and the reader may readily pursue the second and conclude that the solutions are

$$E_S = E_o + \frac{Q + J}{1 + \alpha^2}, \qquad |\varphi\rangle = |\varphi_S\rangle = \frac{1}{\sqrt{2}}(|n\ m\rangle + |m\ n\rangle) \tag{105}$$

(symmetric)

$$E_A = E_o + \frac{Q - J}{1 - \alpha^2}, \qquad |\varphi\rangle = |\varphi_A\rangle = \frac{1}{\sqrt{2}}(|n\ m\rangle - |m\ n\rangle) \tag{106}$$

(antisymmetric)

These state vectors or associated functions account for the orbital motion because φ is taken to depend upon the real coordinates $\mathbf{r}_1$ and $\mathbf{r}_2$, $\varphi = \varphi(\mathbf{r}_1, \mathbf{r}_2)$, and not on the 'spin coordinates'. The complete state is accounted for by indicating the constants of the motion for both the orbital and spin motion as by $|\psi\rangle = |\varphi\rangle|\chi\rangle$, with $|\chi\rangle$ the spin part. Since $\mathcal{H}(\mathbf{r}_1, \mathbf{r}_2)$ does not operate on $|\chi\rangle$ because $|\chi\rangle$ does not depend on $\mathbf{r}_1$ or $\mathbf{r}_2$, $|\varphi\rangle$ could in fact have been replaced by $|\psi\rangle$ since $|\chi\rangle$ would simply carry through as a constant and the states would be $|\varphi_S\rangle|\chi\rangle$ and $|\varphi_A\rangle|\chi\rangle$, with $|\chi\rangle$, in each case, initially undefined. However, there is information on $|\chi\rangle$ because the complete state $|\psi\rangle$ must be antisymmetric and so, formally,

$$\left.\begin{array}{l} |\varphi\rangle \text{ symmetric } \sim |\chi\rangle \text{ antisymmetric} \\ |\varphi\rangle \text{ antisymmetric } \sim |\chi\rangle \text{ symmetric} \end{array}\right\} |\psi\rangle = |\varphi\rangle|\chi\rangle(\text{antisymmetric}) \tag{107}$$

Each electron has spin $S = \frac{1}{2}$ and a single determinate component $M_s = \pm\frac{1}{2}\hbar$ corresponding to $|\frac{1}{2}\ \frac{1}{2}\rangle$ and $|\frac{1}{2}\ -\frac{1}{2}\rangle$ or just $|\frac{1}{2}\rangle$, $|-\frac{1}{2}\rangle$ or $|\alpha\rangle$, $|\beta\rangle$. The four symmetrized product states for the two electrons are

$$\begin{array}{llll} |\alpha\ \alpha\rangle, & |1\ \ 1\rangle, & M_s = 1 & \multirow{3}{*}{} \\ |\alpha\ \beta\rangle + |\beta\alpha\rangle, & |1\ \ 0\rangle, & M_s = 0 & \\ |\beta\ \beta\rangle, & |1\ -1\rangle, & M_s = -1 & \end{array} \left.\right\} \begin{array}{l} S = 1, \text{ symmetric} \\ (2S+1 = 3) \end{array}$$

$$|\alpha\ \beta\rangle - |\beta\ \alpha\rangle, \quad |0\ \ 0\rangle, \quad M_s = 0 \left.\right\} \begin{array}{l} S = 0, \text{ antisymmetric} \\ (2S+1 = 1) \end{array}$$

As will become apparent (see Chapter 3, Sections 1.4, 1.5) the two spins $\frac{1}{2}$ behave much like a single spin 1 with components $\hbar$, 0 and $-\hbar$ when the states are the symmetric $|1\ 1\rangle$, $|1\ 0\rangle$ and $|1\ -1\rangle$ but the total spin is effectively zero for the antisymmetric $|0\ 0\rangle$. The former give a triplet, in accordance with $2S + 1 = 3$ and the latter, of course, a singlet: $2S + 1 = 1$. It will also be shown that these are the four eigenstates obtained for the Heisenberg Hamiltonian:

$$\mathcal{H}_{\mathcal{G}} = -\mathcal{J}\mathbf{S}_1 \cdot \mathbf{S}_2 \tag{108}$$

and that the singlet state gives an energy of $+\frac{3}{4}\hbar^2\mathcal{J}$ while the triplet states are degenerate with energy $-\frac{1}{4}\hbar^2\mathcal{J}$ (so that $\Delta E = \mathcal{J}$). $\mathcal{J}$ may be taken to be positive or negative.

Due to the association of the symmetric (triplet) spin states with the antisymmetric orbital states and vice versa, the spin singlet–triplet splitting in energy is just the difference between E_S and E_A:

$$\Delta E = E_S - E_A = \frac{(Q+J)(1-\alpha^2)-(Q-J)(1+\alpha^2)}{(1+\alpha^2)(1-\alpha^2)} = \frac{2J-2\alpha^2 Q}{1-\alpha^4} \tag{109}$$

and so this may be equated to the Heisenberg exchange parameter $\mathcal{J}$. Although equation (108) is written as though there were a direct 'exchange interaction' between the spins, it is seen that this is only an effective spin–spin interaction resulting from symmetry rules and orbital overlap.

Due to the form of ΔE, or $\mathcal{J}$, it cannot be assumed to be positive or negative. On evaluating the integrals, specifically for H_2 and using the simple 1s functions for atomic hydrogen, Heitler and London [3] found that the singlet state lay below the triplet, i.e. that $\mathcal{J}$ was negative [note the sign in equation (108)] and strongly dependent on the internuclear spacing (Figure 2.9). This is, of course, in line with the diamagnetism of H_2 and the general observation of 'spin-pairing' in bonds, with notable exceptions such as O_2. Calculations of the energies E_S and E_A at varying spacings indicated that for all spacings E_A was above E_0 so that this corresponded to an antibonding state, with E_S corresponding to the bonding state.

When $\mathcal{J}$ is negative and the singlet lowest, and if it could be assumed that similar considerations applied to large arrays or 'hydrogen crystals', then an antiferromagnetic ground state would be approached at low temperatures. However large $\mathcal{J}$ is in magnitude, if this were the only consideration the approach would only be gradual with no magnetic phase transition or critical temperature (T_N). Thus a negative $\mathcal{J}$ can only be considered as *indicative* of antiferromagnetism. Conversely, a positive $\mathcal{J}$ is *indicative* of ferromagnetism, though not alone adequate to explain ferromagnetic ordering.

Although the overlap integral as such has a particular description and a particular significance in relation to orthogonality, the Coulomb and exchange integrals also involve a modified overlap to two orbitals and might be termed 'proximity integrals'. Since the wave functions fall off exponentially from the nucleus the coupling should be very sensitive to the interatomic spacing (though not exceptionally sensitive, with reference to coupling in amorphous materials; see Chapter 4). The s orbitals have spherical symmetry but generally ψ or $|\psi|^2$ extends at high values along particular directions in the crystal structure. Thus

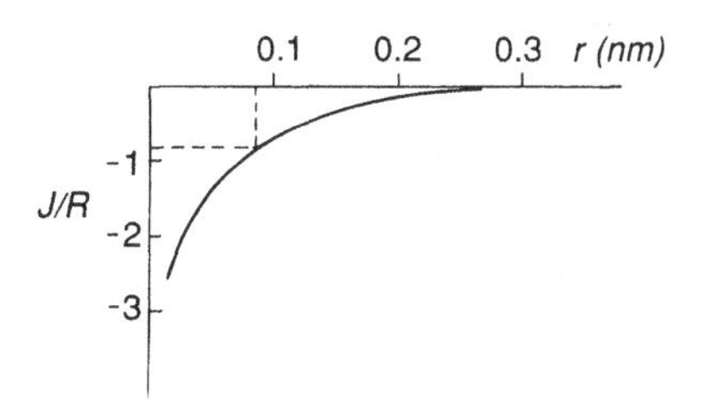

Figure 2.9 Variation of the reduced exchange energy with internuclear spacing. R is the atomic energy unit

it is not surprising that for the commonly occurring orbital singlets in crystals (e.g. MnO, Fe_2O_3), discussed in Section 9, the exchange is isotropic but that in other cases it may be anisotropic and make a contribution to the magnetocrystalline anisotropy (in addition to single-ion anisotropies which exist in the absence of exchange).

For an n-electron system the Hamiltonian should represent all the pair interactions. The appropriate n-particle states might be in the form of Slater determinants, giving automatic antisymmetrization, but according to Dirac [4] simple products of single-particle states, each a product of orbital and spin states or functions, can also be used so long as an appropriate exchange term as such is included in the Hamiltonian. This term would not be present in the original $\mathcal{H}$. The term must be chosen in such a way that the resulting effective Hamiltonian $\mathcal{H}_e$ has matrix elements when represented in the basis of the simple product functions, which are the same as the matrix elements of $\mathcal{H}$ itself when represented in the determinantal states. With no orbital degeneracy, e.g. with orbital singlets lowest in crystal fields (Section 9), $\mathcal{H}_e$ can, following Herring [5] (see also Levy [6, p. 183]), be expressed in a form in which it operates on the initially degenerate set of spin states as

$$\mathcal{H}_e = Q - \sum_{i>j} \mathcal{J}_{ij}(\tfrac{1}{2} + 2\mathbf{S}_i \cdot \mathbf{S}_j) + O \tag{110}$$

The number of spin states is the product of the spin multiplicities $\Pi_{i=1}^{N}(2S_i + 1)$ for N atoms. The sums are taken over the N atoms in the system with $i > j$ to preclude $i = j$. In this,

$$\mathcal{J}_{ij} = \langle \Psi(\cdots \psi_i(m)\psi_j(n)\cdots) \left| \frac{e^2}{4\pi\varepsilon_0 r_{mn}} \right| \Psi(\cdots \psi_i(m)\psi_j(n)\cdots) \rangle \tag{111}$$

where the ψ_i are the single-particle wave functions, each a product of the orbital and spin parts: $R_{nl}(\mathbf{r}_i)\varphi_{lm}(\mathbf{r}_i)\chi_{m_s}$ and Q is the direct Coulomb integral $\langle \Psi | \sum_{i>j} e^2/(4\pi\varepsilon_0 r_{ij}) | \Psi \rangle$. The term O is a small correction depending on $e^{-\alpha R}$, with R the distance between the atoms. $\mathcal{J}_{ij}$ should include overlap terms, but when these are neglected it can be seen that $\mathcal{H}_e$ corresponds to the Heisenberg Hamiltonian. The effective spin Hamiltonian matrix arises as a result of taking the pair interaction elements, in a basis including spin states, and factoring out the orbital integrals. Only spin operators occur in $\mathcal{H}_e$, which thus constitutes a spin Hamiltonian although the interaction is between the orbitals.

In view of the proximity requirement, the separation of two paramagnetic ions by an intervening diamagnetic atom, say, would appear to break the coupling, but exchange effects are exceptionally complex and may be very indirect. It is possible to prepare multilayer films in which ferromagnetic films are separated by non-magnetic layers and if the layers are thin, though still several atoms thick, the films may remain mutually exchange coupled. As further evidence of the complexity involved, if an iron or other 3d atom is situated at low concentration in a non-magnetic metal, the magnetic moment per iron atom may be very small, virtually zero, or may be exceptionally high, depending on the electron number of the host [7]. Giant moments $\sim 10\beta$ or more cannot be ascribed to the atom as such and it appears that these effects are associated with exchange polarization of conduction electrons in the host metals, this interaction being either ferromagnetic or antiferromagnetic in nature. In turn the resistivity of the host metals is expected to be affected. It had long been known that, in some apparently very pure samples of gold or copper, the resistance could increase with decreasing temperature (behaviour more typical of a semiconductor) and this was traced to the presence of minute impurities of iron. This 'Kondo effect' [8] arises when the conduction electrons are antiferromagnetically coupled to those of the magnetic atom and have received very extensive theoretical attention.

The idea that a general indirect exchange interaction between localized moments could arise via the exchange polarization of conduction or itinerant electrons was introduced by Ruderman and Kittel [9], Kasuya [10] and Yosida [11]—hence the 'RKKY' mechanism. This has its lowest order contribution from second-order perturbation of the localized moments by the s–f interactions with the (s) conduction electrons (see, for example, Levy [6]). The mechanism is generally anisotropic. The conduction electrons may also be those in, for example, Si layers, ~ 1.0 nm, separating Fe layers.

A mechanism known as superexchange [12] involves exchange between the ionic moments via outer closed-shell electrons, centred on the same ions, or via an intervening anion, typically O^{2-}. Consider the group Mn^{2+} O^{2-} Mn^{2+} as in Figure 2.10. According to Kramer's theorem the oxygen may be partially in an excited state; a ground state p electron may be excited into a d state and associated with a neighbouring cation. If the 3d shell is half full or more it is expected, by Hund's rule, that the excited spin should be antiparallel to the net spin of the cation, and if the shell is initially less than half full all the d electron spins are expected to become parallel. The spin of the remaining p electron is taken to remain coupled antiparallel to the excited spin and thus either parallel or antiparallel to the d spins of cation 2 in the figure. Since the oxygen now has a net spin, this may be exchange-coupled directly, due to overlap, with cation 1, and this may be expected to be antiferromagnetic in nature. The effect will be to give ferromagnetic or antiferromagnetic coupling between the cations according to whether their d shells are less than half full or at least half full respectively. This accords with the antiferromagnetism of MnO, NiO, CoO, etc., and the ferromagnetism of CrO_2.

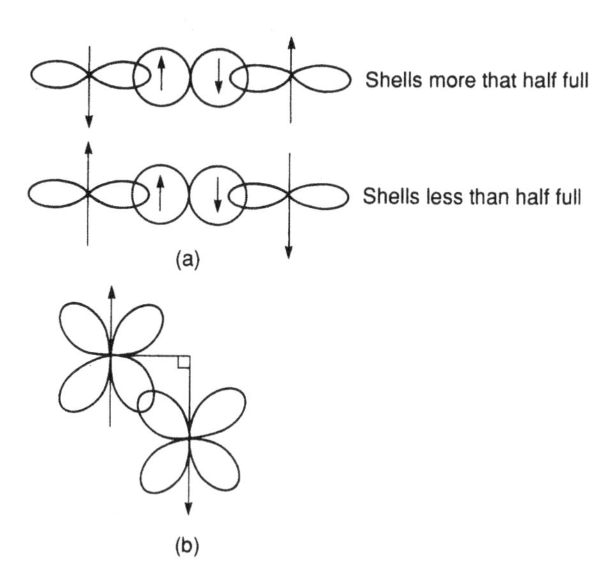

Figure 2.10 (a) This suggests how exchange may be effective via an intervening anion with negligible direct overlap and (b) shows the '90° overlap' of d orbitals which may occur directly. (Reprinted by permission of John Wiley & Sons, Ltd, Magnetic Oxides, ed. D.J. Craik, Wiley 1975)

Zener's double exchange [13] involves the excitation of a d electron, from the cation with the highest number of such, into an overlapping anion orbital, with the transfer of one anion p electron to the second cation. The cations may effectively interchange positions and become ferromagnetically ordered. Differing valencies are specifically involved. The formation of hybrid orbitals by combining d, s and p orbitals to give semi-covalent exchange has been discussed by Goodenough [14].

Some direct overlap giving antiferromagnetic exchange is considered to be possible for octahedral B-site cations in spinel ferrites. The principal interactions in the ferrites are antiferromagnetic, although the result is to produce a spontaneous magnetization. Interactions in ferrites are discussed by Blasse, for example [15].

In Section 9.2 it is suggested that cooperative Jahn–Teller effects, in oxides for example, may lead to some correlation of the orbital angular momentum and influence the exchange coupling.

7 Orbital Angular Momentum Quenching

The field from an electron considered as a particle, $q = -e$, at a nucleus w.r.t. which its position is $\mathbf{r}$, is $B = -[\mu_0/(4\pi)]e\mathbf{r}\times\mathbf{v}/r^3 = -[\mu_0/(4\pi)](e/m)\mathbf{L}/r^3 \sim [\mu_0/(4\pi)]\beta/r^3$ since $\mathbf{r}\times\mathbf{v} = (1/m)\mathbf{r}\times\mathbf{p} = (1/m)\mathbf{L}$ and $L \sim \hbar$ is assumed. For a 2p electron this predicts $B \sim 10$ T. Such fields do not exist in most molecules and crystals due to the quenching of the angular momentum with cancellations due to opposite equivalent components, as noted in Chapter 1. The operators L_k, $k = x, y, z$, are Hermitian so that diagonal matrix elements and mean values are real. The L_k (e.g. $L_z = -i\hbar\, \partial/\partial\phi$) can be written as $L_k = iL'_k$ with L'_k real and with real orbitals appropriate to crystal field environments $\langle L_k\rangle^* = \langle L_k\rangle$ can only apply if $\langle L_k\rangle = 0$. The L_k^2 are real and as an exercise it may be shown that $\langle L_z^2\rangle = \hbar^2$, for p orbitals, and similarly for L_x^2 and L_y^2 using $L_x = \frac{1}{2}(L^+ + L^-)$; the reader may also demonstrate quenching by recourse to the matrices L_k given. At first sight the quenching is complete but contributions do arise as follows.

8 Perturbation Theory, Spin Hamiltonians and g

If $\mathcal{H} = \mathcal{H}_0 + \mathcal{H}_1$ with known eigenstates of $\mathcal{H}_0$, with $\mathcal{H}_0|n\rangle^0 = E^0|n\rangle^0$ and $\mathcal{H}_1$ representing a small effect, then according to perturbation theory the energies for $\mathcal{H}$ are $E_n = E_n^0 + E_n^{(1)} + E_n^{(2)}(+\cdots)$ with $E_n^{(i)}$ the ith-order 'correction'. $E_n^{(1)} = \langle n|\mathcal{H}_1|n\rangle$ and the states remain $|n\rangle^0$. The second-order term is

$$E_n^{(2)} = \sum_{n,k}{}' \frac{\langle k|\mathcal{H}_1|n\rangle\langle n|\mathcal{H}_1|k\rangle}{E_n^0 - E_k^0} \tag{112}$$

and the states are then

$$|n\rangle = |n\rangle^0 + \frac{\sum' \langle k|\mathcal{H}_1|n\rangle|k\rangle}{E_n^{(0)} - E_k^{(0)}} \tag{113}$$

with $k = n$ excluded and the $|n\rangle$, $|k\rangle$ on the right read as the unperturbed states. With $\mathcal{H}_1$ approximated as $-\gamma BS_z$ and the states the products $|n\rangle|m_S\rangle$, then $E^{(1)} = \langle m_S|\langle n| - \gamma BS_z|n\rangle|m_s\rangle = \langle n|n\rangle\langle m_s| - \gamma BS_z|m_s\rangle = -\gamma B\hbar m_s$ with $E^{(2)} = 0$ due to orthogonality of the $|n\rangle$; this is simply consistent since the $|n\rangle|m_S\rangle$ are eigenstates of $\mathcal{H}$. A number of perturbations may be treated in turn but the order in which they are taken may be important. For a p electron in crystal fields (V_1),

$$\begin{aligned}\mathcal{H} &= \left(\frac{p^2}{2m_e} + V_0 + V_1\right) + \lambda\mathbf{L}\cdot\mathbf{S} + \beta\mathbf{B}\cdot(\mathbf{L}+2\mathbf{S}) \\ &= \mathcal{H}_0 + \beta\mathbf{B}\cdot\mathbf{L} + \lambda\mathbf{L}\cdot\mathbf{S} + 2\beta\mathbf{B}\cdot\mathbf{S} \qquad (114)\end{aligned}$$

($\hbar = 1$) for motion in a field, including SO coupling but neglecting the 'diamagnetic term'.

$\mathcal{H}_0$ represents the purely orbital motion in the absence of a field and the three real p orbitals or states $p_x \sim |1\rangle$, $p_y \sim |2\rangle$ and $p_z \sim |3\rangle$ are eigenstates of $\mathcal{H}_0 : \mathcal{H}_0|1\rangle = E_1|1\rangle$ etc., with E_1 the lowest and $|1\rangle$ the ground state. Following Slichter [16] it is taken that the energies are well separated, i.e. that the crystal field term V_1 is large compared to the succeeding terms in equation (114) so that the crystal field states are identifiable as the unperturbed states and the latter terms are perturbations. This is specifically the strong crystal field case and an analogous treatment would apply to a single 3d electron in a 3d transition series ion, but it is not universally applicable.

More fully, the states can be written $|n, m_s\rangle$ or $|n\rangle|m_s\rangle$ to account for spin. The unperturbed states $|n\rangle^0$ are degenerate with respect to spin. Consider first the term

$$\beta\mathbf{B}\cdot\mathbf{L} = \beta(B_xL_x + B_yL_y + B_zL_z)$$

which clearly contains no spin operators. First-order perturbation terms are zero because all the terms

$$\langle m|\langle n|L_i|n\rangle|m\rangle = \langle m|m\rangle\langle n|L_i|n\rangle$$

disappear in correspondence with the quenching for the unperturbed states. However, the states themselves may be taken to be perturbed according to (for the ground state $|1\rangle$)

$$|1\rangle|m\rangle = (|1\rangle|m\rangle)^0 + \beta\sum_{m'}\sum_{i=x,y,z}\sum_{k=2,3}\frac{\langle m'|\langle k|L_i|1\rangle|m\rangle}{E_1 - E_k}B_i(|k\rangle|m'\rangle)^0 \qquad (115)$$

The first summation is over the spin states, e.g. $m' = -\frac{1}{2}, \frac{1}{2}$; the second arises because $\mathbf{B}\cdot\mathbf{L}$ is taken as the sum of the three component products (note the occurrence of B_i) and the third corresponds to the other two orbital states. (Although these are the unperturbed vectors the superscripts may be implied.) Because the L_i do not affect the $|m\rangle$ as such the $\langle m'|m\rangle$ may be factored out of each term to give $\delta_{mm'}$. The first summation disappears and $\langle m'|m\rangle$ selects only those terms with $m' = m$ so that $|m\rangle$ can be written as a factor and equation (115) becomes

$$|1\rangle|m\rangle = \left\{|1\rangle + \beta\sum_{i=x,y,z}\sum_{k=2,3}\frac{\langle k|L_i|1\rangle}{E_1 - E_k}B_i|k\rangle\right\}|m\rangle \qquad (116)$$

Attention is confined to the orbital ground state, i.e. to the set of states $|1, m\rangle$ having $n = 1$ and all permissible m values. The matrix representation of the SO operator in this basis may now be found, expanding $\lambda\mathbf{L}\cdot\mathbf{S}$ in the components and first taking only λL_xS_x for simplicity. Note first the separation

$$\langle m'|\langle 1|\lambda L_xS_x|1\rangle|m\rangle = \lambda\langle m'|S_x|m\rangle\langle 1|L_x|1\rangle$$

$\langle 1|L_x|1\rangle$ is not now zero because this is the mean value in the *perturbed* states. In fact,

$$\langle 1|L_x|1\rangle = \left(\langle 1|^0 + \beta \sum_{i=x,y,z}\sum_{k=2,3} \frac{\langle 1|L_i|k\rangle}{E_1 - E_k} B_i \langle k\right) L_x$$

$$\left(|1\rangle^0 + \beta \sum_i \sum_k \frac{\langle k|L_i|1\rangle}{E_1 - E_k} B_i |k\rangle\right) \tag{117}$$

($L^{i*}_{k1} = L^i_{1k}$). Expansion gives four terms, first, with obvious quenching:

$$\langle 1|^0 L_x |1\rangle^0 = 0$$

Next, consider the term

$$\left(\sum\sum \cdots \langle k|\right) L_x \left(\sum\sum \cdots |k\rangle\right)$$

Since the sums are over $k = 2, 3$ these do not involve the ground state and may be ignored. This leaves

$$\langle 1|^0 L_x| \left(\beta \sum_i \sum_k \frac{\langle k|L_i|1\rangle}{E_1 - E_k} B_i |k\rangle\right) = \beta \sum_i \sum_k \frac{\langle k|L_i|1\rangle\langle 1|L_x|k\rangle}{E_1 - E_k} B_i \tag{118}$$

together with a similar term $\langle \cdots |L_x|1\rangle^0$. To take account of all the components L_x is now replaced by L_j with summation over $j = x, y, z$. Thus, in view of equation (116), a matrix element would be

$$\langle m'|\langle 1|\lambda \mathbf{L}\cdot\mathbf{S}|1\rangle m\rangle = \lambda\beta \sum_{\substack{i=x,y,z\\ j=x,y,z}} \sum_{k=2,3} \frac{\langle k|L_i|1\rangle\langle 1|L_j|k\rangle + \langle 1|L_i|k\rangle\langle k|L_j|1\rangle}{E_1 - E_k} B_i \langle m'|S_j|m\rangle \tag{119}$$

If we postulated an effective Hamiltonian

$$\mathcal{H}_e = \beta \sum_i \sum_j S_i a_{ij} B_j \tag{120}$$

with

$$a_{ij} = \lambda \sum_k \frac{\langle k|L_i|1\rangle\langle 1|L_j|k\rangle + \langle 1|L_i|k\rangle\langle k|L_j|1\rangle}{E_1 - E_k} = a_{ji} \tag{121}$$

then this would clearly give the above matrix elements within the orbital ground state, so the concept (120) is valid. While it contains the spin component operators only, the effects of the terms $\mathbf{B}\cdot\mathbf{L}$ and $\lambda\mathbf{L}\cdot\mathbf{S}$ are incorporated in the a_{ij}, as are the effects of the crystal field splitting itself. The a_{ij} can be shown to constitute a second-rank tensor according to the way in which they are transformed by coordinate rotations. In view of its nature the spin–Zeeman term can be added directly to $\mathcal{H}_e$ to give the total *spin Hamiltonian.*

In certain cases only the a_{ii} will be non-zero: the matrix will be diagonal. For example, recalling that here $|1\rangle \sim xf(r)$, $|2\rangle \sim zf(r)$ and $|3\rangle \sim yf(r)$ and noting that by the chain law

$$\frac{\partial}{\partial x} f(r) = \frac{\partial f}{\partial r}\frac{\partial r}{\partial x} = \frac{\partial f}{\partial r}\frac{\partial}{\partial x}(x^2 + y^2 + z^2)^{1/2} = \frac{\partial f}{\partial r}\frac{x}{r}$$

and similar relations, it is easy to see that $[f \equiv f(r)]$

$$L_x(xf) = -i\left(y\frac{\partial}{\partial z} - z\frac{\partial}{\partial y}\right)xf = -iyx\frac{\partial f}{\partial r}\frac{z}{r} + izx\frac{\partial f}{\partial r}\frac{y}{r} = 0$$

$$L_y(xf) = -i\left(z\frac{\partial}{\partial x} - x\frac{\partial}{\partial z}\right)xf = -izx\frac{\partial f}{\partial r}\frac{x}{r} - izf + ix^2\frac{\partial f}{\partial r}\frac{z}{r}$$
$$= -izf(r)$$

It is left to the reader to fill in, with an obvious notation:

$$\begin{array}{lll} L_x|x\rangle = 0, & L_x|y\rangle = i|z\rangle, & L_x|z\rangle = -i/y\rangle \\ L_y|x\rangle = -i|z\rangle, & L_y|y\rangle = 0, & L_y|z\rangle = i|x\rangle \\ L_z|x\rangle = i|y\rangle, & L_z|y\rangle = -i|x\rangle, & L_z|z\rangle = 0 \end{array}$$

Thus most of the terms in the a_{ij} give zero and it is seen that contributions come from such terms as

$$\langle y|L_z|x\rangle\langle x|L_z|y\rangle = \langle y|i|y\rangle\langle x| - i|x\rangle = 1$$

$$\langle z|L_y|x\rangle\langle x|L_y|z\rangle = \langle z| - i|z\rangle\langle x|i|x\rangle = 1$$

and in particular that the index of $L_{i/j}$ must be common, i.e. $i = j$. Specifically, for this example,

$$\mathcal{H}_e = 2\beta\left\{\frac{\lambda}{E_x - E_z}S_yB_y + \frac{\lambda}{E_x - E_y}S_zB_z\right\} \tag{122}$$

The calculation of the energies in the crystal fields is included in a more general approach in the next section.

$\mathcal{H}_e$ can be regarded as giving the spin energy due to the effect of the applied field on the orbital motion coupled in by the spin–orbit coupling, and to this must be added the direct spin–Zeeman term $2\beta\mathbf{B}\cdot\mathbf{S}$ to give

$$\mathcal{H} = \beta\left[2S_xB_x + \left(2 + \frac{2\lambda}{E_x - E_z}\right)S_yB_y + \left(2 + \frac{2\lambda}{E_x - E_z}\right)S_zB_z\right]$$
$$= \beta(g_{xx}B_xS_x + g_{yy}B_yS_y + g_{zz}B_zS_z)$$
$$= \beta\mathbf{B}\cdot\mathbf{g}\cdot\mathbf{S} \tag{123}$$

where the dyadic $\mathbf{g}$ is given by

$$\mathbf{g} = \mathbf{i}g_{xx}\mathbf{i} + \mathbf{j}g_{yy}\mathbf{j} + \mathbf{k}g_{zz}\mathbf{k} \tag{124}$$

with the usual $\mathbf{i}\cdot\mathbf{i} = 1$, $\mathbf{i}\cdot\mathbf{j} = 0$, etc.

The approximated *spin Hamiltonian* constitutes a direct replacement for the simple spin Hamiltonian for an isolated spin: $g\beta\mathbf{B}\cdot\mathbf{S}$.

The X, Y and Z axes are the principal axes of the tensor a_{ij}, in this case due to the symmetry of the crystal field. The matrix representing $\mathbf{g}$ is diagonal for the same reason.

Taking $g = 2$ for the free spin, an effective field may be defined by the correspondence between

$$2\beta\mathbf{B} \sim \beta\mathbf{B}\cdot\mathbf{g} = \mathbf{i}\beta B_xg_{xx} + \mathbf{j}\beta B_yg_{yy} + \mathbf{k}\beta B_zg_{zz}$$

i.e. the effect of the orbital motion and spin–orbit coupling can be included by replacing the applied field (in the free spin case) by an effective field

$$\mathbf{B}_e = \mathbf{i}\frac{g_{xx}B_x}{2} + \mathbf{j}\frac{g_{yy}B_y}{2} + \mathbf{k}\frac{g_{zz}B_z}{2} \tag{125}$$

In terms of this the spin Hamiltonian is

$$\mathcal{H} = 2\beta\mathbf{B}_e \cdot \mathbf{S} \tag{126}$$

Clearly $\mathbf{B}_e$ differs from $\mathbf{B}$ in direction as well as magnitude and thus the resonant response becomes anisotropic. The spin statics and dynamics are controlled by the Hamiltonian and any theory developed for a free spin applies also to this case with $B \rightarrow B_e$.

Noting that in this section the convention $S_z|\alpha\rangle = \frac{1}{2}|\alpha\rangle$ (rather than $S_z|\alpha\rangle = \frac{1}{2}\hbar|\alpha\rangle$) has been followed, the resonant condition becomes

$$\omega_0 = \frac{2\beta B_e}{\hbar} = -\gamma B_e$$

i.e.

$$\begin{aligned}\hbar\omega_0 &= \beta(g_{xx}^2 B_x^2 + g_{yy}^2 B_y^2 + g_{zz}^2 B_z^2)^{1/2} \\ &= \beta B(\alpha_1^2 g_{xx}^2 + \alpha_2^2 g_{yy}^2 + \alpha_3^2 g_{zz}^2)^{1/2} \\ &= g\beta B \end{aligned} \tag{127}$$

where $\alpha_1 = \cos\theta_1$ with θ_1 the angle between the applied field and the X-axis etc. This defines the *g factor*:

$$g = (\alpha_1^2 g_{xx}^2 + \alpha_2^2 g_{yy}^2 + \alpha_3^2 g_{zz}^2)^{1/2} \tag{128}$$

stressing the anisotropy once more. Clearly $\omega_0 \propto B$ still applies. The g shift δg is defined as the difference between the observed g and the free spin value and the magnitude of δg is an order of magnitude more than the difference between 2.0 and the true free spin value. Recalling that E_x is the lower of the three energies (in this example), $\lambda > 0$ implies that $\delta g = g - 2$ is negative and vice versa. As previously noted, according to Hund's rules, $\lambda > 0$ applies to shells that are less than half full, $\lambda < 0$ for shells more than half full.

The considerations of this section do not apply only to resonance. The spin Hamiltonian also controls such properties as the static magnetic susceptibility and its possible anisotropies. However, g values are much more accurately measured by resonance methods and the development of e.s.r. gave impetus to calculations of states and energies of ions in crystal fields, as follows.

9 General Crystal Field Theory and Magnetic Applications

Oxides are the most intensively studied materials containing ions with magnetic moments and in these the ions usually occupy sites that are variously:

Octahedral	sixfold coordinated
Tetrahedral	fourfold coordinated
Cubal	eightfold coordinated

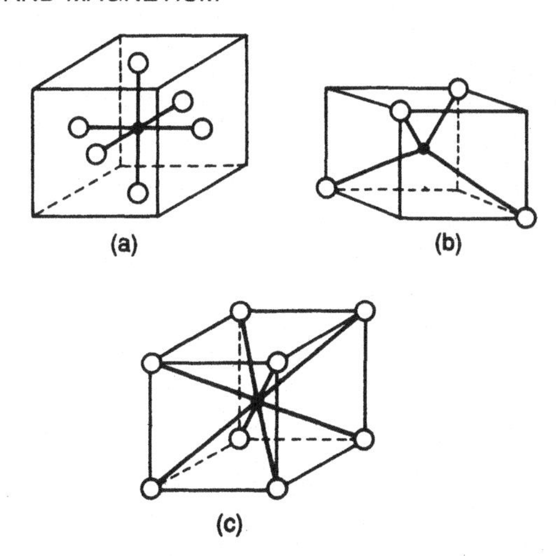

Figure 2.11 Arrangements of oxygen ions (open circles) in a cubic lattice forming (a) an octahedral site, (b) a tetrahedral site and (c) a cubal site. (Reprinted by permission of John Wiley & Sons, Ltd, Magnetic Oxides, ed. D.J. Craik, Wiley 1975)

(see Figure 2.11). The ions most extensively studied are 3d and illustrations will be largely confined to these, although the principles are general and 4f can be treated. These are the important ions in the ferrimagnetic spinels and garnets but the most detailed experimental data comes from e.s.r. or, more properly, paramgnetic resonance, and the ions must then occur at low concentrations in diamagnetic (often oxide) host crystals, e.g. Al_2O_3. Exchange fields, used as a representation of the exchange coupling in the ordered state, are an order of magnitude or more above laboratory e.s.r. fields and possible difficulties introduced by this should be borne in mind.

The electrostatic field due to neighbours must constitute a solution of Laplace's equation and can be expressed as a sum of spherical harmonics [17]. For an ion at the centre of an octahedral array of charges q at $(\pm a, 0, 0)$, etc., the CF potential energy term, to be added to the free ion potential energies for an electron at x, y, z with respect to the nucleus, is

$$\begin{aligned} V = C_0 + C_4(x^4 + y^4 + z^4 - \tfrac{3}{5}r^4) \\ + D_6[x^6 + y^6 + z^6 + \tfrac{15}{4}(x^2y^4 + y^2z^4 + z^2x^4 + y^2x^4 + z^2y^4 + x^2z^4) \\ - \tfrac{15}{4}r^6] + \cdots \end{aligned} \quad (129)$$

The same applies with a tetrahedron of neighbours at $(1/\sqrt{3})(a, a, a)$, etc., and a cube of neighbours at $(1/\sqrt{3})(\pm a, \pm a, \pm a)$, etc., but the coefficients differ, as in Table 2.1. The single-electron energies are to be summed over all the electrons in the ion. Z is chosen as a fourfold axis of rotation for the six- and eightfold coordinations.

Since the cubic CF splitting is usually less than that between the free-ion (L, S) terms but larger than the SO coupling effect $(\lambda \mathbf{L} \cdot \mathbf{S})$, an L, M_L representation may be used to derive the matrix elements and, also, only the ground term needs usually to be considered. For multielectron ions each state $|L\ M_L\rangle$ can be expressed as a determinant state using the 3d states of the n electrons, as given in Chapter 1, and the integrations are then carried

Table 2.1 Values of the coefficients in the cubic crystal fields [18] (Reprinted by permission of John Wiley & Sons Ltd, Magnetic Oxides, ed. D.J. Craik, Wiley 1975)

		Cartesian coordinates C_4	D_6	C_0	D'_4	Spherical polar coordinates D'_6
Coordination number	Units	$\frac{-\lvert e\rvert q}{a^5}$	$\frac{-\lvert e\rvert q}{a^7}$	$\frac{-\lvert e\rvert q}{a}$	$\frac{-\lvert e\rvert q\sqrt{\pi r^4}}{a^5}$	$\frac{-\lvert e\rvert q\sqrt{\pi r^6}}{\sqrt{13}a^7}$
4		$-\frac{35}{9}$	$-\frac{112}{9}$	4	$-\frac{28}{27}$	$\frac{16}{9}$
6		$\frac{35}{4}$	$-\frac{21}{2}$	6	$\frac{7}{3}$	$\frac{3}{2}$
8		$-\frac{70}{9}$	$-\frac{224}{9}$	8	$-\frac{56}{27}$	$\frac{32}{9}$

out. The angular parts are readily integrated and the numbers can be assumed. The radial integrals required,

$$\int_0^\infty [R(r)]^2 r^n r^2 \, dr \equiv \langle r^n \rangle$$

with the $R(r)$ also as given in Section 2.3, can be evaluated separately and the results are given in Table 2.2 together with values of λ (SO) for a few examples. Thus the matrix of the Hamiltonian in the presence of the crystal field can be derived and the eigensystem solved numerically.

A more elegant approach is described by Stevens and Bates [18] on whose paper (SB) this section depends. In general the matrices, or secular equations, are too 'large' to be solved directly but the Wigner–Eckart theorem shows how many of the matrix elements may be interrelated (see for example [1], p. 245) and this suggests that 'smaller' and simpler matrices might serve. Assuming representation in an effectively complete basis, spanning a particular space, if matrices $A = B$ then $A = B$ applies to the operators. If the matrix elements are the same (or proportional) within a certain subspace only, then the operators are not the same, but they can be effectively the same if this subspace only is of interest.

The CF energy terms are homogeneous functions of the coordinates. Since the subspace here is that corresponding to different values of M_L, x, y and z might initially be replaced by L_x, L_y and L_z, but the x, y and z operators commute while the L_i do not and allowance must be made for this; e.g. xy becomes $\frac{1}{2}(L_x L_y + L_y L_x) = yx$. There are also constants of proportionality to be introduced. Table 2.3 gives the equivalent angular momentum operators $O_m^{(n)}$ in terms of L_z and $L^\pm$ (rather than L_x and L_y, for greater ease of evaluation), from SB and references cited there.

The ground terms of the 3d^n have $L = 0, 2$ or 3: S,D,F (Table 2.2). No splitting is expected for S states on purely pictorial grounds with spherical symmetry, though a shift by C_0 occurs. Consider 3d^4, ^{5}D, ignoring the shift C_0. The replacement of the CF operator is

$$C_4\left[x^4 + y^4 + z^4 - \tfrac{3}{5}r^4\right] \to C_4\beta\langle r^4\rangle(L_x^4 + L_y^4 + L_z^4 + \cdots) \tag{130}$$

with $\langle r^4\rangle$ and β as given in Table 2.2. The dots indicate the replacement of r^4, which need not be explicit since its contributions to the matrix elements have been tabulated. On transforming to $L^\pm$ according to Hutchings [17],

$$V = \frac{C_4}{20}\beta\langle r^4\rangle(O_4^0 + 5O_4^4)$$

Table 2.2 Constants appropriate to $3d^n$ free ions [18] (Reprinted by permission of John Wiley & Sons Ltd, Magnetic Oxides, ed. D.J. Craike, Wiley 1975)

Configuration		Ion	Ground state	$\langle r^2 \rangle$ (a.u.)	$\langle r^4 \rangle$ (a.u.)	λ (cm^{-1})	α	β
		Sc^{2+}		4.560	52.360	86		
	→	T_I^{3+}		1.911	7.307	159		
$3d^1$		V^{4+}	^2D	1.389	3.753	255	$-\frac{2}{21}$	$\frac{2}{63}$
		Cr^{5+}		1.080	2.214			
		Mn^{6+}		0.871	1.410			
		Sc^{+}						
		Ti^{2+}		2.400	12.651	61		
$3d^2$	→	V^{3+}	^3F	1.632	5.450	106	$-\frac{2}{105}$	$-\frac{2}{315}$
		Cr^{4+}		1.219	2.883	163		
		Mn^{5+}		0.964	1.756			
		Ti^{+}		3.417	29.570			
$3d^3$		V^{2+}	^4F	2.028	9.112	57	$\frac{2}{105}$	$\frac{2}{315}$
	→	Cr^{3+}		1.434	4.277	91		
		Mn^{4+}		1.102	2.419	135		
		V^{+}		2.767	19.571			
$3d^4$	→	Cr^{2+}	^5D	1.760	7.003	59	$\frac{2}{21}$	$-\frac{2}{63}$
	→	Mn^{3+}		1.278	3.426	87		
		Fe^{4+}		0.995	1.995	125		
		Cr^{+}		2.280	13.211			
$3d^5$	→	Mn^{2+}	^6S	1.528	5.325			
	→	Fe^{3+}		1.141	2.765	—	—	—
		Co^{4+}		0.903	1.658			
		Mn^{+}		2.030	11.025	−64		
$3d^6$	→	Fe^{2+}	^5D	1.391	4.530	−114	$-\frac{2}{21}$	$\frac{2}{63}$
		Co^{3+}		1.052	2.414	(−145)		
		Ni^{4+}		0.839	1.463	(−197)		
		Fe^{+}				−115		
$3d^7$	→	Co^{2+}	^4F	1.262	3.861	−189	$-\frac{2}{105}$	$-\frac{2}{315}$
		Ni^{3+}		0.966	2.071	(−272)		
		Cu^{4+}		0.778	1.273	(−320)		
		Co^{+}		1.582	6.751	−228		
$3d^8$	→	Ni^{2+}	^3F	1.146	3.245	−343	$\frac{2}{105}$	$\frac{2}{315}$
		Cu^{3+}		0.890	1.795	(−438)		
$3d^9$	→	Ni^{+}				−605		
		Cu^{2+}	^2D	1.044′	2.671	−830	$\frac{2}{21}$	$-\frac{2}{63}$

Notes:

1. The commonly found ions in each $3d^n$ configuration are indicated thus: →
2. The values of $\langle r^2 \rangle$ and $\langle r^4 \rangle$ are given in atomic units (1 a.u. = 0.529 Å) and are taken from Michel-Calendini, F.M.O. and Kibler, M.R., Theor. Chim. Acta, *10*, 367 (1968).
3. The values of λ are calculated for free ions and are taken from Dunn, T.M., Trans Faraday Soc., *57*, 1441 (1961). Blume, M and Watson, R.E., Proc. Roy. Soc., A*271*, 565 (1963). For ions in crystals, modified (reduced) values of λ are used. [18]

Table 2.3 The angular momentum tensor operators $O_m^{(n)}$ for $n = 1, 2, 3$ and 4 [18] Reprinted by permission of John Wiley & Sons, Ltd, Magnetic Oxides, ed. D.J. Craike, Wiley 1975

$$O_0^{(1)} = L_z$$
$$O_{\pm 1}^{(1)} = \mp(\tfrac{1}{2})^{1/2} L_\pm$$
$$O_0^{(2)} = \tfrac{1}{2}[3L_z^2 - L(L+1)]$$
$$O_{\pm 1}^{(2)} = \mp(\tfrac{3}{8})^{1/2}(L_z L_\pm + L_\pm L_z)$$
$$O_{\pm 2}^{(2)} = (\tfrac{3}{8})^{1/2} L_\pm^2$$
$$O_0^{(3)} = \tfrac{1}{2}\{5L_z^3 - [3L(L+1) - 1]L_z\}$$
$$O_{\pm 1}^{(3)} = \mp(\tfrac{3}{64})^{1/2}\{[5L_z^2 - L(L+1) - \tfrac{1}{2}]L_\pm + L_\pm(\ldots)\}$$
$$O_{\pm 2}^{(3)} = (\tfrac{15}{32})^{1/2}(L_z L_\pm^2 + L_\pm^2 L_z)$$
$$O_{\pm 3}^{(3)} = \mp(\tfrac{5}{16})^{1/2} L_\pm^3$$
$$O_0^{(4)} = \tfrac{1}{8}\{35L_z^4 - [30L(L+1) - 25]L_z^2 + 3L^2(L+1)^2 - 6L(L+1)\}$$
$$O_{\pm 1}^{(4)} = \mp(\tfrac{5}{64})^{1/2}[\{7L_z^3 - [3L(L+1) + 1]L_z\}L_\pm + L_\pm(\ldots)]$$
$$O_{\pm 2}^{(4)} = (\tfrac{5}{128})^{1/2}\{[7L_z^2 - L(L+1) - 5]L_\pm^2 + L_\pm^2(\ldots)\}$$
$$O_{\pm 3}^{(4)} = \mp(\tfrac{35}{64})^{1/2}(L_z L_\pm^3 + L_\pm^3 L_z)$$
$$O_{\pm 4}^{(4)} = (\tfrac{35}{128})^{1/2} L_\pm^4$$

with

$$\begin{aligned} O_4^0 &= 35L_z^4 - 30L(L+1)L_z^2 + 25L_z^2 - 6L(L+1) + 3L^2(L+1)^2 \\ O_4^4 &= \tfrac{1}{2}(L_+^4 + L_-^4) \end{aligned} \tag{131}$$

O_4^0, containing L_z, does not alter the states and gives diagonal elements. O_4^4 contains $L_+^4 = L^+L^+L^+L^+$, raising M_L four times to give $|2\ 2\rangle$ from $|2\ -2\rangle$ and also L_-^4, giving $|2\ -2\rangle$ from $|2\ 2\rangle$. Inserting an explicit matrix into the SB account for the sake of illustration, it is convenient to order the states ($L = 2$ implied) as

$$|-1\rangle, |0\rangle, |1\rangle, |-2\rangle, |2\rangle \equiv |1\rangle, |2\rangle, |3\rangle, |4\rangle, |5\rangle$$

With $L = 2$, the element $\langle 0|V|0\rangle$ or V_{22} here is just $72(C_4/20)\beta\langle r^4\rangle$, the L_z terms giving zero. $|0\rangle$ does not couple (give elements) with any other state, via the $L_{+/-}^4$, since they are all 'out of reach'. Similar reasoning applies to $|1\rangle$ and $|-1\rangle$ but the L_z give $-48(C_4/20)\beta\langle r^4\rangle$ for both diagonal elements. The states $|-2\rangle$ and $|2\rangle$ give diagonal elements via O_4^0 and couple mutually via the $L_\pm^4$ in O_4^4, so that the matrix is

$$V = \begin{pmatrix} \begin{pmatrix} -48x & 0 & 0 \\ 0 & 72x & 0 \\ 0 & 0 & -48x \end{pmatrix} & \begin{matrix} 0 & 0 \\ 0 & 0 \\ 0 & 0 \end{matrix} \\ \begin{matrix} 0 & 0 & 0 \\ 0 & 0 & 0 \end{matrix} & \begin{pmatrix} V_{44} & V_{45} \\ V_{54} & V_{55} \end{pmatrix} \end{pmatrix}$$

where

$$x = \frac{C_4}{20}\beta\langle r^4\rangle$$

As noted (Section 1.5, Chapter 3), this can be solved as two matrices, the first, 3×3, giving the eigenvalues immediately and the second giving a simple quadratic secular equation which the reader may solve to give:

$$\begin{array}{llllll} \text{Energy:} & -48x & 72x & -48x & 72x & -48x \\ \text{State:} & |-1\rangle & |0\rangle & |1\rangle & \frac{1}{\sqrt{2}}(|2\rangle + |-2\rangle) & \frac{-i}{\sqrt{2}}(|2\rangle - |-2\rangle) \end{array}$$

Since $|-1\rangle$ and $|1\rangle$ are degenerate the two combinations $(i/\sqrt{2})(|1\rangle + |-1\rangle)$ and $-1/\sqrt{2}(|1\rangle - |-1\rangle)$ may be introduced and the list rearranged as:

$$\left.\begin{array}{rl} -\frac{i}{\sqrt{2}}(|2\rangle - |-2\rangle) & \equiv |x\ y\rangle \\ \frac{i}{\sqrt{2}}(|1\rangle + |-1\rangle) & \equiv |y\ z\rangle \\ -\frac{1}{\sqrt{2}}(|1\rangle - |-1\rangle) & \equiv |z\ x\rangle \end{array}\right\} \quad \text{triplet } (T_2)$$

$$\left.\begin{array}{rl} \frac{1}{\sqrt{2}}(|2\rangle + |-2\rangle) & \equiv |x^2 - y^2\rangle \\ |0\rangle & \equiv |3z^2 - r^2\rangle \end{array}\right\} \quad \text{doublet } (E)$$

Thus an algebraic solution has been produced, giving the triplet and doublet shown in Figure 2.12(a). Which multiplet is the lower depends on (a) $\langle r^4\rangle$ is > 0, (b) C_4 alternates between the octahedral, tetrahedral and cubal cases, (c) β alternates throughout the 3d series as indicated in Table 2.2. Referring back to Figure 1.36, in Chapter 1, the corresponding single-electron $|x\ y\rangle$, $|y\ z\rangle$, $|z\ x\rangle$ states give maximum electron densities in between the axes of an octahedron (of negative charges in an oxide), and for $|x^2 - y^2\rangle$ or $|3z^2 - r^2\rangle$ the electron closely approaches the anions so that on pictorial grounds the triplet should be lower. The inversion of the levels as between $3d^1$ and $3d^9$ accords with the convention of considering $3d^9$ as $3d^1$ but with the electron replaced by a positive hole.

If the n electrons were considered to be assigned to the single-electron levels, as shown on the right in Figure 2.12(a) for example, the lowest energy would be obtained by populating the triplet with paired electrons: the low-spin state. However, what may be termed 'Hund's rule' or intra-atomic exchange coupling overcomes (lower) CF splitting, giving the high-spin state in the great majority of cases, even if n is such that the doublet must be occupied. Low-spin compounds, such as some cyanides, are sometimes referred to as covalent complexes. Exceptionally, Gyorgy *et al.* [19] showed that Ni^{3+} ($3d^8$) is low-spin in Al_2O_3 or in yttrium aluminium garnet host crystals. F-state ions, $L = 3$, can be treated similarly by the equivalent operator technique and some results are shown in Figure 2.12(b).

The idea of isomorphism was introduced by Abragam and Pryce [20] and discussed by Griffith [21] and is again developed in SB. The question concerns how an operator may be chosen such that its 3×3 matrix, as if for a p electron problem, is the same as that of an operator in the representation of the CF triplet. Take the lower triplet of $3d^2$, ^3F, and compare:

$$\begin{array}{lll} & \text{cubic CF states} & \text{p states } (|m_l\rangle) \\ |1\rangle: & |0\rangle & \sim |0\rangle \\ |2\rangle: & -\sqrt{\frac{3}{8}}|-1\rangle - \sqrt{\frac{5}{8}}|3\rangle & \sim |-1\rangle \\ |3\rangle: & -\sqrt{\frac{3}{8}}|1\rangle - \sqrt{\frac{5}{8}}|-3\rangle & \sim |1\rangle \end{array}$$

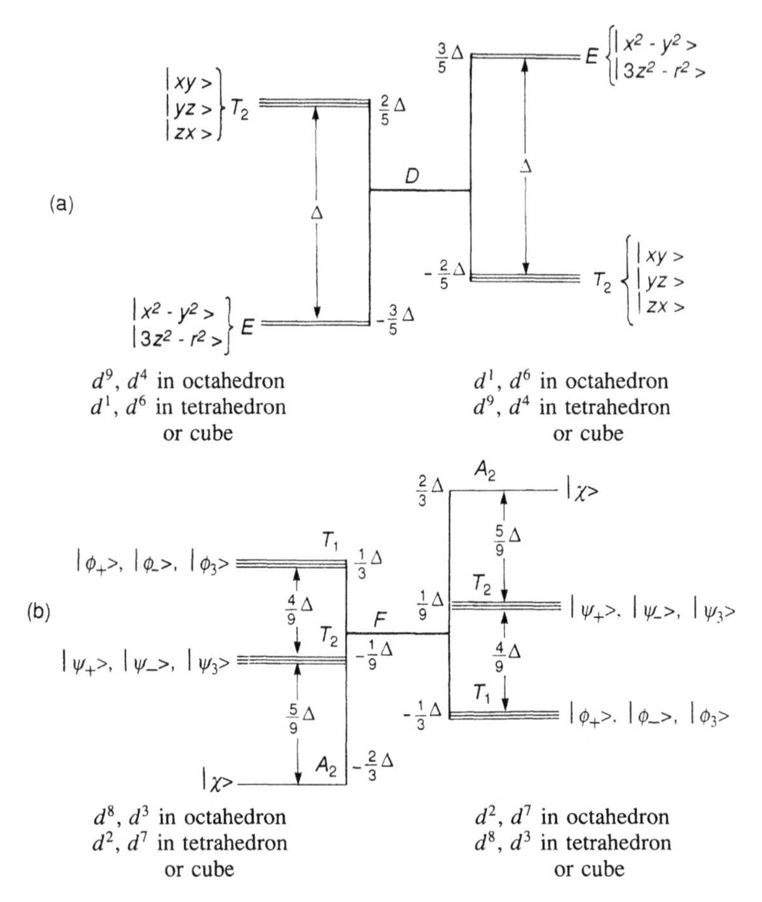

Figure 2.12 Levels for D state (a) and F state (b) ions in cubic environments [18]. (Reprinted by permission of John Wiley & Sons, Ltd, Magnetic Oxides, ed. D.J. Craik, Wiley 1975)

then, for example,

$$\langle 2|L_z|2\rangle = \left(-\sqrt{\frac{3}{8}}\langle -1| - \sqrt{\frac{5}{8}}\langle 3|\right)\left(+\sqrt{\frac{3}{8}}|-1\rangle - 3\sqrt{\frac{5}{8}}|3\rangle\right)$$

$$= -\tfrac{3}{8} + \tfrac{15}{8} = \tfrac{3}{2}$$

and for the p states:

$$\langle 2|l_z|2\rangle = \langle -1|(-1)|-1\rangle = -1$$

(using l for the single-electron p states) so that

$$\underset{\text{(CF)}}{\langle 2|L_z|2\rangle} \equiv \underset{\text{(p)}}{\langle 2| -\tfrac{3}{2}l_z|2\rangle}$$

and this is seen (vitally) to apply to $|3\rangle$ also. The states are described as fictitious, with the fictitious $L' = 1$. States having $L' \neq 1$ might appear necessary for second-order perturbations, but by reference to Messiah [22] and to Stevens [23] it can be seen that these are not, in fact, necessary. For doublets, reduction to 2×2 matrices suggests isomorphism with $S = \frac{1}{2}$, but to avoid confusion when spin as such is introduced, $T = \frac{1}{2}$ is used.

As an example (from SB) consider the effect of the SO coupling on d^6, ^5D with a triplet lowest. $\mathbf{L}$ within the triplet becomes $-\mathbf{L}'$ with $L' = 1$ and the problem becomes that for a free ion with $L' = 1$, $S = 2$:

$$
\begin{array}{ll}
J' = L' + S = 3: & \text{energy } -2\lambda \\
J' = 2: & \text{energy } \lambda \\
J' = 1: & \text{energy } 3\lambda
\end{array}
$$

From Table 2.2, $\lambda = -114\ \text{cm}^{-1}$ for Fe^{2+} and the ground level is a triplet with $J' = 1$ (see Figure 2.13). For the Landé g factor the modifications are

$$\beta\mathbf{B}\cdot(\mathbf{L} + 2\mathbf{S}) = g\beta\mathbf{B}\cdot\mathbf{J} \rightarrow \beta\mathbf{B}\cdot(-\mathbf{L}' + 2\mathbf{S}) = g\beta\mathbf{B}\cdot\mathbf{J}' \tag{132}$$

within $L' = 1$, i.e. specifically for the triplet. With $\mathbf{J}' = \mathbf{L}' + \mathbf{S}$, $\mathbf{L}' + 2\mathbf{S} \equiv \wedge_J \mathbf{J}'$ where

$$\wedge_J = 1 + \frac{J'(J'+1) - L'(L'+1) + S(S+1)}{2J'(J'+1)} \tag{133}$$

and so $(-\mathbf{L}' + 2\mathbf{S})$ becomes $(3\wedge_J - 4)\mathbf{J}'$. For the case $L' = 1$, $S = 2$, $J' = 1$, $\wedge_J = \frac{5}{2}$. It is advisable (Section 9.2) to replace $g = 1$ for the orbital momentum by a flexible $g_{L'}$ which may be < 1.0. The Zeeman Hamiltonian is then

$$\beta\mathbf{B}\cdot(g_{L'}\mathbf{L}' + g_S\mathbf{S}) \equiv g\beta\mathbf{B}\cdot\mathbf{J}' \tag{134}$$

and g becomes

$$g = \tfrac{1}{2}(g_{L'} + g_S) + \tfrac{1}{2}(g_{L'} - g_S)\frac{L'(L'+1) - S(S+1)}{J'(J'+1)} \tag{135}$$

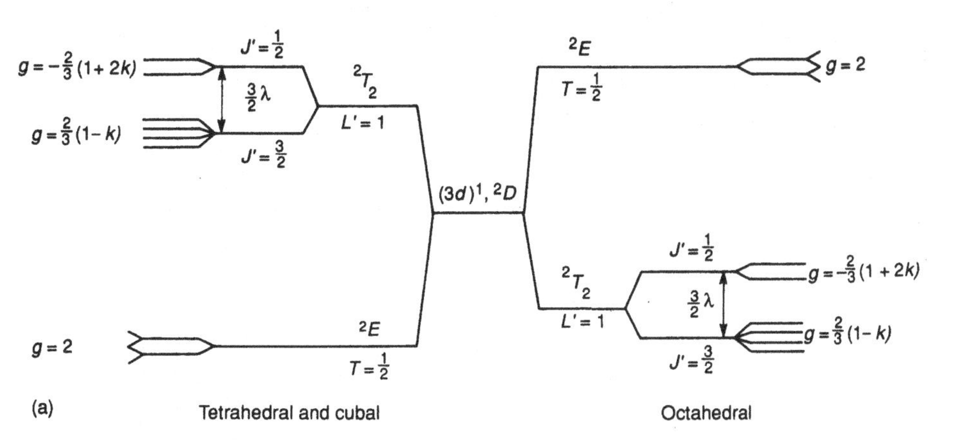

Figure 2.13 Energy levels and g values for $3d^n$ ions with SO coupling $\gg$ distortion; cubic crystal field and SO effects followed by the first-order Zeeman splittings. The SO energy changes are obtained from

$$E_{J'} = \tfrac{1}{2}\lambda\alpha[J'(J'+1) - L'(L'+1) + S'(S'+1)]$$

with J' and L' the fictitious angular momenta. λ is the free ion SO coupling constant and α the factor multiplying L'_z in the equivalence given. The factor k involved in g allows for covalent and Jahn–Teller effects [18]. (Reprinted by permission of John Wiley & Sons, Ltd, Magnetic Oxides, ed. D.J. Craik, Wiley 1975)

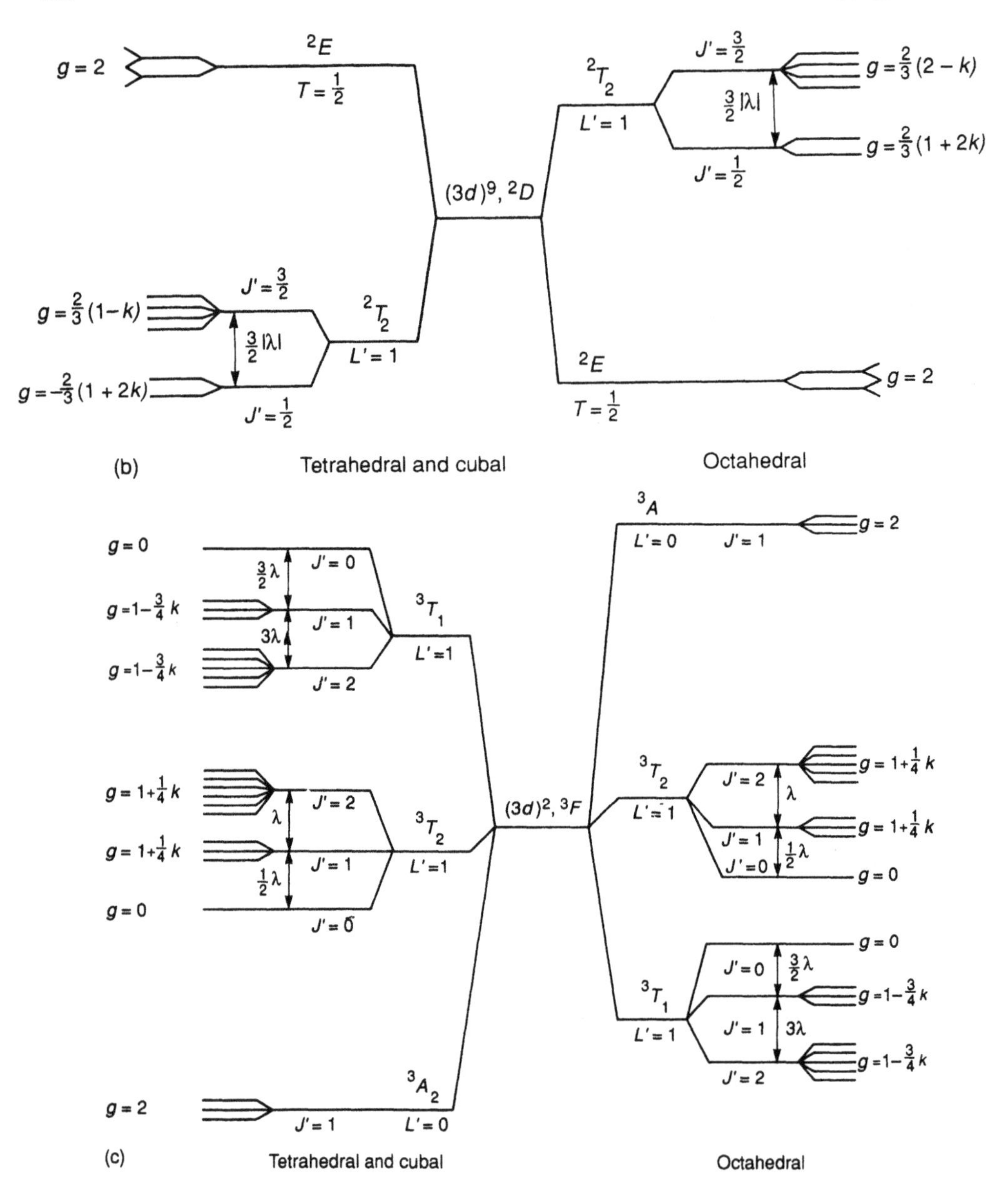

Figure 2.13 *(continued)*

Putting $g_{L'} = 1$ and $g_S = 2$, this has the same form as equation (86), but there J, L and S are the true values, as compared with the fictitious L' and J'. Thus the problem of the SO coupling within the orbital triplet is seen to have been converted to a free-ion problem in which S is preserved but L is replaced by $L' = 1$. The same principles apply to an orbital doublet, but the orbital motion within the doublet is zero and $g = g_S$ simply, as for an orbital singlet.

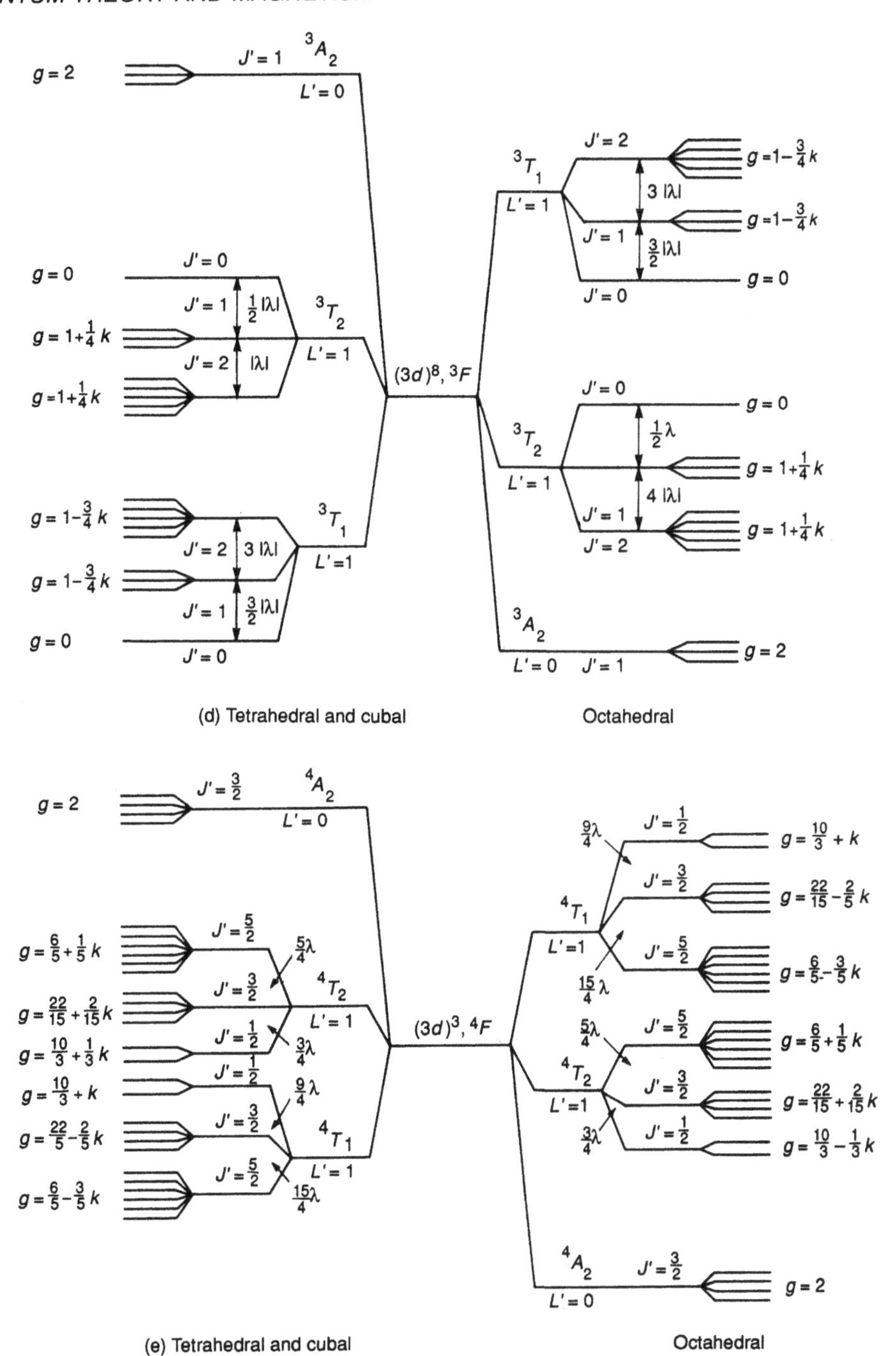

Figure 2.13 (*continued*)

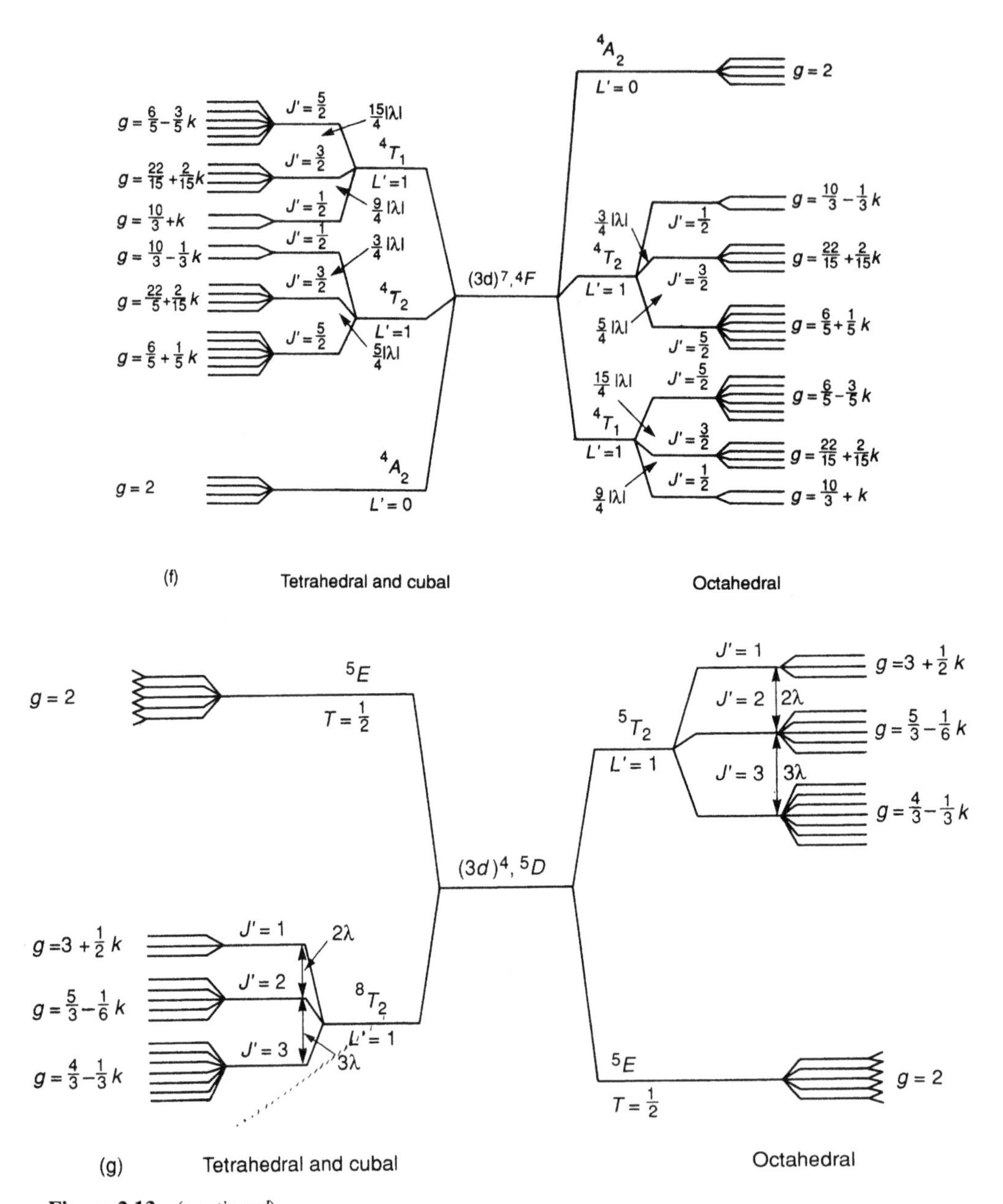

Figure 2.13 *(continued)*

The effect of, for example, a distortion from octahedral symmetry whereby the (negatively charged) neighbours on the z axis are pulled out can be understood pictorially (for $3d^1$) inasmuch as $|y\ z\rangle$ and $|z\ x\rangle$ are affected similarly but differently from $|x\ y\rangle$ (Figure 1.36 in Chapter 1). This tetragonal distortion splits the triplet into a singlet and a doublet: a parameter δ (given by SB) is negative and the doublet will be lower. Conversely, if these

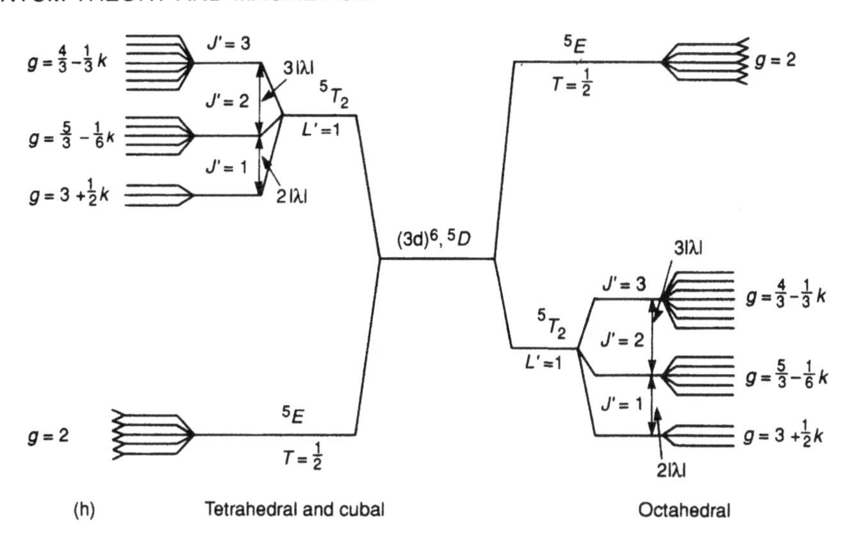

Figure 2.13 *(continued)*

neighbours are pushed in then δ will be positive. For the upper doublet (octahedral case), $|3z^2 - r^2\rangle$ will be favoured over $|x^2 - y^2\rangle$ since the charge is more concentrated along OZ for the former. This is evidenced by the calculated results in Figure 2.14.

That the orbital singlets should behave as simple spin levels is to be expected in first order, involving ground states only. However, if second or higher order approximations are introduced, small splittings are expected. In perfect symmetry, octahedral, cubal or tetragonal, the energy is expected on symmetry grounds to be given by (e.g. for M_n^{2+}, $3d^5$, ^6S: $S = \frac{5}{2}$)

$$W = \tfrac{1}{6}a\left(S_x^4 + S_y^4 + S_z^4 - \tfrac{707}{16}\right) \tag{136}$$

with a constant. With a tetragonal or trigonal distortion, W becomes

$$W = \tfrac{1}{6}a\left(S_x^4 + S_y^4 + S_z^4 - \tfrac{707}{16}\right) + D\left(S_\zeta^2 - \tfrac{35}{12}\right) + \frac{F}{36}\left(7S_\zeta^4 - \tfrac{92}{5}S_\zeta^2 + \tfrac{567}{16}\right) \tag{137}$$

with D and F constants and the ζ axis the z axis for tetragonal distortions, and the ζ axis the trigonal axis for trigonal distortions, giving the shifts and splittings shown in Figure 2.15. The effects are small but make a substantial contribution to the magnetocrystalline anisotropy.

9.1 Paramagnetic Resonance and g Tensor Calculations

Creating an isomorphism or comparison between the zero-field spin doublet and a general doublet which can be assigned a fictitious spin S', the spin Hamiltonian can be written as

$$\mathcal{H}_s = \beta \mathbf{B} \cdot \mathbf{g} \cdot \mathbf{S}', \qquad \text{where } S' = \tfrac{1}{2} \tag{138}$$

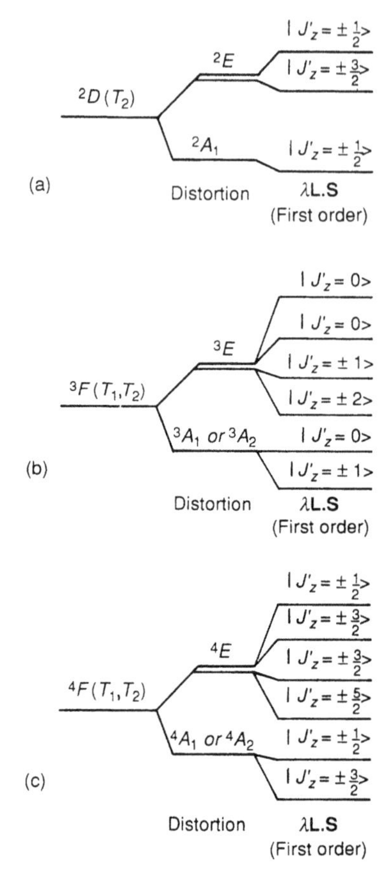

Figure 2.14 Levels for ions in distorted sites including *SO* coupling effects: δ and λ are assumed to be positive [18]. A tetragonal distortion of an octahedron is effected by pulling out the neighbours on the z axis and of a tetrahedron or cube by increasing the magnitude of the z coordinates of all the ligands equally. A trigonal distortion is effected by elongation of the octahedron, tetrahedron or cube along a threefold axis. (Reprinted by permission of John Wiley & Sons, Ltd, Magnetic Oxides, ed. D.J. Craik, Wiley 1975)

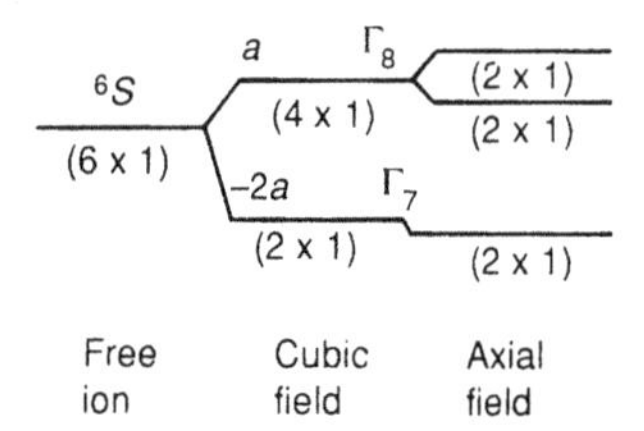

Figure 2.15 Splittings as for a Mn^{2+} ion in a cubic field and with a small axial (tetragonal or trigonal) distortion, affecting the magnetocrystalline anisotropy though usually $< 1\ \text{cm}^{-1}$ [18]. (Reprinted by permission of John Wiley & Sons, Ltd, Magnetic Oxides, ed. D.J. Craik, Wiley 1975)

with **g** a tensor to account for possible anisotropies. S' is defined to the extent that $2S' + 1$ gives the multiplicity and for a triplet $S' = 1$. The prime is often omitted from S' because in writing equation (138) it is apparent that S cannot be the simple spin operator or the accepted scalar g would be involved. With an axis of symmetry defining OZ, and OX and

OY equivalent,

$$\mathcal{H}_s = \beta[g_{\parallel} B_z S_z' + g_{\perp}(B_x S_x' + B_y S_y')] \tag{139}$$

If $\mathbf{B} = \mathbf{k}B$ the splitting is $g_{\parallel}\beta B$ and if $\mathbf{B} = \mathbf{i}B$ the splitting is $g_{\perp}\beta B$. Clearly the term paramagnetic resonance becomes more appropriate than e.s.r. when these concepts are applied, although an alternative would be 'effective e.s.r.'. In particular, the Zeeman splitting is not given by $\beta\mathbf{B}\cdot\mathbf{S}$ but by

$$\beta\mathbf{B}\cdot(\mathbf{L}+2\mathbf{S}) \rightarrow \beta\mathbf{B}\cdot(-\mathbf{L}'+2\mathbf{S}) \tag{140}$$

according to the previous discussion with $L' = 1$ for the doublet initially but subject to substantial modifications. Thus, for example, setting $hf = \Delta E g_{\parallel}\beta B$, $g_{\parallel} = \Delta E/(\beta B)$ and $g_{\parallel}$ is given as the splitting due to $(-L_z' + 2S_z)$ alone. The implication in using a spin Hamiltonian to describe the behaviour of a general doublet is that it is well separated from other levels. If the use of the g components is extended to describe moments in ordered materials it must be recognized that the effective exchange fields, being much greater than those appropriate to paramagnetic resonance, could lead to overlap with other levels.

As an example, $3d^1$, ^2D has an orbital triplet lowest in a tetragonally distorted octahedron of negatively charged neighbours (Figure 2.13). The Zeeman term within the triplet is as in equation (140) and the effective Hamiltonian is

$$\mathcal{H}_e = \delta\left(L_z'^2 - \tfrac{2}{3}\right) - \lambda(\mathbf{L}'\cdot\mathbf{S}) + \beta\mathbf{B}\cdot(-\mathbf{L}'+2\mathbf{S}) \tag{141}$$

where δ characterizes the distortion ($=0$ with no distortion). With $B = 0$, $[\mathcal{H}_e, J_z'] = 0$, with $J_z' = L_z' + S_z$. The energies can thus be characterized by J_z' values. With $L' = 1$, $L_z' = 1, 0, -1$, there are six eigenstates $|L_z'\ S_z\rangle$. It is seen that the pair $|1\ \tfrac{1}{2}\rangle$ and $|-1\ -\tfrac{1}{2}\rangle$ have the same energy, a doublet with $J_z' = \pm\tfrac{3}{2}$. Noting that

$$\beta B(-L_z' + 2S_z)|1\ \tfrac{1}{2}\rangle = \beta B[-1 + 2(\tfrac{1}{2})]|1\ \tfrac{1}{2}\rangle = 0$$

$$\beta B(-L_z' + 2S_z)|-1\ -\tfrac{1}{2}\rangle = \beta B[-(-1) + 2(-\tfrac{1}{2})]|-1\ -\tfrac{1}{2}\rangle = 0$$

it is clear that there is no Zeeman splitting here and this is a 'non-magnetic' doublet. For fields $\parallel OX$ it is necessary to consider the matrix elements $\langle|-L_x' + 2S_x|\rangle$ and it can be seen, for example, on transforming to $L^{\pm}$, $S^{\pm}$ that these are all zero, reinforcing the above. For consistency it may be said that this doublet still corresponds to a fictitious spin $S = \tfrac{1}{2}$, but with $g_{\parallel} = 0 = g_{\perp}$.

Of the remaining four of these states of definite J_z', $|1\ -\tfrac{1}{2}\rangle$ and $|0\ +\tfrac{1}{2}\rangle$ group together, with $J_z' = \tfrac{1}{2}$. To find the zero-field eigenstates it is necessary to solve the secular equation using these two as a basis, and this can be seen to give eigenstates which are

$$\begin{aligned} |1\rangle &= \sin\theta|1\ -\tfrac{1}{2}\rangle + \cos\theta|0\ \tfrac{1}{2}\rangle \\ |2\rangle &= \cos\theta|1 -\ \tfrac{1}{2}\rangle - \sin\theta|0\ \tfrac{1}{2}\rangle \end{aligned} \tag{142}$$

where

$$\tan 2\theta = \frac{\sqrt{2}\lambda}{\delta + \tfrac{1}{2}\lambda} = \left(\frac{2\langle 1\ -\tfrac{1}{2}|\mathcal{H}_e|0\ \tfrac{1}{2}\rangle}{\langle 0\ \tfrac{1}{2}|\mathcal{H}_e|0\ \tfrac{1}{2}\rangle - \langle 1\ -\tfrac{1}{2}|\mathcal{H}_e|1\ -\tfrac{1}{2}\rangle}\right)_{B=0}$$

The eigenvalues are

$$\begin{aligned} E_1 &= (\langle 0\ \tfrac{1}{2}\,|\mathcal{H}_e|\,0\ \tfrac{1}{2}\,\rangle + \langle 0\ \tfrac{1}{2}\,|\mathcal{H}_e|\,1\ -\tfrac{1}{2}\,\rangle \tan\theta)_{B=0} \\ E_2 &= (\langle 0\ \tfrac{1}{2}\,|\mathcal{H}_e|\,0\ \tfrac{1}{2}\rangle\ -\ \langle 0\ \tfrac{1}{2}\,|\mathcal{H}_e|\,1\ -\tfrac{1}{2}\rangle\ \cot\theta)_{B=0} \end{aligned} \tag{143}$$

Finally, it is found that the secular equation using $|\,-1\ \tfrac{1}{2}\,\rangle$ and $|\,0\ -\tfrac{1}{2}\,\rangle$, $J_z' = -\tfrac{1}{2}$, is identical to the above so that the two pairs of levels in fact form two further doublets:

$$\begin{cases} \cos\theta\,|\,1\ -\tfrac{1}{2}\,\rangle - \sin\theta\,|\,0\ \tfrac{1}{2}\,\rangle \\ \cos\theta\,|\,-1\ \tfrac{1}{2}\,\rangle - \sin\theta\,|\,0\ -\tfrac{1}{2}\,\rangle \end{cases}$$

and

$$\begin{cases} \sin\theta\,|\,1\ -\tfrac{1}{2}\,\rangle + \cos\theta\,|\,0\ \tfrac{1}{2}\,\rangle \\ \sin\theta\,|\,-1\ \tfrac{1}{2}\,\rangle + \cos\theta\,|\,0\ -\tfrac{1}{2}\,\rangle \end{cases}$$

(If the reader chooses to write out the appropriate 6×6 matrix it will be apparent why the solutions are obtained in this particular way.)

Recall that $g_{\parallel}$ is given by the splitting corresponding to $\mathcal{H} = -L_z' + 2S_z$ (not the true Hamiltonian) and note that

$$\begin{aligned} (L_z' + 2S_z)|\,1\ -\tfrac{1}{2}\,\rangle &= (-1 - 1)|\,1\ -\tfrac{1}{2}\,\rangle = -2|\,1\ -\tfrac{1}{2}\,\rangle \\ (L_z' + 2S_z)|\,0\ \tfrac{1}{2}\,\rangle &= (1)|\,0\ \tfrac{1}{2}\,\rangle \\ (L_z' + 2S_z)(\cos\theta\,|\,1\ -\tfrac{1}{2}\,\rangle - \sin\theta\,|\,0\ \tfrac{1}{2}\,\rangle) &= -2\cos\theta\,|\,1\ -\tfrac{1}{2}\,\rangle + \sin\theta\,|\,0\ \tfrac{1}{2}\,\rangle \end{aligned}$$

so that

$$\mathcal{H}_{11} = -2\cos^2\theta + \sin^2\theta$$

and similarly

$$\mathcal{H}_{22} = 2\cos^2\theta - \sin^2\theta$$

while obviously $\mathcal{H}_{12} = 0 = \mathcal{H}_{21}$ so that $\mathcal{H}_{11}$ and $\mathcal{H}_{22}$ are the two eigenvalues and thus

$$g_{\parallel} = 2(-2\cos^2\theta + \sin^2\theta) \tag{144}$$

For the second doublet it is only necessary to replace θ by $\theta + \pi/2$, giving

$$g_{\parallel} = 2(-2\sin^2\theta + \cos^2\theta) \tag{145}$$

To calculate $g_{\perp}$ it is convenient to use $-L_x' + 2S_x = -\tfrac{1}{2}(L_+' + L_-') + S^+ + S^-$ and note that, with $L = 1$ and $S = \tfrac{1}{2}$, implied

$$\begin{array}{lll} L_+'|\,1\ S_Z\,\rangle = 0, & L_+'|\,-1\ S_Z\,\rangle = \sqrt{2}|\,0\ S_Z\,\rangle, & L_+'|\,0\ S_z\,\rangle = \sqrt{2}|\,1\ S_Z\,\rangle \\ L_-'|\,1\ S_Z\,\rangle = \sqrt{2}|\,0\ S_Z\,\rangle, & L_-'|\,-1\ S_Z\,\rangle = 0, & L_-'|\,0\ S_z\,\rangle = \sqrt{2}|\,-1\ S_Z\,\rangle \end{array}$$

and

$$\begin{array}{ll} S^+|\,L_Z\ \tfrac{1}{2}\,\rangle = 0, & S^+|\,L_Z\ -\tfrac{1}{2}\,\rangle = |\,L_Z\ \tfrac{1}{2}\,\rangle \\ S^-|\,L_Z\ \tfrac{1}{2}\,\rangle = |\,L_Z\ -\tfrac{1}{2}\,\rangle, & S^-|\,L_Z\ -\tfrac{1}{2}\,\rangle = 0 \end{array}$$

Numbering the states in equation (142) as $|1\rangle$ and $|2\rangle$ the reader may confirm that the matrix and the secular equation are

$$\begin{pmatrix} 0 & \sqrt{2}\sin\theta\cos\theta + \sin^2\theta \\ (S) & 0 \end{pmatrix} \qquad \begin{vmatrix} -\lambda & \mathcal{H}_{12} \\ \mathcal{H}_{21} & -\lambda \end{vmatrix} = 0$$

so that $\lambda^2 = \mathcal{H}_{12}\mathcal{H}_{21} = \mathcal{H}_{12}^2$, $\lambda = \pm\mathcal{H}_{12}$ simply and so

$$g_\perp = 2\mathcal{H}_{12} = 2(\sqrt{2}\sin\theta\cos\theta + \sin^2\theta) \tag{146}$$

Again for the alternative doublet, $\theta \to \theta + \pi/2$. It is noted that $\tan 2\theta = \sqrt{2}\lambda/(\delta + \frac{1}{2}\lambda)$ defines two angles which differ by $\pi/2$.

Thus the elements of **g** can be found for each of the three doublets, regarded as fictitious spins $\frac{1}{2}$, with different $g_\parallel$ and $g_\perp$. Various approximations are involved but the major source of error lies in the assumption that the Zeeman term is simply $\beta\mathbf{B}\cdot(\mathbf{L}+2\mathbf{S})$. It is better to write this as $\beta\mathbf{B}\cdot(g_L\mathbf{L}+2\mathbf{S})$, accepting $g_S = 2$, but acknowledging that g_L may be less (perhaps much less) than unity for reasons discussed below. The g values are brought much closer to the spin value. It can also be shown that a Zeeman splitting does then occur for the $J_z' = \frac{3}{2}$ states. The results may always be obtained by solving the full matrix numerically.

It may be repeated that with cubic symmetry orbital singlets are characterized by $g_\parallel = g_\perp = g_S$ and, with this symmetry, the same applies to orbital doublets also to give the spin-only values, with complete quenching of the orbital angular momentum.

9.2 Covalency and the Jahn–Teller Effect

One reason for introducing g_L is connected with the possibility of covalent effects, the realization that the ion should not always be regarded as an initially isolated ion for which the orbitals are modified by the electrostatic crystal fields: the orbitals may 'spill over' on to the neighbouring ions as suggested in relation to exchange coupling in Section 6.1. This can lead to a reduction in the magnetic moment associated with the orbital. However, the principles of the foregoing treatments still apply since they can be shown to follow from symmetry considerations without necessary reference to the crystal fields as such. On going over to a more general molecular orbital theory all that is necessary is to replace **L** by k**L**, or $g_L = 1$ by $g_L = k$, in the SO and Zeeman terms in equation (141) with k, between 0.8 and 1.0, to be regarded as an experimental parameter (SB, p. 171). Hence the occurrence of k in Figure 2.13, giving the energies for $3d^n$. The following may, however, be more important.

As noted, the doublets and triplets for cubic symmetry can be split by static distortions. Jahn and Teller [24] noted that when, with high symmetry, the lowest level for an ion was degenerate a spontaneous or dynamic distortion might occur which would split the level and produce a lower ground state. According to the population of the levels, this could lead to a net reduction in the total energy of the system, although the elastic energy of the lattice clearly rises. Assuming that this latter depends on Q^2, with Q a displacement parameter (restoring forces depending on Q), while the reduction of the orbital energy is suggested by CF theory to depend on Q:

$$E = AQ^2 - BQ, \qquad \frac{\partial E}{\partial Q} = 2AQ - B$$

with A and B positive constants. A minimum always occurs for some non-zero value of Q. There is an analogy with the reduction of symmetry and energy by Peierls distortions in metals (Section 5, Chapter 1).

In a concentrated material such as a ferrimagnetic oxide, a characteristic ion may be considered to generate a distortion that falls off throughout the lattice. A second ion has a changed energy due to this distortion, giving rise to a mutual energy and a form of coupling via the lattice and a cooperative Jahn-Teller effect. The orbital motion becomes correlated with alignment of the components, but not a perfect alignment any more than that the spins are perfectly aligned by the exchange field, except at $T \to 0$. Excited states analogous to the excited spin states or spin waves (magnons) as discussed in Chapter 3 may be expected. By this analogy, a near-continuous band of excited orbital states would be produced and this is considered to lead to an important contribution to the magnetocrystalline anisotropy at low temperatures. Magnetostriction is also expected to be connected with the distortions.

The Jahn–Teller effect is also indicated, by the work of Stevens and Bates, to lead to reductions in the effective L', which are considerably greater than those due to covalency effects. These might seem apparent simply by considering the changes in degeneracy. With a strong Jahn–Teller effect, $k \sim 0.1$ is to be expected, giving substantial changes in the g values towards the free spin value and reduction of the SO coupling. The observation of such reductions is strong evidence that the Jahn–Teller effects are, in fact, present: as Stevens and Bates point out, they affect nearly everything.

9.3 Magnetocrystalline Anisotropy and Magnetostriction

The effect of magnetostatic or dipolar interactions on the anisotropy (or on the exchange itself, at low temperatures) is only exceptionally taken into account, as by Casimir for hexagonal ferrites [25]. The energy $\mu_0 \sum_{i>j} \mu_i \mu_j (1 - 3\cos^2\theta_{ij})/r_{ij}^3$ does not vary as the common direction of the $\boldsymbol{\mu}_i$, i.e. of $\mathbf{M}$, rotates in a cubic lattice; r_{ij} is the distance between the μ_i, μ_j and θ_{ij} is the angle between their common axis and r_{ij}.

Two major contributions to the anisotropy are recognized. It has been noted that the exchange interaction can, itself, be anisotropic, giving rise to a pair anisotropy which is important when particular pair configurations form in the creation of induced anisotropies (see, for example, Krupička and Závěta [26]. This is also an important contribution in rare earth–iron garnets.

Secondly, a single ion may be anisotropic, even in the absence of exchange coupling, inasmuch as the axis of quantization may become aligned with particular crystallographic directions and the splitting may be such as to indicate that certain components with respect to these axes are favoured. The magnetocrystalline anisotropy then arises when the assembly of anisotropic ions becomes exchange coupled. A distinction may be made in that, when single-ion effects predominate, a particular ion may make a contribution to the resulting anisotropy which is, within limits, directly proportional to its concentration. If pair interactions are important, a quadratic dependence on the concentration is expected.

Single-ion anisotropy is important when 3d ions have the predominant effect. Wolf [27] and Yosida and Tachiki [28] thus ascribed the magnetocrystalline anisotropy of many spinel ferrites principally to the single-ion contribution of the Fe^{3+} ion (^{6}S,$S = \frac{5}{2}$, giving a singlet crystal field ground state) and Krupička *et al.* [29], successfully fitted curves for the theory with experimental results for $Mn_x\,Fe_{3-x}\,O_4$.

Orbitally degenerate CF ground states arise typically for Co^{2+} (^{4}F) in the octahedral sites in spinel ferrites or Fe^{2+} in Si-doped yttrium–iron garnet. A trigonal CF component is superimposed on the octahedral field, with the trigonal axes parallel to the different members of the $\langle 111 \rangle$ set. The crystal field can, of course, be distorted from a cubic symmetry without a lattice distortion. On drawing out the lattice extensively it can be seen that there are four types of octahedral sites, which are not equivalent inasmuch as trigonal axes follow the four different cube diagonals.

The cubic field gives a triplet lowest and since this is $\sim 10^4$ cm^{-1} below the next highest state it can be considered to be isolated, even when exchange splittings are subsequently taken into account. The triplet is thus considered as an effective p state with fictitious angular momentum αL, $L = 1$ and $\alpha = 1$ for Fe^{2+} and $\alpha = \frac{3}{2}$ for Co^{2+}, as indicated in Section 9. In the cubic field alone the degeneracy corresponding to $L_z = 1, 0, -1$ conforms with the lack of an identifiable axis of quantization and there is no anisotropy. Introduction of the uniaxial trigonal field gives a real axis of quantization. The effective Hamiltonian can be approximated as

$$\mathcal{H}_e = D_a L_\zeta^2 + \beta \mathbf{B}_e \cdot \mathbf{g} \cdot \mathbf{S} - \lambda \mathbf{L} \cdot \boldsymbol{\alpha} \cdot \mathbf{S} \tag{147}$$

with D_a representing the strength of the uniaxial CF component, B_e the exchange field giving a Zeeman-like term and $L = 1$. The last term replaces $-\lambda\alpha\mathbf{L} \cdot \mathbf{S}$ in recognition of the tensor nature of $\boldsymbol{\alpha}$ for reduced symmetry. The terms in equation (147) are written in the order of the magnitudes of the splittings produced and, following Slonczewski [30], the latter two can be applied as perturbations once the CF levels have been determined; to first order the splittings are simply added in to give the diagram of Figure 2.16. The trigonal field gives a lower doublet characterized by $\langle L_z \rangle = \pm\alpha$. The exchange is represented by the Zeeman splitting for spin $S = \frac{3}{2}$, in the exchange field. This is followed by the SO splitting

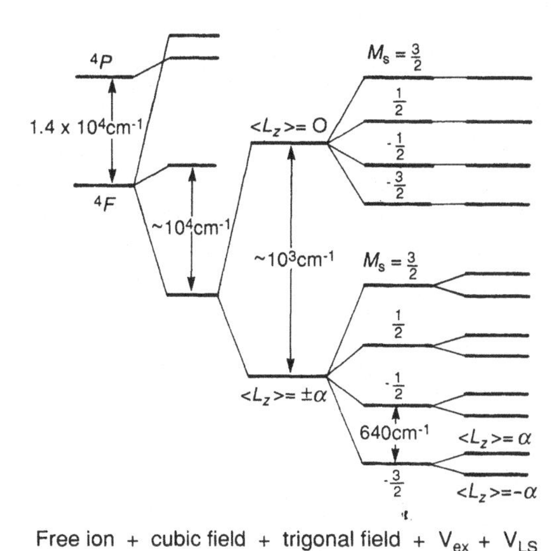

Figure 2.16 Energy levels for a Co^{2+} ion in the case of strong exchange interaction [30]. (Reprinted by permission of John Wiley & Sons, Ltd, Magnetic Oxides, ed. D.J. Craik, Wiley 1975)

and since this is much lower the energies of the lowest pair of levels are approximately

$$E = -2\beta B_e S \pm \alpha\lambda S \cos\theta \tag{148}$$

Slonczewski showed that a statistical distribution over these levels leads to a contribution to the anisotropy energy equal to

$$F_A = -kT \sum_{i=1}^{4} N_i \ln[2\cosh\left(\frac{3\alpha|\lambda|\cos\theta_i}{2\,kT}\right) \tag{149}$$

N_i is the number of Co^{2+} or Fe^{2+} ions in the non-equivalent octahedral positions in the spinel or garnet structures respectively. In the general case a random distribution with all $N_i = N/4$ is expected. θ_i is the angle between the exchange field, and thus **S**, and the axis of quantization or the fictitious orbital moment component.

Equation (148) indicates that if a single ion is indeed considered in isolation the directionally variable part of the reduced energy, $E/(\alpha\lambda S)$, would be minimum and equal to $\varepsilon = -\cos 0° = -1$ with the spins $\parallel$ to [111]. This seems to indicate that the easy axes should be $\langle 111 \rangle$ whereas in fact they are $\langle 100 \rangle$. A simple demonstration appears to resolve this. Recall that there are four inequivalent octahedral sites, each associated with a different choice of the four body diagonals. Thus there is not a unique axis of quantization, but taking four characteristic Co^{2+} ions, one will have [111], another $[\bar{1}\bar{1}\bar{1}]$, etc., as its 'local easy axis'. (In the cubic spinel structure it is impossible to select any of the $\langle 111 \rangle$ as being unique.) Suppose the magnetization in fact lies $\parallel$ to [111]. This will be a local easy axis for one ion, giving $\varepsilon = -1$, but **M** would make an angle of β with the local easy axes for the other three ions. $\beta = \pi - 2\varphi$, where φ is the angle between any $\langle 111 \rangle$ and $\langle 100 \rangle$; $\cos\varphi = 1/\sqrt{3}$, as in Figure 2.17. The total for the four is

$$\varepsilon_4 = -1 - 3\cos(\pi - 2\varphi) = -1 + 6\cos^2\varphi - 3 = -2$$

However, if **M** lies $\parallel$ to [100] the energy is

$$\varepsilon_4 = -4\cos\varphi = -\frac{4}{\sqrt{3}} = -2.31$$

Thus it is energetically favourable for **M** to lie along [100] as opposed to [111]. Co^{2+} is of special interest due to the large magnitudes of the associated effects. Table 2.4 [31] shows how small concentrations can have a great effect on the anisotropy of spinel ferrites.

If a field is applied to rotate **M** to an arbitrary direction the exchange field rotates and the energies of ions on non-equivalent sites differ, according to the θ dependence in

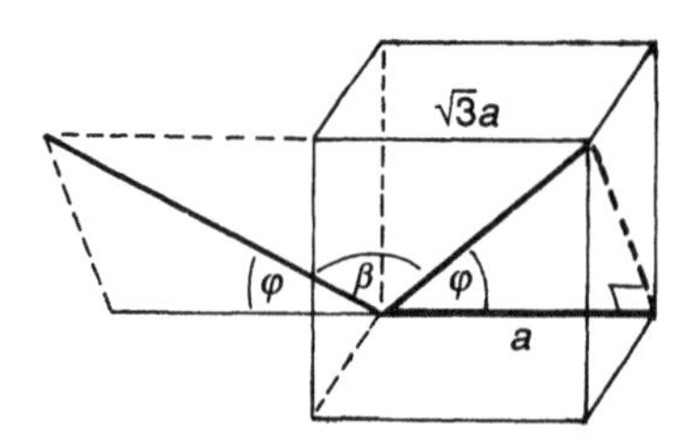

Figure 2.17 The angles for the discussion of magnetocrystalline anisotropy

Table 2.4

	$K_1/10^3\ \mathrm{J\,m^{-3}}$	
	(290 K)	(200 K)
$Mn\ Fe_2O_4$	−4.07	−10.53
$Mn_{0.99}Co_{0.01}Fe_2O_4$	−3.12	−4.68
$Mn_{0.96}Co_{0.04}Fe_2O_4$	−0.70	11.22
$Mn_{0.94}Co_{0.04}Fe_2O_4$	0.63	21.00
$Mn_{0.90}Co_{0.10}Fe_2O_4$	3.73	40.75

The effect of cobalt on the anisotropy constant K_1, of manganese ferrite [31]

equation (148). At a temperature at which the migration of Co^{2+} ions becomes possible, but necessarily below T_c, there will be a tendency towards a preferential occupation of the lower energy sites, ordering the single-ion axes in such a way that an overall anisotropy is induced in the field direction, or rather the magnetization direction during the anneal. The induced anisotropy can, after Néel [32], be decomposed into two terms characterized by constants F and G with different angular dependences (see Krupička and Závěta [26], p. 246). Pennoyer and Bickford [33] made measurements on cobalt-substituted magnetite which showed that while G was proportional to x, as expected for a single-ion effect, the second constant F was quadratically dependent on x, indicating that pairs of ions should be involved, as discussed by Néel [32], introducing a pseudo-dipolar pair interaction.

To the extent that the single-ion picture applies, anisotropy and also magnetostriction constants can be derived in terms of additive contributions from the different atoms or ions included. Iron has a positive K_1 and $K_1 < 0$ for nickel, and it is not surprising that at about 75 per cent nickel both K_1 and λ can pass through zero, giving a pronounced peak in M_s^2/K_T, with K_T the 'total anisotropy' and thus in χ and μ_r. The highest values of μ_r are, however, only obtained for specimens that are quenched from high temperatures, which may be presumed to give a random arrangement of the atoms. If they are annealed below the Curie point, or cooled slowly, long- or short-range ordering occurs, producing anisotropic ion pairs. If cobalt is included very high uniaxial anisotropies may be induced in applied fields, cobalt also having a special role in the metals.

The observation that magnetostriction constants often correspond to an additive law suggests that a single-ion approach is appropriate. The linearity of the effect of Co^{2+} concentration is exemplified by

$$\begin{aligned} \frac{\lambda(Co_{0.04}\ Fe_{2.96}\ O_4)}{0.04} &= -7.0 \times 10^{-4} \\ \frac{\lambda(Co_{0.8}\ Fe_{2.2}\ O_4)}{0.8} &= -7.4 \times 10^{-4} \end{aligned} \tag{150}$$

The properties of garnet films, M_s, K_1 and λ, can be adjusted according to specific requirements by setting up and solving sets of linear equations, as by Giess *et al.*, for example [34]. Many generally observed trends associate magnetostriction with magnetocrystalline anisotropy.

On passing down through T_c the dependence of the (isotropic) exchange energy on lattice deformations gives a volume magnetostriction or exchange striction, distinct from the linear magnetostriction. Conversely, applied pressures influence the exchange energy to give changes in T_c and the piezomagnetic effect.

While anisotropy can be accounted for when all the energy terms are known, microscopic theories of magnetostriction call for a knowledge of strain potentials involving the first derivatives of the various energy density terms. In a single-ion approach the additional electrostatic potential to first order in the strains, the strain potential, may be obtained from $\delta V = \sum_{ij}(\partial V/\partial E_{ij})E_{ij}$, with V the potential energy of the d electrons in the crystal field. The magnetostriction arises from a dependence of the anisotropy energy on the state of strain of the lattice; when $\mathbf{M}_s$ follows an easy direction the energy may be further reduced by the occurrence of strains in the lattice which enhance the trigonal components. The calculation of the magnetoelastic energy calls for the combination of the strain potential and the spin–orbit coupling in a second-order perturbation. Tsuya [35], for example, described extensive calculations on the magnetoelastic effects in spinel ferrites and Slonczewski's studies of the influence of Co^{2+} ions on the anisotropy extended to the effects on the magnetostriction [36], while an example of the calculation of magnetostriction in the antiferromagnetics FeO and CoO is that of Kanamori [37].

If the temperature dependences of the anisotropy and the magnetostriction, falling off in a roughly exponential manner, are compared with the very different behaviour of $M_s(T)$ as the Curie point is approached, it is apparent that the susceptibility should exhibit a peak a little below T_c and this is illustrated later.

References

1. Alonso, M. and Valk, H., *Quantum Mechanics; Principles and Applications*, Addison-Wesley, Reading, Massachusetts, 1973.
2. Dirac, P.A.M., *Principles of Quantum Mechanics*, Oxford University Press, 1947.
3. Heitler, W. and London, F., *Z. Phys.*, **44**, 455 (1927).
4. Dirac, P.A.M., *Proc. Roy. Soc.*, **123A**, 714 (1929).
5. Herring, C., *Magnetism*, Vol. 2B, Eds. G.T. Rado and H. Suhl, Academic Press, New York, p. 1, 1966.
6. Levy, P.M., *Magnetic Oxides*, Ed. D.J. Craik, John Wiley and Sons, London and New York, p. 181, 1975.
7. Friedel, J., T. Phys. Rad. **19**, 573 (1958); see also Clogston, A., Phys. Rev. **125** 541 (1962).
8. Kondo, J., *Prog. Theor. Phys.*, **28**, 846 (1962) and **32**, 37 (1964).
9. Ruderman, M.A. and Kittel, C., *Phys. Rev.*, **96**, 99 (1954).
10. Kasuya, T., *Prog. Theor. Phys.*, **16**, 45 and 58 (1956).
11. Yosida, K., *Phys. Rev.*, **106**, 893 (1957).
12. Anderson, P.W., *Phys. Rev.*, **79**, 350 (1950) and **115**, 2 (1959).
13. Zener, C., *Phys. Rev.*, **81**, 440 (1951), **82**, 403 (1951) and **83**, 299 (1951); see also Zener, C. and Heikes, R.R., *Revs. Mod. Phys.*, **25**, 191 (1953).
14. Goodenough, J.B., *Phys. Rev.*, **100**, 564 (1955); see also Goodenough, J.B. and Loeb, A., *Phys. Rev.*, **98**, 391 (1955).
15. Blasse, G., J. *Phys. Chem. Solids*, **27**, 383 (1966).
16. Slichter, C.P., *Principles of Magnetic Resonance*, Springer-Verlag, Berlin, 1989.
17. Hutchings, M.T., *Solid State Physics*, Vol. 16, Eds. F. Seitz and D. Turnbull, Academic Press, New York, p. 227, 1964.
18. Stevens, K.W.H. and Bates, C.A., *Magnetic Oxide*, Ed. D.J. Craik, John Wiley and Sons, London and New York, p. 141, 1975.
19. Gyorgy, E.M., Sturge, M.D., Fraser, D.B. and LeCraw, R.C., *Phys. Rev. Lett.*, **15**, 19 (1965).
20. Abragam, A. and Price, M.H.L., *Proc. Roy. Soc.*, **A205**, 135 (1951).

21. Griffith, J.S., *Phys. Rev.*, **132**, 316 (1963) and *The Theory of Transition Metal Ions*, Cambridge University Press, 1964.
22. Messiah, A., *Quantum Mechanics*, North Holland, Amsterdam, 1962.
23. Stevens, K.W.H., *J. Phys. C*, **2**, 1934 (1969).
24. Jahn, H.J. and Teller, E., *Proc. Roy. Soc.*, **A161**, 220 (1937).
25. Casimir, H.B.G., *Coll. Int. de Magnétisme*, Grenoble, 1958, CNRS, Paris, p. 296, 1959.
26. Krupička, S. and Závěta, K., *Magnetic Oxides*, Ed. D.J. Craik, John Wiley and Sons, London and New York, p. 235 (1975).
27. Wolf, W.P., *Phys. Rev.*, **108**, 1152 (1957).
28. Yosida, K., Tachiki, M., *Prog. Theor. Phys. Kyoto*, **17**, 331 (1957).
29. Krupička, K., Cervinka, L., Novak, P. and Závěta, K., *Proceedings of International Conference on Magnetism* Nottingham, P., 650 1964; see also Palmer, W., *Phys. Rev.*, **131**, 1057 (1963).
30. Slonczewski, J.C., *Phys. Rev.*, **110**, 1341 (1958) and *Magnetism*, Vol. I, Eds. G.T. Rado and H. Suhl, Academic Press, New York, p. 205 (1963).
31. Pearson, R.F., *J. Appl. Phys.*, **31**, 160S (1960); see also Bozorth, R.M., Tilden, E.F. and Williams, A.J., *Phys. Rev.*, **99**, 1788 (1955).
32. Néel, L., *J. Phys. Rad.*, **15**, 225 (1954).
33. Penoyer, R.F. and Bickford, L.R., Jr, *Phys. Rev.*, **108**, 271 (1957).
34. Giess, E.A., Calhoun, B.A., Klokholm, E., McGuire, T.R. and Rosier, L.L., *Mater. Res. Bull.*, **6**, 317 (1971).
35. Tsuya, N., *Sci. Res. Inst. Tohoku Univ.*, **B8**, 161 (1957) and *J. Appl. Phys.*, **29**, 449 (1958), see also Clark, A.E., Desavage, B.F., Tsuya, N. and Kawakami, S., *J. Appl. Phys.*, **37**, 1324 (1966).
36. Slonczewski, J.C., J. *Phys. Chem. Solids*, **15**, 335 (1960) and *J. Appl. Phys.*, **32**, 253S (1961); see also Callen, E.R. and Callen, H.B., *Phys. Rev.*, **129**, 578 (1963).
37. Kanamori, J., *Prog. Theor. Phys. Kyoto*, **17**, 197 (1957).

CHAPTER 3 Spin and Magnetization, Statics and Dynamics

1 Spin and Magnetization

It has been seen that spin has a central role in magnetism, most obviously in e.s.r. and n.m.r. Due to the common quenching of the orbital angular momentum a wide range of magnetic phenomena is associated primarily with magnetic moments due to spin and orbital effects can be accounted for by the use of suitable g factor or tensor, while for nuclear magnetism and n.m.r. the role of the spins of the protons, for example, is apparent. Technical magnetic materials, at least those devoid of rare earth elements, may be regarded approximately as ordered assemblies of spins; the experimental values of g in most ferro- and ferrimagnetics are very close to 2. Many paramagnetic susceptibilities may be ascribed largely to the polarization of the spins. The strong (non-dipolar) interactions between spins involve orbital effects, but these may be incorporated into a spin Hamiltonian without explicit reference to the orbital motion.

The operators used most generally in this chapter are symbolized by S, S_z, etc., conventionally suggesting application to electron spin, but most of the principles and results apply to intrinsic angular momenta generally and the operators may be considered to be replaced by I, I_z, etc. The Stern–Gerlach experiments described may be considered to be carried out on neutrons, for example.

1.1 The Stern–Gerlach Experiment [1]

A beam of effectively single-electron atoms, such as sodium, which can be regarded as neutralized electrons with anomalous mass and with zero orbital angular momentum in the ground state, is prepared in a symmetrical environment. It is passed through a 'splitter' S_z in which there is a magnetic field **B** having a gradient in the field direction, or through a 'filter' F_z which is a splitter afforded with a stop. With $\mathbf{B} \parallel OZ$, say, the beam with initial intensity I is split into two beams of intensity $I/2$ deflected in a symmetrical manner as in Figure 3.1(a). With the inclusion of a stop as in Figure 3.1(b), half the beam intensity is transmitted.

The emergent beam labelled μ_+ passes completely through an identical filter but μ_- is wholly stopped by such a filter [Figure 3.1(c) and (d)]. S could be rotated to give $\mathbf{B} \parallel OX$, a splitter S_x say, and of course corresponding results would be obtained with emergent beams deflected into or out of the diagram. If the beam μ_+, or the beam μ_- from S_z is passed through S_x it is split into two in exactly the same way as in the initial experiment [Figure 3.1(e)] or equivalently it is reduced to half the intensity by a filter F_x.

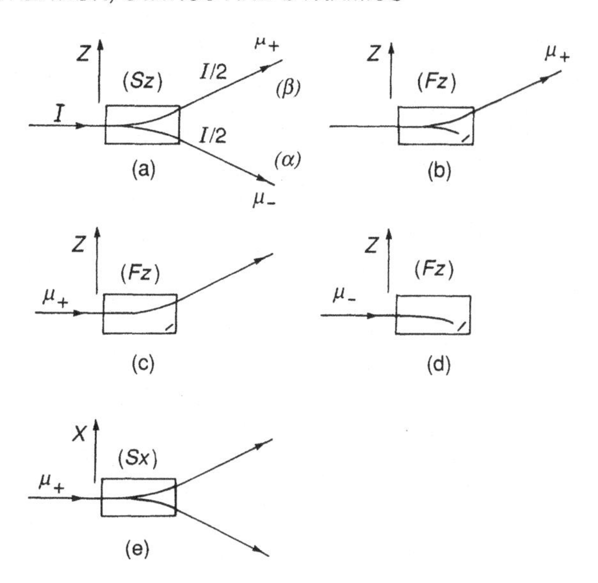

Figure 3.1 The Stern–Gerlach experiments discussed. S indicates a splitter and F a filter. Note the distinction between S_z and S_x

The effect of S_z indicates that the electrons (or atoms) in beam μ_+ are those which have taken a magnetic moment component parallel to the field and gradient and those in μ_- have a moment of equal magnitude antiparallel to the field. Since it is prepared in the absence of any fields that might define specific axes the original beam must have zero polarization, while the subsequent experiments indicate that the emergent beams μ_+ and μ_- are completely polarized. These polarizations are not induced by the field, since this would only give one component.

It can only be concluded that there exist magnetic moments and, by inference, associated angular momenta which are not due to orbital motion ($L = 0$) and are thus of an intrinsic character; for the sake of distinction these are termed the spin moments. Moreover, when an axis OZ is defined by the presence of a field the only components that may arise have the two values $\pm\mu_z$ and $\pm S_z$. The general theory of angular momentum has been seen, in Chapter 2, to show that the permitted values of the components differ by $\hbar$ and are symmetrically distributed about zero. Thus $S_z = \pm\frac{1}{2}\hbar$ unambiguously.

Because either of the beams μ_+ or μ_- is fully transmitted or completely stopped by F_z, according to the rotation of F_z through 180° about the beam axis, it is inferred that each of these beams is completely polarized parallel or antiparallel to OZ, and remains so. Assuming a negative value for γ, as for orbital motion, particles in the beam μ_+ may be assigned the state $|-\frac{1}{2}\rangle$ or $|\beta\rangle$, meaning initially no more than the state for which the value $S_z = -\frac{1}{2}\hbar$, and those in μ_- the state $|\frac{1}{2}\rangle$ or $|\alpha\rangle$. For α, $\mu_z = -g(\mathrm{e}/2m)\frac{1}{2}\hbar$ and for β, $\mu z = +g(\mathrm{e}/2m)\frac{1}{2}\hbar$, since the gyromagnetic ratio must always be of the form $\gamma = -ge/2m$, with g a number. (Recall that e is taken to be the proton charge so that γ is negative, while γ is positive for the proton.) The value of g is, in principle, that obtained from a knowledge of the field gradients and the deflections, but is more easily and accurately determined otherwise.

When the beam now designated α (or the beam β) is passed through S_x the result is as in the original experiment: symmetrical splitting into two beams of equal intensity. Conversely, if a splitter S_x is used first so that the emergent beams could be described by labels α_x and β_x and assigned the states $|x, \frac{1}{2}\rangle$, $|x, -\frac{1}{2}\rangle$, with obvious implications, then beam α_x (or β_x) is split into two by a splitter S_z, as was the original beam. Nevertheless, there is a great difference between the original beam, I, and α_x. If a filter is rotated completely around the beam axis then I is always split into two and half transmitted, for all orientations, but α_x is wholly transmitted or wholly stopped for the two particular orientations corresponding to F_x and is only half transmitted for the orientation corresponding to F_z. The beam I is unpolarized with $P = 0$, whereas the beams α_x, β_x like α and β, are completely polarized parallel or antiparallel to OX, or OZ.

Since beam α_x appears to be in a pure state, with all particles characterized identifically, its description by a single-state symbol, or vector, i.e. as $|x, \frac{1}{2}\rangle$ simply, is justified (and, of course, this applies equally to $|z, \frac{1}{2}\rangle$ or just $|\frac{1}{2}\rangle$ or $|\alpha\rangle$). However, due to the effect of F_z, this state may equally be regarded as one for which there is an equal probability that the state is $|\alpha\rangle$ or is $|\beta\rangle$. (The probability p_1 that the state is $|1\rangle$, for a superposition $|\chi\rangle$ of $|1\rangle$ and $|2\rangle$, is to be read as the extent to which $|\chi\rangle$ resembles $|1\rangle$ or 'consists of' $|1\rangle$: it is not implied that with N particles, $p_1 N$ are in state $|1\rangle$, etc., as for a mixture.) This is consistent with the proposition that the states $|\alpha\rangle$ and $|\beta\rangle$ may be taken as basis states in which the state $|x, \frac{1}{2}\rangle$ may be expressed as a superposition $c_1|\alpha\rangle + c_2|\beta\rangle$ with $c_1 = c_2$ here to indicate equal contributions. This in itself suggests the formulation in which the states are described by generalized vectors with the basis states having a role analogous to unit vectors in the Cartesian space of vector quantities. However, it is the squares of the coefficient that arise on taking scalar self-products, suggesting that, since it is also noted that in the general case coefficients should not be assumed real, the above suggestion should be replaced by $|c_1|^2 = |c_2|^2$ when the probabilities of the two basis states are equal ($|c_1|^2 = c_1^* c_1$).

This certainly circumvents the difficulty that would otherwise arise, i.e. that the equal probabilities of being in $|\alpha\rangle$ or $|\beta\rangle$ should also apply to $|x, -\frac{1}{2}\rangle$ and further that the designation of the axis orthogonal to OZ as OX was arbitrary so that two other states $|y, \frac{1}{2}\rangle$ and $|y, -\frac{1}{2}\rangle$ could similarly be introduced. Now there are three pairs which most simply satisfy $|c_1|^2 = |c_2|^2$, i.e. $1, 1$; $1, -1$; i, i and $i, -i$. The particular associations are arbitrary but the conventional choice between x and y gives

$$\begin{aligned} |x, \tfrac{1}{2}\rangle &= \frac{1}{\sqrt{2}}(|\alpha\rangle + |\beta\rangle), \qquad |x, -\tfrac{1}{2}\rangle = \frac{1}{\sqrt{2}}(|\alpha\rangle - |\beta\rangle) \\ |y, \tfrac{1}{2}\rangle &= \frac{1}{\sqrt{2}}(|\alpha\rangle + i|\beta\rangle), \qquad |y, -\tfrac{1}{2}\rangle = \frac{1}{\sqrt{2}}(|\alpha\rangle - i|\beta\rangle) \end{aligned} \tag{1}$$

The leading factors ensure normalization on the assumption that $|\alpha\rangle$ and $|\beta\rangle$ are normalized according to $\langle\alpha|\alpha\rangle = 1 = \langle\beta|\beta\rangle$; they are orthogonal since they are not degenerate, as may readily be varified. The probabilities must be given by $p_i = |c_i|^2 = c_i^* c_i$ in order to give $\sum p_i = 1$ for a normalized superposition.

The Stern–Gerlach experiment is truly remarkable since it gives a convincing introduction to the concept of spin and, in a sense, to general quantum theory.

1.2 Spin Operators, Matrices and Vectors

In the study of orbital motion, explicit operators can be derived and eigenvalues and eigenfunctions determined by solving differential equations. In the present case it must be assumed that, since spin is inevitably encompassed by quantum theory, the relevant states and operators are those that may be associated with matrices that may be derived on general grounds. Referring to the theory of angular momentum, as in Chapter 2, the two states discussed first in relation to the S–G experiment are more fully denoted $|\,\frac{1}{2}\;\frac{1}{2}\,\rangle$ and $|\,\frac{1}{2}\;-\frac{1}{2}\,\rangle$ in correspondence with the general $|\,j\;m\,\rangle$, since the component m can only range between $+j$ and $-j$. Since $J^+|\,j\;m\,\rangle[(j-m)(j+m+1)]^{1/2}|\,j\;m+1\,\rangle$, and with $j \to S = \frac{1}{2}$, $m \to m_S = \pm\frac{1}{2}$, $J^+ \to S^+$, it is seen that (abbreviating once more and with explicit $\hbar$) $S^+|\frac{1}{2}\rangle = 0$, $S^+|-\frac{1}{2}\rangle = \hbar|\frac{1}{2}\rangle$ and similarly $S^-|\frac{1}{2}\rangle = \hbar|-\frac{1}{2}\rangle$ and $S^-|-\frac{1}{2}\rangle = 0$. The only non-zero matrix elements, in $|\frac{1}{2}\rangle = |1\rangle$ and $|-\frac{1}{2}\rangle = |2\rangle$, are $\langle\frac{1}{2}|S^+|-\frac{1}{2}\rangle$ and $\langle-\frac{1}{2}|S^-|\frac{1}{2}\rangle$, each = $\hbar$. Thus the spin matrices are

$$S^+ = \hbar\begin{pmatrix}0 & 1\\ 0 & 0\end{pmatrix}, \qquad S^- = \hbar\begin{pmatrix}0 & 0\\ 1 & 0\end{pmatrix} \tag{2}$$

$$S_x = \tfrac{1}{2}\hbar\begin{pmatrix}0 & 1\\ 1 & 0\end{pmatrix}, \qquad S_y = \tfrac{1}{2}\hbar\begin{pmatrix}0 & -i\\ i & 0\end{pmatrix}, \qquad S_z = \tfrac{1}{2}\hbar\begin{pmatrix}1 & 0\\ 0 & -1\end{pmatrix} \tag{3}$$

making use of $S^\pm = S_x \pm iS_y$ and with S_z obvious, in the S_z basis. Alternatively, with $S_i = \frac{1}{2}\hbar\sigma_i$

$$\sigma_x = \begin{pmatrix}0 & 1\\ 1 & 0\end{pmatrix}, \qquad \sigma_y = \begin{pmatrix}0 & -i\\ i & 0\end{pmatrix}, \qquad \sigma_z = \begin{pmatrix}1 & 0\\ 0 & -1\end{pmatrix} \tag{4}$$

which are the Pauli matrices [2]. A check that these conform with

$$\sigma_x\sigma_y = i\sigma_z, \qquad \sigma_y, \sigma_z = i\sigma_x, \qquad \sigma_z\sigma_x = i\sigma_y, \qquad \sigma_y\sigma_x = -i\sigma_z, \qquad \text{etc.} \tag{5}$$

(remembered by 'x then y gives z' etc.) and thus with

$$\sigma_x\sigma_y - \sigma_y\sigma_x = 2i\sigma_z, \sigma_y\sigma_z - \sigma_z\sigma_y = 2i\sigma_x, \qquad \sigma_z\sigma_x - \sigma_x\sigma_z = 2i\sigma_y \tag{6}$$

is left as an exercise. Obvious eigenvectors of σ_z are $(\,1\;0)^{\mathrm{T}} \sim |\frac{1}{2}\rangle \equiv |\alpha\rangle$ and $(\,0\;1\,)^{\mathrm{T}} \sim |-\frac{1}{2}\rangle \equiv |\beta\rangle$. Simple matrix algebra indicates that

$$\begin{pmatrix}0 & 1\\ 1 & 0\end{pmatrix}\begin{pmatrix}c_1\\ c_2\end{pmatrix} = \begin{pmatrix}c_2\\ c_1\end{pmatrix} = \pm\begin{pmatrix}c_1\\ c_2\end{pmatrix} \qquad \text{if } c_2 = \pm c_1$$

$$\begin{pmatrix}0 & -i\\ i & 0\end{pmatrix}\begin{pmatrix}c_1\\ c_2\end{pmatrix} = \begin{pmatrix}-ic_2\\ ic_1\end{pmatrix} = \pm\begin{pmatrix}c_1\\ c_2\end{pmatrix} \qquad \text{if } c_2 = \pm ic_1$$

Eigenvectors for $\sigma_x(S_x)$ and $\sigma_y(s_y)$ expressed as general superpositions in the S basis, $|\chi\rangle = C_1|\alpha\rangle + C_2|\beta\rangle$, are thus

$$|x, \tfrac{1}{2}\rangle = \frac{1}{\sqrt{2}}(|\alpha\rangle + |\beta\rangle), \qquad |x, -\tfrac{1}{2}\rangle = \frac{1}{\sqrt{2}}(|\alpha\rangle - |\beta\rangle)$$
$$|y, \tfrac{1}{2}\rangle = \frac{1}{\sqrt{2}}(|\alpha\rangle + i|\beta\rangle), \qquad |y, -\tfrac{1}{2}\rangle = \frac{1}{\sqrt{2}}(|\alpha\rangle - i|\beta\rangle) \tag{7}$$

as inferred previously.

The spin raising and lowering operators $S^{+/-}$ afford a useful expansion for S^2, i.e

$$\begin{aligned} S^2 &= S_x^2 + S_y^2 + S_z^2 = \tfrac{1}{4}(S^+ + S^-)^2 - \tfrac{1}{4}(S^+ - S^-)^2 + S_z^2 \\ &= \tfrac{1}{4}(S^{+^2} + S^{-^2} + S^+S^- + S^-S^+) - \tfrac{1}{4}(S^{+^2} + S^{-^2} - S^+S^- - S^-S^+) + S_z^2 \\ &= S_z^2 + \tfrac{1}{2}(S^+S^- + S^-S^+) \end{aligned} \tag{8}$$

noting that $[S^+, S^-] \neq 0$.

The components of the spin polarization P are defined by $P_i = \bar{\sigma}_i = \langle \chi | \sigma_i | \chi \rangle$. Since

$$\begin{aligned} \sigma_x|\chi\rangle &= c_1|\beta\rangle + c_2|\alpha\rangle \\ \sigma_y|\chi\rangle &= ic_1|\beta\rangle - ic_2|\alpha\rangle \\ \sigma_z|\chi\rangle &= c_1|\alpha\rangle - c_2|\beta\rangle \end{aligned} \tag{9}$$

it is seen that

$$\begin{aligned} P_x &= \overline{\sigma}_x = (c_1^*\langle\alpha| + c_2^*\langle\beta|)\sigma_x|\chi\rangle = c_1c_2^* + c_2c_1^* = \rho_{12} + \rho_{21} \\ P_y &= \overline{\sigma}_y = (c_1^*\langle\alpha| + c_2^*\langle\beta|)\sigma_y|\chi\rangle = i(c_1c_2^* - c_2c_1^*) = i(\rho_{12} - \rho_{21}) \\ P_z &= \overline{\sigma}_z = (c_1^*\langle\alpha| + c_2^*\langle\beta|)\sigma_z|\chi\rangle = c_1c_1^* - c_2c_2^* = \rho_{11} - \rho_{22} \end{aligned} \tag{10}$$

the ρ_{ij} being the elements of the appropriate density matrix:

$$\rho_{ij} = \langle i|\chi\rangle\langle\chi|j\rangle = c_ic_j^*, \qquad |1\rangle \equiv |\alpha\rangle, |2\rangle \equiv |\beta\rangle \tag{11}$$

Making use of $P_i = \bar{\sigma}_i = \mathrm{Tr}\,\rho\sigma_i$, it can be seen that ρ can be expressed in the P_i as

$$\rho = \tfrac{1}{2}\begin{pmatrix} 1 + P_z & P_x - iP_y \\ P_x + iP_y & 1 - P_z \end{pmatrix} \tag{12}$$

and this is clearly consistent with $P_x = \rho_{12} + \rho_{21}$, etc. ρ can also be written as

$$\begin{aligned} \rho &= \tfrac{1}{2}\left\{\begin{pmatrix} 1 & 0 \\ 0 & 1 \end{pmatrix} + \begin{pmatrix} 0 & P_x \\ P_x & 0 \end{pmatrix} + \begin{pmatrix} 0 & -iP_y \\ iP_y & 0 \end{pmatrix} + \begin{pmatrix} P_z & 0 \\ 0 & -P_z \end{pmatrix}\right\} \\ &= \tfrac{1}{2}(P_x\sigma_x + P_y\sigma_y + P_z\sigma_z + I) \end{aligned} \tag{13}$$

exemplifying the following general principle. By taking suitable linear combinations of the matrices σ_i together with the $(n = 2)$ identity matrix it is easy to obtain $\left(\begin{smallmatrix} 1 & 0 \\ 0 & 0 \end{smallmatrix}\right)$ and three other matrices having different single non-zero elements of unity, and it is then apparent that any 2×2 matrix may be obtained as a linear combination of these and thus of the σ_i and I. In this case the coefficients are the P_i.

The spin polarization gives the magnetic polarization as $\boldsymbol{\mu} = -\frac{1}{2}g\beta\mathbf{P}$ or $\mu = -\beta P$, with $g = 2$. For a single electron (proton) in a pure state it is easy to see that $\mu = \frac{1}{2}g\beta$, since

$$P^2 = P_x^2 + P_y^2 + P_z^2 = \cdots = (c_1c_1^* + c_2c_2^*)^2 = 1 \tag{14}$$

The matrix self-products indicate that $\sigma_x^2 = I = \sigma_y^2 = \sigma_z^2$, i.e. $\hat{\sigma}_x^2 = \hat{\sigma}_y^2 = \hat{\sigma}_z^2 = I$, the identity operator. Thus any pure spin state is an eigenstate of σ^2 or S^2 with eigenvalues 3 or

$\frac{3}{4}\hbar^2 = \hbar^2[S(S+1)] : S = \frac{1}{2}$ respectively. It is apparent that $S^2 = \frac{3}{4}\hbar^2 I$ commutes with any of the S_i and thus also with a Hamiltonian, which can be expressed as $\mathcal{H} = -\gamma B S_z$ (with $B \parallel OZ$), but equally apparent that the S_i do not commute with each other. The picture, [Figure 3.2(a)] is one of a spin angular momentum vector of magnitude $\sqrt{3/4}\hbar = 0.87\hbar$ which can, however, never be fully observed in direction due to the uncertainty principle or the failure of the S_i to commute, e.g. for $|\chi\rangle = |\alpha\rangle$, $S_2 = \frac{1}{2}\hbar$ while $S_x = 0 = S_y$ as if the vector was so distributed in space that the components in OXY averaged to zero. A magnetic moment $\sqrt{3/4}g\beta$ may be associated with this vector, using $\gamma = -ge/(2\text{m})$, but again the orientation of this would never be completely known. No field could induce a component $\sqrt{3/4}g\beta$ along its direction.

The magnetic moment components which may be measured or observed must correspond to those that may be calculated, i.e. to the mean values $\bar{S}_i$ or $\bar{\sigma}_i$, i.e. the P_i. The most meaningful picture is that of Figure 3.2(b), in which any state is represented by a particular **P**. An alternative specification of any pure state or magnetic moment consists of the two angles θ and ϕ shown with the relations

$$c_1 = \cos\frac{\theta}{2}, \qquad c_2 = e^{i\phi}\sin\frac{\theta}{2} \tag{15}$$

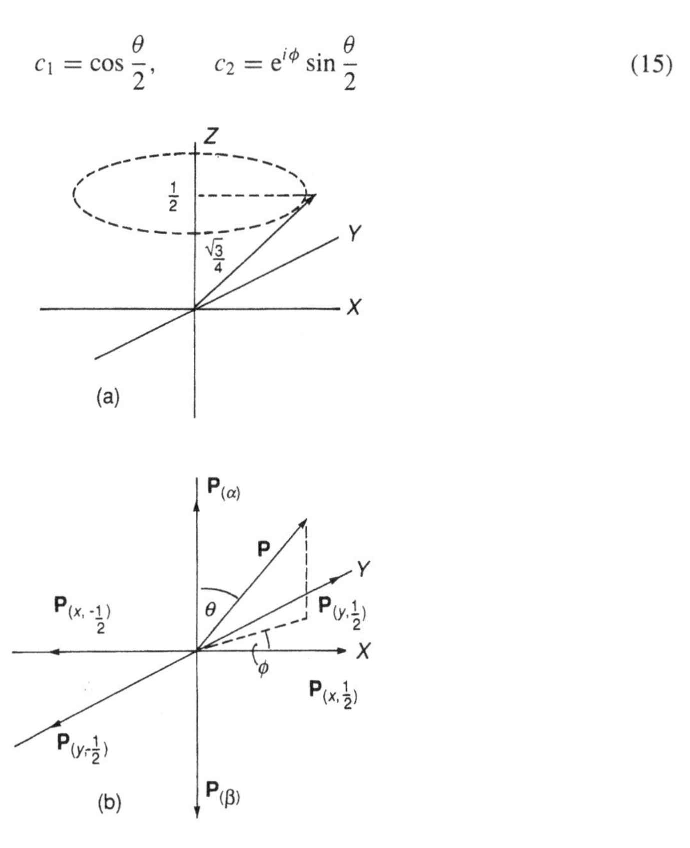

Figure 3.2 (a) For the state $|\alpha\rangle$ the spin angular momentum vector lies indeterminately on the cone, indicated so that the mean components transverse to the chosen axis of quantization are zero (units of $\hbar$). (b) Since $P = 1$ any pure spin state can be specified by the angles shown, with examples for $|\alpha\rangle$, $|x, \frac{1}{2}\rangle$, etc

since the reader may show that these give the appropriate $P_x = \sin\theta\cos\phi$, $P_y = \sin\theta\sin\phi$, $P_z = \cos\theta$ in terms of the c_i. (Since the c_i may be complex, four numbers may appear to be required, but these reduce to two due to normalization and phase indeterminacy.) As an exercises the c_i may be calculated for states for which **P** makes equal angles with OX, OY and OZ, etc.

For a mixed state described as 'W_a in $|\chi_a\rangle$ and W_b in $|\chi_b\rangle$' in terms of probabilities W_i or numbers W_aN and W_bN, the density matrix is $W_a\rho_a + W_b\rho_b$ and each component $P_i = W_aP_i^a + W_bP_i^b$. P^2 for the mixed state is

$$\begin{aligned} P^2 &= W_a^2P_x^{a^2} + W_b^2P_x^{b^2} + 2W_aW_bP_x^aP_x^b + \cdots \\ &= W_a^2(P_x^{a^2} + P_y^{a^2} + P_z^{a^2}) + W_b^2(P_x^{b^2} + P_y^{b^2} + P_z^{b^2}) + 2W_aW_b\mathbf{P}^a \cdot \mathbf{P}^b \\ &= W_a^2 + W_b^2 + 2W_aW_b\mathbf{P}^a \cdot \mathbf{P}^b \end{aligned}$$

since $P^a = 1 = P^b$ for the two pure states. The scalar product of the two unit vectors $\mathbf{P}^a \cdot \mathbf{P}^b = +1$ and $P^2 = (W_a + W_b)^2 = 1$, if $\mathbf{P}^a \parallel \mathbf{P}^b$ so that the nominal mixed state is, in fact, a pure state. Otherwise the $\mathbf{P}^a \cdot \mathbf{P}^b < 1$ and for a true mixed state $P < 1$. If, for example, $|\chi_a\rangle = |\alpha\rangle$, $|\chi_b\rangle = |\beta\rangle$, $P = (W_a - W_b)$ and if $W_a = W_b = 0.5$ then $P = 0$. P ranges from 0 to 1 as M ranges from 0 to $\frac{1}{2}Ng\beta$.

A formal compact review of the algebra of the Pauli operators σ_i may be made (see, for example, Blum [3], p. 18) in the form

$$\sigma_i\sigma_j = \delta_{ij}I + i\sum_k \varepsilon_{ijk}\sigma_k, \qquad \text{for } i, j, k = x, y, z \tag{16}$$

with

$$\varepsilon_{ijk} = \begin{cases} 1 & \text{if } i, j, k = x, y, z \text{ or an even permutation of } x, y, z \\ -1 & \text{if } i, j, k \text{ form an odd permutation of } x, y, z \\ 0 & \text{if two of the } i, j, k \text{ are the same} \end{cases}$$

Thus if $i = j$, $\varepsilon_{ijk} = 0$ and $\sigma_i^2 = I$ is included. If $i = x$ and $j = y$, only $k = z$ gives a non-zero ε_{ijk} and, with $\delta_{ij} = 0$, $\sigma_x\sigma_y = i\sigma_z$. If $i = y$ and $j = x$, $\sum_k \varepsilon_{ijk} = 0 + 0 + \varepsilon_{yxz} = -1$ since a single interchange $x \leftrightarrows y$, an odd permutation, gives yxz from xyz and so

$$\sigma_y\sigma_x = -i\sigma_z, \qquad [\sigma_x, \sigma_y] = 2i\sigma_z$$

using the above $\sigma_x\sigma_y = i\sigma_z$, and so forth.

It is apparent that

$$\begin{aligned} \operatorname{Tr}\sigma_i &= 0, & \operatorname{Tr}\sigma_i^2 &= \operatorname{Tr} I = 2 \\ \operatorname{Tr} S_i &= 0, & \operatorname{Tr} S_i^2 &= \tfrac{1}{2}\hbar^2 \end{aligned} \tag{17}$$

It has been seen that any 2×2 matrix A can be expressed as

$$A = a_x\sigma_x + a_y\sigma_y + a_z\sigma_z + bI = \mathbf{a} \cdot \boldsymbol{\sigma} + bI \tag{18}$$

and it follows that

$$\operatorname{Tr} A = b \operatorname{Tr} I, \qquad b = \operatorname{Tr}\frac{A}{2} \tag{19}$$

Multiplying by σ_z and noting that the product of two σ_i with i different gives a further σ_i with zero trace, $\operatorname{Tr}\sigma_i\sigma_j = 2\delta_{ij}$, which gives $\operatorname{Tr}\sigma_zA = 2a_z$ and generally

$$a_i = \tfrac{1}{2}\operatorname{Tr}\sigma_iA \tag{20}$$

1.3 Equilibrium Magnetization and Magnetic Susceptibility

With a field applied $\parallel OZ$ the magnetization follows from the value of P_z in the equilibrium state. This may be obtained as $P_z = \mathrm{Tr}\,\rho^0\sigma_z$ with ρ^0 the appropriate equilibrium density matrix, as discussed in Section 2.3, Chapter 2. In the present case the Hamiltonian and the energies are

$$\mathcal{H} = -\gamma B S_z = -\tfrac{1}{2}\hbar\gamma B\sigma_z$$

$$E_1 = E_\alpha = -\tfrac{1}{2}\hbar\gamma B, \qquad E_2 = E_\beta = +\tfrac{1}{2}\hbar\gamma B$$

and with $\rho^0 = \mathrm{e}^{-\mathcal{H}/(kT)}/Z$,

$$P_z = \bar{\sigma}_z = \mathrm{Tr}\,\rho^0\sigma_z = \mathrm{Tr}\,\mathrm{e}^{a\sigma_z}\frac{\sigma_z}{Z} \tag{21}$$

with $a = \frac{1}{2}\gamma\hbar B/(kT)$. In the Curie limit for small values of a,

$$P_z = \frac{\mathrm{Tr}\,(1 + a\sigma_z)\sigma_z}{Z} = \frac{\mathrm{Tr}\,(\sigma_z + a\sigma_z^2)}{Z} = \frac{2a}{Z} \tag{22}$$

since $\mathrm{Tr}\,\sigma_z = 0$ and $\mathrm{Tr}\,\sigma_z^2 = 2$. The partition function is

$$Z = \mathrm{Tr}\,\rho^0 = \mathrm{Tr}\,(1 + a\sigma_z) = 2 \tag{23}$$

again with $\mathrm{Tr}\,\sigma_z = 0$ and $\mathrm{Tr}\,1 = 2$ for a dimension of 2. Thus $P_z = a$ and $\mu_z = \frac{1}{2}\hbar\gamma a$. With N spins in unit volume,

$$M = \frac{N\gamma^2\hbar^2 B}{4kT}, \qquad \chi = \frac{M}{H} = \frac{\mu_0 N\gamma^2\hbar^2}{4kT} = \frac{\mu_0 N g^2\beta^2}{4kT} \tag{24}$$

In the general case, since $\sigma_z^2 = I$ then $\sigma_z^4 = I$ also, and clearly any even power gives I, i.e. $\mathrm{Tr}\,\sigma_z^n = 2$ for n even. $\mathrm{Tr}\,\sigma_z = 0$ and $\mathrm{Tr}\,\sigma_z^3 = \mathrm{Tr}\,\sigma_z^2\ \mathrm{Tr}\,\sigma_z = 0$ also, etc. Thus

$$\begin{aligned}
\mathrm{Tr}\,(\mathrm{e}^{a\sigma_z}\sigma_z) &= \mathrm{Tr}\left(1 + a\sigma_z + \frac{a^2\sigma_z^2}{2!} + \cdots\right)\sigma_z \\
&= \mathrm{Tr}\left(\sigma_z + a\sigma_z^2 + \frac{a^2\sigma_z^3}{2!} + \cdots\right) \\
&= 2a + \frac{2a^3}{3!} + \frac{2a^5}{5!} + \cdots = \mathrm{e}^a - \mathrm{e}^{-a}
\end{aligned} \tag{25}$$

as seen by subtracting the series expansions of e^a and e^{-a}. Also

$$\begin{aligned}
Z = \mathrm{Tr}\,\mathrm{e}^{a\sigma_z} &= \mathrm{Tr}\left(1 + a\sigma_z + \frac{a^2\sigma_z^2}{z!} + \cdots\right) \\
&= 2 + \frac{2a^2}{2!} + \frac{2a^4}{4!} + \cdots = \mathrm{e}^a + \mathrm{e}^{-a}
\end{aligned} \tag{26}$$

Thus

$$P_z = \frac{\mathrm{e}^a - \mathrm{e}^{-a}}{e^a + e^{-a}} = \tanh a$$

$$M = \tfrac{1}{2} N\gamma\hbar \tanh a$$

$$\chi = \frac{\partial M}{\partial H} = \frac{\tfrac{1}{4}\mu_0 N\gamma^2\hbar^2}{kT} \operatorname{sech}^2 \frac{\tfrac{1}{2}\gamma\hbar B}{kT}$$

i.e, with $\gamma\hbar = g\beta$,

$$M = \tfrac{1}{2} N g\beta \tanh \frac{\tfrac{1}{2} g\beta B}{kT}, \qquad \chi = \frac{\tfrac{1}{4} N g^2\beta^2\mu_0}{kT} \operatorname{sech}^2 \frac{\tfrac{1}{2} g\beta B}{kT} \tag{27}$$

This rather elaborate derivation may be compared with a much simpler (matrix element) approach, which is to accept that $P_z = \rho_{11} - \rho_{22}$ still when ρ is the equilibrium density matrix. The result is immediate. It is also apparent that since $\rho_{12}^0 = 0 = \rho_{21}^0$, both P_x and P_y are zero at equilibrium. The mixture is then said to be incoherent. It follows that $P = P_z$ and since tanh ( ) varies between 0 and 1 this corresponds to the demonstration that P can vary between 0 and 1.

1.4 Two or More Spins: Direct Products

If each of two spins can be in the two chosen states $|\alpha\rangle$ and $|\beta\rangle$ it is to be expected that the appropriate joint space for the system of the two spins together should be of dimension $n = 4$. As seen most clearly for independent spins, the states should be associated with columns $(c_1\, c_2\, c_3\, c_4)^{\mathrm{T}}$ and the operators with 4×4 matrices. Familiar summation or multiplication of matrices does not change the dimension. Consider, however, the *direct product* of the matrices A or a_{ij} and B or b_{ij} defined, or adequately described, by

$$A \times B = \begin{pmatrix} a_{11}B & a_{12}B \\ a_{21}B & a_{22}B \end{pmatrix} = \begin{pmatrix} a_{11}b_{11} & a_{11}b_{12} & a_{12}b_{11} & a_{12}b_{12} \\ a_{11}b_{21} & a_{11}b_{22} & a_{12}b_{21} & a_{12}b_{22} \\ a_{21}b_{11} & a_{21}b_{12} & a_{22}b_{11} & a_{22}b_{12} \\ a_{21}b_{21} & a_{21}b_{22} & a_{22}b_{21} & a_{22}b_{22} \end{pmatrix} \tag{28}$$

With the associations $|\alpha\rangle \sim (1\ 0)^{\mathrm{T}}$, $|\beta\rangle \sim (0\ 1)^{\mathrm{T}}$, four direct products may be formed:

$$\begin{pmatrix}1\\0\end{pmatrix} \times \begin{pmatrix}1\\0\end{pmatrix} = \begin{pmatrix} 1 \begin{pmatrix}1\\0\end{pmatrix} \\ 0 \begin{pmatrix}1\\0\end{pmatrix} \end{pmatrix} = \begin{pmatrix}1\\0\\0\\0\end{pmatrix}, \qquad \begin{pmatrix}1\\0\end{pmatrix} \times \begin{pmatrix}0\\1\end{pmatrix} = \begin{pmatrix}0\\1\\0\\0\end{pmatrix},$$
$$(|\alpha\rangle \quad |\alpha\rangle \quad = \quad |\alpha\alpha\rangle) \qquad (|\alpha\rangle \quad |\beta\rangle \quad = \quad |\alpha\,\beta\rangle) \tag{29}$$
$$\begin{pmatrix}0\\1\end{pmatrix} \times \begin{pmatrix}1\\0\end{pmatrix} = \begin{pmatrix}0\\0\\1\\0\end{pmatrix}, \qquad \begin{pmatrix}0\\1\end{pmatrix} \times \begin{pmatrix}0\\1\end{pmatrix} = \begin{pmatrix}0\\0\\0\\1\end{pmatrix}$$
$$(|\beta\rangle \quad |\alpha\rangle \quad = \quad |\beta\,\alpha\rangle) \qquad (|\beta\rangle \quad |\beta\rangle \quad = \quad |\beta\,\beta\rangle)$$

Note the distinction achieved between $|\alpha\,\beta\rangle$ and $|\beta\,\alpha\rangle$ which recognizes that, although these are physically equivalent, the state 'one spin in $|\alpha\rangle$ and the other in $|\beta\rangle$' may arise in two different ways. This corresponds to the rule that $A \times B \neq B \times A$ (which applies even if one matrix is replaced by I). Other results which are readily obtained by expansion are

$$\operatorname{Tr} A \times B = \operatorname{Tr} A \operatorname{Tr} B = \operatorname{Tr} B \times A \tag{30}$$

$$(A \times B)(C \times D) = (AC) \times (BD) \tag{31}$$

with products in parentheses to be formed first.

If A and B are matrices representing operators $\hat{A}$ and $\hat{B}$ and C and D are columns representing state vectors $|\chi_c\rangle$ and $|\chi_d\rangle$ then the association is

$$\hat{A}\hat{B}|\chi_c\chi_d\rangle = (\hat{A}|\chi_c\rangle)(\hat{B}|\chi_d\rangle) \tag{32}$$

whereby the operator written first automatically applies to the vector written first and so forth. AB is a direct product of single-particle operators and $|\chi_c\rangle|\chi_d\rangle$ is abbreviated to $|\chi_c\ \chi_d\rangle$, a direct product of states, abbreviated to a product state, in accordance with this association with the matrices. Of course the matrices may be the Pauli matrices and the operators the spin operators. In this way multiparticle operators are derived from single-particle operators.

For example, $\sigma_z\mathfrak{I}|\alpha\ \beta\rangle = \sigma_z|\alpha\rangle\mathfrak{I}|\beta\rangle = |\alpha\ \beta\rangle$ while $\mathfrak{I}\sigma_z|\alpha\ \beta\rangle = \mathfrak{I}|\alpha\rangle\sigma_z|\beta\rangle = -|\alpha\ \beta\rangle$, as illustrated explicitly by ($\mathfrak{I} \sim I$, with $\mathfrak{I}$ the identity operator):

$$\left\{\begin{pmatrix}1 & 0\\ 0 & -1\end{pmatrix} \times \begin{pmatrix}1 & 0\\ 0 & 1\end{pmatrix}\right\}\left\{\begin{pmatrix}1\\ 0\end{pmatrix} \times \begin{pmatrix}0\\ 1\end{pmatrix}\right\} = \begin{pmatrix}1 & 0 & 0 & 0\\ 0 & 1 & 0 & 0\\ 0 & 0 & -1 & 0\\ 0 & 0 & 0 & -1\end{pmatrix}\begin{pmatrix}0\\ 1\\ 0\\ 0\end{pmatrix} = \begin{pmatrix}0\\ 1\\ 0\\ 0\end{pmatrix} \tag{33}$$

$$\left\{\begin{pmatrix}1 & 0\\ 0 & 1\end{pmatrix} \times \begin{pmatrix}1 & 0\\ 0 & -1\end{pmatrix}\right\}\left\{\begin{pmatrix}1\\ 0\end{pmatrix} \times \begin{pmatrix}0\\ 1\end{pmatrix}\right\} = \begin{pmatrix}1 & 0 & 0 & 0\\ 0 & -1 & 0 & 0\\ 0 & 0 & 1 & 0\\ 0 & 0 & 0 & -1\end{pmatrix}\begin{pmatrix}0\\ 1\\ 0\\ 0\end{pmatrix} = -\begin{pmatrix}0\\ 1\\ 0\\ 0\end{pmatrix} \tag{34}$$

The automatic correlation between the order in which the operators are written and the order in which the single-particle states are written is the vital feature. Wave functions for two particles may be distinguished by the convention that 'particle 1' has coordinates $\mathbf{r}_1$ and that operators affecting the $f(\mathbf{r}_1)$ are those exemplified by $\partial/\partial x_1$, but this would not be appropriate here. $|\alpha\ \beta\rangle$ is *not* $|\alpha(\mathbf{r}_1)\rangle|\beta(\mathbf{r}_2)\rangle$. Similarly, an operator S_1^z cannot be taken to operate on 'spin 1' because it depends on the coordinates of that spin. Nevertheless, it is convenient to label operators so that, for example, $S_2^z|\alpha\ \beta\ \alpha\rangle \equiv \mathfrak{I}S_z\mathfrak{I}|\alpha\ \beta\ \alpha\rangle = |\alpha\rangle(S_z|\beta\rangle)|\alpha\rangle$.

As an exercise the product $|x, \frac{1}{2}\rangle|x, \frac{1}{2}\rangle$ may be formed as $\binom{1}{1} \times \binom{1}{1}$ and seen to correspond to $|\chi\rangle = |\alpha\alpha\rangle + |\beta\alpha\rangle + |\alpha\ \beta\rangle + |\beta\ \beta\rangle$. By either of two methods it may be confirmed that $(S_1^x + S_2^x)|\chi\rangle = \hbar|\chi\rangle$ as expected when both spins are polarized $\parallel OX$.

Interactions are most commonly represented by the Heisenberg Hamiltonian:

$$\mathcal{H} = -\mathcal{J}_{ij}\sum_{i,j}{}' \mathbf{S}_i \cdot \mathbf{S}_j \tag{35}$$

the summations, avoiding $i = j$, usually applying to nearest neighbour pairs. It is sometimes, as in the presence of a large field $\parallel OZ$, acceptable to neglect the x and y components and take $\mathcal{H} = -\mathcal{J}\sum' S_i^z S_j^z$, the Ising form, as used in Chapter 1 and discussed later. The Heisenberg interaction will be shown to lead to the existence of coupled states which are characterized by $S = \frac{1}{2}N$ and $M_S = S, S-1, \ldots, -(S-1), -S$. $\mathcal{J} > 0$ may be said to indicate a ferromagnetic interaction inasmuch as if $\mathcal{H}$ were read as a classical expression for the energy, this would be minimal for all the $\mathbf{S}_i$ parallel, suggesting a ground state $|\alpha\ \alpha\ \alpha\ \cdots\ \alpha\ \alpha\rangle$ or $|\beta\ \beta\ \beta\ \cdots\ \beta\ \beta\rangle$, and $\mathcal{J} < 0$ indicates antiferromagnetism with antiparallel neighbours suggestive of $|\alpha\ \beta\ \alpha\ \beta\ \cdots\ \alpha\ \beta\rangle$ as the ground state. The extent to which this actually applies remains to be seen.

A simple treatment may be given for two identical spins in the same field ($B \parallel OZ$), neglecting chemical shifts in the context of n.m.r. It is assumed that the possible product states, $|\alpha\ \alpha\rangle = |1\rangle$, $|\alpha\ \beta\rangle = |2\rangle$, $|\beta\ \alpha\rangle = |3\rangle$, $|\beta\ \beta\rangle = |4\rangle$, must constitute a complete basis so that any state $|\chi\rangle = \sum_i c_i|i\rangle$ and eigenvalues of the matrix $\mathcal{H}_{ij} = \langle i|\mathcal{H}|j\rangle$ are exact. In the present case, expanding $\mathbf{S}_1 \cdot \mathbf{S}_2$,

$$\mathcal{H} = -\gamma B S_1^z - \gamma B S_2^z - \mathcal{J}[S_1^z S_2^z + \tfrac{1}{2}(S_1^+ S_2^- + S_1^- S_2^+)] \tag{36}$$

The effects of the individual operators on the $|i\rangle$ are, for example,

$$\begin{aligned}
S_1^z S_2^z|\alpha\ \beta\rangle &= S_z|\alpha\rangle S_z|\beta\rangle = \tfrac{1}{2}\hbar|\alpha\rangle(-\tfrac{1}{2}\hbar)|\beta\rangle = -\tfrac{1}{4}\hbar^2|\alpha\ \beta\rangle \\
S_1^z S_2^z|\alpha\ \alpha\rangle &= +\tfrac{1}{4}\hbar^2|\alpha\ \alpha\rangle \\
S_1^z S_2^z|\beta\ \beta\rangle &= +\tfrac{1}{4}\hbar^2|\beta\ \beta\rangle \\
S_1^+ S_2^-|\alpha\ \beta\rangle &= S^+|\alpha\rangle S^-|\beta\rangle = 0 \times 0 \\
S_1^+ S_2^-|\beta\ \alpha\rangle &= S^+|\beta\rangle S^-|\alpha\rangle = \hbar|\alpha\rangle\hbar|\beta\rangle = \hbar^2|\alpha\ \beta\rangle \\
S_1^z|\alpha\ \beta\rangle &= \tfrac{1}{2}\hbar|\alpha\ \beta\rangle \\
S_2^z|\alpha\ \beta\rangle &= -\tfrac{1}{2}\hbar|\alpha\ \beta\rangle
\end{aligned}$$

a list which the reader should complete, noting that $S_1^+ S_2^-|\alpha\ \alpha\rangle$ and $S_1^+ S_2^-|\beta\ \beta\rangle$ must give zero due to $S^+|\alpha\rangle = 0$ or $S^-|\beta\rangle = 0$. The matrix elements are then exemplified by

$$\begin{aligned}
\langle\alpha\ \beta|\mathcal{H}|\alpha\ \beta\rangle &= \langle\alpha\ \beta|[-\gamma B(\tfrac{1}{2}\hbar - \tfrac{1}{2}\hbar)|\alpha\ \beta\rangle + \tfrac{1}{4}\hbar\mathcal{J}|\alpha\ \beta\rangle + \cdots] = \tfrac{1}{4}\hbar\mathcal{J} \\
\langle\alpha\ \beta|\mathcal{H}|\beta\ \alpha\rangle &= \langle\alpha\ \beta|[\cdots - \tfrac{1}{2}\hbar^2\mathcal{J}|\alpha\ \beta\rangle] = -\tfrac{1}{2}\hbar^2\mathcal{J}
\end{aligned}$$

noting that the $(\cdots)$ are here irrelevant due to orthogonality of the product states. It is then tedious but simple to complete:

$$\mathcal{H} = \begin{pmatrix} -\hbar\gamma B - \tfrac{1}{4}\hbar^2\mathcal{J} & 0 & 0 & 0 \\ 0 & \tfrac{1}{4}\hbar^2\mathcal{J} & -\tfrac{1}{2}\hbar^2\mathcal{J} & 0 \\ 0 & -\tfrac{1}{2}\hbar^2\mathcal{J} & \tfrac{1}{4}\hbar^2\mathcal{J} & 0 \\ 0 & 0 & 0 & +\hbar\gamma B - \tfrac{1}{4}\hbar^2\mathcal{J} \end{pmatrix} \tag{37}$$

This indicates that the basis states $|1\rangle$ and $|4\rangle$ are eigenstates and that the energies are the matrix elements themselves, as discussed further (Section 1.6). (It is readily confirmed, by inspection, that $|\alpha\ \alpha\rangle$ and $|\beta\ \beta\rangle$ are eigenvectors.) It remains to study the reduced system

$$\begin{pmatrix} \tfrac{1}{4}\hbar^2\mathcal{J} & -\tfrac{1}{2}\hbar^2\mathcal{J} \\ -\tfrac{1}{2}\hbar^2\mathcal{J} & \tfrac{1}{4}\hbar^2\mathcal{J} \end{pmatrix}\begin{pmatrix} c_1 \\ c_2 \end{pmatrix} = E\begin{pmatrix} c_1 \\ c_2 \end{pmatrix}, \qquad \begin{aligned} (\tfrac{1}{4}\hbar^2\mathcal{J} - E)c_1 - \tfrac{1}{2}\hbar^2\mathcal{J}c_2 &= 0 \\ -\tfrac{1}{2}\hbar^2\mathcal{J}c_1 + (\tfrac{1}{4}\hbar^2\mathcal{J} - E)c_2 &= 0 \end{aligned} \tag{38}$$

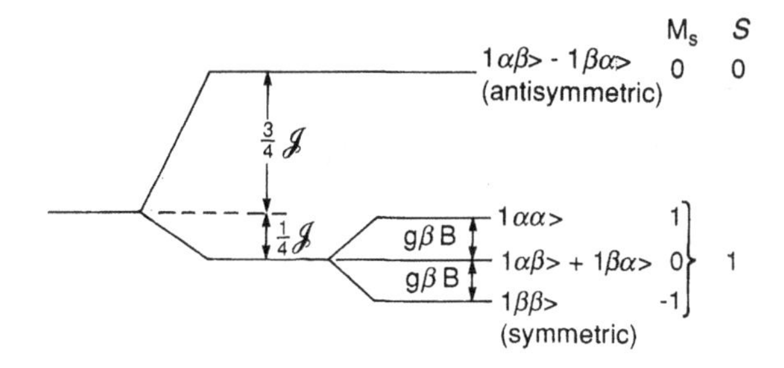

Figure 3.3 Energy levels for two Heisenberg coupled spins $\frac{1}{2}$ with $\mathcal{J} > 0$, with annotation

The secular equation is $E^2 - 2xE - 3x^2 = 0$, where $x = \frac{1}{4}\hbar^2\mathcal{J}$, factorizing to give $E = -x$ and $E = 3x$. When $E = -x$, $2xc_1 - 2xc_2 = 0$ and $c_1 = c_2$, and when $E = 3x$, $-2xc_1 - 2xc_2 = 0$ and $c_1 = -c_2$. The results are collected and illustrated in Figure 3.3 as:

$$
\begin{aligned}
E_4 &= \tfrac{3}{4}\hbar^2\mathcal{J}, & |4\rangle &= \frac{1}{\sqrt{2}}(|\alpha\beta\rangle - |\beta\alpha\rangle),\ M_s = 0, & S &= 0 \text{ antisymmetric} \\
E_3 &= -\hbar\gamma B - \tfrac{1}{4}\hbar^2\mathcal{J}, & |3\rangle &= |\alpha\ \alpha\rangle, \quad M_s = 1 & & \\
E_2 &= -\tfrac{1}{4}\hbar^2\mathcal{J}, & |2\rangle &= \frac{1}{\sqrt{2}}(|\alpha\beta\rangle + |\beta\alpha\rangle),\ M_s = 0 & S &= 1 \text{ symmetric} \\
E_1 &= \hbar\gamma B - \tfrac{1}{4}\hbar^2\mathcal{J}, & |1\rangle &= |\beta\ \beta\rangle, \quad M_s = -1 & &
\end{aligned}
\tag{39}
$$

The eigenvectors are here labelled in order of increasing energy, assuming $\mathcal{J} > 0$ and predominant, and $\gamma < 0$ as for electrons. The information on the right accords with the following.

If the order in which α and β are written in $|2\rangle$ is reversed, the new state remains identical to $|2\rangle$, which is thus symmetric. $|1\rangle$ and $|3\rangle$ are inherently or obviously symmetric. Interchanging α and β in $|4\rangle$ give $|\beta\ \alpha\rangle - |\alpha\ \beta\rangle = -(|\alpha\ \beta\rangle - |\beta\ \alpha\rangle)$ and $|4\rangle$ is thus antisymmetric. The basis states (with the two exceptions) lack symmetry, as do random superpositions. If the state remains the same or changes only in sign on interchange then eigenvalues remain unaltered since, for example, if $A|a\rangle = a|a\rangle$ then $A(-|a\rangle) = -A|a\rangle = -a|a\rangle = a(-|a\rangle)$. Further, for example, if $A|\alpha\ \beta\rangle = |\chi\rangle$ and $A|\beta\ \alpha\rangle = |\varphi\rangle$ the mean values of A for $|4\rangle$ and the permuted state are

$$(\langle\alpha\ \beta| - \langle\beta\ \alpha|)(|\chi\rangle - |\varphi\rangle)$$

and

$$(\langle\beta\ \alpha| - \langle\alpha\ \beta|)(|\varphi\rangle - |\chi\rangle) = \langle 4|A|4\rangle$$

i.e. the mean values are invariant if the states are antisymmetric or symmetric (the latter case being obvious). These are natural requirements for joint states, for indistinguishable particles.

For the system of two spins $\mathbf{S} = \mathbf{S}_1 + \mathbf{S}_2$, i.e.

$$S^2 = (\mathbf{S}_1 + \mathbf{S}_2) \cdot (\mathbf{S}_1 + \mathbf{S}_2) = \mathbf{S}_1 \cdot \mathbf{S}_1 + \mathbf{S}_2 \cdot \mathbf{S}_2 + 2\mathbf{S}_1 \cdot \mathbf{S}_2 = S_1^2 + S_2^2 + 2\mathbf{S}_1 \cdot \mathbf{S}_2 \tag{40}$$

since $\mathbf{S}_1$ and $\mathbf{S}_2$ commute. With the conversion $2\mathbf{S}_1 \cdot \mathbf{S}_2 = 2S_1^z S_2^z + S_1^+ S_2^- + S_1^- S_2^+$ and recalling that $S^2|\alpha\rangle = \frac{3}{4}\hbar^2|\alpha\rangle$, $S^2|\beta\rangle = \frac{3}{4}\hbar^2|\beta\rangle$, it is easy to show that

$$S^2|\alpha\ \alpha\rangle = 2\hbar^2|\alpha\ \alpha\rangle, \qquad S^2|\beta\ \beta\rangle = 2\hbar^2|\beta\ \beta\rangle$$
$$S^2|\alpha\ \beta\rangle = \hbar^2|\alpha\ \beta\rangle + \hbar^2|\beta\ \alpha\rangle, \qquad S^2|\beta\ \alpha\rangle = \hbar^2|\beta\ \alpha\rangle + \hbar^2|\alpha\ \beta\rangle$$

from which

$$\begin{aligned} S^2|1\rangle &= 2\hbar^2|1\rangle, \qquad S^2|2\rangle = 2\hbar^2|2\rangle \\ S^2|3\rangle &= 2\hbar^2|3\rangle, \qquad S^2|4\rangle = 0|4\rangle \end{aligned} \tag{41}$$

The total z components are readily calculated as

$$\begin{aligned} S_z|\alpha\ \alpha\rangle &= (S_1^z + S_2^z)|\alpha\ \alpha\rangle = \tfrac{1}{2}\hbar|\alpha\ \alpha\rangle + \tfrac{1}{2}\hbar|\alpha\ \alpha\rangle = \hbar|\alpha\ \alpha\rangle \\ S_z|\beta\ \beta\rangle &= -\tfrac{1}{2}\hbar|\beta\ \beta\rangle - \tfrac{1}{2}\hbar|\beta\ \beta\rangle = -\hbar|\beta\ \beta\rangle \\ S_z(|\alpha\ \beta\rangle + |\beta\ \alpha\rangle) &= \tfrac{1}{2}\hbar|\alpha\ \beta\rangle - \tfrac{1}{2}\hbar|\alpha\ \beta\rangle - \tfrac{1}{2}\hbar|\beta\ \alpha\rangle + \tfrac{1}{2}\hbar|\beta\ \alpha\rangle = 0(|\alpha\ \beta\rangle + |\beta\ \alpha\rangle) \\ S_z(|\alpha\ \beta\rangle - |\beta\ \alpha\rangle) &= 0(|\alpha\ \beta\rangle - |\beta\ \alpha\rangle) \end{aligned} \tag{42}$$

and in view of their relations to the values of S they may be denoted M_S, where $M = 0$ for $S = 0$ and $M = 1, 0, -1$ associated with $S = 1$ (see Figure 3.3).

1.5 Generalization

Whether interest is in the effects of applied fields or in the coupling itself, account may first be taken of $\mathcal{H}_{\mathcal{J}}(\mathcal{H} = \mathcal{H}_B + \mathcal{H}_{\mathcal{J}})$. All the states are, by choice, eigenstates of the S_i^z and thus of $\mathcal{H}_B$, which contributes only to diagonal elements of $\mathcal{H}$. In a simple demonstration:

$$\text{if } \begin{pmatrix} a_{11} & a_{12} \\ a_{21} & a_{22} \end{pmatrix} \begin{pmatrix} c_1 \\ c_2 \end{pmatrix} = \begin{pmatrix} a_{11}c_1 + a_{12}c_2 \\ a_{21}c_1 + a_{22}c_2 \end{pmatrix} = \lambda \begin{pmatrix} c_1 \\ c_2 \end{pmatrix}$$
$$\text{then } \begin{pmatrix} a_{11} + b & a_{12} \\ a_{21} & a_{22} + b \end{pmatrix} \cdot \begin{pmatrix} c_1 \\ c_2 \end{pmatrix} = \cdots = (\lambda + b) \begin{pmatrix} c_1 \\ c_2 \end{pmatrix} \tag{43}$$

or $(A + bI)C = (\lambda + b)C$ if $AC = \lambda C$. Thus b or $\langle i|\mathcal{H}_B|j\rangle$ can be added to the values obtained for $\mathcal{H}_{\mathcal{J}}$ so long as b is common to each diagonal element, as will be contrived.

Any state may be characterized by the eigenvalue given by the total $S_z = \sum S_i^z$. For the basis states this is just $\frac{1}{2}\hbar$ times the number of α's minus the number of β's. $\mathcal{H}_{\mathcal{J}}$ contains the S_i^z and also the $S_i^+ S_j^-$ and $S_i^- S_j^+$, which either give zero or effectively interchange α's and β's and thus do not change S_z. Thus non-zero matrix elements occur only between product states with the same S_z. If the states are ordered so that the first 'group' contains the single $|\alpha\ \alpha\ \alpha\ \cdots\ \alpha\ \alpha\rangle$, the second contains $|\alpha\ \alpha\ \alpha\ \cdots\ \alpha\ \beta\rangle$, etc., with any one spin in $|\beta\rangle$, the third is $|\alpha\ \alpha\ \cdots\ \alpha\ \beta\ \beta\rangle$, etc., with two in $|\beta\rangle$ and so forth to the final $|\beta\ \beta\ \beta\ \cdots\ \beta\ \beta\rangle$ then the groups will have $S_z = n(\frac{1}{2}\hbar), (n-2)(\frac{1}{2}\hbar), (n-4)(\frac{1}{2}\hbar), \ldots, (n-2n)(\frac{1}{2}\hbar)$. The number of groups is clearly $n+1$. Because non-zero elements only exist between members of each group, the resulting matrix is block diagonal as in the specific example which the reader may verify for $n = 3$:*

$$\mathcal{H} = \begin{pmatrix} -\frac{1}{2}J(+3b) & & & & & & & 0 \\ & 0(+b) & -\frac{1}{2}J & 0 & & & & \\ & -\frac{1}{2}J & \frac{1}{2}J(+b) & -\frac{1}{2}J & & & & \\ & 0 & -\frac{1}{2}J & 0(+b) & & & & \\ & & & & 0(-b) & -\frac{1}{2}J & 0 & \\ & & 0 & & -\frac{1}{2}J & \frac{1}{2}J(-b) & -\frac{1}{2}J & \\ & & & & 0 & -\frac{1}{2}J & 0(-b) & \\ & & & & & & & -\frac{1}{2}J(-3b) \end{pmatrix} \tag{44}$$

In this J stands for $\hbar^2\mathcal{J}$ and $b = -\frac{1}{2}\hbar\gamma B$; b is discounted for the present and is included here only to stress the disposition of the values between the different submatrices.

The matrix $\mathcal{H}$ constitutes the *direct sum* $\mathcal{H} = \mathcal{H}_3 \oplus \mathcal{H}_1 \oplus \mathcal{H}_{-1} \oplus \mathcal{H}_{-3}$, with $\mathcal{H}_1$ the 3×3 matrix over states with $\sigma_z = 1$, etc. A direct sum is described by

$$A \oplus B = \begin{pmatrix} a_{11} & a_{12} & 0 & 0 \\ a_{21} & a_{22} & 0 & 0 \\ 0 & 0 & b_{11} & b_{12} \\ 0 & 0 & b_{21} & b_{22} \end{pmatrix} \tag{45}$$

Consider $Ac = b$ with $A = A^{(1)} \oplus A^{(2)}$, $b = b^{(1)} \oplus b^{(2)}$ and $c = c^{(1)} \oplus c^{(2)}$, e.g.

$$\begin{pmatrix} a_{11}^1 & a_{12}^1 & 0 & 0 \\ a_{21}^1 & a_{22}^1 & 0 & 0 \\ 0 & 0 & a_{11}^2 & a_{12}^2 \\ 0 & 0 & a_{21}^2 & a_{22}^2 \end{pmatrix} \begin{pmatrix} c_1^1 \\ c_2^1 \\ c_1^2 \\ c_2^2 \end{pmatrix} = \begin{pmatrix} b_1^1 \\ b_2^1 \\ b_1^2 \\ b_2^2 \end{pmatrix} \sim \begin{cases} \begin{pmatrix} a_{11}^1 & a_{12}^1 \\ a_{21}^1 & a_{22}^1 \end{pmatrix} \begin{pmatrix} c_1^1 \\ c_2^1 \end{pmatrix} = \begin{pmatrix} b_1^1 \\ b_2^1 \end{pmatrix} \\ \begin{pmatrix} a_{11}^2 & a_{12}^2 \\ a_{21}^2 & a_{22}^2 \end{pmatrix} \begin{pmatrix} c_1^2 \\ c_2^2 \end{pmatrix} = \begin{pmatrix} b_1^2 \\ b_2^2 \end{pmatrix} \end{cases}$$

The equivalence of the systems on the left and on the right is confirmed by expansion. If those on the right are eigensystems, in full:

$$\begin{pmatrix} a_{11}^1 & a_{12}^1 \\ a_{21}^1 & a_{22}^1 \end{pmatrix} \begin{pmatrix} c_{11}^1 & c_{12}^1 \\ c_{21}^1 & c_{22}^1 \end{pmatrix} = \begin{pmatrix} c_{11}^1 & c_{12}^1 \\ c_{21}^1 & c_{22}^1 \end{pmatrix} \begin{pmatrix} \lambda_1^1 & 0 \\ 0 & \lambda_2^1 \end{pmatrix}$$

then $AC = C\lambda$ becomes

$$(A^{(1)} \oplus A^{(2)})(C^{(1)} \oplus C^{(2)}) = (C^{(1)} \oplus C^{(2)})(\lambda^{(1)} \oplus \lambda^{(2)})$$

and thus

$$\begin{pmatrix} \lambda_1 & 0 & 0 & 0 \\ 0 & \lambda_2 & 0 & 0 \\ 0 & 0 & \lambda_3 & 0 \\ 0 & 0 & 0 & \lambda_4 \end{pmatrix} = \begin{pmatrix} \lambda_1^1 & 0 \\ 0 & \lambda_2^1 \end{pmatrix} \oplus \begin{pmatrix} \lambda_1^2 & 0 \\ 0 & \lambda_2^2 \end{pmatrix}$$

i.e. the eigenvalues of the full matrix are the list of those of the submatrices taken together. It is legitimate to solve each subsystem separately and to assemble the list of eigenvalues and vectors for the complete system; this was presumed in the treatment of two spins.

* numbering the basis states $|\alpha\alpha\alpha\rangle$, $|\alpha\alpha\beta\rangle$, $|\alpha\beta\alpha\rangle$, $|\beta\alpha\alpha\rangle$, $|\alpha\beta\beta\rangle$, $|\beta\alpha\beta\rangle$, $|\beta\beta\alpha\rangle$, $|\beta\beta\beta\rangle$, as $|1\rangle$ to $|8\rangle$

To complete the example, the secular equation for $\mathcal{H}_1$ is

$$\begin{vmatrix} 0-E & -x & 0 \\ -x & x-E & -x \\ 0 & -x & 0-E \end{vmatrix} = -E(E^2 - Ex - 2x^2) = 0$$

(where $x = \frac{1}{2}\hbar^2\mathcal{J}$) so $E = 0$, or $E = 2x$ or $E = -x$. When $E = 0$,

$$\begin{pmatrix} 0 & -x & 0 \\ -x & x & -x \\ 0 & -x & 0 \end{pmatrix}\begin{pmatrix} c_1 \\ c_2 \\ c_3 \end{pmatrix} = x\begin{pmatrix} -c_2 \\ -c_1 + c_2 - c_3 \\ -c_2 \end{pmatrix} = 0$$

so $c_2 = 0$, $c_3 = -c_1$ and the vector is $|\,0\ 1\,\rangle = (1/\sqrt{2})(|\,\alpha\ \alpha\ \beta\,\rangle - |\,\beta\ \alpha\ \alpha\,\rangle)$ using labels for the energy, in units of x, and σ_z. Similarly, for $E = -x$, $|\,-1\ 1\,\rangle = (1/\sqrt{3})(|\,\alpha\ \alpha\ \beta\,\rangle + |\,\alpha\ \beta\ \alpha\,\rangle + |\,\beta\ \alpha\ \alpha\,\rangle)$ and for $E = 2x$, $|\,2\ 1\,\rangle = (1/\sqrt{6})(|\,\alpha\ \alpha\ \beta\,\rangle - 2|\,\alpha\ \beta\ \alpha\,\rangle + |\,\beta\ \alpha\ \alpha\,\rangle)$. The eigenvalues of the submatrix $\mathcal{H}_{-1}$ are, obviously, these same three and the eigenvectors correspond if the basis vectors $|2\rangle$, $|3\rangle$, $|4\rangle$ above are replaced by $|5\rangle$, $|6\rangle$, $|7\rangle$. The remaining two energies are clearly both $E = -x$ with $|\,-1\ 3\,\rangle = |\,\alpha\ \alpha\ \alpha\,\rangle$ and $|\,-1\ -3\,\rangle = |\,\beta\ \beta\ \beta\,\rangle$. The contributions due to $\mathcal{H}_B$ may be added appropriately.

The total spin is obtained according to

$$\begin{aligned} S^2 &= (\mathbf{S}_1 + \mathbf{S}_2 + \mathbf{S}_3)^2 \\ &= S_1^2 + S_2^2 + S_3^2 + 2S_1^zS_2^z + 2S_1^zS_3^z + 2S_2^zS_3^z \\ &\quad + S_1^+S_2^- + S_1^-S_2^+ + S_1^+S_3^- + S_1^-S_3^+ + S_2^+S_3^- + S_2^-S_3^+ \end{aligned}$$

so that, for example,

$$\begin{aligned} S^2|\alpha\ \alpha\ \alpha\rangle &= \tfrac{15}{4}\hbar^2|\alpha\ \alpha\ \alpha\rangle \\ S^2|\alpha\ \alpha\ \beta\rangle &= \tfrac{3}{4}\hbar^2|\alpha\ \alpha\ \beta\rangle + \hbar^2|\beta\ \alpha\ \alpha\rangle + \hbar^2|\alpha\ \beta\ \alpha\rangle \\ S^2(|\alpha\ \alpha\ \beta\rangle - |\beta\ \alpha\ \alpha\rangle) &= \tfrac{3}{4}\hbar^2(|\alpha\ \alpha\ \beta\rangle - |\beta\ \alpha\ \alpha\rangle) \\ S^2(|\alpha\ \alpha\ \beta\rangle + |\alpha\ \beta\ \alpha\rangle + |\beta\ \alpha\ \alpha\rangle) &= \tfrac{15}{4}\hbar^2(|\alpha\ \alpha\ \beta\rangle + |\alpha\ \beta\ \alpha\rangle + |\beta\ \alpha\ \alpha\rangle) \end{aligned}$$

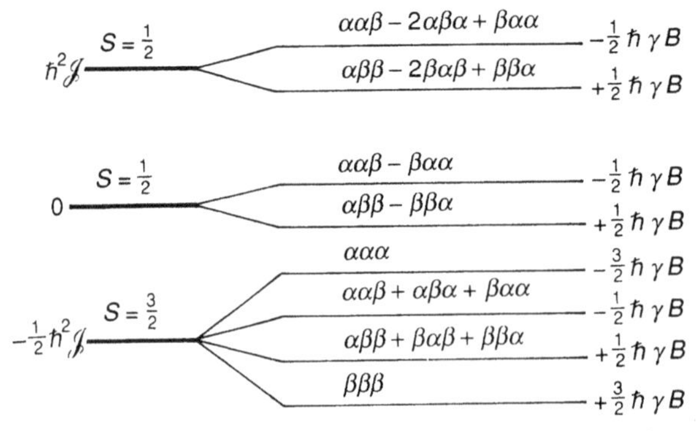

Figure 3.4 Three Heisenberg coupled spins, $\mathcal{J} > 0$. The upper doublet (the ground state for $\mathcal{J} < 0$) is replaced by a singlet for an even number of spins. Recall $\hbar\gamma = -g\beta$. Not to scale

The reader may confirm these and complete the list, noting that the basis vectors are not eigenvectors of S^2 but the solutions are states of definite S. Noting that $S(S+1)\hbar^2 = \frac{15}{4}\hbar^2$ or $\frac{3}{4}\hbar^2$ if $S = \frac{3}{2}$ or $S = \frac{1}{2}$ respectively, for the above examples, the total spin can be designated by the values of S given in Figure 3.4.

The reader may repeat this $n = 3$ demonstration for a rudimentary ring or triangle, rather than for the open-ended line assumed here, by taking account of $\mathcal{J}_{31} = \mathcal{J}$ as well as $\mathcal{J}_{12}$ and $\mathcal{J}_{23}$, and noting the introduction of extra degeneracy.

1.6 Ground States and Excited States

It is noted that the energy $E = -\frac{1}{2}\hbar^2\mathcal{J}$ is common to all four submatrices in the above and, if $\mathcal{J} > 0$, this is the lowest value occurring. The ground state is $(n+1)$-fold degenerate. This feature is general.

Characterize any basis state applying to a particular submatrix by the number m of cases for which the spin of neighbours in the list is the same. Thus $n - m$ pairs of neighbours have unlike spins assuming, for demonstration only, a ring of n spins. Each basis state defines a row in the matrix. It couples to itself, on the diagonal, via $-\mathcal{J}S_i^z S_{i+1}^z$, which gives $-\mathcal{J}(\frac{1}{4}\hbar^2)$, m times and $-\mathcal{J}(-\frac{1}{4}\hbar^2)$, $(n-m)$ times; the diagonal element is $-\frac{1}{4}\hbar\mathcal{J}(2m-n)$. (The $S^{\pm}$ do not contribute here.) For the off-diagonals, the number of states coupling to the chosen state is that number of states which are created by interchanges due to $S_i^+ S_{i+1}^-$ or $S_i^- S_{i+1}^+$, which is the number of unlike neighbours, $n - m$. Each element is of magnitude $-\frac{1}{2}\hbar^2\mathcal{J}$. Thus the sum R of the elements in any one row of the submatrix is $[-2(n-m) - (2m-n)]\frac{1}{4}\hbar^2\mathcal{J} = -n(\frac{1}{4}\hbar^2\mathcal{J})$. Since this is independent of m it applies to all the rows of the submatrix, and indeed to all the rows of all the submatrices. To illustrate this, define a row by $\langle\alpha\ \beta\ \beta\ \alpha\ \alpha\ \alpha|$, $m = 4$, $n - m = 2$. Interchanges give $|\alpha\ \beta\ \beta\ \alpha\ \alpha\ \alpha\rangle$ from the two states $|\beta\ \alpha\ \beta\ \alpha\ \alpha\ \alpha\rangle$ and $|\alpha\ \beta\ \alpha\ \beta\ \alpha\ \alpha\rangle$ and there are two off-diagonal elements in the row, each $-\frac{1}{2}\hbar^2\mathcal{J}$. For the row $\langle\alpha\ \beta\ \alpha\ \beta\ \alpha\ \alpha|$, $m = 2$, $n - m = 4$, the coupling states are the four $|\beta\ \alpha\ \alpha\ \beta\ \alpha\ \alpha\rangle$, $|\alpha\ \beta\ \beta\ \alpha\ \alpha\ \alpha\rangle$, $|\alpha\ \beta\ \alpha\ \alpha\ \beta\ \alpha\rangle$ and $|\alpha\ \alpha\ \beta\ \beta\ \alpha\ \alpha\rangle$, but R remains the same because the diagonal element on this row is $(-4+2)(-\frac{1}{4}\hbar^2\mathcal{J}) = \frac{1}{2}\hbar^2\mathcal{J}$ and in the former case it was $(-2+4)(-\frac{1}{4}\hbar^2\mathcal{J}) = -\frac{1}{2}\hbar^2\mathcal{J}$.

Let R_i be the sum of the elements of row i and consider a vector $\chi = (111\cdots 1)$. For this, $\mathcal{H}\chi = (R_1 R_2 R_3 \cdots R_N) = R(111\cdots 1)$ if all the $R_i = R$. In this case the vector $(111\cdots 1)$ or $|0\rangle = \sum c_i|i\rangle$ with all $c_i = 1$, over the N basis states for that particular submatrix, is an eigenvector and the particular extreme eigenvalue is just R. (The reason for the designation of these states as $|0\rangle$, or $|k = 0\rangle$, will become apparent.) Such a vector, with the same eigenvalue, arises for each submatrix so the ground state is $(n+1)$-fold degenerate. All the ground states will be seen to be characterized by the same value of S, i.e. $S = \frac{1}{2}n$, as seen in the examples $n = 2$, $S = 1$ and $n = 3$, $S = \frac{3}{2}$, but since each is associated with a different submatrix they are distinguished by $S_z/(\frac{1}{2}\hbar) = n, n-1, \ldots, -(n-1), -n$, i.e. by $S_z = M_S\hbar$ with $M_S = S-1, S-2, \ldots, -(S-1), -S$. (See also Section 1.13.1)

This corresponds to the general theory of angular momentum for a general S, i.e. the system behaves like a single spin $S = \frac{1}{2}n$ so long as only the ground states are occupied and the state of magnetic order is then perfect. Associated with each of the degenerate ground states are $N - 1$ excited states, with N applying to the relevant submatrix ($N = 1$ for $|\alpha\ \alpha\ \alpha\ \cdots\ \alpha\rangle$ or $|\beta\ \beta\ \beta\ \cdots\ \beta\rangle$ but $N > 1$ otherwise), with the same S_z values but with lower values of S. Thus, as the excited states become occupied the state of order decreases.

The matrix characterized by $S_z = 0$ (n even) yields the other extreme energy which, for the example of a line of spins, is $+(n-1)(\frac{1}{4}\hbar^2\mathscr{J})$, or $+n(\frac{1}{2}\hbar^2\mathscr{J})$ for a ring. (This would be given by $|\alpha\ \beta\ \alpha\ \beta\ \cdots\ \alpha\ \beta\rangle$ if the $S^{\pm}S^{\mp}$ were ignored, but this is not, in fact, the unique state.) $S = 0$ for the extreme excited state. If $\mathscr{J} < 0$ this becomes the antiferromagnetic singlet ground state and unique occupation of this state again indicates perfect order: a rudimentary illustration consists of Figure 3.4. For $n = 3$ the moment cannot be zero and it is noted that $S = \frac{1}{2}$ for the two antiferromagnetic ground states $|\alpha\ \alpha\ \beta\rangle - 2|\alpha\ \beta\ \alpha\rangle + |\beta\ \alpha\ \alpha\rangle$ and $|\beta\ \beta\ \alpha\rangle - 2|\beta\ \alpha\ \beta\rangle + |\alpha\ \beta\ \beta\rangle$. When n is even the antiferromagnetic ground state is a singlet with $S = 0$.

The energies form an irregular band limited by the two extremes and consideration of the width of the band and the number of levels involved indicates that the density of states increases with n, so that for substantial crystals the population of excited states commences at very low temperatures, with a consequent decrease in the state of order and the spontaneous magnetization. The lower-lying excited states will be seen to constitute spin waves and these will be correlated with the direct solutions for simple models in Section 1.14.2.

1.7 Traces of Spin Matrices

Trace calculations occur frequently in magnetic theory. While $AB \neq BA$ for matrices A and B the diagonals of the commuted products contain the same elements, in different positions, and thus $\mathrm{Tr}\,AB = \mathrm{Tr}\,BA$. Obviously $\mathrm{Tr}\,(A+B) = \mathrm{Tr}\,A + \mathrm{Tr}\,B$. Thus from the basic commutation relations (6) such as

$$\sigma_x\sigma_y - \sigma_y\sigma_x = 2i\sigma_z \tag{46}$$

we have

$$\mathrm{Tr}\,2i\sigma_z = \mathrm{Tr}\,\sigma_x\sigma_y - \mathrm{Tr}\,\sigma_y\sigma_x = 0, \qquad \text{since } \mathrm{Tr}\,\sigma_i = 0 \tag{47}$$

for $i = x, y, z$. The specific low-order examples (Section 1.3) clearly conform to this. Multiplying equation (46) by σ_z on the right,

$$\sigma_x\sigma_y\sigma_z - \sigma_y\sigma_x\sigma_z = 2i\sigma_z^2$$

and on writing out similar relations for σ_x^2 and σ_y^2 it is seen that

$$\mathrm{Tr}\,\sigma_x^2 = \mathrm{Tr}\,\sigma_y^2 = \mathrm{Tr}\,\sigma_z^2 \tag{48}$$

i.e.

$$\mathrm{Tr}\,S_x^2 = \mathrm{Tr}\,S_y^2 = \mathrm{Tr}\,S_z^2 \tag{49}$$

which has also been exemplified. The matrix for S^2 is known to be always $S(S+1)\hbar^2 I$, where I is the unit matrix of appropriate dimension (since only this gives the appropriate result $S(S+1)\hbar^2$ for all eigenstates of S_z with the same S). The dimension is just $2S+1$, i.e. the number of possible values of M_S, so $\mathrm{Tr}\,I = 2S+1$ and $\mathrm{Tr}\,S^2 = S(S+1)(2S+1)$ but

$$\mathrm{Tr}\,S^2 = \mathrm{Tr}\,S_x^2 + \mathrm{Tr}\,S_y^2 + \mathrm{Tr}\,S_z^2$$

and using equation (49),

$$\mathrm{Tr}\,S_x^2 = \mathrm{Tr}\,S_y^2 = \mathrm{Tr}\,S_z^2 = \tfrac{1}{3}S(S+1)(2S+1) \tag{50}$$

and also

$$\mathrm{Tr}\, S^+S^- = \mathrm{Tr}\,(S_x + iS_y)(S_x - iS_y) = \mathrm{Tr}\,(S_x^2 + S_y^2) = \tfrac{2}{3}S(S+1)(2S+1) \qquad (51)$$

For multispin systems, it was seen that S_2^z, for example, was represented by a direct product matrix $\mathfrak{I}S_z\mathfrak{I}(n = 3)$. Over the vector space of spin 2 alone its trace and that of S_z^2 will be as above. However, if it is necessary to calculate the trace over the joint spin space the factor $2S + 1$ must be replaced by the dimension of this space. For example, for 3 spins $\frac{1}{2}$ we would have

$$\mathrm{Tr}\, S_2^{z^2} = (\tfrac{1}{3})(\tfrac{1}{2})(\tfrac{3}{2}) \times 3 \times (2 \times \tfrac{1}{2} + 1)$$

Returning to the commutation relations, we may obtain, for example,

$$(\sigma_y\sigma_x)\sigma_y - (\sigma_z\sigma_y)\sigma_y = (2i\sigma_x)\sigma_y$$

and note that the trace of the left-hand side is zero so that, with two similar relations,

$$\mathrm{Tr}\,\sigma_x\sigma_y = \mathrm{Tr}\,\sigma_y\sigma_z = \mathrm{Tr}\,\sigma_x\sigma_z = 0 \qquad (52)$$

1.8 Energy, Magnetization and Susceptibility for a General Spin

The trace calculations permit the extension of the results to a system with a general S. Use is again made of $\langle E\rangle = \mathrm{Tr}\,(\mathcal{H}\rho_0)/\mathrm{Tr}\,\rho_0$ with $\rho_0 = \exp[-\mathcal{H}/(kT)]$, ρ_0 not being normalized in this case. Expanding the exponentials up to linear terms, the numerator is

$$N = \mathrm{Tr}\,\mathcal{H}\left(1 - \frac{\mathcal{H}}{kT}\right) = \mathrm{Tr}\left(\mathcal{H} - \frac{\mathcal{H}^2}{kT}\right)$$

$$= \mathrm{Tr}\left(-\frac{\gamma BS_z}{kT} - \frac{\gamma^2B^2S_z^2}{kT}\right) \qquad (53)$$

with $\mathcal{H}$ for a spin S in a field B. However, $\mathrm{Tr}\, S_z = 0$ and using equation (50),

$$N = -\frac{\gamma^2B^2}{kT}\tfrac{1}{3}\hbar^2 S(S+1)(2S+1) \qquad (54)$$

By $\mathrm{Tr}\,\rho_0$, in the denominator (D), we mean the trace of the statistical matrix as such in the same representation as that in which S_z^2 is implicitly expressed in order for equation (53) to be applicable, i.e. in the eigenstates $|M_S\rangle$ of S_z. Thus when the exponential is approximated to give

$$D = \mathrm{Tr}\left(\frac{1 - \mathcal{H}}{kT}\right) = \mathrm{Tr}\left(\frac{1 + \gamma BS_z}{kT}\right) = \mathrm{Tr}\, 1$$

(using $\mathrm{Tr}\, S_z = 0$ again), Tr 1 is interpreted as the trace of the representation of the identity operator, which is of course just the identity matrix I of dimension $2S + 1$: $\mathrm{Tr}\, I = 2S + 1$. Thus, finally,

$$\langle E\rangle = -\frac{\gamma^2B^2}{3kT}\hbar^2 S(S+1) \qquad (55)$$

To find the mean magnetic moment component the above might be repeated, substituting the appropriate operator γS_z for $\mathcal{H}$, but the way in which the constants are carried through is apparent and

$$
\begin{aligned}
\langle\mu_z\rangle &= \frac{\operatorname{Tr}\hat{\mu}_z\rho_0}{\operatorname{Tr}\rho_0} \\
&= \frac{\operatorname{Tr}\gamma S_z[1+\gamma BS_z/(kT)]}{\operatorname{Tr}\rho_0} \\
&= \frac{[\gamma^2 B/(kT)]\operatorname{Tr}S_z^2}{\operatorname{Tr}\rho_0} \\
&= \frac{\gamma^2\hbar^2 BS(S+1)}{3kT}
\end{aligned} \tag{56}
$$

Obviously this corresponds to the classical relation between the mean values

$$
\langle E\rangle = -B\langle\mu_z\rangle \tag{57}
$$

Finally, the magnetic susceptibility in this so-called Curie limit or zero-field limit is, with N identical systems per unit volume,

$$
\chi = \frac{N\langle\mu_z\rangle}{H} = \frac{N\mu_0\langle\mu_z\rangle}{B} = \frac{N\mu_0\gamma^2\hbar^2 S(S+1)}{3kT} = \frac{N\mu_0 g^2\beta^2 S(S+1)}{3kT} \tag{58}
$$

using $\gamma = -ge/(2m)$ and $\beta = e\hbar/(2m)$.

A particular example, for $S = 1$, illustrates the role of trace calculations. Use the approximation $\rho_0 = 1 - \mathcal{H}/kT = 1 + \gamma BS_z/(kT)$ and represent this in the $|M_S\rangle = |1\rangle, |0\rangle, |-1\rangle$ which are orthonormal and are eigenvectors of S_z. The matrix is thus diagonal with elements $\langle 1|1+\gamma BS_z/(kT)|1\rangle = \langle 1|1+\gamma B\hbar/(kT)|1\rangle = 1+\gamma B\hbar/(kT)$, etc. Since the basis vectors are also eigenvectors of $\hat{\mu}_z$, this also is diagonal with elements $\gamma M_S\hbar$ and so

$$
\begin{aligned}
\mu_z\rho_0 &= \begin{pmatrix} \gamma\hbar & 0 & 0 \\ 0 & 0 & 0 \\ 0 & 0 & -\gamma\hbar \end{pmatrix}\begin{pmatrix} 1+\gamma B\hbar/(kT) & 0 & 0 \\ 0 & 1 & 0 \\ 0 & 0 & 1-\gamma B\hbar/(kT) \end{pmatrix} \\
&= \begin{pmatrix} \gamma\hbar+\gamma^2 B\hbar^2/(kT) & 0 & 0 \\ 0 & 0 & 0 \\ 0 & 0 & -\gamma\hbar+\gamma^2 B\hbar^2/(kT) \end{pmatrix}
\end{aligned}
$$

$$
\langle\mu_z\rangle = \frac{\operatorname{Tr}\mu_z\rho_0}{\operatorname{Tr}\rho_0} = \frac{2\gamma^2 B\hbar^2/(kT)}{3} = \frac{\gamma^2\hbar^2 BS(S+1)}{3kT} \qquad (S=1)
$$

as predicted generally. The cancellations correspond to the trace results.

Because μ_z and ρ_0 are simultaneously diagonal in the chosen basis the Curie limit may readily be relaxed. Denoting the basis $|i\rangle$, the diagonal elements of μ_z may simply be denoted as μ_{z_i}, the ith eigenvalues of $\hat{\mu}_z$. Similarly, the diagonal elements of ρ_0 are

$$
\langle i|1-\mathcal{H}/(kT)+\mathcal{H}^2/(2k^2T^2)\ldots|i\rangle = \langle i|1-E_i/(kT)+E_i^2/(2k^2T^2)\ldots|i\rangle = \mathrm{e}^{-E_i/(kT)}
$$

and so

$$
\langle\mu_z\rangle = \frac{\sum\mu_{z_i}\mathrm{e}^{-E_i/(kT)}}{\sum\mathrm{e}^{-E_i/(kT)}} \tag{59}
$$

Thus the density matrix method may be considered as a formalization of the procedure in which the characteristic values for each state are multiplied by the Boltzmann factors, normalized by the partition function, and the sum taken.

In practical terms, if a series of states $|i\rangle$ having energies E_i and simultaneously characteristic values of μ_{z_i} arise as the solution of a particular problem, it is justifiable to calculate $\langle\mu_z\rangle$ as the classical Boltzman average over these energy levels. (Due account must be taken of degeneracy by treating a doubly degenerate level as two distinct levels, for example.) This is precisely what has already been assumed in Chapter 1. It is obvious that this applies to a single spin but it also applies to a multispin system if the spin matrices are taken to apply to the space of all the spins.

1.9 Nuclear Spin–Spin Coupling: Different Effective Fields

Although the previous treatments were rather detailed they were basically simple due to the assumption that the field acting on the spins is just the applied field—obviously the same for all. The results cannot be taken to apply generally to the coupling between two protons, say as relevant to n.m.r. spectra, since in the general case the effective field on each is different due to the chemical shifts. The treatment becomes a little more complex and occasions an opportunity to note the relation of the energy levels to the spectra, as influenced by the transition rules. Due to the particular context, I_z, I^+, etc., replace S_z, S^+, etc., in this section.

For two protons 'a' and 'b' subject to different chemical shifts or shielding and with a Heisenberg interaction

$$\begin{aligned}\mathcal{H} &= -\gamma B_a I_a^z - \gamma B_b I_b^z + \frac{J}{\hbar^2}\mathbf{I}_a \cdot \mathbf{I}_b \\ &= -\omega_a I_a^z - \omega_b I_b^z + \frac{J}{\hbar^2}[I_a^z I_b^z + \tfrac{1}{2}(I_a^+ I_b^- + I_a^- I_b^+)] \end{aligned} \tag{60}$$

where

$$B_a = B_0(1-\sigma_a), \qquad B_b = B_0(1-\sigma_b)$$
$$\omega_a = \gamma B_a, \qquad \omega_b = \gamma B_b$$

giving simple solutions only when $\omega_a = \omega_b$, $\sigma_a = \sigma_b$.

The familiar product states are eigenstates of an operator for the total spin components:

$$F_z = I_a^z + I_b^z \tag{61}$$

with

$$F_z|\alpha\,\alpha\rangle = \hbar|\alpha\,\alpha\rangle, \qquad F_z|\alpha\,\beta\rangle = 0|\alpha\,\beta\rangle$$
$$F_z|\beta\,\alpha\rangle = 0|\beta\,\alpha\rangle, \qquad F_z|\beta\beta\rangle = -\hbar|\beta\,\beta\rangle$$

The eigenvalue can be denoted $m_t\hbar$. Now

$$I_z I^+ = I_z I_x + i I_z I_y = i I_y + i(-i I_x) = I^+, \qquad I^+ I_z = -I^+$$

so that

$$[I_z, I^+] = 2I^+$$

(use having been made of the relations $I_x I_y = i I_z$, etc.).

Similarly,

$$[I_z, I^-] = -2I^-$$

and

$$[I_a^z + I_b^z, I_a^+ I_b^-] = 2I_a^+ I_b^- - 2I_a^+ I_b^- = 0$$

since I_a^z necessarily commutes with I_b^+, etc. Thus

$$[F_z, I_a^+ I_b^- + I_a^- I_b^+] = 0$$

and since F_z obviously commutes with the Zeeman terms it commutes with $\mathcal{H}$.

Thus eigenstates of $\mathcal{H}$ are expected to be linear combinations of those of F_z, i.e. of the simple product states.

The states $|\alpha\ \alpha\rangle$ and $|\beta\ \beta\rangle$ are themselves eigenstates of $\mathcal{H}$ because one or other factor of the $I_i^+ I_j^-$ must give zero and these can be 'forgotten', giving simply

$$\mathcal{H}|\alpha\ \alpha\rangle = \left(-\omega_a \times \tfrac{1}{2}\hbar - \omega_b \times \tfrac{1}{2}\hbar + \tfrac{1}{4}J\right)|\alpha\ \alpha\rangle$$

$$\mathcal{H}|\beta\ \beta\rangle = \left(\omega_a \times \tfrac{1}{2}\hbar + \omega_b \times \tfrac{1}{2}\hbar + \tfrac{1}{4}J\right)|\beta\ \beta\rangle$$

This leaves for solution the reduced 2×2 Hamiltonian in $|\alpha\ \beta\rangle \equiv |1\rangle$ and $|\beta\ \alpha\rangle \equiv |2\rangle$:

$$H = \begin{pmatrix} -\omega_a \times \frac{1}{2}\hbar + \omega_b \times \frac{1}{2}\hbar - \frac{1}{4}J & \frac{1}{2}J \\ \frac{1}{2}J & \omega_a \times \frac{1}{2}\hbar - \omega_b \times \frac{1}{2}\hbar - \frac{1}{4}J \end{pmatrix} \tag{62}$$

(lacking the simplicity conferred by $\omega_a = \omega_b$). The secular determinant expands to give $(H_{21} = H_{12})$

$$(H_{11} - E)(H_{22} - E) - H_{12}^2 = 0$$

The solutions of this quadratic equation are easily seen to be

$$\begin{aligned} E_1 &= \tfrac{1}{2}(H_{11} + H_{22}) + \tfrac{1}{2}[(H_{11} - H_{22})^2 + 4H_{12}^2]^{1/2} \\ E_2 &= \tfrac{1}{2}(H_{11} + H_{22}) - \tfrac{1}{2}[(H_{11} - H_{22})^2 + 4H_{12}^2]^{1/2} \end{aligned} \tag{63}$$

Noting that

$$(H_{11} - H_{22})^2 = \hbar^2(\omega_b - \omega_a)^2, \qquad 4H_{12}^2 = J^2$$

a case may be envisaged in which the difference between the two chemical shifts is large and J is small such that

$$\hbar^2(\omega_b - \omega_a)^2 \gg J^2$$

in which case approximately

$$E_1 = H_{11}, \qquad E_2 = H_{22}$$

H is now diagonal in the basis states and

$$E_1 = E_{\alpha\beta} = -\omega_a \times \tfrac{1}{2}\hbar + \omega_b \times \tfrac{1}{2}\hbar - \tfrac{1}{4}J, \ (|\alpha\ \beta\rangle)$$

$$E_2 = E_{\beta\alpha} = \omega_a \times \tfrac{1}{2}\hbar - \omega_b \times \tfrac{1}{2}\hbar - \tfrac{1}{4}J, \quad (|\beta\ \alpha\rangle)$$

as would be obtained by omitting the raising and lowering operators and reducing the interaction term to $(J/\hbar^2)I_a^z I_b^z$, the Ising form. In turn, since the I^+I^- terms were obtained from the $I_i^x I_j^x$ and $I_i^y I_j^y$, this so-called linear case corresponds to the neglect of the transverse components and a little consideration shows that the argument applies to the high-field limit for the e.s.r. spectrum of the hydrogen atom used in Chapter 1, Section 17.1.1, the difference between the chemical shifts being replaced by the large difference between the γ values for the electron and the proton.

In the general case the states giving energies E_1 and E_2 are

$$|1\rangle = c_1^{(1)}|\alpha\ \beta\rangle + c_2^{(1)}|\beta\ \alpha\rangle, \qquad |2\rangle = c_1^{(2)}|\alpha\ \beta\rangle + c_2^{(2)}|\beta\ \alpha\rangle$$

The coefficients are found from

$$(H_{11} - E_i)c_1^{(i)} + H_{12}c_2^{(i)} = 0$$

$$H_{12}c_1^{(i)} + (H_{22} - E_i)c_2^{(i)} = 0$$

together with normalization:

$$c_1^{(i)^2} + c_2^{(i)^2} = 1$$

and orthogonality:

$$(c_1^{(1)}c_2^{(1)})\begin{pmatrix} c_1^{(2)} \\ c_2^{(2)} \end{pmatrix} = c_1^{(1)}c_1^{(2)} + c_2^{(1)}c_2^{(2)} = 0$$

The latter two conditions are satisfied by setting

$$c_1^{(1)} = \cos\theta, \qquad c_2^{(1)} = \sin\theta$$

together with

$$c_1^{(2)} = -\sin\theta, \qquad c_2^{(2)} = \cos\theta$$

as readily verified. It can be confirmed that the equations are satisfied when

$$\cos\theta = \tfrac{1}{2}\left[1 + \frac{\omega_a\hbar - \omega_b\hbar}{D}\right]^{1/2}$$

with

$$D = [(H_{11} - H_{22})^2 + 4H_{12}^2]^{1/2} = [(\omega_b\hbar - \omega_a\hbar)^2 + J^2]^{1/2}$$

It is convenient to note, with $\cos 2\theta = 2\cos^2\theta - 1$, that

$$\cos 2\theta = \frac{\omega_a\hbar - \omega_b\hbar}{D}, \ \sin 2\theta = \frac{J}{D}$$

Re-labelling $|\alpha\ \alpha\rangle = |1\rangle, |\beta\ \beta\rangle = |4\rangle$ for the lowest and highest levels for free spins, Table 3.1 can be drawn up.

The energy level diagrams and conceivable absorptive transitions are as shown in Figure 3.5. This includes (Figure 1.5b), the second limiting or special case of $\omega_a = \omega_b$, i.e. like nucleii in like environments. The equivalent transitions $|1\rangle \to |3\rangle$ and $|3\rangle \to |4\rangle$ give

$$\Delta E = \omega\hbar = g\beta_N B, \qquad f = f_0 = \frac{g\beta_N B}{h}$$

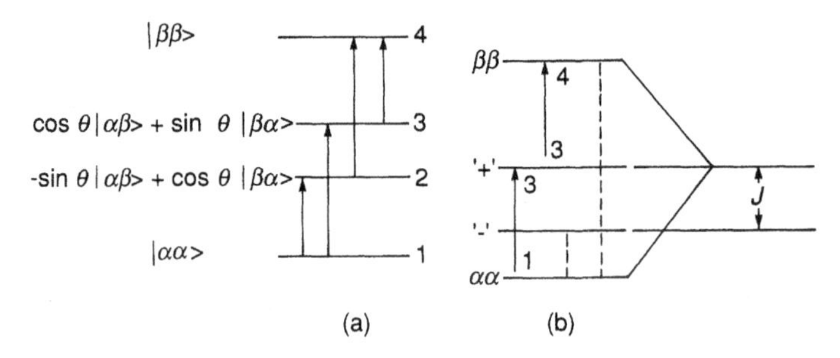

Figure 3.5 Energy levels for two coupled spins in (a) the general case and (b) in the case where $f_0^A = f_0^B$

Table 3.1

State	Energy E	Limit 1, E_f	Limit 2
$\|4\rangle = \|\beta\ \beta\rangle$	$\frac{1}{2}\hbar(\omega_a + \omega_b) + \frac{1}{4}\ J$	$\frac{1}{2}\hbar(\omega_a + \omega_b) + \frac{1}{4}\ J$	$\hbar\omega + \frac{1}{4}\ J$
$\|3\rangle = \cos\theta\|\alpha\ \beta\rangle + \sin\theta\|\beta\ \alpha\rangle$	$\frac{1}{2}D - \frac{1}{4}\ J$	$\frac{1}{2}\hbar(\omega_b - \omega_a) - \frac{1}{4}\ J$	$\frac{1}{4}\ J$
$\|2\rangle = -\sin\theta\|\alpha\ \beta\rangle + \cos\theta\|\beta\ \alpha\rangle$	$-\frac{1}{2}D - \frac{1}{4}\ J$	$\frac{1}{2}\hbar(\omega_a - \omega_b) - \frac{1}{4}\ J$	$-\frac{3}{4}\ J$
$\|1\rangle = \|\alpha\ \alpha\rangle$	$-\frac{1}{2}\hbar(\omega_a + \omega_b) + \frac{1}{4}\ J$	$-\frac{1}{2}\hbar(\omega_a + \omega_b) + \frac{1}{4}\ J$	$-\hbar\omega + \frac{1}{4}\ J$

Energies for two protons subject to different chemical shifts compared to the limit 2, with $\omega_a = \omega_b$, and the limit 1 in which J^2 is neglected in $D = [(\omega_b\hbar - \omega_a\hbar)^2 + J^2]^{1/2}$.

The other transitions shown would give lines of half this height shifted by $\pm J$ in energy, i.e. by $\pm\mathcal{J}$ in frequency with $\mathcal{J} = J/h$, but the transition rules must be considered.

A standard treatment shows that for like nucleii, with the same g or γ, the signal height is proportional to

$$S_{ij} = |\langle j|F^-|i\rangle|^2$$

for $|i\rangle \to |j\rangle$. Since

$$F^-|\alpha\ \alpha\rangle = (I_a^- + I_b^-)|\alpha\ \alpha\rangle = |\alpha\ \beta\rangle + |\beta\ \alpha\rangle$$

then

$$S_{13} = [(\langle\alpha\beta| + \langle\beta\ \alpha|)(|\alpha\ \beta\rangle + |\beta\ \alpha\rangle)]^2 = 4$$

and also

$$F^-(|\alpha\ \beta\rangle + |\beta\ \alpha\rangle) = |\beta\ \beta\rangle + |\beta\ \beta\rangle, \qquad S_{34} = [\langle\beta\beta|2|\beta\ \beta\rangle]^2 = 4$$

However,

$$F^-(|\alpha\ \beta\rangle - |\beta\ \alpha\rangle) = |\beta\ \beta\rangle - |\beta\ \beta\rangle = 0, \qquad S_{24} = 0$$

and also $S_{12} = 0$. Thus in this particular case there is no splitting of the line accompanying the splitting of the levels. In the case where $\sigma_a \doteqdot \sigma_b$ a splitting of $\pm\mathcal{J}$ is expected to be indicated by two extra lines of minimal intensity.

In the other extreme case $\hbar^2(\omega_a - \omega_b)^2 \gg J^2$ (but not $J^2 \to 0$) the levels are as shown in Figure 3.6(a), the states being the basic products. (This is not to scale since the central spacing should be much greater.) All the transitions are allowed with equal probability. Due

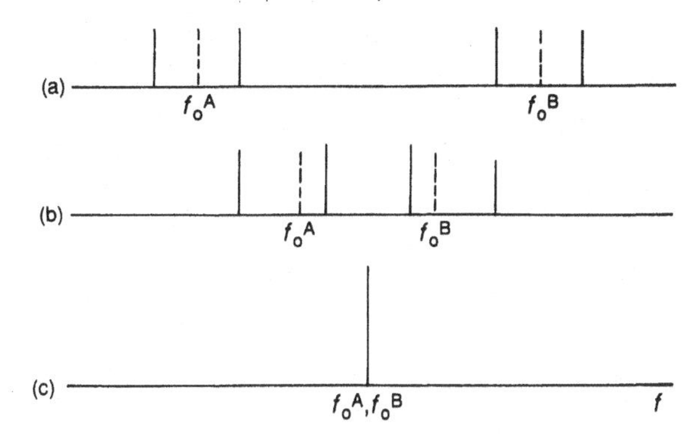

Figure 3.6 The range of spectra corresponding to Figure 3.5, with the limit (*c*) for 3.5*b*

to the symmetry of the levels the spectrum is as shown, the centre of each pair of lines falling at f_a and f_b, the zero-coupling frequencies.

In the general case the spectrum has the appearance of Figure 3.6(b), the pair splitting no longer being symmetrical about f_a and f_b. The heights are obtained by noting that

$$F^-(\cos\theta|\alpha\ \beta\rangle + \sin\theta|\beta\ \alpha\rangle) = (\cos\theta + \sin\theta)|\beta\ \beta\rangle$$

$$F^-(-\sin\theta|\alpha\ \beta\rangle + \cos\theta|\beta\ \alpha\rangle) = (\cos\theta - \sin\theta)|\beta\ \beta\rangle$$

so that

$$S = 1 + \sin 2\theta \qquad (\cos\theta|\alpha\ \beta\rangle + \sin\theta|\beta\ \alpha\rangle)$$

$$S = 1 - \sin 2\theta \qquad (-\sin\theta|\alpha\ \beta\rangle + \cos\theta|\beta\ \alpha\rangle)$$

for transitions to or from the states indicated, using $2\sin\theta\cos\theta = \sin 2\theta$. (The special cases correspond to these with $\theta = \pi/4$, $\cos\theta = \sin\theta = 1/\sqrt{2}$ for the case $\omega_a = \omega_b$ or $\theta = \pi/2$ for the first-order case.)

1.10 Hetero-Spins and the High-Field Approximation

A second source of complexity arises, even for two spins, if the gyromagnetic ratios differ. The electron and proton in isotropic hyperfine coupling will be considered.

$$\mathcal{H} = -\gamma B S_z - \gamma_N B_N I_z + \frac{A}{\hbar^2}\left[S_z I_z + \tfrac{1}{2}(S^+I^- + S^-I^+)\right] \tag{64}$$

B_N is, in general, distinguished from B due to the chemical shift. If the first symbol in the state vector is associated with the electron:

$$\mathcal{H}|\alpha\ \beta\rangle = [-\gamma B(\tfrac{1}{2}\hbar) + \gamma_N B_N(\tfrac{1}{2}\hbar)]|\alpha\ \beta\rangle - \tfrac{1}{4}A|\alpha\ \beta\rangle + \tfrac{1}{2}A|\beta\ \alpha\rangle$$

$$\mathcal{H}|\beta\ \alpha\rangle = [\gamma B(\tfrac{1}{2}\hbar) - \gamma_N B_N(\tfrac{1}{2}\hbar)]|\beta\ \alpha\rangle - \tfrac{1}{4}A|\beta\ \alpha\rangle + \tfrac{1}{2}A|\alpha\ \beta\rangle$$

and the 2×2 matrix for solution is

$$\begin{pmatrix} -\frac{1}{2}g\beta B + \frac{1}{2}g_N\beta_N B_N - \frac{1}{4}A & \frac{1}{2}A \\ \frac{1}{2}A & \frac{1}{2}g\beta B - \frac{1}{2}g_N\beta_N B_N - \frac{1}{4}A \end{pmatrix}$$

(with obvious replacements). The distinction between B_{N} and B will be dropped. With the previous $\gamma_{\mathrm{N}} \equiv \gamma$, the Zeemann terms disappeared and in the general case ($n \geqslant 2$) the Zeeman terms in any matrix block were common and existed on the diagonals only. By subtracting a diagonal matrix and solving the remainder the correct solutions were obtained by superimposing the Zeeman splittings on the exchange or hyperfine splittings, since if

$$\mathcal{H}|n\rangle = \lambda_n|n\rangle \qquad \text{and} \qquad \mathcal{H} = \mathcal{H}' + aI$$

with I the identity, then

$$(\mathcal{H}' + aI)|n\rangle = \mathcal{H}'|n\rangle + aI|n\rangle = \lambda_n'|n\rangle + a|n\rangle = (\lambda_n' + a)|n\rangle$$

and so

$$\lambda_n = \lambda_n' + a$$

This gives the condition for splittings which are linear in the field since the Zeeman terms in the Hamiltonian are linear in the field. The reader might show that these principles follow immediately on writing down the secular equations.

In the present case this essential simplicity is lost (even with $B_{\mathrm{N}} = B$) and we can only write out the secular equation with the solutions of the quadratic, arbitrarily designated E_+ and E_-, readily obtained as

$$E_+ = -\tfrac{1}{4}A + \tfrac{1}{2}[A^2 + (-g\beta + g_{\mathrm{N}}\beta_{\mathrm{N}})^2 B^2]^{1/2}$$
$$E_- = -\tfrac{1}{4}A - \tfrac{1}{2}[A^2 + (-g\beta + g_{\mathrm{N}}\beta_{\mathrm{N}})^2 B^2]^{1/2}$$

which only becomes linear in B for large B/A^2. This is seen by using the second-order approximation $(1 + \delta)^{1/2} = 1 + \delta/2$ to give

$$\begin{aligned} E_+ &= -\tfrac{1}{4}A + \tfrac{1}{2}g\beta B - \tfrac{1}{2}g_{\mathrm{N}}\beta_{\mathrm{N}}B + \frac{A^2}{4(g\beta B - g_{\mathrm{N}}\beta_{\mathrm{N}}B)} \\ E_- &= -\tfrac{1}{4}A - \tfrac{1}{2}g\beta B + \tfrac{1}{2}g_{\mathrm{N}}\beta_{\mathrm{N}}B - \frac{A^2}{4(g\beta B - g_{\mathrm{N}}\beta_{\mathrm{N}}B)} \end{aligned} \tag{65}$$

As B increases, noting that $\beta \gg \beta_{\mathrm{N}}$, the linear high-field approximation, with neglect of the last term, is approached. Moreover $E_+ \to E_{\alpha\beta}$, as obtained from

$$\mathcal{H}|\alpha\ \beta\rangle = E_{\alpha\beta}|\alpha\ \beta\rangle, \qquad \mathcal{H} = \mathcal{H}_B + \left(\frac{A}{\hbar^2}\right) S_z I_z$$

i.e. replacing the Heisenberg term by an Ising term in the spin components only and $E_- \to E_{\beta\alpha}$ as similarly obtained. Descriptively the components S_x and S_y are small and can be neglected. The set of solutions is completed by writing down the remaining

$$\begin{aligned} E_{\alpha\alpha} &= -\tfrac{1}{2}g\beta B - \tfrac{1}{2}g_{\mathrm{N}}\beta_{\mathrm{N}}B + \tfrac{1}{4}A \\ E_{\beta\beta} &= \tfrac{1}{2}g\beta B + \tfrac{1}{2}g_{\mathrm{N}}\beta_{N}B + \tfrac{1}{4}A \end{aligned} \tag{66}$$

for the basic (eigen) states $|\alpha\ \alpha\rangle$ and $|\beta\ \beta\rangle$. These, however, are exact for any A and B so that the curvatures exist only as indicated by Figure 3.7. The origins are established by

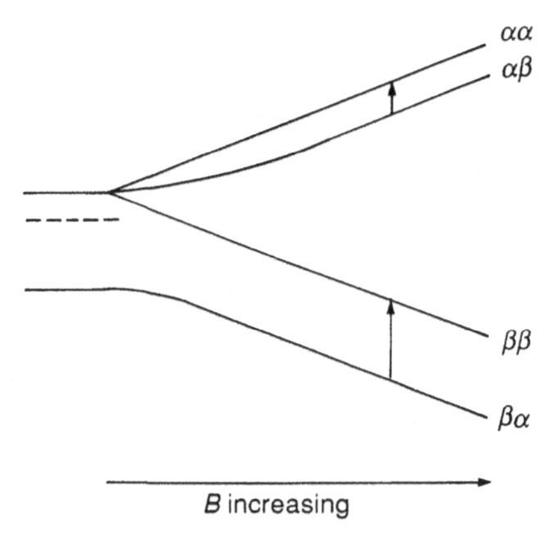

Figure 3.7 Energy levels with increasing field for the H atom

noting that as $B \to 0$ the Zeeman terms may be neglected and the problem is equivalent to solving the interaction Hamiltonian alone as for the homo-spin case. The two ($B = 0$) energies are $E_+ = \frac{1}{4}A$ and $E_- = -\frac{3}{4}A$. Solving for the coefficients with $E = E_+$ readily gives $c_1 = c_2$ and for $E = E_-$, $c_1 = -c_2$, so the states can be designated

$$c_1^+|\alpha\ \beta\rangle + c_2^+|\beta\ \alpha\rangle \xrightarrow{B\to 0} \frac{1}{\sqrt{2}}(|\alpha\ \beta\rangle + |\beta\ \alpha\rangle)$$

$$c_1^-|\alpha\ \beta\rangle + c_2^-|\beta\ \alpha\rangle \xrightarrow{B\to 0} \frac{1}{\sqrt{2}}(|\alpha\ \beta\rangle - |\beta\ \alpha\rangle)$$

The second of these can be described simply by the increasing tendencies of the field to favour the electron spin state $|\beta\rangle$. The coefficients of the general states may be derived in the usual way. It is clear that similar considerations apply to nuclear spin–spin coupling so that the analysis for the hetero-nuclear case is more difficult. For $n > 2$ the difficulty of the analyses increases rapidly but of course it is always easy to obtain numerical solutions for n up to the order of 10 in practice.

1.11 Dipolar Interactions

Consider two identical spins with $\mathcal{H} = \mathcal{H}_0 + \mathcal{H}_1$, $\mathcal{H}_1$ representing the dipolar interaction. $\mathcal{H}_0 = -\gamma B(S_1^z + S_2^z)$ and $\mathcal{H}_1$ is derived from the dipolar energy expression, i.e. $-\boldsymbol{\mu} \cdot \mathbf{B}$ with $\mathbf{B}$ as in equation (87) in Chapter 1, with the replacement $\boldsymbol{\mu} \to \gamma \mathbf{S}$:

$$\mathcal{H} = -\gamma B_0(S_1^z + S_2^z) - \gamma^2\mu_0 \frac{3(\mathbf{S}_1 \cdot \mathbf{r}_0)(\mathbf{S}_2 \cdot \mathbf{r}_0) - \mathbf{S}_1 \cdot \mathbf{S}_2}{r^3} \tag{67}$$

where $\mathbf{r}_0$ is a unit vector along the line of centres of the spins while r is the separation. Using the polar angles θ and φ (Figure 3.8), since $r_0 = 1$,

$$\mathbf{r}_0 = \mathbf{i}\sin\theta\cos\varphi + \mathbf{j}\sin\theta\sin\varphi + \mathbf{k}\cos\theta$$

while

$$\mathbf{S}_1 = \mathbf{i}S_1^x + \mathbf{j}\cdot S_1^y + \mathbf{k}S_1^z$$
$$\mathbf{S}_2 = \mathbf{i}S_2^x + \mathbf{j}\cdot S_2^y + \mathbf{k}S_2^z$$

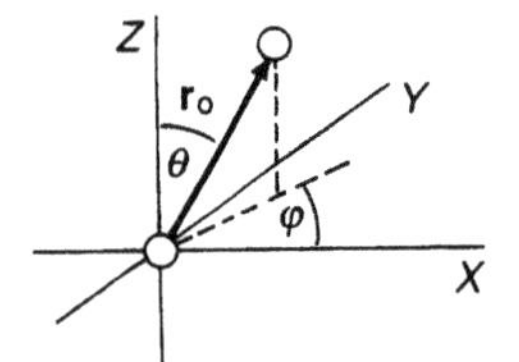

Figure 3.8 Coordinates for two spins in dipolar interaction

Thus since $\mathbf{i}\cdot\mathbf{i}=1$ and $\mathbf{i}\cdot\mathbf{j}=0$, etc.,

$$\mathbf{S}_1\cdot\mathbf{r}_0 = S_1^x\sin\theta\cos\varphi + S_1^y\sin\theta\sin\varphi + S_1^z\cos\theta$$
$$\mathbf{S}_2\cdot\mathbf{r}_0 = S_2^x\sin\theta\cos\varphi + S_2^y\sin\theta\sin\varphi + S_2^z\cos\theta$$

The product $(\mathbf{S}_1\cdot\mathbf{r}_0)(\mathbf{S}_2\cdot\mathbf{r}_0)$ contains nine terms. The term $S_1^zS_2^z\cos^2\theta(\times 3)$ is taken with the corresponding terms from $\mathbf{S}_1\cdot\mathbf{S}_2 = S_1^zS_2^z + \frac{1}{2}(S_1^+S_2^- + S_1^-S_2^+)$ to give $\mu_0\gamma^2S_1^zS_2^z(1-3\cos^2\theta)/r^3$. Taking two further terms, for example, and using $\cos^2\varphi = \frac{1}{4}(e^{2i\varphi}+e^{-2i\varphi}+2)$ and $\sin^2\varphi = -\frac{1}{4}(e^{2i\varphi}+e^{-2i\varphi}-2)$ gives

$$S_1^xS_2^x\sin^2\theta\cos^2\varphi + S_1^yS_2^y\sin^2\theta\sin^2\varphi = \frac{\sin^2\theta}{4}\left[(S_1^xS_2^x - S_1^yS_2^y)e^{2i\varphi}\right.$$
$$\left.+(S_1^xS_2^x - S_1^yS_2^y)e^{-2i\phi} + 2(S_1^xS_2^x + S_1^yS_2^y)\right]$$

The last term gives $\frac{1}{4}\sin^2\theta(S_1^+S_2^-S_1^-S_2^+)$. It is also noted that

$$S_1^+S_2^+ = (S_1^x + iS_1^y)(S_2^x + iS_2^y) = S_1^xS_2^x - S_1^yS_2^y + i(S_1^xS_2^y + S_1^yS_2^x)$$

so that $S_1^+S_2^+$ is also expected to appear in the final expression. The 'trick' is to group all the terms containing $e^{i\varphi}$, $e^{2i\varphi}$ etc., and it is eventually found that $\mathcal{H}_1$ can be expressed in terms of raising and lowering operators and S_1^z and S_2^z as

$$\mathcal{H}_1 = \mu_0\gamma^2\left\{[S_1^zS_2^z - \tfrac{1}{4}(S_1^+S_2^- + S_1^-S_2^+)]Y_0 - \tfrac{3}{2}(S_1^+S_2^z + S_1^zS_2^+)Y_1\right.$$
$$\left.-\tfrac{3}{2}[S_1^-S_2^z + S_1^zS_2^-]Y_1^* - \tfrac{3}{4}S_1^+S_2^+Y_2 - \tfrac{3}{4}S_1^-S_2^-Y_2^*\right\} \quad (68)$$

The functions of the coordinates Y_i are

$$Y_0 = \frac{1-3\cos^2\theta}{r^3}$$

$$Y_1 = \frac{\sin\theta\cos\theta e^{i\varphi}}{r^3}$$

$$Y_2 = \frac{\sin^2\theta e^{-2i\varphi}}{r^3}$$

The reader may readily fill in the complete treatment with no problem other than tedium. The basis states for the two spins may be taken as $|1\rangle = |\alpha\ \alpha\rangle$, $|2\rangle = (1/\sqrt{2})(|\alpha\ \beta\rangle + |\beta\ \alpha\rangle)$, $|3\rangle = |\beta\ \beta\rangle$, $|4\rangle = (1/\sqrt{2})(|\alpha\ \beta\rangle - |\beta\ \alpha\rangle)$. The first three form the triplet, all having $S=1$, and the last is the singlet state with $S = 0$. The matrix of $\mathcal{H}_1$ in this representation is

$$\mathcal{H}_1 = \tfrac{1}{4}\mu_0\gamma^2\hbar^2 \begin{pmatrix} Y_0 & -3\sqrt{2}Y_1 & -3Y_2 & 0 \\ -3\sqrt{2}Y_1^* & -2Y_0 & 3\sqrt{2}Y_1 & 0 \\ -3Y_2^* & 3\sqrt{2}Y_1^* & Y_0 & 0 \\ 0 & 0 & 0 & 0 \end{pmatrix} \tag{69}$$

The diagonal elements can be written down immediately since all the combinations of or with $S_i^{\pm}$ change the state and do not contribute: only $S_1^z S_2^z$ need be considered. The remainder follow routinely. The matrix $\mathcal{H}_0$ is, of course, diagonal with the unperturbed energies $-\gamma\hbar B_0, 0, +\gamma\hbar B_0, 0$ for elements 11, 22, 33 and 44 respectively. It is clear that the singlet state is not affected by the dipole interaction, as is to be expected.

In a rigid lattice with constant r, θ and φ it may be taken as a first approximation that the off-diagonal elements have little effect on the energies. The approximated total matrix is then

$$\mathcal{H} = -\gamma\hbar \begin{pmatrix} B_0 - \frac{1}{4}\mu_0\gamma\hbar Y_0 & 0 & 0 & 0 \\ 0 & \frac{1}{2}\mu_0\gamma\hbar Y_0 & 0 & 0 \\ 0 & 0 & -B_0 - \frac{1}{4}\mu_0\gamma\hbar Y_0 & 0 \\ 0 & 0 & 0 & 0 \end{pmatrix}$$

Note that the energy of state $|2\rangle$ is affected: although $M_s = 0$ the spin and magnetic moment are not zero, as they are for $|4\rangle$. Transitions corresponding to a single spin reversal are those between $|1\rangle$ and $|2\rangle$ or between $|2\rangle$ and $|3\rangle$. The energy changes and ω values involved are (see Figure 3.9)

$$\begin{aligned} \hbar\omega_{12} &= \gamma\hbar\left(B_0 + \tfrac{3}{4}\mu_0\gamma\hbar\frac{3\cos^2\theta - 1}{r^3}\right) \\ \hbar\omega_{23} &= \gamma\hbar\left(B_0 - \tfrac{3}{4}\mu_0\gamma\hbar\frac{3\cos^2\theta - 1}{r^3}\right) \end{aligned} \tag{70}$$

Thus the interaction may be represented by an effective field

$$B_e = \pm\tfrac{3}{4}\mu_0\gamma\hbar\frac{3\cos^2\theta - 1}{r^3} \tag{71}$$

The classical field from one dipole $\boldsymbol{\mu}_1$ acting on a second dipole $\boldsymbol{\mu}_2$, as shown in Figure 3.10, in which the dipoles are specifically parallel to each other, is

$$\mathbf{B} = \mu_0\frac{3(\boldsymbol{\mu}_1 \cdot \mathbf{r})\mathbf{r}}{r^5} - \frac{\mu_0\boldsymbol{\mu}_1}{r^3} = \frac{(3\mu_0\mu_1\cos\theta)\mathbf{r}}{r^4} - \frac{\mu_0\boldsymbol{\mu}_1}{r^3}$$

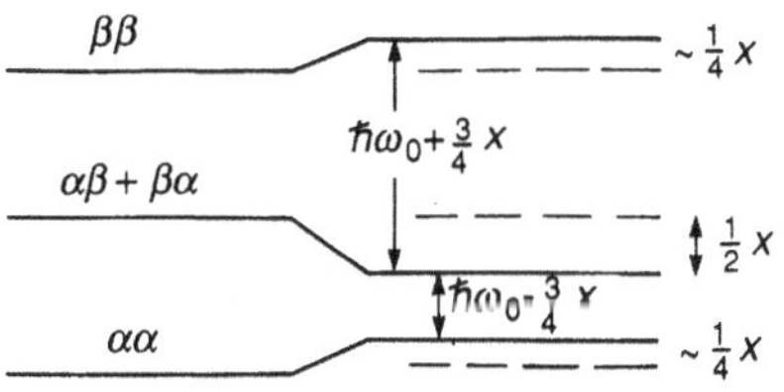

Figure 3.9 Energy levels for dipolar interaction: $x = \frac{1}{4}\mu_0\gamma^2\hbar Y_0$

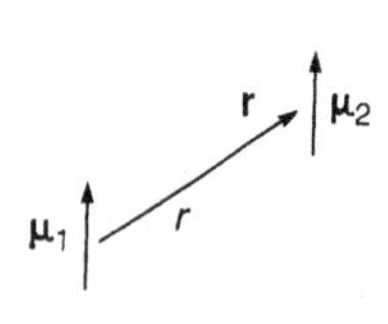

Figure 3.10 Parallel spins for the classical dipolar interactions

The Z component of this is obtained by projecting the first term (the second being already $\parallel OZ$) so that

$$B_z = \pm\frac{\mu_0\mu_1}{r^3}(3\cos^2\theta - 1) = \pm\tfrac{1}{2}\mu_0\hbar\gamma\frac{3\cos^2\theta - 1}{r^3}$$

taking μ_1 to be that of a spin $\frac{1}{2}$ and assuming that $\boldsymbol{\mu}_1$ may be reversed with respect to the orientation shown in the figure.

Thus the dipolar interaction may be accounted for in terms of effective fields which are very similar to those according to the classical calculation described but are greater by a factor of $\frac{3}{2}$.

The general solution of the equation of motion when it contains terms for B, the r.f. field and the dipolar interaction is naturally exceptionally complex, but at a simple level the range of effective fields is considered to give a series of overlapping lines, themselves taken to be very narrow in the absence of dipolar effects. For an order of magnitude the extremes of the fields give

$$\begin{aligned} \Delta B &\sim \frac{\mu_0\mu}{a^3} = \frac{4\pi \times 10^{-7} \times 5 \times 10^{-27}}{8 \times 10^{-30}} = 10^{-3}\ \mathrm{T} \\ \Delta\omega &\sim \gamma\Delta B \sim 10^5\ \mathrm{S}^{-1} \end{aligned} \tag{72}$$

with μ the nuclear magneton, γ for the proton and a spacing $a = 2 \times 10^{-10}$ m. These are enormous values, compared with high-resolution n.m.r. lines, but if r, θ, ϕ and thus Y_0, Y_1, Y_2 vary rapidly in time the fields largely average out, as follows.

1.12 Motional Narrowing and Magic Angle Spinning

The effect of relative motion, as of the protons in water, is dramatic and is vital experimentally. The following treatment is after Kittel [4]. T_2 will correspond to the time taken for a characteristic spin to move out of phase substantially with respect to one precessing at ω_0, say by 1 radian, since at this stage M_t will have fallen greatly. For an effective dipole field B_e the frequency shift is $(\Delta\omega)_0 = \gamma B_e$. When the dipoles are mobile the field can be taken to switch between $\pm B_e$, randomly but with a certain mean time τ at each value. In a time τ the spin advances or lags one which is exposed to B_0 only, by

$$\delta\varphi = \tau(\Delta\omega)_0 = \pm\tau\gamma B_e$$

After n random intervals of mean duration τ the mean square phase angle will be

$$\langle\varphi^2\rangle = n(\delta\varphi)^2 = n\tau^2\gamma^2 B_e^2 \tag{73}$$

This relies on a statistical argument using the random walk theory. Taking n equal steps back and forth gives no net mean displacement, but for n steps of length l in random directions the mean displacement achieved corresponds to $\langle r^2 \rangle = nl^2$. Setting $\langle \varphi^2 \rangle = 1$, for substantial loss of coherence, in equation (73) $n = 1/(\tau^2 \gamma^2 B_e^2)$ and the time taken for this is $n\tau$, i.e.

$$T_2 = n\tau = \frac{1}{\tau \gamma^2 B_e^2}$$

Taking the linewidth to be $1/T_2$ this gives

$$\Delta\omega = \frac{1}{T_2} = \tau(\gamma B_e)^2 = \tau(\Delta\omega)_0^2$$

Recalling that $(\Delta\omega)_0$ is the dipole-controlled linewidth for the rigid lattice, the narrowing or reduction of the dipolar broadening, due to motion, is represented by the factor $F = \tau(\Delta\omega)_0$ or $\tau\gamma B_e$. An appropriate value of τ for water molecules can be derived from dielectric constant results as $\sim 10^{-10}$ s. With B_e as estimated above,

$$F \sim 10^{-10} \times 10^8 \times 10^{-3} = 10^{-5}$$

This very great effect is in line with measured linewidths in water. The importance of this with respect to high-resolution studies is obvious and the implications for the study of a wide range of kinetic problems are equally clear.

In a rigid solid with no motional narrowing as above, motion may be introduced by spinning the specimen about an axis at θ_s to OZ (Figure 3.11). Consider the effects on B_e at a nucleus 'a', which for convenience is taken to be on this axis. For a nucleus 'b' the values of θ and φ, the polar coordinates of nucleus a with nucleus b at the origin, vary rapidly and effective motional narrowing is achieved. However, for a neighbouring nucleus such as c, θ remains constant and equal to θ_s. As noted in Chapter 1, Section 17.2.9, Andrew appreciated the significance of the expression (71) for B_e. If θ_s is chosen as the 'magic angle', i.e. that which sets $3\cos^2\theta_s - 1 = 0$ then the B_e from a nucleus such as c becomes zero. Thus the spinning axis should be at an angle $\cos^{-1} 1/\sqrt{3}$ to B_0. Andrew also overcame the obvious technical problems involved in achieving a mechanical angular frequency of up to 10^4 Hz with the constraints imposed by the available magnets and a great amount of solid state information was obtained in this way. The effects of spinning are to be compared with those obtained by special multiple pulse methods, as also noted in Chapter 1. Spinning at lower frequencies may also be employed to average out applied field inhomogeneities.

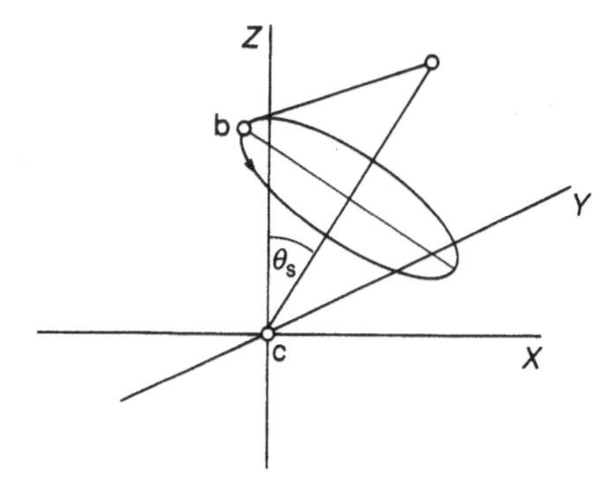

Figure 3.11 The spinning angle θ_s

1.13 Spin Waves

Spin wave theory is elaborate and extensive. The present aims are simplification and unification with other approaches.

The perfectly ferromagnetically ordered state has been seen to exist only as $T \to 0$, or $T/\mathcal{J} \to 0$, when only the ground state $(2S+1)$ multiplet, characterized for N spins $\frac{1}{2}$ by total $S = \frac{1}{2}N$, $M_s = S, S-1, \ldots, -(S-1), -S$, is occupied. At any finite temperature excitations lead to a decrease in the order and a fall in $M(T)$. These may be seen as spin deviations, spin reversals for $S = \frac{1}{2}$, but it is hardly to be expected that the lowest excitations should correspond to one identifiable reversal. The deviations will be seen to be delocalized in a systematic wave-like manner, forming spin waves, and the excitations are called magnons.

1.13.1 Simple treatment of spin waves in a one-dimensional $S = \frac{1}{2}$ chain

The chain of N spins is uniform with spacing a and all $\mathcal{J}_{i,i+1} = \mathcal{J}$, $\mathcal{J} = 0$ otherwise. This includes $\mathcal{J}_{N1} = \mathcal{J}$, meaning that the chain forms a closed ring or that periodic boundary conditions apply equivalently as N becomes very large. Solutions for a Heisenberg interaction, for general N, are sought in terms of the N orthonormal basis states corresponding to a single reversal, which can be designated as

$$
\begin{aligned}
&\overset{\underline{1}\,2\,3\,4 \qquad N}{|1\rangle = |\beta\alpha\alpha\alpha\cdots\alpha\rangle}, \qquad && \overset{1\,\underline{2}\,3\,4 \qquad N}{|2\rangle = |\alpha\beta\alpha\alpha\cdots\alpha\rangle} \\
&\overset{1\,2\,3\,4 \qquad \underline{l} \qquad N}{|l\rangle = |\alpha\alpha\alpha\alpha\cdots\beta\cdots\alpha\rangle}, \qquad && \overset{1\,2\,3\,4 \qquad \underline{N}}{|N\rangle = |\alpha\alpha\alpha\alpha\cdots\beta\rangle}
\end{aligned}
\tag{74}
$$

with β in position l for state $|l\rangle$. The state $|0\rangle = |\alpha\ \alpha\ \alpha\ \alpha\ \cdots\ \alpha\rangle$ for no reversals is akin to the vacuum state in the number representation. The Hamiltonian for nn interactions can be divided as

$$
\mathcal{H} = \mathcal{H}_1 + \mathcal{H}_2 + \mathcal{H}_3 = -2\mathcal{J}\sum_i S_i^z S_{i+1}^z - \mathcal{J}\sum_i S_i^+ S_{i+1}^- - \mathcal{J}\sum_i S_i^- S_{i+1}^+ \tag{75}
$$

with $i = 1, 2, \ldots, N$ and $N+1 \equiv 1$ implied. For the state $|0\rangle$ with no reversals,

$$
\begin{aligned}
\mathcal{H}_1|0\rangle &= -2\mathcal{J}N(\tfrac{1}{4})|0\rangle = -N\frac{\mathcal{J}}{2}|0\rangle \\
\mathcal{H}_2|0\rangle &= 0 \quad \mathcal{H}_3|0\rangle = 0 \\
E_0 &= -N\frac{\mathcal{J}}{2}
\end{aligned}
\tag{76}
$$

since $S^+|\alpha\rangle = 0$, etc.

A general state is

$$
\chi = \sum c_l|l\rangle \tag{77}
$$

With the examples

$$
\begin{aligned}
&S_1^+S_2^-|\beta\ \alpha\ \alpha\ \alpha\ \alpha\ \cdots\rangle = |\alpha\ \beta\ \alpha\ \alpha\ \alpha\ \cdots\rangle, \qquad && S_1^+S_2^-|1\rangle = |2\rangle \quad S_i^+S_{i+1}^-|i\rangle = |i+1\rangle \\
&S_1^-S_2^+|\alpha\ \beta\ \alpha\ \alpha\ \alpha\ \cdots\rangle = |\beta\ \alpha\ \alpha\ \alpha\ \alpha\ \cdots\rangle, \qquad && S_1^-S_2^+|2\rangle = |1\rangle \quad S_i^-S_{i+1}^+|i+1\rangle = |i\rangle
\end{aligned}
$$

it is clear that

$$S_{i-1}^{+} S_i^{-} c_{i-1} |i-1\rangle = c_{i-1} |i\rangle$$

$$S_i^{-} S_{i+1}^{+} c_{i+1} |i+1\rangle = c_{i+1} |i\rangle$$

and that

$$\mathcal{H}_2 |\chi\rangle = -\mathcal{J} \sum_i S_i^{+} S_{i+1}^{-} \sum_l c_l |l\rangle = -\mathcal{J} \sum_{l=1}^{N} c_{l-1} |l\rangle \tag{78}$$

$$\mathcal{H}_3 |\chi\rangle = -\mathcal{J} \sum_i S_i^{-} S_{i+1}^{+} \sum_l c_l |l\rangle = -\mathcal{J} \sum_{l=1}^{N} c_{l+1} |l\rangle \tag{79}$$

since in the double summations the contribution to $\mathcal{H}_2|\chi\rangle$ is zero except when $i = 1$ (N times) and the contribution to $\mathcal{H}_3|\chi\rangle$ is zero except when $i = l - 1$. Specifically for this chain, $\mathcal{H}_2$ and $\mathcal{H}_3$ have the effect of shifting the array of coefficients backwards and forwards through a single step with respect to the array of the basis states. The series remain complete, $l = 1, 2, \ldots, N$, due to the cyclic boundary conditions with $c_{N+1} = c_1$ and $c_{1-1} = c_N$. For the entire Hamiltonian,

$$\mathcal{H}|\chi\rangle = (\mathcal{H}_1 + \mathcal{H}_2 + \mathcal{H}_3)|\chi\rangle = \left(\frac{-N\mathcal{J}}{2} + 2\mathcal{J} \right) |\chi\rangle - \mathcal{J} \sum_l (c_{l+1} + c_{l-1}) |l\rangle$$

$|\chi\rangle$ being an eigenstate of $\mathcal{H}_1$ since $\mathcal{H}_1|\ell\rangle = -2\mathcal{J}(N-4)(\frac{1}{4})|l\rangle$ for each basis state. The requirement for $|\chi\rangle$ to be an eigenstate of $\mathcal{H}$ is clearly that

$$c_{l+1} + c_{l-1} = Bc_l \tag{80}$$

(B constant), which is most obviously met by all $c_l = 1$, with $c_{l+1} + c_{l-1} = 2c_l$, and

$$\mathcal{H}|\chi\rangle = \left(-\frac{N\mathcal{J}}{2} + 2\mathcal{J} \right) |\chi\rangle - \mathcal{J}2|\chi\rangle = -\frac{N\mathcal{J}}{2}|\chi\rangle$$

This is just an alternative proof of an already established principle, i.e. that

$$|\chi\rangle = |\chi_0\rangle = N^{-1/2} \sum |l\rangle$$

is an eigenstate with energy equal to that for $|0\rangle$.

For a more general case, note that equation (80) would correspond, with $B = 2$, to linear interpolation as for $c = la$, but $|\chi\rangle$ could not then be normalized or meet the boundary conditions. Clearly a periodic function is required and it is natural to suggest

$$c_l = \mathrm{e}^{ik(l-1)a}$$

chosen to give $c_1 = 1$ ($|\chi\rangle = \sum c_l |1\rangle$ remaining to be normalized) with the wave vector k restricted by

$$c_{N+1} = \mathrm{e}^{ik[(N+1)-1]a} = \mathrm{e}^{ikNa} = c_1 = 1$$

i.e.

$$k = n\frac{2\pi}{Na} = n\frac{2\pi}{L} \qquad \text{for } n = 0, 1, 2, \ldots$$

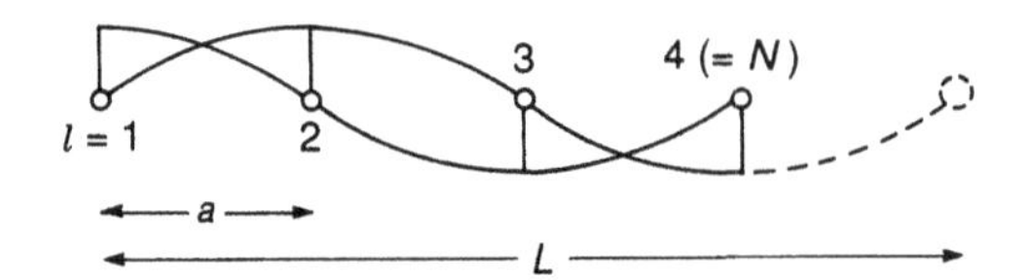

Figure 3.12 A minimal spin wave model, the lowest ($k \neq 0$) wave for $N = 4$, with $\lambda = L/n$ and $n = 1$. $L = 4a$, not $3a$, in accordance with the periodic boundary conditions or by taking the nominal spin on the right to be spin 1. The real and imaginary parts of the c_l are shown: $k = \pi/(2a)$ and $c_l = \cos \pi(l-1)/2 + i \sin \pi(l-1)/2$, for $l - 1, 2, 3, 4$, giving $\sum c_l = 0$

$L = Na$ is taken to be the length of the chain, and the above can alternatively be expressed as $\lambda = L/n$, with the wavelength given by $k = 2\pi/\lambda$: an integral number of wavelengths must fit the chain length. Use will be made of

$$\sum_{l=1}^{N} c_l = 0$$

which is illustrated by Figure 3.12 and exemplified by equations (89) and (90) later. When $a \to 0$ and $N \to \infty$ the sum is replaced by an integral and the result may readily be obtained as an exercise. The choice of the c_l is seen to be appropriate because

$$\begin{aligned} c_{l+1} + c_{l-1} &= \mathrm{e}^{ikla} + \mathrm{e}^{ik(l-2)a} \\ &= \mathrm{e}^{ik(l-1)a+ika} + \mathrm{e}^{ik(l-1)a-ika} \\ &= (\mathrm{e}^{ika} + \mathrm{e}^{-ika})\mathrm{e}^{ik(l-1)a} \\ &= B(k)c_l \end{aligned}$$

(cp Eqn. 80) and eigenstates can now be designated as

$$|k\rangle = \sum c_l |l\rangle = \sum \mathrm{e}^{ik(l-1)a} |l\rangle \tag{81}$$

where the label k is the wave vector designating the state and not an indication of a basis state having β in position k. The spin wave energy is given by

$$\left(-2\mathcal{J}\sum S_i^z S_{i+1}^z - \mathcal{J}\sum S_i^+ S_{i+1}^- - \mathcal{J}\sum S_i^- S_{i+1}^+\right)|k\rangle = -\mathcal{J}\left[\frac{N}{2} + 2 - B(k)\right]|k\rangle$$

and approximating

$$B(k) = \mathrm{e}^{ika} + \mathrm{e}^{-ika} = 2 - k^2a^2$$

gives the dispersion relation for the spin wave energies, as in Figure 3.13:

$$E = E_0 + E(k) = E_0 + \hbar\omega(k) = -\frac{N\mathcal{J}}{2} + \mathcal{J}k^2a^2 \tag{82}$$

which, of course, checks with $E(k = 0) = -N\mathcal{J}/2$ as for the state $|0\rangle$.

To determine the total spin according to $\mathbf{S}^2|k\rangle = S(S+1)|k\rangle$ with $\mathbf{S}^2 = \sum S_i^z S_j^z + \frac{1}{2}\sum S_i^+ S_j^- + \frac{1}{2}\sum S_i^- S_j^+$, first note that

$$\sum_i S_i^z \sum_j S_j^z |k\rangle = \sum_i S_i^z [(N-1)(\tfrac{1}{2}) + (-\tfrac{1}{2})]|k\rangle$$

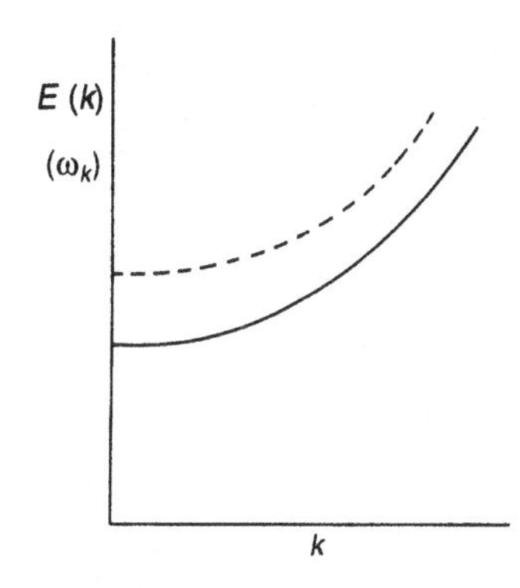

Figure 3.13 ω_k (or energy) versus wave number for spin waves. The 'manifold' indicated by the broken line replaces the plot initially expected due to the involvement of magnetostatic energy which depends, according to certain treatment, on the direction of propagation

$$= \sum_i S_i^z (N-2)(\tfrac{1}{2})|k\rangle$$

$$= \tfrac{1}{4}(N-2)^2|k\rangle \tag{83}$$

Also,

$$\tfrac{1}{2}\sum S_i^- S_j^+ \sum c_l|l\rangle = \tfrac{1}{2}\sum S_i^- \sum c_l|0\rangle$$

$$= \tfrac{1}{2}\left(\sum_l c_l\right)\left(\sum_i |i\rangle\right) = 0 \tag{84}$$

since $\sum c_l = 0$. For the last term,

$$\tfrac{1}{2}\sum S_i^+ \sum S_j^- \sum c_l|l\rangle = \tfrac{1}{2}\sum S_i^+ [c_1(0 + |12\rangle + |13\rangle + |14\rangle + \cdots + |1N\rangle)$$
$$+c_2(|12\rangle + 0 + |23\rangle + \cdots + |2N\rangle) + \cdots + c_N(|1N\rangle + |2N\rangle + |3N\rangle$$
$$+ \cdots + |N-1N\rangle + 0)]$$

where $|12\rangle = |\beta\,\beta\,\alpha\,\alpha\,\alpha\,\cdots\,\alpha\rangle, |13\rangle = |\beta\,\alpha\,\beta\,\alpha\,\alpha\,\cdots\,\alpha\rangle$, etc. Since, for example, $\sum S_i^+|12\rangle = |2\rangle + |1\rangle$, with all $S^+|\alpha\rangle = 0$,

$$\tfrac{1}{2}\sum S_i^+ \sum S_j^- \sum c_l|l\rangle = \tfrac{1}{2}\{c_1[0 + (|1\rangle + |2\rangle) + (|1\rangle + |3\rangle)$$
$$+ \cdots + (|1\rangle + |N\rangle)] + c_2[(|1\rangle + |2\rangle) + 0 + (|2\rangle + |3\rangle)$$
$$+ \cdots + (|2\rangle + |N\rangle)] + \cdots + c_N[(|N\rangle + |1\rangle) + (|N\rangle + |2\rangle)$$
$$+ \cdots + (|N\rangle + |N-1\rangle) + 0]\}$$
$$= \tfrac{1}{2}\left[\left(\sum c_l\right)\left(\sum |i\rangle\right) + (N-2)\sum c_l|l\rangle\right]$$
$$= \tfrac{1}{2}(N-2)\sum c_l|l\rangle \tag{85}$$

Adding the contributions,

$$\mathbf{S}^2 \sum c_l|l\rangle = \left(\frac{N^2}{4} - \frac{N}{2}\right)\sum c_l|l\rangle$$

$$= \left[\left(\frac{N}{2} - 1\right)\right]\left[\left(\frac{N}{2} - 1\right) + 1\right]\sum c_l|l\rangle$$

$$= S_{\mathrm{T}}(S_{\mathrm{T}} + 1)\sum c_l|l\rangle$$

so that the total spin quantum number for the chain is

$$S^{(k)} = (N - 2)s \tag{86}$$

using upper case for the system and lower case $s = \frac{1}{2}$ for the individual spins. The superscript indicates a state with $k \neq 0$ specifically.

For $|k = 0\rangle$ or $\sum|1\rangle$, i.e. all $c_l = 1$, the above may be repeated with the replacement of $\sum c_l = 0$ by $\sum c_l = N$. Then equation (84) gives $\frac{1}{2}N\sum|1\rangle$, equation (85) gives $\frac{1}{2}[N + (N - 2)]\sum|1\rangle$ and summing the contributions

$$S^{(k=0)} = N \times \tfrac{1}{2} = Ns \tag{87}$$

as expected.

Using S_x etc. for values and S^x etc. for the operators the conclusion for the one-dimensional chain, as for the general lattice, that the formation of ($k \neq 0$) spin waves reduces the value of S as compared with that for $|k = 0\rangle$ (itself equal to that for $|0\rangle$), while S_z remains the same, has clear implications with respect to the values of S_x and S_y. The effect of spin wave excitation on S_x and S_y may be illustrated by a simple example. A rudimentary model consists of just four spins $s = \frac{1}{2}$, giving four basis states with the restriction to a single reversal:

$$|1\rangle = |\beta\ \alpha\ \alpha\ \alpha\rangle, \qquad |2\rangle = |\alpha\ \beta\ \alpha\ \alpha\rangle, \qquad |3\rangle = |\alpha\ \alpha\ \beta\ \alpha\rangle, \qquad |4\rangle = |\alpha\ \alpha\ \alpha\ \beta\rangle$$

with the $|k = 0\rangle$ and a typical state $|k\rangle$ being

$$|k = 0\rangle = \tfrac{1}{2}(|1\rangle + |2\rangle + |3\rangle + |4\rangle)$$

$$|k\rangle = \tfrac{1}{2}(|1\rangle + i|2\rangle - |3\rangle - i|4\rangle)$$

Recalling $S^x|\alpha\rangle = \frac{1}{2}|\beta\rangle$, $S^x|\beta\rangle = \frac{1}{2}|\alpha\rangle$, $S^y|\alpha\rangle = \frac{1}{2}i|\beta\rangle$ and $S^y|\beta\rangle = -\frac{1}{2}i|\alpha\rangle$,

$$\sum S_j^x|1\rangle = \tfrac{1}{2}[|\alpha\ \alpha\ \alpha\ \alpha\rangle + |\beta\ \beta\ \alpha\ \alpha\rangle + |\beta\ \alpha\ \beta\ \alpha\rangle + |\beta\ \alpha\ \alpha\ \beta\rangle]$$

$$\sum S_i^x \sum S_j^x|1\rangle = \tfrac{1}{4}[(|1\rangle + |2\rangle + |3\rangle + |4\rangle) + (|2\rangle + |1\rangle + |\rangle + |\rangle)$$
$$+ (|3\rangle + |1\rangle + |\rangle + |\rangle) + (|4\rangle + |1\rangle + |\rangle + |\rangle)]$$

where the 'empty' $|\ \rangle$ indicate states other than the basis states which give zeros on taking mean values and need not be identified. They may be lumped together in single sums σ_i and it is readily confirmed that

$$\sum S_i^x \sum S_j^x|1\rangle = \tfrac{1}{4}(4|1\rangle + 2|2\rangle + 2|3\rangle + 2|4\rangle + \sigma_1)$$

$$\sum S_i^x \sum S_j^x|2\rangle = \tfrac{1}{4}(2|1\rangle + 4|2\rangle + 2|3\rangle + 2|4\rangle + \sigma_2)$$

$$\sum S_i^x \sum S_j^x|3\rangle = \tfrac{1}{4}(2|1\rangle + 2|2\rangle + 4|3\rangle + 2|4\rangle + \sigma_3)$$

$$\sum S_i^x \sum S_j^x|4\rangle = \tfrac{1}{4}(2|1\rangle + 2|2\rangle + 2|3\rangle + 4|4\rangle + \sigma_4)$$

Thus

$$\sum S_i^x \sum S_j^x |k=0\rangle = \tfrac{1}{2}(\tfrac{10}{4})(|1\rangle + |2\rangle + |3\rangle + |4\rangle + \sigma)$$

$\langle S^x \rangle_{k=0} = 0$ obviously and

$$\langle S^{x^2} \rangle_{k=0} = S_x^2 = \tfrac{1}{4}\left(\sum \langle l|\right) \tfrac{10}{4} \left(\sum |l\rangle + \sigma\right) = \tfrac{10}{4}$$

The reader may readily repeat the above, replacing S^x by S^y, and obtain the expected

$$\langle S^y \rangle_{k=0} = 0, \qquad \langle S^{y^2} \rangle_{k=0} = S_y^2 = \tfrac{10}{4}$$

However, for the state $|k\rangle$, while $\langle s^x \rangle_k = 0$ is again apparent,

$$\begin{aligned}\sum S_i^x \sum S_j^x |k\rangle &= \tfrac{1}{2} \times \tfrac{1}{4}(2|1\rangle + i2|2\rangle - 2|3\rangle - i2|4\rangle + \sigma)\\ &= \tfrac{1}{2}(|k\rangle + \sigma)\end{aligned}$$

The nominal sum of 'other states' σ remains though it is conceivable in this case that it could have become zero. Thus again only a mean value can be found, i.e.

$$\langle S^{x^2} \rangle_k = S_x^2 = \tfrac{1}{2}$$

and again the reader may confirm that

$$\langle S^y \rangle_k = 0 \quad \langle S^{y^2} \rangle_k = S_y^2 = \tfrac{1}{2}$$

Naturally these are the values of S_x^2 and S_y^2 that give

$$\begin{aligned} S_{k=0}^2 &= S_x^2 + S_y^2 + S_z^2 = \tfrac{10}{4} + \tfrac{10}{4} + \tfrac{4}{4} = 6 = 2(2+1), \qquad S_{k=0} = 2 \\ S_k^2 &= S_x^2 + S_y^2 + S_z^2 = \tfrac{1}{2} + \tfrac{1}{2} + 1 = 2 = 1(1+1), \qquad S_k = 1 \end{aligned} \tag{88}$$

in line with the earlier conclusions.

1.13.2 Spin waves and direct solutions

There should be direct connections between spin waves and the solutions of the Heisenberg Hamiltonian matrix. The two approaches are simply two different routes to the evaluation of the excitations — the solution of the same Hamiltonian by different means. The connection will be illustrated by a simple example, a closed chain of four spins $\frac{1}{2}$. Only the subspace spanned by the four basis states corresponding to a single spin reversal will be considered, for comparison with one-magnon states, these being denoted

$$|1\rangle = |\beta\ \alpha\ \alpha\ \alpha\rangle,\ |2\rangle = |\alpha\ \beta\ \alpha\ \alpha\rangle,\ |3\rangle = |\alpha\ \alpha\ \beta\ \alpha\rangle \text{ and}$$
$$|4\rangle = |\alpha\ \alpha\ \alpha\ \beta\rangle$$

Noting that, for example,

$$\begin{aligned} -2\mathcal{J} \sum S_i^z S_{i+1}^z |\beta\ \alpha\ \alpha\ \alpha\rangle &= -2\mathcal{J}\left(-\tfrac{1}{4} + \tfrac{1}{4} + \tfrac{1}{4} - \tfrac{1}{4}\right) |\beta\ \alpha\ \alpha\ \alpha\rangle \\ &= 0|\beta\ \alpha\ \alpha\ \alpha\rangle \end{aligned}$$

$$-\mathcal{J} \sum S_i^+ S_{i+1}^- |\beta\ \alpha\ \alpha\ \alpha\rangle = -\mathcal{J}|\alpha\ \beta\ \alpha\ \alpha\rangle$$

it is confirmed that

$$\mathcal{H}|1\rangle = -\mathcal{J}(|2\rangle + |4\rangle), \qquad \mathcal{H}|2\rangle = -\mathcal{J}(|1\rangle + |3\rangle)$$

$$\mathcal{H}|3\rangle = -\mathcal{J}(|2\rangle + |4\rangle), \qquad \mathcal{H}|4\rangle = -\mathcal{J}(|1\rangle + |3\rangle)$$

so that the matrix and the secular equation are

$$\mathcal{H} = \begin{pmatrix} 0 & -\mathcal{J} & 0 & -\mathcal{J} \\ -\mathcal{J} & 0 & -\mathcal{J} & 0 \\ 0 & -\mathcal{J} & 0 & -\mathcal{J} \\ -\mathcal{J} & 0 & -\mathcal{J} & 0 \end{pmatrix}, \qquad \begin{vmatrix} -E & -\mathcal{J} & 0 & -\mathcal{J} \\ -\mathcal{J} & -E & -\mathcal{J} & 0 \\ 0 & -\mathcal{J} & -E & -\mathcal{J} \\ -\mathcal{J} & 0 & -\mathcal{J} & -E \end{vmatrix} = 0$$

the latter evaluating to

$$E^4 - 4E^2\mathcal{J}^2 = (E^2 + 2E\mathcal{J})(E^2 - 2E\mathcal{J}) = 0$$

If $E^2 + 2E\mathcal{J} = 0$ then $E = 0$ or $E = -2\mathcal{J}$ and if $E^2 - 2E\mathcal{J} = 0$ then $E = 0$ once more or $E = +2\mathcal{J}$. When $E = 0$, $\mathbf{Hc} = \mathbf{Ec}$ gives

$$c_2 + c_4 = 0$$

$$c_1 + c_3 = 0$$

satisfied by

$$c_1 = 1, \qquad c_3 = -1, \qquad c_2 = 1, \qquad c_4 = -1$$

but with some ambiguity since the equations do not connect c_1 and c_2 so that the vector

$$(c_1, c_2, c_3, c_4) = (1, 1, -1, -1) \tag{89}$$

would be equally acceptable. $\mathbf{Hc} = 2\mathcal{J}\mathbf{c}$ gives

$$2c_1 + c_2 + c_4 = 0$$

and three other equations satisfied by

$$(c_1, c_2, c_3, c_4) = (1, -1, 1, -1) \tag{90}$$

[or equally, of course, $(-1, 1, -1, 1)$].

The spin wave states within the Brillouin zone are $\sum c_l|l\rangle$; $c_l = \exp[ik(l-1)a]$, $k = 2\pi n/(4a)$, $n = 0, 1, 2$, but it will be deliberately chosen to extend the range to $n = 4$. Summing over nearest neighbours along the chain the spin wave energies are

$$\varepsilon(\mathbf{k}) = 2S\sum_i \mathcal{J}_i(1 - e^{i\mathbf{k}\cdot\mathbf{a}_i}) \tag{91}$$

with the $\mathbf{a}_i$ vectors, all $\mathbf{a}_i = a$, extending from a characteristic spin to each of its two neighbours. Taking all $\mathcal{J}_i = \mathcal{J}$ and $\mathbf{k}$ to be directed along the chain length,

$$\begin{aligned} \varepsilon(\mathbf{k}) &= 2S\mathcal{J}[(1 - e^{ika}) + (1 - e^{-ika})] \\ &= (2S)2\mathcal{J}(1 - \cos ka) \\ &= 2\mathcal{J}(1 - \cos ka) \end{aligned} \tag{92}$$

and with $k = 2\pi/\lambda$, $n\lambda = 4a$, $k = n\pi/(2a)$, the states, spin wave energies and total energies are

n	ka	$\|k\rangle$	$\varepsilon(k)$	E	
0	0	$\|1\rangle + \|2\rangle + \|3\rangle + \|4\rangle$	0	$-2\mathcal{J}$	Direct solutions
1	$\pi/2$	$\|1\rangle + i\|2\rangle - \|3\rangle - i\|4\rangle$	$2\mathcal{J}$	0	
2	π	$\|1\rangle - \|2\rangle + \|3\rangle - \|4\rangle$	$4\mathcal{J}$	$2\mathcal{J}$	Zone boundary
3	$3\pi/2$	$\|1\rangle - i\|2\rangle - \|3\rangle + i\|4\rangle$	$2\mathcal{J}$	0	
4	2π	$\|1\rangle + \|2\rangle + \|3\rangle + \|4\rangle$	0		

Note that $|k = 2\pi/a\rangle \equiv |k = 0\rangle$ and that $|k = 3\pi/(2a)\rangle$ is essentially the same as $|k = \pi/(2a)\rangle$ since the only apparent difference consists of a shift along the line or represents the arbitrary choice of the first basis state $|1\rangle$. The inclusion of $n = 3$ and $n = 4$ simply affords an illustration of the effect of crossing the zone boundary.

It is apparent that direct solution of the Hamiltonian matrix for $N = 4$ with a single spin reversal gives four solutions, the matrix being 4×4. If the resulting degeneracy were ignored the three energies obtained would correlate with the three spin wave energies up to $k = \pi/a[E = E_0 + \varepsilon(k)]$ and the state vectors would correlate inasmuch as the coefficients $(c_1, c_2, c_3, c_4) = (1, i, -1, -i)$ satisfy $\mathbf{Hc} = E\mathbf{c}$ for the state with $E = 0$. The degeneracy in the direct solutions corresponds to the existence of an extra state which can be considered as the first spin wave state which lies outside the Brillouin zone.

1.13.3 Summary of general single-magnon calculations

For N general spins S and with interactions with z nearest neighbours, in a three-dimensional lattice, the perfectly ordered state $|0\rangle$ gives the energy $E_0 = -NS^2 z\mathcal{G}$. States with one spin deviation (not now necessarily a reversal) are described as the state $|0\rangle$ as modified by lowering the lth spin:

$$|l\rangle = \frac{1}{(2S)^{1/2}} S_l^- |0\rangle, \qquad \langle l| = \frac{1}{(2S)^{1/2}} \langle 0| S^+$$

since $(S^-)^\dagger = S^\dagger$. An eigenstate is taken to have the form

$$|\chi\rangle = \sum_l \varphi(\mathbf{R}_l)|l\rangle$$

with the lth spin or atom at position $\mathbf{R}_l$. It is suggested that $\varphi(\mathbf{R}_l) \propto e^{i\mathbf{k}\cdot\mathbf{R}_l}$ and that the exchange constant applying between two different sites is a function of the distance between the sites. If a characteristic spin n is taken to be at the origin, $\mathcal{J}_{nl} = \mathcal{J}(\mathbf{R}_l)$ applies between the site n and the site l, not of course the coupling *at* the site l.

This suggestion is shown to be acceptable because it does in fact yield a solution and the spin wave energy is

$$E(k) = 2S \sum_l \mathcal{J}(\mathbf{R}_l)(1 - e^{i\mathbf{k}\cdot\mathbf{R}_l}) \rightarrow 2S \sum_l \mathcal{J}(\mathbf{R}_l)(1 - \cos \mathbf{k} \cdot \mathbf{R}_l) \tag{93}$$

the second for a crystal with a centre of inversion. If $\mathbf{k}$ is increased by any lattice vector $\mathbf{K}$ $E(\mathbf{k})$ and $|k\rangle$ are unchanged and so $\mathbf{k}$ can be considered to be confined to the Brillouin zone, i.e. for an s.c. crystal, for example, the components k_x, k_y, k_z range from 0 to π/a (and taking values outside this range would simply duplicate results).

For an s.c. lattice all six $(\mathbf{R}_l) = a$ and the $\mathbf{R}_l = \pm a\mathbf{i}, \pm a\mathbf{j}, \pm a\mathbf{k}$, giving

$$\varepsilon(\mathbf{k}) = 2S\mathcal{G}\sum_{l=1}^{6}(1 - \mathrm{e}^{i\mathbf{k}\cdot\mathbf{R}_l}) = 2S\mathcal{G}(6 - \mathrm{e}^{ik_x a}\cdots)$$
$$= 2S\mathcal{G}[6 - 2(\cos k_x a + \cos k_y a + \cos k_z a)]$$

with a maximum of $24S\mathcal{G}(k_x = k_y = k_z = \pi/a)$. The normalized states for N atoms correspond to

$$\varphi(\mathbf{k}, \mathbf{R}_l) = N^{-1/2}\mathrm{e}^{i\mathbf{k}\cdot\mathbf{R}_l} \tag{94}$$

These are clearly eigenstates of $\sum_i S_i^z$ and can be shown to be eigenstates of $\mathbf{S}^2$ for the total spin, giving

$$\mathbf{S}^2|\mathbf{k}\rangle = \sum_{ij}\mathbf{S}_i\cdot\mathbf{S}_j|\mathbf{k}\rangle = (NS-1)[(NS-1)+1]|\mathbf{k}\rangle \tag{95}$$

so that the total spin of the assembly is reduced to $NS - 1$, from the value NS applying to the state of perfect order and maximum possible magnetization.

1.13.4 Generalizations and conclusions

The calculation for the simple, one-dimensional, model and those outlined for the general three-dimensional case refer to single-magnon states, a single spin reversal that can now be seen not to be localized but to extend throughout the crystal in a wave-like manner. The state $|k = 0\rangle$ corresponds to perfect order, since for this the total spin is NS, and to the ideal maximum magnetization. As T increases from zero the occupation of the single-magnon states with total spin $NS - 1$ effects a marginal reduction only, of the order, and of $\mathbf{M}$ (assuming N large). To predict, for example, $M(T)$ realistically it is necessary to study how the single-magnon states combine and this calls for transformations to creation and annihilation operators a and $a^\dagger$ and the number representation. Operators P and Q, expressed in the a and $a^\dagger$, are appropriate for the study of harmonic oscillators and a similarity, which is close so long as the occupation numbers are low, is noted between the matrices of P and Q and those of the spin operators (see, for example [5–7]). In this harmonic oscillator approximation the energy is $E_0 + \sum_k N_k\hbar\omega(k)$ with N_k the number of magnons of energy $E(k) = \hbar\omega(k)$. The eventual result for the magnetization at low temperatures corresponding to the restriction of the theory to low occupation numbers, is

$$\frac{M(T)}{M_0} = 1 - \frac{1}{n^S}\left(\frac{kT}{8\pi S\mathcal{J}}\right)^{3/2}\zeta\left(\tfrac{3}{2}\right) \tag{96}$$

with $n = 1, 2, 4$ for s.c., b.c.c. and f.c.c. crystals and ζ the Riemann zeta function $\zeta(x) = \sum_{n=1}^{\infty} n^{-x}$. When the $T^{3/2}$ dependence is observed values for $\mathcal{G}$ follow. The specific heat due to spin wave excitation is found to be

$$C = Nk_\mathrm{B}\frac{15}{4n}\left(\frac{k_\mathrm{B}T}{8\pi S\mathcal{J}}\right)^{3/2}\zeta\left(\tfrac{5}{2}\right) \tag{97}$$

$C \propto T^3$ for lattice vibrations and the contrast can be associated with $E(k) \sim k^2$ for spin waves and $E(k) \sim k$ for phonons. In ferromagnetic insulators such as EuS spin wave excitation makes the predominant contribution to C at low temperatures and thus, again, values of $\mathcal{G}$ may be determined [8, 9].

2 Spin and Magnetization Dynamics

To a large extent magnetization dynamics can be identified with spin dynamics, the dynamics of the polarization vector. It is only with this association that a reasonably straightforward account can be given. For an electron substantial orbital quenching is assumed, though the g factor should not be necessarily taken to be that for a simple spin but to be suitably modified (Chapter 2). For protons, etc., such problems do not arise. The account is limited to $S = \frac{1}{2}$ or equally to $I = \frac{1}{2}$ for reasons of (relative) simplicity. The results are naturally appropriate to both e.s.r. and n.m.r., though there is the continued annoyance of deciding whether to consider $|\beta\rangle$ (electrons) or $|\alpha\rangle$ (protons) to be the lower-lying state in an applied field. Although the very different magnitudes of γ for the two cases are of great technical significance the frequencies involved are, in both cases, in a range in which stimulated rather than spontaneous transitions are important.

It is easiest, for many demonstrations, to think in terms of a single spin which must be in a pure state with $P = 1$ and $\mu = \frac{1}{2}\hbar\gamma$, $\mu \doteqdot \beta$ for an electron. However, many of the results apply to the magnetization of assemblies in mixed states, e.g. to the equilibrium magnetization M_0 induced by an applied field, particularly when the relaxation or time scales involved are such that the vector $\mathbf{M}_0$ maintains its magnitude during the motion.

2.1 General Principles

All states or all wave functions have a time dependence, $\psi(\mathbf{r}) \to \psi(\mathbf{r}, t)$ or $|\psi\rangle \to |\psi(t)\rangle$, which can often be ignored because the observables or constants of the motion are not affected. The (time-dependent) Schrodinger equation is

$$i\hbar\frac{\partial|\psi(t)\rangle}{\partial t} = \mathcal{H}|\psi(t)\rangle \tag{98}$$

$\mathcal{H}$ is initially assumed to have no time dependence, i.e. not to involve t explicitly or implicitly. The simplest quantized system is an electron free to move in one dimension save that it is confined to a region of length l by potential barriers so high as to render the tunnelling negligible. $\mathcal{H} = -[\hbar^2/(2m)]\mathrm{d}^2/\mathrm{d}x^2$ and obvious eigenfunctions and values are $\psi(x) = N \sin n\pi x/l$, $E_n = [\hbar^2/(2m)](n^2\pi^2/l^2)$. The functions satisfying equation (98) are thus

$$\begin{aligned}\psi(x, t) &= N \sin\frac{n\pi x}{l}\mathrm{e}^{-i\omega_n t}\\ &= N \sin\frac{n\pi x}{l}(\cos\omega_n t - i\sin\omega_n t)\end{aligned}$$

with $\omega_n = E_n/\hbar$, having real and imaginary parts which oscillate, out of phase, in time but with constant energy. In general an eigenstate may be designated as $|\varphi_n(0)\rangle$ or just $|\varphi_n\rangle$ at $t = 0$ and at time t:

$$|\varphi_n(t)\rangle = \mathrm{e}^{-(i/\hbar)E_n t}|\varphi_n\rangle = \mathrm{e}^{-i\omega t}|\varphi_n\rangle \tag{99}$$

Expanding a state in terms of eigenstates,

$$|\psi(t)\rangle = \sum c_n|\varphi_n(t)\rangle \tag{100}$$

Since as noted previously

$$\mathrm{e}^{-(i/\hbar)\mathcal{H}t}|\varphi_n\rangle = \mathrm{e}^{-(i/\hbar)E_n t}|\varphi_n\rangle$$

we may also write

$$|\psi(t)\rangle = \mathrm{e}^{-(i/\hbar)\mathcal{H}t}\sum c_n|\varphi_n\rangle = \mathrm{e}^{-(i/\hbar)\mathcal{H}t}|\psi(0)\rangle \tag{101}$$

For example, if

$$|\psi(0)\rangle = c_1|\alpha\rangle + c_2|\beta\rangle, \qquad \mathcal{H} = -\gamma B S_z = -\omega_0 S_z \tag{102}$$

$(E_\alpha = -\frac{1}{2}\hbar\omega_0,\ E_\beta = \frac{1}{2}\hbar\omega_0)$ then

$$|\psi(t)\rangle = c_1\mathrm{e}^{i\times 1/2\omega_0 t}|\alpha\rangle + c_2\mathrm{e}^{-i\times 1/2\omega_0 t}|\beta\rangle$$

$$\langle\psi(t)| = c_1^*\mathrm{e}^{-i\times 1/2\omega_0 t}\langle\alpha| + c_2^*\mathrm{e}^{i\times 1/2\omega_0 t}\langle\beta|$$

The reader may show that $|\psi(t)\rangle$ satisfies equation (98), with $\mathcal{H} = -\gamma B S_z$. The density matrix is clearly time dependent: $\rho(t) = |\psi(t)\rangle\langle\psi(t)|$. The elements are, for example,

$$\begin{aligned}
\rho_{11} &= \langle\alpha|\rho|\alpha\rangle \\
&= \langle\alpha|\varphi(t)\rangle\langle\varphi(t)|\alpha\rangle \\
&= c_1\mathrm{e}^{i\times(1/2)\omega_0 t}c_1^*\mathrm{e}^{-i\times(1/2)\omega_0 t} = c_1c_1^* \\
\rho_{12} &= \langle\alpha|\rho|\alpha\rangle \\
&= c_1\mathrm{e}^{i\times(1/2)\omega_0 t}c_2^*\mathrm{e}^{i\times(1/2)\omega_0 t} = c_1c_2^*\mathrm{e}^{i\omega_0 t}
\end{aligned}$$

and

$$\rho(t) = \begin{pmatrix} c_1c_1^* & c_1c_2^*\mathrm{e}^{i\omega_0 t} \\ c_2c_1^*\mathrm{e}^{-i\omega_0 t} & c_2c_2^* \end{pmatrix} \tag{103}$$

from which

$$\begin{aligned}
P_x(t) &= \overline{\sigma_x(t)} = c_1c_2^*\mathrm{e}^{i\omega_0 t} + c_2c_1^*\mathrm{e}^{-i\omega_0 t} \\
P_y(t) &= \overline{\sigma_y(t)} = i(c_1c_2^*\mathrm{e}^{i\omega_0 t} - c_2c_1^*\mathrm{e}^{-i\omega_0 t}) \\
P_z(t) &= \overline{\sigma_z(t)} = c_1c_1^* - c_2c_2^*
\end{aligned} \tag{104}$$

2.2 Dynamic Response of a Spin to a Static Magnetic Field: Precession

Since $\rho(t)$ is the density matrix for the time-dependent states that satisfy the Schrodinger equation it is directly associated with the prescribed Hamiltonian, here $\mathcal{H} = -\gamma B S_z$. The associated $P_i(t)$ thus correspond to the motion of $\mathbf{P}$ or of $\boldsymbol{\mu} = \frac{1}{2}\hbar\gamma\mathbf{P}$ in the presence of the field $B \parallel OZ$.

Since the components $\parallel B$ are given by the diagonal elements, they are static; for example, if the initial state is $|\alpha\rangle, c_1 = 1, c_2 = 0$, then

$$\overline{\sigma_z(t)} = c_1c_1^* = 1 = \overline{\sigma_z(0)} \tag{105}$$

The components normal to B are given in magnitude by

$$|P_t| = (\overline{\sigma}_x^2 + \overline{\sigma}_y^2)^{1/2} = (4c_1c_1^*c_2c_2^*)^{1/2} \tag{106}$$

i.e. P_t remains constant. For initial states $|x, \pm\frac{1}{2}\rangle$, $|y, \pm\frac{1}{2}\rangle$, $P_t = 1$ while for $|\chi(0)\rangle = |\pm\frac{1}{2}\rangle$, P_t remains zero (one of the $c_i = 0$).

For $|\chi(0)\rangle = |x, \frac{1}{2}\rangle$, $c_1 = 1/\sqrt{2} = c_2$,

$$\begin{aligned} P_x(t) &= \overline{\sigma_x(t)} = \tfrac{1}{2}(e^{i\omega_0 t} + e^{-i\omega_0 t}) = \cos\omega_0 t \\ P_y(t) &= \frac{i}{2}(e^{i\omega_0 t} - e^{-i\omega_0 t}) = -\sin\omega_0 t \\ P_z(t) &= 0 \end{aligned} \tag{107}$$

and P is a vector of unit magnitude rotating at $\omega_0 = \gamma B$ as in Figure 3.14(a).

The use of the density matrix constitutes a convenience here and as an exercise the reader may study how these results could have been obtained otherwise.

The angular coordinates of **P**, and of $\boldsymbol{\mu}$ may be introduced (see Section 1.3). By choice of phase, $c_1 = c_1^* = \cos\theta/2$, $c_2 = e^{i\phi}\sin\theta/2$, and $c_2^* = e^{-i\phi}\sin\theta/2$. This facilitates the specification of any particular initial state, e.g. that defined by **P** lying at equal angles with the three coordinate axes as in Figure 3.14(b): $\cos\theta = 1/\sqrt{3}$. The reader may show that

$$c_1 = \frac{1}{\sqrt{2}}\left(1 + \frac{1}{\sqrt{3}}\right)^{1/2}, \qquad c_2 = \tfrac{1}{2}(1+i)\left(1 - \frac{1}{\sqrt{3}}\right)^{1/2}$$

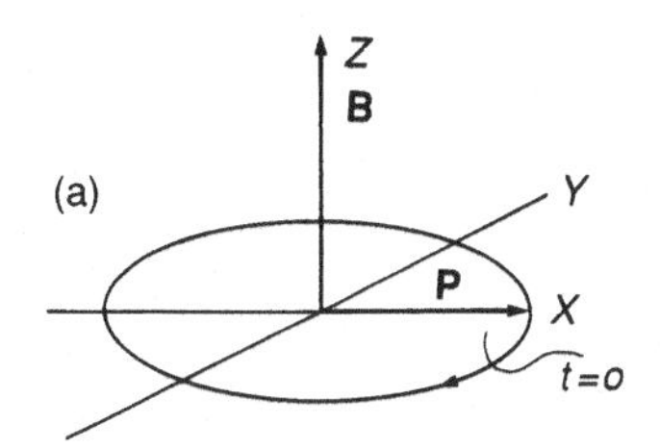

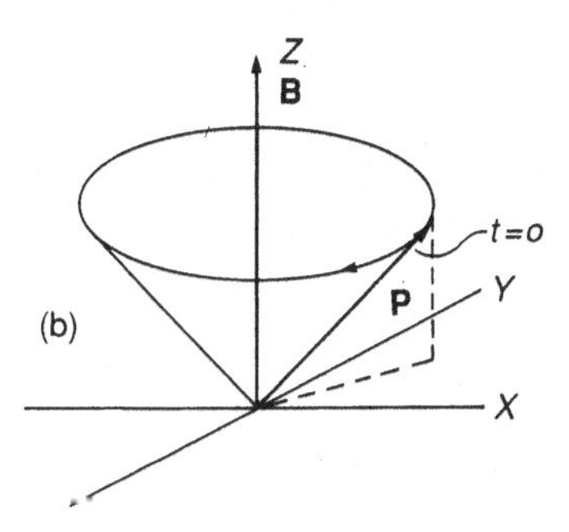

Figure 3.14 The motion of **P** according to $|(0)\rangle = |x, \frac{1}{2}\rangle$ (a) or a more general initial state (b). The magnetic moment is parallel or antiparallel to **P**

giving

$$P_x(t) = \frac{1}{\sqrt{3}}(\cos\omega_0 t + \sin\omega_0 t)$$
$$P_y(t) = \frac{1}{\sqrt{3}}(\cos\omega_0 t - \sin\omega_0 t) \tag{108}$$
$$P_z(t) = \frac{1}{\sqrt{3}}$$

$P_x(t)$ and $P_y(t)$ are thus the components that would be given by two vectors of length $1/\sqrt{3}$ and initially $\parallel$ OX and OY, rotating at ω_0, projected on to OX and OY, i.e. they are the projections of a vector $\mathbf{P}_t$ of length $\sqrt{2/3}$ initially at 45° to OX and OY and rotating in OXY. P_t is the projection of $\mathbf{P}$ on OXY and as noted above is of constant magnitude.

Thus the response of $\mathbf{P}$ or $\boldsymbol{\mu}$ to an applied field $\mathbf{B}$ consists of rotation on a cone at a constant angle to $\mathbf{B}$, i.e. to classical precession. For an assembly of non-interacting spins no magnetization is induced and the energy remains constant.

Generalizing, $\rho(0)$ in θ, ϕ is

$$\rho(0) = \begin{pmatrix} \cos^2\dfrac{\theta}{2} & e^{i\phi}\sin\dfrac{\theta}{2}\cos\dfrac{\theta}{2} \\ e^{i\phi}\sin\dfrac{\theta}{2}\cos\dfrac{\theta}{2} & \sin^2\dfrac{\theta}{2} \end{pmatrix} = \begin{pmatrix} \frac{1}{2}(1+\cos\theta) & \frac{1}{2}e^{-i\phi}\sin\theta \\ \frac{1}{2}e^{i\phi}\sin\theta & \frac{1}{2}(1-\cos\theta) \end{pmatrix} \tag{109}$$

and the representation in $|\alpha\rangle$ and $|\beta\rangle$ of $\rho(t) = |\chi(t)\rangle\langle\chi(t)|$, with the replacements for the c_i giving $|\chi(t)\rangle$ as

$$|\chi(t)\rangle = \cos\frac{\theta}{2}e^{i\times(1/2)\omega_0 t}|\alpha\rangle + \sin\frac{\theta}{2}e^{i(\phi-(1/2)\omega_0 t)}|\beta\rangle$$

is easily seen to be

$$\rho(t) = \begin{pmatrix} \frac{1}{2}(1+\cos\theta) & \frac{1}{2}e^{i(\omega_0 t-\phi)}\sin\theta \\ \frac{1}{2}e^{-i(\omega_0 t-\phi)}\sin\theta & \frac{1}{2}(1-\cos\theta) \end{pmatrix} \tag{110}$$

from which

$$P_x = \cos(\omega_0 t - \phi)\sin\theta$$
$$P_y = -\sin(\omega_0 t - \phi)\sin\theta \tag{111}$$
$$P_z = \cos\theta$$

This constitutes a general illustration of precession for any state characterized, initially, by particular values of θ and ϕ. The simplicity of the treatment depends on the assumption that $\mathcal{H}$ does not depend on time.

2.3 Time Evolution Operators

If a state is $|\psi(0)\rangle$ at an arbitrary $t = 0$ and becomes $|\psi(t)\rangle$ at time t there must be some operator, a time evolution operator U or $U(t)$, say, that transforms one state into the other as

$$|\psi(t)\rangle = U|\psi(0)\rangle, \qquad \langle\psi(t)| = \langle\psi(0)|U^\dagger \tag{112}$$

In the example $|\psi(t)\rangle = c_1 e^{-iE_1t/\hbar}|1\rangle + c_2 e^{-iE_2t/\hbar}|2\rangle$ it has been seen that although the exponential is not common it can be replaced by $e^{-i\mathcal{H}t/\hbar}$ if $|i\rangle = |E_i\rangle$, since this 'picks out' each different E_i and factorization is then possible:

$$|\psi(t)\rangle = e^{-i\mathcal{H}t/\hbar}\sum c_n|\varphi_n\rangle = e^{-i\mathcal{H}t/\hbar}|\psi(0)\rangle$$

so that the time evolution operator is, with the restriction noted,

$$U(t) = e^{-i\mathcal{H}t/\hbar}, \qquad \text{when } \mathcal{H} \neq \mathcal{H}(t) \tag{113}$$

The Schrodinger equation becomes

$$i\hbar\frac{\partial U(t)}{\partial t}|\psi(0)\rangle = \mathcal{H}U(t)|\psi(0)\rangle \tag{114}$$

and since $|\psi(0)\rangle$ is general the operator equation is

$$i\hbar\frac{\partial U(t)}{\partial t} = \mathcal{H}U(t), \qquad -i\hbar\frac{\partial U(t)^\dagger}{\partial t} = U(t)^\dagger\mathcal{H}^\dagger = U(t)^\dagger\mathcal{H} \tag{115}$$

Multiplying by $U^\dagger$ or by U:

$$i\hbar U^\dagger\frac{\partial u}{\partial t} = U^\dagger\mathcal{H}U, \qquad -i\hbar\frac{\partial U^\dagger}{\partial t}U = U^\dagger\mathcal{H}U$$

and so

$$i\hbar\left(U^\dagger\frac{\partial U}{\partial t} + \frac{\partial U^\dagger}{\partial t}U\right) = i\hbar\frac{\partial}{\partial t}(U^\dagger U) = 0$$

Thus $U^\dagger U$ is constant in time, as is suggested by $U = e^{-i\mathcal{H}t/\hbar}$, $U^\dagger = e^{+i\mathcal{H}t/\hbar}$. A state $U^\dagger|\psi(t)\rangle = U^\dagger U|\psi(0)\rangle$ ceases to be dependent on time but remains as $|\psi(0)\rangle$, and so it can be taken that $U^\dagger$ 'counteracts' U, i.e. that $U^\dagger U = 1$. The inverse of an operator or matrix is defined by $U^{-1}U = 1$ and so $U^\dagger = U^{-1}$ and U is said to be unitary.

2.4 The Liouville Equation

For generality ρ may be taken to be for a mixed state so that $\rho(0) = \sum W_n|\psi_n(0)\rangle\langle\psi_n(0)|$ and

$$\begin{aligned}\rho(t) &= \sum W_n|\psi_n(t)\rangle\langle\psi_n(t)| \\ &= \sum W_n U|\psi_n(0)\rangle\langle\psi_n(0)|U^\dagger \\ &= U\rho(0)U^\dagger\end{aligned} \tag{116}$$

Differentiating by parts with $\partial\rho(0)/\partial t = 0$,

$$\begin{aligned}i\hbar\frac{\partial\rho(t)}{\partial t} &= i\hbar\frac{\partial U}{\partial t}\rho(0)U^+ + i\hbar U\rho(0)\frac{\partial U^\dagger}{\partial t} \\ &= \mathcal{H}U\rho(0)U^\dagger - U\rho(0)U^\dagger\mathcal{H} = \mathcal{H}\rho(t) - \rho(t)\mathcal{H}\end{aligned}$$

where $i\hbar\partial U/\partial t = \mathcal{H}u$ has been used. Thus

$$i\hbar\frac{\partial\rho(t)}{\partial t} = [\mathcal{H}, \rho(t)] \tag{117}$$

which is the Liouville equation.

For example, using matrices in $|\alpha\rangle$ and $|\beta\rangle$ and with ρ that for a general spin state and $\mathcal{H} = -\gamma B\frac{1}{2}\hbar\left(\begin{smallmatrix}1 & 0\\ 0 & -1\end{smallmatrix}\right)$:

$$\mathcal{H}\rho(t) = -\gamma B(\tfrac{1}{2}\hbar)\begin{pmatrix}1 & 0\\ 0 & -1\end{pmatrix}\begin{pmatrix}\frac{1}{2}(1+\cos\theta) & e^{i(\omega t-\phi)}\times\frac{1}{2}\sin\theta\\ e^{-i(\omega t-\phi)}\times\frac{1}{2}\sin\theta & \frac{1}{2}(1-\cos\theta)\end{pmatrix}$$

and it is readily demonstrated that

$$\frac{\partial\rho(t)}{\partial t} = \frac{1}{i\hbar}[\mathcal{H}, \rho(t)] = \frac{-\gamma B}{2i}\begin{pmatrix}0 & e^{i(\omega t-\phi)}\\ e^{-i(\omega t-\phi)} & 0\end{pmatrix}$$

giving, with $\omega_0 = \gamma B$,

$$\begin{aligned}\overline{\dot{\sigma}_x(t)} &= -\omega_0\sin(\omega_0 t-\phi)\sin\theta\\ \overline{\dot{\sigma}_y(t)} &= -\omega_0\cos(\omega_0 t-\phi)\sin\theta\\ \overline{\dot{\sigma}_z(t)} &= 0\end{aligned} \tag{118}$$

as an alternative demonstration of precession, the field being implicitly $\parallel OZ$.

For a general field direction

$$\mathcal{H} = -\boldsymbol{\mu}\cdot\mathbf{B} = -\tfrac{1}{2}\hbar\gamma\sum\sigma_j B_j \qquad \text{for } j = x, y, z \tag{119}$$

Using $\overline{\sigma_i} = \operatorname{Tr}\rho\sigma_i$,

$$\begin{aligned}i\hbar\frac{\partial\overline{\sigma_i}}{\partial t} &= i\hbar\frac{\partial}{\partial t}\operatorname{Tr}\rho\sigma_i = \operatorname{Tr}[\mathcal{H},\rho]\sigma_i\\ &= \operatorname{Tr}(\mathcal{H}\rho\sigma_i - \rho\mathcal{H}\sigma_i) = \operatorname{Tr}[\sigma_i,\mathcal{H}]\rho\\ &= -\tfrac{1}{2}\hbar\gamma\sum_j B_j\operatorname{Tr}[\sigma_i,\sigma_j]\rho\end{aligned} \tag{120}$$

Thus

$$\dot{\overline{\sigma}}_x = -\frac{\gamma}{2i}B_y\operatorname{Tr}[\sigma_x,\sigma_y]\rho + B_z\operatorname{Tr}[\sigma_x,\sigma_z]\rho$$

With $[\sigma_x,\sigma_y] = 2i\sigma_z$, etc.,

$$\begin{aligned}\dot{\overline{\sigma}}_x &= -\gamma(B_y\operatorname{Tr}\sigma_z\rho - B_z\operatorname{Tr}\sigma_y\rho)\\ &= -\gamma(B_y\overline{\sigma_z} - B_z\overline{\sigma_y})\end{aligned}$$

With similar results which the reader may derive for $\dot{P}_y = \dot{\overline{\sigma}}_y$ and $\dot{P}_z = \dot{\overline{\sigma}}_z$ the three component equations combine to

$$\frac{d\mathbf{P}}{dt} = \gamma\mathbf{P}\times\mathbf{B}, \qquad \frac{d\boldsymbol{\mu}}{dt} = \gamma\boldsymbol{\mu}\times\mathbf{B}, \qquad \frac{dM}{dt} = \gamma\mathbf{M}\times\mathbf{B} \tag{121}$$

the second corresponding to an implicit factor and the third following when it is recalled that ρ may be that for a mixed state. This is the general equation of motion of the magnetization in an applied field in the absence of damping (because $\mathcal{H}$ contains no terms that might be associated with damping or relaxation).

2.5 Time-Dependent Hamiltonian

In a common situation

$$\mathcal{H} = \mathcal{H}(t) = \mathcal{H}_0 + V(t) \tag{122}$$

as, for example, $\mathcal{H}_0 = -\gamma B S_z$ with $V(t)$ representing the effect of an oscillating or r.f. field. Assume that $|\psi_n(t)\rangle$ may be expanded in the eigenstates of $\mathcal{H}_0$, $\mathcal{H}_0|\varphi_n\rangle = E_n|\varphi_n\rangle$, but not that the coefficients have the simple time dependence obtained previously, i.e. that

$$|\psi_n(t)\rangle = \sum_n c_n(t)|\varphi_n(t)\rangle = \sum_n c_n(t)\mathrm{e}^{(-i/\hbar)E_n t}|\varphi_n\rangle \tag{123}$$

where $(|\varphi_n\rangle \equiv |\varphi_n(0)\rangle$. To study the evolution of ρ in the $|\varphi_n\rangle$ basis, note that

$$\mathcal{H}_{ij} = \langle\varphi_i|\mathcal{H}_0|\varphi_j\rangle + \langle\varphi_i|V|\varphi_j\rangle = E_j^0 S_{ij} + \langle\varphi_i|V|\varphi_j\rangle$$

and bracketing the Liouville equation between $\langle i|$ and $|j\rangle$,

$$\begin{aligned} i\hbar\frac{\partial\rho_{ij}}{\partial t} &= \sum_n \left(E_i^0\delta_{in}\rho_{nj} + \langle\varphi_i|V|\varphi_n\rangle\rho_{nj} - \rho_{in}E_j\delta_{nj} - \rho_{in}\langle\varphi_n|V|\varphi_j\rangle\right) \\ &= (E_i - E_j)\rho_{ij} + \sum_n \left(\langle\varphi_i|V|\varphi_n\rangle\rho_{nj} - \rho_{ij}\langle\varphi_n|V|\varphi_n\rangle\right) \end{aligned}$$

or, equivalently,

$$i\hbar\frac{\partial\rho_{ij}}{\partial t} = (E_i - E_j)\rho_{ij} + \langle\varphi_i|[V,\rho]|\varphi_j\rangle \tag{124}$$

For $n = 2$ with the matrix $\mathcal{H}_0 = \begin{pmatrix} E_1 & 0 \\ 0 & E_2 \end{pmatrix}$

$$\mathcal{H}_0\rho = \begin{pmatrix} E_1 & 0 \\ 0 & E_2 \end{pmatrix}\begin{pmatrix} \rho_{11} & \rho_{12} \\ \rho_{21} & \rho_{22} \end{pmatrix} = \begin{pmatrix} E_1\rho_{11} & E_1\rho_{12} \\ E_2\rho_{21} & E_2\rho_{22} \end{pmatrix}$$

and

$$[\mathcal{H}_0, \rho] = \begin{pmatrix} 0 & |E_1^0 - E_2^0)\rho_{12} \\ (E_2^0 - E_1^0)\rho_{21} & 0 \end{pmatrix} \tag{125}$$

Since $i\hbar\partial\rho/\partial t = [\mathcal{H}_0, \rho] + [V, \rho]$, the last term remains to be studied.

2.6 Interactions

Considering $\mathcal{H} = \mathcal{H}_0 + V$, with typically $\mathcal{H}_0 = -\gamma B S_z$ and $V \equiv V(t)$ corresponding to the interaction of the system, e.g. spin, with an r.f. field, if V were zero the time evolution operator would be

$$U_0 = \mathrm{e}^{-i\mathcal{H}_0 t/\hbar}, \qquad U_0^\dagger = \mathrm{e}^{+i\mathcal{H}_0 t/\hbar}, \qquad \text{with } U_0^\dagger U_0 = 1 \tag{126}$$

defining U_0. Note that U_0 is not the time evolution operator in the general case $V \neq 0$. If the state $|\psi(t)\rangle$ of equation (123) is alternatively written as

$$|\psi(t)\rangle = \mathrm{e}^{-i\mathcal{H}_0 t/\hbar} \sum c_n(t)|\varphi_n\rangle \tag{127}$$

for the case that the $|\varphi_n\rangle$ are eigenstates of $\mathcal{H}_0$, then this can be expressed as

$$|\psi(t)\rangle = \mathrm{e}^{-i\mathcal{H}_0 t/\hbar}|\psi_\mathrm{I}(t)\rangle, \qquad \text{where } |\psi_\mathrm{I}(t)\rangle = \sum c_n(t)|\varphi_n\rangle$$

i.e.

$$\begin{aligned} |\psi(t)\rangle &= U_0|\psi_\mathrm{I}(t)\rangle, \qquad U_0 = \mathrm{e}^{-i\mathcal{H}_0 t/\hbar} \\ |\psi_\mathrm{I}(t)\rangle &= U_0^\dagger|\psi(t)\rangle, \qquad U_0^\dagger = \mathrm{e}^{i\mathcal{H}_0 t/\hbar} \end{aligned} \tag{128}$$

The second follows because $U_0^\dagger U = 1$ (operate on the first with $U_0^\dagger$). Note that with $t = 0$, $U_0 = 1$ so that $|\psi(0)\rangle = |\psi_\mathrm{I}(0)\rangle$. $|\psi_\mathrm{I}(t)\rangle$ is said to be the state in the interaction picture and for consistency $|\psi(t)\rangle$ is said to be that in the Schrodinger picture.

2.6.1 Interaction picture: $V = 0$

The interaction picture is one in which the evolution of the state in time differs from that in the original, Schrodinger, picture. With this, generalization is a valid concept even when the interaction disappears. In fact, when $V = 0$, with each $|\varphi_n(t)\rangle = \mathrm{e}^{-iE_n t/\hbar}|\varphi_n\rangle = \mathrm{e}^{-i\mathcal{H}_0 t/\hbar}|\varphi_n\rangle$, $|\varphi_n\rangle \equiv |\varphi_n(0)\rangle$ and $|\psi(t)\rangle = \mathrm{e}^{-i\mathcal{H}_0 t/\hbar}\sum c_n|\varphi_n\rangle$ with the c_n independent of time, i.e. $|\psi(t)\rangle = U_0|\psi(0)\rangle$:

$$|\psi_\mathrm{I}(t)\rangle = U_0^\dagger|\psi(t)\rangle = U_0^\dagger U_0|\psi(0)\rangle = |\psi(0)\rangle \tag{129}$$

and the time evolution would be totally suppressed. More briefly, when $\mathcal{H} \equiv \mathcal{H}_0$, U_0 becomes the time evolution operator and $U_0^\dagger$ suppresses the evolution since $U_0^\dagger U_0 = 1$.

Suppose, for example, $\mathcal{H} = \mathcal{H}_0 = -\gamma B_0 S_z = -\omega_0 S_z$, $|\chi(0)\rangle = c_1|\alpha\rangle + c_2|\beta\rangle$ so that

$$\begin{aligned} |\chi(t)\rangle &= U_0|\chi(0)\rangle = \mathrm{e}^{-i\mathcal{H}_0 t/\hbar}|\chi(0)\rangle \\ &= c_1\mathrm{e}^{i(\omega_0/2)t}|\alpha\rangle + c_2\mathrm{e}^{-i(\omega_0/2)t}|\beta\rangle \end{aligned} \tag{130}$$

and

$$\begin{aligned} P_x(t) &= \overline{\sigma_x(t)} = \langle\chi(t)|\sigma_x|\chi(t)\rangle \\ &= c_1^* c_2 \mathrm{e}^{-i\omega_0 t} + c_2^* c_1 \mathrm{e}^{i\omega_0 t} \end{aligned} \tag{131}$$

recalling $\sigma_x|\alpha\rangle = |\beta\rangle$, $\sigma_x|\beta\rangle = |\alpha\rangle$ and that $\mathrm{e}^{ix} \to \mathrm{e}^{-ix}$ in the adjoint. If, for example, $|\chi(0)\rangle = |x, \frac{1}{2}\rangle$, $P_x(t) = \cos\omega_0 t$, $P_y(t)$ and $P_z(t)$ follow similarly and precession in $\mathbf{B}_0$ is readily demonstrated by use of the time evolution operator. In the interaction picture for this case $U_0^\dagger|\chi(t)\rangle$ simply remains as $|\chi(0)\rangle$ and the P^i are

$$P_\mathrm{I}^i = \langle\chi_\mathrm{I}(t)|\sigma_i|\chi_\mathrm{I}(t)\rangle = \langle\chi(0)|\sigma_i|\chi(0)\rangle \tag{132}$$

i.e. simply the values assigned to the initial state. Thus the literal interaction picture is one in which $\mathbf{P}$ remains stationary (for $\mathcal{H} \equiv \mathcal{H}_0$). The precession in the Schrodinger picture, occurs at angular frequency $\omega_0 = \gamma B_0$, where B_0 is introduced specifically via the Hamiltonian and the absence of this motion corresponds to setting $B_0 = 0$ in $\mathcal{H}_0$. Thus it may reasonably be said that in the interaction picture the field B_0 is zero whatever other fields may be introduced via V.

The observation that **P** remains stationary is that which would be made by an observer located in a coordinate frame $X'Y'Z'$ rotating at ω_0 about OZ. Further, this observer would infer that no field existed, since no precession was observed. Thus the transformation to the interaction picture may be taken to be illustrated by the introduction of a rotating frame, one rotating at ω_0 about OZ. U_0 might be redesignated U_{ω_0} and thought of as introducing this rotation of the coordinates.

2.6.2 Interaction picture: $V(t) \neq 0$

Recall that in writing $|\psi(t)\rangle = \mathrm{e}^{-i\mathcal{H}_0 t/\hbar}\sum c_n(t)|\varphi_n\rangle$ it was only acceptable to take the c_n as time independent $[c_n(t) \to c_n]$ because this could be specifically justified for the case $V = 0$. Assuming that in the general case the coefficients may be functions of time,

$$|\chi_{\mathrm{I}}(t)\rangle = U_0^{\dagger}|\chi(t)\rangle = \sum c_n(t)|\varphi_n\rangle \tag{133}$$

remains time dependent.

For the case $\mathcal{H} = \mathcal{H}_0 + V$,

$$\begin{aligned} i\hbar\frac{\partial|\chi(t)\rangle}{\partial t} &= i\hbar\frac{\partial}{\partial t}\mathrm{e}^{-i\mathcal{H}_0 t/\hbar}|\chi_{\mathrm{I}}(t)\rangle \\ &= i\hbar\left(\frac{-i\mathcal{H}_0}{\hbar}\mathrm{e}^{-i\mathcal{H}_0 t/\hbar}|\chi_{\mathrm{I}}(t)\rangle + \mathrm{e}^{-i\mathcal{H}_0 t/\hbar}\frac{\partial|\chi_{\mathrm{I}}(t)\rangle}{\partial t}\right) \\ &= (\mathcal{H}_0 + V)|\chi(t)\rangle = (\mathcal{H}_0 + V)\mathrm{e}^{-i\mathcal{H}_0 t/\hbar}|\chi_{\mathrm{I}}(t)\rangle \end{aligned} \tag{134}$$

the last line in accordance with the Schrodinger equation. Thus, from the last two lines,

$$i\hbar\mathrm{e}^{-i\mathcal{H}_0 t/\hbar}\frac{\partial|\chi_{\mathrm{I}}(t)\rangle}{\partial t} = V\mathrm{e}^{-i\mathcal{H}_0 t/\hbar}|\chi_{\mathrm{I}}(t)\rangle \tag{135}$$

or

$$i\hbar U_0\frac{\partial|\chi_{\mathrm{I}}(t)\rangle}{\partial t} = VU_0|\chi_{\mathrm{I}}(t)\rangle$$

Multiplying by $U_0^{\dagger}$,

$$i\hbar U_0^{\dagger}U_0\frac{\partial|\chi_{\mathrm{I}}(t)\rangle}{\partial t} = U_0^{\dagger}VU_0|\chi_{\mathrm{I}}(t)\rangle$$

i.e

$$i\hbar\frac{\partial|\chi_{\mathrm{I}}(t)\rangle}{\partial t} = V_{\mathrm{I}}|\chi_{\mathrm{I}}(t)\rangle, \qquad \text{with } V_{\mathrm{I}} = U_0^{\dagger}VU_0 \tag{136}$$

Thus the time dependence of $|\chi_{\mathrm{I}}(t)\rangle$ depends wholly on V and is governed by a particular form of the Schrodinger equation, which does not involve V as such but the transform $U_0^{\dagger}VU_0$.

The density operator in the interaction representation is

$$\rho_{\mathrm{I}}(t) = \sum W_n|\chi_{\mathrm{I}}^n(t)\rangle\langle\chi_{\mathrm{I}}^n(t)| = \sum W_n U_0^{\dagger}|\chi_n(t)\rangle\langle\chi_n(t)|U_0 = U_0^{\dagger}\rho(t)U_0 \tag{137}$$

as for other operators. Also $\rho(t) = \sum W_n U|\chi(0)\rangle\langle\chi(0)|U^\dagger$, noting the occurrence of $U = e^{-i\mathcal{H}t/\hbar}$, with $\mathcal{H} = (t)$, not U_0, and so

$$\rho_I(t) = U_0^\dagger U\rho(0)U^\dagger U_0$$

Taking

$$U_I = e^{-i\mathcal{H}_0 t/\hbar}U = U_0^\dagger U, \qquad U_I^\dagger = U^\dagger U_0$$

gives

$$\rho_I(t) = U_I\rho_I(0)U_I^\dagger \tag{138}$$

since $\rho_I(0) = \rho(0)$.

Differentiating $\rho(t) = U\rho(0)U^\dagger$ [equation (116)],

$$\begin{aligned} i\hbar\frac{\partial\rho(t)}{\partial t} &= i\hbar\frac{\partial U}{\partial t}\rho(0)U^\dagger + i\hbar U\rho(0)\frac{\partial U^\dagger}{\partial t} \\ &= \mathcal{H}U\rho(0)U^\dagger - U\rho(0)U^\dagger\mathcal{H} \\ &= \mathcal{H}\rho(t) - \rho(t)\mathcal{H} = [\mathcal{H}, \rho] \end{aligned} \tag{139}$$

obtaining the Liouville equation by utilizing the time evolution operator. (Note that $[\mathcal{H}, U] = 0 = [\mathcal{H}, U^\dagger]$ since U, $U^\dagger$ can be expanded as polynomials in $\mathcal{H}$.) Substituting $\rho = U_0\rho_I U$,

$$i\hbar\frac{\partial}{\partial t}U_0\rho_I(t)U_0^\dagger = \mathcal{H}U_0\rho_I(t)U_0^\dagger - U_0\rho_I(t)U_0^\dagger\mathcal{H}$$

and also

$$\begin{aligned} i\hbar\frac{\partial}{\partial t}U_0\rho_I(t)U_0^\dagger &= i\hbar\left[\frac{-i\mathcal{H}_0}{\hbar}U_0\rho_I(t)U^\dagger + U_0\frac{\partial}{\partial t}\rho_I(t)U_0^\dagger + U_0\rho_I(t)\left(\frac{i\mathcal{H}_0}{\hbar}\right)U^\dagger\right] \\ &= \mathcal{H}_0\rho(t) + i\hbar U_0\frac{\partial\rho_I(t)}{\partial t} + \rho(t)\mathcal{H}_0 \end{aligned}$$

This is also equal to $(\mathcal{H}_0 + V)\rho - \rho(\mathcal{H}_0 + V)$ and so

$$i\hbar U_0\frac{\partial\rho_I(t)}{\partial t}U_0^\dagger = V\rho - \rho V$$

Multiplying by $U_0^\dagger$ and by U and inserting $U_0U_0^\dagger = 1$,

$$\begin{aligned} i\hbar\frac{\partial\rho_I(t)}{\partial t} &= U_0^\dagger V\rho U - U_0^\dagger\rho VU = (U_0^\dagger VU_0)(U_0^\dagger\rho U_0) - (U_0^\dagger\rho U_0)(U_0^\dagger VU) \\ &= [V_I, \rho_I(t)] \end{aligned} \tag{140}$$

which might be considered as a natural result in view of the transformed Schrodinger equation.

In the situation of current interest $V = V(t)$ introduces an oscillating or, initially, a rotating field transverse to B_0:

$$\mathcal{H} = -\gamma B_0S_z - \gamma B_1\cos\omega_0tS_x - \gamma B_1\sin\omega_0tS_y \tag{141}$$

The magnitude of the field is B_1 and $\mathbf{B}_1$ rotates in OXY with angular frequency ω_0. Take

$$|\chi_I\rangle = \sum c_n(t)|\varphi_n\rangle = c_1(t)|\alpha\rangle + c_2(t)|\beta\rangle$$

so that the equation of motion is

$$i\hbar\frac{\partial c_1}{\partial t}|\alpha\rangle + i\hbar\frac{\partial c_2}{\partial t}|\beta\rangle = V_I c_1|\alpha\rangle + V_I c_2|\beta\rangle$$

Taking products with $\langle\alpha|$ and with $\langle\beta|$,

$$i\hbar\frac{\partial c_1}{\partial t} = c_1\langle\alpha|V_I|\alpha\rangle + c_2\langle\alpha|V_I|\beta\rangle$$

$$i\hbar\frac{\partial c_2}{\partial t} = c_1\langle\beta|V_I|\alpha\rangle + c_2\langle\beta|V_I|\beta\rangle$$

Noting that $U_0|\alpha\rangle = e^{i\omega_0 t/2}|\alpha\rangle$, $U_0|\beta\rangle = e^{-i\omega_0 t/2}|\beta\rangle$, as before,

$$\begin{aligned} V_I|\alpha\rangle &= e^{i\mathcal{H}_0 t/\hbar} V U_0|\alpha\rangle = e^{i\mathcal{H}_0 t/\hbar}[-\gamma B_1 \cos\omega_0 t(\tfrac{1}{2}\hbar) - \gamma B_1 \sin\omega_0 t(\tfrac{1}{2}i\hbar)]e^{i\omega_0 t/2}|\beta\rangle \\ &= -\tfrac{1}{2}\hbar\gamma B_1 e^{i\omega_0 t/2}(\cos\omega_0 t + i\sin\omega_0 t)e^{i\omega_0 t/2}|\beta\rangle \\ &= -\tfrac{1}{2}\hbar\gamma B_1 e^{i\times 2\omega_0 t}|\beta\rangle \end{aligned} \tag{142}$$

This corresponds to one choice of V corresponding to a particular sense of rotation for $\mathbf{B}_1$. If this is reversed, i.e. $\sin\omega_0 t \to -\sin\omega_0 t$, $\cos\omega_0 t - i\sin\omega_0 t = e^{-i\omega_0 t}$ so that the exponentials are reciprocal and the time variable disappears,

$$\begin{aligned} V_I|\alpha\rangle &= -\tfrac{1}{2}\hbar\gamma B_1|\beta\rangle = -\tfrac{1}{2}\hbar\omega_1|\beta\rangle \\ V_I|\beta\rangle &= -\tfrac{1}{2}\hbar\gamma B_1|\alpha\rangle = -\tfrac{1}{2}\hbar\omega_1|\alpha\rangle \qquad \text{with } \omega_1 = \gamma B_1 \end{aligned} \tag{143}$$

the second following similarly. The matrix is thus

$$V_I = \begin{pmatrix} 0 & -\frac{1}{2}\hbar\omega_1 \\ -\frac{1}{2}\hbar\omega_1 & 0 \end{pmatrix} \tag{144}$$

and

$$i\hbar\frac{\partial c_1}{\partial t} = -\tfrac{1}{2}\hbar\omega_1 c_2$$

$$i\hbar\frac{\partial c_2}{\partial t} = -\tfrac{1}{2}\hbar\omega_1 c_1$$

$$\frac{d^2 c_1}{dt^2} = -\left(\frac{\omega_1}{2}\right)^2 c_1$$

so the solutions are

$$c_1 = b_1\cos\frac{\omega_1 t}{2} + b_2\sin\frac{\omega_1 t}{2}$$

$$c_2 = ib_1\sin\frac{\omega_1 t}{2} - b_2\cos\frac{\omega_1 t}{2}$$

$$[b_1 = c_1(0), \qquad b_2 = c_2(0)]$$

For example, for $|\chi(0)\rangle = |\chi_I(0)\rangle = |\alpha\rangle$, $b_1 = 1$, $b_2 = 0$:

$$|\chi_I\rangle = \cos\frac{\omega_1 t}{2}|\alpha\rangle + i\sin\frac{\omega_1 t}{2}|\beta\rangle = c_1|\alpha\rangle + c_2|\beta\rangle$$

$$\begin{aligned} P_{x'}(t) &= c_1 c_2^* + c_2 c_1^* = 0 \\ P_{y'}(t) &= i(c_1 c_2^* - c_2 c_1^*) = \sin\omega_1 t \\ P_{z'}(t) &= c_1 c_1^* - c_2 c_2^* = \cos^2\frac{\omega_1 t}{2} - \sin^2\frac{\omega_1 t}{2} = \cos\omega_1 t \end{aligned} \tag{145}$$

(The primes on the subscripts stress that these correspond to the time development of $|\chi_I\rangle$, not $|\chi\rangle$.) The reader may check that these are as given by $P_i = \langle\chi_I|\sigma_i|\chi_I\rangle$.

The motion in the interaction picture is a rotation at $\omega_1 = \gamma B_1$ in $OY'Z$, as in Figure 3.15(a). The prime on Y' is a reminder that this is not the true motion in the coordinate frame defined by the Hamiltonian but in the simplified picture illustrated by a rotating frame. The true motion is that which accords with the time development of $|\chi(t)\rangle$. For the example,

$$|\chi\rangle = U_0|\chi_I\rangle = e^{-i\mathcal{H}_0 t/\hbar}|\chi_I\rangle = e^{i\omega_0 t/2}\cos\frac{\omega_1 t}{2}|\alpha\rangle + ie^{-i\omega_0 t/2}\sin\frac{\omega_1 t}{2}|\beta\rangle$$

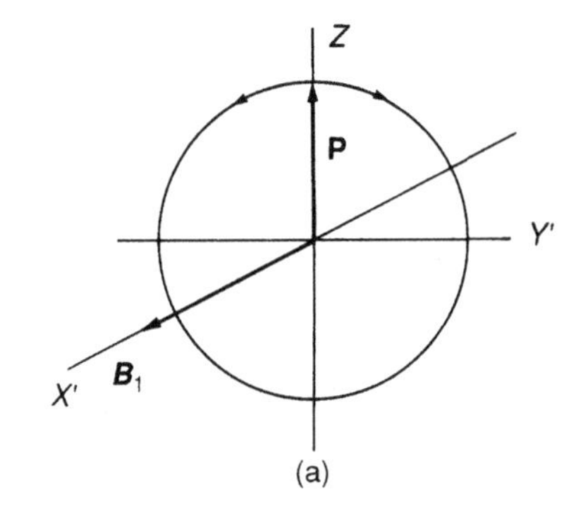

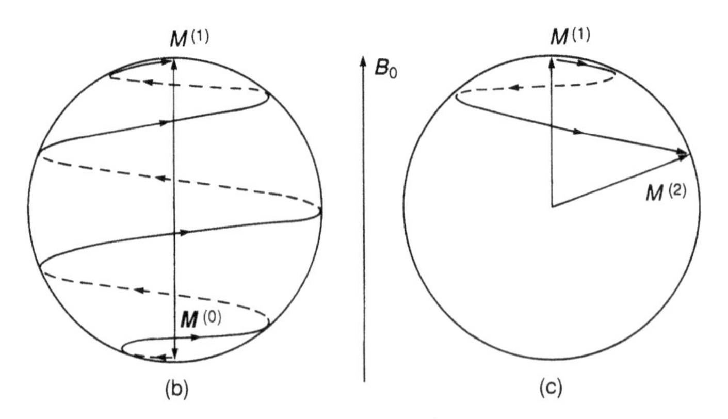

Figure 3.15 (a) Motion of the polarization or magnetization in the interaction picture, effectively in a field $\mathbf{B}_1$ along OX'. (b) Undamped motion in the real frame; in appropriate fields **M** may approach the static field direction but in the absence of damping it then proceeds towards$-$**B**, as in (c), and so forth. The sense accords with the sign of γ

$$
\begin{aligned}
P_x &= \langle \chi | \sigma_x | \chi \rangle \\
&= \left(e^{-i\omega_0 t/2} \cos\frac{\omega_1 t}{2} \langle\alpha| - i e^{i\omega_0 t/2} \sin\frac{\omega_1 t}{2} \langle\beta| \right) \left(e^{i\omega_0 t/2} \cos\frac{\omega_1 t}{2} |\beta\rangle - i e^{-i\omega_0 t/2} \sin\frac{\omega_1 t}{2} |\alpha\rangle \right) \\
&= i(e^{-i\omega_0 t} - e^{i\omega_0 t})(\tfrac{1}{2}) \sin\omega_1 t \\
&= \sin\omega_0 t \sin\omega_1 t
\end{aligned} \tag{146}
$$

This is equivalent to redefining c_1 and c_2:

$$
c_1 = e^{i\omega_0 t/2} \cos\frac{\omega_1 t}{2}, \qquad c_2 = i e^{-i\omega_0 t/2} \sin\frac{\omega_1 t}{2}
$$

$$
P_x = c_1 c_2^* + c_2 c_1^*
$$

Similarly,

$$
\begin{aligned}
P_y &= \cos\omega_0 t \sin\omega_1 t \\
P_z &= \cos\omega_1 t \\
|P_t| &= (P_x^2 + P_y^2)^{1/2} = \sin\omega_1 t
\end{aligned} \tag{147}
$$

In the physical situation of interest it is taken that $B_1 < B_0$, i.e. $\omega_1 < \omega_0$. In fact, $\omega_1 \ll \omega_0$ applies in a resonance experiment with B_0 a large uniform field (the larger the better to increase splittings and enhance resolution and sensitivity) and B_1 limited technically since the frequencies involved are in the megahertz or the microwave regions and are also limited to avoid saturation of the signals, as will be seen later. Purely for the sake of illustration, assume that $\omega_0 = 4\omega_1 \cdot P_t$, the transverse component, goes from 0 to 1 in magnitude as $\omega_1 t$ goes from 0 to $\pi/2$, and when $\omega_1 t = \pi/2$, $\omega_0 t = 2\pi$, $P_x = 0$ and $P_y = 1$. Continuing, $P_z = -1$ at $\omega_1 t = \pi$ and the motion is as shown in Figure 3.15(b). At $\omega_1 t = 2\pi$, **P** is directed once more in its original direction, and so forth.

These results may equally be obtained using the density matrix. For $|\chi(0)\rangle = |\alpha\rangle$ the coefficients in the expansion $|\chi(t)\rangle = c_1(t)|\alpha\rangle + c_2(t)|\beta\rangle$ are $c_1 = \cos\omega_1 t/2$ and $c_2 = i\sin\omega_1 t/2$, so that

$$
\rho_\mathrm{I}(t) = \begin{pmatrix} \frac{1}{2}(1+\cos\omega_1 t) & -\frac{1}{2} i \sin\omega_1 t \\ \frac{1}{2} i \sin\omega_1 t & \frac{1}{2}(1-\cos\omega_1 t) \end{pmatrix} \tag{148}
$$

giving immediately, for the mean values in the interaction picture,

$$
\overline{\sigma_\mathrm{I}^x} = 0, \qquad \overline{\sigma_\mathrm{I}^y} = \sin\omega_1 t, \qquad \overline{\sigma_\mathrm{I}^z} = \cos\omega_1 t \tag{149}
$$

For the Schrodinger picture, with $|\chi(t)\rangle = U_0|\chi_\mathrm{I}(t)\rangle = e^{i\omega_0 t/2}\cos\omega_1 t/2|\alpha\rangle + i e^{-i\omega_0 t/2}\sin\omega_1 t|\beta\rangle$,

$$
\begin{aligned}
\rho(t) &= \begin{pmatrix} \cos^2\frac{\omega_1 t}{2} & -i e^{i\omega_0 t}\sin\frac{\omega_1 t}{2}\cos\frac{\omega_1 t}{2} \\ i e^{-i\omega_0 t}\sin\frac{\omega_1 t}{2}\cos\frac{\omega_1 t}{2} & \sin^2\frac{\omega_1 t}{2} \end{pmatrix} \\
&= \begin{pmatrix} \frac{1}{2}(1+\cos\omega_1 t) & -\frac{1}{2} i e^{i\omega_0 t}\sin\omega_1 t \\ \frac{1}{2} i e^{-i\omega_0 t}\sin\omega_1 t & \frac{1}{2}(1-\cos\omega_1 t) \end{pmatrix}
\end{aligned} \tag{150}
$$

and the $P_i(t)$ follow as before.

As μ is either parallel (proton) or antiparallel (electron) to $\mathbf{P}$ the motion of $\mathbf{P}$ thus represents the response of a magnetic moment to the combination of the two fields, i.e. of a single spin moment or of the magnetic moment or magnetization $\mathbf{M}$ of an assembly. In the latter case, as discussed later, some spin–lattice relaxation mechanisms must be present so that, in the presence of B_0, an equilibrium magnetization $\parallel B_0$ may be established at a rate characterized by a rate constant, which is $1/T_1$ with T_1 the spin–lattice relaxation time. However, the motion is specifically undamped or free of relaxation and it must be assumed that the relaxations are arbitrarily 'switched off' during the study or, more realistically, that the total time over which the motion is studied is much less than T_1 (since the rate of the destruction of the magnetization vector is similarly controlled by T_1).

The motion in the interaction picture is again seen to be that obtained by taking the real motion, in the Schrodinger picture, and observing it in a frame rotating at ω_0.

If $B_1(t)$ is applied at $t = 0$ and removed at $t = \pi/(2\omega_1)$ then $P_z = 0$ and $P_t = 1$, i.e. $\mathbf{P}$ and μ or $\mathbf{M}$ remain in the transverse plane, rotating at a frequency $f_0 = \omega_0/(2\pi)$ and inducing a signal in an appropriate pickup coil, such as the coil used to generate $B_1(t)$. Such a '90° pulse' gives rise to a free induction signal (i.e. one recorded free of B_1) at this frequency f_0 and apparently, so far, with no attenuation.

In the real case, as noted, the spins in the system interact with other spins or with other degrees of freedom of the surroundings or 'lattice' so as to come into thermal equilibrium in relation to the energy levels in the field B_0, the distribution in the system then matching that in the lattice (i.e. the second system or reservoir with which energy is exchanged). At this stage the system is said to have a spin temperature that is equal to that of the lattice. If an initial state is random, $\mathbf{P} = 0$ and $\mathbf{M} = 0$, then at the instant the field is applied it is as if the spin temperature were very high since only for $T \to \infty$ would $\mathbf{P} = 0$ be expected in the presence of B_0; the approach to equilibrium then corresponds to cooling to the temperature of the lattice, as the energy in the field is reduced. For $T_1 \to 0$ equilibrium is established almost instantaneously and, conversely, the absence of relaxation or any approach to equilibrium (or decay of an initial equilibrium magnetization) corresponds to $T_1 \to \infty$. The transverse magnetization, $M_\mathrm{t} = M_0$, constitutes a high-energy state and with a short T_1, M_t should break up while a new equilibrium $\mathbf{M}_0 \parallel \mathbf{B}_0$ becomes established. The equilibrium density matrix is diagonal so that $P_x = 0 = P_y$, i.e. there are no transverse components at this stage and the state is said to be incoherent. Even if T_1 is very long, spin–spin relaxations characterized by a second relaxation time T_2 lead to loss of coherence, the decay of M_t and thus the decay of the free induction signal.

2.6.3 The interaction picture and the rotating frame

To review, in the case that $\mathcal{H} = \mathcal{H}_0 + V(t) \to \mathcal{H}_0$, independent of time, the time-evolution operator is $U = e^{-i\mathcal{H}_0 t/\hbar}$: $[|\chi(t)\rangle = U|\chi(0)\rangle]$. It is then the same as U_0 which gives the transform to the interaction picture as $U_0^\dagger|\chi(t)\rangle = |\chi_\mathrm{I}(t)\rangle$, where $U_0 = e^{-i\mathcal{H}_0 t/\hbar}$. Operating on $|\chi(t)\rangle = U|\chi(0)\rangle$ by $U_0^\dagger$ gives $U^\dagger|\chi(t)\rangle = |\chi_\mathrm{I}(t)\rangle = U_0^\dagger U|\chi(0)\rangle$ so that $|\chi_\mathrm{I}(t)\rangle$ ceases to evolve in time so long as $U = U_0$ and $U_0^\dagger U = U_0^\dagger U_0 = 1$. The $P_i = \langle\chi_\mathrm{I}|\sigma_i|\chi_\mathrm{I}\rangle = \langle\chi(0)|\sigma_i|\chi(0)\rangle$ maintain the values corresponding to the initial state and $\mathbf{P}^\mathrm{I}$ remains static. In fact P_t, transverse to the implied field direction ($\mathcal{H}_0 = -\gamma B_0 S_z = -\omega_0 S_z$), rotates at ω_0 so the transformation to the interaction picture is equivalent to the introduction of a rotation of the coordinates around OZ, at ω_0. In view of the form of $\mathcal{H}$ the transform could be designated $U_{\omega_0}^{\dagger}$ and formally:

$$U_0^{\dagger} = U_{\omega_0}^{\dagger} = e^{-i\omega_0 S_z t/\hbar} \sim \begin{cases} \text{coordinate rotation at } \omega_0, \text{ about } OZ, \text{ with} \\ \mathbf{P} \text{ stationary in } X'Y'Z \end{cases}$$

A transform may equally be defined as $U_{\omega}^{\dagger} = e^{-i\omega S_z t/\hbar}$ with $\omega \neq \omega_0$ necessarily. In the presence of B_0 the time evolution of the transformed states is now

$$\begin{aligned} |\chi_{\omega}(t)\rangle &= UU_{\omega}^{\dagger}|\chi(0)\rangle = e^{i\omega_0 S_z t/\hbar} e^{-i\omega S_z t/\hbar}|\chi(0)\rangle \\ &= e^{i(\omega_0-\omega)S_z t/\hbar}|\chi(0)\rangle \\ &= e^{i\gamma b_0 S_z t/\hbar}|\chi(0)\rangle \end{aligned} \tag{151}$$

The effective time evolution operator for the transformed states is $e^{i\gamma b_0 S_z}$ so the effective Hamiltonian governing the motion is $\mathcal{H} = -\gamma b_0 S_z$ and the effective field is $b_0 = (\omega_0 - \omega)/\gamma$, i.e.

$$b_0 = B_0 - \frac{\omega}{\gamma} = \begin{cases} B_0 & (\omega = 0) \\ 0 & (\omega = \omega_0) \\ -B_0 & (\omega = 2\omega_0) \end{cases} \tag{152}$$

When $V(t) \neq 0$ but depends on a rotating field $B_1(t)$, the time-evolution operator does not have the above simple expression. Nevertheless, the analysis has shown that the evolution of the states transformed by $U_0^{\dagger} = U_{\omega_0}^{\dagger} = e^{i\mathcal{H}_0 t/\hbar}$ ceases to depend on $\mathcal{H}_0$, as if $B_0 \to 0$ once more and it can be seen that $U_{\omega}^{\dagger}$ effectively reduces the z field to b_0. However, these are not the only effects of the transformations. It has been seen that development of the transformed states remains dependent on $V(t)$, or more specifically is controlled by $V_{\mathrm{I}} = U_0^{\dagger} V U_0$. In the case of present interest the reader may show that $V(t)$ can be written as $-(\omega_1/2)(S^+ e^{i\omega_0 t} + S^- e^{-i\omega_0 t})$ and thus, for example, that $e^{-i\omega_0 S_z t/\hbar} V(t) e^{i\omega_0 S_z t/\hbar}|\alpha\rangle = -\hbar\omega_1/2|\beta\rangle$ (commutation is not to be assumed). Thus, in effect $V_{\mathrm{I}} = -\omega_1 S_x = -\gamma B_1 S_x$ is independent of time and corresponds to the presence of a field $\mathbf{B}_1$ which is static and directed along OX, or rather OX' in the rotating frame since the field vector in fact rotates at ω_0. If ω_0 is replaced by a general ω it is clear that it is $U_{\omega}^{\dagger} V(t) U_{\omega}$ that represents an effectively static field and thus corresponds to a rotation of the coordinates at ω. Thus, formally again,

$$U_{\omega}^{\dagger} \sim \begin{cases} \text{coordinate rotation about } OZ, \text{ at } \omega \\ \text{field } b_0 = B_0 - \omega/\gamma \parallel OZ \\ \text{field } B_1 \parallel OX \end{cases}$$

The motion of $\mathbf{P}$ or $\mathbf{M}$ in the 'ω frame' is thus precession about the total field $\mathbf{B} = \mathbf{i}B_1 + \mathbf{k}b_0$, as in Figure 3.16. Motion in the static frame is now, clearly, rather complicated.

If ω is varied from 0 to $2\omega_0$ then $\mathbf{b}$ rotates from $+OZ$ to $-OZ$. If ω is varied very slowly then the precession occurs over a cone, around $\mathbf{b}$, with a very small angle, and this can cause $\mathbf{M}$ to reverse adiabatically, with no exchange of energy. This remarkable adiabatic reversal contrasts with that caused by reversing $\mathbf{B}_0$ so as to destroy $\mathbf{M}$ and in due course, due to relaxations, recreate $\mathbf{M}$ along $-OZ$.

It is only when $\omega = \omega_0$ that $\mathbf{M}$ reverses completely at a rate given by $\omega_1 = \gamma B_1$, with a large periodic change in the energy—$\mathbf{M} \cdot \mathbf{B}$; otherwise the motion is more restricted. In the complete absence of relaxations such considerations cannot affect the (non-existent) energy flow, but if it is assumed that weak spin–lattice interactions only cause limited modifications

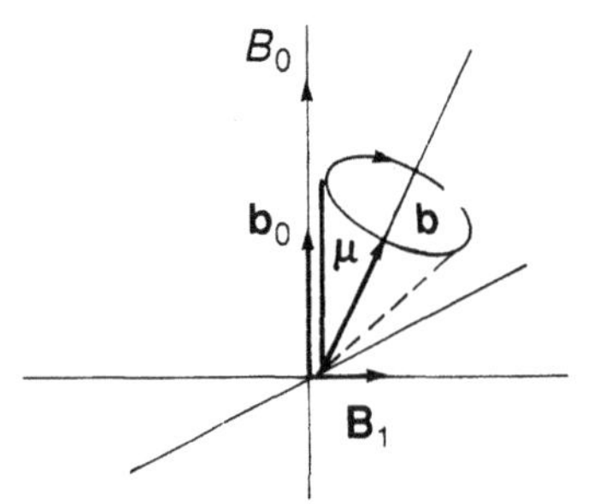

Figure 3.16 Motion in the interaction picture, in the total field

of the free motion then it is to be expected that the absorption of energy should be greatest at $\omega = \omega_0$. This gives a description of the occurrence of resonance peaks, corresponding to continuous wave spectroscopy, which is validated by more convincing but elaborate treatments, as later.

It has been noted that spin–lattice relaxations (also known as longitudinal relaxations since they most directly affect the creation of the magnetization along the B direction) are expected to attenuate the free induction signal since they simultaneously lead to the destruction of the transverse magnetization components. This free induction decay is related to the absorption signal by Fourier transformation and a rapid decay corresponds to a large absorption linewidth and vice versa (see Section 2.8). However, other effects remain to be considered, as follows.

2.7 Spin–Spin or Transverse Relaxation: T_2

If B_0 is changed to some value B then, following a 90° pulse, say, **M** would remain static in a picture corresponding to the transformation $U_\omega^\dagger = e^{-i\omega S_z t/\hbar}$, i.e. in a frame rotating at $\omega = \gamma B$. Thus in the ω_0 frame, **M** would rotate in $OX'Y'$ at an angular frequency $\omega_0 - \omega$. Suppose the static field is nominally B_0 but that experimental field inhomogeneities exist. For the sake of illustration take B_0 to apply over one-third of the spins constituting a magnetization $M_1 = M_0/3$ while $B = B_0 + \delta B$ over another third, M_2, and $B = B_0 - \delta B$ otherwise (M_3). Now, following a 90° pulse (taken to apply to M_0 although it cannot apply precisely to all three groups), only M_1 remains static in the ω_0 frame while M_2 and M_3 rotate in opposite senses at angular velocity $\omega = \gamma \delta B$ as in Figure 3.17d. Recalling that all these vectors are, together, rotating at ω_0 in the static frame containing the axis of a pickup coil, the signals from the different vectors lose their phase coherency and the signal falls. Alternatively expressed, the signal falls as the net transverse magnetization falls [Figure 3.17(d)]. The fall can be shown to be approximately exponential and is characterized by a rate constant $1/T_2$ or relaxation time T_2 (or T_2^*, to distinguish this particular contribution to the transverse relaxation). In reality a continuous distribution of magnetic moments should replace the three $\mathbf{M}_i$ in the illustration.

Next suppose that, at a time τ after the 90° pulse, a 180° pulse is applied, i.e. a short but intense pulse with $\mathbf{B}_1$ (at ω_0) along OY', such that the three $\mathbf{M}_i$ are rotated through 180° about OY'. The remarkable result is that, after the pulse, the three $\mathbf{M}_i$ vectors, assumed to remain in their same local fields and thus rotating at their original ω's, are now approaching each other rather than diverging. Since the rate of approach to a common direction is equal to the rate of the original divergence, they should coalesce to fully restore the original signal

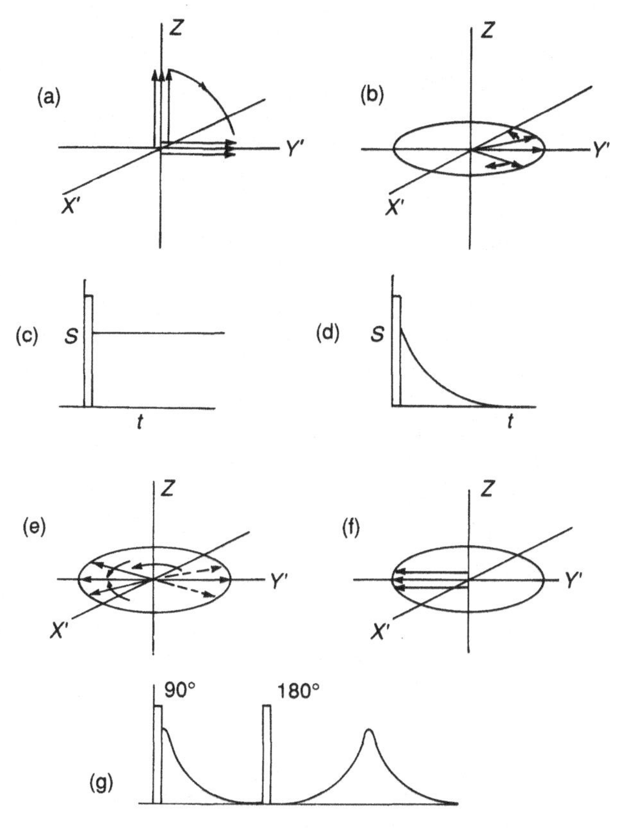

Figure 3.17 (a) A 90° pulse, $B_1 \parallel OX'$, (b) the fanning out in the interaction picture due to different local fields, (c) the ideal (uniform field) signal and (d) the actual signal as the net transverse M decays, (e) the 180° pulse leading to coalescence (f), and (g) the reappearance of the (echo) signal (cf. Figure 3.18)

after a further time interval τ. This so-called echo signal is also illustrated by Figure 3.17. Following the echo the magnetizations fan out once more but can again be re-clustered by a further 180° pulse and so forth. Initially the initial and echo signals may be expected to be of equal height, but a decrease is generally observed, according to the following discussion. Spin echoes are widely utilized in e.s.r. and particularly in n.m.r., though the interest has yet to be made clear.

Even if $\mathbf{B}_0$ were ideally uniform there would still be a decay of the signal (generally slower than that occurring in practice due to the field inhomogeneities) associated with magnetic fields varying in both magnitude and direction from the moments of neighbouring atoms or molecules undergoing Brownian or different types of motion. Due to the rapid rates of motion and consequent field variations these can be viewed equivalently in the static and rotating frames. They cause the spin vectors constituting the transverse magnetization to move in the rotating frame in such a way that their points of intersection diffuse out over a sphere in a random walk manner. (It is recalled that if a number of initially coincident points undergo repeated random displacements there is no net translation of the assembly but in time the points spread out in space.) According to a simplified model the local field h at a given nucleus changes abruptly once every τ_c seconds to a new randomly selected magnitude and direction, while having a certain average value. The local fields at different nucleii vary similarly but without any correlation. It was shown by Carr and Purcell [10] that the transverse magnetization should then decay exponentially:

$$M_{y'}(t) = M_0 e^{-\gamma^2 \langle h^2 \rangle \tau_c t/3} \tag{153}$$

with $\langle \quad \rangle$ denoting the average. This is now taken to apply to the three arbitrary magnetizations M_i in the three different static fields representing the inhomogeneity, so that these decrease exponentially while they fan out and coalesce to give the echo [Figure 3.18(a)]. The random divergence due to the spin–spin interactions is not reversed to give coalescence but continues. Thus if an echo is recorded at $2\tau_1$ and in new experiments at $2 \times (2\tau_1)$, $2 \times (3\tau_1)$, etc., the echo heights obtained should decrease exponentially at a rate given by $1/T_2$. This is sometimes called the natural T_2 to distinguish it from the apparent spin–spin (T_2) effects, which are in fact due to static field inhomogeneities (though still validly regarded as transverse relaxation effects). This corresponded to observations on viscous samples but not to those on water. In water the spins do not reside in regions of particular d.c. field strength but diffuse through these regions; Hahn showed that an e^{-kt^3} dependence was then to be

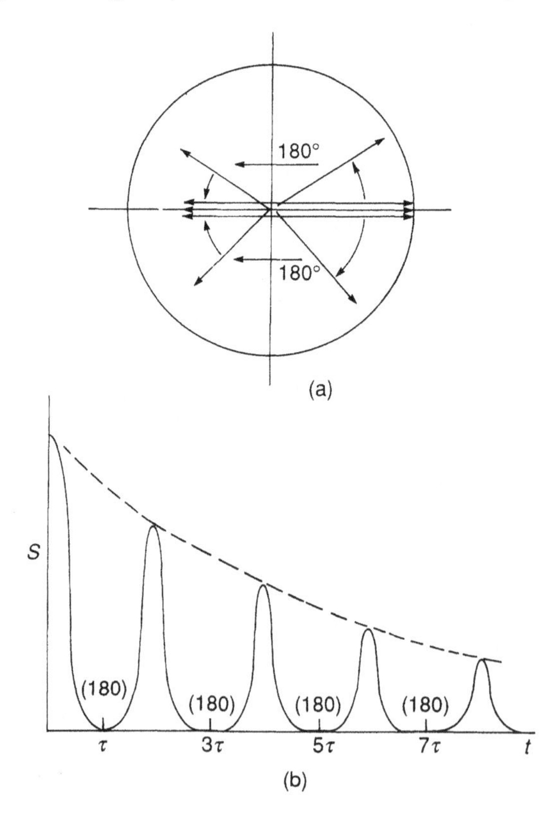

Figure 3.18 Due to spin–spin interactions the arbitrarily separated magnetization vectors decay (a) and successive echoes decrease in height (b)

expected [11]. Carr and Purcell [10] developed the echo method in which, in a single experiment, a 90° pulse was applied at $t = 0$, a 180° pulse at $t = \tau$ with the echo at 2τ, a further 180° pulse at 3τ giving an echo at 4τ and so forth, as in Figure 3.18b. It was shown that these (n) repeated pulses tended to counteract the effects associated with diffusion and the transverse magnetization decay was given by

$$M_{y'}(t) = \mathrm{e}^{-t/T_2 + [-\gamma^2 G^2 D t^3/(12n^2)]} \tag{154}$$

with D the diffusion coefficient and G the gradient of $\mathbf{B}$, in the field direction in which the spins move. With n sufficiently great the exponential decay follows, giving the natural T_2 as 2.0 s for protons in water compared with 0.2 s derived from k^{-1} as above.

If the accidental field gradients are minimized and swamped by deliberate substantial linear field gradients it becomes possible to measure the self-diffusion coefficients. The importance of such measurements in relation to general molecular kinetics or dynamids is apparent. Carr and Purcell [10] calculated the transient response and the absorption line shape (Fourier transform of the transient) expected for particular geometrical homogeneous distributions of spins, the converse procedure to that constituting magnetic resonance imaging (see Section 7, Chapter 6).

2.8 Fourier Transforms

While a Fourier series gives the frequency domain representation of a periodic function, the frequency domain representations of non-periodic functions are given by Fourier transforms with the Fourier transform pair

$$\begin{aligned} F(\nu) &= \int_R f(t)\mathrm{e}^{-2\pi i \nu t}\,\mathrm{d}t, \qquad & F(\nu) &= \mathcal{F}[f(t)] \\ f(t) &= \int_R F(\nu)\mathrm{e}^{2\pi i \nu t}\,\mathrm{d}t, \qquad & f(t) &= \mathcal{F}^{-1}[F(\nu)] \end{aligned} \tag{155}$$

(using ν for frequency). The functions must exist over the entire real line $R = (-\infty, \infty)$. For example, for a one-sided exponential,

$$f(t) = \begin{cases} \mathrm{e}^{-kt}, & t \geqslant 0 \\ 0, & t < 0 \end{cases} \tag{156}$$

it follows that

$$\begin{aligned} F(\nu) &= \int_0^\infty \mathrm{e}^{-(k+2\pi i\nu)t}\,\mathrm{d}t = \frac{-1}{k+2\pi i\nu}\mathrm{e}^{-(k+2\pi i\nu)t}\Big|_0^\infty \\ &= \frac{1}{k+2\pi i\nu} = \frac{k-2\pi i\nu}{k^2+4\pi^2\nu^2} = \frac{k-i\omega}{\omega^2+k^2} \end{aligned} \tag{157}$$

The real and imaginary parts are sketched in Figure 3.19. When the exponential is the envelope of a signal having the frequency ω_0 the effect is to replace ω by $\omega_0 - \omega$ in $F(\omega)$ and it will be seen in Section 2.10 that the real part corresponds to the absorption signal, with the replacement $k = 1/T_2$. The absorption signals or spectra and the f.i.d. signals are interrelated by Fourier transformation and it is technically advantageous to record the f.i.d.'s repetitively and then transform by automatic numerical means, as first realized by

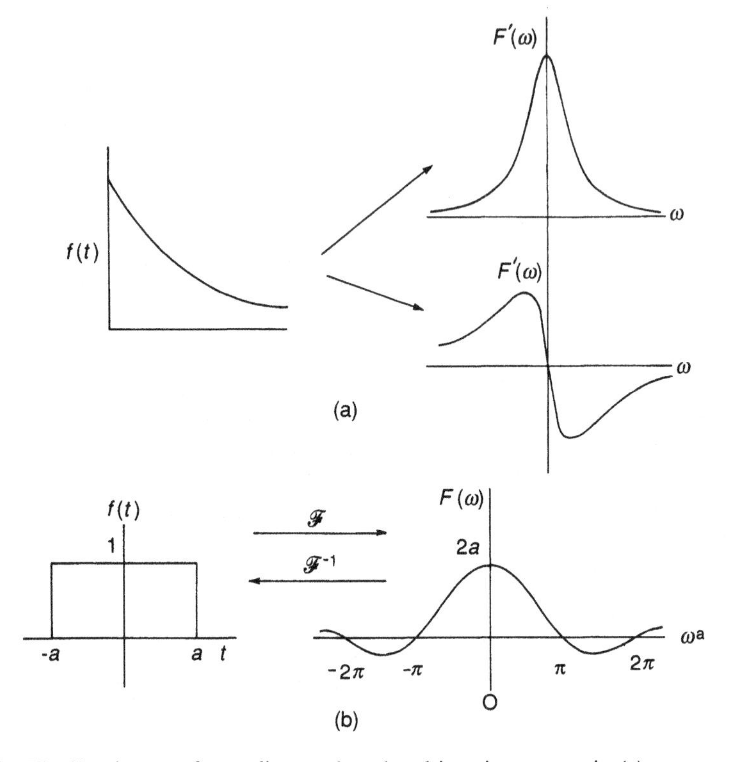

Figure 3.19 The Fourier transforms discussed, real and imaginary parts in (a)

Ernst and Anderson [12]. The real and imaginary parts of $F(\omega) = F' - iF''$ are interrelated by a Hilbert transformation, giving the Kramers–Kronig relations:

$$F'(\omega_1) = \frac{2}{\pi}\int_0^{\infty} \frac{\omega F''(\omega)}{\omega^2 - \omega_1^2}\, d\omega + F'(\infty) \tag{158}$$

$$F''(\omega_1) = -\frac{2}{\pi}\int_0^{\infty} \frac{\omega_1 F'(\omega)}{\omega^2 - \omega_1^2}\, d\omega \tag{159}$$

where ω_1 is any chosen frequency. Usually $F'(\infty) = 0$ is assumed.

In a second example, for a pulse function

$$f(t) = \begin{cases} 1, & -a \leqslant t \leqslant a \\ 0, & \text{otherwise} \end{cases} \tag{160}$$

it follows that

$$F(\nu) = \int_{-a}^{a} e^{-2\pi i \nu t}\, dt = -\left.\frac{e^{-2\pi i \nu t}}{2\pi i \nu}\right|_{-a}^{a}$$

$$= \frac{i(e^{2\pi i \nu a} - e^{-2\pi i \nu a})}{i(2\pi i \nu)} = \frac{\sin 2\pi \nu a}{\pi \nu}$$

i.e.

$$F(\omega) = \frac{2a \sin \omega a}{\omega a} = 2a \operatorname{sinc} \omega a \tag{161}$$

with $\operatorname{sinc} x = \sin x / x$, as in Figure 3.19b. Now $F(\omega)$ is real. A short and sharply defined pulse of monochromatic radiation is equivalent to a wide range of frequencies in the frequency domain. If the pulse width is $\sim 10\ \mu s$ the spread of frequencies is $\sim 10^5$ Hz, which is adequate to cover all the chemical shifts likely to be encountered in n.m.r. so that, for example, all the protons in a sample may be simultaneously excited. This leads to a regime that is much more efficient than continuous-wave operation. Conversely, a sinc-shaped pulse of r.f. in the time domain is equivalent to the application of a rectangular band of frequencies with a continuous distribution and equal intensities, since the Fourier transform and its inverse are uniquely related. This will be seen to be of particular value in magnetic resonance imaging.

2.9 Transitions and Relaxation

For a single spin initially in $|\alpha\rangle$ and $\mathcal{H} = \mathcal{H}_0 + \mathcal{H}_1(t) + \mathcal{H}_2(t) = -\gamma B_0 S_z - \gamma B_1 \cos \omega_0 t S_x + \gamma B_1 \sin \omega_0 t S_y$ the state at time t is $|\chi(t)\rangle = e^{i\omega_0 t/2}(\cos \omega_1 t/2)|\alpha\rangle + i e^{-i\omega_0 t/2}(\sin \omega_1 t/2)|\beta\rangle$ and it is left as an exercise to show that

$$\langle \chi(t)|\mathcal{H}_0|\chi(t)\rangle = -\gamma B_0 \frac{\hbar}{2} \cos \omega_1 t$$

$$\langle \mathcal{H}_1(t)\rangle = -\gamma B_1 \cos \omega_0 t \cos \frac{\omega_1 t}{2} \sin \frac{\omega_1 t}{2} \sin \omega_0 t$$

$$\langle \mathcal{H}_2(t)\rangle = -\langle \mathcal{H}_1(t)\rangle$$

and that the energy oscillates as $E = \langle \mathcal{H}\rangle = \langle \mathcal{H}_0\rangle = -\frac{1}{2} g\beta B_0 \cos \omega_1 t$ and to note that this is the result obtained by accepting the motion already demonstrated and applying the classical $E = -\boldsymbol{\mu} \cdot \mathbf{B}$ with $\mathbf{B}$ the total field. The spin undergoes continuous transitions between the states $|\alpha\rangle$ and $|\beta\rangle$. The transitions $\alpha \to \beta$ and $\beta \to \alpha$ are induced equally by the r.f. field and there is no net exchange of energy.

During the motion, as particularly seen classically as $\mathbf{P}$ passes through the transverse plane, the precessing spin generates fields of frequency ω_0 and these may induce transitions in neighbouring spins or in other systems. A system with which the system of interest S is in contact is known as a reservoir R or lattice. R is taken to be relatively massive and to interchange energy with S while remaining in equilibrium at an unchanged temperature. For energy exchange to occur there must be some interaction with R and this should be reflected by appropriate terms in the Hamiltonian for S, or rather for $S + R$ together. The relative simplicity of the account of dynamics given above corresponded to the absence of such terms, which also indicated that the motion must be undamped.

Direct solutions covering the dynamic behaviour of R as well as S cannot be obtained. In indirect approaches R is regarded as unobserved and solutions obtained for the behaviour of S alone, though as modified by the presence of R. These are complex and lengthy and will only be described in brief; particular reference may be made to the work by Blum on density matrix theory for a complete account [3].

Since nothing further has been defined energy must be conserved within $R + S$ together. Each 'up' transition in S must be accompanied by a 'down' transition in R, which is

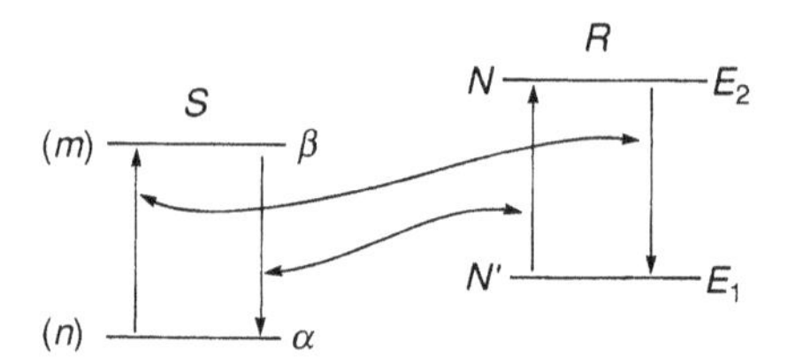

Figure 3.20 Transferring the effect of the greater population of the level E_1 in the reservoir R (at equilibrium) to the system S

assumed to have a great number of degrees of freedom. Suppose the populations n_α and n_β of the levels of Figure 3.20 were set equal. The stimulated up and down transitions of equal probability might appear to occur at equal rates, maintaining the equal populations. However, with R at equilibrium, the population of the E_1 levels indicated are greater than those of the E_2 levels and when the contact is taken into account it is seen that down transitions in S together with up transitions in R are favoured: the effect of the difference in populations of R is transferred to S. This applies until the relative populations of the α and β levels match those of E_1 and E_2 in R, i.e. the spin temperature matches the lattice or reservoir temperature.

If the interaction could be cut off, with $n_\alpha/n_\beta = e^{\Delta E/(kT)}$, then the r.f. field would cause a preponderance of $\alpha \to \beta$ transitions until the populations became equal. However, with appropriate interactions giving a short T_1, equilibrium between α and β may be maintained. The excess of absorptive, radiative, transitions is maintained, but this is accompanied by continuous spin–lattice or relaxation $\beta \to \alpha$ transitions, for a constant total population of the two levels, these latter corresponding to an increase in energy of R. Energy is absorbed from the field and passed to the reservoir. The induced transition rate increases with B_1 so this should be large to give a large absorptive signal, but if B_1 is too large, in view of the relaxation rate, the relaxation transitions may fail to 'keep up' with the induced transitions so that the populations approach equality and the signal decreases, an effect known as saturation.

Formally, the probability that a transition $|n\rangle \to |m\rangle$ occurs in an atom or for a spin together with $|N'\rangle \to |N\rangle$ in R is $\langle m\ N|V|n\ N'\rangle$, the arrow being a reminder of the direction. V represents the excitation. First-order perturbation theory gives the transition probabilities for absorption and transmission:

$$\begin{aligned} W_{fi} &= \frac{2\pi}{\hbar}\delta(E_f - E_i - \hbar\omega)|\langle f|\mathcal{H}|i\rangle|^2 \qquad \text{(absorption)} \\ W_{fi} &= \frac{2\pi}{\hbar}\delta(E_f - E_i + \hbar\omega)|\langle f|\mathcal{H}^\dagger|i\rangle|^2 \qquad \text{(emission)} \end{aligned} \tag{162}$$

applying for a harmonic interaction of the form

$$\mathcal{H} = \mathcal{H}_1 e^{-i\omega t} + \mathcal{H}_1^\dagger e^{i\omega t} \tag{163}$$

in which $\mathcal{H}_1$ is itself taken to be independent of time $\delta(E_f - E_i - \hbar\omega)$ is non-zero only if $E_f - E_i - \hbar\omega = 0$, i.e. $E_f - E_i - \hbar\omega$, the Bohr condition. For example,

$$\mathcal{H} = V = -\gamma B_1 \cos\omega_0 t S_x + \gamma B_1 \sin\omega_0 t S_y$$

$$= -\frac{\gamma B_1}{2}(e^{i\omega_0 t} + e^{-i\omega_0 t})S_x - \frac{\gamma B_1}{2} i(e^{i\omega_0 t} - e^{-i\omega_0 t})S_y$$

$$= -\frac{\gamma B_1}{2}[e^{i\omega_0 t}(S_x + iS_y) + e^{-i\omega_0 t}(S_x - iS_y)]$$

$$= -\frac{\gamma B_1}{2} S^+ e^{i\omega_0 t} - \frac{\gamma B_1}{2} S^- e^{-i\omega_0 t} \tag{164}$$

has the appropriate form since $S^- = (S^+)^*$. Using either expression the reader may show that

$$|\langle\beta|V|\alpha\rangle|^2 = \tfrac{1}{4}\hbar^2\gamma^2 B_1^2 = |\langle\alpha|V|\beta\rangle|^2 \tag{165}$$

This illustrates two results already referred to and a dependence of the signals on γ^2 is also indicated, a feature favouring nucleii such as that of hydrogen, in n.m.r.

For a general spin $S : M_s = S, S-1, \cdots (S-1), -S$, the only non-zero matrix elements of S^+ and S^- are

$$\langle m'|S^+|m'-1\rangle, \qquad \langle m'|S^-|m'+1\rangle$$

and the selection rule is thus

$$\Delta M_s = \pm 1 \tag{166}$$

In the presence of the reservoir the probabilities can be shown to become, with V_{SR} representing interaction between S and R, $\mathcal{H} = \mathcal{H}_S + \mathcal{H}_R + V_{SR}$:

$$W_{mn} = \frac{2\pi}{\hbar}\sum_{N,N'} |\langle m\ N|V_{SR}|n\ N'\rangle|^2 \langle N'|\rho_R|N'\rangle \delta(E_{N'} - E_N - \hbar\omega_{mn})$$

(for $n \to m$) recalling that $\langle m\ N|$ signifies S in $\langle m|$ and R in $\langle N|$. With R in equilibrium and ρ_R relating to R,

$$\langle N'|\rho_R|N'\rangle = \frac{e^{-E_{N'}/(kT)}}{Z}$$

and

$$W_{mn} = \frac{2\pi}{\hbar Z}\sum_{N,N'} |\langle m\ N|V_{SR}|n\ N'\rangle|^2 e^{-E_{N'}/(kT)} \delta(E_{N'} - E_N - \hbar\omega_{mn})$$

$$W_{nm} = \frac{2\pi}{\hbar Z}\sum_{N,N'} |\langle n\ N'|V_{SR}|m\ N\rangle|^2 e^{-E_N/(kT)} \delta(E_N - E_{N'} - \hbar\omega_{mn}) \tag{167}$$

The squared term is equal in these and $\omega_{mn} = \omega_{nm}$, and for energy conservation $E_N - E_{N'} = E_n - E_m$ (Figure 3.20), i.e. $E_N = E_{N'} + E_n - E_m$ so that $e^{-E_N/(kT)} = e^{-E_{N'}/(kT)} e^{-(E_n - E_m)/(kT)}$ and

$$\frac{W_{mn}}{W_{nm}} = e^{(E_n - E_m)/(kT)} = \frac{e^{-E_m/(kT)}}{e^{-E_n/(kT)}} \tag{168}$$

Thus if $E_m < E_n$, $e^{-E_m/kT} < e^{-E_n/(kT)}$, then $W_{mn} > W_{nm}$. This contrasts with $W_{mn} = W_{nm}$ in the absence of relaxation and conforms with the foregoing general arguments since with R at equilibrium the population of $|N'\rangle$ is greater than that of $|N\rangle$, and since the transitions are paired it is as if the 'effective population' of $|n\rangle$ were greater than that of $|m\rangle$. (Recall that $W_{mn} \sim n \to m$.)

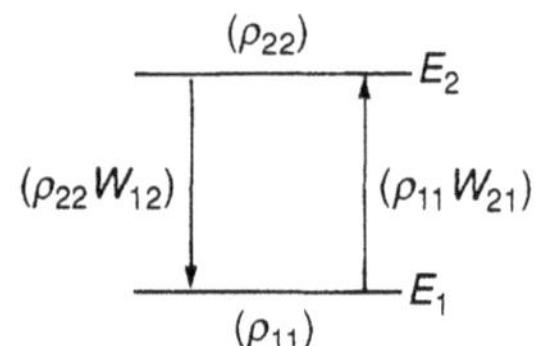

Figure 3.21 ρ_{11} and ρ_{22} give the populations of the two levels

The diagonal elements of ρ, $\rho_{mm} = c_m c_m^*$ are the probabilities or populations of $|m\rangle$ and the rate of change of the populations are $\dot{\rho}_{mm}(t)$. In terms of the W_{mn} and the populations these are given by

$$\dot{\rho}_{mm}(t) = \sum_{n \neq m} \rho_{nn}(t) W_{mn} - \rho_{mm}(t) \sum_{n \neq m} W_{nm} \tag{169}$$

e.g. $\dot{\rho}_{11}(t) = \rho_{22}(t)W_{12} - \rho_{11}(t)W_{21}$ (Figure 3.21). In the steady state, $\dot{\rho}_{11}(t) = 0 = \dot{\rho}_{22}(t)$, using equation (168):

$$\frac{\rho_{11}}{\rho_{22}} = \frac{\mathrm{e}^{-E_1/(kT)}}{\mathrm{e}^{-E_2/(kT)}} \tag{170}$$

consistent with the steady state corresponding to equilibrium.

It may be recalled here that while it has been shown that the r.f. field stimulates emission of energy as well as absorption, and spontaneous emission has been ignored, this is in fact relevant to the Herzian range of frequencies up to 10^{12} Hz (r.f. of 10^4 to 10^9 and microwaves 10^9 to 10^{12} Hz) although in the optical range spontaneous emission becomes important. If it can be contrived that the population of the higher level is greater than that of the lower, e.g. by establishing equilibrium in a field $\parallel OZ$ and rapidly reversing the field to $-OZ$, the high-energy configuration being maintained for a time substantially less than T_1, then exposure to a high-frequency field leads for a time to a net emission of radiation. This is one basis of maser action, (microwave amplification by stimulated emission of radiation). Three-level masers may also be devised.

It may also be recalled that while T_2 is most directly associated with the transverse or spin–spin relaxations and T_1 with the longitudinal or spin–lattice relaxations discussed here, the two are associated in that the transverse components are given by the off-diagonal elements of the density matrix, and these go to zero as the system approaches equilibrium. Thus any theory of spin–lattice relaxation should also relate to T_2, even if spin–spin interactions are not invoked. Clearly $T_2 \leqslant T_1$ necessarily and often $T_2 \ll T_1$ applies.

2.10 Stimulated Transitions and Steady State Absorption

An r.f. field is introduced according to $V(t) = -\gamma B_1 \cos \omega t S_x = -\frac{1}{2}\gamma B_1(\mathrm{e}^{i\omega t} + \mathrm{e}^{-i\omega t})S_x$, i.e. initially a linearly oscillating rather than a rotating field.

The Liouville equation for the density matrix elements is modified by adding the relaxation terms

$$\dot{\rho}_{m'm} = -i\omega_{m'm}\rho_{m'm} - \frac{i}{\hbar}\langle m'|[V, \rho]|m\rangle + \sum_{n,n'} R_{m'mn'n}\rho_{n'n} \tag{171}$$

with $\omega_{mm'} = (E_{m'} - E_m)/\hbar$, as discussed, for example, by Blum [3, p 177]. In the absence of V, i.e. due to relaxation only, the rates of change of the populations are, for example,

$$\dot{\rho}_{11} = W_{12}\rho_{22} - W_{21}\rho_{11}, \qquad \dot{\rho}_{22} = W_{21}\rho_{11} - W_{12}\rho_{22} \tag{172}$$

The rates of change specifically due to V are $\dot{\rho}_{mm} = -i/\hbar\langle m|[V,\rho]|m\rangle$, and summing the two contributions

$$\begin{aligned} \dot{\rho}_{11} &= -\frac{i}{\hbar}\langle 1|[V,\rho]|1\rangle + W_{12}\rho_{22} - W_{21}\rho_{11} \\ \dot{\rho}_{22} &= -\frac{i}{\hbar}\langle 2|[V,\rho]|2\rangle + W_{21}\rho_{11} - W_{12}\rho_{22} \end{aligned} \tag{173}$$

For the off-diagonals the generalized master equation for irreversible motion may be quoted (again after Blum) as

$$\dot{\rho}_{mm'} = -\frac{i}{\hbar}\langle m'|[\mathcal{H},\rho]|m\rangle + \delta_{mm'}\sum_n \langle n|\rho|n\rangle W_{mn} - \gamma_{m'm}\langle m'|\rho|m\rangle \tag{174}$$

with $\mathcal{H} = \mathcal{H}_0 + V(t)$. For example,

$$\begin{aligned} \dot{\rho}_{21} &= -\frac{i}{\hbar}\langle 2|\mathcal{H}\rho - \rho\mathcal{H}|1\rangle - \gamma_{21}\langle 2|\rho|1\rangle \qquad (= \dot{\rho}^*_{12}) \\ &= -\frac{i}{\hbar}(E_2 - E_1)\rho_{21} - \frac{i}{\hbar}\langle 2|[V(t),\rho]|1\rangle - \gamma_{21}\rho_{21} \end{aligned} \tag{175}$$

In this case take $V(t) = V\cos\omega t = (\frac{1}{2})V(\mathrm{e}^{i\omega t} + \mathrm{e}^{-i\omega t})$ which, with V appropriate, is for a linearly oscillating rather than a rotating field:

$$\dot{\rho}_{21} = -i(\omega_{21} - i\gamma_{21})\rho_{21} - \frac{i}{2\hbar}\langle 2|V|1\rangle(\mathrm{e}^{i\omega t} + \mathrm{e}^{-i\omega t})(\rho_{11} - \rho_{22}) \tag{176}$$

on evaluating the commutator $[V(t),\rho]$ and taking element 2,1. In the steady state $\dot{\rho}_{11} = 0 = \dot{\rho}_{22}$, stimulated emission and absorption being balanced by relaxation. In the interaction representation $\rho_{21} = \mathrm{e}^{-i\omega_{21}t}\rho^{\mathrm{I}}_{21}$ and

$$\begin{aligned} \dot{\rho}^{\mathrm{I}}_{21} &= -\gamma_{21}\rho^{\mathrm{I}}_{21} - \frac{i}{2\hbar}\langle 2|V|1\rangle[\mathrm{e}^{i(\omega_{21}+\omega)t} + \mathrm{e}^{i(\omega_{21}-w)t}](\rho_{11} - \rho_{22}) \\ &\to -\gamma_{21}\rho^{\mathrm{I}}_{21} - \frac{i}{2\hbar}\langle 2|V|1\rangle\mathrm{e}^{i(\omega_{21}-\omega)t}(\rho_{11} - \rho_{22}) \end{aligned} \tag{177}$$

The second line corresponds to neglecting the very high frequency term and constitutes the 'rotating wave approximation'. This appears to be universally accepted and is the justification for the use of a rotating field operator in the previous treatment, although it is a linearly oscillating field which is in fact applied. The reader may confirm that the approximation is equivalent to replacing $V(t)$ specifically by that for a rotating field.

In the steady state the density matrix elements have ceased evolving in time [note that the time dependence is implicit in the foregoing, $\dot{\rho}^{\mathrm{I}}_{21} \equiv \dot{\rho}^{\mathrm{I}}_{21}(t)$, etc.] but do depend on the frequency. Postulate

$$\rho^{\mathrm{I}}_{21}(t) = \mathrm{e}^{i(\omega_{21}-\omega)t}\rho_{21}(\omega), \qquad \rho_{21}(t) = \mathrm{e}^{-i\omega t}\rho_{21}(\omega) \tag{178}$$

the second returning to the Schrodinger picture, because on substituting in equation (177) the exponential functions of time then cancel as required:

$$\rho_{21}(\omega) = -\frac{i}{2\hbar}\langle 2|V|1\rangle\frac{\rho_{11} - \rho_{22}}{i(\omega_{21} - \omega) + \gamma_{21}} = \frac{1}{2\hbar}\langle 2|V|1\rangle\frac{\rho_{11} - \rho_{22}}{\omega - \omega_{21} + i\gamma_{21}} \tag{179}$$

Postulating $\rho_{12}(t) = e^{+i\omega t}\rho_{12}(\omega)$, it is found that

$$\rho_{12}(\omega) = \frac{1}{2\hbar}\langle 2|V|1\rangle^* \frac{\rho_{11} - \rho_{22}}{\omega - \omega_{21} - i\gamma_{21}^*} \tag{180}$$

with $\gamma_{ij} = \gamma_{ji}^*$, as seen later when the significance of γ is described. Note that the rotating wave approximation is equivalent to setting $\langle 2|V(t)|1\rangle = (\frac{1}{2})\langle 2|V|1\rangle e^{-i\omega t}$ and $\langle 1|V(t)|2\rangle = (\frac{1}{2})\langle 1|V|2\rangle e^{i\omega t}$.

For the diagonal elements,

$$\begin{aligned}\dot{\rho}_{11} &= -\frac{i}{\hbar}\langle 1|[V(t), \rho]|1\rangle + W_{12}\rho_{22} - W_{21}\rho_{11} \\ \dot{\rho}_{22} &= -\frac{i}{\hbar}\langle 2|[V(t), \rho]|2\rangle + W_{21}\rho_{11} - W_{12}\rho_{22}\end{aligned} \tag{181}$$

Noting that the matrix of V is $\begin{pmatrix} 0 & V_{12} \\ V_{21} & 0\end{pmatrix}$ and taking element 2,2 in the commutator of the matrices of V and ρ, the first term in the second of these equations, the 'radiation term', becomes

$$-\frac{i}{\hbar}V_{12}\rho_{21} + \frac{i}{\hbar}V_{21}\rho_{12} \tag{182}$$

and substituting for ρ_{21} and ρ_{22},

$$\begin{aligned}\dot{\rho}_{22}^{\text{rad}} &= -\frac{i}{2\hbar^2}|\langle 2|V|1\rangle|^2(\rho_{11} - \rho_{22})\left(\frac{1}{\omega - \omega_{21} - i\gamma_{21}^*} - \frac{1}{\omega - \omega_{21} + i\gamma_{21}}\right) \\ &= \frac{1}{\hbar^2}|V_{21}|^2(\rho_{11} - \rho_{22})\frac{\gamma_{21}'}{(\omega - \omega_{21} - \gamma_{21}'')^2 + \gamma_{21}'^2}, \qquad \text{where } \gamma_{21} = \gamma_{21}' + i\gamma_{21}''\end{aligned} \tag{183}$$

Similarly,

$$\dot{\rho}_{11}^{\text{rad}} = \frac{1}{\hbar^2}|V_{21}|^2\frac{\gamma_{21}'}{(\omega - \omega_{21} - \gamma_{21}'')^2 + \gamma_{21}'^2}(\rho_{22} - \rho_{11}) \tag{184}$$

which is equal to $-\dot{\rho}_{22}^{\text{rad}}$ as necessary to conserve numbers. Writing

$$\frac{1}{\hbar^2}|V_{21}|^2\frac{\gamma_{21}'}{(\omega - \omega_{21} - \gamma_{21}'')^2 + \gamma_{21}'^2} = W_{21}(\omega) = W_{12}(\omega) \tag{185}$$

and adding the other terms to the radiation term:

$$\begin{aligned}\dot{\rho}_{11} &= [W_{12} + W_{12}(\omega)]\rho_{22} - [W_{21} + W_{21}(\omega)]\rho_{11} = 0 \\ \dot{\rho}_{22} &= [W_{21} + W_{21}(\omega)]\rho_{11} - [W_{12} + W_{12}(\omega)]\rho_{22} = 0\end{aligned} \tag{186}$$

equating to zero for the steady state. These combine the effects of relaxation (W_{21} and W_{12}) and of the r.f. field [$W_{21}(\omega)$ and $W_{12}(\omega)$]. If B_1 is small, i.e. $|V_{21}|$ and $W_{12}(\omega)$ are small, these are consistent with $\rho_{11} \sim \rho_{11}^0$ and $\rho_{22} \sim \rho_{22}^0$, the equilibrium values, but if B_1 is large this no longer applies and the system is said to be pumped.

The power absorbed by the system from the r.f. field and transferred to the lattice is

$$\frac{dE}{dt} = E_1\dot{\rho}_{11}^{\text{rad}} + E_2\dot{\rho}_{22}^{\text{rad}} = (E_1 - E_2)\dot{\rho}_{11}^{\text{rad}} \tag{187}$$

with E_1 the lower. In the usual situation, close to equilibrium, $\rho_{11} > \rho_{22}$ and by equation (186), $\dot{\rho}_{11} < 0$ and thus $\mathrm{d}E/\mathrm{d}t > 0$. If population inversion can be achieved, $\rho_{22} > \rho_{11}$, it follows that $\mathrm{d}E/\mathrm{d}t < 0$ with net emission, as inferred previously, leading to maser and analogous laser effects.

Substituting for $\dot{\rho}_{11}^{\mathrm{rad}}$:

$$\frac{\mathrm{d}E}{\mathrm{d}t} = \frac{1}{\hbar^2}(E_2 - E_1)|V_{21}|^2 \frac{\gamma'_{21}}{(\omega - \omega_{21} - \gamma''_{21})^2 + \gamma'^2_{21}}(\rho_{11} - \rho_{22}) \tag{188}$$

with the last factor usually approximating $\rho_{11}^0 - \rho_{22}^0$. Note that γ''_{21} is very small and is usually neglected, while γ'_{21} is set equal to $1/T_2$. In the rotating wave approximation $V(t) = -\gamma B_1 \cos\omega t S_x + \gamma B_1 \sin\omega t S_y$ and the matrix in $|\alpha\rangle$ and $|\beta\rangle$ follows from

$$V(t)|\alpha\rangle = -\frac{\gamma B_1}{2}\mathrm{e}^{-i\omega t}\hbar|\beta\rangle, \qquad V(t)|\beta\rangle = -\frac{\gamma B_1}{2}\mathrm{e}^{i\omega t}\hbar|\alpha\rangle$$

i.e $V_{12} = -\frac{1}{2}\hbar\gamma B_1 \mathrm{e}^{i\omega t}$, $V_{21} = V_{12}^*$. Substituting for V_{12} and noting that

$$\rho_{11} - \rho_{22} = 2\bar{\sigma}_z = \frac{4}{\hbar}\bar{S}_z = \frac{4\mu_z}{\gamma\hbar} \tag{189}$$

which becomes $4M_z/(\gamma\hbar)$ with implicit multiplication by the number of spins per unit volume, it is easy to see that the power absorption is

$$P = \frac{\mathrm{d}E}{\mathrm{d}t} = \omega_0 M_z \gamma B_1^2 T_2 \frac{1}{(\omega - \omega_0)^2 T_2^2 + 1} \tag{190}$$

This peaks at $\omega = \omega_0$, the resonance frequency. Writing $P = Af(\omega)$, $P = A$ at $\omega = \omega_0$ and when $P = A/2$, $(\delta\omega)^2 T_2^2 = 1$ where $\delta\omega = \omega - \omega_0$. Thus the half-power level occurs at $\delta\omega = 1/T_2$ or $\delta f = 1/(2\pi T_2)$ (Figure 3.22) and the linewidth as thus defined is proportional to $1/T_2$. It can be seen by inspection that if γ''_{21} were not neglected it would give a (very small) shift as opposed to the broadening.

This section has substantially followed the approach of Blum (3), to which direct reference is recommended for a more complete account.

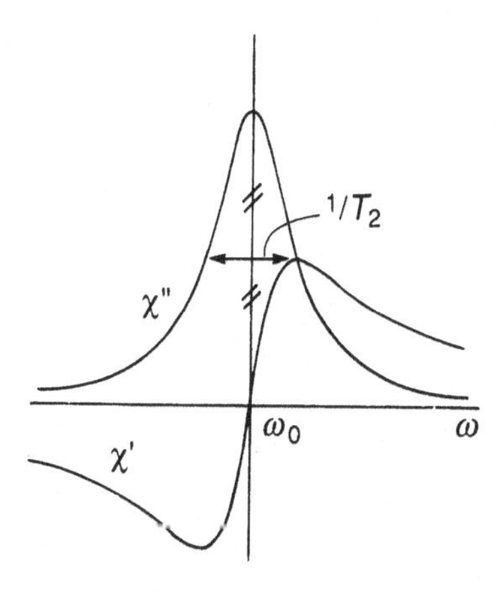

Figure 3.22 Absorption and dispersion plots, relating the linewidth to T_2

2.11 The Bloch Equations

With $V(t) = 0$ initially and noting that the sum of occupation probabilities $\rho_{11} + \rho_{22} = 1$,

$$\begin{aligned}\dot{\rho}_{11} &= W_{12}\rho_{22} - W_{21}\rho_{11} = W_{12}\rho_{22}\ (+W_{12}\rho_{11}) - W_{21}\rho_{11}\ (-W_{12}\rho_{11}) \\ &= W_{12} - (W_{21} + W_{12})\rho_{11} \\ \dot{\rho}_{22} &= W_{21} - (W_{12} + W_{21})\rho_{22}\end{aligned} \tag{191}$$

In the steady state $\rho_{ii} = \rho_{ii}^0$ for the field B_0 and $\dot{\rho}_{ii} = 0$:

$$\rho_{11}^0 - \rho_{22}^0 = \frac{W_{12} - W_{21}}{W_{12} + W_{21}} \tag{192}$$

In the general case, equation (191) gives

$$\dot{\rho}_{11} - \dot{\rho}_{22} = \frac{\rho_{11}^0 - \rho_{22}^0}{1/(W_{12} + W_{21})} - \frac{\rho_{11} - \rho_{22}}{1/(W_{12} + W_{21})} = \frac{\rho_{11}^0 - \rho_{22}^0 - (\rho_{11} - \rho_{22})}{T_1} \tag{193}$$

if we set

$$T_1 = \frac{1}{W_{12} + W_{21}} \tag{194}$$

In the absence of relaxation $W_{12} = -W_{21}$, consistent with $T_1 \to \infty$. Associating ρ_{11}, ρ_{22} with populations and with an implied factor:

$$\frac{\mathrm{d}M_z}{\mathrm{d}t} = \frac{M_z^0 - M_z(t)}{T_1} \tag{195}$$

Introducing the rotating field,

$$\begin{aligned}\dot{\rho}_{11} - \dot{\rho}_{22} &= \frac{2i}{\hbar}(V_{12}\rho_{21} - V_{21}\rho_{12}) + \frac{M_z^0 - M_z}{T_1} \\ &= i\gamma B_1(\mathrm{e}^{i\omega t}\rho_{21} - \mathrm{e}^{-i\omega t}\rho_{12}) + \frac{M_z^0 - M_z}{T_1}\end{aligned} \tag{196}$$

Consider $B_1(t) = B_1 \cos \omega t - B_1 \sin \omega t = B_x + B_y$, i.e.

$$B_x = B_1(\tfrac{1}{2})(\mathrm{e}^{i\omega t} + \mathrm{e}^{-i\omega t}), \qquad B_y = -B_1\left(\frac{1}{2i}\right)(\mathrm{e}^{i\omega t} - \mathrm{e}^{-i\omega t})$$

Then

$$\begin{aligned}\gamma[\mathbf{M}\times\mathbf{B}_1]_z &= \gamma[M_x B_y - M_y B_x] = N\gamma(\tfrac{1}{2}\hbar\gamma\bar{\sigma}_x B_y - \tfrac{1}{2}\hbar\gamma\bar{\sigma}_y B_x) \\ &= \tfrac{1}{2}\hbar^2\gamma^2[(\rho_{12} + \rho_{21})B_y - i(\rho_{12} - \rho_{21})B_x]\end{aligned} \tag{197}$$

Comparing this with equation (196) it is clear that

$$\frac{\mathrm{d}M_z}{\mathrm{d}t} = \gamma[\mathbf{M}\times\mathbf{B}_1(t)]_z + \frac{M_z^0 - M_z}{T_1} \tag{198}$$

Also

$$\dot{P}_x = \dot{\rho}_{21} + \dot{\rho}_{12} = \frac{-\rho_{12} + \rho_{21}}{T_2} + \frac{i}{\hbar}(i\hbar\gamma B_y)(\rho_{11} - \rho_{22}) + i\omega_{12}(\rho_{12} - \rho_{21}) \tag{199}$$

$$\dot{P}_y = i(\dot{\rho}_{12} - \dot{\rho}_{21})$$

$$= i(i\omega_{12})(\rho_{12} + \rho_{21}) + i\left[\frac{i}{\hbar}(-\hbar\gamma B_x)\right](\rho_{11} - \rho_{22})\frac{-i}{T_2}(\rho_{12} - \rho_{21}) \tag{200}$$

($\gamma'_{12} = 1/T_2$) i.e., with $\omega_{21} = \gamma B_0$,

$$\dot{P}_x = \gamma B_0 P_y - \gamma B_y P_z - \frac{P_x}{T_2}$$

$$\dot{P}_y = -\gamma B_0 P_x + \gamma B_x P_z - \frac{P_y}{T_2}$$

or $(\times\frac{1}{2}\hbar\gamma N)$

$$\dot{M}_x = \gamma[\mathbf{M}\times\mathbf{B}]_x - \frac{M_x}{T_2}$$

$$\dot{M}_y = \gamma[\mathbf{M}\times\mathbf{B}]_y - \frac{M_y}{T_2} \tag{201}$$

$$\dot{M}_z = \gamma[\mathbf{M}\times\mathbf{B}]_z + \frac{M_z^0 - M_z}{T_1}$$

which are the famous Bloch equations [13], with $\mathbf{B}$ the total field, $\mathbf{B}_0 + \mathbf{B}_1(t)$. These can be solved but there is little point in doing so since the main feature of such solutions, the nature of the resonant absorption, has already been demonstrated.

2.12 The Relaxation Terms

It is recalled that $T_1 = 1/(W_{12} + W_{21})$ and $T_2 = 1/\gamma'_{12}$. In the full treatment (e.g. [3]) these are expressed using

$$W_{21} = \Gamma^+_{1221} + \Gamma^-_{1221}, \qquad W_{12} = \Gamma^+_{2112} + \Gamma^-_{2112} \tag{202}$$

which are shown to be real since $(\Gamma^-_{mnkl})^* = \Gamma^+_{lknm}$ and the complex γ_{12} is given by

$$\gamma_{12} = \sum_k (\Gamma^+_{1kk1} + \Gamma^-_{2kk2}) - \Gamma^+_{2211} - \Gamma^-_{2211} \tag{203}$$

(so that $\gamma_{12} = \gamma^*_{21}$). In these,

$$\Gamma^+_{mkln} = \left(\frac{1}{\hbar}\right)^2 \sum_{ij} \langle m|Q_i|k\rangle\langle l|Q_i|n\rangle \int_0^\infty e^{-i\omega_{ln}t''}\langle F_i(t'')\ F_j\rangle\, dt''$$

$$\Gamma^-_{mkln} = \left(\frac{1}{\hbar}\right)^2 \sum_{ij} \langle m|Q_j|k\rangle\langle l|Q_i|n\rangle \int_0^\infty e^{-i\omega_{mk}t''}\langle F_j\ F_i(t'')\rangle\, dt'' \tag{204}$$

The interaction term V_{SR} in the Hamiltonian $\mathcal{H} = \mathcal{H}_S + \mathcal{H}_R + V_{SR}$ is treated as a combination of operators Q_i which themselves operate on the system states $|m\rangle$, $|n\rangle$, and operators F_i on the reservoir states $|N\rangle$ or $|R_i\rangle$ according to

$$V_{SR} = \sum_i Q_i F_i \tag{205}$$

It is the density matrix for the system $\rho_S(t)$ that is of interest, and by implication it is this 'reduced density matrix' that has been studied in the foregoing. The density matrix for the system and reservoir taken together, $\rho(t)$, to which the basic theory applies, must be approximated by making assumptions on the nature of the reservoir. No information on the reservoir is to be obtained: it is said to be unobserved. The reservoir clearly affects the system but the (massive) reservoir is negligibly affected by interactions with the system and is taken to remain in thermal equilibrium at constant temperature. The system $\rho_S(t)$ can be shown to be obtained by taking the trace of $\rho(t)$ over the states of R (denoted Tr_R), i.e. for the matrix elements

$$\langle m|\rho_S(t)|n\rangle = \sum_i \langle R_i m|\rho(t)|R_i n\rangle \tag{206}$$

The approximated $\rho_S(t)$ in the interaction picture is introduced according to

$$\rho_{\mathrm{I}}(t) = \rho_{\mathrm{I}}^S(t)\rho_R(0) = \rho_{\mathrm{I}}^S(t)\rho_R^0 \;:\; \rho_R^0 = \frac{\mathrm{e}^{-\mathcal{H}_R/kT}}{Z} \tag{207}$$

The full development of the relaxation theory introduces certain mean values for those parts of the interaction Hamiltonian that operate on R:

$$\langle F_i(t)\rangle = \mathrm{Tr}_R F_i(t)\rho_R^0 = \sum_N \langle N|F_i(t)|N\rangle\langle N|\rho_R^0|N\rangle \tag{208}$$

The $|N\rangle$ are eigenstates of $\mathcal{H}_R$ and it may be assumed that the F_i have no diagonal elements in this basis (since otherwise they would in fact be part of $\mathcal{H}_R$) and so the $\langle F_i(t)\rangle = 0$.

Also introduced are the mean values for products

$$\langle F_i(t)\; F_j(t')\rangle = \mathrm{Tr}_R F_i(t)F_j(t)\rho_R^0 \tag{209}$$

where the t and t' indicate that the interactions may occur at different times. If $t - t' \gg \tau$, a characteristic correlation time for the reservoir during which it is expected that the reservoir should have recovered from the effect of any interaction, then the interactions are not correlated so that $\langle F_i(t)\; F_j(t')\rangle = \langle F_i(t)\rangle\langle F_j(t')\rangle = 0$. The functions are maximum at $t - t' = 0$ and decrease as $t - t'$ increases above τ. In n.m.r. τ may be related to the time during which a nucleus is close to another nucleus or to a paramagnetic centre, etc. The reservoir is specified by specifying $\mathcal{H}_R$ and the F_i and these occur, meaningfully, only in the correlation functions. Note that the integrals in the expressions for the $\Gamma^{\pm}$ are taken over $t'' = t - t$ and $\langle F_i(t)\; F_j(t')\rangle \to \langle F_i(t'')\; F_j\rangle$. It is taken that the F_i do not, in general, commute. If they did then $\Gamma^{\pm}_{nmmn} = \Gamma^{\pm}_{mnnm}$ and thus $W_{mn} = W_{nm}$ would apply with obvious consequences. This implies that no classical treatment of relaxation could be valid.

Calculations of the Γ values and thus of T_1 and T_2 in realistic cases are too extensive to be included here, being more relevant to works on resonance as such. The objective has been to elucidate the more basic aspects of the spin and magnetization dynamics underlying spin resonance techniques.

3 The Magnetization Dynamics of Ordered Materials

This represents something of a break in the chapter, since the spins are taken to be coupled to give a macroscopic magnetization which may be treated classically, as by Kittel and others [14, 15].

3.1 Ferromagnetic Resonance (F.M.R.)

Initially, damping is ignored and solutions sought for $\mathrm{d}\mathbf{M}/\mathrm{d}t = \gamma\mathbf{M}\times\mathbf{B} = \mu_0\gamma\mathbf{M}\times\mathbf{H}$ with $\mathbf{B}$ comprising a static $\mathbf{B}_0 \parallel OZ$ and an oscillating $B_1\mathrm{e}^{i\omega t}$. With B_1 assumed small (ω in the microwave range) and B_0 large, so as to cause saturation, the magnetization may be taken to be given by adding small oscillating components to M_s (lying $\parallel OZ$): $\mathbf{M} = \mathbf{k}M_\mathrm{s}+M(t) = \mathbf{i}M_x(t)+\mathbf{j}M_y(t)+\mathbf{k}[M_z(t)+M_\mathrm{s}]$. It is also assumed that $\mathbf{M}(t) = \mathbf{M}\mathrm{e}^{i\omega t}$. Demagnetizing fields must also be taken into account, consideration of magnetocrystalline anisotropy being postponed. For an ellipsoidal specimen with principal axes coinciding with the coordinate axes, a component of $M \parallel OX$ gives demagnetizing fields $\parallel OX$ only, etc. The demagnetizing tensor is diagonal and

$$\begin{pmatrix} B_x^\mathrm{d} \\ B_y^\mathrm{d} \\ B_z^\mathrm{d} \end{pmatrix} = -\mu_0 \begin{pmatrix} D_x & 0 & 0 \\ 0 & D_y & 0 \\ 0 & 0 & D_z \end{pmatrix} \begin{pmatrix} M_x \\ M_y \\ M_z \end{pmatrix} = -\mu_0 \begin{pmatrix} D_x M_x \\ D_y M_y \\ D_z M_z \end{pmatrix} \tag{210}$$

or

$$\mathbf{B}_\mathrm{d} = -\mu_0\mathbf{D}\cdot\mathbf{M} = -\mu_0[\mathbf{i}D_xM_x + \mathbf{j}D_yM_y + \mathbf{k}D_z(M_z + M_\mathrm{s})] \tag{211}$$

The equation of motion including demagnetizing effects is

$$\begin{aligned} \frac{\mathrm{d}M}{\mathrm{d}t} &= \gamma[\mathbf{M}_\mathrm{s} + \mathbf{M}(t)]\times\{\mathbf{B}_0 + \mathbf{B}(t) - \mu_0\mathbf{D}\cdot[\mathbf{M}_\mathrm{s} + \mathbf{M}(t)]\} \\ &= \gamma[\mathbf{M}_\mathrm{s}\times\mathbf{B}(t) + \mathbf{M}(t)\times\mathbf{B}_0] - \gamma\mu_0\{[\mathbf{M}_\mathrm{s} + \mathbf{M}(t)]\times\mathbf{D}\cdot[\mathbf{M}_\mathrm{s} + \mathbf{M}(t)]\} \end{aligned} \tag{212}$$

noting that $\mathbf{M}_\mathrm{s}\times\mathbf{B}_0 = 0 = \mathbf{M}(t)\times\mathbf{B}(t)$ with $\mathbf{M}_\mathrm{s} \parallel \mathbf{B}_0$ and $\mathbf{M}(t) \parallel \mathbf{B}(t)$. The last product is $-\gamma\mu_0\mathbf{M}\times\mathbf{D}\cdot\mathbf{M}$ and, with equation (211),

$$\begin{aligned} \mathbf{M}\times\mathbf{D}\cdot\mathbf{M} = \mathbf{i}&[M_yD_z(M_z + M_\mathrm{s}) - (M_z + M_\mathrm{s})D_yM_y] \\ &+ \mathbf{j}[(M_z + M_\mathrm{s})D_xM_x - M_xD_z(M_z + M_s)] \\ &+ \mathbf{k}[M_xD_yM_y - M_yD_xM_x] \\ &\to \mathbf{i}M_y(D_z - D_y)M_\mathrm{s} + \mathbf{j}M_x(D_x - D_z)M_\mathrm{s} + \mathbf{k}(0) \end{aligned} \tag{213}$$

neglecting products of two (small) oscillating components for the last line. The left-hand side of equation (212) is $i\omega(\mathbf{i}M_x + \mathbf{j}M_y + \mathbf{k}M_z)$ and recalling that $\mathbf{M}_\mathrm{s} = \mathbf{k}M_\mathrm{s}$, $\mathbf{B}_\mathrm{a} = \mathbf{k}B_\mathrm{a}$, equating components gives

$$\begin{aligned} i\omega M_x &= -\gamma M_\mathrm{s}B_y + \gamma M_yB_0 + \mu_0\gamma(D_y - D_z)M_yM_\mathrm{s} \\ i\omega M_y &= \gamma M_\mathrm{s}B_x - \gamma M_xB_0 + \mu_0\gamma(D_z - D_x)M_xM_\mathrm{s} \end{aligned} \tag{214}$$

Setting $B_x = 0 = B_y$, $B(t) \parallel OZ$ and rearranging, the secular equation (condition for non-zero solutions) is

$$\begin{vmatrix} i\omega & \gamma[-B_0 - \mu_0(D_y - D_z)M_s] \\ \gamma[B_0 - \mu_0(D_z - D_x)M_s] & i\omega \end{vmatrix} = 0$$

From this the resonance frequency in the field B_0 is

$$\omega_0 = \gamma\{[B_0 + \mu_0(D_y - D_z)M_s][B_0 + \mu_0(D_x - D_z)M_s]\}^{1/2} \tag{215}$$

For a sphere with all components of D equal,

$$\omega_0 = \gamma B_0 \tag{216}$$

and thus ω_0 is specifically not that obtained by simply matching the precession frequency in B_0 as modified for demagnetizing effects to give $B = B_i = B_0 - \mu_0 N M_s$. This latter would only be obtained by ignoring D_x and D_y in equation (215), and as the magnetization responds to the combined field components $M_x(t)$ and $M_y(t)$ are generated and D_x and D_y must in fact affect the motion—except coincidentally for a thin sheet with surface $\perp OZ$ with $D_x = 0 = D_y$ and, with $D_z = 1$, for which

$$\omega_0 = \gamma(B_0 - \mu_0 M_s) \tag{217}$$

Other simple expressions are easily seen to be

$$\omega_0^{(s)} = \gamma[B_0(B_0 + \mu_0 M_s)]^{1/2}, \qquad \omega_0^{(c)} = \gamma(B_0 + \tfrac{1}{2}\mu_0 M_s) \tag{218}$$

for a thin sheet with its surface $\parallel OZ$ (s) and a long cylinder (c) with its axis $\parallel OZ$. Special techniques exist for grinding and polishing single-crystal spheres and these are widely employed.

The foregoing applies to specimens that are isotropic apart from shape effects. To account for magnetocrystalline anisotropy [14] a term may be added to the total field:

$$\mathbf{B} = \mathbf{B}_0 + \mathbf{B}(t) - \mu_0 \mathbf{D}\cdot\mathbf{M} + \mu_0 \mathbf{H}_K, \qquad \text{where } \mathbf{H}_K = -\mathbf{D}_K\cdot\mathbf{M} \tag{219}$$

where, if the easy directions coincide with coordinate axes and $\mathbf{D}_K$ is diagonal,

$$\mathbf{H}_K = -\mathbf{D}_K\cdot\mathbf{M} = -\mathbf{i}D_x^K M_x - \mathbf{j}D_y^K M_y - \mathbf{k}D_z^K(M_z + M_s) \tag{220}$$

It is easy to see that the elements D_x^K, etc., appear in the general treatment in the same way as do the D_x, etc., so that for a sphere

$$\begin{aligned} \omega_0 &= \gamma\{[B_0 + \mu_0(D_y^K - D_z^K)M_s][B_0 + \mu_0(D_x^K - D_z^K)M_s]\} \\ &\to \gamma\left[B_0 + \mu_0\left(\frac{2K_1}{M_s}\right)\right] \end{aligned} \tag{221}$$

the latter for a uniaxial crystal with B_0 along the easy axis. Even with $B_0 = 0$, resonance occurs at $\omega_0 = \gamma\mu_0(2K_1/M_s)$ if the crystal remains saturated, as for a single-domain particle or high-coercivity material. For a cubic crystal with $B_0 \parallel$ a [100] easy direction, Kittel showed that equation (221) still applied, but with $B_0 \parallel$ [110]:

$$\omega_0 = \gamma\left[\left(B_0 - \mu_0\frac{2K_1}{M_s}\right)\left(B_0 + \mu_0\frac{K_1}{M_s} + \mu_0\frac{K_2}{2M_s}\right)\right] \tag{222}$$

and for $B_0 \parallel [111]$:

$$\omega_0 = \gamma \left(B_0 - \mu_0 \frac{4K_1}{3M_s} - \mu_0 \frac{4K_2}{9M_s} \right) \tag{223}$$

Thus f.m.r. can give both M_s and the K_i as well as γ or g, whence $g = 2.10$, 2.18 and 2.21 for metallic iron, cobalt and nickel respectively ($\gamma = g\beta/\hbar$).

For a random polycrystalline specimen in a field that saturates each crystallite, the resonance condition differs throughout (even if the field is adequate to cause rotations to overall saturation) and this leads to a broadening of the resonance absorption peak which is additional to that caused by relaxations, as below. Porosity, common in oxide polycrystals, disturbs the field uniformity and can have a similar effect, as well as influencing relaxation as such, as studied by Schlömann for example [16]. Resonance and relaxation becomes very complex in polycrystals, as discussed, for example, by Patton [17].

For an isotropic sphere equation (212) can be solved to give $M_x(t)$ and $M_y(t)$ in terms of the fields and the results expressed as $\mathbf{M}(t) = \boldsymbol{\chi}\mathbf{H}(t)$ with $\boldsymbol{\chi}$ the susceptibility tensor. The corresponding permeability tensor $[\mathbf{B}(t) = \mu_0\boldsymbol{\mu}_r\mathbf{H}(t) = \boldsymbol{\mu}H(t)]$ is

$$\boldsymbol{\mu} = \begin{pmatrix} \mu_0^2\gamma^2 B_0 M_s + \mu_0 & \dfrac{-i\mu_0^2\omega\gamma M_s}{\omega_0^2 - \omega^2} & 0 \\ \dfrac{i\mu_0^2\omega\gamma M_s}{\omega_0^2 - \omega^2} & \dfrac{\mu_0^2\gamma^2 B_0 M_s}{\omega_0^2 - \omega^2} + \mu_0 & 0 \\ 0 & 0 & \mu_0 \end{pmatrix} \tag{224}$$

Ferromagnetic resonance occurs typically at very high frequencies. With $B_0 \sim 1.0$ T for saturation, or as the anisotropy field in barium ferrite for example, $g = 2$, $\gamma = g\beta/\hbar \sim 2 \times 10^{-23}/10^{-34}$: $\omega \sim 2 \times 10^{10}$ or $f \sim 2 \times 10^9$ Hz, $\lambda \sim 10$ cm or less. This is in the microwave or wave propagation region and the foregoing theory applies only so long as the specimen is much smaller than a wavelength so that the instantaneous field can be taken to be uniform.

If the fields are deliberately non-uniform then well-defined non-uniform modes of oscillation may be excited, magnetostatic or Walker modes [18] giving multiple absorption peaks for the different modes. The spins precess with phases which break up the surface pole densities and reduce the magnetostatic energy below that for the uniform mode. The 'wavelengths', or rather periods since plane waves are not involved, are long, comparable to the specimen dimensions, so the exchange energies involved are low. For spin waves, as below, the wavelengths are much shorter and exchange effects predominate, while the pole densities produced on surfaces $\perp B_0$ are so finely subdivided that magnetostatic energies are minimal.

At frequencies so high that the fields can only be regarded as propagating waves, the **H** vectors must be associated with **E** vectors. However, up to and including microwaves the response is gyromagnetic rather than electrical in nature, and this has been a tacit assumption above. Conversely, if ω is increased by a few orders above ω ($\lambda \sim 1$ μm, say) then $\boldsymbol{\mu} \rightarrow \mathbf{I}$ and gyromagnetic effects are minimal in the optical region. Thus it is the permeability tensor that controls the propagation of microwaves through magnetic materials and is of vital importance in the design of numerous devices.

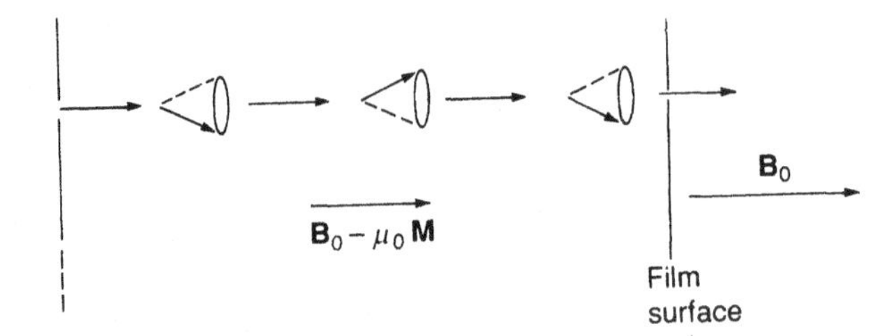

Figure 3.23 A line of spins through a film, with pinning at the surfaces

In metals the penetration or skin depths of $\sim 10^{-5}$ m for microwaves indicate the use of thin-film specimens. Poorly conducting ferrimagnetic oxides (which exhibit f.m.r. so long as the coupling is preserved; see Section 3.5) are not subject to such restrictions.

In certain cases the spins may become pinned at the film surfaces, due to anomalous anisotropies at the surface. In this case the response consists of spin wave generation as in Figure 3.23, as opposed to the uniform or Kittel mode. Kittel [19] gave

$$\omega_0 = \gamma(B_0 - \mu_0 M) + \frac{D}{\hbar}k^2 = \gamma(B_0 - \mu_0 M) + \frac{D}{\hbar}\left(\frac{n}{\pi L}\right)^2 \tag{225}$$

with $D = \mathcal{J}a^2$ for $S = \frac{1}{2}$, exchange energy being of obvious importance since the film thicknesses approximate domain wall widths; n is the order number related to the spin wave vector by $k = 2\pi/\lambda = n\pi/L$ since $n\lambda/2 = L$, the film thickness. Spin wave generation is not restricted to films in this particular way.

3.2 Relaxation and Damping

The Bloch–Bloembergen approach to damping is similar to that given in the foregoing sections, involving the relaxation times T_1 and T_2. Phenomenological approaches, adding relaxation terms to the undamped equation of motion, were made by Landau and Lifshitz [20] and by Gilbert [21], the latter according to

$$\frac{\mathrm{d}M}{\mathrm{d}t} = \gamma \mathbf{M} \times \mathbf{B} - \frac{\alpha}{M}\mathbf{M} \times \frac{\mathrm{d}\mathbf{M}}{\mathrm{d}t} \tag{226}$$

Using this the elements of χ become, for an isotropic sphere,

$$\begin{aligned} \chi_{11} = \chi_{22} &= \frac{\gamma\mu_0 M_s\left(\gamma B_0 + i\omega\alpha + \frac{1}{3}\mu_0\gamma M_s\right)}{(\omega_0 - i\omega\alpha)^2 - \omega^2} \\ \chi_{21} = -\chi_{12} &= \frac{i\omega\mu_0\gamma M_s}{(\omega_0 + i\omega\alpha)^2 - \omega^2} \end{aligned} \tag{227}$$

The complex elements imply phase differences between the field and magnetization components and thus indicate power absorption, over a certain range of frequencies $\Delta\omega$, or of fields at fixed frequency to give the linewidths as ΔB. Damping and line broadening can be associated with eddy currents in metals and in some of the more highly conducting oxides, which contain both ferrous and ferric ions, the thermally activated conductivity involving interchange of electrons between these. When this interchange is restricted to localized 'hopping' the anisotropy may be affected by the ordering of ferrous and ferric ions on octahedral sites so that, with a particular orientation of $M(t)$, the energy falls if the ordering occurs in such a way as to make that orientation a favourable direction. The motion of $M(t)$ may then induce hopping, leading to an energy dissipation which should be greatest when the frequency matches the reciprocal of the characteristic time period for hopping [22]. Distinctions between this damping mechanism and eddy current damping may be made by studying whether there is a dependence on crystal size. Ferrimagnetic garnets such as YIG, $Y_3Fe_5O_{12}$ or GdIG, $Gd_3Fe_5O_{12}$ (see Section 3, Chapter 5) ideally contain no ferrous ions.

ΔB for cobalt-substituted nickel ferrite rises by 0.7 mT for each atomic per cent of cobalt added [23] in line with calculations by Haas and Callen [24] based on the large associated spin–orbit coupling and anisotropic effects. Similarly, an increase of about 2 mT per atomic per cent of Sm in YIG may be associated with the unquenched orbital angular momentum.

Spin waves afford a relaxation mechanism when their (magnon) energies match those of the lattice vibrations (phonons), and general theories of relaxation and damping are

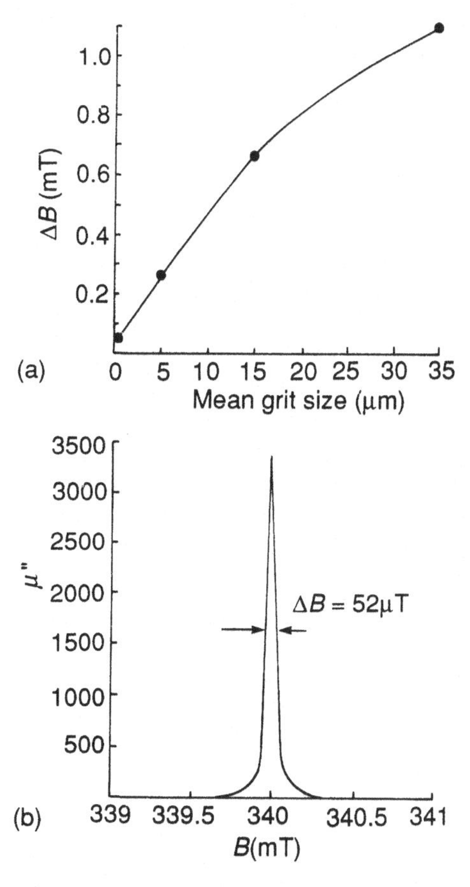

Figure 3.24 (a) The effects of surface roughness as indicated by the grit size in the device used to polish single-crystal spheres of YIG and (b) a line profile (9300 MHz) for a highly polished sphere [25]. (Reprinted by permission of John Wiley & Sons, Ltd, Magnetic Materials, R.S. Tebble & D.J. Craik, Wiley 1969)

developed in terms of spin wave generation and magnon scattering (see, for example, Patton [17]). The progressive grinding and polishing of single-crystal YIG spheres [25] reduces ΔB progressively, as in Figure 3.24, suggesting that the existence of field disturbances associated with minute surface pits may facilitate appropriate spin wave generation, although the formation and subsequent removal of strained layers has also been suggested to be important here.

Spin waves, rather than the uniform mode, are excited at high field amplitudes, as in high-power devices, and the threshold fields may be increased by introducing polycrystals with grain diameters ($\sim 10^{-8}$ m) below the spin wavelengths of importance.

In an ellipsoid, say with D_a relating to the axis of the applied field (OZ), the spin waves may be shown to propagate more easily in directions at certain angles θ to OZ, giving rise to the dispersion (with D as in equation 225)

$$\omega_k^2 = \gamma^2 \left(B_0 - \mu_0 D_a M_s + \frac{DK^2}{\gamma\hbar}\right)\left(B_0 - \mu_0 D_a M_s + \frac{DK^2}{\gamma\hbar} + \mu_0 M_s \sin^2\theta\right) \qquad (228)$$

The plots of ω_k versus k for $\theta = 0$ and for $\theta = \pi/2$ enclose the spin wave manifold. For general ellipsoids there may be a large number of waves, with different k values, which overlap the $k = 0$ mode. Some of these may be degenerate with thermal magnons providing for energy transfer from the uniform mode to the lattice. The damping and ΔB may thus depend on specimen shape and in fact thin discs are favoured.

3.3 Domain Structures and Saturation

For the simple principles to apply, B_0 must be adequate to drive out domains and give saturation. The effects of domains on the resonance were considered by Polder [26] and Polder and Smit [27] and discussed by LeCraw and Spencer [28] for example.

With domain widths $\ll$ the specimen dimensions and demagnetization assumed, the demagnetizing fields for the static magnetization may be neglected. The coherent precessions in each set of domains may be driven by oscillating fields (a) normal or (b) $\parallel$ the walls, for the sake of illustration (Figure 3.25). In (a) the dynamic components of M produce no pole densities on the walls, only on the surface; m_x is continuous across the walls and m_y alternates in sign throughout the structure. Due to these phases and since the specimen is demagnetized so that the pole densities due to m_y average to zero, the effective demagnetizing factors governing the dynamics are

$$N_x^e = N_x, \qquad N_y^e = 0 \qquad (N_z^e = 0)$$

In the limiting case of a thin sheet with $N_x \to 0$ the demagnetizing effects disappear from the resonance condition and $\omega_\perp = 0$ if anisotropy is neglected.

In (b) the phases are such that pole densities due to m_x appear on the walls and m_y is continuous so that

$$N_y^e = N_y, \qquad N_x^e = 1$$

since the domains are plate shaped. The maximum resonance frequency thus corresponds to a thin plate with $N_y \to 1$, and including anisotropy the overall extremes are

$$\omega_{\min} = \mu_0 \gamma H_K, \qquad \omega_{\max} = \mu_0 \gamma (H_K + M_s) \qquad (229)$$

indicating $\omega_{\max} \sim 5$ GHz for YIG.

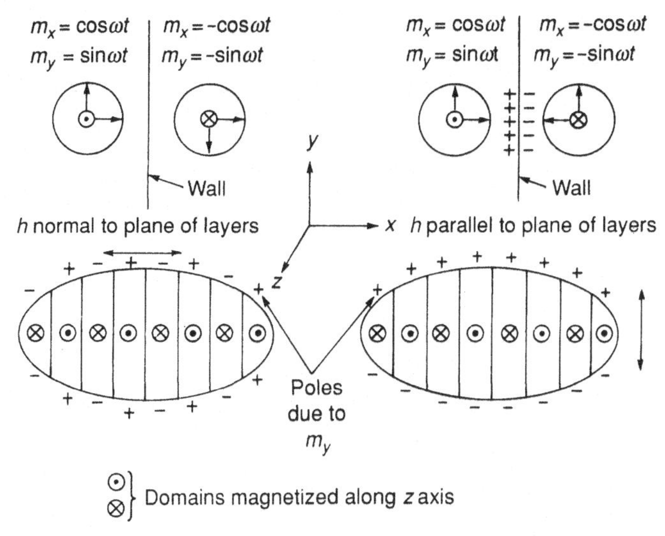

Figure 3.25 The influence of a domain structure on the resonance — the Polder–Smit model (Reprinted by permission of John Wiley & Sons Ltd, Magnetic Oxides, ed. D.J. Craik, Wiley 1975)

In general it is clear that in demagnetized specimens resonance losses may spread from very low frequencies, depending on H_K, up to microwave frequencies, and to avoid losses in remanent devices $\omega > \omega_{max}$ must be satisfied.

If it is required that the resonance frequency should be relatively low but that losses associated with the domain structures should be absent, then reduction of the bias field must be accompanied by a relatively low M_s so that saturation may still be achieved. Thus considerable attention has been devoted to, for example, the reduction of M_s for YIG by incorporating Al or Ga.

Losses peaking at relatively low frequencies are associated with resonant responses of the domain walls as noted in Section 14, Chapter 1.

3.4 Dynamic Response to Static or Pulsed Fields

Taking the undamped equation of motion [(226) with $\alpha = 0$] with $\mathbf{B} = \mathbf{k}B$, in component form and taking $\mathrm{d}/\mathrm{d}t(\mathrm{d}M_x/\mathrm{d}t)$ and substituting, it is readily seen that

$$\frac{\mathrm{d}^2 M_x}{\mathrm{d}t^2} = -\omega_0^2 M_x, \qquad \omega_0 = \gamma^{\mathrm{B}} \tag{230}$$

An obvious solution is $M_x = M_t \mathrm{e}^{i\omega_0 t}$ with M_t the component transverse to $\mathbf{B}$. It then follows simply that the equation for M_y is satisfied by $M_y = iM_t\mathrm{e}^{i\omega_0 t} = M_t \mathrm{e}^{i(\omega_0 t + \pi/2)}$. With θ the polar angle between $\mathbf{M}$ and OZ (constant here, with $\mathrm{d}M_z/\mathrm{d}t = 0$),

$$\begin{aligned} M_x &= M_s \sin\theta \mathrm{e}^{i\omega_0 t} \\ M_y &= M_s \sin\theta \mathrm{e}^{i(\omega_0 t + \pi/2)} \\ M_z &= M_s \cos\theta \end{aligned} \tag{231}$$

The physical components are the real parts. On introducing the (Gilbert) damping, the products may be expanded to give equations for $\mathrm{d}M_x/\mathrm{d}t$, $\mathrm{d}M_y/\mathrm{d}t$, $\mathrm{d}M_z/\mathrm{d}t \cdot M_z$ may be eliminated from the first two, using the third, and it is left as an exercise to show that

$$\begin{aligned}
\frac{\mathrm{d}M_x}{\mathrm{d}t} &= \frac{\omega_0}{1+\alpha^2}M_y + \frac{\omega_0\alpha}{1+\alpha^2}\frac{M_y M_z}{M_s} \\
\frac{\mathrm{d}M_y}{\mathrm{d}t} &= -\frac{\omega_0}{1+\alpha^2}M_x + \frac{\omega_0\alpha}{1+\alpha^2}\frac{M_y M_z}{M_s} \\
\frac{\mathrm{d}M_z}{\mathrm{d}t} &= -\frac{\omega_0}{1+\alpha^2}M_s + \frac{\omega_0\alpha}{1+\alpha^2}\frac{M_z^2}{M_s}
\end{aligned} \tag{232}$$

Now it can be shown that equations (231) still constitute solutions except in that $\omega = \omega_0$ cannot be assumed and M_z cannot be assumed constant so that $\theta \to \theta(t)$. Differentiating,

$$\frac{\mathrm{d}M_x}{\mathrm{d}t} = i\omega M_s \sin\theta \mathrm{e}^{i\omega t} + M_s \mathrm{e}^{i\omega t}\cos\theta\frac{\mathrm{d}\theta}{\mathrm{d}t}$$

The first term indicates that $\omega = \omega_0/(1+\alpha^2)$. The second is appropriate if $\cos\theta\,\mathrm{d}\theta/\mathrm{d}t$ is proportional to $\cos\theta\sin\theta$, which is the case if

$$\begin{aligned}
\tan\frac{\theta}{2} &= c\mathrm{e}^{at} \\
\frac{\mathrm{d}\left(\dfrac{\theta}{2}\right)}{\cos^2\dfrac{\theta}{2}} &= ac\mathrm{e}^{at} = a\tan\frac{\theta}{2}\mathrm{d}t \\
\frac{\mathrm{d}\theta}{\mathrm{d}t} &= 2a\cos\frac{\theta}{2}\sin\frac{\theta}{2} = a\sin\theta
\end{aligned}$$

and it is clearly required that $a = \alpha\omega/(1+\alpha^2)$. The reader may then demonstrate consistency throughout. Thus $c = \tan\theta_0/2$ with $\theta = \theta_0$ at $t = 0$ for the initial conditions when the field is applied and $a \sim (\text{time})^{-1}$ is replaced by $1/\tau$.

Thus equations (231) are solutions of the Gilbert equation for a static field only, so long as $\omega_0 \to \omega = \omega_0/(1+\alpha^2)$ and also

$$\tan\frac{\theta}{2} = \tan\frac{\theta_0}{2}\mathrm{e}^{t/\tau}, \qquad \text{where } \tau = \frac{1+\alpha^2}{\alpha\omega_0} \tag{233}$$

When α is small the motion is a modified precession and reversal, for example, involves several periods, but with α large the motion is more directly from the initial direction to the field direction. Taking $\theta_0 = \pi/8$ and $B \parallel -OZ$ for example, it can be calculated that for $t/\tau = 0, 1, 2, 3, 4$ and 5; $M_z/M_s = 0.926, 0.556, -0.335, -0.879, -0.993$, and -0.998, and switching can be taken to be achieved for $t \sim 5\tau$. In both the limits $\alpha = 0$ and $\alpha \to \infty$ there is no switching (rotation w.r.t. OZ) and Kikuchi [29] showed that the most rapid switching should occur for $\alpha = 1$, with a time of $t = 2/\gamma B$, e.g. $t = 10$ ns with $B = 1$ mT, $g = 2$, $|\gamma| = 1.8 \times 10^{11}\ \mathrm{s^{-1}\ T^{-1}}$.

Current pulses with rise-times < 1 ns can be produced and transmitted along strip lines (conductors carefully positioned w.r.t. earthed sheets) so as to cause a virtually instantaneous

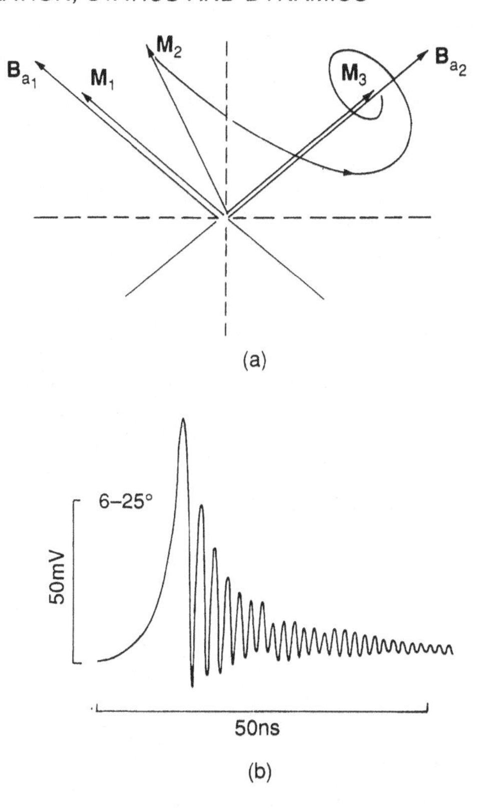

Figure 3.26 (a) The predicted motion of **M** in a garnet crystal when the field rotates slowly, between easy directions, from $\mathbf{B}_{a_1}$ to $\mathbf{B}_{a_2}$ with a processional discontinuity between positions $\mathbf{M}_2$ and $\mathbf{M}_3$ and (b) a recorded signal in a neighbouring strip line [S.J. Blundell, Thesis, Nottingham, 1975]

effective rotation in the direction of a bias field applied to a thin sheet of, say, Al–YIG. The dynamic response is detected via the stray fields intersecting a second orthogonal strip line and the precessional nature of the response is readily demonstrated.

In a more interesting situation, for a (110) single-crystal slice of YIG there are four directions of minimum anisotropy energy and, between these, four directions of maximum energy. For a very thin disc a small bias field gives saturation along an easy direction 1 in Figure 3.26(a). A slow pulse field effectively rotates B_a to B_{a_2} in a second easy direction. **M** rotates 'uphill' to a direction $\mathbf{M}_2$ where there is a point of inflection in the total energy. At this point it is as if the anisotropy field were instantaneously switched because a different easy direction becomes effective. Figure 3.26(b) shows voltages induced in a neighbouring strip line. It is stressed that in this case the rise-time of the drive field was much greater than the precession time period. The material was $Y_3Ga_{0.95}Fe_{4.05}O_{12}$, with $M_s = 33.5 \times 10^3$ A m^{-1}.

3.6 Antiferromagnetic and Ferrimagnetic Resonance

In an antiferromagnetic, following Kittel [30], each sublattice may be supposed to be subjected to an exchange field proportional to the magnetization of the other sublattice as well as to anisotropy fields B_K:

$$\mathbf{B}_1 = -\mu_0 \lambda \mathbf{M}_2 + \mathbf{k} B_K$$
$$\mathbf{B}_2 = -\mu_0 \lambda \mathbf{M}_1 - \mathbf{k} B_K$$

(with colinear ordering $\parallel OZ$ and no external fields applied). The equations of motion are

$$\frac{d\mathbf{M}_1(t)}{dt} = \gamma \mathbf{M}_1(t) \times \mathbf{B}_1$$
$$\frac{d\mathbf{M}_2(t)}{dt} = \gamma \mathbf{M}_2(t) \times \mathbf{B}_2$$

so that $d\mathbf{M}_1(t)/dt$ depends upon $\mathbf{M}_2(t)$ via $\mathbf{B}_1$. Assuming $\mathbf{M}_1(t) = \mathbf{M}_1 e^{i\omega t}$, etc., it is possible, as for f.m.r., to derive four equations for M_1^x, M_1^y, M_2^x and M_2^y. A key step is the reduction of these to two equations in $M_1^+ = M_1^x + iM_1^y$ and $M_2^+ = M_2^x + iM_2^y$. The secular equation then indicates that

$$\omega_0^2 = \gamma^2 (B_K^2 + 2B_e B_K)$$
$$\omega_0 \doteqdot \pm\gamma (2B_e B_K)^{1/2} \tag{234}$$

the second on the assumption that $B_e \gg B_K$ with the exchange field $B_e = \mu_0 \lambda M$ common.

Antiferromagnetic resonance (a.f.m.r.) occurs at this (extremely high) frequency, even in the absence of applied fields. With applied fields the modified calculation leads to

$$\omega_0 = \gamma B_0 \pm \gamma (B_K^2 + 2B_e B_K)^{1/2} \tag{235}$$

If the calculations introduce $M^- = M_x - iM_y$ in place of M^+, it can be seen on introducing the expressions for M_1^- and M_2^- that the condition for non-zero solutions corresponds to the existence of components in OXY which rotate oppositely to the original case and in general

$$\omega_0 = \pm\gamma B_0 \pm \gamma (B_K^2 + 2B_e B_K)^{1/2} \tag{236}$$

The magnitude of ω_0 has two values according to whether the terms are of the same or of different sign. In fact there is a field for which one value of ω_0 is zero, i.e.

$$B_a = (B_K^2 + 2B_e B_K)^{1/2} \doteqdot (2B_e B_K)^{1/2} = B_f \tag{237}$$

The approximation is the field that can be shown to cause spin flopping or reordering of the spins in a direction normal to the original axis.

The two resonance modes have been identified, for example, for Cr_2O_3 around 165 GHz $\sim 3 \times 10^8/(165 \times 10^9) = 0.002$ m, i.e. millimetre waves [31], with lower frequencies applying when the Néel points are lower. Measurements in applied fields can give B_e and B_K separately and a great deal of information on antiferromagnetics has been obtained in this way. Large applied fields reduce the frequency for one of the modes and this is of value when the Néel point is high.

For ferrimagnetics with two non-equivalent sublattices with magnetization M_A and M_B, the different γ_A and γ_B are introduced as well as B_K^A and B_K^B. The resonance condition becomes (expanding the determinant)

$$\left[\frac{\omega_0^\pm}{\gamma_A} \pm (B_0 + \mu_0 \lambda M_B + B_K^A)\right]\left[\frac{\omega_0^\pm}{\gamma_B} \pm (B_0 - \mu_0 \lambda M_A - B_K^B)\right] + \mu_0^2 \lambda^2 M_A M_B = 0 \tag{238}$$

There are two solutions for ω_0^+ and two for ω_0^-. Those for ω_0^+ are equal in magnitude and opposite in sign and those for ω_0^- have this same magnitude and the opposite signs to those of the corresponding ω_0^+. In one approximation only products containing an exchange field are preserved. It is then found that λ drops out of the equation and

$$\omega_0^{\pm} = \pm\gamma_e(B_0 + B_K) \tag{239}$$

with the effective γ and anisotropy field as

$$\gamma_e = \frac{M_A - M_B}{M_A/\gamma_A - M_B/\gamma_B}, \qquad B_K = \frac{B_K^A M_A + B_K^B M_B}{M_A - M_B} \tag{240}$$

This is the effective f.m.r. condition, the approximation being that the frequencies are much smaller than those given by resonance in the exchange fields; it corresponds to the motion of $\mathbf{M}$ with M_A remaining antiparallel to M_B.

In a second approximation no assumptions are made about the magnitude of ω_0 but the applied and anisotropy fields are neglected, in relation to the exchange fields. In this case equation (238) becomes

$$\left(\frac{\omega_0^{\pm}}{\gamma_A} \pm \mu_0\lambda M_B\right)\left(\frac{\omega_0^{\pm}}{\gamma_B} \mp \mu_0\lambda M_A\right) + \mu_0^2\lambda^2 M_A M_B = 0$$

so that

$$\omega_0^{\pm} = \pm\mu_0\lambda(\gamma_B M_A - \gamma_A M_B) = \pm\mu_0\gamma_A\gamma_B\lambda\left(\frac{M_A}{\gamma_A} - \frac{M_B}{\gamma_B}\right) \tag{241}$$

This specifically involves the exchange field and the response is known as exchange resonance, occurring from millimetre waves to the infra-red.

References

1. Gerlach, W. and Stern, O., *Ann. Phys. (Leipzig)*, **74**, 673 (1924).
2. Pauli, W., *Z. Phys.*, **31**, 765 (1925).
3. Blum, K., *Density Matrix Theory and Applications*, Plenum Press, New York and London, 1989.
4. Kittel, C., *Solid State Physics*, John Wiley and Sons, New York and London, 1963.
5. Callaway, J., *Quantum Theory of the Solid State*, Academic Press, New York, 1976.
6. Mattis, D.C., *The Theory of Magnetism*, Springer-Verlag, Berlin, 1981.
7. Kittel C., *Quantum Theory of Solids*, John Wiley and Sons, New York and London, 1963.
8. McCollum, D.C. and Callaway, J., *Phys. Rev. Lett.*, **9**, 376 (1962).
9. Charap, S.H. and Boyd, E.L., *Phys. Rev.*, **133**, A811 (1964).
10. Carr, H.Y. and Purcell, E.M., *Phys. Rev.*, **94**, 630 (1954).
11. Hahn, E.L., *Phys. Rev.*, **80**, 580 (1950) and **94**, 630 (1950).
12. Ernst, R.R. and Anderson, W.A., *Rev. Sci. Instr.*, **37**, 93 (1966).
13. Bloch, F., *Phys. Rev.*, **70**, 460 (1946).
14. Kittel, C., *Phys. Rev.*, **73**, 155 (1948) and *Phys. Rev.*, **76**, 743 (1949).
15. Luttinger, J.M. and Kittel, C., *Helv. Phys. Acta*, **21**, 480 (1958); Richardson, J.M., *Phys. Rev.*, **75**, 1630 (1949); Polder, D., *Phil. Mag.*, **40**, 99 (1949); Van Vleck, J.H., *Phys. Rev.*, **78**, 266 (1950).
16. Schlömann, E., *Proceedings of Conference on Magnetism and Magnetic Materials*, Boston, p. 600, 1956 and Raytheon Technical Report R-15, 1956.
17. Patton, C.E., *Magnetic Oxides*, Ed. D.J. Craik, John Wiley and Sons, New York and London, p. 575, 1975.

18. Walker, L.R., *Phys. Rev.*, **105**, 390 (1957).
19. Kittel, C., *Phys. Rev.*, **110**, 1295 (1958); see also Searle, C.W., Morrish, A.H. and Prosen, R.J., *Physica*, **29**, 1219 (1963).
20. Landau, L. and Lifshitz, E., *Phys. Z. Sowjet.*, **8**, 153 (1935).
21. Gilbert, T.L., *Phys. Rev.*, **100**, 1243 (1955).
22. Wijn, H.P.J., Thesis, Leiden, 1953.
23. Schlomann, E., Green, J.J. and Milano, U., *J. Appl. Phys.*, **31**, 386S (1960).
24. Haas, C.W. and Callen, H.B., *J. Appl. Phys.*, **32**, 157 (1961).
25. LeCraw, R.C., Spencer, E.G. and Porter, C.S., *Phys. Rev.*, **110**, 1311 (1958).
26. Polder, D., *J. Phys. Rad.*, **12**, 337 (1951).
27. Polder, D. and Smit, J., *Rev. Mod. Phys.*, **25**, 89 (1953).
28. LeCraw, R.C. and Spencer, E.G., *J. Appl. Phys.*, **28**, 399 (1957).
29. Kikuchi, R., *J. Appl. Phys.*, **27**, 1352 (1956).
30. Kittel, C., *Phys. Rev.*, **82**, 565 (1951); see also Nagamiya, T., *Prog. Theor. Phys. (Kyoto)*, **6**, 342 (1951).
31. Heller, G.S., Stickler, J.J and Thaxter, J.B., *J. Appl. Phys.*, **32**, 307S (1961).

CHAPTER

4 Magnetostatics, Magnetic Domains and Magnetic Design

1 Magnetostatics

To review the material introduced in Chapter 1, currents or circuits generate fields $\mathbf{H}_0$ or induction fields $\mathbf{B}_0$ in free space, which are connected by $\mathbf{B}_0 = \mu_0\mathbf{H}_0$ and with μ_0 constant must have the same attributes; in particular both are solenoidal: $\nabla \cdot \mathbf{H}_0 = 0 = \nabla \cdot \mathbf{B}_0$. If the whole of space is occupied by a medium, with relative permeability μ_r representing its magnetic response, then $\mathbf{H}$ is defined by the Biot–Savart law exactly as is $\mathbf{H}_0$ (it does not depend on the medium) whereas the definition of $\mathbf{B}$ does involve the medium, i.e. the forces in the defining relation are proportional to μ_r and $\mathbf{B} = \mu_r\mathbf{B}_0$ or $\mathbf{B} = \mu_r\mu_0\mathbf{H}_0 = \mu_r\mu_0\mathbf{H} = \mu\mathbf{H}$. With μ_r constant everywhere, $\mathbf{B}$ and $\mathbf{H}$ remain essentially similar and the introduction of two differently designated fields is not necessary since $\mathbf{B}_0$ could be used as the field dependent on the circuits alone and the response of the medium described by $\mu_r = \mathbf{B}/\mathbf{B}_0$. (Some reference to the detailed nature of the medium is necessary, whether $\mathbf{B}_0$ or $\mathbf{H}$ is used, to see that while $\mu_r \to \infty$ may arise $\mathbf{B} = \mu_r\mathbf{B}_0$ may only rise up to a certain limit.)

It is when μ differs in different regions of space that it becomes, at the least, very difficult to develop a reasonable account without the designation of a solenoidal field $\mathbf{B}$ and a separate field $\mathbf{H}$, which may be conservative inasmuch as it is convenient to calculate contributions to $\mathbf{H}$ as if they were derived from sources described as surface (σ) or volume pole densities ρ, with $\sigma = \mathbf{M} \cdot \mathbf{n}$ and $\rho = -\nabla \cdot \mathbf{M}$, with $\mathbf{M}$ defined in terms of dipole densities. (This concept is introduced systematically in Section 1.1.) It was seen that $\mathbf{B} = \mu_0(\mathbf{H} + \mathbf{M})$ and that $\nabla \cdot \mathbf{B} = 0$ is consistent with $\nabla \cdot \mathbf{H} \neq 0$ because $\nabla \cdot (\mathbf{H} + \mathbf{M}) = 0$, i.e.

$$\nabla \cdot \mathbf{H} = -\nabla \cdot \mathbf{M} = \rho, \qquad \nabla^2\varphi = -\rho$$

($\mathbf{H} = -\nabla\varphi$). If it were practicable to calculate $\mathbf{B}$ due to all the individual dipoles constituting $\mathbf{M}$, account being taken of the contributions within the quasi-point dipoles, no conservative field would necessarily be introduced.

$\mathbf{H}$ and φ may be considered as quantities contrived, in association with poles, to facilitate calculations, with $\mathbf{B}$ and $\mathbf{A}$ the more fundamental or essential. The force on a charged particle (electron) passing through a magnetized material (thin foil, as in Lorentz microscopy) is certainly given by the cross-product not with **H**, since **H** in the example may well be zero, but with **B** which may be considered as the 'total field including **M**'. In a particular picture or example the trajectory must be seen to sample the internal fields within the dipoles considered as minute cylindrical current sheets and not just the fields between the dipoles which, in general, average to zero.

Equally, a filamentary current must be seen as a line cutting through the dipoles so that the average field experienced is not just any applied field, as modified perhaps by demagnetizing fields (which have been seen to be the average of the external dipole fields), but includes the average of the internal dipole fields over the specimen volume which has been seen to be equal to the magnetization. This, of course, corresponds to involving $\mathbf{B} = \mu_0(\mathbf{H} + \mathbf{M})$ and not just $\mathbf{B} = \mu_0\mathbf{H}$.

It is not surprising that completely inappropriate results would be obtained by taking the dipoles to be pairs of poles, in suitable limits, the field contribution within these quasi-point dipoles clearly being inappropriately directed.

In considering the energy of, or torques on, an individual dipole $\boldsymbol{\mu}$ surrounded by further dipoles constituting a magnetization, the situation is different because here the internal dipole fields are not sampled. The energy is $-\boldsymbol{\mu} \cdot \mathbf{B}_0$ with $\mathbf{B}_0$ or $\mu_0\mathbf{H}_0$ the applied field, perhaps modified by demagnetizing fields, i.e. including the average of the external dipole fields, but it is not $-\boldsymbol{\mu} \cdot \mathbf{B}$. If it were the latter it would not be possible to calculate susceptibilities. For an assembly of spins $\frac{1}{2}$ in the Curie limit, the energy expression leads to a magnetization $M = Ng^2\beta^2B/(4kT)$, i.e. $M = cB$ with c constant at a given T. Setting $B = \mu_0(H_0 + M) = B_0 + \mu_0 M$ gives $M = c(B_0 + \mu_0 M)$, i.e. $M = cB_0/(1 - \mu_0 c)$, which is not acceptable since $\mu_0 c = 1$ applies for some value of T. B must be taken here as B_0 or $\mu_0 H_0$.

1.1 Fields from Magnetized Bodies

For a uniformly magnetized body, the expression for the potential from a dipole $\mathbf{M}\,\mathrm{d}v$ (see Chapter 1, equation 89) together with

$$\nabla_2\left(\frac{1}{r}\right) = -\nabla_1\left(\frac{1}{r}\right) \qquad \text{for } r = |\mathbf{r}_2 - \mathbf{r}_1|$$

which can easily be established as a generalization of $\partial f(r)/\partial x_1 = -\partial f(r)/\partial x_2$ ($\nabla_1 = \mathbf{i}\partial/\partial x_1 + \cdots$, $\nabla_2 = \mathbf{i}\partial/\partial x_2 + \cdots$) lead to

$$\varphi(\mathbf{r}_2) = -\frac{1}{4\pi}\mathbf{M} \cdot \int \nabla_2 \frac{1}{r}\,\mathrm{d}v \tag{1}$$

with integration over the magnetized body, i.e.

$$\varphi = \mathbf{M} \cdot (-\nabla V) = \mathbf{M} \cdot \mathbf{F}, \qquad \text{where } \mathbf{F} = -\nabla V, \qquad V = \frac{1}{4\pi}\int \frac{1}{r}\,\mathrm{d}v \tag{2}$$

V has the form of a potential from a uniformly charged body with unit charge density, according to $V = (1/4\pi)q/r$ for a point charge q. F is a conservative field of no particular designation.

The outwards flux of F across a closed surface is $N_F = \int_S \mathbf{F} \cdot \mathrm{d}\mathbf{S}$ with $\mathrm{d}\mathbf{S}$ a vector surface element: $\mathrm{d}\mathbf{S} = \mathbf{n}\,\mathrm{d}S$, with $\mathbf{n}$ the outwardly directed unit vector normal to the surface. For a sphere of radius a with unit charge density, $\mathbf{F}$ must by symmetry be radially directed and depend only on r so that $(r > a)$

$$N_F = \int_S \mathbf{F} \cdot d\mathbf{S} = 4\pi r^2 F = \tfrac{4}{3}\pi a^3$$

in accordance with Gauss, the latter term being the total charge enclosed by the surface. Thus

$$\mathbf{F} = \frac{1}{4\pi}\left(\tfrac{4}{3}\pi a^3\right)\frac{\mathbf{r}}{r^3}$$

and

$$\varphi = \mathbf{M} \cdot \mathbf{F} = \frac{1}{4\pi}\frac{\boldsymbol{\mu} \cdot \mathbf{r}}{r^3}, \qquad \text{where } \boldsymbol{\mu} = \tfrac{4}{3}\pi a^3 \mathbf{M} \tag{3}$$

The potential and field $\mathbf{H} = -\nabla\varphi$ external to a uniformly magnetized sphere is thus that of a point dipole equal to the total moment of the sphere at the centre of the sphere.

If the variable surface is taken to lie within the sphere, $r < a$, the volume enclosed is $\frac{4}{3}\pi r^3$ and by Gauss's theorem and symmetry,

$$\mathbf{F}_i = \frac{1}{4\pi}\tfrac{4}{3}\pi r^3 \frac{\mathbf{r}}{r^3} = \tfrac{1}{3}\mathbf{r} \tag{4}$$

With $\mathbf{M} \parallel OZ$, $\varphi_i = \mathbf{M} \cdot \mathbf{r}/3 = \frac{1}{3}Mz$ and $\mathbf{H}_i = -\nabla\varphi_i = -\frac{1}{3}\mathbf{k}M$. Thus the internal (demagnetizing) field is uniform and can be written as $\mathbf{H} = -N\mathbf{M}$, with the demagnetizing factor $N = \frac{1}{3}$. At the surface $r \to a$ and $\varphi_i = \frac{1}{3}\mathbf{M} \cdot \mathbf{a} = \varphi$, exemplifying a general principle for φ at an interface.

The method can be seen to be equivalent to making the calculations for two nearly superimposed spheres of volume pole density $\pm\rho$ displaced through an infinitesimal distance $\boldsymbol{\delta}$ according with $\rho\boldsymbol{\delta} = \mathbf{M}$, $\rho = M/\delta$, the notional fields $\mathbf{F}$ being those from spheres of (notional) magnetic pole density. This is, in turn, equivalent by simple geometry to the production of a surface pole density $\sigma = \mathbf{M} \cdot \mathbf{n}$ on the surface, i.e.

$$\varphi = \frac{1}{4\pi}\int \frac{\mathbf{M} \cdot \mathbf{n}}{r}\, dS = \frac{1}{4\pi}\int \frac{\mathbf{M} \cdot d\mathbf{S}}{r} \tag{5}$$

Referring to Figure 4.1, the reader may show that the fields external to a circular cylinder of great length and sectional area $\mathcal{A}$, with uniform $\mathbf{M}$ normal to the axis, are those of two lines of pole density $\pm\mathcal{A}M/\delta$ per unit length, or equally those of oppositely directed pairs

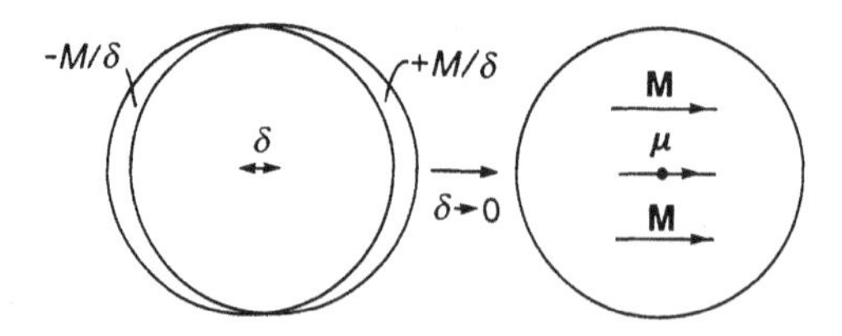

Figure 4.1 Two nearly overlapping cylinders of mythical volume pole density $\rho = \pm M/\delta$ may be seen to become equivalent to a transversely magnetized cylinder, $\mathbf{M} \parallel \boldsymbol{\delta}$, as $\delta \to 0$, so that the external fields are those of a 'linear dipole'

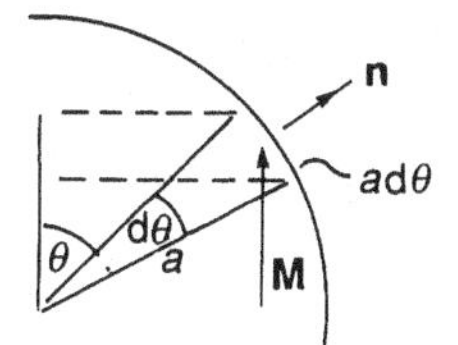

Figure 4.2 The pole representation for a uniformly magnetized sphere

of appropriate filamentary currents, i.e. of a 'line dipole' extending along the cylinder axis. This may be confirmed by the general theory.

It is left as a further simple exercise, referring to Figure 4.2, to write down (a) the contribution to the field specifically at the centre of a uniformly magnetized sphere from an element of surface $a^2 \sin\theta \, d\theta \, d\phi$ bearing pole density $\sigma = \mathbf{M} \cdot \mathbf{n}$ and (b) the resolved contribution dH_z obtained by trivial integration w.r.t. $\phi(d\phi \to 2\pi)$ and to complete the simple integration w.r.t. θ to show that the field is $-\frac{1}{3}M$. There can be no other components along the axis $\parallel$ $\mathbf{M}$ and this must thus be the field at all points along this axis, and indeed at all points within the sphere, both conclusions being the only ones compatible with $\nabla \cdot \mathbf{H} = 0$, where $\mathbf{M}$ is uniform. Thus the explicit pole model can be shown to give the correct results.

The sphere and long cylinder are special examples of ellipsoids and it is only for these that the internal fields are uniform (with $\mathbf{M}$ uniform) and are readily calculable in terms of the demagnetization factors N. Attention is usually confined to oblate or prolate spheroids generated by rotating an ellipse about a principal axis and specified by the polar semi-axis a and equatorial semi-axis b or by $q = a/b \cdot \mathbf{H}_d = -N(q)\mathbf{M}$ when $\mathbf{M}$ lies along a principal axis, with N_a relating to the polar axis and N_b to the equatorial axis. For prolate spheroids $(q > 1)N_a$ is as given in equation (216) of Chapter 1 and $N_b = \frac{1}{2}(1 - N_a)$. For oblate spheroids $(q < 1)$, $M \perp$ the circular section:

$$N_a = \frac{1}{1-q^2}\left[1 - \frac{q}{(1-q^2)^{1/2}} \cos^{-1} q\right] \tag{6}$$

and an approximation for $q > 2$ is $N_b = (2/q)(3-1/q)$. Some values are given in Table 4.1. It is noted that $N(\text{c.g.s.}) = 4\pi N(\text{SI})$, e.g. $N = 4\pi/3$ for a sphere in the unrationalized c.g.s., in correspondence with $d\varphi = (1/4\pi)\mathbf{M} \cdot \mathbf{n} \, ds/r$ in the SI.

It will be seen that the field from a line of pole density $\sigma \, dx$ is $H = (1/2\pi)\sigma \, dx \sin\alpha/z$, with z and α as in Figure 4.3 and the demonstration indicated by the caption may be treated as a simple example.

Table 4.1

a/b	Na	a/b	Na
0	0.1	2.0	0.1736
0.01	0.9845	5.0	0.0558
0.1	0.8608	10.0	0.0203
0.5	0.5272	50	0.0014
1.0	0.3333	100	0.0004

Examples of N_a for ellipsoids of revolution with polar axis a, equatorial axis b. $N_b = (1 - N_a)/2$. Recall $N_a(\text{e.g.s}) = 4\pi N_a$ and $N_a + 2N_b = 4\pi$ in the c.g.s. system.

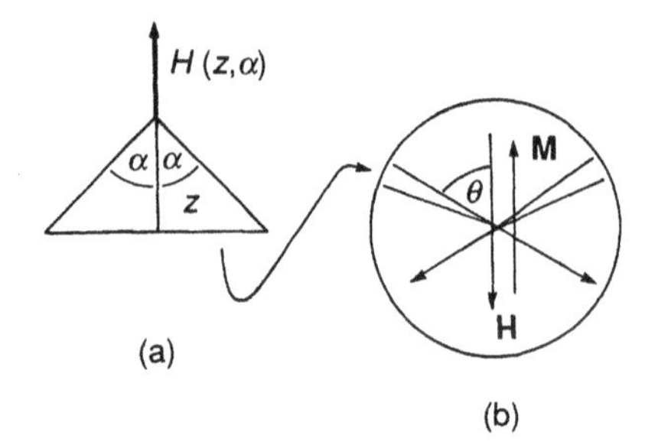

Figure 4.3 Applying the expression for the field from a line of pole density (a) to the surface element indicated in (b), the field at the centre of the cylinder may be found by simple integration

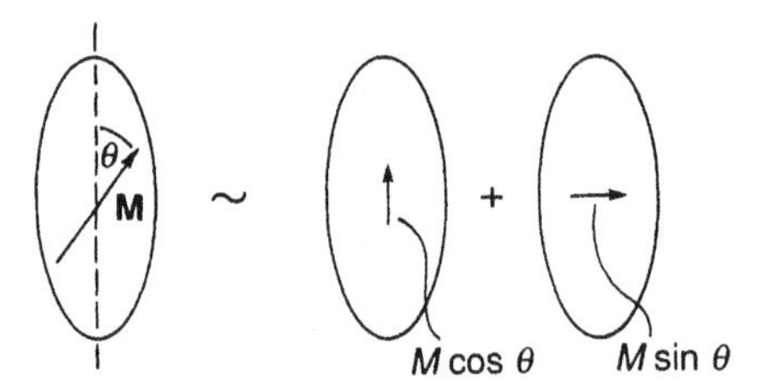

Figure 4.4 The fields within the ellipsoid with uniform **M** on the left are the sum of those within the two ellipsoids on the right, and must thus be uniform (though not antiparallel to **M**)

If **M** makes an angle θ with the a axis, $\mathbf{H}_M$ will still be uniform by the superposition principle, as illustrated in Figure 4.4:

$$\mathbf{H}_M = -N_a M \cos\theta \mathbf{e}_z - N_b M \sin\theta \mathbf{e}_r \tag{7}$$

or with an alternative notation, and for a general ellipsoid,

$$\begin{pmatrix} H_x \\ H_y \\ H_z \end{pmatrix} = -\begin{pmatrix} N_{xx} & 0 & 0 \\ 0 & N_{yy} & 0 \\ 0 & 0 & N_{zz} \end{pmatrix} \begin{pmatrix} M_x \\ M_y \\ M_z \end{pmatrix}, \qquad \mathbf{H} = -\mathbf{NM} \tag{8}$$

but it is clear that $\mathbf{H}_M \parallel \mathbf{M}$ does not apply generally.

For non-ellipsoidal bodies there is no uniform demagnetizing field but it may be useful to derive effective or approximate 'demagnetizing factors' giving the mean field component antiparallel to M, or the field at the centre of a body. The field on the axis of a disc (radius a) with $\sigma = M$ has been given (Section 1, Chapter 1) and setting $z = a$ and with a factor of 2 it is seen that the demagnetizing field at the centre of an axially magnetized cylinder with length = diameter is $H = -(1 - 1/\sqrt{2})M = -0.293M$, comparable with $N = 0.333$ for the sphere. Making use of relations given in Section 1.17, the field at the centre of a uniformly magnetized square-section rectangular block ($M \perp$ the section) is

$$H_M = -\frac{2\mathbf{M}}{\pi} \sin^{-1} \frac{1}{1+q} \tag{9}$$

where q is the dimensional ratio (Figure 4.5) and for a cube $\mathbf{H} = -\frac{1}{3}\mathbf{M}$ exactly as for the sphere. Mean demagnetizing fields can be calculated from energies of formation which, as

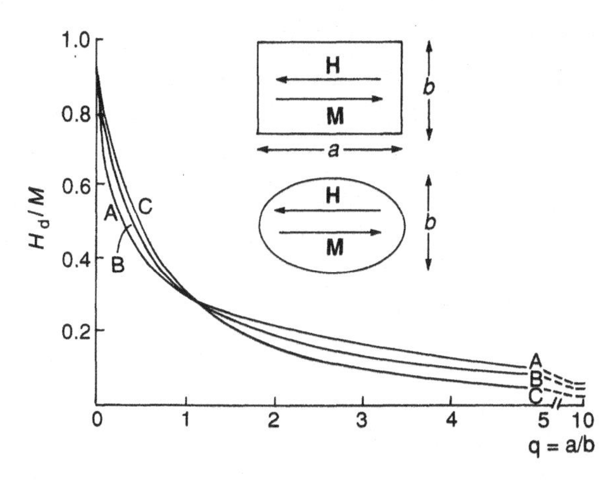

Figure 4.5 Demagnetizing fields at the centre of the square-section block (A) compared with those at all points within the inscribed ellipsoids (C) and those corresponding to the calculated magnetostatic self-energies (B)

discussed later, can be equated to $\frac{1}{2}\mu_0 H_d M$, with H_d the mean field, giving curve B in the figure. Again, specifically for the cube, H_d is equal to the true demagnetizing field for the sphere; from the centre, H_d rises at points along the axis $\parallel M$ but falls at points in the transverse plane to average to the central value.

Within an ellipsoid $\mathbf{H}_M$ is uniform when $\mathbf{M}$ is uniform and it is scarcely surprising that the induced $\mathbf{M}$ in an ellipsoid in a uniform applied field is uniform, as demonstrated in Section 1.3.2. The internal field with H_a along a principal axis is $\mathbf{H} = \mathbf{H}_i = \mathbf{H}_a - N\mathbf{M}$; rearranging, $M = \chi H$ gives

$$\mathbf{M} = \frac{\chi \mathbf{H}_a}{1 + N\chi} \tag{10}$$

For disordered materials the demagnetizing effects are very small, but sufficient to give measurable field shifts in intrinsically accurate n.m.r. measurements on two-phase specimens such as colloids or emulsions, so that n.m.r. can be used to determine χ values. If χ is very large, as for soft ferromagnetics, $M \to H_a/N$ (unless $N \to 0$) and the response is dominated by shape effects with an effective $\chi = 1/N$.

For a sphere, for example, $M = \chi H_a/(1+\frac{1}{3}\chi)$ and the internal field is $H_i = H_a - \frac{1}{3}M = H_a - \frac{1}{3}H_a/(1+\frac{1}{3}\chi) = 3H_a/(2+\mu_r)$. The induction is

$$\mathbf{B} = \mu_0(\mathbf{H}_i + \mathbf{M}) = \mu_0(\mathbf{H}_i + \chi\mathbf{H}_i) = \mu_0\mu_r H_i$$
$$= \mu_0\mu_r \frac{3\mathbf{H}_a}{2+\mu_r} = \mu \frac{3\mathbf{H}_a}{2+\mu_r}$$

By considering each component of the magnetization to produce a separate pole distribution it is easy to see that even if the field is not applied along a principal axis the induced magnetization is uniform, but is no longer $\parallel H$.

1.1.1 Generalization: non-uniform magnetization

M cannot be taken outside the integral [equation (1)]. Expanding $\nabla \cdot (\mathbf{M}r^{-1})$ and grouping alternate terms:

$$\nabla \cdot (\mathbf{M}r^{-1}) = r^{-1}\frac{\partial M_x}{\partial x} + M_x \frac{\partial}{\partial x} r^{-1} + r^{-1}\frac{\partial M_y}{\partial y} + \cdots$$

$$= r^{-1}\nabla \cdot \mathbf{M} + \mathbf{M} \cdot \nabla r^{-1}$$

Thus equation (1) becomes

$$\varphi(\mathbf{r}_2) = \frac{1}{4\pi}\int_V \nabla_1 \cdot (\mathbf{M}r^{-1})\,\mathrm{d}v - \frac{1}{4\pi}\int_V r^{-1}\nabla_1 \cdot \mathbf{M}\,\mathrm{d}v$$

The first term can be transformed to an integral over the surface S of the magnetized body to give

$$\varphi(\mathbf{r}_2) = \frac{1}{4\pi}\int_S r^{-1}\mathbf{M} \cdot \mathrm{d}\mathbf{S} - \frac{1}{4\pi}\int r^{-1}\nabla_1 \cdot \mathbf{M}\,\mathrm{d}v$$

$$= \frac{1}{4\pi}\int_S \frac{\mathrm{d}\sigma}{r} + \frac{1}{4\pi}\int_V \frac{\rho}{r}\,\mathrm{d}v, \qquad \text{where } \rho = -\nabla \cdot \mathbf{M} \tag{11}$$

and the potential is obtained as contributions from surface poles, as suggested above, plus those from the volume poles ρ now that $\nabla \cdot \mathbf{M} \neq 0$. Since

$$\int_V \nabla \cdot \mathbf{M}\,\mathrm{d}v = \int_S \mathbf{M} \cdot \mathrm{d}\mathbf{S}, \qquad \int_S \sigma\,\mathrm{d}S + \int_V \rho\,\mathrm{d}v = 0$$

it is clear that the total pole strength is always zero, in line with the formal existence of the poles as constituents of dipoles.

Volume pole densities arise, for example, in cylinders in which **M** is uniform in magnitude but radially directed w.r.t. the cylinder axis. The flux of M for a cylindrical shell bounded by r and $r + \mathrm{d}r$ is $M \times [2\pi(r + \mathrm{d}r) - 2\pi r]$ per unit length along the axis and the volume of the shell is $\pi[(r + \mathrm{d}r)^2 - r^2]$ so that, neglecting $\mathrm{d}r^2$, in the limit the divergence is the flux per unit volume and

$$\rho(r) = -\frac{M}{r} \tag{12}$$

For a hollow cylinder the sum of the appropriate surface and volume integrals is then readily seen to be zero.

1.2 Equations of Poisson and Laplace

Recall that the point divergence is given by the flux across a closed surface (nominally closed in the above example by annuli across which the components of **M** were zero) divided by the volume V enclosed, in the limit $V \to 0$. By Gauss' theorem for a conservative field the flux is equal to the total source enclosed; further in a sufficiently small element the pole density can be taken to be constant and since $\nabla \cdot \mathbf{B} = 0$, $\nabla \cdot \mathbf{M} = -\rho$, $\nabla \cdot \mathbf{H} = \rho$ so that

$$\nabla^2\varphi = -\rho \tag{13}$$

($\mathbf{H} = -\nabla\varphi$). This is Poisson's equation for the scalar potential, reducing to Laplace's equation when $\mathbf{M}$ is uniform or zero: $\nabla \cdot (\mathbf{H} + \mathbf{M}) = 0 = -\nabla^2\varphi + \nabla \cdot \mathbf{M} = -\nabla^2\varphi$.

It has been seen that $\nabla\times\mathbf{H} = (1/\mu_0)\nabla\times\mathbf{B} = \mathbf{J}$, the current density, and $\nabla\cdot\mathbf{A} = 0$ due to the simple relation of $\mathbf{A}$ to current distributions or circuits. Using a general identity which the reader may confirm by expansion, it follows that

$$\nabla\times\mathbf{B} = \nabla\times(\nabla\times\mathbf{A}) = \nabla(\nabla \cdot \mathbf{A}) - \nabla^2\mathbf{A} = -\nabla^2\mathbf{A}$$

i.e.

$$\left(\frac{1}{\mu_0}\right)\nabla^2\mathbf{A} = -\mathbf{J}, \qquad \left(\frac{1}{\mu}\right)\nabla^2\mathbf{A} = R\nabla^2\mathbf{A} = -\mathbf{J} \tag{14}$$

which is Poisson's equation for the vector potential, where R is the reluctance. In regions where no current densities exist $\nabla\times\mathbf{B} = 0$ and since $\mathbf{B}$ is then irrotational it can be taken to be given by the gradient of a scalar potential: $\mathbf{B} = -\mu_0\nabla\varphi$. With $\nabla\cdot\mathbf{B} = 0$ this indicates that $\nabla^2\varphi = 0$ always, for $J = 0$, but Poisson's equation does apply when the potentials and fields are those arising from magnetized material according to the models discussed above. The apparent anomaly is simply a consequence of the arbitrary introduction of conservative fields arising from the magnetization as such, rather than directly from current distributions.

The most general magnetostatic calculations consist of the solution of these equations, together with appropriate boundary conditions and numerical approaches are surveyed later.

1.3 Boundary Conditions

A boundary is a surface at which μ or $\mathbf{M}$ changes, e.g. the surface of a magnetic material, or a surface occupied by a current sheet. Figure 4.6 shows a region of space of 'pill-box' shape spanning the surface. The volume is so small that the surface enclosed can be considered flat. The ends are parallel to the surface and the curved side is normal to the surface. As $\delta h \to 0$ the flux across the side becomes negligible and the total outward flux is $\mathbf{B}_2 \cdot \mathbf{n}_2 - \mathbf{B}_1 \cdot \mathbf{n}_1$. Also, both $\mathbf{n}_1$ and $\mathbf{n}_2$ become $\mathbf{n}$ at the surface, and since $\nabla \cdot \mathbf{B} = 0$,

$$(\mathbf{B}_2 - \mathbf{B}_1) \cdot \mathbf{n} = 0, \qquad \mathbf{B}_1 \cdot \mathbf{n} = \mathbf{B}_2 \cdot \mathbf{n} \tag{15}$$

Thus there is no change in the normal component of $\mathbf{B}$ on crossing a surface. (The normal component of $\mathbf{B}$ at points contiguous to a surface, on either side, is the same.)

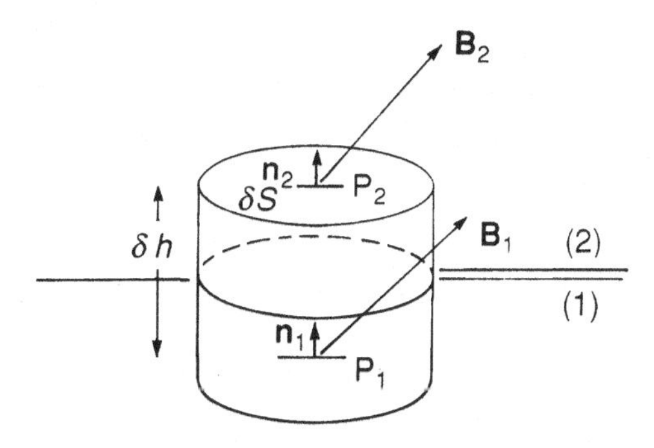

Figure 4.6 The continuity of the normal component of **B** across an interface

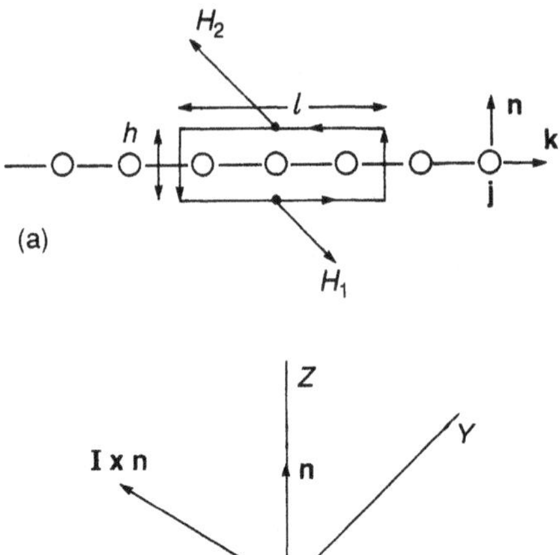

Figure 4.7 The change in the tangential component of **H** on crossing an interface carrying a surface current: **I**×**n** is in *OXY* (b)

It will be seen below that there is no change in the tangential component of **H** across a surface in the absence of currents. Thus with $\mathbf{B} = \mu_0(\mathbf{H} + \mathbf{M})$ there *are* changes in the tangential components of **B** given by $\mu_0\times$ the change in the component of **M** or, equivalently, the corresponding surface current density representing **M**.

Figure 4.7(a) shows a rectangular circuit crossing a surface carrying a surface current **I**. The circuit is, analogously to the preceding case, small enough for the surface to be considered flat and for the fields on either side to be considered to be uniform within the circuit. Here **n** is the unit normal vector, **k** a unit vector in the surface and **j** a unit vector out of the diagram: $\mathbf{j} = \mathbf{k}\times\mathbf{n}$. The direction of **I**, in the surface, is arbitrary but its component normal to the circuit is $\mathbf{I}\cdot\mathbf{j}$ and the current enclosed by the circuit is $(\mathbf{I}\cdot\mathbf{j})l$. As $h \to 0$ the line integral becomes

$$\int_C \mathbf{H}\cdot d\mathbf{r} = l\mathbf{H}_1\cdot\mathbf{k} - l\mathbf{H}_2\cdot\mathbf{k} = (\mathbf{I}\cdot\mathbf{j})l \tag{16}$$

according to the circuital law. Since $\mathbf{j} = \mathbf{k}\times\mathbf{n}$, $\mathbf{I}\cdot\mathbf{j} = -(\mathbf{I}\times\mathbf{n})\cdot\mathbf{k}$ by a general rule or according to the diagram of Figure 4.7(b) (with $|\mathbf{n}| = 1$). Thus equation (16) gives

$$\mathbf{H}_2 - \mathbf{H}_1 = \mathbf{I}\times\mathbf{n} \tag{17}$$

If **I** is in fact normal to the plane of the diagram $\mathbf{I}\cdot\mathbf{j} = I$ and from equation (16),

$$(\mathbf{H}_1 - \mathbf{H}_2)\cdot\mathbf{k} = I \tag{18}$$

i.e. the tangential components, parallel to the surface and normal to the current, change by I on crossing the surface. This is, of course, exemplified by the case of a long solenoidal with

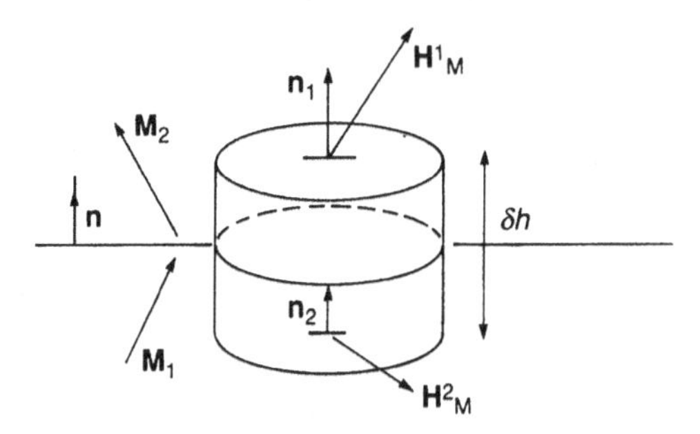

Figure 4.8 The change in the normal component of **H** on crossing an interface at which **M** changes

surface current density $\mathbf{I} = I\mathbf{e}_\varphi$ and $\mathbf{H} = I\mathbf{e}_z$ within the solenoidal while $\mathbf{H} = 0$, outside. If there are no real surface currents, $\mathbf{I} = 0$, there is clearly no change in the tangential component of H on crossing the interface.

Turning to $\mathbf{H}_M$, from the magnetization, i.e. from ρ and σ, the volume element in Figure 4.8 can be taken to be so small that $\mathbf{M}$ is effectively uniform within it and $\rho = 0$ but

$$\sigma = (\mathbf{M}_1 - \mathbf{M}_2) \cdot \mathbf{n} \tag{19}$$

An analogous treatment to that referring to Figure 4.6 readily gives

$$(\mathbf{H}_{M_2} - \mathbf{H}_{M_1}) \cdot \mathbf{n} = \sigma \tag{20}$$

i.e. the normal component of **H** changes by σ on crossing the surface. Moreover, as $\delta h \to 0$ the fields $\mathbf{H}_M$ become normal to the surface so that no tangential component is introduced. Thus the relation (18) still applies if **H** is taken to be $\mathbf{H}_C + \mathbf{H}_M$. Conversely, the magnetization does not affect $\mathbf{H}_C$ and thus equation (20) can be written for the total field:

$$\mathbf{H}_2 \cdot \mathbf{n} - \mathbf{H}_1 \cdot \mathbf{n} = \sigma = \mathbf{M}_1 \cdot \mathbf{n} - \mathbf{M}_2 \cdot \mathbf{n} \tag{21}$$

This gives

$$\mathbf{B}_2 \cdot \mathbf{n} = \mu_0(\mathbf{H}_2 \cdot \mathbf{n} + \mathbf{M}_2 \cdot \mathbf{n}) = \mu_0(\mathbf{H}_1 \cdot \mathbf{n} + \mathbf{M}_1 \cdot \mathbf{n}) = \mathbf{B}_1 \cdot \mathbf{n} \tag{22}$$

In accordance with equation (15).

This applies for permanent or induced magnetization, for which latter $B = \mu H$ also. Thus with $H_M = -\nabla\varphi$ and $\partial\varphi/\partial n$ the gradient normal to the surface,

$$\mu_1 \frac{\partial \varphi_1}{\partial n} = \mu_2 \frac{\partial \varphi_2}{\partial n} \tag{23}$$

The existence of this relation indicates that φ is continuous at the surface and so

$$\varphi_1 = \varphi_2 \tag{24}$$

is the basic boundary condition for the scalar potential. φ may not be calculable on the interface and this should be read in terms of the potentials at points closely approaching from each side of the interface.

The following two sections exemplify the use of the boundary conditions in simple situations.

1.3.1 Magnetic images

Applications of the boundary conditions show, very simply, how fields and forces can, in certain circumstances, be accounted for in terms of magnetic images. In Figure 4.9(a), a dipole $\boldsymbol{\mu}$ is at a distance d from the plane surface of a block of material characterized by $\mu_r \to \infty$ and is directed normal to the surface. All dimensions of the block are $\gg d$. In the material, whatever magnetization is induced, $\mathbf{H} = \mathbf{M}/\chi = 0$, $(\chi \to \infty)$, and so, by the continuity of the tangential field component, at the point P in space adjacent to the surface, $H_t = 0$. This is obviously achieved by adding fields from a notional image dipole $\boldsymbol{\mu}' = \boldsymbol{\mu}$, as indicated in the figure. Indeed, only such a dipole field can have this effect at all such points P. Alternatively, the potential corresponding to each dipole field must obey Laplace's equation and since ∇^2 is linear, φ for the two dipoles constitutes a solution of the equation which also fulfills the boundary conditions, so the solution is acceptable: the reader may readily show that the continuity of φ across the interface may be invoked to give the same result.

It is easy to see, similarly, that if $\boldsymbol{\mu}$ is parallel to the surface the image is as shown in Figure 4.9(b), with $\boldsymbol{\mu}' = -\boldsymbol{\mu}$. The dipoles with these two particular orientations can be regarded as giving the two components of a generally oriented dipole so that the general case is easily envisaged.

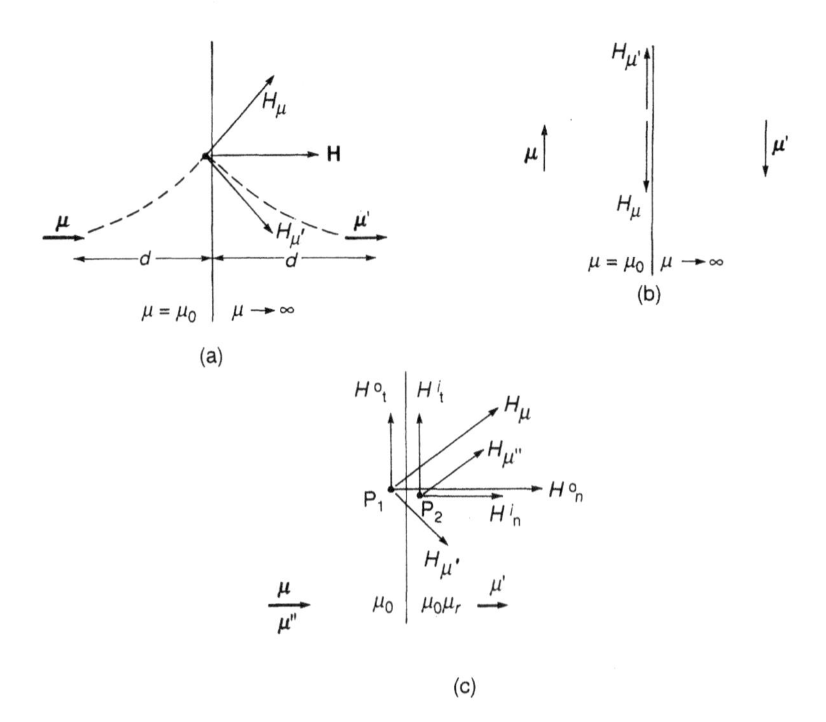

Figure 4.9 The introduction of imaginary image dipoles which ensure that H_t is continuous for the two components in (a) and (b) and the construction for finite permeability (c)

There is no real point dipole in the material, but a magnetization distribution is created which gives the potentials and fields that would be generated by such an image dipole. This particular case suggests that more generally, with μ_r finite, a solution might be achieved in which the fields at a point P_1, close to the surface in space, are those from $\boldsymbol{\mu}$ together with an image $\boldsymbol{\mu}'$ ($\neq \boldsymbol{\mu}$, necessarily), while those at P_2 in the material adjacent to P_1 are those of a modified dipole $\boldsymbol{\mu}''$ replacing $\boldsymbol{\mu}$ [Figure 4.9(c)]. (Whether or not this seems reasonable, the important question is whether such a system can be shown to conform with the boundary conditions.) Since P_1 and P_2 differ infinitesimally in position, the appropriate field components are proportional to the dipole magnitudes so that it is unnecessary to write the explicit expressions and it is apparent by reference to the figure that

$$\left.\begin{array}{l}(H_t^0 = H_t^i)\ \boldsymbol{\mu} - \boldsymbol{\mu}' = \boldsymbol{\mu}'' \\ (B_n^0 = B_n^i)\ \boldsymbol{\mu} + \boldsymbol{\mu}' = \mu_r \boldsymbol{\mu}''\end{array}\right\} \boldsymbol{\mu}'' = \frac{2\boldsymbol{\mu}}{1+\mu_r},\ \boldsymbol{\mu}' = \frac{\mu_r - 1}{\mu_r + 1}\boldsymbol{\mu} \tag{25}$$

reducing to the foregoing, $\boldsymbol{\mu}' = \boldsymbol{\mu}$ and $\boldsymbol{\mu}'' = 0$, as $\mu_r \to \infty$ and also showing that the image disappears as $\mu_r \to 1$. Again the existence of these images, or modifications, purely as affording a method of calculation can be considered to have been justified.

Calculations based on images are genuinely useful as simple approximations since it can be demonstrated, by calculating the actual magnetization distributions in finite blocks and the consequent fields and forces, that the approximations are reasonable so long as the block dimensions are two or three times d (by the method of Section 4.1 for example).

An interesting situation arises when $\mu_r = 0$, $\boldsymbol{\mu}' = -\boldsymbol{\mu}$, which is pursued in Chapter 6, Section 3.

1.3.2 Permeable sphere in a uniform applied field

If $\mathbf{H}_a = H_a \mathbf{e}_z$ the corresponding potential is

$$\varphi = -H_a r \cos\theta$$

because (see Figure 4.10)

$$\begin{aligned}-\nabla\varphi &= \mathbf{e}_r \frac{\partial\varphi}{\partial r} + \mathbf{e}_\theta \frac{1}{r}\frac{\partial\varphi}{\partial\theta} + \mathbf{e}_\phi \frac{1}{r\sin\theta}\frac{\partial\varphi}{\partial\phi} \\ &= \mathbf{e}_r H\cos\theta - \mathbf{e}_\theta H\sin\theta\end{aligned}$$

In the presence of the sphere trial solutions are

$$\varphi_1 = Ar\cos\theta, \qquad \varphi_2 = -H_a r\cos\theta + \frac{B\cos\theta}{r^2} \tag{26}$$

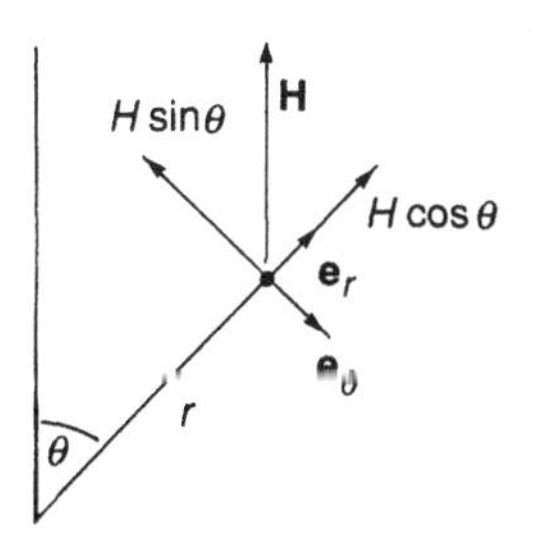

Figure 4.10 Coordinates and unit vectors referred to in the text

with φ_1 and φ_2 the potentials inside and outside the sphere, based on the assumptions that the internal fields may be expected to be uniform and the external fields due to the sphere should then be dipole fields. A and B are to be determined. It is clear that these satisfy Laplace's equation. Setting $r = a$, the condition $\varphi_1 = \varphi_2$ gives

$$Aa\cos\theta = -H_a a\cos\theta + \frac{B\cos\theta}{a^2} \tag{27}$$

and noting that for the sphere $\partial\varphi/\partial n = \partial\varphi/\partial r$, equation (23) gives, with equation (26),

$$\mu_r \left.\frac{\partial\varphi_1}{\partial r}\right|_{r=a} = \left.\frac{\partial\varphi_2}{\partial r}\right|_{r=a}, \qquad \mu_r A\cos\theta = -H_a\cos\theta - \frac{2B\cos\theta}{a^3} \tag{28}$$

Multiplying equation (27) by $2/a$ and adding to equation (28) gives $A = -3H_a/(\mu_r + 2)$. Multiplying equation (27) by μ_r/a and equating right-hand sides with equation (28) gives $B = a^3 H_a(\mu_r - 1)/(\mu_r + 2)$. Thus,

$$\varphi_1 = -\frac{3H_a}{\mu_r + 2} r\cos\theta, \qquad \varphi_2 = -H_a r\cos\theta + a^3 H_a \frac{\mu_r - 1}{\mu_r + 2}\frac{\cos\theta}{r^2}$$

The internal field is $\mathbf{H}_1 = -\nabla\varphi_1 = [3H_a/(\mu_r + 2)]\mathbf{k}$ using $r\cos = z$ for convenience. Writing $\varphi_2 = \varphi_a + \varphi_b$, $-\nabla\varphi_a = \mathbf{H}_a$. Also,

$$\varphi_b = \tfrac{4}{3}\pi a^3 H_1 \chi \frac{\cos\theta}{r^2}\frac{1}{4\pi} = \frac{1}{4\pi}(VM)\frac{\cos\theta}{r^2}$$

on substituting for H_a, using $\mu_r - 1 = \chi$ and $\chi H_1 = M$, and with V the volume of the sphere. This is the expression for the potential due to a dipole of strength equal to the total moment, at the centre of the sphere, as expected from the permanent magnet case. The total external field is the sum of this dipole field and the applied field. Also, the internal field is readily seen to be given by

$$H_1(\mu_r + 2) = H_1(\chi + 3) = M + 3H_1 = 3H_a$$

i.e.

$$H_1 = H_a - \tfrac{1}{3}M \tag{29}$$

Noting that the normal component of $\mathbf{B}$ at the surface is $-\mu_0\mu_r\,\partial\varphi/\partial r$ at $r = a$, inside the sphere

$$B_{n_1} = \frac{3\mu_0\mu_r H_a\cos\theta}{\mu_r + 2}$$

and outside, at $r = a$, $B_n = -\mu_0\,\partial\varphi_2/\partial r$,

$$B_{n_2} = \mu_0 H_a\cos\theta + \mu_0 a^3 H_a \frac{\mu_r - 1}{\mu_r + 2}\frac{2\cos\theta}{a^3} = \cdots = B_{n_1}$$

as required. The tangential component of $\mathbf{H}$ approaching the surface from within is $H_{t_1} = H_1\sin\theta = (H_a - \frac{1}{3}M)\sin\theta$ and at neighbouring points outside

$$H_{t_2} = H_a\sin\theta - \frac{1}{4\pi}\tfrac{4}{3}a^3\frac{M\sin\theta}{a^3} = \cdots = H_{t_1}$$

again as required. The components of $\mathbf{H}$ normal to the surface at $r = a$ are

$$H_{n_1} = -\frac{\partial \varphi_1}{\partial r} = H_1 \cos\theta = (H_a - \tfrac{1}{3}M)\cos\theta$$

$$H_{n_2} = -\frac{\partial \varphi_2}{\partial r} = H_a \cos\theta + \frac{1}{4\pi} VM \frac{2\cos\theta}{a^3} = H_a \cos\theta + \tfrac{2}{3}M\cos\theta$$

Thus

$$H_{n_2} - H_{n_1} = M\cos\theta = \mathbf{M}\cdot\mathbf{n}$$

$$\frac{\partial \varphi_1}{\partial n} - \frac{\partial \varphi_2}{\partial n} = \mathbf{M}\cdot\mathbf{n}$$

and the value of this simple model is apparent. In fact, an alternative calculation may be based purely on the boundary conditions and this is left as an exercise.

1.4 Vector Potential and Surface Currents

By commencing with the vector potential from an element $\mathbf{M}\,\mathrm{d}v$ treated as a point dipole,

$$\mathrm{d}\mathbf{A}(\mathbf{r}_2) = \frac{\mu_0}{4\pi}[\mathbf{M}(\mathbf{r}_1)\,\mathrm{d}v]\frac{\mathbf{r}}{r^3} \tag{30}$$

(where $\mathbf{r} = \mathbf{r}_2 - \mathbf{r}_1$), a similar treatment to the above gives

$$\mathbf{A}(\mathbf{r}_2) = \frac{\mu_0}{4\pi}\int_S \frac{\mathbf{M}(\mathbf{r}_1)\times\mathrm{d}\mathbf{s}}{r} + \frac{\mu_0}{4\pi}\int_V \frac{\nabla_1\times\mathbf{M}(\mathbf{r}_1)}{r}\,\mathrm{d}v \tag{31}$$

with the second term zero for uniform magnetization. The first term is loosely comparable to the foregoing surface pole term but its detailed interpretation is completely different. For a cylinder with uniform axial $\mathbf{M}$ (Figure 4.11), for example, $\mathrm{d}s = a\,\mathrm{d}\phi\,\mathrm{d}z$ and $|\mathbf{M}\times\mathbf{n}| = M$ with the direction indicated. For the strip,

$$\mathrm{d}\mathbf{A} = \frac{a\mu_0}{2}\frac{\mathbf{M}\mathrm{d}z}{r}$$

which is the expression giving $\mathrm{d}\mathbf{A}$ from a ring current of magnitude $M\,\mathrm{d}z$. Integrating formally,

$$\mathbf{A}(\mathbf{r}_2) = \frac{\mu_0}{4\pi}\int_S \frac{\mathbf{I}_M(\mathbf{r}_1)\mathrm{d}S}{r} \tag{32}$$

where

$$\mathbf{I}_M = \mathbf{M}\times\mathbf{n}, \qquad I_M = M \tag{33}$$

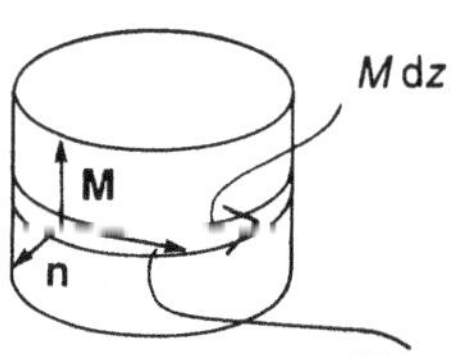

Figure 4.11 The direction of the effective surface current vector

i.e. the potential, and $\mathbf{B} = \nabla \times \mathbf{A}$, is that which would be given by a surface current of magnitude M (or $|\mathbf{M} \times \mathbf{n}|$ in the general case with $\mathbf{M}$ not $\parallel$ the surface) flowing around the cylinder surface.

This (undetectable) surface current has no greater reality than the alternative surface poles. Each may be considered as a representation of the magnetization to facilitate calculations which could scarcely be made for Avogadro's number of dipoles. However, it must be stressed that the $\mathbf{H}$ fields calculated from σ or ρ are specifically the conservative or 'source' fields while the fields calculated from $\mathbf{I}$ are obviously solenoidal and can be identified with B as below. It is convenient at this stage to introduce the concept of the magnetic shell.

1.5 Magnetic Shell or Dipole Sheet

Consider a sheet of material with uniform magnetization $\mathbf{M}$ normal to the surface and thickness t, in the form of a disc of radius a. Suppose that a/t is so great that the thickness can be considered negligible in any calculation (without assuming $t \to 0$ specifically). Since $\mathbf{M}$ is the magnetic dipole moment per unit volume the dipole moment per unit surface area of the sheet is $\mathbf{m} = \mathbf{M}t$. Such a sheet of dipole density, of negligible thickness, is sometimes called a magnetic shell (without these particular restrictions on the geometry).

The scalar potential from a dipole $\boldsymbol{\mu}$ at the origin at P given by $\mathbf{R}$ is $(1/4\pi)\boldsymbol{\mu} \cdot \mathbf{R}/R^3$. In Figure 4.12(a), $\mathrm{d}\boldsymbol{\mu} = \mathbf{m}\,\mathrm{d}\mathcal{A}$ and due to the symmetry $\mathrm{d}\mathcal{A}$ can be taken as $2\pi r\,\mathrm{d}r$. Then $\mathrm{d}\boldsymbol{\mu} \cdot \mathbf{R} = R\,\mathrm{d}\mu \cos\theta$ and the potential for the ring is

$$\mathrm{d}\varphi = \frac{1}{4\pi}(m 2\pi r\,\mathrm{d}r)\frac{R\cos\theta}{R^3} = \frac{m}{2}\sin\theta\,\mathrm{d}\theta$$

since $R = z/\cos\theta$ and $\mathrm{d}R = (z/\cos^2\theta)\sin\theta\,\mathrm{d}\theta$. For the whole sheet,

$$\varphi = -\frac{m}{2}[\cos\theta]_0^\alpha = \frac{m}{2}(1 - \cos\alpha) = \frac{m}{2}\left[1 - \frac{z}{(z^2 + a^2)^{1/2}}\right] \tag{34}$$

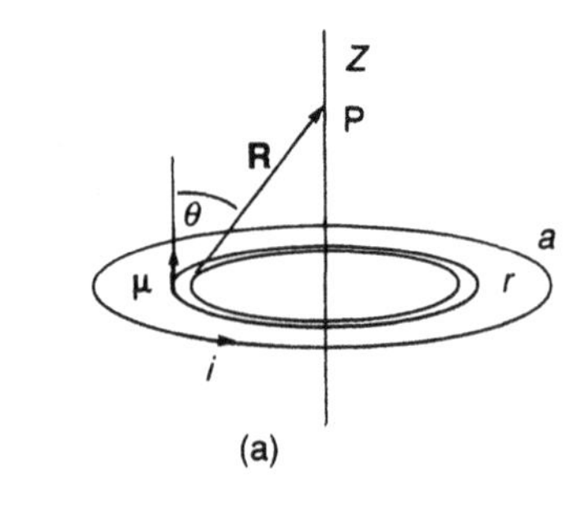

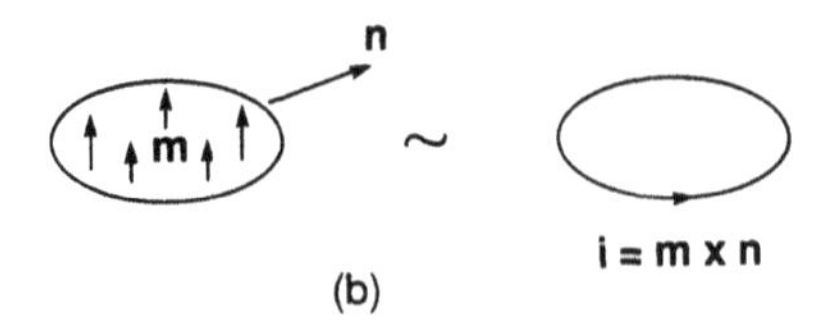

Figure 4.12 A magnetic shell or sheet of dipoles, equivalent to a bounding current loop (b)

(where $\alpha = \tan^{-1} a/z$) and the axial field is given by

$$H_z = -\frac{\partial\varphi}{\partial z} = \frac{m}{2}\left[\frac{1}{(z^2+a^2)^{1/2}} - \frac{z^2}{(z^2+a^2)^{3/2}}\right]$$

$$= \frac{m}{2}\frac{a^2}{(z^2+a^2)^{3/2}} = \frac{i}{2}\frac{a^2}{(z^2+a^2)^{3/2}} \tag{35}$$

where $i = m = Mt$.

This is a striking result because equation (35) shows that the fields from the dipole sheet are exactly those generated by an equivalent line current flowing around the perimeter [Figure 4.12(b)]. The magnitude of the current is simply m or Mt and the direction given by $\mathbf{i} = \mathbf{m} \times \mathbf{n}$ with $\mathbf{n}$ the outward normal at the perimeter as shown. (It would be too great a coincidence for this to apply on the axis only and indeed it can be shown to apply to H_x and H_z at general points.)

In the corresponding physical picture it is supposed that, in taking the dipoles to form a continuum, the individual microscopic current loops merge in such a way that they cancel at all internal points and leave a net current on the perimeter only. Particularly with this in mind, it can be accepted that the conclusions are not restricted to the particular geometry used for the demonstration.

1.6 Scalar Potential and Solid Angle

The potential (34) can be written as

$$\varphi = m\frac{1}{4\pi}[2\pi(1-\cos\alpha)] \tag{36}$$

i.e.

$$\frac{\varphi}{m} = \frac{1}{4\pi}\omega_a \tag{37}$$

where ω_a is the solid angle subtended by the disc of radius a at the point P. The Gaussian image Q′ of any point Q on the perimeter of the disc, i.e. on C, is the point of intersection of PQ with a unit sphere centred on P. The image of C is C′, the locus of the points Q′. The solid angle is the area of the portion of the unit sphere bounded by C′. According to Figure 4.13(a), this area is

$$\omega_a = \mathcal{A} = \int_0^\alpha 2\pi(1)\sin\theta(1)\,\mathrm{d}\theta$$

$$= 2\pi(1-\cos\alpha) \tag{38}$$

and so equation (37) follows.

In Figure 4.13(b), S is a spherical surface of radius R and S_1 an arbitrary surface. Then $\mathbf{r}\cdot\mathrm{d}\mathbf{S}_1/r$ projects the element $\mathrm{d}\mathbf{S}_1$ parallel to $\mathrm{d}\mathbf{S}_1'$ on S. Thus, by simple geometry,

$$\frac{\mathbf{r}\cdot\mathrm{d}\mathbf{S}_1}{r^3} = \frac{\mathbf{R}\cdot\mathrm{d}\mathbf{S}_1'}{R^3} = \frac{\mathrm{d}S_1}{R^2}$$

This is the element of solid angle subtended by $\mathrm{d}S_1$ at 0. Integrating over the surface S_1,

$$\omega = \int_{S_1}\frac{\mathbf{r}\cdot\mathrm{d}\mathbf{S}_1}{r^3} \tag{39}$$

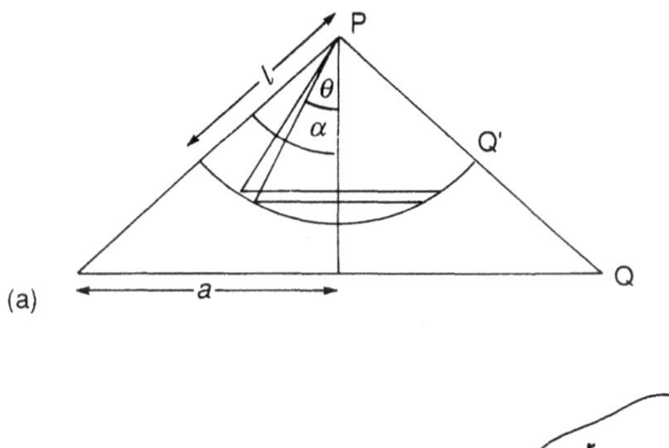

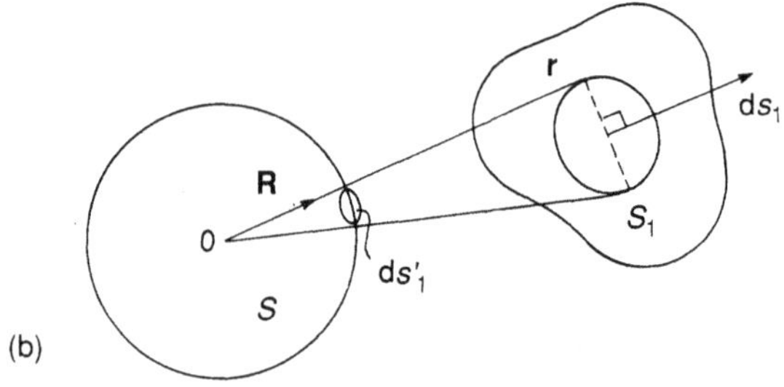

Figure 4.13 Scalar potential and solid angle

If S is now replaced by a 'dipole sheet' or magnetic shell of strength m or $m\mathbf{n}$, then the elementary dipole associated with $\mathbf{dS}$ is $\mathrm{d}\boldsymbol{\mu} = m\mathbf{n}\,\mathrm{d}S = m\,\mathbf{dS}$ and

$$\mathrm{d}\varphi = \frac{1}{4\pi}m\frac{\mathbf{r}\cdot\mathbf{dS}}{r^3}$$

For the whole shell,

$$\varphi = \frac{m}{4\pi}\int_S \frac{\mathbf{r}\cdot\mathbf{dS}}{r^3} = \frac{m}{4\pi}\omega_S \tag{40}$$

as follows by comparison with equation (39). According to the foregoing we could equally write

$$\varphi = \frac{I}{4\pi}\omega_S \tag{41}$$

where I is the line current flowing around the perimeter of S and this applies whether I is in fact a real current or the effective current $I = m = Mt$ representing the magnetization of the shell. Obviously there is no restriction on the shape of the shell or circuit. Generally we should write $\varphi = \pm(M/4\pi)\omega_S$ or $\varphi = \pm(I/4\pi)\omega_S$ to account for the direction of the magnetization or the current. Referring to Figure 4.13, $\partial\omega/\partial z$ is negative and $H_z = -\partial\varphi/\partial z$ is positive if the positive sign is taken.

Referring to Figure 4.14(a), it is postulated that the fields from a short cylindrical magnet with $\mathbf{M}$ as shown are those from a disc (1) of surface poles of density $\sigma = \mathbf{M}\cdot\mathbf{n} = M$ at $z = 0$ together with a disc (2) of density $\sigma = -M$ at $z = -t$. These fields can themselves be expressed in terms of solid angles because $H_{z_1} = \frac{1}{2}M[1 - z/(z^2 + a^2)^{1/2}]$ is also

$$H_{z_1} = \left(\frac{M}{4\pi}\right)[2\pi(1 - \cos\alpha_1)] = \left(\frac{M}{4\pi}\right)\omega_1$$

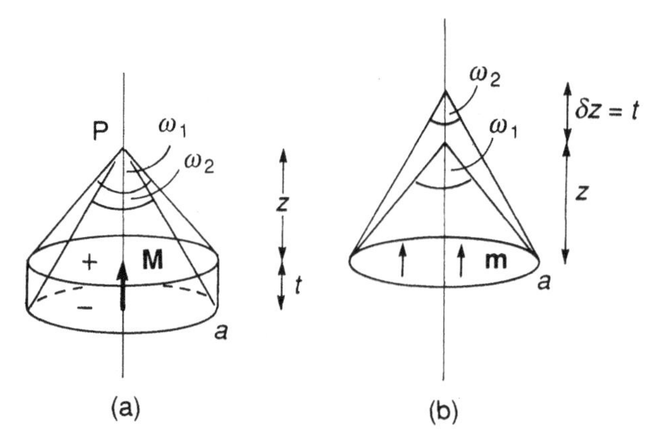

Figure 4.14 A magnet represented by two pole sheets (a) and a magnetic shell (b)

and similarly $H_{z_2} = (-M/4\pi)\omega_2$ with ω_1 and ω_2 the solid angles subtended by disc 1 and disc 2 respectively. The total field at P is thus

$$H_z = H_{z_1} + H_{z_2} = \left(\frac{M}{4\pi}\right)(\omega_1 - \omega_2) \tag{42}$$

For the circular magnetic shell with the same value of **M** giving $\mathbf{m} = \mathbf{M}t$ and same radius a [Figure 4.14(b)] the *potentials* at points z and $z + \delta z$ designated φ_1 and φ_2 are

$$\varphi_1 = \left(\frac{m}{4\pi}\right)\omega_1, \qquad \varphi_2 = \left(\frac{m}{4\pi}\right)\omega_2 \tag{43}$$

and the field is thus

$$H_z = -\frac{\partial\varphi}{\partial z} \doteqdot \frac{\varphi_1 - \varphi_2}{\delta z} = \frac{m}{4\pi}\frac{\omega_1 - \omega_2}{\delta z} \tag{44}$$

If δz is set equal to t these are the same solid angles as those in equation (43). Moreover, as t becomes very small equation (44) becomes exact, and with $m = Mt$ and $\delta z = t$ the fields given by equations (42) and (44) become equal. Thus the magnetic shell may indeed be regarded as two pole sheets so far as the external fields are concerned, as well as being represented by a peripheral current.

1.7 The Current Sheet and Pole Sheet Conventions

The axial field from a disc of pole density σ reverses as the disc is crossed and with the origin of z on the disc:

$$H_z = \mathrm{Sg}(z)\frac{\sigma}{2}\left[1 - \frac{|z|}{(z^2 + a^2)^{1/2}}\right] \tag{45}$$

with $\mathrm{Sg}(z) = \pm 1$ for $z \gtrless 0$. Applying this to the two discs representing a cylindrical magnet with uniform axial **M**, $\sigma = \pm M$, of length l and with the origin at the centre of the cylinder:

$$\begin{aligned} H_z = {} & \frac{M}{2}\mathrm{Sg}\left(z - \frac{l}{2}\right)\left\{1 - \frac{|z - l/2|}{[(z - l/2)^2 + a^2]^{1/2}}\right\} \\ & - \frac{M}{2}\mathrm{Sg}\left(z + \frac{l}{2}\right)\left\{1 - \frac{|z + l/2|}{[(z + l/2)^2 + a^2]^{1/2}}\right\} \end{aligned}$$

$$
= \begin{cases} \dfrac{M}{2}\left\{\dfrac{|z+l/2|}{[(z+l/2)^2+a^2]^{1/2}} - \dfrac{|z-l/2|}{[(z-l/2)^2+a^2]^{1/2}}\right\}, & z > \dfrac{l}{2} \\ \dfrac{M}{2}\left\{\dfrac{|z-l/2|}{[(z+l/2)^2+a^2]^{1/2}} + \dfrac{|z+l/2|}{[(z-l/2)^2+a^2]^{1/2}}\right\} - M, & -\dfrac{l}{2} < z < \dfrac{l}{2} \\ \dfrac{M}{2}\left\{\dfrac{|z-l/2|}{[(z-l/2)^2+a^2]^{1/2}} - \dfrac{|z+l/2|}{[(z+l/2)^2+a^2]^{1/2}}\right\}, & z < -\dfrac{l}{2} \end{cases}
$$

However, when $z > l/2$ both $(z + l/2)$ and $(z - l/2)$ are > 0 and the $|\;|$ symbols are superfluous. When $z < -l/2$ both $(z + l/2)$ and $(z - l/2)$ are < 0 so again the $|\;|$ symbols can be omitted, but in this case the signs of both terms must be reversed. For $-l/2 < z < l/2$, $(z - l/2) < 0$ and $(z + l/2) > 0$ so the $|\;|$ symbols can be omitted if the sign of the first term is reversed. Thus we arrive at an expression that applies to all points on the axis apart from the extra term for internal points:

$$
H_z = \begin{cases} \dfrac{M}{2}\left\{\dfrac{z+l/2}{[(z+l/2)^2+a^2]^{1/2}} - \dfrac{z-l/2}{[(z-l/2)^2+a^2]^{1/2}}\right\} & \text{(external)} \\ \dfrac{M}{2}\left\{\dfrac{z+l/2}{[(z+l/2)^2+a^2]^{1/2}} - \dfrac{z-l/2}{[(z-l/2)^2+a^2]^{1/2}}\right\} - M & \text{(internal)} \end{cases} \tag{46}
$$

Obviously this may be simplified to the second case only, since where $M = 0$ the 'external' expression follows automatically. This may also be obtained, as an exercise by integrating for the two discs together.

A comparative calculation can be made assuming a surface current I to flow round the cylinder surface, $\mathbf{I} \perp OZ$. For the circular ring current $I\,\mathrm{d}z'$ (OZ' coincident with OZ affording the variable of integration),

$$
\mathrm{d}H = \frac{Ia^2}{2}\frac{\mathrm{d}z'}{[(z-z')^2+a^2]^{3/2}}
$$

Using the standard form $\int \mathrm{d}x/(x^2+a^2)^{3/2} = (1/a^2)(x^2+a^2)^{1/2}$ and $\mathrm{d}(z-z') = -\mathrm{d}z'$, the field from the whole cylinder is

$$
H_z = \frac{I}{2}\left\{\frac{z+l/2}{[(z+l/2)^2+a^2]^{1/2}} - \frac{z-l/2}{[(z-l/2)^2+a^2]^{1/2}}\right\} \tag{47}
$$

(It is rather interesting to show that integration of the potentials from magnetic shells gives this same expression, but this will not be pursued.)

On setting $I = M$ (or $\mathbf{I} = \mathbf{M}\times\mathbf{n}$ generally; see Figure 4.15) and redesignating the field in equation (46), $\mathbf{H}_M$, and that in equation (47), $\mathbf{H}_I$, it is seen that

$$
\mathbf{H}_I = \mathbf{H}_M + \mathbf{M}, \qquad \mu_0\mathbf{H}_I = \mu_0(\mathbf{H}_M + \mathbf{M}) = \mathbf{B} \tag{48}
$$

so that the current sheet and pole sheet models give fields that vary in the same way save for the constant difference M (within the material, of course, the fields being the same externally where $M = 0$). Equally, since H_I is necessarily solenoidal, this can be considered as an alternative demonstration of the relation $\mathbf{B} = \mu_0(\mathbf{H}+\mathbf{M})$ (including in $\mathbf{H}$ any applied fields which by definition have no sources in the region of interest and are thus solenoidal) with the contrasting natures of $\mathbf{B}$ and of $\mathbf{H}$ apparent.

The demonstration relates to points on the axis. This is an axisymmetric, quasi-two-dimensional, system and the space is bounded by the axis plus lines at infinity, where the

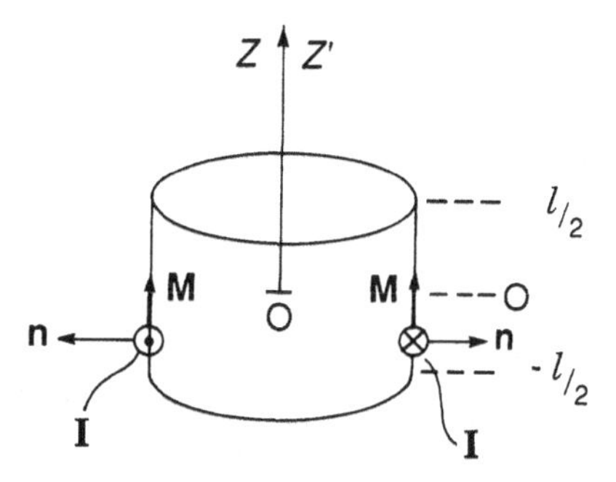

Figure 4.15 The origins of z for the point in space and of z' for integration are at the centre of the magnet

fields and potentials are zero. The different fields may be taken to be implicitly defined over all space since the potentials are defined along the boundaries and Laplace's equation applies. Thus relations between the fields are general. Other simple models could be invoked and full generality may be assumed.

Practical implications of this current–pole sheet equivalence will be illustrated later (Section 4.5), but it is apparent that calculations for magnets and for (real) coils are largely equivalent.

1.8 Magnetic Energy

Expressions for magnetic energies have been introduced in a simple way. To summarize more general treatments, a current distribution $\mathbf{J}_2(\mathbf{r}_2)$ gives a potential $\mathbf{A}_2(\mathbf{r}_1)$ at $\mathbf{r}_1$ and induction $\mathbf{B}_2(\mathbf{r}_1) = \nabla_1 \times \mathbf{A}_2(\mathbf{r}_1)$. A rather lengthy treatment in which $\mathbf{F}_1 \cdot \delta\mathbf{r}_1$ is integrated as $\mathbf{J}_1$ is brought from infinity to a general $\mathbf{r}_1$ gives the work term or mutual energy of the two circuits or distributions (see, e.g. [1]) as

$$W_{12} = W_{21} = \frac{\mu_0}{4\pi} \int \int \frac{\mathbf{J}_1(\mathbf{r}_1) \cdot \mathbf{J}_2(\mathbf{r}_2)}{r} \, \mathrm{d}v_1 \, \mathrm{d}v_2 \tag{49}$$

and using $(\mu_0/4\pi) \int (\mathbf{J}_2/r) \, \mathrm{d}v = \mathbf{A}_2$, etc.,

$$W_{12} = \int \mathbf{J}_1(\mathbf{r}_1) \cdot \mathbf{A}_2(\mathbf{r}_1) \, \mathrm{d}v_1 = \int \mathbf{J}_2(\mathbf{r}_2) \cdot \mathbf{A}_1(\mathbf{r}_2) \, \mathrm{d}v_2 \tag{50}$$

With the replacement $\nabla \times \mathbf{H}_1 = \mathbf{J}_1$ other expressions for the magnetic interaction energies are seen to be

$$W_{12} = \int \mathbf{H}_1 \cdot \mathbf{B}_2 \, \mathrm{d}v = \int \mathbf{H}_2 \cdot \mathbf{B}_1 \, \mathrm{d}v \tag{51}$$

or for linear media with $\mathbf{B} = \mu\mathbf{H} = \mu_r\mu_0\mathbf{H}$,

$$W_{12} = \int \mu \mathbf{H}_1 \cdot \mathbf{H}_2 \, \mathrm{d}v \tag{52}$$

In spherical coordinates the volume element for either integration in equation (49) is $r \sin\theta r \, \mathrm{d}r \, \mathrm{d}\theta \, \mathrm{d}\phi$ and $\mathrm{d}v/r = 0$ even when $r = 0$, so there is no singularity even if the distributions overlap: the case that both distributions have the same configuration in space is not precluded. Suppose that a fraction f of the current in a particular configuration exists, $0 < f < 1$, the distribution being designated $f\mathbf{J}(\mathbf{r})$. Increasing the current density to $(f + \delta f)\mathbf{J}(\mathbf{r})$ is equivalent to bringing up a distribution $\delta f \mathbf{J}(\mathbf{r})$ to the $f\mathbf{J}(\mathbf{r})$ and the interaction energy of the two is

$$\delta W = \frac{\mu_0}{4\pi}\int\int [f\mathbf{J}(\mathbf{r})]\cdot[\delta f\mathbf{J}(\mathbf{r}')]\,\mathrm{d}v\,\mathrm{d}v' \tag{53}$$

Increasing f from 0 to 1 gives the energy of formation or self-energy of the distribution as

$$W = \frac{\mu_0}{4\pi}\int_0^1 f\,\mathrm{d}f \int\int \frac{\mathbf{J}(\mathbf{r})\cdot\mathbf{J}(\mathbf{r}')}{r}\,\mathrm{d}v\,\mathrm{d}v'$$

$$= \frac{1}{2}\frac{\mu_0}{4\pi}\int\int \frac{\mathbf{J}(\mathbf{r})\cdot\mathbf{J}(\mathbf{r}')}{r}\,\mathrm{d}v\,\mathrm{d}v' \tag{54}$$

$\mathbf{J}(\mathbf{r})$ and $\mathbf{J}(\mathbf{r}')$ are the same, single distribution with $\mathbf{r}$ and $\mathbf{r}'$ required for the double integration. As before, it follows also that

$$W = \tfrac{1}{2}\int \mathbf{J}\cdot\mathbf{A}\,\mathrm{d}v \tag{55}$$

or alternatively

$$W = \tfrac{1}{2}\int \mathbf{B}\cdot\mathbf{H}\,\mathrm{d}v \tag{56}$$

and with $B = \mu H$,

$$W = \tfrac{1}{2}\int (\mu\mathbf{H})\cdot\mathbf{H}\,\mathrm{d}v = \tfrac{1}{2}\int \mu H^2\,\mathrm{d}v \tag{57}$$

This can be regarded as the energy of creation of a magnetic field in a permeable medium.

As an alternative for self-energies, set $\mathbf{J}_1$, to some particular $\mathbf{J}'$ with $\mathbf{A}_1 = \mathbf{A}'$ and take a second distribution to be identical to the first, i.e. $\mathbf{J}_2 = \mathbf{J}'$ and $\mathbf{A}_2 = \mathbf{A}'$ also. A third distribution consists of $\mathbf{J}_1$ and $\mathbf{J}_2$ superimposed so that $\mathbf{J}_3 = 2\mathbf{J}'$ and by linearity $\mathbf{A}_3 = 2\mathbf{A}'$ at a point where $\mathbf{A}_1 = \mathbf{A}' = \mathbf{A}_2$. The self-energies must have the same form as the mutual energies and $W = k\int \mathbf{J}\cdot\mathbf{A}\,\mathrm{d}v$, with k to be determined, can be assumed. Thus $W_1 = W_2 = k\int \mathbf{J}'\cdot\mathbf{A}'\,\mathrm{d}v$ and $W_3 = k\int 2\mathbf{J}'\cdot 2\mathbf{A}'\,\mathrm{d}v = 4kW_1$. However, $W_3 = W_1 + W_2$ plus the mutual energy W_{12}, i.e.

$$4k\int \mathbf{J}'\cdot\mathbf{A}'\,\mathrm{d}v = 2k\int \mathbf{J}'\cdot\mathbf{A}'\,\mathrm{d}v + \int \mathbf{J}'\cdot\mathbf{A}'\,\mathrm{d}v$$

and thus $k = \frac{1}{2}$. For n identical systems the number N of interactions can be deduced since for $n = 2$, $N = 1$; for $n = 3$ the interactions are listed as 12, 13; 23, for $n = 4$ as 12, 13, 14; 23, 24; 34, etc., and generally $N = (n-1) + (n-2) + \cdots + [n - (n-1)]$ with $n - 1$ terms so that

$$N = (n-1)n - \sum_{i=1}^{n-1} i = \tfrac{1}{2}(n^2 - n)$$

The condition for k is

$$n^2 k\int \mathbf{J}'\cdot\mathbf{A}'\,\mathrm{d}v = \tfrac{1}{2}(n^2 - n)k\int \mathbf{J}'\cdot\mathbf{A}'\,\mathrm{d}v + nk\int \mathbf{J}'\cdot\mathbf{A}'\,\mathrm{d}v$$

giving $k = \frac{1}{2}$ in general.

Equation (56) states that the interaction energy can be calculated as the integral of the field produced by one circuit over B for the second circuit. Considering the second circuit as a dipole or volume element of magnetization, the potential energy of the dipole in the field becomes

$$W = -\mathbf{H} \cdot \mu_0 \mathbf{M}\,\mathrm{d}v = -\boldsymbol{\mu} \cdot \mathbf{B} \tag{58}$$

and for a body with general magnetization M,

$$W = -\int \mathbf{B} \cdot \mathbf{M}\,\mathrm{d}v \tag{59}$$

However, the self-energy of a magnetized body involves the factor of $\frac{1}{2}$ and

$$W_s = -\tfrac{1}{2} \int \mathbf{M} \cdot \mathbf{B}\,\mathrm{d}v \tag{60}$$

Here the B field is that produced by the magnetization itself and is proportional to M so that the energy required to add an increment of magnetization $(\delta f)M$ when the magnetization is fM is

$$\delta W_s = -\int (\delta f)\mathbf{M} \cdot f\mathbf{B}\,\mathrm{d}v \tag{61}$$

and on integrating as f goes from 0 to 1 the required result is obtained. When equation (60) is applied to the energy of formation of a permanently magnetized body, B must be taken as $\mu_0 H_M$, with H_M the field produced by the magnetization.

1.8.1 Self-energy of ellipsoids: shape anisotropy

For an ellipsoid of revolution with $\mathbf{M}$ uniform and at θ to the a axis (OZ), with $\mathbf{M} = \mathbf{k}M\cos\theta + \mathbf{i}M\sin\theta$, the internal field is $H_M = -\mathbf{k}N_a(M\cos\theta) - \mathbf{i}N_b(M\sin\theta)$ and since this is also uniform the energy *density* is $-\frac{1}{2}\mathbf{B}\cdot\mathbf{M}$, i.e.

$$E_s = -\tfrac{1}{2}\mu_0 \mathbf{H}_M \cdot \mathbf{M} = \tfrac{1}{2}\mu_0 M^2 (N_a \cos^2\theta + N_b \sin^2\theta) = \tfrac{1}{2}\mu_0 M^2 (N_b - N_a)\sin^2\theta + C$$

The constant term is ignored when only the θ dependence is of interest. With a single constant the magnetocrystalline energy is $E_K = K_1 \sin^2\theta$ and E_s corresponds to this with $\frac{1}{2}\mu_0(N_b - N_a)M_s^2 \sim K_1$. In particular the shape anisotropy field is [cf. $2K_1/(\mu_0 M_s)$]

$$H_s = 2\frac{[\frac{1}{2}\mu_0 M_s^2 (N_b - N_a)]}{\mu_0 M_s} = (N_b - N_a)M_s \tag{62}$$

Thus, as noted in Chapter 1, the behaviour of uniaxial single-domain particles is described in the same way whether the anisotropy is magnetocrystalline or is due to the shape, e.g. in a field $\perp$ the easy or long (a) axis, the magnetization curve is linear up to saturation at $H_a = H_s$ and in a field $\parallel$ the a axis switching is expected at $H_a = H_s$.

1.8.2 Alternative calculations using H^2

The self-energy may be equated to the energy of the field created by the magnetization. For a transversely magnetized thin sheet $E_s = \frac{1}{2}\mu_0 N M^2 = \frac{1}{2}\mu_0 M^2$ with $N = 1$ and equally

$E_H = \frac{1}{2}\mu_0 H^2 = \frac{1}{2}\mu_0 M^2$ with $H = -NM$, in a direct treatment dependent on $H = 0$ externally.

For a uniformly magnetized sphere the external (dipole) field H_0 is given in spherical coordinates as $\mathbf{e}_r \mu \cos\theta/(2\pi r^3) + \mathbf{e}_\theta \mu \sin\theta/(4\pi r^3)$ with $\mu = \frac{4}{3}\pi a^3 M$. Since this is independent of ϕ a volume element is taken as $2\pi(r \sin\theta) r \, dr \, d\theta$. It is left as an exercise to show that

$$W_H^0 = \frac{1}{2}\mu_0 \int H_0^2 \, dv = \mu_0 \frac{1}{9}(\tfrac{4}{3}\pi a^3) M^2$$

and that on adding W_H^i for the uniform field within the sphere the result neatly demonstrates the equivalence to $\frac{1}{2}\mu_0 \int \mathbf{H}_\mathrm{d} \cdot \mathbf{M} \, dv$.

1.8.3 Magnetic energy and surface distributions

Using $\mathbf{H} = -\nabla\varphi$,

$$W_\mathrm{s} = -\frac{1}{2}\int \mathbf{B} \cdot \mathbf{M} \, dv = \frac{1}{2}\mu_0 \int (\nabla\varphi) \cdot \mathbf{M} \, dv$$

It can be confirmed by expansion that $\nabla \cdot (\varphi\mathbf{M}) = \varphi\nabla \cdot \mathbf{M} + \mathbf{M} \cdot \nabla\varphi$ so that

$$W_\mathrm{s} = \frac{1}{2}\mu_0 \int \nabla \cdot \varphi\mathbf{M} \, dv - \frac{1}{2}\mu_0 \int \varphi\nabla \cdot \mathbf{M} \, dv \tag{63}$$

Transforming the first integral by Gauss' theorem and with $\mathbf{M} \cdot \mathbf{ds} = \sigma \, ds$, $\nabla \cdot \mathbf{M} = -\rho$,

$$\begin{aligned} W_\mathrm{s} &= \frac{1}{2}\mu_0 \int_S \varphi\mathbf{M} \cdot \mathbf{ds} + \frac{1}{2}\mu_0 \int_V \varphi\nabla \cdot \mathbf{M} \, dv \\ &= \frac{1}{2}\mu_0 \int_S \varphi\sigma \, ds + \frac{1}{2}\mu_0 \int_V \varphi\rho \, dv \end{aligned} \tag{64}$$

In the presence of a domain structure σ may be expressed as a Fourier series and φ as a general solution of Laplace's equation ($\rho = 0$) with coefficients identified according to the boundary conditions and, with considerable effort, the integrations may be achieved (Section 2).

The simplest illustration is inevitably that of Figure 4.16 with $\varphi = \frac{1}{3}Ma\cos\theta$ and $\sigma = M\cos\theta$. The area of the strip is $ds = 2\pi(a \sin\theta) a \, d\theta$ and the reader may show that integration of $\frac{1}{2}\mu_0\varphi\mathbf{M} \cdot \mathbf{n} \, ds$ gives the expected result.

In a second example, on the axis of a disc with pole density σ, at height z, $\varphi = \frac{1}{2}[(z^2 + a^2)^{1/2} - z] \rightarrow \frac{1}{2}\sigma(z - a)$ for $z \ll a$. If a disc of unit area is taken in the centre of a disc with a very great then the expression can be considered general, over the central unit disc. For a corresponding unit surface area of a transversely magnetized sheet of thickness $t \ll a$, $\varphi = \frac{1}{2}\sigma(z - t)$ gives the potential at one surface due to the other. The energy may be regarded as the self-energies of the two discs W_s plus the interaction energy W_i. With

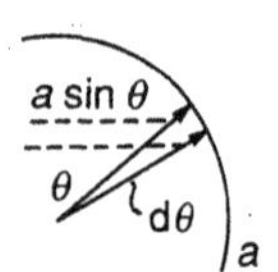

Figure 4.16 For calculating the magnetostatic self-energy of a uniform sphere

$z = 0$, each $W_s = \frac{1}{2}\mu_0\varphi\sigma = \frac{1}{4}\mu_0 M^2 a$ and $W_i = \mu_0\varphi\sigma = -\frac{1}{2}\mu_0 M^2(a - t)$ so the total is $W = W_i + 2W_s = \frac{1}{2}\mu_0 M^2 t$, as expected for the energy per unit surface area, $E = \frac{1}{2}\mu_0 M^2$. It is easy to see that the same result is obtained by taking the product of σ and φ from one surface, at the other, for the two surfaces in turn, effectively integrating $\mu_0\sigma\varphi$ over all the surface.

1.9 Energy, Flux and Inductance

The mutual potential energy E_{12} of two circuits, i.e. two filamentary current loops, is obtained by replacing $\mathbf{J}(\mathbf{r})\,dv$ for a general distribution in space by $i\,\mathbf{dr}$ where $\mathbf{dr}$ gives the local direction [Figure 4.17(a)]. Thus

$$E_{12} = -W_{12} = -\frac{\mu_0 i_1 i_2}{4\pi}\int_{C_1}\int_{C_2}\frac{\mathbf{dr}_1\cdot\mathbf{dr}_2}{r} = -i_1 i_2 L_{12} \tag{65}$$

where

$$L_{12} = \frac{\mu_0}{4\pi}\int_{C_1}\int_{C_2}\frac{\mathbf{dr}_1\cdot\mathbf{dr}_2}{r}$$

is called the mutual inductance of the two circuits. L_{12} is a function of the geometry of the circuits and due to the presence of μ_0 [or from equation (65)] the units are JA^{-2} ($\mu_0 \sim \mathrm{JA}^{-2}\mathrm{m}^{-1}$).
With

$$\mathbf{A}_2(\mathbf{r}_1) = \frac{\mu_0}{4\pi}\int_{C_2}\frac{i_2\,\mathbf{dr}_2}{r} \tag{66}$$

we have

$$E_{12} = i_1\int_{C_1}\mathbf{A}_2(\mathbf{r}_1)\cdot\mathbf{dr}_1 \tag{67}$$

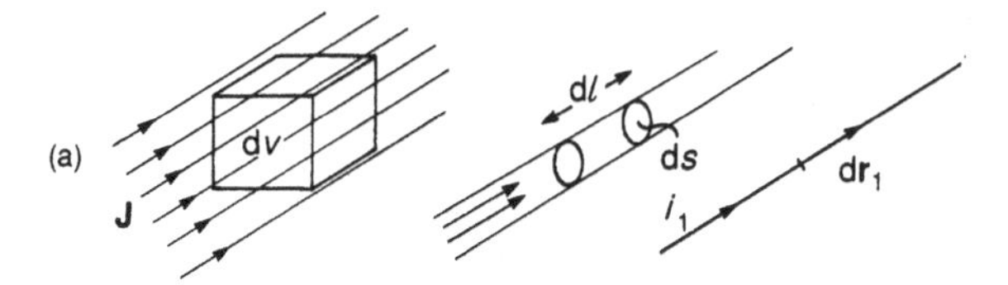

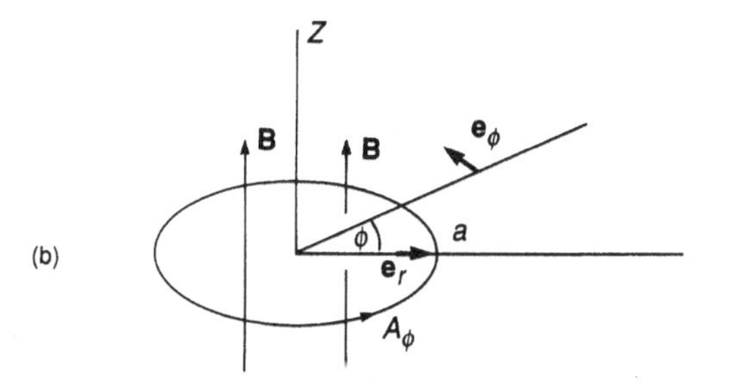

Figure 4.17 (a) **J** is a vector current density and i a scalar current magnitude; (b) shows the unit vectors and the directions of **A** and **B**

Since $\mathbf{B} = \nabla\times\mathbf{A}$ and by Stoke's theorem the surface integral of $\nabla\times\mathbf{v}$ over a surface spanning a circuit C is equal to the line integral of $\mathbf{v}\cdot d\mathbf{r}$ around C, it follows that

$$E_{12} = i_1 \int_{S_1} \nabla\times\mathbf{A}_2 \cdot d\mathbf{s}_1 = i_1 \int_{S_1} \mathbf{B}_2 \cdot d\mathbf{s}_1 \tag{68}$$

For example, if $\mathbf{A} = \mathbf{e}_\phi \frac{1}{2} rB$ [Figure 4.17(b)] in a cylindrical coordinate system then taking $\nabla\times\mathbf{A}$ in cylindrical form (Appendix 2) and picking out the only appropriate term,

$$\mathbf{B} = \nabla\times\mathbf{A} = \frac{1}{r}\frac{\partial}{\partial r} rA_\phi \mathbf{e}_z = \frac{1}{r} rB\mathbf{e}_z = \mathbf{e}_z B \tag{69}$$

This potential corresponds to a uniform induction for which the surface integral spanning the ring of radius a is $\pi a^2 B$. Since $\mathbf{A}$ is directed around the ring its line integral is simply $2\pi a A = \pi a^2 B$, with $A = \frac{1}{2} rB$ and $r = a$.

The integral of $\mathbf{B}_2$ across the surface spanned by C_1 is the flux of $\mathbf{B}_2$ (due to C_2) through C_1, i.e. $N_{C_1}^{B_2}$. Thus

$$E_{12} = i_1 N_{C_1}^{B_2} = i_2 N_{C_2}^{B_1}$$

(since $\mathbf{A}_1$ could be introduced instead of $\mathbf{A}_2$ and the treatment repeated). Comparing this with E_{12} in equation (65),

$$i_1 i_2 L_{12} = i_1 N_{C_1}^{B_2} = i_2 N_{C_2}^{B_1}$$

i.e.

$$L_{12} = \frac{N_{C_1}^{B_2}}{i_2} = \frac{N_{C_2}^{B_1}}{i_1} \tag{70}$$

This can be illustrated by taking two coaxial ring currents constituting C_1 and C_2, as shown in Figure 4.18 in which b is to be taken as $\gg a_1$ or a_2. At large distances the field from C_1 is equivalent to that of a point dipole $\mu_1 = \mathcal{A}_1 i_1$ and $C_2 \sim \mu_2 \doteq \mathcal{A}_2 i_2$. In this same approximation $\mathbf{B}$ is uniform over each surface and due to the geometry $\mathbf{B}$ is normal to each surface. The value of B_1, generated by C_1, at C_2, is $(\mu/4\pi)2(\pi a_1^2)i_1/b^3$ so that

$$N_{C_2}^{B_i} = \pi a_2^2 \frac{\mu}{4\pi} \frac{2\pi a_1^2 i_1}{b^3} \tag{71}$$

and similarly

$$N_{C_1}^{B_2} = \pi a_1^2 \frac{\mu}{4\pi} \frac{2\pi a_2^2 i_2}{b^3} \tag{72}$$

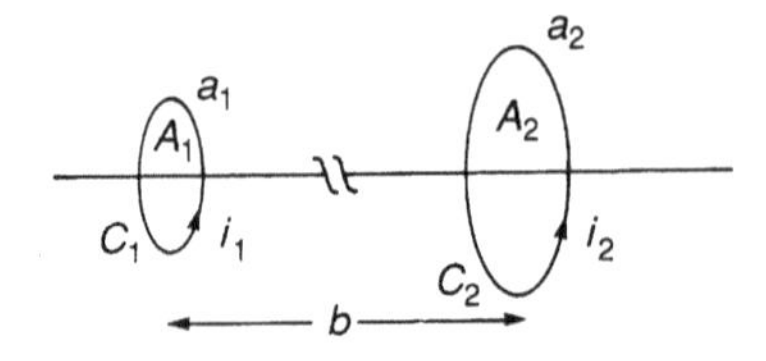

Figure 4.18 Simple model of circuits C_1 and C_2 for inductance calculations

so that clearly $N_{C_1}^{B_2}/I_2 = N_{C_2}^{B_1}/I_1$. A little consideration shows that so long as this approximation is assumed the same principle is illustrated whatever the mutual disposition of the two coils.

The approximate mutual inductance of the two coils follows from either of equation (71) or (72) with (70), and the true mutual inductance is found simply by recourse to the exact relevant field expression, with (in the general case) numerical integration over the appropriate areas. One simple model which gives an approach to an exact result is a long solenoid with $B_1 = \mu H = \mu I$, where I is the surface current density: $I = n_1 i_1$ with n_1 turns per unit length carrying a current i_1. (The wires are assumed to be filamentary and closely spaced.) The mutual inductance with a single coil or circuit C_2 wound tightly on top of the primary is $N_{C_2}^{B_1}/i_1 = \mu n_1 i_1 \mathcal{A}/i_1 = \mu n_1 \mathcal{A}$. If the secondary circuit consists of n_2 single circuits or turns then the inductance can be taken to be the sum of the contribution over all the turns, in view of the manner in which the basic relation is derived (the mutual energies adding), and so

$$L_{12} = \mu n_1 n_2 \mathcal{A} = \mu_r \mu_0 n_1 n_2 \mathcal{A} \tag{73}$$

For this to be exact, the primary current must be taken to constitute an infinitely thin sheet and the solenoid must be of infinite length. The turns of the secondary coil are not subject to such restrictions.

A ring of permeable material with a close-fitting primary coil may be taken to give a result corresponding to the above if the cross-sectional radius of the material is much less than the radius of the ring. The approximation depends on the extent to which the extreme circumferences taken within the ring may be considered to be equal.

1.9.1 Self-inductance

The field, induction and flux associated with a single circuit or a coil of wire must always be proportional to the current in the circuit: $N \propto i$. Just as $N_{C_1}^{B_2} = L_{12} i_2$, the self-inductance L can be described by $N = Li$:

$$L = \frac{N}{i} \tag{74}$$

The coil need not, of course, be a single turn but by definition the current is now common to all the turns.

The energy relation $E_{12} = L_{12} i_1 i_2$ is replaced by $E = \frac{1}{2} L i^2$ since the self-energy $\frac{1}{2}\int \mathbf{B}\cdot\mathbf{H}\,dv$ is invoked. Thus the energy of a single circuit is also $E = \frac{1}{2} i N$. The definition of the self-inductance following from the energy relation is

$$L = \frac{\mu_0}{4\pi i^2} \int_{V_1} \int_{V_2} \frac{\mathbf{J}(\mathbf{r}_1)\cdot\mathbf{J}(\mathbf{r}_2)}{r} dv_1\, dv_2 \tag{75}$$

with $r = |\mathbf{r}_1 - \mathbf{r}_2|$ and $\mathbf{r}_1$ and $\mathbf{r}_2$ simply indicating different points in space at which the single distribution has the values $\mathbf{J}(\mathbf{r}_1)$ and $\mathbf{J}(\mathbf{r}_2)$. The apparent singularity does not exist when the elements for the integration are taken into account so long as the conductors are not infinitely thin; i is the surface integral of $\mathbf{J}$ over the section of the conductor.

Suppose two coils are taken to be virtually coincident, say C_1 with n_1 turns and current i_1 and C_2 with n_2 turns and current i_2. The induction through both coils must be the same and the flux, being the integrated induction accounting for all the turns in each circuit, must obey $N_1 \propto n_1$ and $N_2 \propto n_2$: $N_1/N_2 = n_1/n_2 \cdot N_2$ can be considered as the flux due to i_1, i.e. $N_2 = L_{12}i_1$. Thus

$$\frac{N_1}{N_2} = \frac{L_1 i_1}{L_{12} i_1} = \frac{L_1}{L_{12}} = \frac{n_1}{n_2}$$

Similarly for the current i_2:

$$\frac{N_2}{N_1} = \frac{L_2 i_2}{L_{12} i_2} = \frac{L_2}{L_{12}} = \frac{n_2}{n_1}$$

Combining these,

$$\frac{L_1}{L_{12}} = \frac{L_{12}}{L_2} = \frac{n_1}{n_2}$$

and also

$$L_{12}^2 = L_1 L_2 \tag{76}$$

It is recalled that this is a special case. More generally all the flux generated by C_1 will not be enclosed by C_2 or vice versa and then

$$L_{12} = k\sqrt{L_1 L_2} \tag{77}$$

$(0 \leqslant k \leqslant 1)$, where k, called the coupling coefficient, is a function of the particular geometry.

Assuming that $B = \mu n_1 i$ for a very long solenoid, the flux through each turn is $N' = \mu n_1 i \mathcal{A}$. The self-inductance per unit length is thus $L' = n_1 N'/i$ and if the total length is l then

$$L \doteqdot \mu n_1^2 \mathcal{A} l = \mu_r \mu_0 n_1^2 \mathcal{A} l \tag{78}$$

The exact result for a real solenoid is readily computed.

Although it appears to be stressing the obvious, attention must be drawn to the presence of μ, or μ_r, in equations (73) and (78). The inductance of a coil is very greatly increased by filling it with a material with high relative permeability (as compared with the air-filled coil with $\mu_r \doteqdot 1$) and this is indeed one of the principal applications of magnetic materials.

1.10 Electromagnetic Induction

When a conductor moves in such a way that it 'cuts through the field lines,' i.e. lines giving the direction of **B**, an electric field is set up in and around the conductor. Referring to Figure 4.19(a), as the motion commences a mobile charge carrier q moving within the rod experiences a force $\mathbf{F} = q\mathbf{v}\times\mathbf{B}$ as shown. (The opposite direction for **F** might be considered more appropriate for metals with electrons as the carriers.) The charge is accelerated but eventually a steady state must be achieved in which the magnetic forces are balanced by electrical forces due to the uneven distribution of charge along the rod: $q\mathbf{E} = -\mathbf{F}$. Thus the electric fields and electromotive force between the ends of the rod are a consequence of the magnetic forces.

Electromotive forces (e.m.f.'s) may originate from the chemical potentials in voltaic cells but more generally include any effect that causes a charge to circulate around a circuit. In this case the e.m.f. is

$$\varepsilon = \frac{1}{q}\int \mathbf{F}\cdot d\mathbf{r} = \int (\mathbf{v}\times\mathbf{B})\cdot d\mathbf{r} \tag{79}$$

and the corresponding field is $\mathbf{E} = \mathbf{F}/q = \mathbf{v}\times\mathbf{B}$.

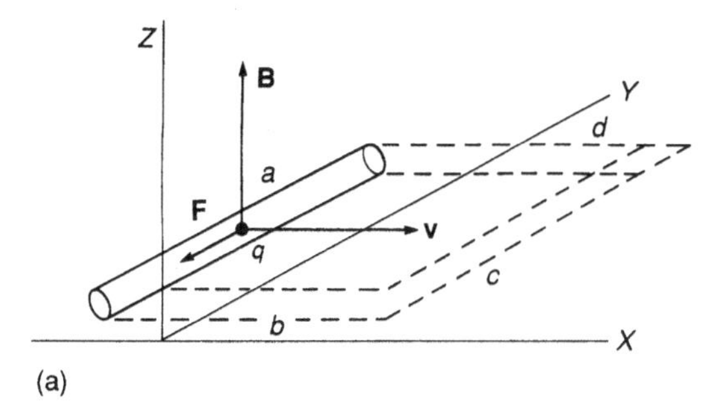

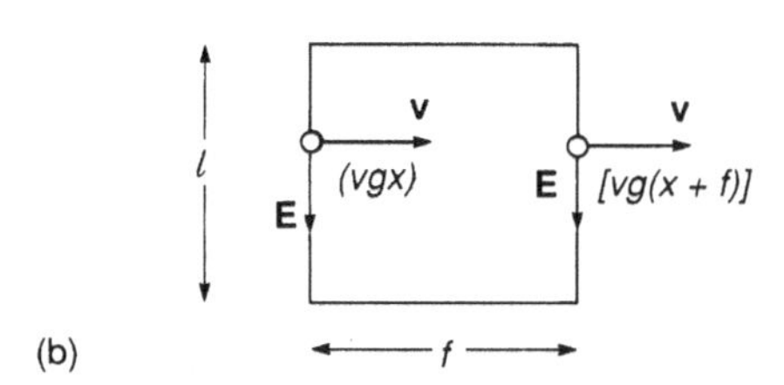

Figure 4.19 (a) The force on the charge creates an e.m.f. between the ends of the conductor. (b) An illustration for Faraday's law of induction

Suppose three further conductors b, c and d are added to a to form the rectangular circuit C [Figure 4.19]. Whatever field $\mathbf{E}$ exists in a is equivalently induced in c since $\mathbf{v}\times\mathbf{B}$ is the same and clearly no fields are induced along the axes of b or d. The line integral of $\mathbf{E}\cdot\mathbf{dr}$ around C is zero. Note that since $\mathbf{B}$ is assumed constant the flux N through C is constant as C moves: $\mathrm{d}N/\mathrm{d}t = 0$.

Now let $\mathbf{B}$ vary linearly in magnitude with x, i.e. $\mathbf{B} = \mathbf{k}gx$. If the centre of conductor a is x, o, z the centre of c is $x + f$, o, z. Along a, $\mathbf{B} = \mathbf{k}gx$ and along c, $\mathbf{B} = \mathbf{k}g(x + f)$. Thus in a, $E_{\mathrm{a}} = |\mathbf{v}\times\mathbf{B}| = \nu gx$ and in c, $E_{\mathrm{c}} = \nu g(x + f)$ with the directions shown in Figure 4.19(b). The line integral of $\mathbf{E}\cdot\mathbf{dr}$ taken in an anticlockwise direction is

$$\begin{aligned}\int_c \mathbf{E}\cdot\mathbf{dr} &= (\nu gx)e - [\nu g(x + f)]e \\ &= -\nu gfe \end{aligned} \tag{80}$$

The flux here is just the product of the mean value of B and the area, i.e. $N = g(x+f/2)\times fe$ and $\mathrm{d}N/\mathrm{d}t = (\mathrm{d}N/\mathrm{d}x)(\mathrm{d}x/\mathrm{d}t) = gfe \times \nu$. Thus

$$\int_c \mathbf{E}\cdot\mathbf{dr} = -\frac{\mathrm{d}N}{\mathrm{d}t}, \qquad \text{where } N = \int_s \mathbf{B}\cdot\mathrm{d}\mathbf{s} \tag{81}$$

This relation can be seen to be completely general and to constitute Faraday's law of induction.

The relation (81) applies whether N changes due to the motion of the circuit or due to a change of $\mathbf{B}$ through a stationary circuit. The sign in equation (81) may seem arbitrary. The direction of E can be explained as follows. Suppose $\mathbf{B}$ traversing a closed conducting loop increases by $\Delta\mathbf{B}$ over a certain time interval. A current i is induced in one of two

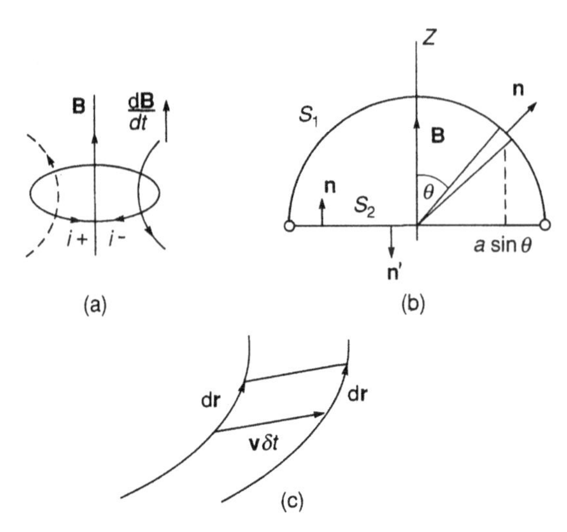

Figure 4.20 (b) Illustration of the arbitrary nature of the surface bounded by the circuit and (c) the area swept out by an element dr moving through $v\delta t$

directions or senses, say i^+ or i^-. Suppose i^+ gives a contribution $\mathbf{B}^+$ in the same direction as **B** (Figure 4.20). In this case $\mathrm{d}N/\mathrm{d}t$ is increased by i^+ and so i^+ is itself increased and with minimal resistance i^+ will increase indefinitely. This is contrary to experience and the direction of **E** and of I must be that which opposes **B** or tends to reduce $\mathrm{d}N/\mathrm{d}t$. This is known as Lenz's law and gives the classical analogy of diamagnetism.

In Figure 4.20(a), **B** is taken to be uniform over the loop and $\parallel OZ$. B increases in magnitude so $\mathrm{d}\mathbf{B}/\mathrm{d}t$ and $\mathrm{d}N/\mathrm{d}t$ are > 0 w.r.t. positive OZ. By the right-hand screw rule the circulation corresponding to i^+ is given the positive sign inasmuch as the screw with the head rotated in this direction would move along $+OZ$ and vice versa. Thus i^- corresponds to the negative sign in equation (81).

A finite $\mathrm{d}N/\mathrm{d}t$ and e.m.f. would also arise if **B** were constant in magnitude but rotated in direction since it is specifically $\mathbf{B}\cdot\mathbf{ds}$ that contributes to the flux: this is, of course, the dynamo effect. It may also be noted that while it is natural to think of N as $\mathbf{B}\cdot\mathbf{ds}$, integrated specifically over a plane surface spanning a plane circuit, any surface can be taken. If $\mathbf{B} = \mathbf{k}B$ in Figure 4.20(b) it is apparent that $N = \int_0^{\pi/2} 2\pi a \sin\theta B \cos\theta a\, \mathrm{d}\theta = 2\pi a^2[\frac{1}{2}\sin^2\theta]_0^{\pi/2} = \pi a^2 B$ when evaluated over the hemispherical surface S_1, as it is over the plane surface S_2. This also corresponds to Gauss's theorem (with div $\mathbf{B} = 0$) since the flux over the closed surface formed by S_1 and S_2 together must be zero (in which case the flux over the portion S_2 would be taken to be negative using $\mathbf{n}'$ as the specifically outward normal: $\mathbf{B}\cdot\mathbf{n}' = -B$). Clearly any surface may be taken, **B** need not be uniform and the circuit itself need not be planar. Also, of course, if the ring in Figure 4.20 is replaced by a coil of n circular turns effectively superimposed, $N = n\pi a^2 B$.

Formally, if in general **B** changes by $\delta\mathbf{B}$,

$$\delta N = \int_S \delta\mathbf{B}\cdot\mathbf{ds}$$

If the circuit also moves (not necessarily rigidly) so that v is the velocity of an element $\mathbf{dr}$ of C, the area swept out by $\mathbf{dr}$ in time δ_t is [Figure 4.20(c)]

$$\mathbf{dS} = \mathbf{v}\delta t \times \mathbf{dr}$$

Thus the total change of flux due to the motion is

$$\delta N_V = \int_C (\mathbf{v}\delta t \times \mathbf{dr}) \cdot \mathbf{B} = -\delta t \int (\mathbf{B} \times \mathbf{v}) \cdot \mathbf{dr}$$

with a conversion used previously. Adding δN due to $\partial B/\partial t$ and dividing by δt,

$$\varepsilon = \int_C \mathbf{E} \cdot \mathbf{dr} = -\frac{\delta N}{\delta t} = -\int_S \frac{\partial \mathbf{B}}{\partial t} \cdot \mathbf{dS} - \int_C (\mathbf{B} \times \mathbf{v}) \cdot \mathbf{dr} \tag{82}$$

This is the general form of Faraday's law for the electromotive force due to both motion of the circuit, in which ε is induced, and the rate of change of B linking the circuit.

1.11 Inductance and E.M.F.

There is clearly a close connection between inductance, depending on N, and e.m.f.'s depending on $\mathrm{d}N/\mathrm{d}t$. For a single circuit $C, \varepsilon = -\mathrm{d}N/\mathrm{d}t$ where N is the flux due to the current i in C, taken across C. Since the self-inductance is $L = N/i$, $N = Li$ and $\varepsilon = -L\,\mathrm{d}i/\mathrm{d}t$. Figure 4.21 shows a torus uniformly wound by a total of n turns carrying a current i, one of which is illustrated—effectively a solenoid closed on itself. By symmetry the field $\mathbf{H}$ produced can be taken to be circumferential and to have the same magnitude around the torus, at any given r. We make use of the equivalence of $\int_C \mathbf{H} \cdot \mathbf{dr}$ to the total current enclosed by C.

The circuit chosen is that marked C_H in the figure, comprising the edges of a cylinder with an infinitesimal longitudinal slot. By symmetry and since $\mathbf{H} \cdot \mathbf{dr}$ is non-zero only along the ring formed by the base of the cylinder:

$$\int_{C_H} \mathbf{H} \cdot \mathbf{dr} = 2\pi r H = ni, \qquad H = \frac{ni}{2\pi r}, \qquad B = \frac{\mu n i}{2\pi r}$$

This involves the assumption that $\mathbf{H} = 0$ along the circular portion of C_H at the top of the cylinder. A circuit consisting of this circle itself encloses no current and though this does not in general mean that $\mathbf{H}$ is zero, $\mathbf{H} = 0$ here because of the symmetry. It is made

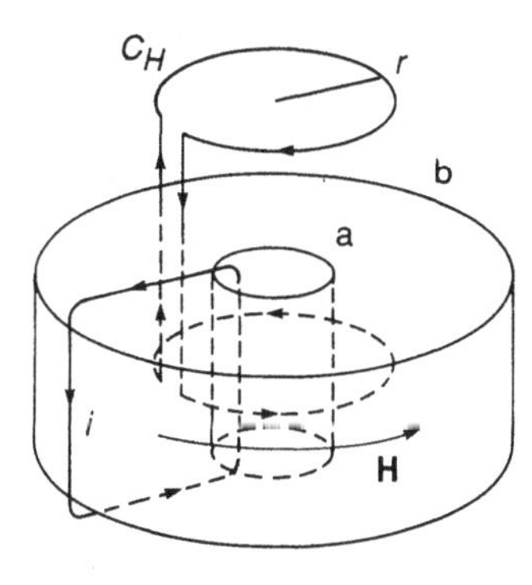

Figure 4.21 A uniformly wound torus (one turn shown) and a circuit chosen for integration

clear that C_H encloses the 'total current' nI by imagining the torus itself to be slit and opened out to form a solenoid.

Considering one turn, the flux N^1 would be given by

$$N^1 = l\int_a^b \frac{\mu ni}{2\pi r}\,\mathrm{d}r = \frac{\mu nil}{2\pi}\ln\frac{b}{a}$$

since $B = B(r)$ only. The total flux N through all the turns constituting a single circuit is $n \times N^1$. Thus the e.m.f. is

$$\varepsilon = -\frac{\mathrm{d}N}{\mathrm{d}t} = -\frac{\mu n^2 l}{2\pi}\ln\frac{b}{a}\frac{\mathrm{d}i}{\mathrm{d}t}$$

while the inductance is

$$L = \frac{N}{i} = \frac{\mu n^2 l}{2\pi}\ln\frac{b}{a}$$

It is again to be stressed that both ε and L are proportional to μ and filling an otherwise air-filled coil with high-permeability magnetic material can have a dramatic effect on the magnitudes.

If a battery (ε_0) is connected across an inductor L and resistance R in series (Figure 4.22) the current rises exponentially towards its Ohm's law value. By elementary circuit theory i is governed by $\varepsilon_0 - L\,\mathrm{d}i/\mathrm{d}t = Ri$, the solution to which is readily seen to be

$$i = \left(\frac{\varepsilon_0}{R}\right)(1 - \mathrm{e}^{-(R/L)t})$$

If S_1 is opened so that the current ceases instantaneously, the indication is that $L\,\mathrm{d}i/\mathrm{d}t \to \infty$. In practice sparking or breakdown in the insulation of the coil would tend to occur. It is most inadvisable to switch off a large current passing through a large electromagnet (with its massive inductance). The current should always be reduced slowly to zero.

Suppose now that, having attained the steady current $i = i_0 = \varepsilon_0/R$, $(t \to \infty)$, S_1 is opened but also S_2 is simultaneously closed. The relevant equation is now $0 = L(\mathrm{d}i/\mathrm{d}t) + Ri$ for which the solution is immediately

$$i = i_0\mathrm{e}^{-(R/L)t}$$

(with a new time zero). As this current flows though the resistor energy is dissipated at the rate Ri^2, neglecting the resistance of the coil constituting L.

The establishment of i_0, during which more energy is drawn from the battery than is dissipated in R, can be thought of as pumping up the field in L, with its associated energy. When the circuit is shorted the fields collapse and the field energy appears as heat in the

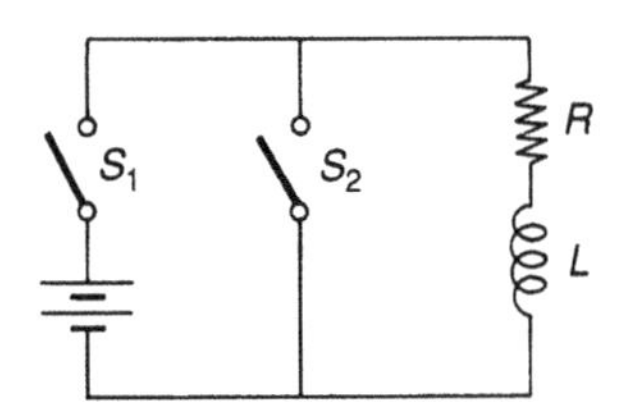

Figure 4.22 The R–L circuit discussed

resistors. The total energy dissipated is

$$E = \int_0^\infty Ri^2\,\mathrm{d}t = \int_0^\infty Ri_0^2 \mathrm{e}^{-2(R/L)t}\,\mathrm{d}t$$
$$= Ri_0^2\left(-\frac{L}{2R}\right)\left[\mathrm{e}^{-2(R/L)t}\right]_0^\infty = \tfrac{1}{2}Li_0^2$$

The power required to build up the current is

$$\frac{\mathrm{d}E}{\mathrm{d}t} = i\varepsilon = iL\frac{\mathrm{d}i}{\mathrm{d}t} = \tfrac{1}{2}\frac{\mathrm{d}}{\mathrm{d}t}Li^2$$

This suggests that the instantaneous stored energy should be

$$E = \tfrac{1}{2}Li^2 \tag{83}$$

This is readily confirmed, by example at least, using the case of the torus for which $L = [\mu n^2 l/(2\pi)]\ln(b/a)$. The magnetic stored energy should be $\frac{1}{2}\int \mathbf{B}\cdot\mathbf{H}\,\mathrm{d}v$, i.e.

$$E = \tfrac{1}{2}\int_a^b \left(\frac{\mu n i}{2\pi r}\right)\left(\frac{ni}{2\pi r}\right)(l2\pi r\,\mathrm{d}r)$$
$$= \tfrac{1}{2}\frac{\mu n^2 i^2 l}{2\pi}\ln\frac{b}{a}$$
$$= \tfrac{1}{2}Li^2 \tag{84}$$

assuming a linear material ($B = \mu H$) which is also isotropic so that $\mathbf{B}\cdot\mathbf{H} = BH$. A similar demonstration may also be made for a long solenoid and for a single circular current loop. This approach is sometimes taken to constitute a derivation of the basic relation for magnetic energy.

When the current loop is considered it is realized that this should not be taken to be a filamentary current with cross-section $\to 0$ or infinities arise. The line currents or effective surface current sheets should be considered as closely restricted current densities. They should be considered sufficiently thin to render trivial any ambiguities over the positioning of the limits of integration for field calculations.

1.12 Electromagnetic Waves: Maxwell's Equations

Recall that the magnetic flux $N = \int_S \mathbf{B}\cdot\mathrm{d}\mathbf{s}$ over a surface S and if S is bounded by a conductor a current may be induced in accordance with an e.m.f. $= -\mathrm{d}N/\mathrm{d}t$. The e.m.f. $= \int_C \mathbf{E}\cdot\mathrm{d}\mathbf{l}$ and the presence of the conductor is immaterial:

$$\int_C \mathbf{E}\cdot\mathrm{d}\mathbf{l} = -\frac{\mathrm{d}}{\mathrm{d}t}\int_S \mathbf{B}\cdot\mathrm{d}\mathbf{s} = -\int_S \frac{\mathrm{d}\mathbf{B}}{\mathrm{d}t}\cdot\mathrm{d}\mathbf{s} \tag{85}$$

where the circuit C encloses S. The second equality applies when S and C are fixed. The electric fields are associated with time-varying magnetic fields, as well as with charges according to Coulomb's law:

$$\mathbf{E}(\mathbf{r}) = q\frac{\hat{\mathbf{R}}}{4\pi\varepsilon_0 R^2} \tag{86}$$

(for the field at $\mathbf{r}$ from charge q at $\mathbf{r}_1$, with $\mathbf{R} = \mathbf{r} - \mathbf{r}_1$ and $\hat{\mathbf{R}} = \mathbf{R}/R$). This applies in space with $\varepsilon_0 = 8.8542 \times 10^{-12}\,\mathrm{C}^2\ \mathrm{N}^{-1}\ \mathrm{m}^{-2}$, the electric permittivity of free space. Gauss's electrostatic law is

$$\int_S \mathbf{E} \cdot \mathrm{d}\mathbf{s} = \frac{1}{\varepsilon} \int_V \rho \, \mathrm{d}v \tag{87}$$

with ε the permittivity of the medium. When a field is applied to an insulator or dielectric there is no movement of charge across a boundary of the material, but a dipole moment $\mathbf{P}$, per unit volume, may be induced—the electric polarization. As for a magnetized body, effective surface and volume charge densities $-\nabla \cdot \mathbf{P}$ and $\mathbf{P} \cdot \mathbf{n}$ may be associated with $\mathbf{P}$. Gauss's law becomes

$$\varepsilon_0 \int_S \mathbf{E} \cdot \mathrm{d}\mathbf{s} = \int_V \rho \, \mathrm{d}v + \int_V (-\nabla \cdot \mathbf{P}) \, \mathrm{d}v$$

or

$$\int_V \nabla \cdot (\varepsilon_0 \mathbf{E} + \mathbf{P}) \, \mathrm{d}v = \int_V \rho \, \mathrm{d}v \tag{88}$$

The volume is arbitrary and so

$$\nabla \cdot (\varepsilon_0 \mathbf{E} + \mathbf{P}) = \rho$$

i.e.

$$\nabla \cdot \mathbf{D} = \rho, \qquad \mathbf{D} = \varepsilon_0 \mathbf{E} + \mathbf{P} \tag{89}$$

defining the displacement $\mathbf{D}$. In space $\mathbf{P} = 0$ and $\varepsilon_0 \nabla \cdot \mathbf{E} = \rho$. Assuming linearity it is convenient to write

$$\mathbf{P} = \varepsilon_0 \chi_\mathrm{e} \mathbf{E} \tag{90}$$

where the electric susceptibility χ_e is positive. Thus

$$\mathbf{D} = \varepsilon_0 \mathbf{E} + \varepsilon_0 \chi_\mathrm{e} \mathbf{E} = \varepsilon_0 (1 + \chi_\mathrm{e}) \mathbf{E} = \varepsilon_0 K \mathbf{E}, \qquad \text{where } K = 1 + \chi_\mathrm{e} \tag{91}$$

K is the dielectric constant and

$$\varepsilon = K \varepsilon_0 \tag{92}$$

the permittivity of the material. Since $\mathbf{D} = \varepsilon \mathbf{E}$ and $\nabla \cdot \mathrm{D} = \rho$,

$$\nabla \cdot (\varepsilon \mathbf{E}) = \rho \tag{93}$$

which is the equivalent, in differential form, of equation (88). From the force relations the units are

$$E \sim \mathrm{NC}^{-1}$$
$$P \sim (\mathrm{Cm})\mathrm{m}^{-3} = \mathrm{C\ m}^{-2}$$
$$\varepsilon_0 \sim \mathrm{C}^2\mathrm{N}^{-1}\mathrm{m}^{-2}$$

therefore,

$$\chi_\mathrm{e} \sim 1, \qquad K \sim 1$$

The correspondence with the magnetic ratios is sometimes stressed by using K_e, for K as above, and K_m in place of μ_r, the relative permeability. It is worth comparing

$$\mathbf{B} = \mu_0(\mathbf{H} + \mathbf{M}) = \mu_0(\mathbf{H} + \chi\mathbf{H}) = \mu_0\mu_r\mathbf{H} = \mu\mathbf{H}$$

$$\mathbf{D} = \varepsilon_0\mathbf{E} + \mathbf{P} = \varepsilon_0\mathbf{E} + \varepsilon_0\chi_e\mathbf{E} = \varepsilon_0(1 + \chi_e)\mathbf{E} = \varepsilon_0 K\mathbf{E} = \varepsilon\mathbf{E}$$

Dimensionally $H = M$ but $E \neq P$.

In crystals **D** and **E** are related by a permittivity tensor:

$$\mathbf{D} = \boldsymbol{\varepsilon}\mathbf{E}$$

$$\begin{pmatrix} D_x \\ D_y \\ D_z \end{pmatrix} = \begin{pmatrix} \varepsilon_{xx} & \varepsilon_{xy} & \varepsilon_{xz} \\ \varepsilon_{yx} & \varepsilon_{yy} & \varepsilon_{yz} \\ \varepsilon_{zx} & \varepsilon_{zy} & \varepsilon_{zz} \end{pmatrix} \begin{pmatrix} E_x \\ E_y \\ E_z \end{pmatrix}$$

The magnetic Gauss's law is $\int_S B\,\mathrm{d}s = 0$ and Ampere's circuital law can be written as

$$\int_C \mathbf{B} \cdot \mathrm{d}\mathbf{l} = \mu \int_S \mathbf{J} \cdot \mathrm{d}\mathbf{s} = \mu i \tag{94}$$

for the integral of **B** tangential to the closed curve C bounding S (open); i is the total current across C. There is no current density **J** through a dielectric, e.g. between the plates of a condenser. Transient currents flow in the leads to charge the condenser and set up the field **E**. These generate B fields and B fields also arise around and within the dielectric. When the charge is Q,

$$E = \frac{Q}{\varepsilon A}$$

approximately, A being the plate area. Thus

$$\varepsilon\frac{\partial E}{\partial t} = \frac{1}{A}\frac{\partial Q}{\partial t} = J_D$$

where J_D is the displacement current. Maxwell showed that equation (94) should be extended to account for this, as

$$\int_C \mathbf{B} \cdot \mathrm{d}\mathbf{l} = \mu \int_S \left(\mathbf{J} + \varepsilon\frac{\partial \mathbf{E}}{\partial t}\right) \cdot \mathrm{d}\mathbf{s} = \mu \int_S \left(\mathbf{J} + \frac{\partial \mathbf{D}}{\partial t}\right) \cdot \mathrm{d}\mathbf{s} \tag{95}$$

As with the example of the current loop, the specific picture of the condenser may be ignored and it is taken that B fields can arise from real current densities or from displacement current densities, in effect from time-varying E fields (cf. above). With $\mathbf{B} = \mu\mathbf{H}$, equation (95) also gives

$$\int_C \mathbf{H} \cdot \mathrm{d}\mathbf{l} = \int_S \left(\mathbf{J} + \frac{\partial \mathbf{D}}{\partial t}\right) \cdot \mathrm{d}\mathbf{s} \tag{96}$$

The equations governing time-varying electromagnetic fields, emphasizing the interdependence of electrical and magnetic effects, i.e. Maxwell's equations, can be collected as

$$\nabla \times \mathbf{B} = \mu\left(\mathbf{J} + \frac{\partial \mathbf{D}}{\partial t}\right) = \mu\left(\mathbf{J} + \varepsilon\frac{\partial \mathbf{E}}{\partial t}\right) \tag{97a}$$

$$\left(\nabla\times\mathbf{H} = \mathbf{J} + \frac{\partial\mathbf{D}}{\partial t} = \mathbf{J} + \varepsilon\frac{\partial\mathbf{E}}{\partial t}\right) \tag{97b}$$

$$\nabla\times\mathbf{E} = -\frac{\partial\mathbf{B}}{\partial t} \tag{97c}$$

$$\nabla\cdot\mathbf{B} = 0 \tag{97d}$$

$$\nabla\cdot\mathbf{D} = \rho \tag{97e}$$

or, in integral form,

$$\int_C \mathbf{B}\cdot\mathrm{d}\mathbf{l} = \mu\int_S\left(\mathbf{J} + \varepsilon\frac{\partial\mathbf{E}}{\partial t}\right)\cdot\mathrm{d}\mathbf{s} = \mu\int_S\left(\mathbf{J} + \frac{\partial\mathbf{D}}{\partial\mathbf{t}}\right)\cdot\mathrm{d}\mathbf{s} \tag{98a}$$

$$\int_C \mathbf{E}\cdot\mathrm{d}\mathbf{l} = -\int_S\frac{\partial\mathbf{B}}{\partial t}\cdot\mathrm{d}\mathbf{s} \tag{98b}$$

$$\int_S \mathbf{B}\cdot\mathrm{d}\mathbf{s} = 0 \tag{98c}$$

$$\int_S \mathbf{E}\cdot\mathrm{d}\mathbf{s} = \frac{1}{\varepsilon}\int_S \mathbf{D}\cdot\mathrm{d}\mathbf{s} = \int_V \rho\,\mathrm{d}v \tag{98d}$$

1.12.1 The wave equation

Assuming that, by Ohm's law, **J** is replaced as

$$\mathbf{J} = \sigma\mathbf{E}$$

with σ, the conductivity, and operating with $\nabla\times$ on equation (97a),

$$\nabla\times(\nabla\times\mathbf{B}) = \mu\sigma\nabla\times\mathbf{E} + \mu\varepsilon\nabla\times\frac{\partial\mathbf{E}}{\partial t}$$

Using

$$\nabla\times(\nabla\times\mathbf{B}) = \nabla(\nabla\cdot\mathbf{B}) - \nabla^2\mathbf{B} = -\nabla^2 B$$

(where $\nabla\cdot\mathbf{B} = 0$) and assuming $\partial^2\mathbf{E}/\partial x\partial t = \partial^2\mathbf{E}/\partial t\partial x$, etc., so that, since $\nabla\times\mathbf{E} = -\partial\mathbf{B}/\partial t$, $\nabla\times\partial\mathbf{E}/\partial t = -\partial^2\mathbf{B}/\partial t^2$, then

$$\nabla^2\mathbf{B} = \mu\sigma\frac{\partial\mathbf{B}}{\partial t} + \mu\varepsilon\frac{\partial^2\mathbf{B}}{\partial t^2}$$

Alternatively, with $B = \mu H$,

$$\nabla^2\mathbf{H} = \mu\sigma\frac{\partial\mathbf{H}}{\partial t} + \mu\varepsilon\frac{\partial^2\mathbf{H}}{\partial t^2}$$

Similarly,

$$\nabla^2\mathbf{E} = \mu\sigma\frac{\partial\mathbf{E}}{\partial t} + \mu\varepsilon\frac{\partial^2\mathbf{E}}{\partial t^2} - \nabla\left(\frac{\rho}{\varepsilon}\right)$$

(using $\nabla\cdot\mathbf{E} = \rho/\varepsilon$).

Considering a region containing no charge or current densities (but not necessarily free space),

$$\nabla^2 \mathbf{H} = \mu\varepsilon \frac{\partial^2 \mathbf{H}}{\partial t^2} \tag{99}$$

$$\nabla^2 \mathbf{E} = \mu\varepsilon \frac{\partial^2 \mathbf{E}}{\partial t^2} \tag{100}$$

The first is, in Cartesian coordinates,

$$\mathbf{i}\nabla^2 H_x + \mathbf{j}\nabla^2 H_y + \mathbf{k}\nabla^2 H_z = \mu\varepsilon \left(\mathbf{i}\frac{\partial^2 H_x}{\partial t^2} + \mathbf{j}\frac{\partial^2 H_y}{\partial t^2} + \mathbf{k}\frac{\partial^2 H_z}{\partial t^2} \right)$$

Thus there is a total of six differential equations for the components of **H** and **E**, each having the form

$$\nabla^2 f - \mu\varepsilon \frac{\partial^2 f}{\partial t^2} = 0, \qquad \text{cf. } \nabla^2 f - \frac{1}{v^2}\frac{\partial^2 f}{\partial t^2} = 0 \tag{101}$$

The equation on the right is the familiar wave equation. [In one dimension, if $f = A\exp(ikx + \omega t)$ then $\partial^2 f/\partial x^2 = -k^2 f$ and $\partial^2 f/\partial t^2 = -\omega^2 f$ and f is seen to be a solution of the wave equation so long as $k^2/\omega^2 = (4\pi^2/\lambda^2)/(4\pi^2\nu^2) = 1/(\lambda^2\nu^2) = 1/v^2$ using ν for frequency.] Thus the **H** and **E** components propagate in the form of a wave with velocity $v = 1/\sqrt{\mu\varepsilon}$ or, *in vacuo*, $1/\sqrt{\mu_0\varepsilon_0}$. As noted, it is chosen that $\mu_0 = 4\pi \times 10^{-7}$ exactly, in magnitude, and ε_0 then takes the value corresponding to the measured velocity of approximately 3×10^8 ms^{-1}, *in vacuo*. Conventionally $c = 1/\sqrt{\mu\varepsilon}$ and $c_0 = 1/\sqrt{\mu_0\varepsilon_0}$ (but c is often used for c_0).

The orientation of the **E** and **H** or **B** vectors with respect to the direction of propagation is seen simply for a plane wave, propagating along *OX*: $\mathbf{E} = \mathbf{E}(x, t)$, constant in planes $\parallel$ *OYZ*. With $\rho = 0$, $\nabla \cdot \mathbf{E} = 0$ becomes $\partial E_x/\partial x = 0$. Either E_x is zero or E_x is non-zero but constant with respect to x. The latter case would not correspond to propagation along *OX*, so E_x is zero.

When there is a single component normal to *OX* the light is plane polarized or linearly polarized. For example, if $\mathbf{E} = \mathbf{j}E(x, t)$ and only E_y exists equation (97c) gives

$$\frac{\partial B_z}{\partial t} = -\frac{\partial E_y}{\partial x} \tag{102}$$

and also

$$\frac{\partial B_x}{\partial t} = 0 = \frac{\partial B_y}{\partial t}$$

The only time-varying component of **B** is B_z. For a harmonic wave, which is a solution of the wave equation

$$E_y = E_0 e^{i[\omega(t - x/c) + \delta]}$$

using $\omega/c = 2\pi/\lambda = k$. From equation (102),

$$\frac{\partial B_z}{\partial t} = -\frac{\partial E_y}{\partial x} = i\frac{\omega}{c}E_y$$

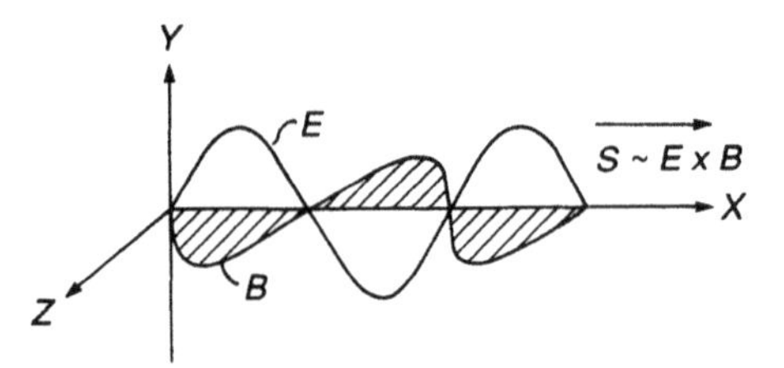

Figure 4.23 Orthogonal **E** and **H** or **B** vectors for an electromagnetic wave

and integrating w.r.t. time, using the equation for E_y above,

$$B_z = \frac{1}{c} E_y \tag{103}$$

Thus **E** and **B** are orthogonal and in phase and may be pictured as in Figure 4.23.

The energy density of a magnetic field has been seen to be $\frac{1}{2}\mu H^2 = \frac{1}{2}BH = \frac{1}{2}B^2/\mu$ and it can be shown that the energy of an electric field is $\frac{1}{2}\varepsilon E^2 = \frac{1}{2}DE$. Using $E = cB$ with no restriction to plane waves,

$$U_M = \frac{1}{2}\frac{B^2}{\mu} = \frac{1}{2}\frac{E^2}{c^2\mu} = \frac{1}{2}\varepsilon E^2 \frac{1}{c^2(\mu\varepsilon)} = U_E \tag{104}$$

since $c = 1/\sqrt{\mu\varepsilon}$. The total energy density of the wave is thus

$$U = \varepsilon E^2 = \frac{1}{\mu} B^2 \tag{105}$$

Using this it can be seen that

$$\begin{aligned}
\frac{\partial U}{\partial t} &= \frac{\partial}{\partial t}\left(\tfrac{1}{2}\varepsilon E^2 + \tfrac{1}{2}\mu H^2\right) = \varepsilon E \frac{\partial E}{\partial t} + \mu H \frac{\partial H}{\partial t} \\
&= \varepsilon E \frac{\partial E}{\partial t} + H \frac{\partial B}{\partial t} = \mathbf{E} \cdot \frac{\partial \mathbf{D}}{\partial t} + \mathbf{H} \cdot \frac{\partial \mathbf{B}}{\partial t} \\
&= \mathbf{E} \cdot (\nabla \times \mathbf{H} - \mathbf{J}) - \mathbf{H} \cdot (\nabla \times \mathbf{E}) \\
&= -\nabla \cdot (\mathbf{E} \times \mathbf{H}) - \mathbf{E} \cdot \mathbf{J}
\end{aligned} \tag{106}$$

Integrating over any fixed volume,

$$\frac{\mathrm{d}}{\mathrm{d}t}\int_V U\,\mathrm{d}v + \int_V \nabla \cdot (\mathbf{E} \times \mathbf{H})\,\mathrm{d}v + \int_V \mathbf{E} \cdot \mathbf{J}\,\mathrm{d}v = 0$$

and using Gauss's theorem for the second term, with S enclosing V,

$$\frac{\mathrm{d}}{\mathrm{d}t}\int U\,\mathrm{d}v + \int_S (\mathbf{E} \times \mathbf{H}) \cdot \mathrm{d}\mathbf{s} + \int_V \mathbf{E} \cdot \mathbf{J}\,\mathrm{d}v = 0 \tag{107}$$

The third term gives the energy dissipated as heat and is zero in space. Due to the conservation of energy the second term must represent the rate at which electromagnetic energy flows across the surface, since it must balance the rate of change of energy in the arbitrary volume which it surrounds. The original treatment was given by Poynting and Heaviside and the vector

$$\mathbf{E}\times\mathbf{H} = \mathbf{S} \qquad (= \mathbf{E}\times\mathbf{B}/\mu) \tag{108}$$

is called the Poynting vector.

The picture is particularly clear for plane waves, as above, of uniform intensity passing through unit area normal to OX. Recalling the directions of $\mathbf{E}$ and $\mathbf{H}$, $\mathbf{E}\times\mathbf{H} = \mathbf{i}EH$ and

$$\begin{aligned}\int_S (\mathbf{E}\times\mathbf{H})\cdot \mathbf{ds} = EH &= \frac{EB}{\mu} \\ &= \frac{E^2}{\mu c} = \frac{(\varepsilon E)^2}{(\mu\varepsilon)c} = \varepsilon E^2 c\end{aligned}$$

(using $B = E/c$ and $\mu\varepsilon = 1/c^2$). This is simply the total electromagnetic energy in a cylinder of length c and unit sectional area normal to the direction of propagation. The magnitude of $\mathbf{S}$ gives the power per unit area normal to $\mathbf{S}$.

$\mathbf{E}$, $\mathbf{B}$ and $\mathbf{S}$ vary in time. Assuming harmonic oscillations, $\langle E^2\rangle = \frac{1}{2}E_0^2$, averaged over time and

$$I = \langle S\rangle = \tfrac{1}{2}c\varepsilon E_0^2 = \tfrac{1}{2}\frac{c}{\mu}B_0^2 \tag{109}$$

where I is the intensity, or irradiance in an optical context. In this context it is common to use $c = 1/\sqrt{\mu_0\varepsilon_0}$ for the velocity in space (as c_0 above) and $v = 1/\sqrt{\mu\varepsilon}$ for the velocity in a general dielectric.

The ratio of c and v gives the refractive index $n = c/v$. In the great majority of dielectrics, *optical frequencies* are too high for the B field to induce any appreciable magnetization and $\mu = \mu_0$, $\mu_r = 1$ very nearly. The interactions with the radiation are *gyro-electric* rather than gyro-magnetic. To this extent,

$$v = \frac{1}{\sqrt{\varepsilon\mu_0}}, \qquad n = \frac{c}{v} = \sqrt{\frac{\varepsilon}{\varepsilon_0}} = \sqrt{K}$$

1.13 General Calculations of Fields from Magnets and Coils

By the preceding sections, calculation of the fields from coils and magnets are virtually equivalent. It may well be convenient to replace a real current distribution by an equivalent magnet, or set of magnets, for currents over finite areas normal to the currents. Compensations by the addition of $\mathbf{M}$ must be made for the fields within the coils. The first objective is to generalize the analysis of cylindrical systems, with the following preliminary note.

1.13.1 Elliptic integrals and Bessel functions

Elliptic integrals of the first and second kind may be denoted $K(\alpha, \varphi)$ and $E(\alpha, \varphi)$ or more usually $K(k, \varphi)$ and $E(k, \varphi)$ with $k = \sin\alpha$. (Unfortunate variations in notation exist, e.g.

F for the first.) In the notation of Heumann [2],

$$K(k, \varphi) = \int_0^{\varphi} \frac{d\theta}{\Delta(\theta)} = \int_0^{\varphi} \frac{d\theta}{\sqrt{1 - k^2 \sin^2 \theta}} \qquad \text{(first)} \tag{110}$$

$$E(k, \varphi) = \int_0^{\varphi} \Delta(\theta)\, d\theta = \int_0^{\varphi} \sqrt{1 - k^2 \sin^2 \theta}\, d\theta \qquad \text{(second)} \tag{111}$$

(Recall that a definite integral is not a function of the variable of integration but is a function of the limits.) When the upper limit is $\pi/2$ the *complete* first and second elliptic integrals are defined respectively as

$$K_0(k) = \int_0^{\pi/2} \frac{d\theta}{\Delta(\theta)}, \qquad E_0(k) = \int_0^{\pi/2} \Delta(\theta)\, d\theta \tag{112}$$

While these cannot be evaluated as explicit functions they are of such common occurrence (as below) that values have been tabulated for reference and efficient subroutines (e.g. NAG FBOIA) have been developed for their evaluation in programs. Thus any calculation giving an expression in terms of elliptic integrals may be considered to have attained a reasonable conclusion. Moreover certain relations between the elliptic integrals may be developed [3], and they are readily differentiated to give

$$\frac{dK}{dk} = \frac{E}{k(1 - k^2)} - \frac{K}{k}, \qquad \frac{dE}{dk} = \frac{E}{k} - \frac{K}{k} \tag{113}$$

Legendre's normal elliptical integral of the third kind is defined by

$$\Pi(\alpha, p, \varphi) = \int_0^{\varphi} \frac{d\theta}{(1 - p \sin^2 \theta)\Delta(\theta)} \tag{114}$$

with $\Delta(\theta)$ as in equation (110) and $k = \sin\alpha$. Introducing an angle β such that

$$\sin^2 \beta = \frac{p - k^2}{pk'^2}, \qquad \cos^2 \beta = \frac{k^2(1 - p)}{pk'^2} \qquad (k'^2 = 1 - k^2) \tag{115}$$

(for $k^2 \leqslant p \leqslant 1$) the lambda function is defined by

$$\Lambda(\alpha, \beta, \varphi) = (1 - p)^{1/2}(1 - k^2/p)^{1/2}\Pi(\alpha, p, \varphi) \tag{116}$$

The complete elliptic integral corresponding to this is

$$\Lambda\left(\alpha, \beta, \frac{\pi}{2}\right) \equiv \Lambda(\alpha, \beta) \tag{117}$$

A paper that is important in the present context [4] makes use of

$$K_0(k) = \frac{2}{\pi} K(k), \qquad E_0(k) = \frac{2}{\pi} E(k), \qquad \Lambda_0(\alpha, \beta) = \frac{2}{\pi} \Lambda(\alpha, \beta)$$

(though in fact using F in place of K): $K_0(0) = E_0(0) = 1$.

According to Heumann [2] the lambda function can be evaluated as

$$\Lambda_0(\alpha, \beta) = \Lambda_0(k, \beta) = K_0(k)E(k', \beta) - [K_0(k) - E_0(k)]K(k', \beta) \tag{118}$$

($k = \sin\alpha$, $k'^2 = 1 - k^2$; see [4] p. 540). Note that the factors $2/\pi$ are not included in $K(k', \beta)$ or $E(k', \beta)$. When the abbreviations K, K_0, E, E_0 are used these are to be taken as the complete integrals $K(k)$ or $K(\alpha)$, etc.

General solutions of Laplace's equation $\nabla^2\varphi = 0$ in cylindrical coordinates, for systems with cylindrical symmetry so that the coordinates are effectively just the pair (ρ, z) or (x, z), having the property that $\varphi \to 0$ as $z \to \infty$, can be obtained in the form

$$\varphi = \frac{1}{4\pi}\int_0^\infty J_m(\varepsilon a)J_n(\varepsilon x)\varepsilon^\lambda e^{-\varepsilon z}\,d_\varepsilon$$

$$= \frac{1}{4\pi}I(m, n, \lambda; a, x, z) = \frac{1}{4\pi}I(m, n, \lambda) \tag{119}$$

with a, x and z implied in the second abbreviation.

$J_n(x)$ is the nth-order Bessel function of the first kind:

$$J_n(x) = \frac{x^n}{2^n}\sum_{k=0}^{\infty}(-1)^k\frac{x^{2k}}{2^{2k}k!(n+k)!} \tag{120}$$

These themselves constitute the solutions of Bessel's differential equation

$$y'' + \frac{y'}{x} + \left(1 - \frac{n^2}{x^2}\right)y = 0$$

and are discussed at length by Watson [5]. For example, the derivatives are

$$J_n'(x) = \frac{n}{x}J_n(x) - J_{n+1}(x) \tag{121}$$

A basic relation between Bessel functions is

$$J_{n-1}(x) + J_{n+1}(x) = \frac{2n}{x}J_n(x) \tag{122}$$

so it is easy to see that the derivative can also be written as

$$J_n'(x) = J_{n-1}(x) - \frac{n}{x}J_n(x) \tag{123}$$

Putting $n = 0$, equation (122) also shows that

$$J_{-1}(x) + J_1(x) = 0, \qquad J_{-1}(x) = -J_1(x) \tag{124}$$

Differentiating by parts and making use of equation (121), it is easy to see that

$$\frac{d}{dx}[x^{-n}J_n(x)] = -x^{-n}J_{n+1}(x) \tag{125}$$

e.g. with $n = -1$,

$$\frac{d}{dx}xJ_{-1}(x) = -xJ_0(x) \tag{126}$$

Clearly the integrals $I(m, n, \lambda)$ can be evaluated numerically and if m, n and λ can be determined and a specified it may be considered that a solution has been found for $\varphi(x, z)$.

To facilitate the evaluation and enable evaluation by recourse to tabulated functions, Eason *et al.* [4] showed how the $I(m, n, \lambda)$ could be expressed in terms of elliptic integrals of the first, second and third kind; the relevant examples will be quoted below. In certain cases the usefulness of these exceeds the context of evaluation.

1.13.2 Field from a circular current loop

By the axial symmetry [Figure 4.24(a)] $\mathbf{A}$ is independent of ϕ and p can be taken to be the point x, z; ϕ is used as the variable of integration around the loop in evaluating $\mathbf{A} = (\mu_0/4\pi)\int_C i\,\mathrm{d}\mathbf{r}/R$, with R as indicated. Temporarily designate $\mathbf{A}' = (4\pi/\mu_0)\mathbf{A} = \int_C i\,\mathrm{d}\mathbf{r}/R$.

According to Figure 4.24(b), on pairing elements in each semicircle it becomes apparent that there is no component $\mathrm{d}A_x$ (and obviously there is no $\mathrm{d}A_z$), i.e. $\mathbf{A} = A\mathbf{e}_\phi$. Further, we may integrate around the semicircle and double the result, and thus use the scalar relation

$$A'_\phi = 2\int_0^\pi i\frac{\cos\phi\,\mathrm{d}r}{R} \tag{127}$$

Using $\mathrm{d}r = a\,\mathrm{d}\phi$, $R^2 = z^2 + b^2$ [Figure 4.24(b)] and $b^2 = a^2 + x^2 - 2ax\cos\phi$,

$$A'_\phi = 2i\int_0^\pi \frac{a\cos\phi\,\mathrm{d}\phi}{(a^2 + x^2 + z^2 - 2ax\cos\phi)^1/2} \tag{128}$$

On introducing an angle related to ϕ by $\phi = \pi + 2\theta$, the appropriate limits for complete elliptic integrals are obtained. Noting that $\cos\phi = \cos(\pi+2\theta) = \cos\pi\cos 2\theta - \sin\pi\sin 2\theta = -\cos 2\theta = -(2\cos^2\theta - 1) = -(1 - 2\sin^2\theta)$, setting $\mathrm{d}\phi = 2\,\mathrm{d}\theta$ and adjusting the limits (with a double change of sign),

$$A'_\phi = 4ai\int_0^{\pi/2} \frac{(2\sin^2\theta - 1)\mathrm{d}\theta}{[a^2 + x^2 + z^2 - 2ax(2\sin^2\theta - 1)]^{1/2}} \tag{129}$$

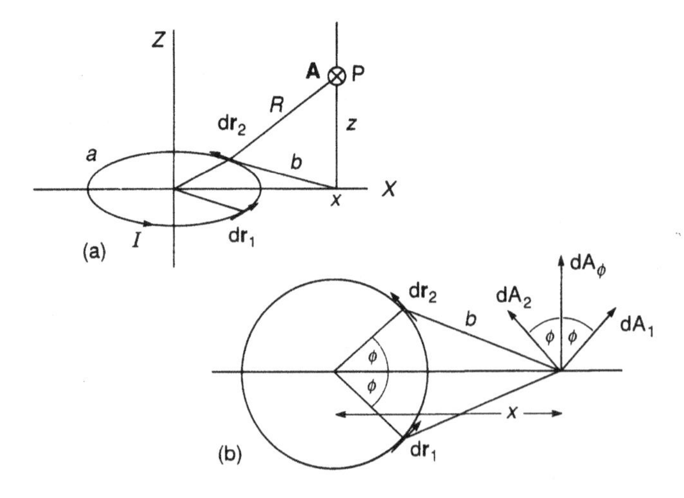

Figure 4.24 Vector potential of a circular current loop. Note the pairing of the elements. (b) A projection looking down OZ

The denominator contains

$$(a+x)^2 + z^2 - 4ax\sin^2\theta = [(a+x)^2 + z^2](1 - k^2\sin^2\theta)$$

with

$$k^2 = \frac{4ax}{(a+x)^2 + z^2} \tag{130}$$

so that

$$A'_\phi = \frac{4ai}{[(a+x)^2 + z^2]^{1/2}} \int_0^{\pi/z} \frac{(2\sin^2\theta - 1)\mathrm{d}\theta}{(1 - k^2\sin^2\theta)^{1/2}} \tag{131}$$

In view of the final form desired, note that

$$\frac{1}{(1-k^2\sin^2\theta)^{1/2}} - (1-k^2\sin^2\theta)^{1/2} = \frac{1-(1-k^2\sin^2\theta)}{(1-k^2\sin^2\theta)^{1/2}}$$

so that

$$\frac{\sin^2\theta}{(1-k^2\sin^2\theta)^{1/2}} = \frac{1}{k^2}\frac{1}{(1-k^2\sin^2\theta)^{1/2}} - \frac{1}{k^2}(1-k^2\sin^2\theta)^{1/2}$$

Using this equation (131) becomes, for A_ϕ itself,

$$\begin{aligned} A_\phi &= \frac{\mu_0 i}{2\pi}\left(\frac{a}{x}\right)^{1/2} k \left[\left(\frac{2}{k^2} - 1\right)\int_0^{\pi/2} \frac{\mathrm{d}\theta}{(1-k^2\sin^2\theta)^{1/2}} - \frac{2}{k^2}\int_0^{\pi/2}(1-k^2\sin^2\theta)^{1/2}\,\mathrm{d}\theta\right] \\ &= \frac{\mu_0 i}{2\pi}\left(\frac{a}{x}\right)^{1/2} k \left[\left(\frac{2}{k^2} - 1\right)K(k) - \frac{2}{k^2}E(k)\right] \end{aligned} \tag{132}$$

and the conversion to elliptic integrals is achieved.

Since A_ϕ is the only component and $\mathbf{B} = \nabla\times\mathbf{A}$ there can be no ϕ component of $\mathbf{B}$. Picking out the non-zero terms in the curl in cylindrical coordinates,

$$B_x = -\frac{1}{x}\frac{\partial}{\partial z}(xA_\phi) + \frac{1}{x}\frac{\partial}{\partial\phi}A_z \rightarrow -\frac{\partial}{\partial z}A_\phi$$

$$B_z = -\frac{1}{x}\frac{\partial}{\partial\phi}A_x + \frac{1}{x}\frac{\partial}{\partial x}xA_\phi \rightarrow \frac{1}{x}\frac{\partial}{\partial x}(xA_\phi)$$

Noting that

$$\frac{\partial k}{\partial z} = -\frac{zk^3}{4ax}, \qquad \frac{\partial k}{\partial x} = (a^2 - x^2 + z^2)\frac{k^3}{8ax^2}$$

and using equation (113) with $\partial K/\partial x = (\partial K/\partial k)(\partial k/\partial x)$, etc.,

$$B_x = \frac{\mu_0 i}{2\pi}\frac{z}{x}\frac{1}{[(a+x)^2 + z^2]^{1/2}}\left[E\frac{a^2+x^2+z^2}{(a-x)^2+z^2} - K\right] \tag{133}$$

$$B_z = \frac{\mu_0 i}{2\pi}\frac{1}{[(a+x)^2 + z^2]^{1/2}}\left[E\frac{a^2-x^2-z^2}{(a-x)^2+z^2} + K\right] \tag{134}$$

Note, for example, that when $x = 0$, $k^2 = 0$ and $K = E = \pi/2$ and,

$$B_z = \frac{\mu_0 i}{2\pi} \frac{1}{(a^2 + z^2)^{1/2}} \frac{\pi}{2} \left(\frac{a^2 - z^2}{a^2 + z^2} + 1 \right)$$

$$= \frac{\mu_0 i}{4} \frac{2a^2}{(a^2 + z^2)^{3/2}} \rightarrow \frac{\mu_0}{4\pi} \frac{2(i\pi a^2)}{Z^3} \tag{135}$$

the second for $a \ll z$, as expected for the dipole.

For a cylindrical current sheet, $I\,\mathrm{Am}^{-1}$, extending between $z' = \pm l/2$, the field or induction components at (x, z) are given by

$$B_x = \frac{\mu_0 I}{2\pi} \int_{-l/2}^{l/2} \frac{z - z'}{x[(a + x)^2 + (z - z')^2]^{1/2}} \left[\frac{a^2 + x^2 + (z - z')^2}{(a - x)^2 + (z - z')^2} E(k) - K(k) \right] \mathrm{d}z'$$

$$B_z = \frac{\mu_0 I}{2\pi} \int_{-l/2}^{l/2} \frac{1}{[(a + x)^2 + (z - z')^2]^{1/2}} \left[\frac{a^2 - x^2 - (z - z')^2}{(a - x)^2 + (z - z')^2} E(k) + K(k) \right] \mathrm{d}z' \tag{136}$$

replacing i by $I\,\mathrm{d}z'$ and noting that now, in $E(k)$ and $K(k)$,

$$k^2 = \frac{4ax}{(a + x)^2 + (z - z')^2}$$

1.13.3 Circular pole density ring

A ring of pole density ρ per unit length may arise in association with the surface of a cylindrical magnet $\perp$ **M**, $\rho = \sigma\,\mathrm{d}r = \pm M\,\mathrm{d}r$, where $\mathrm{d}r$ is the width of an annulus taken on the surface. Such a ring may also be used in association with div **M**, where div $\mathbf{M} \neq 0$.

With $\mathrm{d}l$ an element of the ring length the contribution to the scalar magnetic potential is $d\varphi = (1/4\pi)\rho(\mathrm{d}l/R)$ with R as shown in Figure 4.25. $R^2 = a^2 + x^2 + z^2 - 2ax\cos\phi$ and $\mathrm{d}l = a\,\mathrm{d}\phi$. Making use of the cylindrical symmetry again to integrate (twice) around the semicircle,

$$\varphi = \frac{1}{4\pi} 2\rho \int_0^{\pi} \frac{a\,\mathrm{d}\phi}{(a^2 + x^2 + z^2 - 2ax\cos\phi)^{1/2}} \tag{137}$$

Again introduce θ by $\phi = \pi + 2\theta$, $\mathrm{d}\phi = 2\,\mathrm{d}\theta$, giving

$$\varphi = \frac{1}{4\pi} 4a\rho \int_0^{\pi/2} \frac{\mathrm{d}\theta}{[a^2 + x^2 + z^2 - 2ax(2\sin^2\theta - 1)]^{1/2}}$$

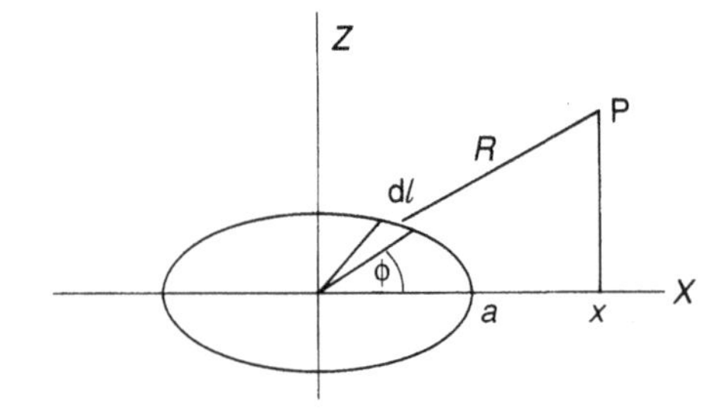

Figure 4.25 For the scalar potential at general points from a circular ring of uniform pole density

which is rearranged in a similar way to that in the preceding section to

$$\varphi = \frac{a\rho}{\pi} \frac{1}{[(a+x)^2 + z^2]^{1/2}} \int_0^{\pi/2} \frac{\mathrm{d}\theta}{(1 - k^2 \sin^2 \theta)^{1/2}}$$

i.e.

$$\varphi = \frac{a\rho}{\pi[(a+x)^2 + z^2]^{1/2}} K(k), \qquad k^2 = \frac{4ax}{(a+x)^2 + z^2} \tag{138}$$

This checks against $\varphi = \rho a/[2(a^2 + z^2)^{1/2}]$ for $x = 0$, $K(0) = \pi/2$.
More compactly,

$$\varphi = \frac{ak\rho}{2\pi\sqrt{ax}} K(k) \tag{139}$$

Eason *et al.* [4] give

$$I(0, 0; 0) = \frac{k}{2\sqrt{ax}} K_0(k) = \frac{k}{\pi\sqrt{ax}} K(k)$$

Thus we may also write

$$\varphi = \frac{1}{4\pi}(2\pi a\rho) I(0, 0; 0) = \frac{1}{4\pi}(2\pi a\rho) \int_0^\infty J_0(\varepsilon a) J_0(\varepsilon x) \mathrm{e}^{-\varepsilon z} \, \mathrm{d}\varepsilon \tag{140}$$

which corresponds to the solution of Laplace's equation by Bateman [6] for the potential due to a ring of radius a carrying a total charge Q as $QI(0, 0; 0)$, setting $2\pi a\rho \to Q$ and rationalizing. As $a \to 0$, $J_0(\varepsilon a) \to 1$ according to equation (120) and Watson [5, p. 384] gives

$$\int_0^\infty J_0(\varepsilon x) \mathrm{e}^{-\varepsilon Z} \, \mathrm{d}\varepsilon = \frac{1}{\sqrt{z^2 + x^2}} \tag{141}$$

In this limit,

$$\varphi = \frac{1}{4\pi} \frac{2\pi a\rho}{\sqrt{z^2 + x^2}} = \frac{1}{4\pi} \frac{m}{r}$$

where m is the total pole strength and r the distance from the ring to (x, z).

The above conversion is useful because it follows immediately that

$$H_z = +\frac{a\rho}{2} \int_0^\infty J_0(\varepsilon a) J_0(\varepsilon x) \mathrm{e}^{-\varepsilon z} \varepsilon \, \mathrm{d}\varepsilon = \frac{a}{2} I(0, 0; 1)$$

and recourse to Eason *et al.* [4] gives

$$H_z = \frac{a\rho}{2} I(0, 0; 1) = \frac{a\rho}{2} \frac{zk^3 E(k)}{4\pi k'^2 (ax)^{3/2}} \tag{142}$$

and also since, from equation (121), $\partial/\partial x \; J_0(\varepsilon x) = -\varepsilon J_1(\varepsilon x)$,

$$\begin{aligned} H_x &= \frac{a\rho}{2} \int_0^\infty J_0(\varepsilon a) J_1(\varepsilon x) \mathrm{e}^{-\varepsilon z} \varepsilon \, \mathrm{d}\varepsilon \\ &= \frac{a\rho}{2} I(0, 1; 1) \\ &= \frac{a\rho}{2} \left[\frac{k^3(x^2 - a^2 - z^2) E(k)}{8\pi x k'^2 \sqrt{ax}} + \frac{k}{2\pi x \sqrt{ax}} K(k) \right] \end{aligned} \tag{143}$$

It is noted that Eason gives $I(1,0;1)$ and not $I(0,1;1)$, but it is only necessary to interchange a and x for the required result since k and k' are symmetric in a and x. Inevitably the same results are obtained by differentiating φ in the form given by equation (138), as the reader may confirm as an exercise.

The utility of these expressions is apparent since they give approximate values for the fields from any magnet with axial symmetry by suitable schemes of subdivision and condensation of the surface distributions on to rings, as well as the following developments.

1.13.4 Fields from surface distributions on discs

For the scalar potential from a disc of radius a, having a uniform pole density σ, and an annulus of radius r, width dr gives a potential, from equation (140), of

$$\mathrm{d}\varphi = \frac{\sigma}{2}\int_0^\infty r J_0(\varepsilon r)\,\mathrm{d}r\, J_0(\varepsilon x)\mathrm{e}^{-\varepsilon z}\,\mathrm{d}\varepsilon \tag{144}$$

replacing a by the variable r and ρ by $\sigma\,\mathrm{d}r$. Integrating over the disc,

$$\varphi = \frac{\sigma}{2}\int_0^\infty \left[\int_0^a r J_0(\varepsilon r)\,\mathrm{d}r\right] J_0(\varepsilon x)\mathrm{e}^{-\varepsilon z}\,\mathrm{d}\varepsilon \tag{145}$$

By equation (123),

$$\int_0^a r J_0(\varepsilon r)\,\mathrm{d}r = \varepsilon^{-1} a J_1(\varepsilon a) \tag{146}$$

and so

$$\varphi = \frac{\sigma a}{2}\int_0^\infty J_1(\varepsilon a) J_0(\varepsilon x)\mathrm{e}^{-\varepsilon z}\varepsilon^{-1}\,\mathrm{d}\varepsilon \tag{147}$$

Differentiating as before, the field components are

$$H_x = -\frac{\partial\varphi}{\partial x} = \frac{a\sigma}{2}\int_0^\infty J_1(\varepsilon a) J_1(\varepsilon x)\mathrm{e}^{-\varepsilon z}\,\mathrm{d}\varepsilon \tag{148}$$

$$H_z = -\frac{\partial\varphi}{\partial z} = \frac{a\sigma}{2}\int_0^\infty J_1(\varepsilon a) J_0(\varepsilon x)\mathrm{e}^{-\varepsilon z}\,\mathrm{d}\varepsilon \tag{149}$$

or, by Eason *et al.* [4],

$$H_x = \frac{a\sigma}{\pi k\sqrt{ax}}[(1-\tfrac{1}{2}k^2)K(k) - E(k)] \tag{150}$$

$$H_z = \frac{a\sigma}{\pi}\begin{cases}\left[-\dfrac{kz}{4a\sqrt{ax}}K(k) - \dfrac{1}{2a}\Lambda(\alpha,\beta) + \dfrac{1}{a}\right], & a > x\\[2ex] \left[-\dfrac{kz}{4a^2}K(k) + \dfrac{1}{2a}\right], & a = x\\[2ex] \left[-\dfrac{kz}{4a\sqrt{ax}}K(k) + \dfrac{1}{2a}\Lambda(\alpha,\beta)\right], & a < x\end{cases} \tag{151}$$

Applying these to two discs with $\sigma = \pm M$, the fields from an axially magnetized right cylindrical magnet may readily be calculated. For a hollow magnet, of internal and external radii a and b, the integral w.r.t. r is to be taken from a to b or the principle of superposition used:

$$\varphi = \frac{\sigma}{2} \int_0^\infty [bJ_1(\varepsilon b) - aJ_1(\varepsilon a)] J_0(\varepsilon x) \mathrm{e}^{-\varepsilon z} \varepsilon^{-1} \, \mathrm{d}\varepsilon \tag{152}$$

with the fields following.

The potential for the disc may also be obtained [7] by noting that the general solution of Laplace's equation, with axial symmetry and having the property that $\varphi \to 0$ as $z \to \infty$, can be expressed as

$$\varphi(x, z) = \int_0^\infty A(\varepsilon) J_0(\varepsilon x) \mathrm{e}^{-\varepsilon z} \, \mathrm{d}\varepsilon \tag{153}$$

with A to be determined. The boundary conditions for the disc are

$$-\left.\frac{\partial \varphi}{\partial z}\right|_{z=0} = \int_0^\infty \varepsilon A(\varepsilon) J_0(\varepsilon x) \, \mathrm{d}\varepsilon = \begin{cases} 0, & x > a \\ \dfrac{\sigma}{2}, & a > x > 0 \end{cases} \tag{154}$$

By the Hankel inversion theorem (Hankel transforms), if

$$f(x) = \int_0^\infty \varepsilon A(\varepsilon) J_0(\varepsilon x) \, \mathrm{d}\varepsilon \tag{155}$$

then

$$A(\varepsilon) = \int_0^\infty x f(x) J_0(\varepsilon x) dx \tag{156}$$

Thus, using equations (154) with $f(x) = \sigma/2$ from 0 to a and zero for $x > a$,

$$\begin{aligned} A(\varepsilon) &= \int_0^a x \frac{\sigma}{2} J_0(\varepsilon x) \, \mathrm{d}x \\ &= \frac{\sigma}{2} \left[\tfrac{1}{2} x J_1(\varepsilon x)\right]_0^a \\ &= \frac{\sigma}{2} \varepsilon^{-1} a J_1(\varepsilon a) \end{aligned} \tag{157}$$

and substitution for $A(\varepsilon)$ in equation (153) gives φ as before.

1.13.5 Surface distributions on cylindrical surfaces

If a cylinder has a surface pole density σ, due to a 'uniform' radial component of the magnetization, on the cylindrical surface, an element $\mathrm{d}z'$ (Figure 4.26) can be regarded as a ring with $\rho = \sigma \, \mathrm{d}z' (2\pi a \rho = 2\pi a dz' \sigma)$. Writing down the contribution $\mathrm{d}\varphi$ to the potential from this ring and integrating over the cylinder surface

$$\begin{aligned} \varphi &= \frac{a\sigma}{2} \int_0^\infty J_0(\varepsilon a) J_0(\varepsilon x) \left(\int_{-l/2}^{l/2} e^{-\varepsilon(z-z')} \, \mathrm{d}z' \right) \mathrm{d}\varepsilon \\ &= \frac{a\sigma}{2} \int_0^\infty J_0(\varepsilon a) J_0(\varepsilon x) \mathrm{e}^{-\varepsilon(z-l/2)} \varepsilon^{-1} \, \mathrm{d}\varepsilon - \frac{a\sigma}{2} \int_0^\infty J_0(\varepsilon a) J_0(\varepsilon x) \mathrm{e}^{-\varepsilon(z+l/2)} \varepsilon^{-1} \, \mathrm{d}\varepsilon \end{aligned}$$

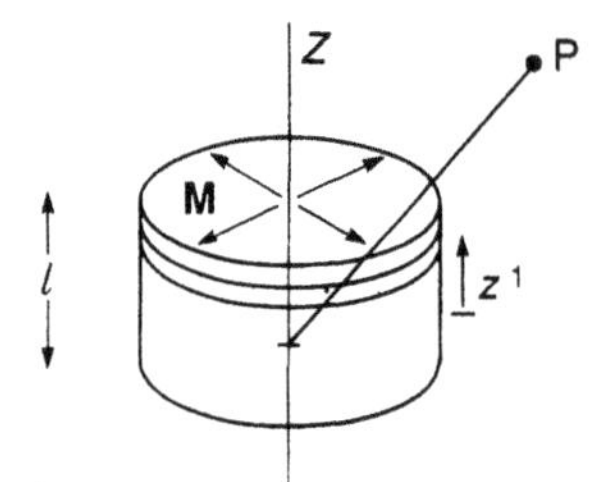

Figure 4.26 A radial magnet

Again using $\partial/\partial x\ J_0(\varepsilon x) = -\varepsilon J_1(\varepsilon x)$, the fields are

$$H_x = \frac{a\sigma}{2}\left[\int_0^\infty J_0(\varepsilon a)J_1(\varepsilon x)\mathrm{e}^{-\varepsilon(z-l/2)}\,\mathrm{d}\varepsilon - \int_0^\infty J_0(\varepsilon a)J_1(\varepsilon x)\mathrm{e}^{-\varepsilon(z+l/2)}\,\mathrm{d}\varepsilon\right] \quad (158)$$

$$H_z = \frac{a\sigma}{2}\left[\int_0^\infty J_0(\varepsilon a)J_0(\varepsilon x)\mathrm{e}^{-\varepsilon(z-l/2)}\,\mathrm{d}\varepsilon - \int_0^\infty J_0(\varepsilon a)J_0(\varepsilon x)\mathrm{e}^{-\varepsilon(z+l/2)}\,\mathrm{d}\varepsilon\right] \quad (159)$$

(nothing the double sign change in each case).

Since the variables cannot now be implied the integrals should be denoted $I(m, n, \lambda; a, x, z)$. Thus

$$H_x = \frac{a\sigma}{2}\left[I\left(1, 0, 0, x, a, z - \frac{l}{2}\right) - I\left(1, 0, 0, x, a, z + \frac{l}{2}\right)\right] \quad (160)$$

noting the interchange of a and x in order to utilize the expression for $I(1, 0, 0)$ for the conversion to elliptic integrals; similarly,

$$H_z = \frac{a\sigma}{2}\left[I\left(0, 0, 0, a, x, z - \frac{l}{2}\right) - I\left(0, 0, 0, a, x, z + \frac{l}{2}\right)\right]$$

$$= \frac{a\sigma}{2}\frac{k}{\pi\sqrt{ax}}[K(k_1) - K(k_2)] \quad (161)$$

where

$$k_1^2 = \frac{4ax}{(a+x)^2 + (z - l/2)^2}$$

$$k_2^2 = \frac{4ax}{(a+x)^2 + (z + l/2)^2}$$

Here the sign of $(z \pm l/2)$ is irrelevant. The sign of H_z reverses on going from $z = z_1$, say, to $z = -z_1$ and this is achieved because H_z is the difference of two terms. If $H_1 = f(z_1 - c)^2 - f(z_1 + c)^2 = f(z_1^2 - 2cz_1 + z_1^2) - f(z_1^2 + 2cz_1 + c^2)$ then $H_2 = f(-z_1 - c)^2 - f(-z_1 + c)^2 = -H_1$.

It is possible to produce ring-shaped or hollow cylindrical magnets with radial magnetization. To calculate the fields from these the first requirement is the application of equations (160), (161) for both the inner and outer cylindrical surfaces. Then div M must also be taken into account for this configuration, as previously noted; $\nabla \cdot \mathbf{M} = -M/r$. An obvious way to approximate the effect of div $\mathbf{M}$ is to divide the magnet into a number of cylindrical shells, calculate the volume pole density $\rho = -\nabla \cdot \mathbf{M}$ within each shell and condense this on to a cylindrical surface at the centre of each shell. This is readily programmed and it is easy to judge the degree of subdivision for the requisite convergence.

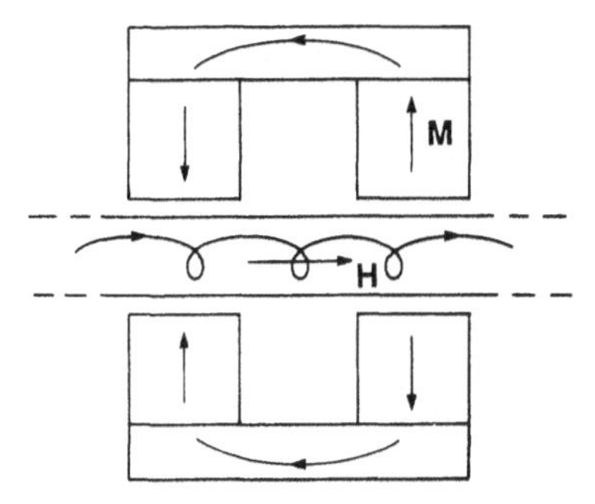

Figure 4.27 Radial magnet assemblies give good axial access, e.g. for microwave generators, and **H** may be contrived to be uniform over a substantial space

An example of the way in which radial magnets are of value is indicated by Figure 4.27. The pair of oppositely magnetized rings were studied as a means of providing the axial magnetic field required for the operation of microwave tubes or generators. It was found that with an appropriate geometry and using a high-coercivity material such as $SmCo_5$ so that the assumption of uniform magnetization was justified, an adequately intense field which was sufficiently uniform over a 'working volume' between the rings, as indicated, could be produced. The biasing magnets constitute a replacement for the conventional solenoid and indeed the fields equivalent to those in a massive solenoid of high current density were achieved. The particular relevance was to satellite-borne radar for earth resources studies: the absence of the power unit required to drive the solenoid was a paramount consideration.

1.14 Magnetic Self-energies of Cylindrical Magnets

A uniformly (axially) magnetized right circular cylinder (radius a, length $l \parallel OZ$) $\mathbf{M} \parallel OZ$, can be represented by two discs with surface pole densities $\sigma = \pm M$. In this representation the magnetic energy, which is also $\frac{1}{2}\int \mathbf{B}\cdot\mathbf{M}\,\mathrm{d}v$, can be calculated as the self-energies of the two discs plus the interaction energy between them. Each term is given by $(\mu_0 X)$ the integral of $\varphi\sigma$ over the surfaces, with a factor of $\frac{1}{2}$ for the self-energies.

For the (identical) self-energies we require the potential on the disc surface

$$\varphi = \frac{a\sigma}{2}\int_0^\infty J_1(\varepsilon a)J_0(\varepsilon x)\varepsilon^{-1}\,\mathrm{d}\varepsilon$$

The energy of an annulus between x and $x + dx$ with pole strength $2\pi x M\,\mathrm{d}x$ is

$$\mathrm{d}W_\mathrm{s} = \tfrac{1}{2}\mu_0\frac{aM^2}{2}\int_0^\infty J_1(\varepsilon a)2\pi[xJ_0(\varepsilon x)\,\mathrm{d}x]\varepsilon^{-1}\,\mathrm{d}\varepsilon$$

For the whole disc, integration within the brackets from 0 to a gives

$$\begin{aligned} W_\mathrm{s} &= \mu_0\frac{\pi a M^2}{2}\int_0^\infty J_1(\varepsilon a)[\varepsilon^{-1}aJ_1(\varepsilon a)]\varepsilon^{-1}\,\mathrm{d}\varepsilon \\ &= \mu_0\frac{\pi a^2M^2}{2}\int_0^\infty J_1^2(\varepsilon a)\varepsilon^{-2}\,\mathrm{d}\varepsilon \end{aligned} \qquad (162)$$

This is the same for both discs since it involves M^2; both φ and σ reverse in sign.

The mutual energy is given by integrating the product of the pole density on one disc and the potential at that disc produced by the second. (The choice is, of course, arbitrary but certainly the integration should not be taken over both pairs: the procedure corresponds to bringing one disc from infinity to a distance l from the other.)

The calculation is virtually that for W_s but (a) the potential now includes $e^{-\varepsilon z}$ with $z = l$, (b) if $\sigma = M$ in the expression for φ, $\sigma = -M$ in the integration and (c) the factor $\frac{1}{2}$ is not included. Thus the result can be written down as

$$W_M = -\mu_0 \pi a^2 M^2 \int_0^\infty J_1^2(\varepsilon a) e^{-\varepsilon l} \varepsilon^{-2} \, d\varepsilon \tag{163}$$

The total magnetic energy or energy of formation of the magnet is thus

$$\begin{aligned} W &= W_M + 2W_s \\ &= \mu_0 \pi a^2 M^2 \int_0^\infty J_1^2(\varepsilon a)(1 - e^{-\varepsilon l}) \varepsilon^{-2} \, d\varepsilon \end{aligned} \tag{164}$$

As an exercise the reader may readily show that the same result is obtained by using equation (149) for H_z and evaluating:

$$W = \frac{1}{2} \int_V \mu_0 \mathbf{H} \cdot \mathbf{M} \, dv$$

over the volume of the cylinder.

Suppose, in equation (164), that l becomes sufficiently small for $e^{-\varepsilon l} = 1 - \varepsilon l$ to be accepted. Then

$$\begin{aligned} W &\to \mu_0 \pi a^2 M^2 \int_0^\infty J_1^2(\varepsilon a) \varepsilon l \varepsilon^{-2} \, d\varepsilon \\ &= \mu_0 M^2 V \int_0^\infty J_1^2(\varepsilon a) \varepsilon^{-1} \, d\varepsilon \end{aligned}$$

where $V = \pi a^2 l$ is the volume of the magnet. The integral is now $I(1, 1, -1)$ which is given for the particular case $a = x$ applying here as

$$I(1, 1, -1) = \frac{z}{2ka} E_0 - \frac{kz}{4a^3}(2a^2 + \tfrac{1}{2} z^2) K_0 + \tfrac{1}{2}$$

but in the present case $z \to 0$ and so in the limit

$$W = \tfrac{1}{2} \mu_0 M^2 V$$

This is just what would be obtained from $\frac{1}{2} \int \mathbf{B} \cdot \mathbf{M} \, dv$ for a thin sheet with a demagnetizing field equal in magnitude to M, i.e. $N = 1$, and $\mathbf{B} = \mu_0 \mathbf{H}$.

1.15 Forces between Magnets

For simplicity, taking two coaxial cylindrical magnets as in Figure 4.28, the force is given by $F_z = -\partial W / \partial z$, where W is now the energy of interaction of the two magnets. W may be calculated as the integral of the pole density on surfaces 3 and 4 times the potential from

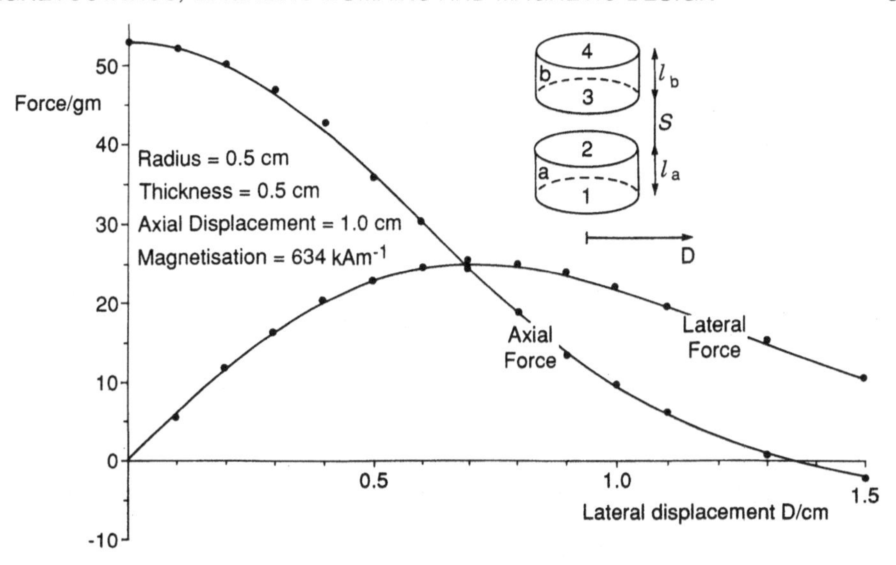

Figure 4.28 The calculated forces between the ideal magnets, as inset, corresponding to measurements on $SmCo_5$ magnets within graphical accuracy (A.J. Harrison, Thesis, Nottingham, 1976)

magnet a. Since this latter is the sum of the potentials from faces 1 and 2 the procedure is equivalent to summing the surface interaction energies $W_{13} + W_{14} + W_{23} + W_{24}$ in an obvious notation.

If φ_3^a is the potential from magnet a at face 3 and φ_4^a that at face 4, then

$$W = \mu_0 \int_{S_3} (-M_b)\varphi_3^a \, dS + \mu_0 \int_{S_4} M_b \varphi_4^a \, dS$$

and

$$F = \mu_0 \int_{S_3} M_b \frac{\partial \varphi_3^a}{\partial z} \, dS - \mu_0 \int_{S_4} M_b \frac{\partial \varphi_4^a}{\partial z} \, dS$$

$$= \mu_0 \int_{S_4} M_b H_z \, dS - \mu_0 \int_{S_3} M_b H_z \, dS \tag{165}$$

with $H_z = -\partial\varphi^a/\partial z$ at S_3 or at S_4. Thus the force may equally be calculated by integrating the products of the relevant $\sigma's$ and field components.

For example, H_z at S_3 is (S and l_a as in Figure 4.28)

$$H_{z_3} = \frac{aM_a}{2} \int_0^\infty J_1(\varepsilon a) J_0(\varepsilon x)(e^{-\varepsilon S} - e^{-\varepsilon(S+l_a)}) \, d\varepsilon$$

Taking H_{z_4} similarly and making use again of $\int_0^b J_0(\varepsilon x)\, dx = bJ_1(\varepsilon b)\varepsilon^{-1}$ it is straightforward to assemble the full expression as

$$F_z = \mu_0 \pi ab M_a M_b \int_0^\infty J_1(\varepsilon a) J_1(\varepsilon b)[e^{-\varepsilon(S+l_a)} + e^{-\varepsilon(S+l_b)} - e^{-\varepsilon S} - e^{-\varepsilon(S+l_a+l_b)}]\varepsilon^{-1} \, d\varepsilon$$

$$= \mu_0 \pi ab M_a M_b [I(1, 1, -1; a, b, s + l_a) + I(\ldots) - I(\ldots) - I(\ldots)] \tag{166}$$

$$I(1,1,-1;a,x,z)=\begin{cases}\dfrac{zE}{\pi k\sqrt{ax}}-\dfrac{kz}{2\pi ax\sqrt{ax}}(a^2+x^2+\tfrac{1}{2}z^2)K+\dfrac{a^2-x^2}{2\pi ax}\Lambda+\dfrac{x}{2a}, & a>b\\[2ex] \dfrac{z}{\pi ka}E-\dfrac{kz}{2\pi a^3}(2a^2+\tfrac{1}{2}z^2)K+\tfrac{1}{2}, & a=b\\[2ex] \dfrac{zE}{\pi k\sqrt{ax}}-\dfrac{kz}{2\pi ax\sqrt{ax}}(a^2+x^2+\tfrac{1}{2}z^2)K+\dfrac{x^2-a^2}{2\pi ax}\Lambda+\dfrac{a}{2x}, & a<b\end{cases}$$

If the magnets are not coaxial a lateral component F_x arises and since the overall axial symmetry of the pair is lost no such explicit expressions can be derived. It is necessary to carry out numerical integrations using incompletely circular annuli around which the field components are constant, and this calls for some careful geometry. Examples of such forces for laterally displaced magnets are shown in Figure 4.28. These were calculated for shapes corresponding to available $SmCo_5$ specimens (kindly supplied by Brown-Boveri of Zurich) on which careful measurements were made. The experimental points lie on the curves within graphical accuracy. Clearly the assumption of uniform **M** or 'ideality' is valid for such materials (see also Section 4.1).

The forces can also be calculated using the surface current model in an obvious way, integrating the cross-product of the fields from one magnet and line current elements on the second magnet, but this is the more difficult approach.

1.16 Fields from Transversely Magnetized Cylinders

Knowing the potential from a disc with surface pole density σ, the potential from a cylinder with a volume pole density of ρ can be evaluated by taking an elementary disc of thickness dz' to have a surface density $\sigma = \rho\, dz'$ and integrating over the cylinder:

$$\varphi = \frac{a\rho}{2}\int_0^\infty J_1(\varepsilon a)J_0(\varepsilon x)(e^{-\varepsilon(z-l/2)} - e^{-\varepsilon(z+l/2)})\varepsilon^{-2}\, d\varepsilon \tag{167}$$

Again with $\partial/\partial x\ J_0(\varepsilon x) = -J_1(\varepsilon x)\varepsilon$,

$$H_x = \frac{a\rho}{2}\int_0^\infty J_1(\varepsilon a)J_1(\varepsilon x)(e^{-\varepsilon(z-l/2)} - e^{-\varepsilon(z+l/2)})\varepsilon^{-1}\, d\varepsilon \tag{168}$$

and clearly

$$H_z = \frac{a\rho}{2}\int_0^\infty J_1(\varepsilon a)J_0(\varepsilon x)(e^{-\varepsilon(z-l/2)} - e^{-\varepsilon(z+l/2)})\varepsilon^{-1}\, d\varepsilon \tag{169}$$

Since a body with a uniform volume pole density cannot exist, since the net pole density must always be zero, these may appear to be singularly valueless expressions. Consider, however, the effect of taking two superimposed identical cylinders with $\rho = +M$ and $\rho = -M$ and displacing the positive cylinder through a small distance δ along a transverse axis OX, as in Figure 4.29(a). From the triangle in Figure 4.29(b) $(l\cos\theta + \delta)^2 + l^2\sin^2\theta = a^2$, noting that OQ is a radius. Thus

$$l = \frac{-2\delta\cos\theta + \sqrt{4\delta^2\cos^2\theta + 4(a^2-\delta^2)}}{2}$$

as $\delta \to 0$ terms in δ^2 can be ignored and $l = a - \delta\cos\theta$. However, if PR is the radius of the displaced cylinder such that $QR = a - l = \delta\cos\theta$ then the thickness of the shell where

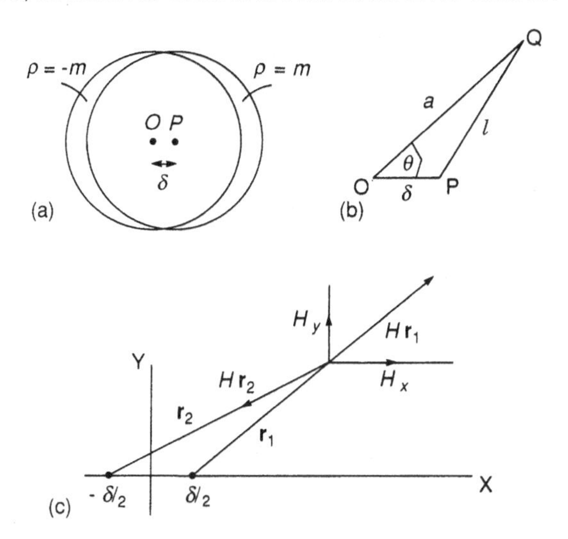

Figure 4.29 Cross-section of two cylinders of uniform pole density, nearly superimposed, becoming equivalent to a transversely magnetized cylinder

the two cylinders do not overlap is $\delta\cos\theta$ in the limit $\delta \to 0$. Where the cylinders overlap the total $\rho = 0$ and thus the only pole distribution is one on the surfaces equal to $\pm\rho\delta\cos\theta$. This is just δ times the pole density $\mathbf{M}\cdot\mathbf{n}$ for $\mathbf{M} \parallel OX$, i.e. to the direction defined by $\boldsymbol{\delta}$, if ρ is set equal to M. Thus if the fields from the two 'solid' cylinders are computed, with due regard to the sign of ρ and with δ set to an adequately small fraction of the distance to the point of calculation, the sum of these fields divided by δ gives the field for the transversely magnetized cylinder.

Confining attention, for simplicity, to points on OX only and defining $F(x, z) = H_x^s/\sigma$, where H_x^s refers to the 'solid' cylinders, we have

$$H_x = \frac{MF(x-\delta) - MF(x)}{\delta} \to \frac{\partial\varphi}{\partial x} \tag{170}$$

with $\varphi = \mathbf{M}.\mathbf{H}^s$ and $\mathbf{H}^s$ the intensity due to a uniform distribution of charge of unit density. Thus it may be said that this section has merely achieved an illustration of the principle developed in Section 1.1.

The transverse cylinder does not have axial symmetry. Replacing x in equation (168) by the radial coordinate r, the H_r components are as illustrated in Figure 4.29(c) and these give rise to components H_x and H_y in addition to H_z. Practical calculations may make use of the effective numerical differentiation as above or, of course, differentiation of H_x^s to give H_x, H_y and H_z, but in the latter case there is no easy conversion to elliptic integrals.

1.17 Potentials and Fields from Orthorhombic Magnets and Coils

Orthorhombic magnets $2a \times 2b \times l$, as in Figure 4.30(a), can be represented by two rectangles of surface pole density $\sigma = \pm M$. Calculations of both the scalar potentials and the fields are required for different purposes. Although $\mathbf{H} = -\operatorname{grad}\varphi$, it will be seen to be easy to calculate $\mathbf{H}$ directly. The calculations implicitly apply to coils of corresponding geometry.

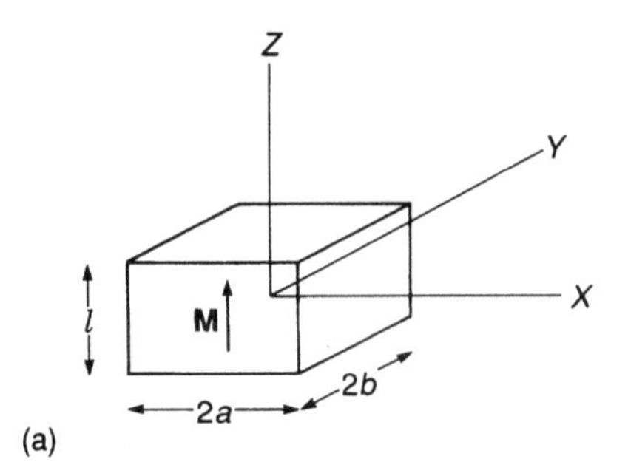

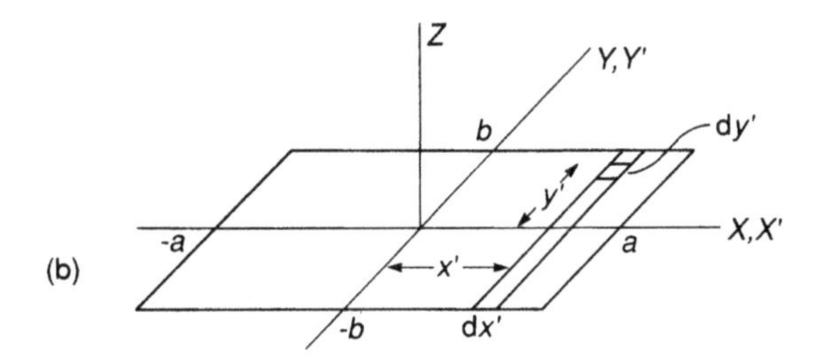

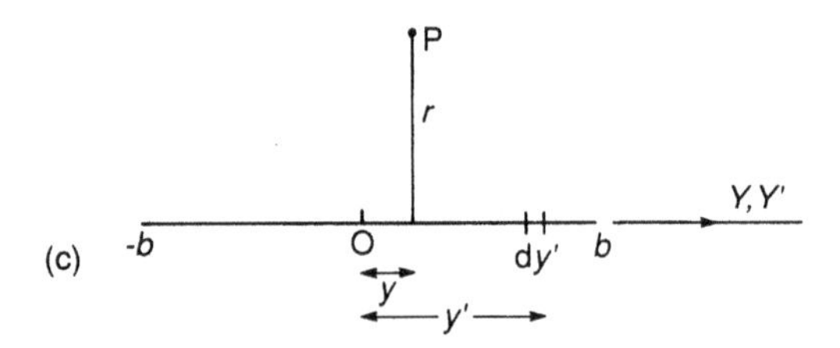

Figure 4.30 For the calculation of scalar potentials from rectangular section magnets as in (a)

Referring to Figure 4.30(b), to find $\varphi(x, y, z)$ at any point P we first require the contribution φ_s from the strip of width dx' at x', at x, y, z. Consider the plane containing the strip and the point P [Figure 4.30(c)] in which $r^2 = (x - x')^2 + z^2$. The potential from the element dy' is

$$d\varphi_s = \frac{\sigma dx'}{4\pi} \frac{dy'}{[(y - y')^2 + r^2]^{1/2}}$$

Using the standard form $\int da/\sqrt{u^2 + a^2} = \ln(u + \sqrt{u^2 + a^2}) + C = \ln a + \ln[u/a + \sqrt{(u/a)^2 + 1}] + C$, we obtain

$$\begin{aligned}\varphi_s = \int d\varphi &= \frac{-\sigma\, dx'}{4\pi} \int_{y'=-b}^{y'=b} \frac{du}{\sqrt{u^2 + r^2}} \\ &= \frac{\sigma\, dx'}{4\pi} \left(\sinh^{-1} \frac{y + b}{r} - \sinh^{-1} \frac{y - b}{r} \right) \qquad (171)\end{aligned}$$

(Note the sign change since $du = -dy'$.) Here use has been made of the relation

$$\sinh^{-1} x = \ln(x + \sqrt{x^2 + 1}) \tag{172}$$

to provide an abbreviation.

For the whole sheet, integrating w.r.t. x' and using φ_s as $d\varphi$,

$$\varphi(x, y, z) = \frac{\sigma}{4\pi} \int_{-a}^{a} \left\{ \sinh^{-1} \frac{y + b}{[(x - x')^2 + z^2]^{1/2}} - \sinh^{-1} \frac{y - b}{[(x - x')^2 + z^2]^{1/2}} \right\} dx'$$
$$= \frac{\sigma}{4\pi} \int_{x-a}^{x+a} \left(\sinh^{-1} \frac{y + b}{\sqrt{t^2 + z^2}} - \sinh^{-1} \frac{y - b}{\sqrt{t^2 + z^2}} \right) dt \tag{173}$$

introducing $t = x - x'$.

Aziz [8] quotes the integral

$$\int \sinh^{-1} \frac{a}{\sqrt{t^2 + z^2}} dt = a \sinh^{-1} \frac{t}{\sqrt{a^2 + z^2}} + t \sinh^{-1} \frac{a}{\sqrt{t^2 + z^2}} - z \tan^{-1} \frac{at}{z\sqrt{a^2 + z^2 + t^2}}$$

and on making use of this the required result follows. It does not seem necessary to write out the sum of the twelve terms at length. For repetitive computation, making use of equation (172), the expression should be evaluated as the single logarithmic function of the appropriate product of terms rather than the sum of the logs (so far as the $\sinh^{-1}$ terms are concerned).

To calculate the fields directly the simplest procedure is to initially carry out a restricted calculation for the field components above one corner of a rectangle $a \times b$ disposed as in Figure 4.31(a). The coordinates x and y can then be used for the integration. For the element $dx\,dy$ the field is $dH_r = (\sigma/4\pi)\,dx\,dy/r^2$ along r. Since $\tan\theta = y/p$, $dy = p\,d\theta/\cos^2\theta$ and also $1/r^2 = \cos^2\theta/p^2$:

$$dH_r = \frac{\sigma}{4\pi}\frac{dx d\theta}{p}$$

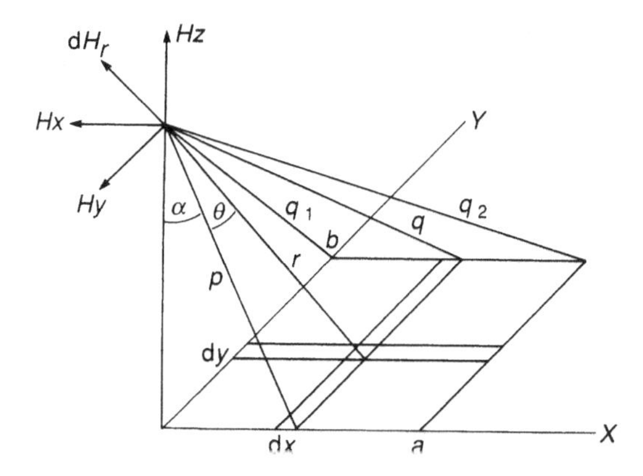

Figure 4.31 For the fields from a rectangle of uniform pole density at a point above one corner with generalizations achieved by superimposing sets of rectangles. $\alpha_1 = \tan^{-1} a/z$, $\theta_1 = \sin^{-1} b/q$

The resolute along p is

$$\mathrm{d}H_p = \frac{\sigma}{4\pi}\frac{\mathrm{d}x\cos\theta\,\mathrm{d}\theta}{p}$$

Integrating to give the contribution from the strip,

$$\mathrm{d}H_p^{(s)} = \frac{\sigma}{4\pi}\frac{\mathrm{d}x}{p}\sin\theta_1 = \frac{\sigma}{4\pi}\mathrm{d}x\frac{b}{pq} \tag{174}$$

and the contribution of the strip to H_x is

$$\mathrm{d}H_x^{(s)} = -\frac{\sigma}{4\pi}\frac{b}{pq}\sin\alpha\,\mathrm{d}x$$

Changing the variable to q, $q^2 = b^2 + p^2 = b^2 + x^2 + z^2$, $q\,\mathrm{d}q = x\,\mathrm{d}x$ and

$$\mathrm{d}H_x^{(s)} = -\frac{\sigma b}{4\pi}\frac{q\,\mathrm{d}q}{q^2 - b^2}$$

Using the standard form $\int \mathrm{d}u/(a^2 - u^2) = (1/2a)\ln[(u+a)/(u-a)]$ and integrating w.r.t. q for the whole rectangle,

$$H_x = -\tfrac{1}{2}\frac{\sigma}{4\pi}\left[\ln\frac{q-b}{q+b}\right]_{q_1}^{q_2} = \frac{\sigma}{4\pi}\left[\ln\frac{p}{q+b}\right]_{q_1}^{q_2}$$

since $p^2/(q+b)^2 = (q^2 - b^2)/(q+b)^2 = (q+b)(q-b)/(q+b)^2$.
Finally, substituting for p and q,

$$H_x = \frac{-\sigma}{4\pi}\ln\left[\frac{(a^2+z^2)^{1/2}}{z}\frac{(b^2+z^2)^{1/2}+b}{(a^2+b^2+z^2)^{1/2}+b}\right] \tag{175}$$

(As illustrated, the component H_x for $\sigma > 0$ is along $-OX$.) By symmetry the component H_y is

$$H_y = \frac{-\sigma}{4\pi}\ln\left[\frac{(b^2+z^2)^{1/2}}{z}\frac{(a^2+z^2)^{1/2}+a}{(a^2+b^2+z^2)^{1/2}+a}\right] \tag{176}$$

From equation (174) the contribution to H_z due to the strip is

$$\mathrm{d}H_z^{(s)} = \frac{\sigma}{4\pi}\frac{b\cos\alpha\,\mathrm{d}x}{pq}$$

Using α as the variable of integration for the whole sheet, $x = z\tan\alpha$ and $\mathrm{d}x = (z/\cos^2\alpha)\,\mathrm{d}\alpha$,

$$\begin{aligned}
H_z &= \frac{\sigma}{4\pi}\int_0^{\alpha_1}\frac{b\,\mathrm{d}\alpha}{q} = \frac{\sigma}{4\pi}\int_0^{\alpha_1}\frac{b\,\mathrm{d}\alpha}{(b^2+z^2/\cos^2\alpha)^{1/2}} \\
&= \frac{\sigma}{4\pi}\int_0^{\alpha_1}\frac{b\cos\alpha\,\mathrm{d}\alpha}{[b^2(1-\sin^2\alpha)+z^2]^{1/2}} = \frac{\sigma}{4\pi}\int_0^{\alpha_1}\frac{b\,\mathrm{d}(\sin\alpha)}{[(z^2+b^2-b^2)\sin^2\alpha]^{1/2}} \\
&= \frac{\sigma}{4\pi}\left[\sin^{-1}\frac{b\sin\alpha}{(b^2+z^2)^{1/2}}\right]_0^{\alpha_1}
\end{aligned}$$

with $\tan\alpha_1 = a/z$ so that

$$H_z = \frac{\sigma}{4\pi}\sin^{-1}\frac{ab}{[(a^2+z^2)(b^2+z^2)]^{1/2}} \tag{177}$$

Although it is possible to derive more general expressions than these, they are in fact all that is required for most purposes. Rather than changing the limits of integration to remove the above restrictions the fields at a general point P or (x, y, z) can be found by superposition.

1.18 Magnetic Energies of Uniform Orthorhombic (Rectangular) Blocks

Recourse to the potential expression for a rectangular pole sheet permits the calculation of these magnetic energies by numerical integration. Rhodes and Rowlands [9] considered the basic problem of evaluating

$$E_m = \mu_0\sigma_2\int_{S_2}\varphi(x_2, y_2, c)\,dx_2\,dy_2$$

where

$$\varphi(x_2, y_2, c) = \frac{\sigma_1}{4\pi}\int_{S_1}[(x_2-x_1)^2+(y_2-y_1)^2+c^2]^{-1/2}\,dx_1\,dy_1$$

for plane parallel surfaces with separation c. This is the mutual energy ($\sigma_1 = -\sigma_2$) and also, with $\sigma_1 = \sigma_2$ and $S_1 \equiv S_2 (c = 0)$ and with a factor of $\frac{1}{2}$, the expression for the self-energy of a plane surface distribution. Results were obtained in a form permitting ready computation, one that was particularly applicable to the analysis of domain structures as illustrated by Figure 4.32. The treatment involved standard integrals and also several rather complex and unfamiliar integrals. With the restrictions implied all the energies can be given in terms of a lengthy but simple function $F(p, q)$ with the reduced parameters $p = b/a$, $q = c/a$:

$$E_m = 2\mu_0 a^3\sigma^2 F(p, q)$$

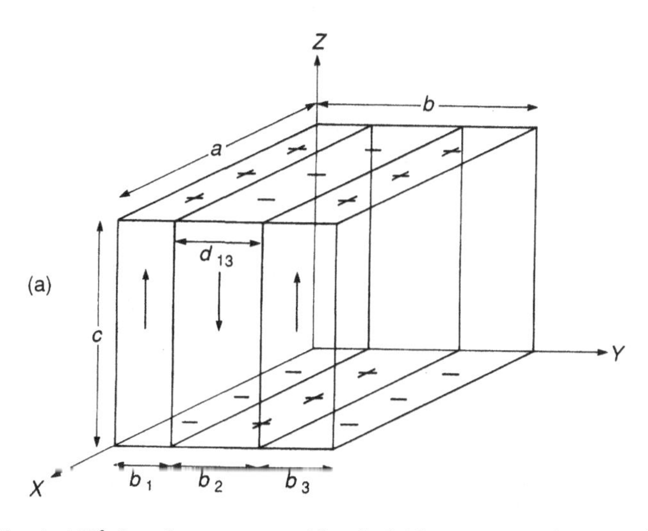

Figure 4.32 Simple 180° domain structures with calculable magnetostatic energy [9]

for two identical rectangles $a \times b$ opposite each other in parallel planes with separation c,

$$E_s = \mu_0 a^3 \sigma^2 F(p, 0)$$

for the self-energy of a rectangle $a \times b$ and

$$E_m = \mu_0 a^3 \sigma^2 [F(p_i + p_j + r, q) + F(r, q) - F(p_i + r, q) - F(p_j + r, q)] \tag{178}$$

with

$$p_i = \frac{b_i}{a}, \qquad q = \frac{c}{a}, \qquad r = \frac{d_{ij}}{a}$$

for two similarly oriented strips of length a and widths b_i and b_j separated by d_{ij} in the y direction and in parallel planes with spacing c. For brevity, setting

$$f(x) = \sinh^{-1}(x) = \ln(x + \sqrt{x^2 + 1})$$

the function F is given by

$$\begin{aligned} 4\pi F(p, q) = {} & (p^2 - q^2) f \left[\frac{1}{(p^2 + q^2)^{1/2}}\right] \\ & + p(1 - q^2) f \left[\frac{p}{(1 + q^2)^{1/2}}\right] + pq^2 f \left(\frac{p}{q}\right) \\ & + 2pq \tan^{-1} \left[\frac{q(1 + p^2 + q^2)^{1/2}}{p}\right] - \pi pq \\ & - \tfrac{1}{3}(1 + p^2 - 2q^2)(1 + p^2 + q^2)^{1/2} + \tfrac{1}{3}(1 - 2q^2)(1 + q^2)^{1/2} \\ & + \tfrac{1}{3}(p^2 - 2q^2)(p^2 + q^2)^{1/2} + \tfrac{2}{3} q^3 \end{aligned} \tag{179}$$

The expressions given clearly permit the evaluation of the total magnetic energies of uniaxial domain structures as illustrated (see Section 2.1, 2.2).

The simplest special case is simply the uniformly magnetized cube, $\mathbf{M} = \mathbf{k}M$, with $p = q = 1$. The total energy is $E = 2E_s + E_m = 2\mu_0 a^3 M^2 F(1, 0) - 2\mu_0 a^3 M^2 F(1, 1)$. From equation (179), $F(1, 0) - F(1, 1) = (\frac{1}{3}\pi)/(4\pi)$, giving $E = \frac{1}{2} \times \frac{1}{3}\mu_0 a^3 M^2$. The energy density is thus identical to that for a sphere : $\frac{1}{2}\mu_0(NM)M$ with $N = \frac{1}{3}$, although for the cube the 'demagnetizing field' is certainly not uniform at the value $-\frac{1}{3}M$. Rhodes and Rowlands [9] also showed that the energy was independent of the direction of the magnetization in the cube.

An analogous coincidence has already been noted, i.e. that the field at the centre of a cube is given by $-N\mathbf{M}$ with $N = \frac{1}{3}$. The reader may demonstrate that this applies whatever the direction of $\mathbf{M}$.

2 Magnetostatics and Domain Structures

The main task is to calculate the magnetostatic energy for a particular structure since terms in the total energy due to the presence of the walls, anisotropies, etc., usually follow by simple geometry. The variation of the energy w.r.t. a restricted number of variables such as

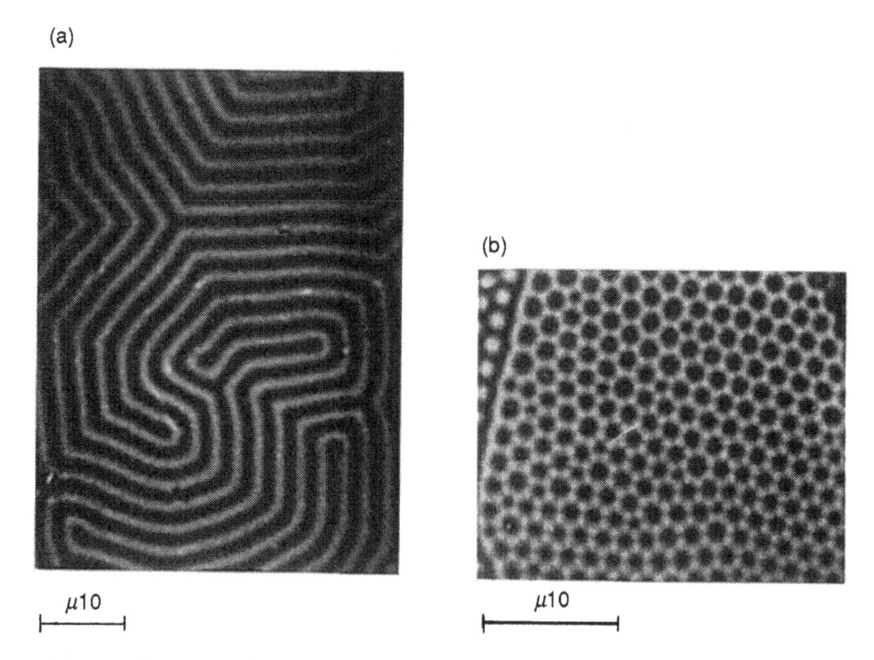

Figure 4.33 180° strip and bubble domains, as revealed by Faraday domain contrast, in thin barium ferrite crystals, (b) showing the bubbles after applying large fields ⊥ the easy axis, i.e. ∥ the surface (G. Myers, Thesis, Nottingham, 1975)

the domain width or the widths of alternate domains, for simple parallel-sided or strip 180° domains or bubble domain radii and spacings, then indicates (numerically if necessary) the equilibrium energy. Magnetization curves follow when an additional term for the energy in an applied field is included. Comparing the results for alternative probable structures then indicates which one may be considered as characteristic, in the hope that all reasonable alternatives have been taken into account. To an extent, discounting metastable cases, the characteristic structure may simply be taken as that observed. If it were feasible for any but simple and restricted cases to apply the calculus of variations with **M** as the variable together with the dependent exchange, anisotropy and magnetostatic terms (not the wall energy as such), then this micromagnetic approach would be more direct and satisfactory. (The combination of micromagnetic principles with numerical methods does permit some progress.)

Among the few structures that may be analysed reasonably simply are regularly periodic 180° domains as expected in sheets or films with easy axis ⊥ the surface, and H_K moderate or high, and which are 'magnetically thin' i.e. $c \sim \lambda$, the characteristic length (Chapter 5, Section 2). Typical relevant specimens are thin sheets that are readily detached from crystals of barium ferrite (Figure 4.33), certain metallic films and epitaxially grown ferrimagnetic garnet layers with strain-induced anisotropy due to controlled lattice mismatch with the substrate crystals (see Figure 1.16).

2.1 180° Strip and Cylindrical (Bubble) Domains

Although methods using the vector potentials and effective surface currents are feasible, the most usual procedure consists, as noted previously, of the integration over the surfaces

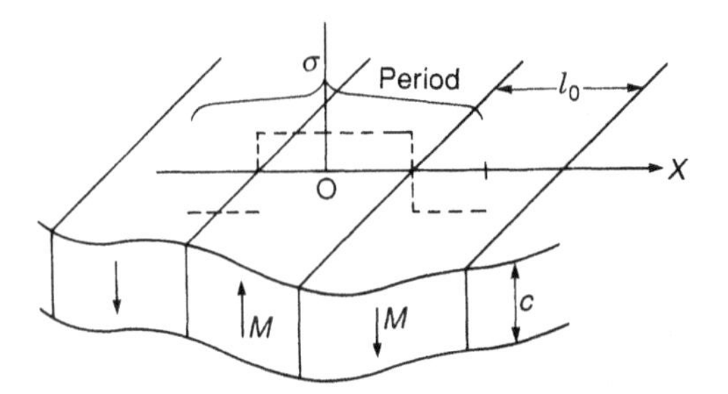

Figure 4.34 Domains with a one-dimensional periodicity, the period being as indicated

of the products of the pole densities σ and the scalar potentials ($\nabla \cdot \mathbf{M} = 0$ assumed), as pioneered by Landau and Lifshitz [10] and by Kittel [11]. Strip domains can be taken to be singly periodic (w.r.t. OX; see Figure 4.34) with σ Fourier expanded as

$$\sigma(x) = \tfrac{1}{2}a_0 + \sum_{n=1}^{\infty} a_n \cos\frac{n\pi x}{l} \tag{180}$$

φ can be expressed as a general solution of Laplace's equation, restricted on account of symmetry:

$$\varphi = \begin{cases} b_0 + \sum b_n \cos\left(\dfrac{n\pi x}{l}\right) e^{n\pi z/l}, & z > \dfrac{c}{2} \\ c_0 z + \sum c_n \cos\left(\dfrac{n\pi x}{l}\right) \sinh\left(\dfrac{n\pi z}{l}\right), & \dfrac{c}{2} > z > -\dfrac{c}{2} \\ d_0 + \sum d_n \cos\left(\dfrac{n\pi x}{l}\right) e^{n\pi z/l}, & z < \dfrac{c}{2} \end{cases}$$

A lengthy and detailed consideration of the boundary conditions eventually leads to the (c.g.s) expression for the energy per unit surface area (for any singly periodic structure):

$$W_{\text{m}} = \frac{1}{2l}\int_{-l}^{l}\left[\left(\frac{a_0}{2} + \sum a_n \cos px\right)\left(\pi a_0 c + \sum \frac{4l}{n} a_n \sinh q \cos px\, e^{-q}\right)\right] dx \tag{181}$$

with coefficients that correspond to the particular structure of concern; $p = n\pi/l$ with l the period, $q = pc/2$.

Specifically for domains of uniform width l_0, i.e. the demagnetized state, the a_n follow from the simplified Fourier expansion and can be shown to give

$$W_{\text{m}} = \frac{16M_{\text{s}}^2 l_0}{\pi^2}\sum_{n\ \text{odd}} n^{-3}(1 - e^{-pc}) \tag{182}$$

as first obtained by Malek and Kambersky [12]. For $c \gg 1$ the exponential can be ignored, i.e. the c-dependence disappears so the interaction between the two surfaces is implicitly ignored, and

$$W_{\text{m}} = \frac{16M_{\text{s}}^2 l_0}{\pi^2}\sum_{n\ \text{odd}} n^{-3} = 1.71\ M_{\text{s}}^2 l_0 \tag{183}$$

since $\text{Lt}_{n\to\infty}\sum n^{-3} = 1.0518\ldots$, as obtained directly by Kittel on evaluating the energy for a single surface.

When fields are applied normal to the surface and ∥ the easy axis to give strips of alternate widths l_1 and l_2, with a period l_1+l_2, and thus a net magnetization $M = M_s(l_1-l_2)/(l_1+l_2)$, it was shown by Kooy and Enz [13] that

$$W_m = M_s^2\left\{2\pi c\left(\frac{M}{M_s}\right)^2 + \frac{8c}{\pi^2 g}\sum_{n=1}^{\infty} n^{-3}\sin^2\left[\tfrac{1}{2}n\pi\left(1+\frac{M}{M_s}\right)\right](1-e^{-2n\pi g})\right. \tag{184}$$

$[g = c/(l_1+l_2)]$.

For any doubly periodic structure with the period for $\sigma(x,y)$ a rectangle $L_x \times L_y$, the Fourier expansion is

$$\sigma(x,y) = \sum_{m=-\infty}^{\infty}\sum_{n=-\infty}^{\infty} a_{mn}e^{i(m\alpha+n\beta)} \tag{185}$$

where $\alpha = 2\pi x/L_x$, $\beta = 2\pi y/L_y$. An appropriate general solution for φ giving $\varphi = 0$ at $z = 0$ (origin central to the sheet) and φ constant at $z \to \pm\infty$ is

$$\varphi = \begin{cases} b_{00} + \sum\limits_m\sum\limits_n b_{mn}e^{ir}e^{-k_{mn}z}[= \varphi(1)], & z > \dfrac{c}{2} \\ c_{00}z + \sum\limits_m\sum\limits_n c_{mn}e^{ir}\sinh(-k_{mn}z), & \dfrac{c}{2} > z > -\dfrac{c}{2} \\ -\varphi(1), & z < \dfrac{c}{2} \end{cases}$$

[all sums from $-\infty$ to $+\infty$ with $m = n = 0$ omitted; $r = m\alpha + n\beta$, $k_{mn} = (\alpha^2+\beta^2)^{1/2}$]. Again boundary conditions such as $\varphi(1) = \varphi(2)$ at $z = c/2$ and those relating the gradients of φ, i.e. the stray fields, at $z = \pm c/2$ introduce relations between the b_{mn}, c_{mn} and a_{mn}, and on carrying out the integration of $\sigma \times \varphi$ it is eventually found that

$$W_m = 2\pi a_{00}^2 c + 2\pi\sum_m\sum_n \frac{a_{mn}a_{-m-n}}{k_{mn}}(1 - e^{-k_{mn}\frac{c}{2}}) \tag{186}$$

as by Goodenough [14] and by Ignatchenko and Sakharov [15]. With $c \to \infty$ and specifically for the demagnetized state when $a_{00} = 0$,

$$E_m(c\to\infty, a_{00}=0) = \frac{2\pi}{c}\sum_m\sum_n \frac{a_{mn}a_{-m-n}}{k_{mn}} \tag{187}$$

which is twice the value found for a single sheet by Kittel [11] and by Spaček [16].

Again application to a particular structure calls for the appropriate evaluation of the Fourier coefficients, and for a rectangular array ($b_1 \times b_2$) of domains of elliptical cross-section ($a_1 \times a_2$) it is found that

$$W_m = M_s^2(2\pi R_e^2 c + 2b_1 L) \tag{188}$$

with the reduced magnetization

$$R_e = \frac{a_1 a_2}{b_1 b_2}\frac{\pi}{2} - 1$$

and

$$L = \sum_m \sum_n \frac{J_1^2 \left\{ \frac{a_1 a_2}{b_1 b_2} \pi \left[\left(\frac{m b_2}{a_2} \right)^2 + \left(\frac{n b_1}{a_1} \right)^2 \right]^{1/2} \right\} \left\{ 1 - e^{-c\pi/b_1} \left[m^2 + \left(\frac{n b_1}{b_2} \right)^2 \right]^{1/2} \right\}}{\left[\left(\frac{m b_2}{a_2} \right)^2 + \left(\frac{n b_1}{a_1} \right)^2 \right] \left[m^2 + \left(\frac{n b_1}{b_2} \right)^2 \right]^{1/2}} \tag{189}$$

as by the author and McIntyre [17], in relation to domain observations on silicon iron (see also Charap and Nemchik [18]). The periodic unit is a right parallelopiped $2b_1 \times 2b_2 \times c$ with an elliptical section cylinder $2a_1 \times 2a_2 \times c$ at its centre. With $a_1 = a_2 = a$, $b_1 = b_2 = b$, i.e. a square array of circular cylinders,

$$W_m = M_s^2 \left(2\pi R_c^2 c + \frac{2a^2}{b} F \right) \tag{190}$$

where

$$F = \sum_m \sum_n J_1^2 \frac{[(a\pi/b)(m^2+n^2)^{1/2}](1 - e^{-(c\pi/b)(m^2+n^2)^{1/2}})}{(m^2+n^2)^{3/2}}$$

as obtained by Craik, Cooper and Druyvestyn [19]. The summations are as above. R_c is again $M/M_s = \pi a^2/2b^2 - 1$.

For an hexagonal array of bubble domains,

$$W_m = M_s^2 \left[2\pi R_h^2 c + \frac{2a^2}{b} \left(\mathcal{H} + \frac{2}{3\pi} G \right) \right] \tag{191}$$

where

$$\mathcal{H} = \sum_{\substack{-\infty \\ m \text{ even}}}^{\infty} \left[\frac{J_1^2 \left(\frac{a\pi m}{b} \right) (1 - e^{-c\pi m/b})}{3m^2} + \frac{\sqrt{3} J_1^2 \left(\frac{a\pi m}{\sqrt{3} b} \right) \left(1 - e^{-c\pi^2/(\sqrt{3}b)} \right)}{m^3} \right]$$

$$G = \sum_m \sum_n \frac{J_1^2 \left[\frac{a\pi}{b} \left(m^2 + \frac{n^2}{3} \right)^{1/2} \right] \left(1 - e^{-(c\pi/b)(m^2+n^2/3)^{1/2}} \right)}{(m^2 + n^2/3)^{3/2}}$$

as given by Druyvesteyn and Dorleyn [20].

Strips can be shown to give the lower energy generally in a zero field but bubble arrays may form on removing saturating in-plane fields and remain in metastable equilibrium (see Figure 4.33). Bubble arrays may be stable in applied fields causing substantial magnetization, with the bubbles well spaced. If a film is saturated by an axial field which is then reduced somewhat, stable bubbles may be formed by pulsing locally applied reverse fields, and these may be caused to move round tracks formed by depositing specially shaped Permalloy elements on the surface. Motion along one period of a loop may be achieved by rotating an in-plane field and the nucleation–non-nucleation sequence in a subsidiary loop that feeds the main loop is coordinated with this field. The bubbles pass a device (magnetoresistive, Hall effect, etc.) that detects their presence, or conversely the absence of a bubble, during one period, and thus reads the information that has been written. The bias field is provided by permanent magnets, as noted in Section 4.1, and the store is thus non-volatile (information is preserved in the absence of power).

In general, numerical minimization is necessary to give equilibrium domain geometries in zero or applied fields and thus to predict anhysteretic magnetization curves such as that compared with experimental curves in Figure 4.35. Appropriate analyses, together with subsidiary observations in in-plane fields, then give M_s, the wall energy, anisotropy and exchange parameters.

The foregoing calculations were given in the original e.g.s., conversions corresponding simply to $W_m(\text{SI}) = (\mu_0/4\pi)W_m$ (c.g.s.). The expressions for $E_m = W_m/c$ then have the form $E_m = \frac{1}{2}\mu_0 M^2$, as for homogeneous magnetization, plus further terms. Using approximations such as $W_m = 1.71(\mu_0/4\pi)M_s^2 l$ for strips, the condition $\partial/\partial l(W_m + \gamma c/l)$ with $\gamma c/l$ the wall energy per unit surface area gives the equilibrium width and energy as

$$l_0 = \left(\frac{4\pi\gamma c}{1.71\mu_0}\right)^{1/2} \frac{1}{M_s}, \qquad W_0 = \left(\frac{1.71\mu_0\gamma c}{4\pi}\right)^{1/2} M_s \tag{192}$$

or $l_0 = \sqrt{4\pi/1.71}\sqrt{\lambda}\sqrt{c}$ in terms of the characteristic length $\lambda = \gamma/\mu_0 M_s^2$. For barium ferrite, for example, Kooy and Enz [13] (using the full expression) estimated $\gamma = 3.0$ erg cm^{-2} or 3.0×10^{-3} Jm^{-2} and Kojima and Goto [21] gave 6.7 erg cm^{-2}. Averaging, with $M = 3.8 \times 10^5$ A m^{-1}, gives $\lambda = 2.8 \times 10^{-8}$ m. Comparing $\frac{1}{2}\mu_0 M_s^2 c$, for the energy per unit surface area in the saturated state with the approximation, it is seen that the multidomain state has the lower energy for $c' > 2.18$. Craik *et al.* [19] evaluated approximations, with

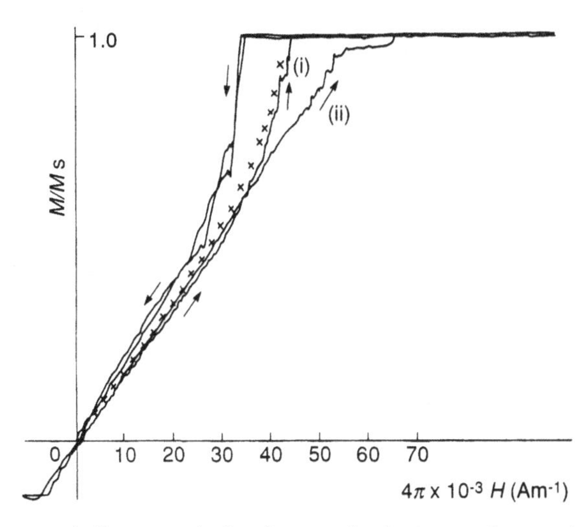

Figure 4.35 The crosses indicate a calculated magnetization curve for a strip structure, which is closely followed by results for an epitaxially grown garnet film after conventional demagnetization (i). The hysteretic loop (ii) was recorded after applying a 3 MHz, 58 Oe field which accelerates the walls strongly and induces a complex wall structure involving Bloch lines or periodic reversals of the sense of rotation ([28] of Chapter 1, see Figure 1.61)

c/b large, for bubble arrays:

$$E_{\mathrm{m}}^{(1)} = \frac{1.506}{4\pi}\mu_0\frac{b}{c}M_{\mathrm{s}}^2, \qquad E_{\mathrm{m}}^{(2)} = \frac{1.345}{4\pi}\mu_0\frac{b}{c}M_{\mathrm{s}}^2 \tag{193}$$

the first for square and the second for hexagonal arrays. Then

$$b_0 = \left(\frac{4\pi(2\pi/\sqrt{3})}{1.345\mu_0}\right)^{1/2}\frac{\sqrt{\gamma c}}{M_{\mathrm{s}}} = 3.9\sqrt{\lambda}\sqrt{c}, \qquad a_0 = \left(\frac{\sqrt{3}}{\pi}\right)^{1/2} b_0 \tag{194}$$

for the demagnetized state, although it is apparent that due to a wall 'surface tension' effect there should always be a small spontaneous remanence antiparallel to $\mathbf{M}_{\mathrm{s}}$ in the bubbles. For the example of barium ferrite, taking $c = 10$ μm, $b_0 = 2.2$ μm and $a_0 = 1.6$ μm. Bubble diameters and strip widths are similar in the same specimen.

2.2 Discrete Structures: Single Bubble Domains

The account given here [7], restricted to circular section bubbles in cylindrical crystals, can be extended to investigate stability against deformation, e.g. to an ellipse as a precursor to a strip domain. For the axial symmetry and $\varphi \to 0$ as $z \to \infty$ the general solution of Laplace's equation in cylindrical coordinates x, z is

$$\varphi(x, z) = \int_0^\infty \varepsilon^{-1} A(\varepsilon) J_0(\varepsilon x)\mathrm{e}^{-\varepsilon z}\,\mathrm{d}\varepsilon \tag{195}$$

with J_0 a Bessel function of the first kind and $A(\varepsilon)$ to be determined. The derivative boundary conditions for a single disc of radius a, pole density σ, at $z = 0$, are

$$-\left.\frac{\partial\varphi}{\partial z}\right|_{z=0} = +\int_0^\infty A(\varepsilon) J_0(\varepsilon x)\,\mathrm{d}\varepsilon = \begin{cases} 0 & x > a \\ \dfrac{\sigma}{2}, & a > x \geqslant 0 \end{cases} \tag{196}$$

In general, when the integral here is equal to $\sigma(x)$, $\sigma(x)$ is the Hankel transform of $A(\varepsilon)$ and by the Hankel inversion theorem

$$A(\varepsilon) = \varepsilon\int_0^\infty x\sigma(x) J_0(\varepsilon x)\,\mathrm{d}x \tag{197}$$

For the concentric discs defined by

$$\sigma(x) = \begin{cases} M_{\mathrm{s}}, & a > x \geqslant 0 \\ -M_{\mathrm{s}}, & b > x > a \\ 0, & x > b \end{cases} \tag{198}$$

it follows that

$$\varphi = \frac{M_{\mathrm{s}}}{2}\int_0^\infty [2aJ_1(\varepsilon a) - bJ_1(\varepsilon b)]J_0(\varepsilon x)\mathrm{e}^{-\varepsilon z}\varepsilon^{-1}\,\mathrm{d}\varepsilon \tag{199}$$

and on integrating the products $\mu_0\sigma\varphi$ over concentric discs the interaction energy is

$$W_{\mathrm{i}} = \mu_0\pi M_{\mathrm{s}}^2\int_0^\infty \left[4a^2J_1^2(\varepsilon a) - 4abJ_1(\varepsilon a)J_1(\varepsilon b) + b^2J_1^2(\varepsilon b)\right]\mathrm{e}^{-\varepsilon z}\varepsilon^{-2}\,\mathrm{d}\varepsilon \tag{200}$$

with $z \to c$ for a cylinder of length c as in Figure 4.36(a), and with a related expression for the self-energy W_s of one pair of coplanar discs ($\times\frac{1}{2}$ and $z = 0$). The energy for the structure is the interaction energy plus two identical self-energies, equivalent to integrating overall.

Introduce $k = a/b$ and also

$$W' = \frac{W}{(\frac{4}{3})\mu_0 a^3 M_s^2} \tag{201}$$

since this is dimensionless and the limiting value of W' as $k \to 0$ is just $\frac{1}{2}$, as the reader may demonstrate by reference to Figure 4.36(b). As $k \to 0$ and $c \to 0$, for $\sigma(x, c) = -\sigma(x, 0)$, the total energy approaches zero so the limiting value of $|W_i'|$ approaches unity. For small values of c/b (Figure 4.37) the magnetostatic energy is only slightly dependent on the presence of the structure, remaining close to that for a uniformly magnetized sheet, and since alternative structures would give a lower energy the bubble would not be stable as such in a zero field. From Figure 4.37, the magnetostatic energy appears to be minimum for the demagnetized state $k = 1/\sqrt{2}$, and this can be confirmed. Since the domain wall energy falls continuously as the bubble contracts, equilibrium in the zero field corresponds to a spontaneous (anhysteretic) remanence. There is a point of inflection in the curve in Figure 4.37 and if a bubble is brought to the corresponding radius by an applied field then both energy terms correspond to pressures contracting the bubble, which indicates a collapse field as observed experimentally in, for example, $EuFeO_3$ crystals (Figure 4.38). Conversely, the increasing gradients in Figure 4.37 at high k indicate an effective repulsion of the wall at the specimen surface, which is also observed. There is extensive literature on

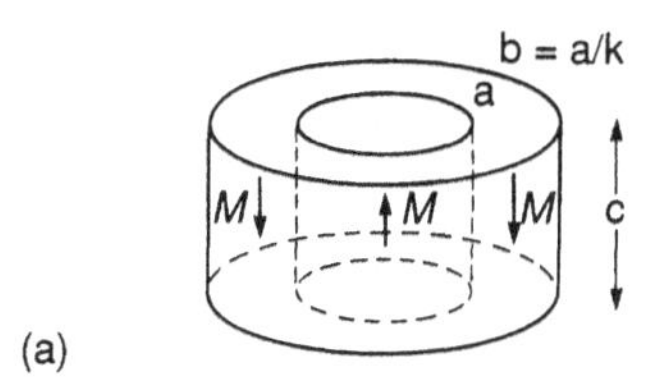

(a)

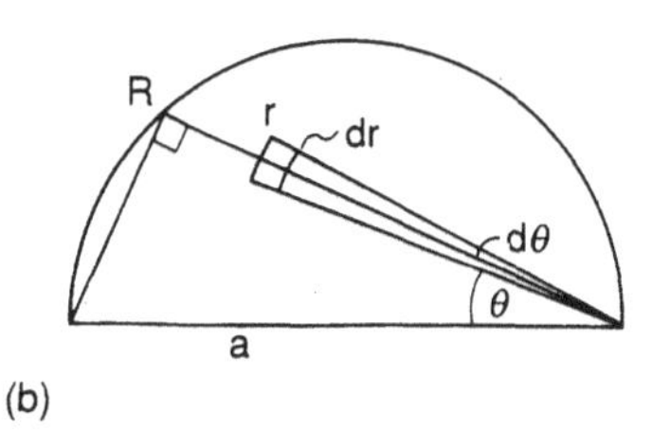

(b)

Figure 4.36 (a) A readily analysed discrete domain structure. (b) As an exercise the scalar potential at a point immediately adjoining the edge of a disc of surface density σ may be found by integrating $d\varphi$ from the element $rd\theta dr$, $r = 0$ to R and $\theta = 0$ to $\pi/2$. The self-energy of the disc follows by integrating $\mu_0\varphi\, d\sigma$ as the disc is created by expansion from a point since φ is constant over a bounding annulus

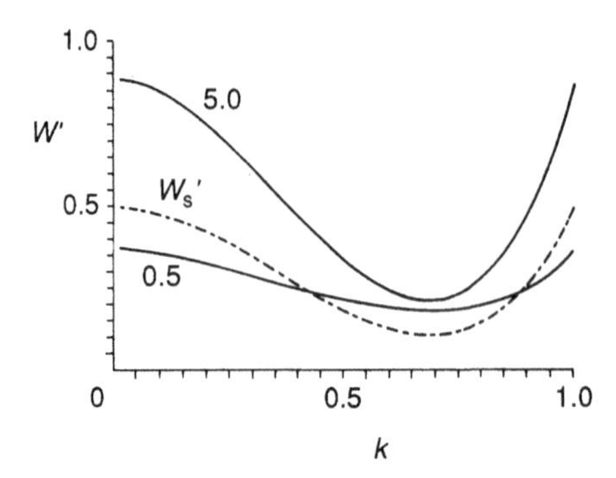

Figure 4.37 The variation with k of the reduced self-energy W_s' as indicated and of the total energies $2W_s' - |W_i|$ for two dimensional ratios c/b, for the structure of Figure 4.36(a). The minimum corresponds to demagnetization and thus an anhysteretic remanence and the points of inflection at lower values of k to bubble collapse fields (when the wall energy is taken into account, in both cases)

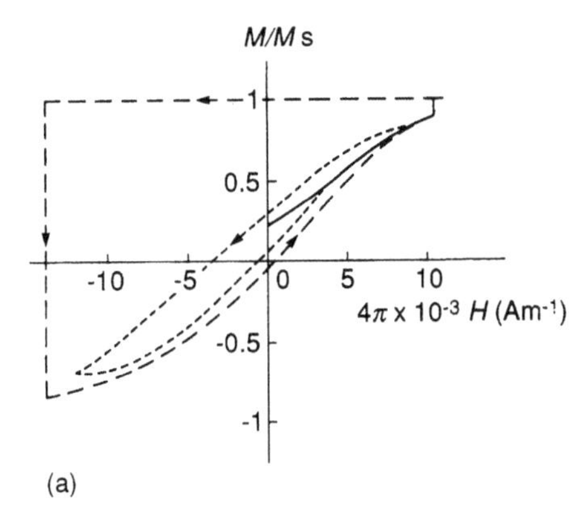

Figure 4.38 A measured magnetization curve for a crystal of europium orthoferrite with the domain structure as in Figure 4.36(a), exhibiting spontaneous remanence and bubble collapse

the magnetostatics and dynamics of bubbles in the storage context, following early work by Bobeck [22], Thiele [23] and others.

3 Energy Product and Permanent Magnets

Before proceeding to general considerations of magnetic design a concept frequently referred to in relation to permanent magnet properties may be mentioned (see, for example, Parker [24]). The product BH is called the energy product and the maximum value of this, as a loop is plotted in a closed flux configuration, i.e. with the specimen in a yoke, is the maximum energy product $(BH)_{max}$ which is still used as an indication of the value of a material. (It has been claimed that, despite the much lower cost per unit volume of other materials, NdFeB is the cheapest in terms of energy product.) It is arguable that the concept is outmoded in relation to modern materials. Most of the rather dubious discussions of this seem to be based on models of ring magnets containing narrow gaps, which are scarcely of topical interest.

Consider an isolated ellipsoid magnetized uniformly along its principal axis and thus with energy of formation $W = \frac{1}{2}\mu_0 H_d MV = \frac{1}{2}\mu_0 NM^2V$. Let W_H be the useful or external field energy. The energy of the internal field is $\frac{1}{2}\mu_0 H^2V = \frac{1}{2}\mu_0 N^2M^2V$. Equating the energy of creation of the total field to the energy of formation of the magnets,

$$W_H = \frac{1}{2}\mu_0 NM^2V - \frac{1}{2}\mu_0 N^2M^2V \tag{202}$$

which is zero for the cases $N = 0$ (needle) and $N = 1$ (transversely magnetized thin sheet) in which no external field is produced. Thus it is maximum when stationary: $\partial W_H/\partial N = 0$, i.e. clearly, when $1 - 2N = 0$, for $N = \frac{1}{2}$. The common shapes for modern magnets are right parallelopipeds or axially magnetized cylinders and N must then be regarded as an effective demagnetizing factor. Assuming uniform magnetization, which is now reasonable, the shape that is most effective in the production of external field energy is that for which the dimension in the direction of M is, coincidentally, approximately half that in the transverse direction. Again with this assumption, the external field energy is $\frac{1}{2}\mu_0 M^2V \times (\frac{1}{4})$ and the value of a particular material is well represented by M^2 for that material.

The convention, however, is to consider the demagnetizing curve as referred to above, in which case the demagnetizing fields are zero and the reverse applied field is implicitly uniform, as in Figure 4.39. In terms of the reverse field as such (the coercivity being conventionally a positive quantity), $B = \mu_0(M - H)$, $BH = \mu_0(MH - H^2)$ and $\partial(BH)/\partial H = 0$ when $H = M/2$, as is apparent from the geometry of Figure 4.39. Of course this corresponds to the demagnetizing factor of $\frac{1}{2}$ for the isolated magnet. Due to this coincidence it remains common to value a magnet according to this maximum value of BH as such, i.e. according to its $(BH)_{max}$. The units of BH or $(BH)_{max}$ are, of course, J m^{-3}, and it is more apparent that this indicates the field energy produced per unit volume of material. The c.g.s. MGO or 'mega gauss oersteds' are not, sadly, equated to 10^6 erg cm^{-3} (see Appendix 1). When the demagnetizing B–H curve is not linear, M is not constant and $(BH)_{max}$ does not occur at a simply specified point, but it may be argued that the most effective shape is one indicated by the effective demagnetizing field corresponding to the field at which the observed $(BH)_{max}$ occurs; more elongated specimens become appropriate.

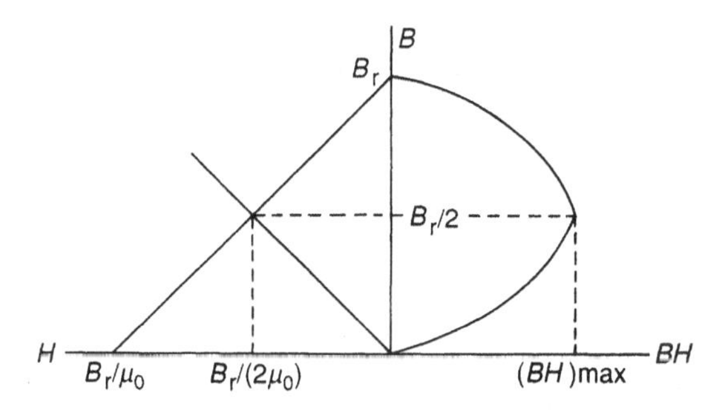

Figure 4.39 $(BH)_{max}$ occurring at $B = B_r/2$ for a linear demagnetizing plot

4 Numerical Methods and Magnetic Design

The general problem is to calculate the field, magnetization and induction, with secondary forces, inductance, etc., over a space containing magnets, current-carrying coils and structures of specified permeability. The magnets and coils may be implicit, giving applied fields, and may even be outside the space. If $\mu = \mu'$ within an ellipsoidal region with $\mu = \mu_0$ elsewhere then an immediate analysis is possible but any departure from this situation calls for analysis by numerical methods or approximations, three of which will be described.

4.1 A Direct Method for Magnetization Distributions

This is a generalization of the relations $\mathbf{M} = \chi\mathbf{H}$ and $\mathbf{H} = \mathbf{H}_a - N\mathbf{M}$, which together give the solutions noted above, a simple-minded approach appropriately developed by the author [25]. In a two-dimensional example a long bar with square section is divided into $n = 4$ elements, in each of which $\mathbf{M}$ is assumed uniform (Figure 4.40). H_a is necessarily in OXY for the two-dimensional example. The internal field components at the centre of each element are

$$\begin{aligned} H_i^x &= H_{a_i}^x + \sum_{j=1}^{n} N_{ij}^{xx} M_j^x + \sum_{j=1}^{n} N_{ij}^{xy} M_j^y, \qquad i = 1, 2, \ldots, n \\ H_i^y &= H_{a_i}^y + \sum_{j=1}^{n} N_{ij}^{yx} M_j^x + \sum_{j=1}^{n} N_{ij}^{yy} M_j^y \qquad i = 1, 2, \ldots, n \end{aligned} \tag{203}$$

where N_{ij}^{kl} is calculated as the component k of the field at the centre of the element i due to a magnetization component l of unit magnitude in element j. Subroutines giving such results for elements of general (principally quadrilateral) shape are based on the foregoing expressions. (The example could also be taken to be *effectively* two-dimensional and to apply to cylinders.) The reader may readily set each $H_i^x = M_i^x/\chi_i^x$ and $H_i^y = M_i^y/\chi_i^y$, make an obvious extension to three dimensions and with a little rearrangement see that the magnetization components are given by $(\mathbf{N} - \boldsymbol{\chi})\mathbf{M} = -\mathbf{H}_a$, i.e.

$$\left\{ \begin{pmatrix} (N_{ij}^{xx}) & (N_{ij}^{xy}) & (N_{ij}^{xz}) \\ (N_{ij}^{yx}) & (N_{ij}^{yy}) & (N_{ij}^{yz}) \\ (N_{ij}^{zx}) & (N_{ij}^{zy}) & (N_{ij}^{zz}) \end{pmatrix} - \begin{pmatrix} (\chi_x) & 0 & 0 \\ 0 & (\chi_y) & 0 \\ 0 & 0 & (\chi_z) \end{pmatrix} \right\} \begin{pmatrix} (M_x) \\ (M_y) \\ (M_z) \end{pmatrix} = - \begin{pmatrix} (H_a^x) \\ (H_a^y) \\ (H_a^z) \end{pmatrix} \tag{204}$$

with, for example,

$$(\chi_x) = \begin{pmatrix} \chi_1^x & 0 & \cdots \\ 0 & \chi_2^x & \cdots \\ \cdots & \cdots & \cdots \end{pmatrix}, \qquad (M_x) = \begin{pmatrix} M_1^x \\ M_2^x \\ \cdots \end{pmatrix}, \qquad (H_a^x) = \begin{pmatrix} H_{a_1}^x \\ H_{a_2}^x \\ \cdots \end{pmatrix} \tag{205}$$

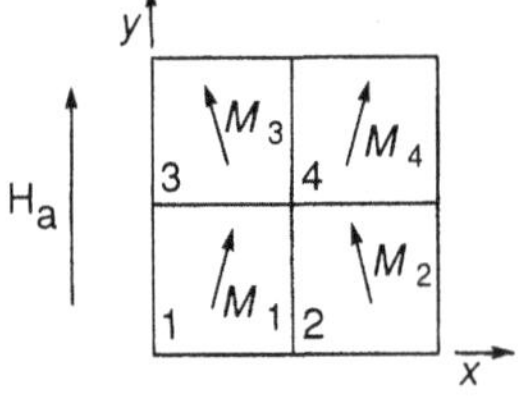

Figure 4.40 The element numbering for a simple two-dimensional example

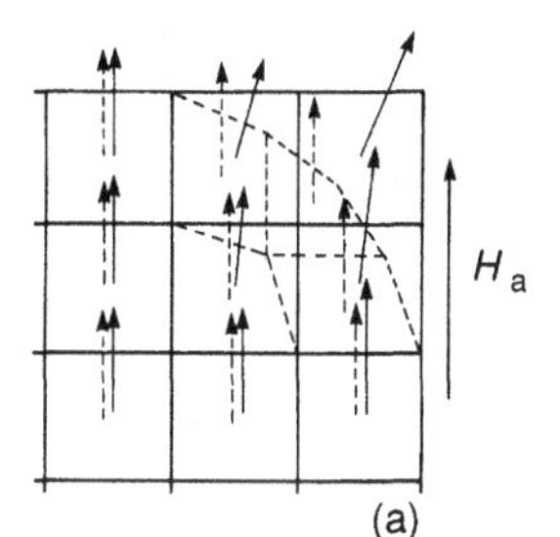

Figure 4.41 The $\mathbf{M}_i$ vectors in a square section bar in a transverse field (one quadrant shown, the calculation itself taking advantage of the symmetry) and (broken arrows) in an approximation to a circular cross-section bar which is expected to become uniformly magnetized

The solution is standard (e.g. [26]). Internal fields follow as, for example,

$$H_i^x = H_{a_i}^x + \sum_{j=1}^{n} N_{ij}^{xx} M_j^x + \sum_{j=1}^{n} N_{ij}^{xy} M_j^y + \sum_{j=1}^{n} N_{ij}^{xz} M_j^z, \qquad i = 1, 2, \ldots, n \tag{206}$$

and external fields are found by recourse to the subroutines used for the N values so that $\mathbf{B}$ follows throughout all space.

An obvious example is given in Figure 4.41 showing a distribution in a square-section bar, which is much as expected according to the demagnetizing fields that would exist if $\mathbf{M}$ were uniform throughout and also the remarkably close approach to the predicted uniform $\mathbf{M} = \mathbf{H}_a/N$ (for $\chi \to \infty$) achieved by adjusting a few coordinates so as to approximate a long cylinder. A technical application is illustrated by Figure 4.42. The way in which the field profiles between pairs of magnets could be evened out, by applying thin sheets of

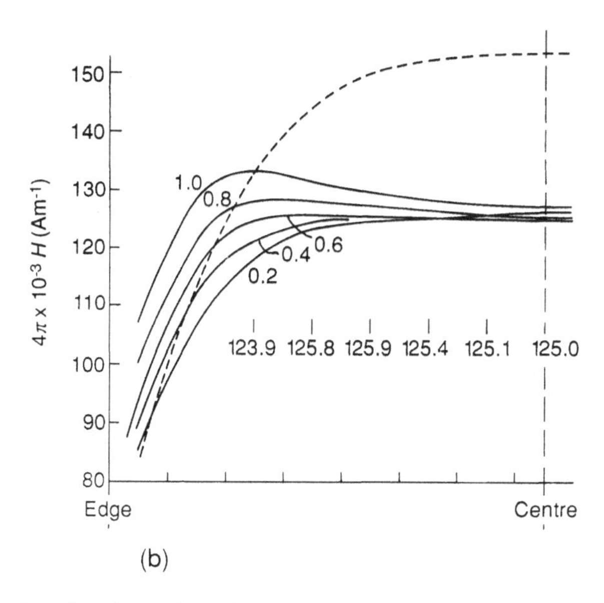

Figure 4.42 The broken line shows field plots between the pair of magnets alone and the remainder indicate how the calculated fields, from the magnets and the magnetization induced in permeable facing sheets of different thicknesses, become more uniform; different sheet thicknesses are indicated with the numbers for the optimum [27]

material with specified permeability to the magnet faces, could be illustrated in a program of value in the design of the devices, which afford the bias fields for bubble domain stores [27]. The components of **M** in the ferrite sheets, which were transverse to **M** in the magnets, were found to be of most importance.

Anisotropy and non-linearity are readily encompassed. Using iterative methods a check is made on elements for which the predicted M exceeds a pre-set M_s and a switch made to an alternative procedure in which M is set equal to M_s and the direction becomes that of the total field. A very simple program is one in which all the $M_i = M$ and the directions of the $\mathbf{M}_i$ follow the total field which includes anisotropy field contributions, applicable to a permanent magnet in which some rotations but no reversals are expected. Applying this to two face-to-face magnets of $SmCo_5$, the least favourable configuration, M_r, differed from M_s in the fifth place only, although reduced forces were obtained as 5.06 compared to the ideal 5.10. For two cubes of barium ferrite, also in opposing contact, $M_r = 0.9988 M_s$ was indicated with a reduced force of 4.81 (ideal of 5.10), but for a single cube $M_r = 0.9995 M_s$. Thus $SmCo_5$ can be considered near-ideal in all circumstances but BaF may depart substantially from the ideal.

4.2 Finite Difference Approximation to Laplace's Equation

On replacing differential coefficients by finite differences the relation between the potentials at the nodes in Figure 4.43, according to Laplace's equation, can be shown to be

$$\frac{2}{1+r}\varphi_1 + \varphi_2 - 4\varphi_0 + \frac{2r}{1+r}\varphi_3 + \varphi_4 = 0 \tag{207}$$

where $r = \mu_a/\mu_b$. This applies also to components of **A** on replacing r by $R = \mu_b/\mu_a$, and three-dimensional equations are readily derived [28]. Such equations are 'centred' on each node in turn and the set solved simultaneously. At external boundaries extraneous potentials are eliminated by, for example, assuming zero or constant fields there. If the problem is the typical one of determining the induction in a permeable body in the presence of fields from a magnet or coil, φ may be divided into a contribution from the magnet, an 'applied potential' φ^0, which is constant in the solution process, and φ^m from the magnetization of the body, the variable to be determined initially; thus $\varphi = \varphi^0 + \varphi^m$, since ∇^2 is a linear operator. Each equation then becomes in effect the relevant equation in φ^m with a right-hand side obtained as the negative of that expression written in the φ_i^0. In one dimension with node 2 on an interface, 1 in a space with $\mu = 1 (\mu \equiv \mu_r$ since μ_0 cancels) and 3 in a

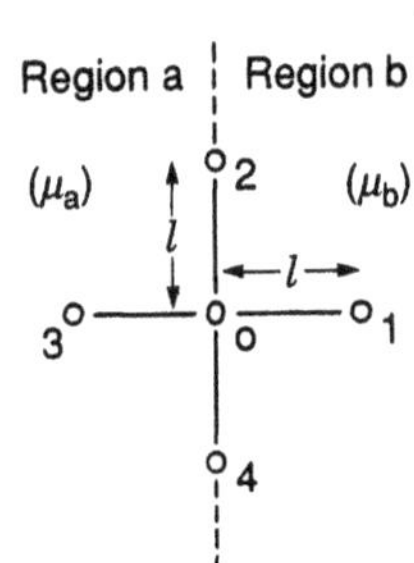

Figure 4.43 Nodes 2, 0 and 4 lying on an interface between region a, $\mu = \mu_a$, and region b, $\mu = \mu_b$

material with $\mu \to \infty$, the appropriate equation is

$$\frac{2}{1+\mu}\varphi_1 - 2\varphi_2 + \frac{2\mu}{1+\mu}\varphi_3 = 0 \to -2\varphi_2 + 2\varphi_3 = 0, \qquad \varphi_3 - \varphi_2 = 0 \tag{208}$$

or

$$\varphi_3^{\mathrm{m}} - \varphi_2^{\mathrm{m}} = \varphi_2^0 - \varphi_3^0 = l H_{\mathrm{a}} \tag{209}$$

with H_a the field from the magnet. If the central node is not on an interface the right-hand side must clearly be zero, so it is only necessary to calculate the applied fields or corresponding potentials at the interfaces, naturally because a field defined over a closed surface is defined at all points in space. This method is of value for much demonstrations and, arguably, only in this way.

4.3 Finite Element Approximations

Some reference should be made to the calculus of variations according to which the solution of a differential equation corresponds to the condition

$$\delta\mathcal{F} = 0, \qquad \mathcal{F} = \int_0^\infty F(\varphi, \varphi^x, \varphi^y, \varphi^z, x, y, z)\,\mathrm{d}x\,\mathrm{d}y\,\mathrm{d}z \tag{210}$$

($\varphi^x = \partial\varphi/\partial x$, etc.) where the functional $\mathcal{F}$ is identified to the extent that the Euler equation for F,

$$\frac{\partial}{\partial x}\frac{\partial F}{\partial \varphi^x} + \frac{\partial}{\partial y}\frac{\partial F}{\partial \varphi^y} + \frac{\partial}{\partial z}\frac{\partial F}{\partial \varphi^z} - \frac{\partial F}{\partial \varphi} = 0 \tag{211}$$

constitutes the differential equation. The derivation of F may be intuitive but such intuition may always be verified. For Poisson's equation in Cartesian coordinates,

$$F = \frac{\mu}{2}(\varphi^{x^2} + \varphi^{y^2} + \varphi^{z^2}) - \rho\varphi(x, y, z) = \frac{\mu}{2}(\nabla\varphi)^2 - \rho\varphi$$

since it is readily checked that this gives the Euler equations

$$\mu\nabla^2\varphi = -\rho, \qquad R\nabla^2\mathbf{A} = -\mu_0\mathbf{J} \tag{212}$$

the second being obtained by replacing, in F, φ^x by A^x, φ by $\mathbf{A}$, ρ by $\mu_0\mathbf{J}$ and μ by the reluctance $R = 1/\mu$. Poisson's axisymmetric equation can be written symmetrically as

$$\mu\left[\frac{1}{x}\frac{\partial}{\partial x}\left(x\frac{\partial\varphi}{\partial x}\right) + \frac{1}{x}\frac{\partial}{\partial z}\left(x\frac{\partial\varphi}{\partial z}\right)\right] + \rho = 0 \tag{213}$$

and the relevant F is seen (cf. Appendix 2) to be

$$F = \frac{\mu}{2}\left[x\left(\frac{\partial\varphi}{\partial x}\right)^2 + x\left(\frac{\partial\varphi}{\partial z}\right)^2\right] - \rho x\varphi = 0 \tag{214}$$

Note that $\nabla \cdot \mathbf{B} = 0, \mathbf{B} = \mu\mathbf{H}$ and $\mathbf{H} = -\nabla\varphi$ together indicate that $\mu\nabla^2\varphi = 0$ and division by μ gives $\nabla^2\varphi = 0$. However, if $\mu = \mu_1$ in a region R_1 and $\mu = \mu_2$ in R_2 and these are to be considered jointly, such division is not justified.

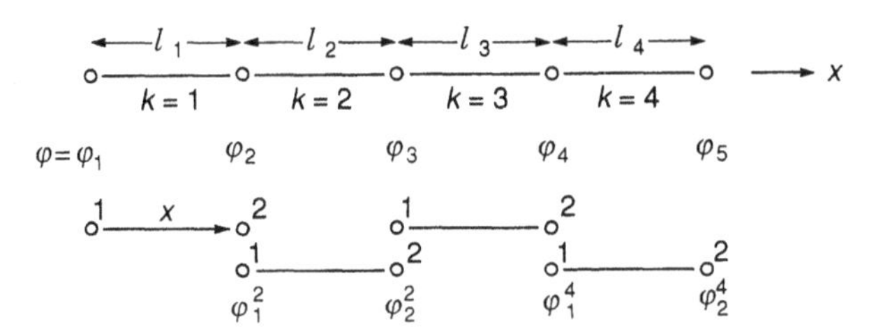

Figure 4.44 Each element numbered or indexed k has two nodes indexed 1 and 2 at which $\varphi = \varphi_1^{(k)}$ and $\varphi = \varphi_2^{(k)}$ locally, but when the elements are contiguous the nodes and nodal values are indexed globally, as shown, with φ_2 (global) = $\varphi_2^{(1)}$ (local) for element 1 and $\varphi_1^{(2)}$ (local) for element 2, etc. The local normal coordinate X exists within each element

In a one-dimensional illustration a line is formed from a set of elements numbered $1, 2, \ldots, k, \ldots, n$ each bound by a node numbered locally as 1 on the left and 2 on the right (Figure 4.44). The elements are contiguous so the whole set of nodes may also be numbered sequentially or 'globally' and if φ has the value $\varphi_1^{(k)}$ and $\varphi_2^{(k)}$ at the extremes of element k, it is seen that $\varphi_1 = \varphi_1^{(1)}, \varphi_2 = \varphi_2^{(1)} = \varphi_1^{(2)}, \varphi_3 = \varphi_1^{(3)}$, etc.

The integral in equation (210) is over all space (the line here) and is the sum of integrals over all the elements: $\mathcal{F} = \mathcal{F}^{(1)} + \mathcal{F}^{(2)} + \cdots + \mathcal{F}^{(k)} + \cdots$. Thus φ must be interpolated within each element, at the simplest (linear) case as $\varphi^k = f_1\varphi_1^k + f_2\varphi_2^k$ with shape functions $f_1 = 1 - X, f_2 = X$, with X ranging from 0 to 1 in each element. These clearly give $\varphi = \varphi_1^k$ at $X = 0, \varphi = \varphi_2^k$ at $X = 1$. X is a local normal coordinate. The coordinate x itself may be interpolated similarly and it is clear that $\mathrm{d}x = l_k\,\mathrm{d}X$ if the element length is l_k; $(\varphi^x)^2\,\mathrm{d}x = (1/l_k^2)(\varphi^x)^2 l_k\,\mathrm{d}X$ and a transformed integral for any element has the limits 0 and 1.

$\delta\mathcal{F} = 0$ corresponds to $\partial\mathcal{F}/\partial\varphi_i = 0$ for all variables φ_i, but effectively only for φ_1^k and φ_2^k in each $\mathcal{F}^k$, since all other differential coefficients are zero, associating two equations with each element. Noting that

$$\frac{\partial}{\partial\varphi_1^k}\left[\frac{\mathrm{d}}{\mathrm{d}X}(f_1\varphi_1^k + f_2\varphi_2^k)\right]^2 = 2[f_1^X\varphi_1^k + f_2^X\varphi_2^k]f_1^X \tag{215}$$

(where $f_1^X = \mathrm{d}f_1/\mathrm{d}X$) and that $\partial/\partial\varphi_1^k(f_1\varphi_1^k + f_2\varphi_2^k) = f_1$, the equations combine as

$$\int_0^1 \frac{\mu_k}{l_k}\begin{pmatrix} f_1^X f_1^X & f_1^X f_2^X \\ f_2^X f_1^X & f_2^X f_2^X \end{pmatrix}\mathrm{d}X\begin{pmatrix}\varphi_1^k \\ \varphi_2^k\end{pmatrix} = \int_0^1 \rho_k \begin{pmatrix} f_1 \\ f_2 \end{pmatrix} l_k\,\mathrm{d}X,$$

$$\begin{pmatrix} a_{11}^k & a_{12}^k \\ a_{21}^k & a_{22}^k \end{pmatrix}\begin{pmatrix}\varphi_1^k \\ \varphi_2^k\end{pmatrix} = \begin{pmatrix} b_1^k \\ b_2^k \end{pmatrix} \tag{216}$$

However, all the pairs of equations for all the elements apply simultaneously and an overlap arises because equations associated with, for example, elements $k = 1$ and $k = 2$ both refer to the globally numbered node 2. The global equations for $n = 3$ are formally

$$\left(\begin{pmatrix} - & - \\ - & - \end{pmatrix}_{(k=1)} \begin{pmatrix} - & - \\ - & - \end{pmatrix}_{(k=2)} \begin{pmatrix} - & - \\ - & - \end{pmatrix}_{(k=3)}\right)\begin{pmatrix} \varphi_1 \\ \varphi_2 \\ \varphi_3 \\ \varphi_4 \end{pmatrix} = \left(\begin{pmatrix} - \\ k=1 \\ - \end{pmatrix}\begin{pmatrix} - \\ k=2 \\ - \end{pmatrix}\begin{pmatrix} - \\ k=3 \\ - \end{pmatrix}\right) = \begin{pmatrix} b_1^1 \\ b_2^1 + b_1^2 \\ b_2^2 + b_1^3 \\ b_2^3 \end{pmatrix} \tag{217}$$

with overlap indicating element addition, the diagonal on the left being of the form $a_{11}^1, a_{22}^1 + a_{11}^2, a_{22}^2 + a_{11}^3, a_{22}^3$ and the right-hand side as indicated. Completing the simple integrations in equation (216), the element equations are

$$\frac{\mu_k}{l_k}\begin{pmatrix} 1 & -1 \\ -1 & 1 \end{pmatrix}\begin{pmatrix} \varphi_1^k \\ \varphi_2^k \end{pmatrix} = \frac{\rho_k l_k}{2}\begin{pmatrix} 1 \\ 1 \end{pmatrix} \tag{218}$$

The reader may find it helpful to write down the pairs of equations for $k = 1, 2, 3, \ldots$ setting $\varphi_1 = \varphi_1^1$, $\varphi_2 = \varphi_2^1 = \varphi_1^2$, etc., and to note the association of these with equation (217). Furthermore, setting all $\rho_k = 0$ (Laplace) and $l_k = 1$ it is seen that the first equation becomes $\varphi_1 = \varphi_2$ so that, for the method as formulated here, there is an implicit boundary condition with $H = 0$ at the external boundary, a condition that can be overwritten if this is appropriate. Setting all $\mu_k = 1$, the second equation is seen to be $-\varphi_1 + 2\varphi_2 - \varphi_3 = 0$, that applying in the finite difference (FD) scheme and if $\mu_k = \mu$ for certain elements then equations such as $\varphi_3 - (1 + \mu)\varphi_4 + \mu\varphi_5 = 0$ arise, again, in this example, as for the FD scheme. (This correspondence should not be overstressed since the methods are utterly distinct and the finite element (FE) method is much the easier to understand and to use in practice.) Further, by introducing certain values of ρ_k, or equally of J_k (one component) with obvious modifications, the reader may readily obtain model results such as that in Figure 4.45 by sequential solution of the equations. (Any one φ_i or A_i can be set arbitrarily.) The model illustrated is wholly artificial since a single pole sheet is introduced as a device to generate a field and it is necessary to consider that the elements extend

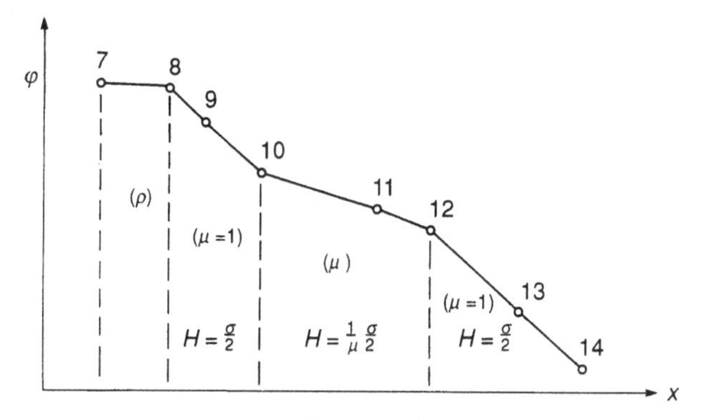

Figure 4.45 Minimal FE example results in one dimension for Poisson's equation. Each node may be taken to be a characteristic node in a two-dimensional sheet normal to the diagram, with all nodal values equal in the sheet. Element 7 (nodes 7 and 8) constituted a 'source' and elements 10 and 11 a permeable sheet. Such results (exact in this case) may be obtained by hand solution as an exercise, with elements 1 to 6 implied to give a model that is symmetrical but unrealistic at this level ($\rho \neq 0$ in element 7 only and $\sigma = \rho l_7$)

indefinitely in the one-dimensional case. The external boundary conditions may be set by replacing the first equation by $\varphi_2 = \varphi_1 - l_1 H_a$ and then it is easily seen that H remains at H_a whenever $\mu = 1$ but $H = H_a/\mu$ in the 'material', which is implicitly a sheet of infinite extent $\perp OX$, but this is incompatible with the introduction of sources here.

4.3.1 Two-dimensional systems

It is clear that

$$\tfrac{1}{4}(1+X)(1+Y) = \begin{cases} 1, & X = Y = 1 \\ 0, & X = -1 \text{ or } Y = -1 \end{cases} \tag{219}$$

and suitable shape functions are

$$\begin{aligned} f_1 &= \tfrac{1}{4}(1-X)(1-Y) && = 1 \text{ at } -1, -1 \\ f_2 &= \tfrac{1}{4}(1+X)(1-Y) && = 1 \text{ at } 1, -1 \\ f_3 &= \tfrac{1}{4}(1-X)(1+Y) && = 1 \text{ at } -1, 1 \\ f_4 &= \tfrac{1}{4}(1+X)(1+Y) && = 1 \text{ at } 1, 1 \end{aligned} \tag{220}$$

the local coordinates now being zero at the element centres: $\varphi^k = \sum_{i=1}^{4} f_i \varphi_i^k$. In relation to Figure 4.46, $(\partial x/\partial X)^k = a_k$ and $(\partial y/\partial Y)^k = b_k$ so the development from one to two dimensions is illustrated by

$$\frac{1}{l_k^2}\frac{\mathrm{d}\varphi^k}{\mathrm{d}X} \rightarrow \frac{1}{a_k^2}\left(\frac{\partial \varphi^k}{\partial X}\right)^2 + \frac{1}{b_k^2}\left(\frac{\partial \varphi^k}{\partial Y}\right)^2, \qquad l_k\,\mathrm{d}X \rightarrow a_k b_k\,\mathrm{d}X\,\mathrm{d}Y$$

and

$$\mathcal{F}^k = \int_{-1}^{1}\int_{-1}^{1} \frac{\mu_k}{2}\left[\frac{b_k}{a_k}\left(\frac{\partial \varphi^k}{\partial X}\right)^2 + \frac{a_k}{b_k}\left(\frac{\partial \varphi^k}{\partial Y}\right)^2\right]\mathrm{d}X\,\mathrm{d}Y - \int_{-1}^{1}\int_{-1}^{1} \rho_k \varphi^k a_k b_k\,\mathrm{d}X\,\mathrm{d}Y \tag{221}$$

It can be seen that the four equations for a characteristic element can be written as

$$\begin{aligned} &\int_{-1}^{1}\int_{-1}^{1} [c_k f_i^X f_j^X + d_k f_i^Y f_j^Y]\,\mathrm{d}X\,\mathrm{d}Y\,(\varphi_i^k) \\ &= \int_{-1}^{1}\int_{-1}^{1} \rho_k a_k b_k (f_i)\,\mathrm{d}X\,\mathrm{d}Y, \quad \text{for } j = 1, 2, 3, 4;\ i = 1, 2, 3, 4 \end{aligned} \tag{222}$$

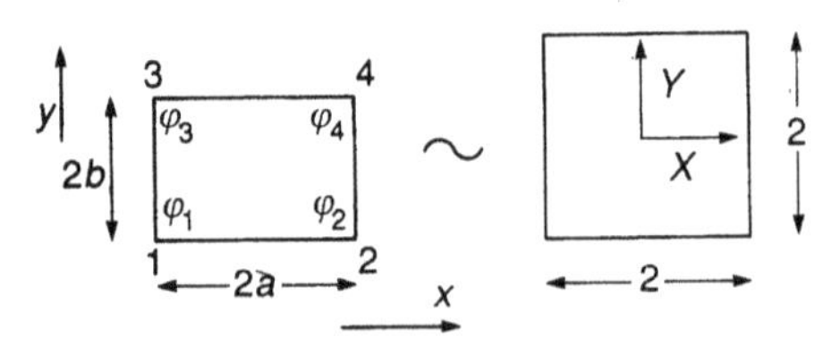

Figure 4.46 A simple rectangular two-dimensional element and associated square element with two local normal coordinates

(where $c_k = b_k/a_k$, $d_k = 1/c_k$). With simple integrations this becomes for $a_k = b_k = 1$, for example,

$$\frac{\mu_k}{6}\begin{pmatrix} 4 & -1 & -1 & 2 \\ -1 & 4 & -2 & -1 \\ -1 & -2 & 4 & -1 \\ 2 & -1 & -1 & 4 \end{pmatrix}\begin{pmatrix} \varphi_1^k \\ \varphi_2^k \\ \varphi_3^k \\ \varphi_4^k \end{pmatrix} = \begin{pmatrix} \rho_k \\ \rho_k \\ \rho_k \\ \rho_k \end{pmatrix}, \qquad \mathbf{A}^k\boldsymbol{\varphi}^k = \mathbf{B}^k \tag{223}$$

The global matrices must again be assembled to give $\mathbf{A}\boldsymbol{\varphi} = \mathbf{B}$ over all the variables φ_i. Referring to Figure 4.47, $\varphi_1^1 = \varphi_1$, $\varphi_2^1 = \varphi_2$, $\varphi_3^1 = \varphi_4$, $\varphi_4^1 = \varphi_5$. The matrix $\mathbf{A}^1$ may be considered to be $n \times n (9 \times 9)$ relating only nominally to all but four of the variables, i.e. with non-zero elements only where the rows and columns 1, 2, 4, 5 intersect. Similarly $\mathbf{B}^1$ can be regarded as the column $(\rho_1\ \rho_1\ 0\ \rho_1\ \rho_1\ 0\ 0\ 0\ 0)^T$ and the other $\mathbf{A}^k$ and $\mathbf{B}^k$ similarly 'expanded'. The global $\mathbf{A}$ and $\mathbf{B}$ are then simply the sums of these, and $\boldsymbol{\varphi}$ contains the list of variables φ_i, $i = 1, 2, \ldots, n$. In a program null matrices are designated and as the element matrices are formed the indices of the elements are adjusted appropriately and the elements are then added in to the null matrix.

The restriction to rectangular elements may be lifted in two ways. The first is by introducing general triangular elements. A rather lengthy development shows that the 3×3 element equations are all

$$\frac{\mu_k}{4T_k}[b_i b_j + c_i c_j](\varphi_i^k) = \frac{\rho_k T_k}{3}(1) \quad i, j = 1, 2, 3 \tag{224}$$

with the unit column on the right. T_k is the area of the triangle k and

$$\begin{aligned} b_1 &= y_2 - y_3, & b_2 &= y_3 - y_1, & b_3 &= y_1 - y_2 \\ c_1 &= -(x_2 - x_3), & c_2 &= -(x_3 - x_1), & c_3 &= -(x_1 - x_2) \end{aligned} \tag{225}$$

where the triangle is defined by the points $x_i y_i$ with the superscript k implied. Similarly, for each component of the vector potential, or when only one component exists,

$$\frac{R_k}{4T_k}[b_i b_j + c_i c_j](A^k) = \frac{J_k T_k}{3}(1) \tag{226}$$

A couple of simple examples of results are shown in Figure 4.48, the direction of A being normal to the diagram. It is recalled that equipotentials of A are flux lines. Figure 4.48(b) illustrates the ease with which the shapes of permeable structures may now be indicated.

An alternative is the introduction of isoparametric rectangular elements. A rectangle may, first, be sheared to form a parallelogram, enclosing a surface S', as in Figure 4.49. Associate

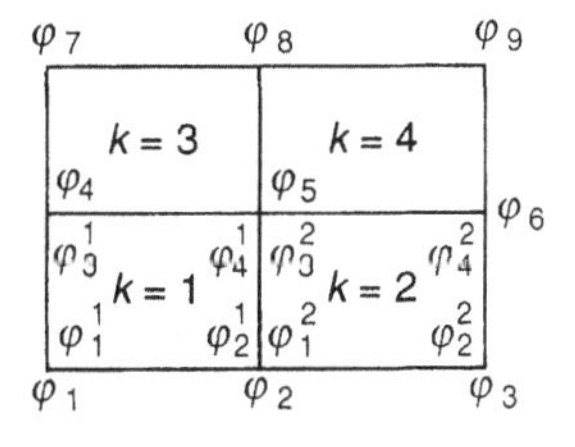

Figure 4.47 The relation between the local and global indexing illustrated by an array of four elements

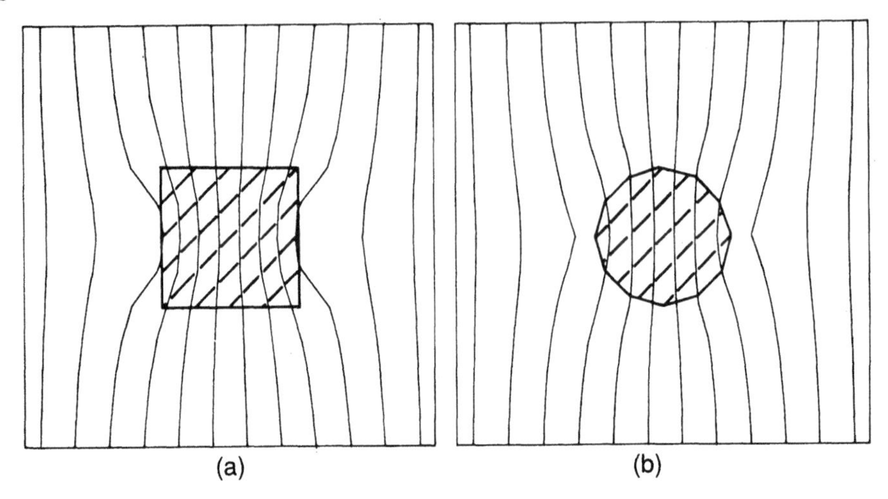

Figure 4.48 (a) Equipotentials (of **A**), which are the lines of induction, for a permeable square-section bar and (b) the way in which a few coordinate adjustments effectively convert this to a long cylinder in which **B** is uniform

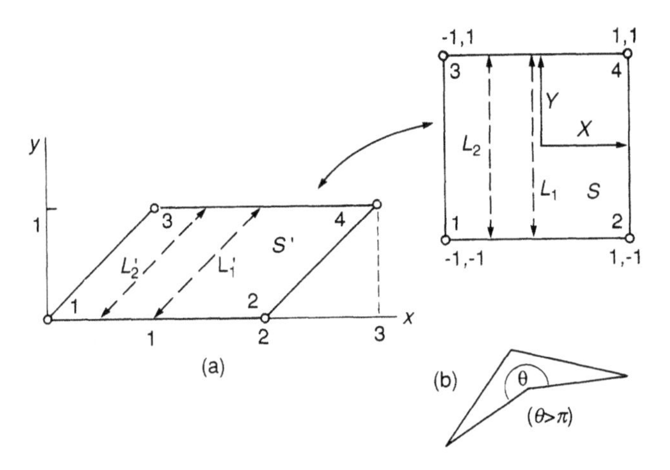

Figure 4.49 A quadrilateral element, a parallelogram for the sake of illustration, and its associated normalized square element. (b) An impermissible element

with this a square 2×2 element with the coordinates X, Y enclosing S, and take a line L_1 with $X = 0$ in S. The shape functions can be used to interpolate the coordinates as well as the functions, where $x = \sum_{i=1}^{4} f_i x_i$ and $y = \sum_{i=1}^{4} f_i y_i$ with $f_1 = \frac{1}{4}(1 - X)(1 - Y)$, $f_2 = \frac{1}{4}(1 + X)(1 - Y)$, $f_3 = \frac{1}{4}(1 - X)(1 + Y)$, $f_4 = \frac{1}{4}(1 + X)(1 + Y)$. It is easy to see that, for the line L_1,

$$x = \tfrac{3}{2} + \tfrac{1}{2}Y, \qquad y = \tfrac{1}{2} + \tfrac{1}{2}Y$$

When $Y = -1$, $x = 1$ and $y = 0$ and when $Y = 1$, $x = 2$ and $y = 1$. The f_i connect points on L_1 in S with points on L_1' in S^1 in a linear manner. For every point P in S there is a unique point P′ in S', so long as no internal angle $> 180°$ — a basic limitation. Since x is a function of X and Y, $\mathrm{d}x = (\partial x/\partial X)\,\mathrm{d}X + (\partial x/\partial Y)\,\mathrm{d}Y$ and, for example, if S is simply a rectangle $2a \times 2b$, $\mathrm{d}x = a\,\mathrm{d}X$ and $\mathrm{d}y = b\,\mathrm{d}Y$, as before. For a general quadrilateral it may be seen that

$$\int_{S^1} \mathrm{d}x\,\mathrm{d}y = \int_{-1}^{1}\int_{-1}^{1} |\det \mathbf{J}|\,\mathrm{d}X\,\mathrm{d}Y, \qquad \text{where } \mathbf{J} = \begin{pmatrix} \dfrac{\partial x}{\partial X} & \dfrac{\partial y}{\partial X} \\ \dfrac{\partial x}{\partial Y} & \dfrac{\partial y}{\partial Y} \end{pmatrix} \tag{227}$$

$\mathbf{J}$ being the Jacobian matrix. In the elementary example $\mathbf{J} = \left(\begin{smallmatrix} a & 0 \\ 0 & b \end{smallmatrix}\right)$ and $\det \mathbf{J} = ab$. Expressing x and y in X and Y via the f_i given, it is seen, on forming the products, that

$$\mathbf{J} = \tfrac{1}{4}\begin{pmatrix} -(1-Y) & 1-Y & -(1+Y) & 1+Y \\ -(1-X) & -(1+X) & 1-X & 1+X \end{pmatrix}\begin{pmatrix} x_1 & y_1 \\ x_2 & y_2 \\ x_3 & y_3 \\ x_4 & y_4 \end{pmatrix} \tag{228}$$

The significance of this development is that integrals such as

$$\int_{S_k} \mathcal{G}_k f_i(X,Y) f_j(X,Y)\,\mathrm{d}S = \int_{-1}^{1}\int_{-1}^{1} \mathcal{G}_k f_i(X,Y) f_j(X,Y) |\det \mathbf{J}|\,\mathrm{d}X\,\mathrm{d}Y \tag{229}$$

can now be evaluated for general quadrilaterals because while that on the left is formally 'over S_k' that on the right is explicit and is a simple polynomial, with $\mathcal{G}_k$ a constant current density in the element (see Figure 4.50 and the exercise noted in the caption). It is also necessary to evaluate integrals of the form

$$d_{ij} = \int_{S_k} \left[\frac{\partial f_i}{\partial x}\frac{\partial f_j}{\partial x} + \frac{\partial f_i}{\partial y}\frac{\partial f_j}{\partial y}\right] \mathrm{d}S, \qquad i, j = 1, 2, 3, 4 \tag{230}$$

noting that it is the $\partial f_i/\partial X$, etc., that are immediately available. This can be written as

$$d_{ij} = \int_{S^k} \mathbf{L}^T f_i \mathbf{L} f_j\,\mathrm{d}x\,\mathrm{d}y \tag{231}$$

with $\mathbf{L}^{\mathrm{T}} = (\partial/\partial x \;\; \partial/\partial y)$. Now

$$\frac{\partial}{\partial X} = \frac{\partial x}{\partial X}\frac{\partial}{\partial x} + \frac{\partial y}{\partial X}\frac{\partial}{\partial y}, \qquad \frac{\partial}{\partial Y} = \frac{\partial x}{\partial Y}\frac{\partial}{\partial x} + \frac{\partial y}{\partial Y}\frac{\partial}{\partial y}$$

which can be combined as

$$\begin{pmatrix} \dfrac{\partial}{\partial X} \\ \dfrac{\partial}{\partial Y} \end{pmatrix} = \begin{pmatrix} \dfrac{\partial x}{\partial X} & \dfrac{\partial y}{\partial X} \\ \dfrac{\partial x}{\partial Y} & \dfrac{\partial y}{\partial Y} \end{pmatrix}\begin{pmatrix} \dfrac{\partial}{\partial x} \\ \dfrac{\partial}{\partial y} \end{pmatrix} = \mathbf{J}\begin{pmatrix} \dfrac{\partial}{\partial x} \\ \dfrac{\partial}{\partial y} \end{pmatrix} = \mathbf{JL} \tag{232}$$

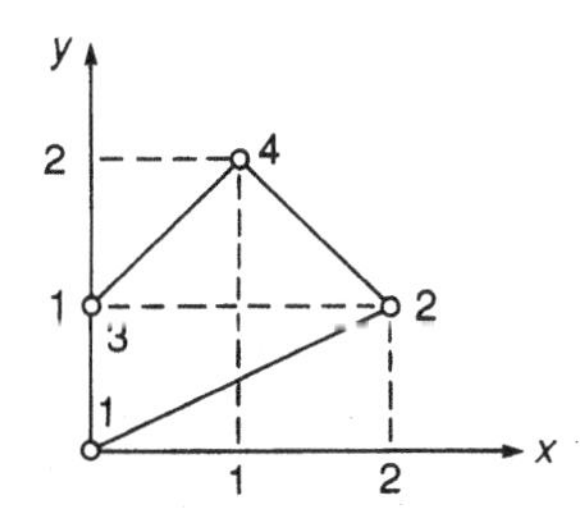

Figure 4.50 As an exercise the area of the quadrilateral may be verified by double integration using the Jacobian. (A natural error may give the value 8. How is $\det k\mathbf{A}$ related to $\det \mathbf{A}$?)

i.e.

$$\mathbf{G} = \mathbf{J}\mathbf{L} \tag{233}$$

with $\mathbf{G} = (\partial/\partial X \ \partial/\partial Y)^{\mathrm{T}}$. Thus, multiplying by $\mathbf{J}^{-1}$,

$$\begin{aligned} \mathbf{L} &= \mathbf{J}^{-1}\mathbf{G} \\ \mathbf{L}^{\mathrm{T}} &= \mathbf{G}^{\mathrm{T}}(\mathbf{J}^{-1})^{\mathrm{T}} \end{aligned} \tag{234}$$

and d_{ij} becomes

$$d_{ij} = \int_{S^k} \mathbf{G}^{\mathrm{T}}(\mathbf{J}^{-1})^{\mathrm{T}} f_i \mathbf{J}^{-1}\mathbf{G} f_j \, \mathrm{d}x \, \mathrm{d}y \tag{235}$$

and using equation (227),

$$d_{ij} = \int_{-1}^{1}\int_{-1}^{1} \mathbf{G}^{\mathrm{T}} f_i (\mathbf{J}^{-1})^{\mathrm{T}}\mathbf{J}^{-1}\mathbf{G} f_j |\det \mathbf{J}| \, \mathrm{d}X \, \mathrm{d}Y \tag{236}$$

Replacing $\mathbf{G}^{\mathrm{T}} f_i$ by $(f_i^X f_i^Y)$ and $\mathbf{G} f_j$ by $(f_j^X f_j^Y)^T$, using Q_{mn} for the elements of $\mathbf{J}^{-1}$ and evaluating the matrix products,

$$\begin{aligned} d_{ij} &= \int_{-1}^{1}\int_{-1}^{1} \left[(f_i^X f_i^Y) \begin{pmatrix} Q_{11} & Q_{21} \\ Q_{12} & Q_{22} \end{pmatrix} \right] \left[\begin{pmatrix} Q_{11} & Q_{12} \\ Q_{21} & Q_{22} \end{pmatrix} \begin{pmatrix} f_j^X \\ f_j^Y \end{pmatrix} \right] |\det \mathbf{J}| \, \mathrm{d}X \, \mathrm{d}Y \\ &= \int_{-1}^{1}\int_{-1}^{1} \{ (Q_{11} f_i^X + Q_{12} f_i^Y)(Q_{11} f_j^X + Q_{12} f_j^Y) \\ &\quad + (Q_{21} f_i^X + Q_{22} f_i^Y)(Q_{21} f_j^X + Q_{22} f_j^Y) \} |\det \mathbf{J}| \, \mathrm{d}X \, \mathrm{d}Y \end{aligned} \tag{237}$$

It is to be stressed that $\partial x/\partial X$, etc., and thus $\mathbf{J}$ are readily found when x and y are expressed in the $f_i(X, Y)$. The Q_{ij} follow by inverting $\mathbf{J}$. Thus it is easy to specify the integrand for numerical, e.g. Gaussian, quadrature.

The element equations may now be expressed as, for example,

$$R_k d_{ij} A_j^k = \mathcal{G}_k c_i, \qquad i = 1, 2, 3, 4 \tag{238}$$

with implied addition over the repeated index, $j = 1, 2, 3, 4$, and

$$c_i = \int_{-1}^{1}\int_{-1}^{1} f_i(X, Y) |\det \mathbf{J}| \, \mathrm{d}X \, \mathrm{d}Y \tag{239}$$

For the elementary example of the rectangular $2a \times 2b$ element with the Jacobian as given, $\mathbf{J}^{-1}$ can be found by taking the transpose of the matrix of cofactors and dividing by the determinant, $\det \mathbf{J}$:

$$\mathbf{J}^{-1} = \frac{\begin{pmatrix} J_{22} & -J_{21} \\ -J_{12} & J_{11} \end{pmatrix}^{\mathrm{T}}}{\det \mathbf{J}} = \frac{\begin{pmatrix} b & 0 \\ 0 & a \end{pmatrix}}{ab} \tag{240}$$

(In a general case inversion is achieved by a numerical algorithm.) Thus $Q_{11} = 1/a$, $Q_{12} = 0$, $Q_{21} = 0$, $Q_{22} = 1/b$ and equation (238) can, on using equation (237), be seen to correspond to the equation derived directly for this special case.

4.3.2 Three-dimensional isoparametric elements

A three-dimensional isoparametric element is shown in Figure 4.51. The function $\varphi^k(x, y, z)$ is linearly interpolated as

$$\varphi^k = \sum_{i=1}^{8} f_i(X, Y, Z)\varphi_i^k \tag{241}$$

where the shape functions such as

$$f_1 = \tfrac{1}{8}(1 - X)(1 - Y)(1 - Z)$$

$$f_8 = \tfrac{1}{8}(1 + X)(1 + Y)(1 + Z)$$

clearly give $\varphi^k = \varphi_1^k$ and $\varphi^k = \varphi_8^k$ respectively at $(-1, -1, -1)$ and $(1, 1, 1)$. It can be seen, or verified, that all the f_i can be expressed compactly as

$$f_i = \tfrac{1}{8}(1 + X_iX)(1 + Y_iY)(1 + Z_iZ) \tag{242}$$

with the X_i, Y_i and Z_i as indicated in the figure. The numbering scheme is chosen for a convenient match with the global numbering and again the most obvious procedure is to generate the coordinates of an array of rectangular-block elements and then introduce appropriate adjustments.

The development of the equations follows that in the preceding section with the linear operators now

$$L^T = \left(\frac{\partial}{\partial x}\frac{\partial}{\partial y}\frac{\partial}{\partial z}\right) \tag{243}$$

$$G^T = \left(\frac{\partial}{\partial X}\frac{\partial}{\partial Y}\frac{\partial}{\partial Z}\right) \tag{244}$$

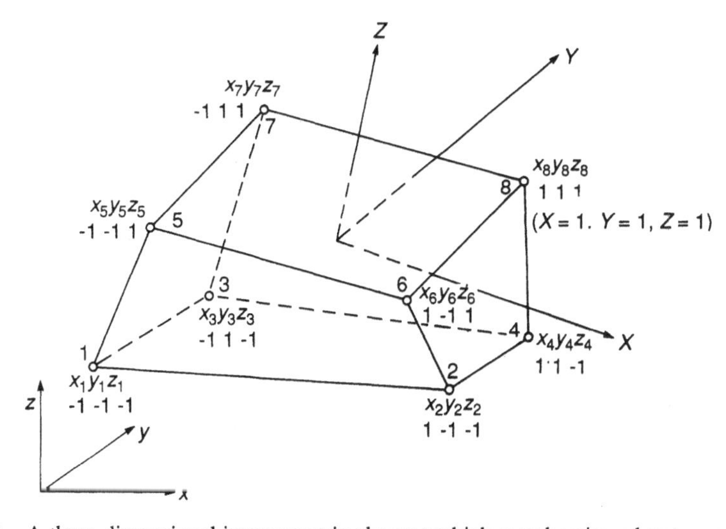

Figure 4.51 A three-dimensional isoparametric element which may be viewed as two corresponding elements (the second being a cube; cf. Figure 4.49)

and the Jacobian

$$
\mathbf{J} = \begin{pmatrix} \dfrac{\partial x}{\partial X} & \dfrac{\partial y}{\partial X} & \dfrac{\partial z}{\partial X} \\ \dfrac{\partial x}{\partial Y} & \dfrac{\partial y}{\partial Y} & \dfrac{\partial z}{\partial Y} \\ \dfrac{\partial x}{\partial Z} & \dfrac{\partial y}{\partial Y} & \dfrac{\partial z}{\partial Z} \end{pmatrix}
= \begin{pmatrix} f_1^X & f_2^X & f_3^X & f_4^X & f_5^X & f_6^X & f_7^X & f_8^X \\ f_1^Y & f_2^Y & f_3^Y & f_4^Y & f_5^Y & f_6^Y & f_7^Y & f_8^Y \\ f_1^Z & f_2^Z & f_3^Z & f_4^Z & f_5^Z & f_6^Z & f_7^Z & f_8^Z \end{pmatrix} \begin{pmatrix} x_1 & y_1 & z_1 \\ x_2 & y_2 & z_2 \\ \cdots & \cdots & \cdots \\ x_8 & y_8 & z_8 \end{pmatrix} \tag{245}
$$

expressing x, y, z in the shape functions and coordinates. The matrix d_{ij} becomes

$$
\begin{aligned}
d_{ij} = \int_{-1}^{1}\int_{-1}^{1}\int_{-1}^{1} &\{(Q_{11}f_i^X + Q_{12}f_i^Y + Q_{13}f_i^Z)(Q_{11}f_j^X + Q_{12}f_j^Y + Q_{13}f_j^Z) \\
&+ (Q_{21}f_i^X + Q_{22}f_i^Y + Q_{23}f_i^Z)(Q_{21}f_j^X + Q_{22}f_j^Y + Q_{23}f_j^Z) \\
&+ (Q_{31}f_i^X + Q_{32}f_i^Y + Q_{33}f_i^Z)(Q_{31}f_j^X + Q_{32}f_j^Y + Q_{33}f_j^Z)\}|\det \mathbf{J}|\mathrm{d}X\,\mathrm{d}Y\,\mathrm{d}Z, \\
&i, j = 1, 2, \ldots, 8
\end{aligned} \tag{246}
$$

and the remainder of the treatment is obvious.

4.3.3 Elements of higher order

For quadratic interpolation (in one dimension) a central node is introduced and $\varphi = f_1\varphi_1 + f_2\varphi_2 + f_3\varphi_3$ with $f_1 = -\frac{1}{2}(X - X^2)$, $f_2 = 1 - X^2$, $f_3 = \frac{1}{2}(X + X^2)$, X ranging from -1 to 1 in each element. For two-dimensional quadratic elements (Figure 4.52) the shape functions are

$$
\begin{aligned}
f_1 &= \tfrac{1}{4}(1-X)(1-Y)(-X-Y-1), & f_2 &= \tfrac{1}{2}(1-Y)(1-X^2) \\
f_3 &= \tfrac{1}{4}(1+X)(1-Y)(X-Y-1), & f_4 &= \tfrac{1}{2}(1-X)(1-Y^2) \\
f_6 &= \tfrac{1}{4}(1-X)(1+Y)(-X+Y-1), & f_5 &= \tfrac{1}{2}(1+X)(1-Y^2) \\
f_8 &= \tfrac{1}{4}(1+X)(1+Y)(X+Y-1), & f_7 &= \tfrac{1}{2}(1+Y)(1-X^2)
\end{aligned} \tag{247}
$$

Three-dimensional quadratic elements have 32 nodes. Shape functions for these and for cubic interpolations are given, for example, by Zienkiewicz [29]. The remainder of the development is as before, save for the increase in dimension for the element matrices. For the same total number of nodes there is a substantial increase in the accuracy on going to the quadratic case, but improvements on going to higher orders are limited. Colloquially, the introduction of some 'curvature' is very beneficial.

On increasing the order the number of elements must be limited and they should not be wastefully used to introduce fixed fields from magnets or currents. Instead, the concept of the 'applied potentials' described in relation to the FD method should be utilized.

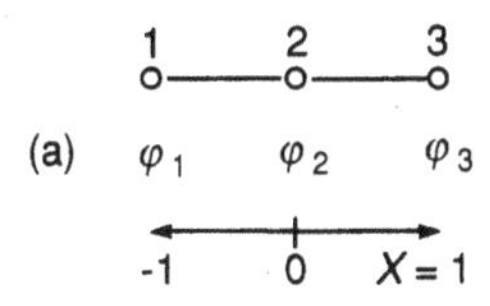

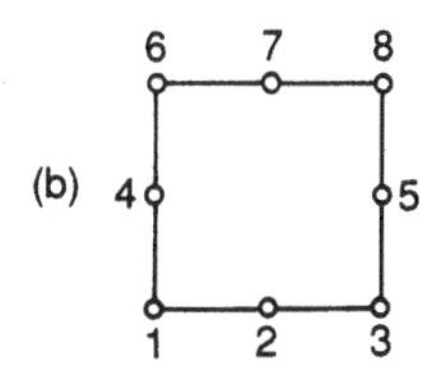

Figure 4.52 A single one-dimensional element containing a central node to permit quadratic interpolation and a two-dimensional quadratic element [see equations (247)]

4.4 Survey of the Numerical Methods

The limiting factor for practical solutions is n, the number of nodes or equations or dimension of the matrix: the solution time goes as n^3. In the FD or FE methods the nodes must span not only the permeable members but also a region of surrounding space adequate (assisted by various devices) for the potential variations to fall to low values. In the direct method described there is no such problem but the disadvantage is that solutions are sought for the three components of **M** rather than the scalar potentials. This also favours FE solutions for φ rather than **A** generally. The direct and the FD methods are useful for general demonstrations and for simple design work but it is hoped that the flexibility of the FE method has been demonstrated to an extent which indicates that it is usually the method of choice. It can be developed much further than may be described here and is of very wide scope: the author has even found that, with a different functional and only a simply modified strategy, phenomena such as the quantum mechanical tunnelling or reflection of wave packets may be beautifully demonstrated.

4.5 A Design Study: Uniform Field Environments

As one example of practical design studies, a very common requirement is a magnetic field that is uniform over a prescribed space. The magnitude may be low or may be as high as can reasonably be achieved. The space may be very large: $\sim$ 1 m^3, as for magnetic resonance imaging, (MRI). The space is taken to be finite, precluding long solenoids or wide slits. Attention is confined to methods involving simple principles or intuition rather than complex analysis.

Without regard for practicality, the most obvious approach is to take four magnets with magnetization closure as in Figure 4.53(a) with a spherical or elliptical cavity in one magnet. Assuming, as in the following, ideal magnets with **M** uniform and ideal construction, $\mathbf{H} = +N\mathbf{M}$ is precisely uniform. There is no access, but the enclosed cavity could, in a compromise, be replaced by a long open cylinder with the useful space around the centre. The magnet enclosing the cavity could be expanded relatively, maintaining closure by using magnets with differing **M** as in the later models.

For a space of any shape, only exemplified by a cube, the question is whether any arrangement of magnets or corresponding current distributions can be devised which form a hollow cube and give a uniform field throughout the internal cubic cavity. Firstly, consider

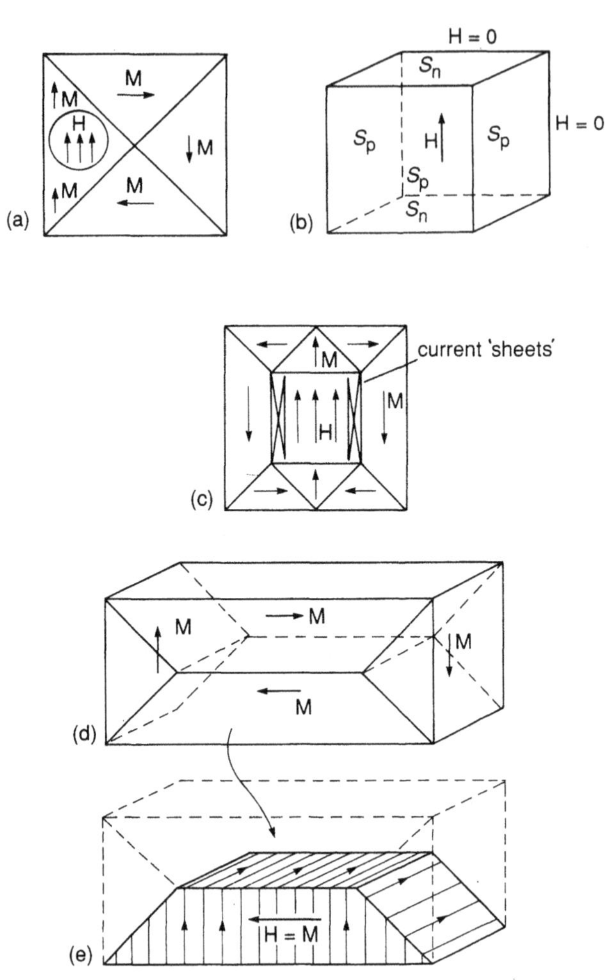

Figure 4.53 (a) The circle may be viewed as a sphere giving an enclosed uniform field environment or as a cylinder giving a uniform field with access in the obvious limit. (b) To 'enclose' a uniform field by reference to the boundary conditions current sheets must be located at surfaces S_p and pole sheets at surfaces S_n. (c) This shows how this may be achieved in practice, the current sheets being seen as continuing parallel to the diagram at the front and back to enclose the uniform field space. In an equivalent approach one or more (or all) of the magnets in any array that gives a zero field everywhere (d) is replaced by its equivalent coil(s) to give an ideally uniform $H = M$ within the coil(s) with a zero field externally as in (e) [Patent G B 9323905.1]

the question of shielding, using a cubic box of soft material and in an applied field H_a. With $\chi \to \infty$ the field in the material must be zero or $\mathbf{M}$ would respond accordingly. Thus the tangential field component at all points adjacent to the internal surface must be zero and φ is constant at such points. The only solution of Laplace's equation in an enclosed space with these boundary conditions consists of φ constant throughout. Thus the shielding is perfect and $H_i = 0$ within the cavity. This means that a magnetization distribution has been created, governed solely by shape effects in this limit, which generates fields $\mathbf{H}_m$ which are equal and opposite to the applied field $\mathbf{H}_a$. This particular distribution may be calculated by one of the methods outlined. To the extent that it can be duplicated by an assembly of permanent magnets, and in the absence of $\mathbf{H}_a$ of course, a uniform field equal in magnitude to that of H_a in the calculation is produced throughout the cavity. External fields are also generated. (For the shielding as such the implications of the constant potential at the external surfaces are completely different.) Note that external (dipole) fields arise in the model 4.53a.

4.5.1 Field uniform within and zero outside the working space

A large superconducting solenoid, with $B \sim 1$ T say, or the gap in a large permanent or electromagnet of simple design may be regarded as a massive dipole producing large and extensive fields and gradients and substantial hazards due to the forces on ferrous objects. It is desirable, for large MRI installations for example, that the fields should not only be uniform within the working space but should be zero outside that space.

Considering the boundary conditions and referring to Figure 4.53(b), a simply described space V is bound by surfaces parallel (S_p) and normal (S_n) to the field. The tangential component of $\mathbf{H}$ must change at S_p so this must coincide with a current sheet. H_n must change at the S_n, which must be pole sheets.

It has been stressed that a magnet, e.g. cylindrical and uniformly axially magnetized, may be represented either by a surface current $\mathbf{I} = \mathbf{M}\times\mathbf{n}$ flowing round the cylindrical surface or by a pair of pole sheets with $\sigma = \pm\mathbf{M}\cdot\mathbf{n}$ for field calculations. (The principle is not restricted to the example.) If the current gives $\mathbf{H}_I$ and the poles give $\mathbf{H}_M$ then the two differ by the constant $\mathbf{M}$ throughout the magnet:

$$\mathbf{H}_I - \mathbf{H}_M = \mathbf{M}$$

Thus if $\mathbf{H}_I$ is the field of a real surface current, i.e. a thin-walled coil, in the limit and the pole field is reversed by locating pole sheets with signs that are opposite to those of the nominal magnet, at the ends of the coil, it follows that the field is ideally uniform at the value $M = I$ at all points occupied by the nominal magnet and identically zero outside that space. The space involved is clearly that enclosed by the coil and the pole sheets. An example is shown by Figure 4.53(c), the field having the value $\mathbf{M}_s$.

Alternatively, the ideally thin-walled coil can be considered equivalent to a magnet occupying the space enclosed by the coil if the surface current is set equal to $\mathbf{M}\times\mathbf{n}$, inasmuch as the same external fields are produced. The fields within the equivalent coil and the magnet differ by the constant $\mathbf{M}$. Thus, commencing with the closed configuration of Figure 4.53(d), which clearly produces no fields, if one or more of the magnets is replaced by its equivalent coil then the fields external to the assembly must remain zero, but ideally uniform fields equal to the magnetization of the replaced magnet(s) exist within the coil(s), as in Figure 4.53(e). For coils of the shape shown the correct current densities are achieved automatically with uniform winding on the front face shown.

In an obvious development, all of the magnets are replaced to give a system of coils alone, with freedom to adjust the field magnitudes. To the author's knowledge these are the only finite coils that give ideally uniform fields with an abrupt transition to zero externally. (The direction changes between the different coils, of course.)

Assemblies such as that suggested by Figure 4.53(c) are proposed as practical possibilities for full-body MRI with fields between 0.1 and 1.0 T. Low fields may be achieved by laminating the magnet frameworks, rather than commissioning unusually low-M materials.

When the currents are present there are no demagnetizing fields acting on the magnets and only recoil effects need be taken into account. Considerations of power consumption for resistive coils and of real current densities J (Am^{-2}) in relation to the possible utilization of high-T_c superconductors, recalling that the coils must be thin walled in relation to the dimensions of the space, indicate that the design is the more favoured the larger the assembly. With no stringent permanent magnet specifications (favouring low-cost barium ferrite) the potential cost might permit 'walk-in' facilities for continuous monitoring of invasive techniques to be considered.

Access to the working space must be contrived by compromise (continuous) or by considering part of the coil to be demountable or hinged, with appropriate contacts (intermittent). By considering the boundary conditions once more it may be concluded that, for these particular requirements, the space must, in principle, always be enclosed.

References

1. Chambers Ll.G., *Mathematics of Electricity and Magnetism*, Chapman and Hall, London, 1973
2. Heumann, C., *J. Math. Phys.* **20**, 127 and 336 (1941)
3. Hancock H., *Elliptic Integrals*, Dover, New York, 1958; Dwight, H.B., *Tables of Integrals and Other Mathematical Data*, Macmillan, London, 1961
4. Eason, G. Noble, L. and Sneddon, I.N., *Phil. Trans. Roy. Soc.* **247**, 529 (1955).
5. Watson, G.N., *A Treatise on the Theory of Bessel Functions*, Cambridge University Press, 1922
6. Bateman, H., *Partial Differential Equations of Mathematical Physics*, Cambridge University Press, 1932
7. Craik, D.J., *J. Phys. D*, **7**, 1566 (1974).
8. Aziz, Z., Thesis, Manchester University 1978.
9. Rhodes, P. and Rowlands, G., *Proc. Leeds Phil. Soc.*, **6**, 191 (1954).
10. Landau, L. and Lifshitz, E., *Physik. Z. Sowjet.*, **8**, 153 (1935).
11. Kittel, C., *Phys. Rev.*, **71**, 270 (1947) and **73**, 155 (1948); *Rev. Mod. Phys.*, **21**, 541 (1949).
12. Malek, Z. and Kambersky, V., *Czech. J. Phys.*, **8**, 416 (1958).
13. Kooy, C. and Enz, U., *Philips Res. Repts*, **15**, 7 (1960).
14. Goodenough, J.B., *Phys. Rev.*, **102**, 356 (1956)
15. Ignatchenko, V.A. and Sakharov, Y.V., *Bull. Acad. Sci., USSR (Phys. Ser.)*, **28**, 475 (1964).
16. Spacek, L., *Czech. J. Phys.*, **9**, 186 (1959).
17. Craik, D.J. and McIntyre, D.A., *IEEE Trans. Magn.*, **MAG-5**, 378 (1969).
18. Charap, S.H. and Nemchik, J.M., *IEEE Trans. Mag.*, **5**, 566 (1969).
19. Craik, D.J. Cooper, P.V. and Druyvesteyn, W.F., *Phys. Lett.*, **34a**, 244 (1971).
20. Druyvesteyn, W.F. and Dorleyn, J.W.F., *Philips Res. Repts*, **26**, 11 (1971).
21. Kojima, H. and Goto, K., *Proceedings International Conference on Magnetism*, Nottingham, p. 727, 1964, and *J. Appl. Phys.*, **36**, 538 (1965).
22. Bobeck, A.H., *Bell Syst. Tech. J.*, **46**, 1901 (1967); Bobeck, A.H., Fischer, R.F., Perneski, A.J., Remeika, J.P. and van Uitert, L.G., *IEEE Trans. Magn.*, **MAG-5**, 544 (1969); Bobeck, A.H., *Magnetic Oxides*, John Wiley and Sons, London and New York, p. 743, 1975.
23. Thiele, A.A., *Bell Syst. Tech. J.*, **48**, 3287 (1969); **50**, 725 (1971) and *J. Appl. Phys.*, **41**, 1139 (1970).
24. Parker, R.J., *Advances in Permanent Magnetism*, John Wiley and Sons, New York and Chichester, 1990.
25. Craik, D.J., *Phil. Mag.*, **B41**, 485 (1980).
26. Young, D.M., *Iterative Solution of Large Linear Systems*, Academic Press, New York, 1971.
27. Craik, D.J. and Cooper, P.V., *IEEE Trans. Magn.*, **MAG-14**, 306 (1978).
28. Binns, K.J. and Lawrenson, P.J., *Analysis and Computation of Electric and Magnetic Field Problems*, Pergamon Press, New York, 1963.
29. Zienkiewicz, O.C., *The Finite Element Method*, McGraw-Hill, London, 1977.

CHAPTER

5 Survey of Magnetic Materials

1 Introduction

The technical properties of magnetic materials depend on both the intrinsic properties and the microstructure. Following a brief survey of hard and soft materials an account of intrinsic properties will be presented simply as a collection of data.

2 Technical Properties: Microstructure and Magnetic Behaviour

It is convenient to consider three particular lengths: (a) a typical lattice spacing a, (b) an exchange length L_e, the order of a domain wall width, and (c) a characteristic length λ for domain studies (wall energy γ):

$$L_e = \sqrt{\frac{A}{K_1}}, \qquad \lambda = \frac{\gamma}{\mu_0 M_s^2} \text{ (SI)}, \qquad \lambda - \frac{\gamma}{4\pi M_s^2} \text{ (c.g.s.)} \tag{1}$$

A further particular length may be associated with a volume V (for an approximately spherical particle) such that, at any chosen T such as room temperature,

$$CV = 2kT$$

$C = 2K_1$, $\mu_0(N_b - N_a)M_s^2$ or $3\lambda_s\sigma$ according to the predominance of magnetocrystalline, shape or stress anisotropies. The maximum energy as M rotates, θ passing through $\pi/2$, is $\frac{1}{2}CV \sin\theta = \frac{1}{2}CV$ and if $CV < kT$ substantially it may be expected that $\mathbf{M}$ should be subject to reversals due to thermal fluctuations (Section 2.2).

If the particle diameter $d \sim a$ then the occurrence of an odd number of layers of cations in antiferromagnetics such as NiO or Cr_2O_3 may give a significant net moment in relation to the volume and, as below, detectable superparamagnetism. Departures from the properties and behaviour expected for bulk material are expected to be common and this applies also to films with thickness $\sim a$.

A critical size may be defined as that for which the magnetostatic energy for a uniform-$\mathbf{M}$, single-domain particle, $\frac{1}{2} \times \frac{1}{3}\mu_0 M_s^2(\frac{4}{3}\pi r^3)$ for a sphere, is equal to that for a two-domain particle with a central wall, approximately half the above, plus the energy of the wall. This immediately gives

$$r_c = 9\lambda, \qquad a_c = 12\lambda \tag{2}$$

the second applying to a cubic particle and following from the expressions given by Rhodes and Rowlands (see Section 1.19, Chapter 4). Due to the dependence of the energy terms on r^3 or on r^2 the single-domain state is favoured for $r < r_c$. This is one context in which λ is a useful parameter: it can also be shown that for films, thickness c, with easy axis $\perp$ the surface, the multidomain state has the lower energy when $c > 2.18\lambda$ and that $c \sim \lambda$ is the rough condition for the domains to maintain the form of simple strips or cylinders.

2.1 High-Coercivity Materials: Recording and Permanent Magnets

In grain-oriented specimens the magnetostatic energy term giving equation 2 is reduced (to zero for perfect orientation) and the critical diameter raised to, for example, 10^{-6} m for barium ferrite, according to observation. The coercivities of permanent magnet specimens are substantially lower than those expected for coherent rotation (Table 5.1) and it has been proposed that the coercivities be associated with the pinning of walls, running extensively through the material, at grain boundaries. Such pinning could result if these relatively disordered regions are characterized by anomalously low magnetocrystalline anisotropy constants, and this could account for the coercivity of other high-anisotropy materials [1].

Coercivities which would be considered low in terms of permanent magnet behaviour but are much higher than those for soft materials, i.e. range from 20 to 40 kA m^{-1}, are required

Table 5.1 Examples of commercial barium ferrite magnets, by Swift-Levick Magnets Ltd, Sheffield. $d = 4.8$–5 g cm^{-3}, temperature coefficient of remanence = -0.19% K^{-1}, maximum operating temperature = 350°C. ${}_BH_c$ is the reverse field giving $B = 0$

	B_r (T)	${}_BH_c$ (kA m^{-1})	$(BH)_{max}$ (kJ m^{-3})	Recoil permeability
Feroba 1 (isotropic)	0.22	136	7.9	1.2
Feroba 2 (anisotropic)	0.39	176	28	1.15
Feroba 3 (anisotropic)	0.37	240	26	1.15

for conventional, longitudinal, magnetic recording [2–4]. In this, minute regions of a tape or film on a disc that is initially uniformly magnetized are switched by a recorder head as in Figure 5.1 [5]. Too high a value of H_c would render the switching difficult and with too low a value the reversals would not be stable against their own demagnetizing fields or be capable of production with an acceptably high packing. The presence, or absence, of the reversals may be detected by the signals induced in a similar (or, indeed, the same) head as the track passes and this constitutes digital information. Suitable H_c values are obtained for acicular particles of magnetite or, most commonly, γ-Fe_2O_3 (maghaemite), typically some 0.1 μm by 0.5 μm; Table 5.2 compares other materials in use.

α-(FeO)OH, goethite, is precipitated in colloidal form, providing seeds for the growth of acicular particles in a ferrous sulphate solution containing Fe metal, with heating and

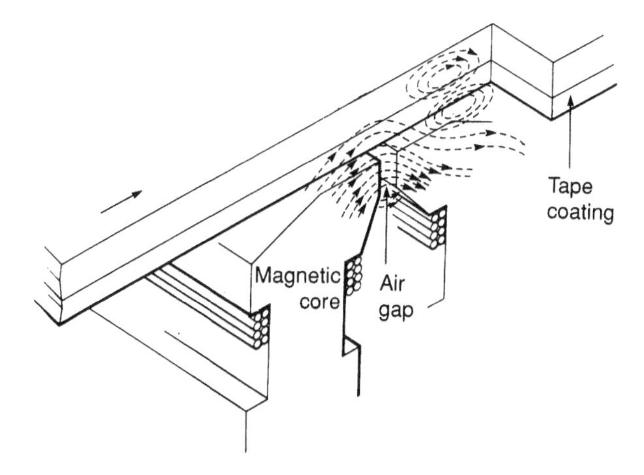

Figure 5.1 The conventional magnetic recording process [5]. (Reprinted by permission of John Wiley & Sons Ltd, Magnetic Oxides, ed. D.J. Craik, Wiley 1975)

Table 5.2 An approximate guide to materials for conventional recording (after Bate [15]). Shape: C = cubic, E = equiaxed, A = acicular, P = plates, r = length/width or width/thickness

	σ_s (A m^2 kg^{-1})	T_c (°C)	K_1 (kJ m^{-3})	d (g cm^{-3})	H_c (kA m^{-1})	M_r/M_s	Shape	Size (μm)
γ-Fe_2O_3	74	590	−4.64	5.07	6–12	0.46	E	0.05–0.3
					20–29	0.80	A, $r = 7$	$l = 0.2$–0.7
					26–30	0.75	P	
Fe_3O_4	92	575	−11	5.20	~24	0.52	E	
					24–27	0.7	A, $r = 7$	
γ-$Co_xFe_{2-x}O_3$					27–70	0.47–0.61	A, $r = 5$	
$x = 0.04$	50		+100	4.67	32	0.7	C	0.05–0.08
$x = 0.06$	44				41–48	0.7	C	0.05–0.08
$Co_{0.04}Fe_{2.96}O_4$	80	570	+30	5.19	48–64		C	~0.2
$CoFe_2O_4$	65–70	520	+60	5.33	60–78	0.65	C	0.004–0.016
$BaO{\cdot}6Fe_2O_3$	68	450	330	5.28	426	0.5	P, $r = 15$	$d = 0.08$–0.15
CrO_2	89–92	117	25	4.83	4–12		E	$l = 0.2$–1.5
CrO_2+Sb	77–85	117		4.9	12–48	0.5–0.9	A	$w = 0.03$–$.1$

aeration. This is followed by the transformations

$$\alpha\text{-(FeO)OH} \rightarrow \alpha\text{-Fe}_2\text{O}_3 \rightarrow \text{Fe}_3\text{O}_4 \rightarrow \gamma\text{-Fe}_2\text{O}_3$$

which are pseudomorphic, particle sizes and shapes being retained. The process may stop at the penultimate stage and magnetite has been utilized to some extent; the relatively high M_s enhances the signal but introduces certain difficulties. Co-substituted γ-Fe_2O_3, even in the form of equi-axed particles, may give high H_c, permitting high packing densities. Alternatively, epitaxial layers of Co ferrite may be formed on the surface of γ-Fe_2O_3 particles, using solutions of $CoCl_2$ and NaOH, increasing H_c to 80 $\mathrm{KA\,m^{-1}}$. Such Co-modified materials have, according to Mallinson [2], been extensively utilized in professional and consumer video tapes and computer discs, one advantage being a relatively low temperature coefficient of H_c. The Co ferrite enhances the shape anisotropy and thus appears to adopt a uniaxial anisotropy during formation. CrO_2, which is ferromagnetic with $T_c = 121°\mathrm{C}(\mathrm{Cr}^{4+} \sim 3\,\mathrm{d}^2 \sim 2\beta)$, can be formed from CrO_3 by catalysed reactions under pressure at 400–500°C, in the form of particles typically 2.5×0.5 μm, having fewer pores and protuberances than those of γ-Fe_2O_3 and, when aligned in tapes, giving H_c to 52 $\mathrm{KA\,m^{-1}}$ and good reduced remanences of 0.9. CrO_2 is substantially utilized, as is modified barium ferrite, e.g. $BaCo_{0.8}Ti_{0.8}Fe_{10.4}O_{19}$, which has virtually unlimited H_c in the present context and has been reported to perform well. As indicated by comparing particles of Fe_3O_4 and γ-Fe_2O_3 with the same shape and size (produced by transforming one into the other) but with substantially different M_s [6] and by studying temperature dependences of H_c, noting that over certain regions M_s changes slowly while K_1 changes very rapidly [7], the coercivity is largely associated with shape anisotropy here.

The particles are aligned in the tapes and with easy axes ∥ the field direction $H_c = (N_b - N_a)M_s$ predicts $H_c \sim 200\ \mathrm{kA\,m^{-1}}$ with axial ratio ~ 5 while $H_c \sim 25$–$30\ \mathrm{kA\,m^{-1}}$ is typically observed. To account for this and related observations, Jacobs and Bean [8] suggested that the elongated particle might be treated as a chain of spheres, each single sphere being isotropic although the assembly is clearly anisotropic [Figure 5.2(a)]. For this, $H_c = a\pi M_s$ (c.g.s.) with a from 0.5 for two spheres to unity for an infinite chain [c.f. $2\pi M_s$ (c.g.s.) for a long cylinder]. This does not give a size-dependent H_c but for the model of Figure 5.2(b) some exchange energy is involved on reversal and a size dependence arises, as for the buckling mode of Figure 5.2(c) [9], where again the surface pole densities break up to reduce the maximum magnetostatic energies during reversal, at the expense of the exchange energy introduced. A further incoherent reversal mode, curling as in Figure 5.2(d), put forward by Frei *et al.* [9] also reduces pole formation. Wohlfarth [10] gave H_c for a prolate spheroid, with minor semi-axis b and negligible magnetocrystalline or strain anisotropy, as

$$H_c \geqslant \frac{2\pi k A}{b^2 M_s} - N_a M_s \tag{3}$$

with N_a corresponding to the field direction, with k from 1.38 for a sphere to 1.08 for a long cylinder, $a \gg b$. For the latter the equality sign applies (and $N_a = 0$). The effect of crystal anisotropy is to add $2K_1/M_s$ to the above so long as the effect is smaller than that of the shape. Incoherent rotation does not apply to the rotational response of the magnetization to small applied fields and initial susceptibilities are unaffected.

A second critical size can be defined by the value of b for which curling is replaced by coherent rotation, i.e. $H_c = (N_b - N_a)M_s$ giving, with the above,

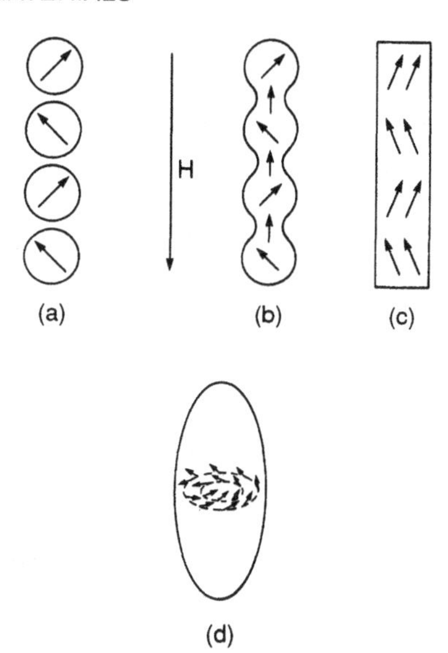

Figure 5.2 The fanning reversal mechanism in a chain of spheres (a) with a modification (b) introducing some exchange energy as for the buckling (c) and curling modes (d) [8, 9]

$$b_c = \left(\frac{2\pi k A}{N_b M_s^2} \right)^{1/2} \tag{4}$$

The size dependence of H_c predicted for curling is to be stressed, in relation to the experimental data in Figure 5.3 [11]. Reference may also be made, for example to the particle size dependence of H_c for γ-Fe_2O_3 used in recording, noted by Corradi *et al.* [12].

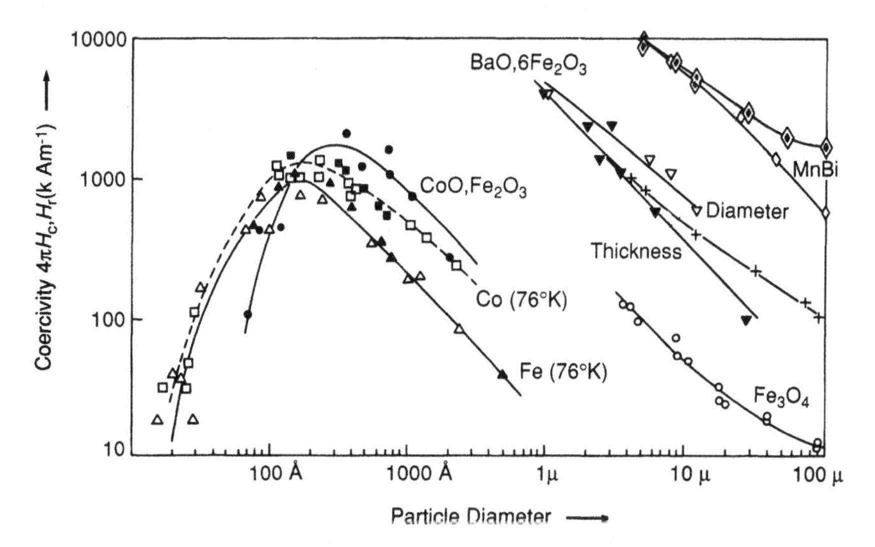

Figure 5.3 Data connecting coercivity with particle size, collected by Luborsky [11]. (Reprinted by permission of John Wiley & Sons Ltd, Magnetic Materials, R.S. Tebble & D.J. Craik, Wiley 1969)

For coherent rotation H_c decreases regularly with increasing angle between the field and easy axis, but for incoherent rotation H_c initially varies very slowly and may rise as the angle increases; the observations by Bate in Figure 5.4 [13] are of clear significance.

Strong magnetostatic interactions are to be expected when the particles are embedded at high density in a plastic, for recording, and additional indirect exchange interactions may arise between fine particles produced as precipitates in permanent magnets as such. Iron-rich particles in an Al-rich matrix are formed on appropriate heat treatment of 'precipitation-hardened' permanent magnet materials: Alnico or Alcomax with typical compositions 8% Al, 13% Ni, 24% Co, 3% Cu. H_c is high only when the precipitation particles have sizes as indicated by Figure 5.5 [14, 15]. It may be that the optimum properties arise not only when the particle size is appropriate but when the heat treatment is such as to give the clearest distinction between the magnetic particles and the non-magnetic matrix. However, it is clear from studies on multilayer films, e.g. Fe–Si–Fe–Si..., that exchange effects can bridge or penetrate non-magnetic layers, presumably due to the RKKY mechanism, with polarization of conduction electrons in the Si layers. Evidence of strong interactions of some type, leading to cooperative reversals, is afforded by the observation of large-scale domains in the Alnico materials. As shown in Figure 5.6, the domains are not so clearly

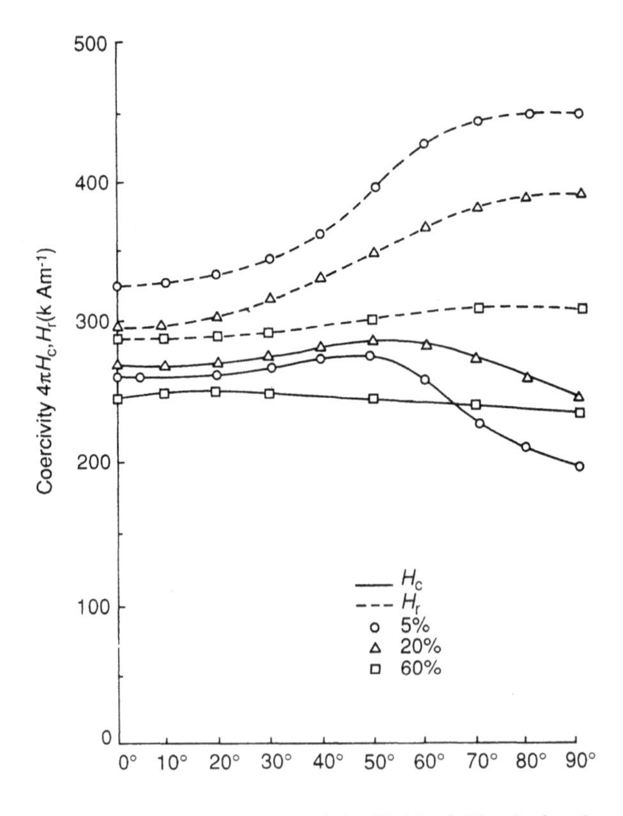

Figure 5.4 Dependence of H_c and remanence coercivity H_r (the field reducing the remanence to zero) on the angle between the directions of orientation and of measurement for partially oriented acicular γ-Fe_2O_3 particles [13]. (Reprinted by permission of John Wiley & Sons Ltd, Magnetic Oxides, ed. D.J. Craik, Wiley 1975)

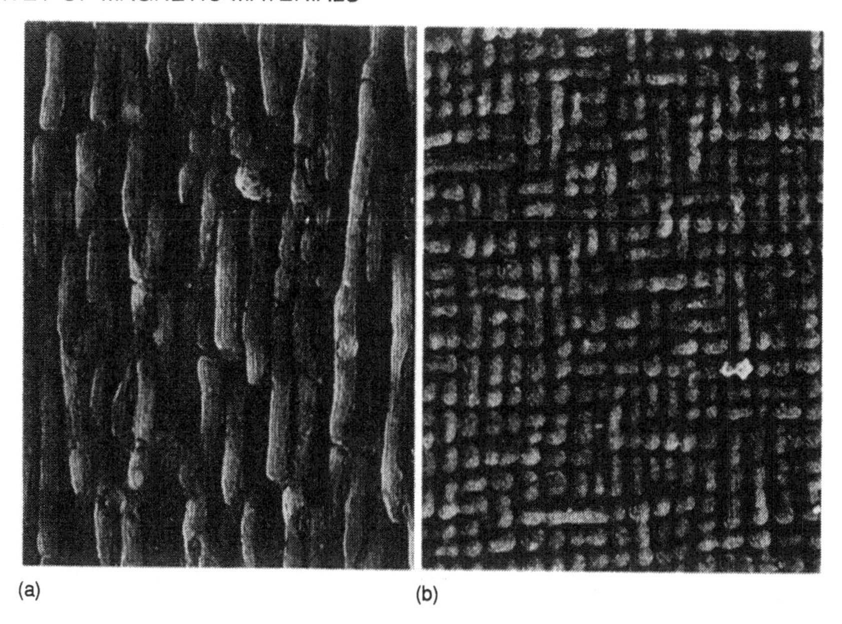

Figure 5.5 Electron micrographs of replicas of etched surfaces of Ticonal X (a) ∥ and (b) ⊥ to the direction of the aligning field, by de Jong *et al.* [14], to be compared with an oxide replica technique of de Vos [15]

defined as those in homogeneous materials. Rather ragged 180° domains even form during the reversal of the magnetization in ESD specimens [16], which are rather like synthetic Alnicos, with elongated singe-domain iron particles, produced by electrodeposition into mercury, embedded in a non-magnetic matrix [11].

Equi-atomic PtCo can be produced with a cubic crystal structure, with low anisotropy and coercivity, which converts on heat treatment to an ordered tetragonal structure which is uniaxial with very high anisotropy (Table 5.3). H_c rises to a peak of about 400 kA m^{-1} [17] when about half the material is ordered and occurs in the form of exceptionally fine lamellae, as shown in Figure 5.7(a). The magnetic integrity of the finely divided structure here depends

Table 5.3 Equiatomic PtCo [17], with typical commercial data for permanent magnets (e.g. General Electric, Johnson Matthey)

	σ_s (A m^2 kg^{-1})	K_1 (kJ m^{-3})	K_2 (kJ m^{-3})	$K_1 + K_2$ (kJ m^{-3})	Easy axes
Disordered, cubic	44.7	−70	160	—	⟨1 1 1⟩
Ordered, tetragonal	37.7	—	—	1720	⟨0 0 1⟩
	σ_r	σ_s	σ_r/σ_s	H_c (kA m^{-1})	
Partially ordered	21.7	—	0.56	239	(along [110])
	31.0	38.5	0.80	279	(along [111])
Commercial (partially ordered)	$B_r = 0.6$–0.64 T	${}_BH_c = 334$–392 kA m^{-1}		$(BH)_{max} = 60$–73 kJ m^{-3}	

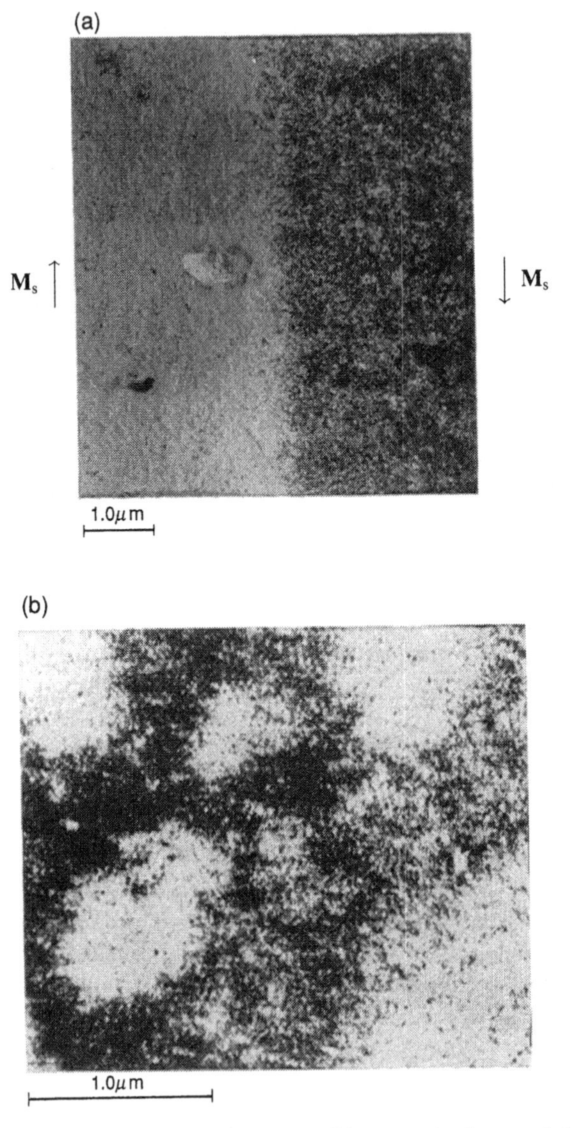

Figure 5.6 Domains in Alcomax, in its optimum condition, are clearly revealed but are unlike those in high-anisotropy magnets: (a) $\parallel$ and (b) $\perp$ to the alignment direction, with some evidence in (b) of the fine structure of Figure 5.5

on the enforced changes in the directions of **M**, but the interactions are obviously strong and again domains with a fine structure matching the microstructure can be seen to form as in Figure 5.7(b). When the lamellae grow a little, domains are seen within them. Despite obvious disadvantages this material is of value when the quantity employed is minimal, as in thin films for magneto-optical recording.

The outstanding permanent magnet materials typified by $SmCo_5$, as developed by Strnat *et al.* [18] (see also [19]), and Sm_2Co_{17} are high-uniaxial-anisotropy two-sublattice materials

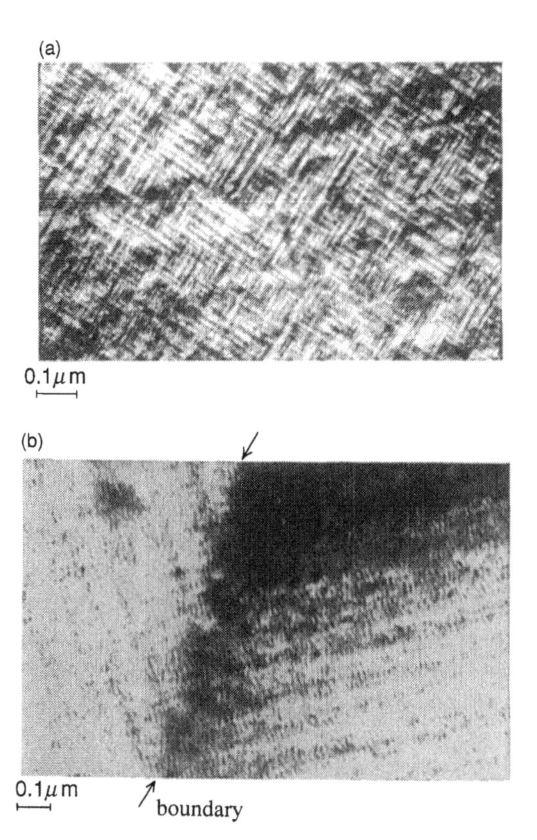

Figure 5.7 (a) An electron micrograph of PtCo in its optimum condition indicating incipient precipitation of the ordered phase, the streaks being ∥ (110) planes (A. J. Harrison, Thesis, Nottingham, 1976). (b) Domains in the optimum material showing some corresponding fine structure

Table 5.4 Properties of RCo_5 alloys [20]. MM = mischmetal, here 54.5% Ce, 26.6% La, 13% Nd, 5% Pt. $(BH)^{ideal}_{max}$ is $\frac{1}{4}\mu_0 M_s^2$. H_K = anisotropy field by extrapolation of hard-axis magnetization curves. $K_u \sim H_k = 2K_u/(\mu_0 M_s)$

RCo_5 : R =	Y	La	Ce	Pr	Sm	MM
T_c(°C)	648	567	374	612	724	~ 520
M_s (kA m^{-1})	845	723	613	955	768	~ 708
$(BH)^{ideal}_{max}$ (kJ m^{-3})	224	164	118	286	183	158
H_K (kA m^{-1})	10000	14000	13000–17000	12000–17000	17000–23000	14000–16000
K_u (MJ m^{-3})	5.5	6.3	5.2–6.4	6.9–10.0	8.1–11.2	6.4–6.9
d (10^3 kg m^{-3})	7.69	8.03	8.55	8.34	8.60	8.35

with hexagonal crystal structures (Table 5.4). For $SmCo_5$, M_s = 760 kA m^{-1} and B_r > 0.9 T can be achieved in aligned compacts of fine particles, formed by liquid-phase sintering, while H_K has the remarkable value of 23 000 kA m^{-1} and $\mu_0 H_c > B_r$ is readily achieved. Sm_2Co_{17} has M_s = 920 kA m^{-1} and potentially the higher energy product [20, 21]. $SmCo_5$ is distinctive in that, as the particle size is reduced by milling, H_c for other RCo_5's rises and then falls once more before reaching the very high values specifically applying to $SmCo_5$ (see, for example, Becker [22]). Materials with compositions containing copper, such as $SmCo_{5-x}Cu_x$, and $Sm_2Co_{17-x}Cu_x$ [23], develop a two-phase structure, with a scale of 50 nm, of Co-rich and Cu-rich phases. With properties dependent on heat treatment they are, in a sense, akin to the Alnicos; they are bulk-hardened and need not have a low crystallite size to develop a high H_c. An important distinction between the two classes of materials can be made by commencing with thermally demagnetized specimens which are presumed to contain domain walls. In single-phase materials saturation is approached in moderate fields and the initial susceptibility is high, as in Figure 5.8. For bulk (precipitation)-hardened materials very little magnetization is induced until the fields are near to the magnitude of H_c, and then M rises steeply (Figure 5.8), as expected if the fields causing reversal and those initiating the induction of the magnetization are equally the fields required to cause rotations in the precipitate particles or to move the 'domain' walls separating the interacting regions.

Arguably the most impressive permanent magnet materials of all are, colloquially, Nd–Fe–B. Sagawa *et al.* [24] found that the tetragonal $Nd_2Fe_{14}B$ had the remarkable uniaxial anisotropy constant of 3500 kJ m^{-3} and obtained $B_r = 1.23$ T in aligned compacts, with $H_c = 960$ kA m^{-1}. (It is recalled that the power of a magnet may be described by B_r^2 if H_c is very great.) The material had the composition $Nd_{15}Fe_{77}B_8$ but the principal phase consisted of $Nd_2Fe_{14}B$ crystallites $\sim$ 10 μm across. The grains are originally larger than this and post sintering Nd-rich material gives the fine-grain structure with non-magnetic grain boundary layers (Figure 5.9) rich in Nd and oxygen but low in iron [25].

Coinciding with the above, a group at General Motors [26] found that melt-spun ribbons of Nd–Fe–B compositions were virtually amorphous (according to the quench rate) and that exceptional coercivities developed on subsequent heat treatment during which spherical crystallites some 20–100 nm in diameter formed. (This, also remarkably, predated work on nanocrystalline soft materials, as noted later.) In a striking further development Lee [27] found that crystallite alignment, and very high B_r, could be produced by die-upsetting, which is a special hot pressing technique involving deformation. This work has been reviewed by Croat [28].

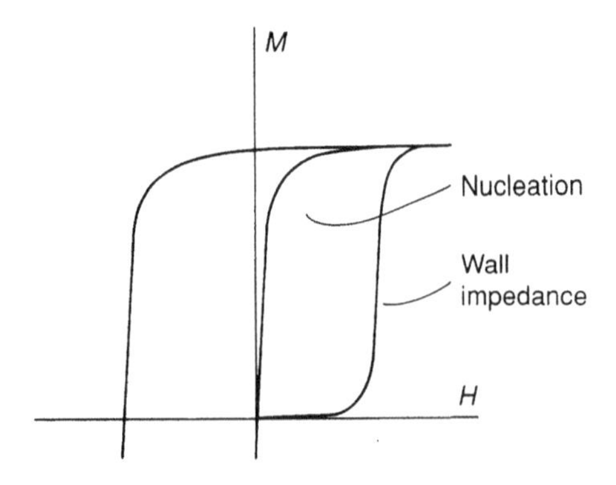

Figure 5.8 Contrasting the virgin magnetization curves expected if the coercivity is controlled by nucleation or by domain wall impedance

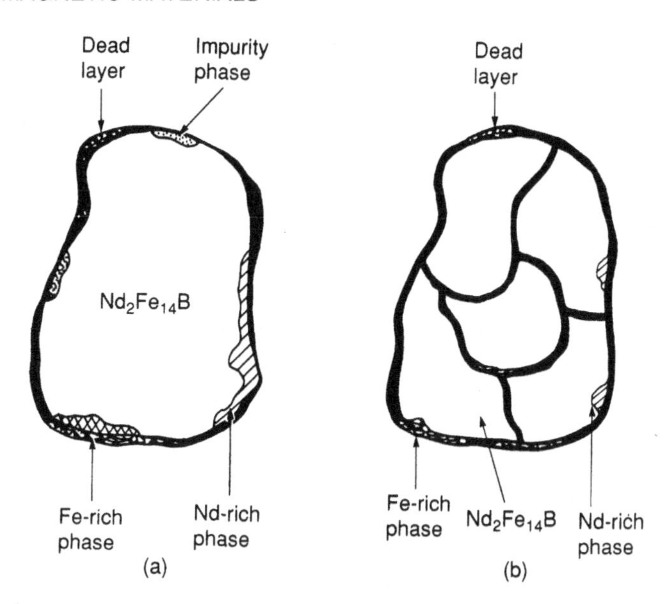

Figure 5.9 (a) A typical grain in a Nd–Fe–B magnet prior to post-sintering, containing Fe-rich and Nd-rich regions together with non-magnetic impurities such as Nd_2O_3. The dead layer is a Nd-rich film, taken to have no magnetic order. (b) This shows how the large grain breaks up on post-sintering while surplus Nd diffuses into the Fe-rich regions forming $Nd_2Fe_{14}B$. Remaining non-magnetic Nd-rich inclusions do not impair the coercivity [25]

Both cobalt and samarium are costly and the current and potential economical advantages of Nd–Fe–B are substantial. It has been claimed that in cost per unit energy product it is competitive even with barium ferrite, which has a very low cost per unit volume.

Particle size management is clearly vital. Knowing that rare earth alloys could absorb large volumes of hydrogen at high pressures and temperatures, and that on subsequent removal of the hydrogen the material may be reduced to a fine powder or at least riven by microcracks which facilitate its pulverization, Harris *et al.* [29], (see also McGuiness *et al.* [30], Harris and McGuiness [31] and Takeshita and Nakayama [32]) made detailed and productive studies of this hydrogen decrepitation [or hydrogenation, disproportionation, desorption and recombination (HDDR)] in relation to the newer materials.

The set of Tables 5.5 to 5.8 indicates currently available specifications for the more recently developed materials.

Returning to Figure 5.3, it is noted that as extremely small particle diameters are reached the trend is reversed and H_c tends to fall. In fact H_c may become zero for the reasons outlined below.

Table 5.5 Two grades of bonded $SmCo_5$ magnet, by Swift-Levick Magnets Ltd ('predominantly Samarium'). H_a is the field required for magnetization

	B_r (T)	${}_BH_c$ ($kA\,m^{-1}$)	H_c ($kA\,m^{-1}$)	$(BH)_{max}$ ($kJ\,m^{-3}$)	H_a ($kA\,m^{-1}$)
'B2/50'	0.55–0.58	360–380	720	48–56	2000
'B3/60'	0.58–0.59	400–420	800	56–64	2000

Table 5.6 Two grades of sintered $SmCo_5$ magnet by Swift-Levick Magnets Ltd. The approach to ideal behaviour is indicated by $\mu_0\,{}_BH_c \approx B_r$ or by $\frac{1}{4}B_r^2/\mu_0 \approx (BH)_{max}$

	B_r (T)	${}_BH_c$ ($kA\,m^{-1}$)	$\mu_{0B}H_c$ (T)	H_c ($kA\,m^{-1}$)	$(BH)_{max}$ ($kJ\,m^{-3}$)	$\frac{1}{4}B_r^2/\mu_0$ ($kJ\,m^{-3}$)	H_a ($kA\,m^{-1}$)
'S2/150'	0.87–0.93	640–715	0.80–0.90	1500	160–185	150–170	2000
'S2/175'	0.92–0.98	680–750	0.85–0.94	1500	185–200	170–190	2000

Table 5.7 Two grades of Sm_2Co_{17} by Swift-Levick Magnets Ltd (see notes for Tables 5.6, 5.8) Materials with $(BH)_{max}$ to 240 $kJ\,m^{-3}$ and $B_r \sim 1.1$ T with ${}_BH_c \sim 800$ $kA\,m^{-1}$ ($\mu_{0B}H_c \sim 1.0$ T) are under development

	B_r (T)	${}_BH_c$ ($kA\,m^{-1}$)	H_c ($kA\,m^{-1}$)	$(BH)_{max}$ ($kJ\,m^{-3}$)	H_a ($kA\,m^{-1}$)	$\mu_{0B}H_c$ (T)	$\frac{1}{4}B_r^2/\mu_0$ ($kJ\,m^{-3}$)
'S3/195'	1.0 –1.04	715–770	1500–2000	185–200	4400	0.90–0.97	200–220
'S3/205'	1.02–1.06	730–785	1500–2000	190–210	4400	0.92–0.99	210–220

Table 5.8 Two out of a wide range of Nd–Fe–B magnets, trade name Neorem, by Sumitomo Special Metals Ltd, produced by Outokumpu Magnets *OY*, courtesy of Swift-Levick Magnets Ltd. Note that $\mu_0 H_c \gtrsim B_r$, a specification only achieved in recent years [and that $\mu_0\,{}_BH_c \doteqdot B_r$, $(BH)_{max} \doteqdot (BH)_{max}^{ideal}$]

	B_r (T)	${}_BH_c$ ($kA\,m^{-1}$)	H_c ($kA\,m^{-1}$)	$(BH)_{max}$ ($kJ\,m^{-3}$)	$\mu_{0B}H_c$ (T)	$\frac{1}{4}B_r^2/\mu_0$ ($kJ\,m^{-3}$)	$\mu_0 H_c$ (T)
'400i'	1.30	985	1000	320	1.24	336	1.25
'490i'	1.15	870	1900	250	1.09	263	2.39

2.2 Superparamagnetism

McNab *et al.* [33] showed by Mossbauer spectroscopy that there was no stable magnetization in particles of magnetite with diameters ranging from 10 to 16 nm. Assemblies of such particles, which are typical of those in magnetic colloids, known as ferrofluids when concentrated, are readily saturated in fields of the order of 1 T and have M_s values characteristic for the material, so the particles have characteristic magnetic moments. The remanence is found to be zero. The remanent state could thus be taken as that in which the moment orientations are randomly distributed in space. However, the Mossbauer studies show that in this particular case the moment orientations are also fluctuating continuously in time.

For a particle of volume v the anisotropy energy tending to define the direction of the moment in the absence of an applied field can be written as

$$E_A = \tfrac{1}{2} C v \sin^2\theta$$

where

$$C = \begin{cases} 2K_1 & \text{(crystal)} \\ \mu_0(N_b - N_a)M_s^2 & \text{(shape)} \\ 3\lambda\sigma & \text{(strain)} \end{cases} \tag{5}$$

This has a maximum of $\frac{1}{2}Cv$ as θ goes through $\pi/2$ for a uniaxial particle, presenting a barrier that must be overcome by thermal activation as for an activated chemical reaction, etc. The probability per second, designated $1/\tau$, that the magnetization will reverse is thus given by the Boltzmann relation

$$\frac{1}{\tau} = f e^{-Cv/(2kT)} \tag{6}$$

where τ is the relaxation time for the decay of the remanence:

$$M(t) = M(0)e^{-t/\tau} \tag{7}$$

In the complete treatments by Néel [34] and Brown [35] a typical value for the frequency factor f in equation (6) is shown to be 10^9 s^{-1} using arguments in which the frequency of approach to the energy barrier is associated with precession.

Assuming that sufficient time is allowed for equilibrium to be attained, the assembly of particles must behave just like an assembly of very large spins in a magnetic field, i.e. exhibit classical Langevin paramagnetism. The magnetization is reversible with respect to the field with zero remanence and the zero-field susceptibility corresponds to a Curie law. For obvious reasons this behaviour is known as superparamagnetism.

Effective superparamagnetism can only be defined in relation to an arbitrary time constant τ. For $\tau = 100$ s, $Cv/2 = 50\,kT$, and this may be used to give a blocking volume V_b separating the stable and superparamagnetic regions of behaviour at a given temperature. The significance of V_b follows from the rapid variation of the exponential function as indicated by calculating τ at room temperature for magnetite spheres with only crystal anisotropy effective:

d (nm)	44	48.5	50.5	56.5
τ	2.42 s	9 min	28 days	19 years

The blocking volumes, at room temperature, for spheres of iron and cobalt can similarly be identified as corresponding to diameters of 10 and 5 nm respectively. For α-Fe_2O_3 at room temperature $d = 20$ nm [36].

At a fixed volume τ changes very rapidly with temperature and a blocking temperature may be similarly defined. If this is above T_c, particles of that arbitrary volume will never behave superparamagnetically.

Remarkably, antiferromagnetics can exhibit superparamagnetism [37]. This arises because, with a simple collinear ordering, in crystallites that contain an odd number of layers of ions (counted in an appropriate direction) the moments cannot sum to zero. This has been shown to apply to NiO and Cr_2O_3 [38].

Clearly superparamagnetism is an important topic in its own right and is particularly relevant to studies of ferrofluids. In the present context its role is negative, inasmuch as the particles used for recording must have stable remanence and are thus subject to limiting minimum sizes.

2.3 High-Permeability Materials

It might be supposed that as the particle size rises and H_c falls, the susceptibilities or permeabilities should, conversely, rise. It is, in fact, observed that the χ's or μ's rise as r or d increases as in Figure 5.10 by Herzer [39, 40] who suggested $\mu_i \propto d$ approximately at the larger sizes; the effects at the lower sizes will be considered later (Section 2.3.1). The interest here is in high-μ materials which may be typified by three materials or classes.

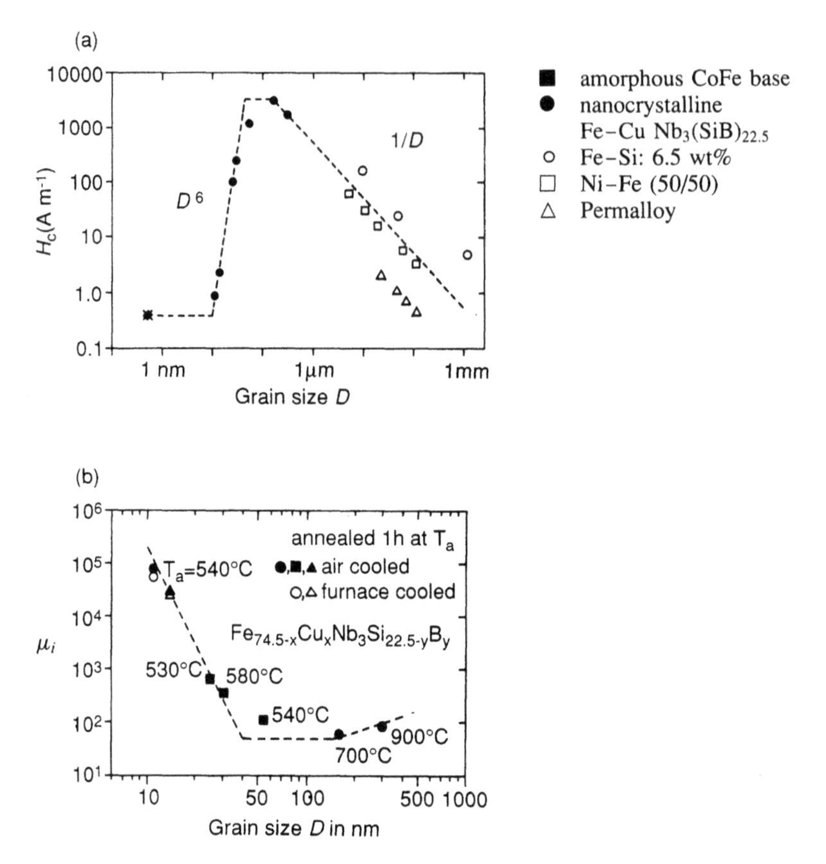

Figure 5.10 Coercivity and initial permeability versus grain size for the materials indicated: $H_c \propto 1/D$ and $\mu_i \propto D$ above 150 nm and $H_c \propto D^6$, $\mu_i \propto D^{-6}$ below 150 nm [39]. The fall in H_c is, in this case, not to be ascribed to superparamagnetism (which would not accord with the high μ_i)

The first, appropriate for high-power devices, e.g. the large transformers at low or moderate frequencies which link the consumer to the mains and involve substantial potential losses or wastage of total power resources, is silicon iron. The call is specifically for high maximum permeability (see Figure 5.11) and this is favoured by the high M_s of iron, assuming that H_c may be minimized. As opposed to the Permalloys, following, Si–Fe is a moderate-anisotropy material in which well-defined and mainly 180° domain structures occur and it is essential that wall pinning should be low. Of the common impurities those that give rise to inclusions such as carbides or nitrides are most important in this respect. Levels of impurities giving, for example, $\mu_m = 200\,000$ and $H_c \sim 1.0\ \mathrm{A\,m^{-1}}$ are only achieved by annealing in moist hydrogen, with great difficulty for pure iron, the use of which is restricted to, for example, special shielding materials. The inclusion of Si at 1–4 per cent weight favours the precipitation of carbon as graphite rather than cementite Fe_3C and immobilizes N as Si_3N_4, both of which are advantageous, so that lower purity levels may be tolerated than for Fe as such. Si also reduces the crystal anisotropy, with

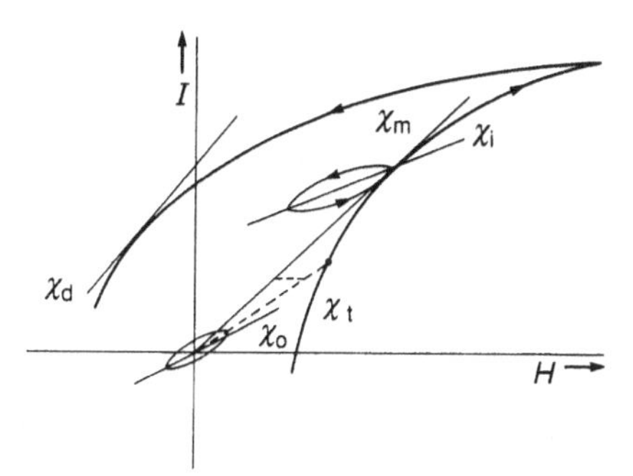

Figure 5.11 The slopes of the relevant lines give the initial (χ_i), total (χ_t), maximum (χ_m), incremental (χ_{inc}) and differential (χ_d) susceptibilities (or relative permeabilities since the fields are generally very low for soft materials). (Reprinted by permission of John Wiley & Sons, Ltd, Magnetic Materials, R.S. Tebble & D.J. Craik, Wiley 1969)

a minimum at 6.4 per cent, and Si–Fe responds to magnetic annealing. Moreover, Si can reduce the conductivity ($\times\frac{1}{4}$ for 3.5 per cent) and eddy current losses and is found to favour the extent to which grain-oriented sheets can be produced by successive rolling and annealing. The grains then have common [001] directions and when strips are cut and assembled into frames appropriately, the magnetization processes are scarcely dependent on rotation; saturation may be approached in fields $\sim H_c$, giving high values of μ_m. Cube texture with two easy directions in the sheet surface may also be produced. It is to be expected that the walls should move more readily in larger grain-size specimens since, with imperfect texture, the grain boundaries influence the magnetostatic energies involved.

An appropriate alternative specification, to μ_m, is the induction in a chosen field of, for example, 800 $\mathrm{A\,m^{-1}}$: B_8. This does not correspond to saturation along the easy axes because a misorientation angle of 3° can be achieved for single, (110) [001], texture and while $\cos 3^\circ = 0.999$ a value of $B_r = 1.97$ T $= 0.97B_s$ is high; the demagnetizing effect due to components of M_s across the sheet surfaces are presumably also involved.

When a very good grain orientation is achieved it becomes apparent that surface roughness can have a significant effect on the hysteresis since any magnetic irregularity impedes the motion of domain walls. Thus surface polishing has been found to be beneficial in reducing hysteresis.

Sheet thickness is the most obvious factor influencing eddy current losses, but thickness reduction tends to conflict with the production of a good texture. However, by employing a complex process in which strips were successively rolled and annealed and a final extended annealing process at > 1150°C eventually gave a tertiary recrystallization, Arai *et al.* [41] obtained the outstanding results illustrated by Figure 5.12. The growth in B_8 towards $0.97B_s$ mirrors the production of the very good orientation by the recrystallization. The dramatic fall in H_c may be connected with the large size of the newly formed crystallites. As shown by Figure 5.12(b) the orientation is maintained down to very low rolled thicknesses.

Figure 5.12(c) shows the losses for a tertiary recrystallized sheet 75 μm thick, $B_8 =$ 1.98 T, $H_c = 2.8$ $\mathrm{A\,m^{-1}}$. The 'iron loss', i.e. total loss, is only 20 per cent lower than that for conventional grain-oriented material and this is because (a) it is almost wholly eddy current loss and (b) the measures taken to decrease the hysteresis loss actually increase the eddy current loss because the domain walls become very widely spaced due to the

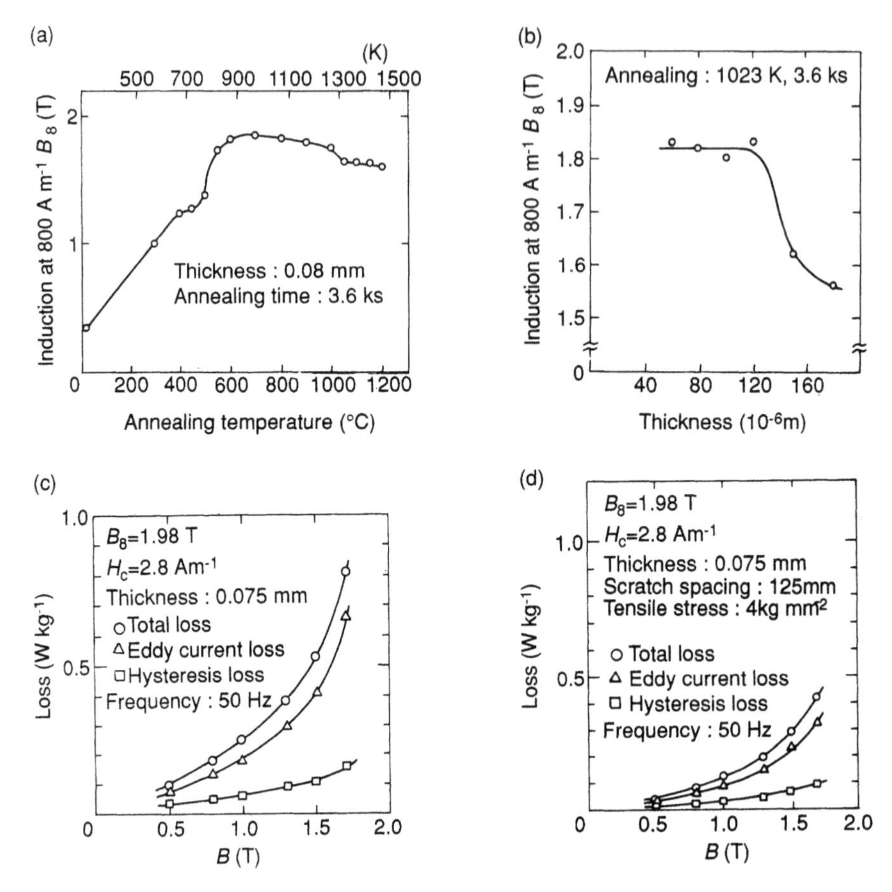

Figure 5.12 (a) Induction 'B_8' as a function of annealing temperature for recrystallized SiFe. (b) Dependence of B_8 on sheet thickness for primary recrystallized sheets. (c) Losses for tertiary recrystallized sheets. (d) Losses after domain refinement, deliberately increasing the number of walls and minimizing the eddy current anomaly [41]

good orientation and large grain size. This, in fact, provides an excellent example of the anomalous eddy current losses associated with the domain walls which must move rapidly, even at 50 Hz frequencies, due to spacings of $\sim$ 1.0 mm.

This effect was found to be counteracted by introducing 'domain refinement' by chemically etching grooves in the surface, 2.5 μm deep and 40 μm wide, some mm in spacing, oriented across the rolling direction. (Scratching the surface gave a similar effect, but this was due to the stresses set up and the effect was lost during the annealing necessarily involved in core fabrication.) The consequent great effect on the losses is shown by Figure 5.12(d).

The so-called electronic devices, high-frequency transformers, pulse transformers, magnetic amplifiers, etc., usually involve low power levels and the call is for high initial permeability, μ_i, remaining constant over substantial frequency ranges (low losses), as achieved by Ni–Fe alloys or Permalloys. The key feature here is that the anisotropy constants of Ni and Fe are of opposite sign and since the single-ion additive principle applies at least approximately, K_1 and K_2 for the alloy become zero close to 75 per cent Ni. This suggests that the wall widths should increase indefinitely and rotational processes should apply to give $\chi \rightarrow \infty$. However, magnetostriction is not zero at this composition and several observations, e.g. of eddy current loss anomalies [42], implicate domain wall motion. As the anisotropies become small the walls become very wide in relation to any inclusions or other imperfections, and can then be shown to interact weakly with the inclusions. Bozorth [43] suggested that the peak in μ_i close to the zero-λ_{111} composition corresponded to the motion of walls bisecting $\langle 1\ 1\ 1 \rangle$ directions since $K_1 < 0$ in this region with $\mathbf{M}_s \parallel \langle 1\ 1\ 1 \rangle$; μ_i passes through a peak between the zero-K and zero-λ compositions (Figure 5.13). The second peak occurs around the zero-λ_{100} composition; it is lower because K_1 is higher here. The 45–50 per cent Ni alloys ('45-Permalloy') give the higher induction and are also of technical importance. It is necessary to avoid the formation of ordered Ni_3Fe by a second heat treatment following stress-reduction annealing. The ordering is less pronounced if 4–5 per cent Mo is included, the anisotropy at the zero-λ composition is further reduced and $\mu_i \rightarrow 100\,000$ may be achieved without double heat treatments. The inclusion of Cu, to give 'Mumetal', improves malleability which favours the construction of magnetic shields for example. These materials are used in the form of thin strips or ribbons, wound to form toroids, and it must be noted that when thin strips of high-purity Si–Fe are taken into account the classification employed here becomes blurred. Furthermore, uniaxial anisotropies may be induced in the NiFe alloys by field annealing, the production of texture or by deformation (roll-induced anisotropy wherein an anisotropic distribution of atom pairs is produced by slip occurring in an otherwise isotropically ordered material [44]) and then the *maximum* permeabilities of these alloys may be high. Alloys containing cobalt, e.g. 43 per cent Ni 23 per cent Co, or Perminvar, naturally respond well to magnetic annealing with the induced anisotropy overwhelming the cubic component to give uniaxial behaviour, with the easy axis running around the toroidal core and giving rise to almost ideal square-loop behaviour (Figure 5.14). If the annealing occurs in the absence of an applied field, and necessarily below the Curie point, a unique domain structure is 'frozen in' because the induced easy axes

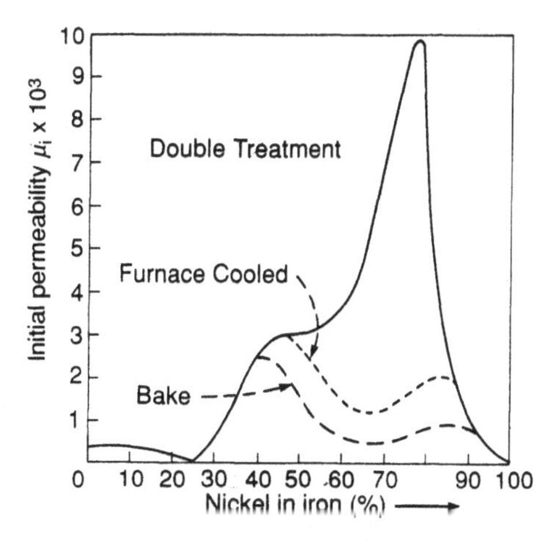

Figure 5.13 Initial permeability for NiFe alloys, after Bozorth [43]. (Reprinted by permission of John Wiley & Sons, Ltd, Magnetic Materials, R.S. Tebble & D.J. Craik, Wiley 1969)

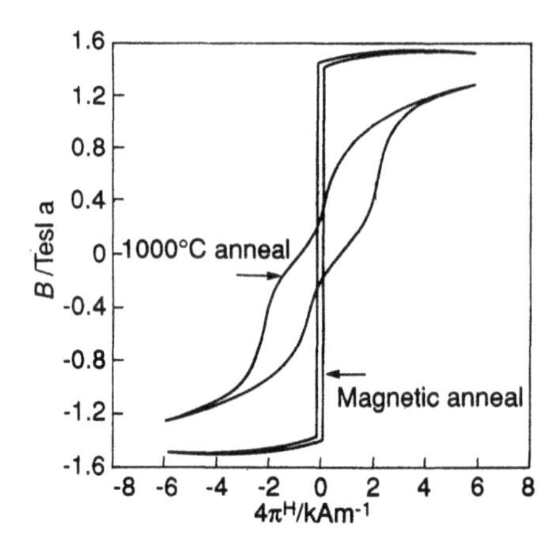

Figure 5.14 A constricted loop for a Perminvar (43% Ni, 23% Co) ring slowly cooled after annealing, compared with the loop for the magnetically annealed specimen [45]. (Reprinted by permission of John Wiley & Sons, Ltd, Magnetic Materials, R.S. Tebble & D.J. Craik, Wiley 1969)

follow the direction of M within the existing domain walls, which then have an anomalously low energy so long as they maintain these particular positions. As a loop is traversed the walls are effectively attracted to these positions, which correspond to demagnetization, and it is thus easy to see how the unusual constricted loops [45] of Figure 5.14 arise.

Losses, and the corresponding fall in the real or 'useful' part of the permeability with increasing frequency, consist of hysteresis loss at low frequencies with additional eddy current losses of increasing importance as f increases for the metals. Tapes must be very thin for use at high frequencies (e.g. up to 10 kHz for which $t \sim 0.1$ mm), but eventually the relatively insulating spinel ferrites become appropriate. It is, however, useful to stress the wide range of the relevant resistivities by the display:

Iron	$\rho = 10^{-7}\ \Omega\text{m}$
Iron alloys	ρ to $10^{-6}\ \Omega\text{m}$
Ferrous ferrite, magnetite	$\rho = 4 \times 10^{-5}\ \Omega\text{m}$
Nickel/manganese ferrite	$\rho \sim 10^{2}\ \Omega\text{m}$
Magnesium ferrite	$\rho = 10^{5}\ \Omega\text{m}$

Magnetite is scarcely an insulator. It is a semiconductor up to the Verwey temperature, 119 K, below which there is a slight tetragonal distortion due to ordering of Fe^{2+} and Fe^{3+} ions on octahedral sites, as shown by neutron diffraction [46] and by Mossbauer spectroscopy [47]; above 119 K the behaviour is near-metallic, ρ rising slowly with temperature. The absence of order facilitates the electron transfer: $Fe^{2+}\ O^{2-}\ Fe^{3+} \rightarrow Fe^{3+}\ O^{2-}\ Fe^{2+}$. The incorporation of ferrous ions in, for example, Mn ferrite, which is of value in reducing the crystal anisotropy by the single-ion effect because Mn and Ni ferrites have $\langle 1\ 1\ 1 \rangle$ easy directions while Fe^{2+} and Co^{2+} ions make a positive contribution to K_1, rapidly reduces ρ due to the low activation energy required for the transfer of electrons between Fe^{2+} and Fe^{3+} ions on equivalent sites. Thus eddy currents may be important in high-permeability MnFe ferrites and special attention to grain boundaries to mutually insulate the grains can reduce the losses.

For $Ni_{1-y}Fe_{2+y}O_4$ the metallic behaviour typifying magnetite disappears at $y = 0.8$, i.e. for Fe^{3+} $[Ni^{2+}_{0.2}\ Fe^{2+}_{0.8}\ Fe^{3+}]\ O_4$, the square brackets indicating octahedral sites, and the semiconductivity corresponds to $\sigma = Ae^{-B/(kT)}/T^{\beta-1}$, with B increasing from 0.06 eV for $y = 0.8$ to 0.12 eV for $y = 0.1$ and $\beta \doteqdot 2$, which is consistent with a hopping mechanism.

Mn ferrite has a relatively high M_s, which is further increased by zinc substitution ($Mn_{1-x}Zn_x Fe_2 O_4$) since the Zn^{2+} ions with zero moment replace ions in the minor sublattice (Figure 5.15) [48]. For $Mn_{0.5}Zn_{0.4}Fe^{2+}_{0.1}Fe^{3+}_2O_4$, K_1 becomes zero at room temperature: the temperature dependences of the contributions to K_1 due to the different ions differ. λ is also very low for this composition since Fe^{2+} ions make positive contributions to the magnetostriction so the low-f permeabilities can be high, as indicated [49] by Figure 5.16. The lower peak corresponds to the passage through zero of K_1 and the higher to the general observation that $K_1 \to 0$ more rapidly than $M_s \to 0$ as T_c is approached. The increase of μ_i with particle size is here very clear, as in Figure 5.17, so long as the porosity is confined to the grain boundaries as indicated [50]. Such materials are formed by counter-diffusion of the cations, which are much smaller than the O^{2-} ions, on heating intimate mixtures of appropriate single oxides, followed or accompanied by sintering. A variety of special methods involving co-precipitation or the formation of mixed crystals of compounds, from which intimately mixed oxides may be produced, also exists and these may be employed to produce particularly dense ferrites, but the 'dry' method is economic and common. In the course of sintering there is a tendency for certain grains to grow rapidly at the expense of

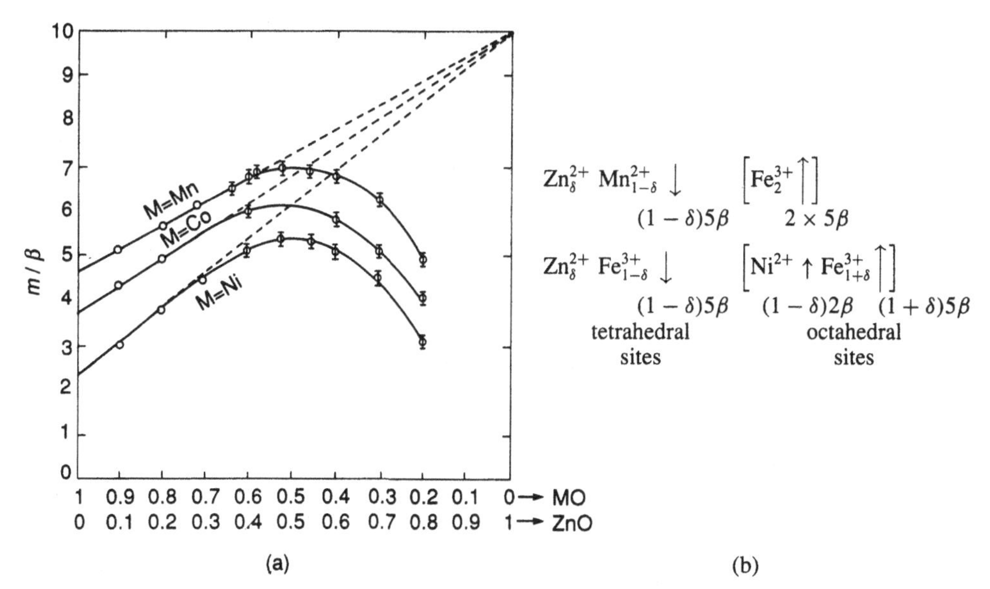

Figure 5.15 The increase in M_s for Mn ferrite due to substitution by Zn ions with zero moment, as indicated by the site occupations in (b). The extrapolated 10_β is not achieved because there would then no longer be a second sublattice with which the principally antiferromagnetic coupling could be effective [48]. (Reprinted by permission of John Wiley & Sons, Ltd, Magnetic Oxides, ed. D.J. Craik, Wiley 1975)

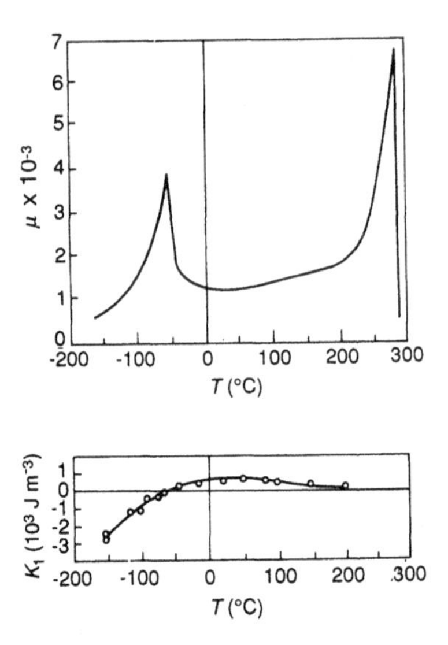

Figure 5.16 $\mu(T)$ and $K_1(T)$ for 31MnO 11ZnO 58Fe_2O_3 crystals [49]. (Reprinted by permission of John Wiley & Sons, Ltd, Magnetic Oxides, ed. D.J. Craik, Wiley 1975)

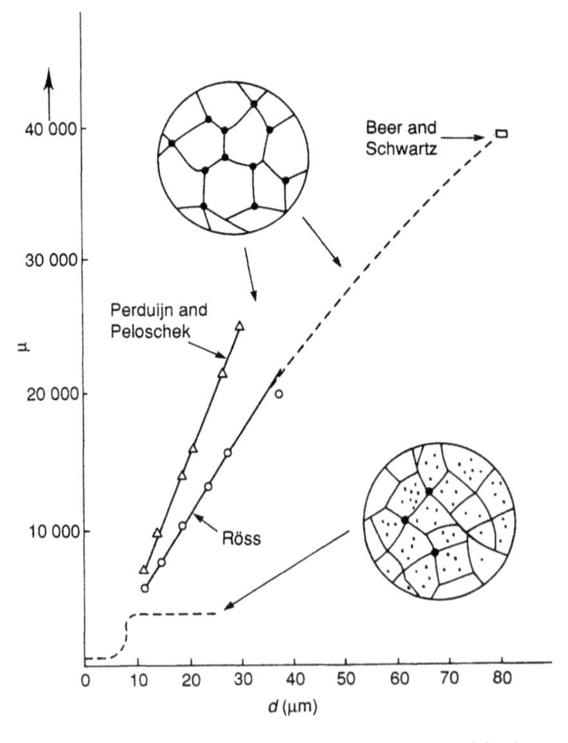

Figure 5.17 Linear increase of permeability for MnZn ferrites with the particular microstructure indicated [50]. (Reprinted by permission of John Wiley & Sons, Ltd, Magnetic Oxides, ed. D.J. Craik, Wiley 1975)

others, entrapping the pores between them, and this can be avoided by attention to purity and sintering conditions [51]. An appropriate choice of composition and sintering conditions (which may also affect the ferrous content) can reduce the large temperature coefficients that might otherwise be expected. It may be assumed that the walls are pinned by the pores at the grain boundaries and bow in small a.c. fields. Considering both changes in wall energy and in the magnetostatic energies arising, it can be seen that the more extensive walls in large crystallites respond more readily to applied fields.

For operation at high frequencies eddy currents must be avoided and the inclusion of substantial ferrous concentrations is precluded since these lead to damping and losses, an early fall in $\mu_i'(f)$, even if they are inadequate to lead to extensive motion of the electrons and only introduce a local hopping. In any case the high-frequency losses will consist of resonance peaks, grossly broadened in the demagnetized specimens, and it is fruitless to attempt to optimize the low-frequency μ_i due to a basic conflict which emerges on taking the limiting frequency of operation f_c to be the resonance frequency in the anisotropy field and μ_i to be that for rotation:

$$f_c = \frac{\gamma}{2\pi}\mu_0 H_k = \frac{\gamma\mu_0}{2\pi}\frac{4k}{3M_s}, \qquad \mu_i = 1 + \chi_i = 1 + \frac{\mu_0 M_s^2}{2k} = 1 + \frac{\gamma M_s}{3\pi}\frac{1}{f_c} \tag{8}$$

Thus, as pointed out by Snoek [52], $\mu_i \propto 1/f_c$ approximately and this applies to an extent to the set of curves in Figure 5.18 [53].

Some of the losses at frequencies below f_c may be connected with domain wall resonance or damped wall motion [54]. This may be suppressed by stabilizing the wall positions using compositions, including Co^{2+}, for which the anisotropy is induced at room temperature, so that the effect is not lost if the walls are displaced from a unique low-energy configuration, or by deliberately introducing some intragranular porosity. Alternatively, the grains may be reduced below the critical, single-domain, size, but then special hot-pressing methods are

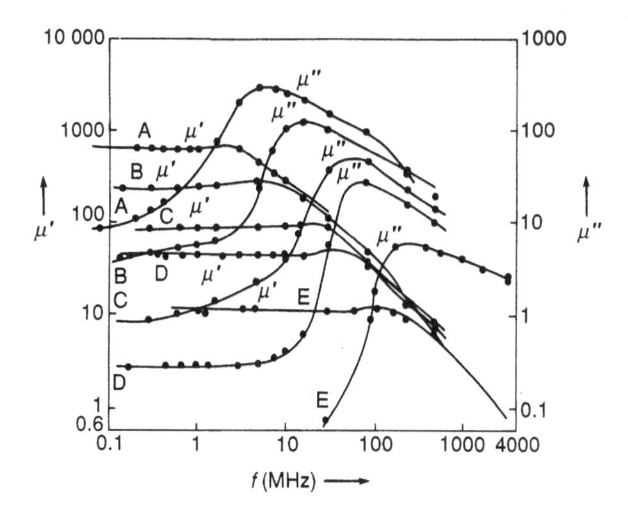

Figure 5.18 Permeability spectra for NiZn ferrites with contents NiO:ZnO of A, 17.5:33.2; B, 24.9:24.9; C, 31.7:16.5; D, 39.0:9.4; E, 48.2:0.7; balance Fe_2O_3 [53]. The higher the value of μ' the lower is the frequency for the onset of resonance losses. (Reprinted by permission of John Wiley & Sons, Ltd, Magnetic Oxides, ed. D.J. Craik, Wiley 1975)

required to give densification without undue grain growth, as by de Lau [55], who obtained $Ni_{0.31}$ $Zn_{0.64}$ Fe_2 O_4 at 99.7 per cent density with a grain diameter of 0.5 μm.

To an extent the conflict noted above is circumvented by introducing easy-plane materials, Ferroxplana's [56], for which μ_i is controlled by the lower in-plane anisotropy, but f_c is also influenced by the higher anisotropy favouring the easy plane.

2.3.1 Disappearing crystallites: amorphous and nanocrystalline materials

Exchange depends primarily on proximity and the regularity of a crystal lattice is not essential: ferromagnetism might well be expected in the amorphous state if the mean interatomic spacing resembles that of a ferromagnetic crystal. Significantly, however, magnetocrystalline anisotropy, though not volume magnetostriction and anisotropy associated with any stresses present, would disappear.

When a common glass or specifically SiO_2 is cooled from an apparent liquid state its viscosity η changes by orders of magnitude over a limited range around a temperature T_g to give an apparent solid in that η/G, with G the shear modulus, is greater than the time over which a typical observation is made and molecular motion virtually ceases. When a metal in the liquid state is cooled, crystalline nucleii form and grow, but this takes a certain time and it is conceivable that with very rapid supercooling T_g may be reached and crystallization completely inhibited to give the amorphous state. This is found to be the case for compositions typically TM–M, with TM (transition metal) a combination of Fe, Co and Ni and M (metalloid) usually boron or phosphorus. An enormous range of compositions has been studied, such as Fe_{78} Si_{12} B_{10}, with $M_s = 1270\ \mathrm{kA\,m^{-1}}$ [57], and another general class is FeNiCo–rare earth, or TM–RE, from which amorphous films may be produced by sputtering. The common methods, however, are splat cooling, or the direction of a jet of liquid on to a spinning chilled disc or cylinder, to give (continuously) thin narrow ribbons with thicknesses appropriate for eddy current suppression (10–100 μm or more). Widths may be increased to ~ 10 cm.

For crystalline alloys the so-called Slater–Pauling plots [58] in Figure 5.19 suggest that m, the moment per atom ($m = n_e\beta$), is given by a simple average for two constituents over substantial compositional ratios. Consider a common band for the alloys, which is to be expected when the atomic numbers differ by only one or two. The case proposed

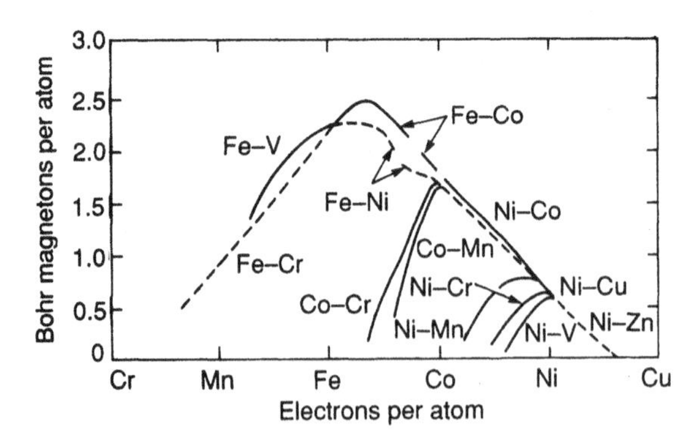

Figure 5.19 Slater–Pauling plots: dependence of M on electron concentration, after Bozorth [43], comparing closely with such plots for amorphous materials, i.e. Figure 25 of [59]

by Stoner, in which one of the sub-bands (for 'spin-up' and 'spin-down' electrons, shifted relatively by the exchange field) is filled and the other partly filled, is called the strong magnetic case. The Fermi level exists in one sub-band only. The alternative with a smaller coupling and shift is called the weak case and here both sub-bands are partially filled and are crossed by the Fermi level. Suppose that the strong case applies to the alloys, though not necessarily to the pure metals. The depiction of the finite number of levels in Figure 5.20 is unrealistic but adequate as a model. According to this, iron itself has $n_e = 5 - 2 = 3$, cobalt $n_e = 2$ and nickel $n_e = 1$, the integral values simply being consequential on the illustration. What matters is that on going from iron to cobalt or cobalt to nickel, n_e changes by -1 and from iron to nickel, n_e changes by -2, independently of the shape of the band. The numbers of the outer electrons $n = 8, 9$ and 10 may be associated with Fe, Co, Ni on the Slater–Pauling plots, indicating $\partial n_e/\partial n = -1$ for the FeCoNi alloys. The argument gives no real information on the n_e, only on the $\partial n_e/\partial n$. For x close to unity in $Fe_x\ Co_{1-x}$ and $Fe_x\ Ni_{1-x}$ the values of n_e indicated are greater than those for pure iron, suggesting that the band structure changes to give the weak case for pure iron. The indication that $n_e \sim 2.5$ for about 30 per cent Co in FeCo does correspond to the exceptional M_s of such alloys.

The most important aspect of this discussion, at the present, is the observation [59] that when the experimental n_e is plotted as a function of composition for amorphous alloys the results are virtually equivalent to the above, save for the reductions due to the non-magnetic elements in, for example $(FeCo)_{80}\ B_{20}$ [60]. The slope of -1 is preserved as the Fe/Co ratio is changed and the increased M due to small Co concentrations in (FeCo)M is included. So far as these trends are concerned it is of no consequence whether the state is crystalline or amorphous. For Co_3B and Co_2B, M is the same for both states [61], but when a wide range of alloys is considered it is impossible to generalize simply. When a change in M_s does occur on crystallizing an amorphous material, it is usually small.

The metalloids reduce the moment of, for example, Ni to the extent that $Ni_{80}\ M_{20}$ is non-magnetic; in $Ni_{100-x}\ P_x$, $x < 19$ is necessary for magnetization to exist, recalling the effect of Al in crystalline NiAl.

In the absence of a lattice, and thus of magnetocrystalline anisotropy, only isotropic magnetostriction and volume magnetostriction $w = \delta v/v$ exist: δv is the change in volume on passing through T_c and not that on saturating the specimen. For Fe-based alloys $\lambda_s \sim 30 - 50 \times 10^{-6}$. For the CoNi base λ_s is small and negative and zero-λ_s compositions can be obtained for $(Fe, Co, M)_{78}Si_8B_{14}$, where M = Mn, etc. [62]. The temperature dependence of λ_s corresponds to the single-ion theory of Callen and Callen, developed for crystalline materials [63]. Substantial non-uniform stresses arise due to non-uniform freezing and give rise to stress anisotropies which may be partially removed by annealing. For certain Fe-rich compositions the temperature dependence of the volume magnetostriction compensates the thermal expansion, over certain ranges, giving the Invar effect [64]. A pressure dependence of T_c is the converse of the Invar effect.

The low-λ compositions compare with the Permalloys and give high permeabilities. The ready production of sheet thicknesses of $\sim$ 10 μm is relevant to operation at MHz frequencies, and the resistivities are generally relatively high; comparison with ferrite cores is claimed, with high inductions and volume advantage. High saturation inductions may be obtained, e.g. about 1.6 T for $Fe_{78}\ Si_{12}\ B_{10}$, and the inclusion of Si again improves stability and reduces H_c to give high μ_m, so that such materials are serious contenders for high power amplifiers. It may be noted that the crystalline materials that give the best performance have required intensive and prolonged development and that the procedures required to produce such materials may be elaborate.

Nanocrystalline materials. On heating amorphous $Fe_{73.5}Cu_1Nb_3Si_{13.5}B_9$ above a crystallization temperature of 510 °C Yoshiwaza and Yamauchi [65] formed an ultrafine structure of α-FeSi crystallites with $d \sim 10$–20 nm. Herzer [40] compared an amorphous ribbon for which $M_s(T) = M_s(0)(1 - T/T_c)^{1/3}$ applied, with a recrystallized (1 h at 520°C) specimen which showed a change of slope at 320°C compatible with the existence of two phases, one consisting of α-FeSi with $\mu_0 M = 1.3$ T (75 per cent) and the other 25 per cent, constituting an amorphous grain boundary phase, $\mu_0 M = 1.16$ T, $T_c = 320$°C, with thickness 1.2 nm or 4–5 atomic layers. The crystallites are supposed to be exchange-coupled via the amorphous layers below T_c for these, explaining the change in slope at 320°C in Figure 5.20.

A return may now be made to Figure 5.10 which, at the lower sizes, referred to this material and, of course, appeared completely anomalous: these are low-H_c, high-μ materials. It is also recalled that the exchange length indicates the minimum distance over which $\mathbf{M}_s$ may change direction without involving very high exchange energy densities, since it resembles a domain wall width. In crystalline materials changes in direction of $\mathbf{M}_s$ occur at grain boundaries where the easy direction changes, and these are opposed by the exchange coupling if it is assumed, referring to the general properties of amorphous materials, that regions of disorder at grain boundaries do not substantially affect the exchange coupling. Rotations of $\mathbf{M}_s$ to smooth out the change in direction should apply over distances of the order of L_e and would, in general, have negligible effects. In the nanocrystalline materials the effects are far from negligible since $d \sim L_e$ and any substantial rotation from one crystallite to another is precluded: the exchange overwhelms the anisotropy and $\mathbf{M}_s$ tends to a common direction throughout the assembly of crystallites. If the easy directions are randomly distributed no axis is favoured and the assembly behaves cooperatively and isotropically (see Figure 5.22). This is a limiting case conclusion for $d \ll L_e$ and, in practice, referring to the work of Alben *et al.* [66], Herzer concluded that for $d < L$ a mean anisotropy of

$$\langle K \rangle = \frac{K_1^4 d^6}{A^3}$$

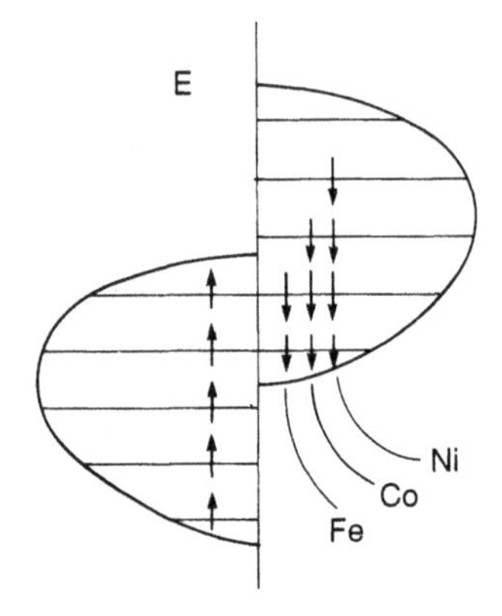

Figure 5.20 Formalized band structure in relation to Figure 5.19

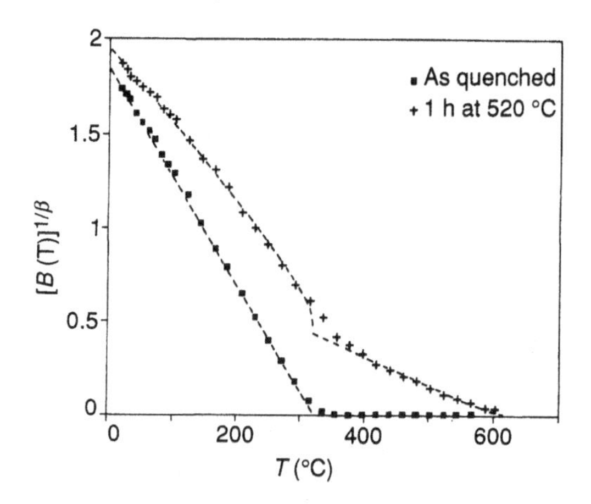

Figure 5.21 Temperature dependence of the saturation induction, i.e. of $B^{1/\beta}$ with $\beta \approx 0.36$, for amorphous and noncrystalline $Fe_{73.5}Cu_1Nb_3Si_{13.5}B_9$, with a change of slope at 320 °C [40]. This is taken to correspond to Figure 5.22(a) up to 320 °C and Figure 5.22(b) above 320 °C

should remain (using $K_1 = 8$ kJ m^{-3}, $A = 10^{-11}$ J m^{-1}, $L_e = 35$ nm.) For coherent rotation $H_c \propto \langle K \rangle / M_s$ so that the observed dependence, in Figure 5.22, of H_c on d^6 follows. The predictions that $H_c \rightarrow 0$ and $\mu_i \rightarrow \infty$ as $d \rightarrow 0$ are modified in practice by the occurrence of magnetoelastic anisotropies, surface roughness, etc., as for amorphous materials. These materials can have $\mu_0 M = 1.2$–1.3 T, far higher than for Permalloy, and $\mu_m = 500\,000$, with approach to saturation in $H = 10$ A m^{-1} and $B(10$ A m$^{-1}) = 1.18$ T and $H_c \sim 1.0$ A m^{-1}.

A striking comparison may be made with the production of Nd–Fe–B magnets with exceptionally high H_c by crystallizing amorphous ribbons — the General Motors method (as in Section 2.1). There is some significant difference in particle size but the major difference is that the intergranular phase in the Nd–Fe–B alloy is non-magnetic and serves specifically to break the exchange coupling between the grains.

This section may be concluded, overall, by referring to Tables 5.9 and 5.10, which compare some properties of amorphous and crystalline materials.

Table 5.9 Comparison of properties, relevant to high-power applications, of Fe-based amorphous sheet with conventional Si–Fe and with tertiary recrystallized Si–Fe with domain refinement [41]. W13/50 and W17/50 are the losses at 50 Hz with peak inductions of 1.3 and 1.7 T. The final column indicates the effects of tensile stress of 4 kg mm^{-2}. A competition is suggested between the introduction of new materials and the improvement of established materials

Sample	Grain-oriented Silicon steel (With coating)	Fe-based Amorphous sheet (Field annealed)	Tertiary recrystallized silicon Steel sheet (domain refined) Mechanical	Chemical	
Thickness (nm)	0.3	0.02 ~ 0.04	0.075	0.071	0.032
B_s(T)	2.03	1.5 ~ 1.6	2.03	2.03	2.03
W13/50 (W kg^{-1})	0.6	0.15 ~ 0.30	0.19	0.17	0.13
W17/50 (W kg^{-1})	1.02	—	0.42	0.35	0.21

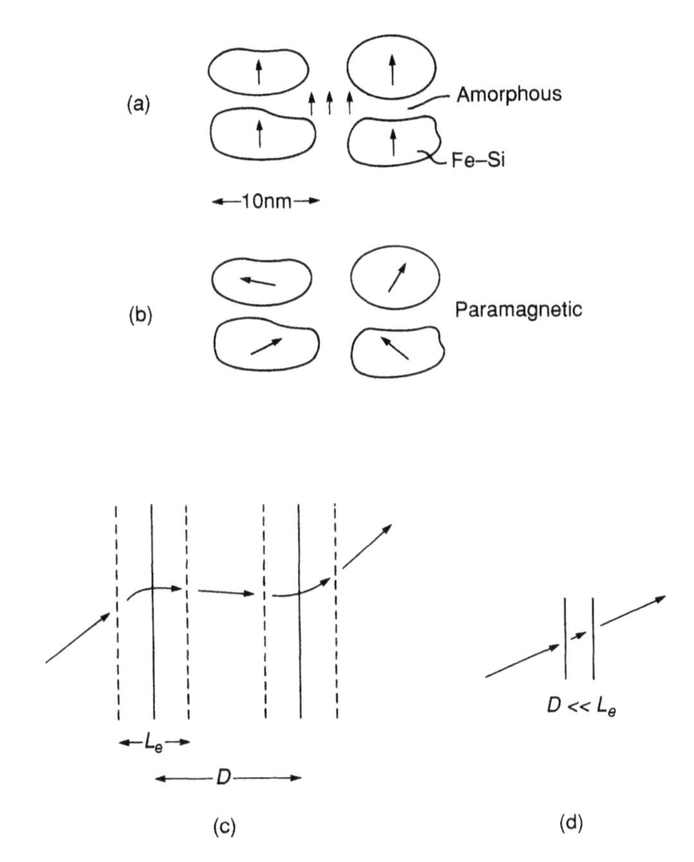

Figure 5.22 (a,b) An indication of the decoupling between the SiFe crystallites when the intervening layer becomes paramagnetic. With coupling the directions of $\mathbf{M}_s$ may vary from one microcrystallite to another if the diameters $D > L_e$, the exchange length (c), but with $D \ll L_e$ as in (d) this would involve very great exchange energy densities and the magnetocrystalline anisotropies are effectively suppressed

Table 5.10 Core loss per cycle of ultrathin amorphous cores compared with other core materials used in the high-frequency range for $B_m = 0.1$ T [67]

		Core loss W_c for $B_m = 0.1$ T (J/m^{-3})				
	Core materials	100 kHz	1 MHz	2 MHz	5 MHz	10 MHz
Amorphous cores	3.8 μm $((FeCo)Cr)_{75}(SiB)_{25}$	0.23	1.0	1.5	4.0	7.3
	5.5 μm $((FeCo)Cr)_{75}(SiB)_{25}$	0.16	1.0	2.0	5.2	8.8
	6.4 μm $(FeCo)_{71}(SiB)_{29}$	0.40	1.8	3.2	7.3	13.0
	14.1 μm $(FeCo)_{75}(SiB)_{25}$	1.0	5.3	—	—	—
Crystalline cores	Mn–Zn high-frequency ferrite	0.88	2.5	—	—	—
	5 μm Supermalloy tape	0.8	2.8	4.9	9.7	16.5

3 Intrinsic Properties

This does not constitute a systematic treatment but is rather a brief collection of data, which may be amplified by reference to particular works on materials (e.g. [68]). The general principles underlying the data have, it is hoped, been covered in the text and the collection here should facilitate reference. Intrinsic data are given on the transition metals iron, cobalt and nickel, on the rare earth metals and on a few of the alloys, followed by the most important oxide materials [with a single Figure (5.23) showing the interesting $M_s(T)$ plots for ferrimagnetic garnets]. The properties are principally the spontaneous magnetization at room temperature, M_s, and as $T \rightarrow 0$ K, $M_0 \equiv M_s(0)$; σ and σ_0 or $\sigma(0)$, the specific magnetizations; the first magnetocrystalline anisotropy constants K_1 and the corresponding anisotropy fields H_K; the saturation magnetostriction λ_s; the Curie temperature T_c and the Weiss temperature θ; and the Curie constant in the paramagnetic region.

Tables 5.11 to 5.18 give pure numbers x. The quantities correspond to the list:

$$T = x \text{ K}$$

$$\theta = x \text{ K}$$

$$M = x \text{ kA m}^{-1} = x \text{ gauss } = x\mathcal{A} \text{ cm}^{-1}(1\mathcal{A} = 1 \text{ abamp } = 10 \text{ A})$$

$$\sigma = x \text{ A m}^2\text{kg}^{-1} = x\mathcal{A} \text{ cm}^2\text{g}^{-1}$$

$$H_K = x \text{ kA m}^{-1} = (x \times 4\pi)\mathcal{A} \text{ cm}^{-1} = (x \times 4\pi)\text{Oe}$$

$$K_1 = x \text{ kJ m}^{-3} = (x \times 10^4) \text{ erg cm}^{-3}$$

$$C_m = x \times 10^{-6} \text{ m}^3\text{mol}^{-1}\text{K} = \left(\frac{x}{4\pi}\right) \text{cm}^3\text{mol}^{-1}\text{K}$$

$$\lambda_s = x \times 10^{-6}$$

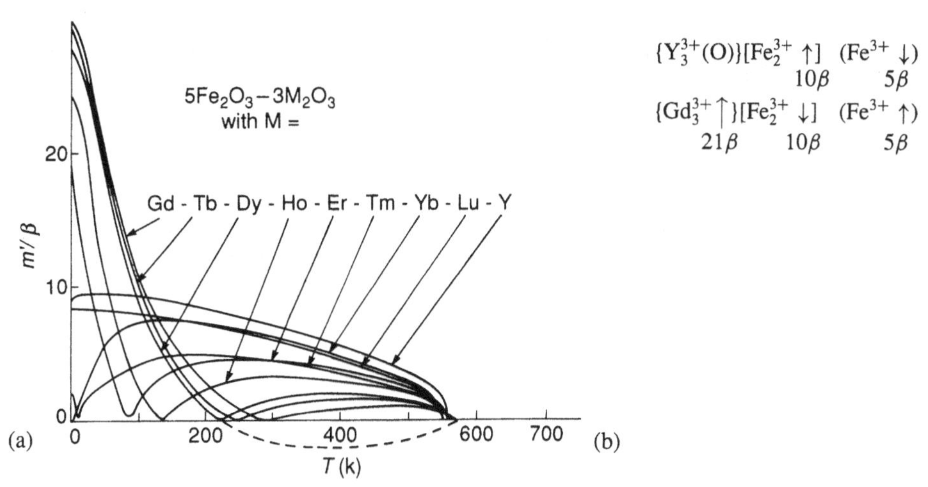

Figure 5.23 (a) $M_s(T)$ in Bohr magnetons per formula unit for the ferrimagnetic RIGs showing how the rare earths affect the compensation points but not the Curie points, since the latter are controlled by the coupling between the Fe^{3+} lattices [85]. Compensation points also occur for transition metal–rare earth alloys, for example. (Reprinted by permission of John Wiley & Sons, Ltd, Magnetic Oxides, ed. D.J. Craik, Wiley 1975) (b) Order at $T \rightarrow 0$ K, { }, [ ] and ( ) indicating dodecahedral, octahedral and tetrahedral sites

Table 5.11 Iron, cobalt and nickel metals [69–74]. Cobalt is h.c.p. at low and room temperatures, f.c.c. above 400°C with smooth $\sigma(T)$ curve applying to both. $\sigma = 155.2$ at 412°C for f.c.c.

	M_0	σ_0	n_β	T_c	M_s	θ	C	K_1	H_K	g	λ_s
Fe	1752	221.7	2.216	1043	1710	1100	15.9	48	44.6	2.094	−7
Co (h.c.p.)	1446	162.5	1.716		1431			530	589	2.17	−50
Co (f.c.c.)			1.75	1394		1408	15.5				
Ni	519	58.57	0.616	631	494	650	4.05	−4.5	14.7	2.185	−36

Table 5.12 The anisotropy and saturation magnetostriction of several Ni–Fe crystals at room temperature [75]. Where two states of order were examined, 'a' denotes the more nearly disordered state and 'b' the more nearly ordered state. All 'b' data were taken on crystals cooled at 1.2°C h^{-1}. In the 'a' state, the Ni–Fe crystals were given a normal furnace cool

Composition (weight %)	Anisotropy constant, K_1 (kJ m^{-3})		Magnetostriction $\lambda_{100} \times 10^6$		Magnetostriction $\lambda_{111} \times 10^6$	
	a	b	a	b	a	b
Ni	−5.6		−58.3		−24.3	
80.4 Ni–Fe	−1.0	−1.3	7.9	8.2		2.1
75.4 Ni–Fe	−0.7	−3.0	16.2	12.0	6.1	9.2
70.1 Ni–Fe	−0.1	−0.95	22.7	20.9		10.8
59.9 Ni–Fe	0.1	−0.35	21.9	21.1		20.2
49.4 Ni–Fe	0.5	0.3	6.4	7.4		27.5
39.8 Ni–Fe	0.5	0.5	−9.5	−10.7		27.4
Fe	4.6		20.0		−15.7	

Table 5.13 Section of the rare earth metal series. n_β^c and p_e^c are calculated according to the ground states given. n_β is the first value from saturation, second from neutron diffraction. All save Gd behave formally as 0 K $|$ferro$|T_c|$antiferro$|T_N|$ paramagnetic (Curie–Weiss). All h.c.p. structures [76, 77]

	d	T_c	T_N	θ	σ_0	n_β	p_e	n_β^c	p_e^c	State
Gd	7.866	293.2	—	317	268.4	7.55	7.98	7.0	7.94	$^8S_{7/2}$
Tb	8.253	221	229	236	328	9.34 (9.5)	9.77	9.0	9.7	$^7F_6$
Dy	8.567	85	178.5	151	350	10.2 (9.0)	10.67	10.0	10.6	$^6H_{15/2}$
Ho	8.799	20	132	87.7	350.2	10.3 (10.0)	10.8	10.0	10.6	$^5I_8$
Er	9.058	30	85	47.2	267	8.0 (9.0)	9.8	9.0	9.6	$^4I_{15/2}$

Table 5.14 Examples of rare earth–transition metal compounds, many more of which exist [78]. The GdM_2 are cubic ($MgCu_2$) and the GdM_5 are hexagonal ($CaCu_5$)

	Order	T_c	n_β	M_0	M_s
$GdMn_2$	Ferri	575	1.0	159	
$GdFe_2$	Ferri	782	3.4	605	406
$GdCo_2$	Ferri	408	4.8	692	400
$GdNi_2$	Ferro	85	6.7	1350	
$GdMn_5$	Ferri	465	6.2	684	318
$GdFe_5$	Ferri	455	5.0	550	358
$GdCo_5$	Ferri	1030	1.4	111	151
$GdNi_5$	Ferro	27	6.1	597	

Table 5.15 Spinel ferrites M Fe_2O_4 [53, 79]. Cu is slow cooled, with properties dependent on heat treatment and ordering and 'Li' is $Li_{0.5}Fe_{2.5}O_4$. $\lambda_{100} = -25, -20, -46$ and $\lambda_{111} = +4.5, +78, +22$ for M = Mn, Fe, Ni with $\lambda_{110} = +57$ for M = Fe, all $\times 10^{-6}$

M	d	ρ	n_β^{cal}	n_β	σ_0	M_0	σ_s	M_s	T_c	K_1	λ_s	H_K
Mn	5.00	10^2	5	4.55	112	560	80	400	573	−4.0	−5	16
Fe	5.24	4×10^{-5}	4	4.1	98	510	92	480	858	−13	+4	43
Co	5.29	10^5	3	3.94	93.9	496	80	425	793	+200	−110	750
Ni	5.38	10–100	2	2.3	56	300	50	270	858	−6.9	−17	41
Cu	5.35		1	1.3	30	160	25	135	726	−6.0		71
Mg	4.52	10^5	0	1.1	31	140	27	110	713	−4.0	−6	58
'Li'	4.75	10	2.5	2.6	69	330	65	310	943	−8.3	−8	43

Table 5.16 Rare earth orthoferrites [80]. σ_0^M is the molar spontaneous magnetization in A m^2 mol^{-1}; α, canting angle in mrad

	La	Sm	Eu	Gd	Tb	Dy	Ho	Er	Tm	Yb	Lu	(Y)
n_β	0.047	0.041	0.040	0.049	0.039	0.040	0.041	0.041	0.040	0.045	0.054	0.045
σ_0^M	0.225	0.230	0.225	0.273	0.230	0.225	0.230	0.228	0.225	0.250	0.300	0.250
T_c	740	674	664	663	650	646	643	640	631	631	623	643
α	9.1	8.2	8.0	9.8	7.8	8.0	8.2	8.1	8.0	8.9	10.7	8.9

Table 5.17 Hexagonal oxides: M, W, Y and Z compounds [81]. Y (Ferroxplana): **M** lies along the c axis below 215 K, in the hexagonal plane above 215 K. Co_2Z: moments on the cone below 220 K, in the hexagonal plane to 480 K, along the c axis above 480 K

	d	σ_0	M_s	T_c	n_β	K_1 or $K_1 + 2K_2$	H_K
BaM	5.28	100	380	723	20	330	1350
PbM	5.63	80	320	724	18.6	220	1090
SrM	5.11	108	370	733	20.6	350	1590
Fe_2W	5.31	98	320	728	27.4	300	1510
ZnFeW		108	380	793	30.7	240	995
Mg_2Y	5.14	29	119	553	6.9	−60	796
Co_2Y	5.40	39	185	613	9.8	−260	2230
Ni_2Y	5.40	25	127	663	25	−90	1110
Co_2Z	5.88	69	267	683	31.2	−180	103
Zn_2Z	5.37		310	633		60	

Table 5.18 Rare earth and yttrium iron garnets [82–84]. Data in brackets by extrapolation from substituted compounds. T_{comp}: compensation temperature

R=	(Y)	La	Pr	Nd	Sm	Eu	Gd	Tb	Dy	Ho	Er	Tm	Yb	Lu
d	5.17	(5.67)	(5.87)	(6.00)	6.23	6.31	6.46	6.55	6.61	6.77	6.87	6.94	7.06	7.14
n_β	5.01	(5.0)	(9.8)	(8.7)	5.43	2.78	16.0	18.2	16.9	15.2	10.2	1.2	0.0	5.07
M_s					135	93	135	4	43	78	103	110	130	140
T_{comp}							286	246	226	137	83	—	0–6	
T_c	553				578	566	561	568	563	567	556	549	548	549

References

1. Craik, D.J. and Hill, E.W., *IEEE Trans. Magn.* **MAG-11**, 1379 (1975); *J. de Physique*, **38** (Colloque C1, Suppl. 4), C1-39 (1977); *Physica*, 86–88B, 1486 (1977).
2. Mallinson, J.C., *The Foundations of Magnetic Recording*, Academic Press Inc., San Diego, 1987.
3. Mee, C.D., *The Physics of Magnetic Recording*, North Holland, Amsterdam, 1964.
4. Mee, C.D. and Daniels, E. (Eds.), *Magnetic Recording*, Vol. 1, *Technology*, McGraw-Hill, New York, 1981.
5. Bate, G., *Magnetic Oxides*, Part 2, Ed. D.J. Craik, John Wiley and Sons, London and New York, p. 689, 1975.
6. Eagle, D.F. and Mallinson, J.C., *J. Appl. Phys.*, **38**, 995 1963.
7. Speliotis, D.E., Morrison, J.R. and Bate, G., *Proceedings of International Conference on Magnetism*, Nottingham, p. 623, 1965.
8. Jacobs, I.S. and Bean, C.P., *J. Appl. Phys.*, **29**, 537 (1955); *Phys. Rev.* **100**, 1060 (1955).
9. Frei, E.H., Shtrikman, S. and Treves, D., *Phys. Rev.*, **106**, 446 (1957).
10. Wohlfarth, E.P., *Magnetism* Vol.III, Eds. G.T. Rado and H. Suhl, Academic Press, New York, p. 351, 1963.
11. Luborsky, F.E., *J. Appl. Phys.*, **32**, 171S (1961).
12. Corradi, A.R., Green, W.B., Price, T.C., Bottoni, G. Candolfo, D. Cecchetti, A., Masoli, F. and Molesini, L., *IEEE Trans. Magn.*, **26**, 237 (1990).
13. Bate, G., *J. Appl. Phys.*, **32**, 361S (1961).
14. de Jong, J. Smeets, J.M.G. and Haanstra, H.B., *J. Appl. Phys.*, **29**, 297 (1958).
15. de Vos, K.J., *Philips Res. Repts*, **18**, 405 (1963); and *Proceedings of International Conference on Magnetism*, Nottingham, p. 772 1964 and Thesis, Eindhoven, 1966.
16. Craik, D.J. and Isaac, E.D., *Proc. Phys. Soc.*, **76**, 160 (1960).
17. McCurrie, P.A. and Gaunt, P., *Proceedings of International Conference on Magnetism*, Nottingham, 1964 (Institute of Physics and Physics Society, London, p. 780, 1965; Newkirk, J.B., Smoluchowski, R., Geisler, A.H. and Martin, D.L., *J. Appl. Phys.* **22**, 290 (1951).
18. Strnat, K.J., Hoffer, G., Olson, J., Ostertag, W. and Becker, J.J., *J. Appl. Phys.* **38**, 1001 (1967).
19. Nesbitt, E.A., Williams, R.H., Sherwood, R.C., Buehler, E. and Wernick, J.H., *Appl. Phys. Lett.*, **12**, 361 (1968); Nesbitt, E.A., *J. Appl. Phys.* **40**, 1259 (1969).
20. Nesbitt, E.A. and Wernick, J.H., *Rare Earth Permanent Magnets*, Academic Press, New York and London, 1973.
21. Nesbitt, E.A., Williams, H.J, Wernick, J.H. and Sherwood, R.C., *J. Appl. Phys.*, **32**, 342S (1961) and **33**, 1674 (1962).
22. Becker, J.J., *J. Appl. Phys.*, **41**, 1055 (1970).
23. Tawara, Y. and Senno, H., *Jap. J. Appl. Phys.*, **7**, 966 (1968); Senno, H. and Tawara, Y., *Jap. J. Appl. Phys.*, **8**, 118 (1968).
24. Sagawa, M., Fujimura, S., Togawa, N., Yamamoto, H. and Matsuura, Y., *J. Appl. Phys.*, **55**, 2083 (1984); see also Sagawa, M., Hirosawa, S., Yamamoto, H., Fujimura, S. and Matsuura, Y., *Jap. J. Appl. Phys.*, **26**, 785 (1987).
25. Elbicki, J.M., Wallace, W.H., and Korahlev, *IEEE Trans. Magn.*, **25**, 3567 (1989).
26. Croat, J.J., Herbst, J.F., Lee, R.W. and Pinkerton, F.E., *J. Appl. Phys.*, **55**, 2075 (1984); see also Lee, R.W., Brewer, E.G. and Schaffel, N.A., *IEEE Trans. Magn.*, **MAG-21**, 1958 (1985).
27. Lee, R.W., *Appl. Phys. Lett.*, **46**, 790 (1985).
28. Croat, J.J., *IEEE Trans. Magn.* **25**, 3550 (1989).
29. Harris, I.R., Evans, J. and Nyholm, P., Br. Patent 1 554 384, (1979).
30. McGuiness, P.J., Devlin, E., Harris, I.R., Rozendaal, E. and Ormerod, J., *J. Mater. Sci.*, **24**, 2541 (1989).
31. Harris, I.R. and McGuiness, P.J., *J. Less-Common Metals*, **172–174**, 1273 (1991).
32. Takeshita, T. and Nakayama, R.N., *Proceedings of 10th International Workshop on Rare Earth Magnets and Their Applications*, Kyoto, p. 551, 1989.
33. McNab, T.K., Fox, R.A. and Boyle, A.J.F., *J. Appl. Phys.*, **39**, 5703 (1968).
34. Néel, L., *Ann. de Geophysique*, **7**, 8 (1951).

35. Brown, W.F., Jr, *J. Appl. Phys.* **30**, 1303 (1959); see also *Micromagnetics*, Interscience-John Wiley and Sons, New York and London, 1963; *Magnetostatic Principles in Ferromagnetism*, North-Holland, Amsterdam, 1962.
36. Hedley, I.G., *Phys. Earth Planet. Int.*, **1**, 103 (1968).
37. Néel, L., *Compt. Rend.*, **252**, 4075, **253**, 9 and **253**, 203 (1961).
38. Cohen, J., Creer, K.M., Pauthenet, R. and Srivastava, K., *J. Phys. Soc. Japan*, **17**, Suppl. B-1, 685 (1962).
39. Herzer, G., *IEEE Trans. Magn.*, **26**, 1397 (1990).
40. Herzer, G., *IEEE Trans. Magn.*, **25**, 3327 (1989).
41. Arai, K.I., Ishiyama, K. and Mogi, H., *IEEE Trans. Magn.* **25**, 3949 (1989).
42. Jackson, R.C. and Lee, E.W., *J. Mater. Sci.*, **1**, 362 (1966).
43. Bozorth, R.M., *Rev. Mod. Phys.*, **25**, 42 (1953).
44. Chikazumi, S., Suzuki, K. and Iwata, H., *J. Phys. Soc. Japan*, **15**, 250 (1960); Tamagawa, N., Nakagawa, Y. and Chikazumi, S., *J. Phys. Soc. Japan*, **17**, 1256 (1962).
45. Williams, H.J. and Goertz, M., *J. Appl. Phys.*, **23**, 316 (1952).
46. Hamilton, W.C., *Phys. Rev.*, **110**, 1059 (1958).
47. Kundig, W. and Hargrove, R.S., *Solid State Commun.*, **7**, 223 (1969).
48. Guillaud, C., *J. Phys. Rad.*, **12**, 239 (1951); Guillaud, C. and Roux, M., *Compt. Rend.*, **229**, 1133 (1949); see also Pauthenet, R., *Ann. Phys.*, **7**, 710 (1952).
49. Ohta, K., *J. Phys. Soc. Japan*, **18**, 684 (1963).
50. Perduijn, D.J. and Peloschek, H.P., *Proc. Br. Ceram. Soc.*, **10**, 263 (1968); Ross, E., Hanke, I. and Moser, E., *Z. Angew. Phys.*, **17**, 504 (1964); Beer, A. and Schwartz, J., *IEEE Trans. Magn.*, **MAG-3**, 470 (1966).
51. Peloschek, H.P. and Perduijn, D.J., *IEEE Trans. Magn.*, **MAG-4**, 453 (1968).
52. Snoek, J.L., *Physica*, **14**, 207 (1948).
53. Smit, J. and Wijn, H.P.J., *Ferrites*, Philips Technical Library, Philips, Eindhoven, 1959; *Adv. Electr. and Electr. Phys.*, **6**, 64 (1954).
54. Rado, G.T., *Rev. Mod. Phys.*, **25**, 81 (1953); Rado, G.T., Wright, R.W. and Emerson, W.H., *Phys. Rev.*, **80**, 273 (1950).
55. de Lau, J.G.M., *Proc. Brit. Ceram. Soc.*, **10**, 275 (1968); de Lau, J.G.M. and Stuijtes, A.L., *Philips Res. Repts*, **21**, 104 (1966).
56. Jonker, G.H., Wijn, H.P.J. and Braun, P.B., *Philips Tech. Rev.*, **18**, 145 (1956); *Proc. IEE*, **104B**, 249 (1957)
57. Luborsky, F.E., Becker, J.J., Frischman, P.J. and Johnson, L.A., *J. Appl. Phys.*, **49**, 1769 (1978); see also Liebermann, H.H. and Graham, C.D. Jr, *IEEE Trans. Magn.*, **MAG-12**, 921 (1976); Hagiwara, M., Inoue, A. and Masumoto, T., *Metall. Trans.*, **13**, 373 (1982).
58. Slater, J.C., *J. Appl. Phys.*, **8**, 385 (1937); Pauling, L., *Phys. Rev.*, **54**, 899 (1938).
59. Egami, T., *Repts Prog. Phys.*, **47**, 1601 (1984).
60. O'Handley, R.C., Hasegawa, R., Ray, R. and Chou, C.P., *Appl. Phys. Lett.*, **29**, 330 (1977); O'Handley, R.C., Mendelsohn, L.I. and Nesbitt, E.A., *IEEE Trans. Magn.*, **MAG-12**, 942 (1976).
61. Hasegawa, R. and Ray, R., *J. Appl. Phys.*, **50**, 1586 (1979).
62. Jagielinski, T., Arai, K.I., Tsuya, N., Ohnuma, S. and Masumoto, T., *IEEE Trans. Magn.*, **MAG-13**, 1553 (1977); Luborsky, F.E., *Amorphous Magnetism*, Vol III, Plenum, New York, p. 345, 1977; Graham, C.D. Jr and Egami, T., *Ann. Rev. Mater. Sci.*, **8**, 423 (1978).
63. Callen, E.R. and Callen, H.B., *Phys. Rev.*, **120**, 578 (1963) and **A139**, 455 (1965).
64. Fukamichi, K., Kikuchi, M., Arakawa, S. and Masumoto, T., *Solid State Commun.*, **23**, 955 (1977); Fukamichi, K., *Amorphous Metallic Alloys*, Ed. F.E. Luborsky, Butterworth, London, p. 317 1983.
65. Yoshizawa, Y. and Yamauchi, K., *IEEE Trans. Magn.*, **25**, 3324 (1989).
66. Alben, R., Becker, J.J. and Chi, M.C., *J. Appl. Phys.*, **49**, 1653 (1978).
67. Yagi, M. and Sawa, T., *IEEE Trans. Magn.*, **26**, 1409 (1990).
68. Tebble, R.S. and Craik, D.J., *Magnetic Materials*, John Wiley and Sons, London and New York, 1969.
69. Danan, H., Herr, A. and Meyer, A.J.P., *J. Appl. Phys.*, **39**, 669 (1968).
70. Williams, G.M. and Pavlovic, A.S., *J. Appl. Phys.*, **39**, 571 (1968).
71. Benninger, G.N. and Pavlovic, A.S., *J. Appl. Phys.*, **38**, 1325 (1967).

72. Bozorth, R.M. and Sherwood, R.C., *Phys. Rev.*, **94**, 1439 (1954).
73. Meyer, A.J.P. and Asch, G., *J. Appl. Phys.*, **32**, 330S (1961).
74. Tatsumoto, E., Okamoto, T., Iwata, N., and Kadena, Y., *J. Phys. Soc. Japan*, **20**, 1541 (1965).
75. Hall, R.C., *J. Appl. Phys.*, **30**, 816 (1959).
76. Koehler, W.C., *J. Appl. Phys.*, **36**, 1078 (1965).
77. Gibson, J.A. and Harvey, G.S., Air Force Materials Laboratory, Wright-Patterson AFB, Ohio, (1966).
78. Methfessel, S., *IEEE Trans. Magn.*, **1**, 144 (1965).
79. Von Aulock, W.H., *Handbook of Microwave Ferrite* Materials. Academic Press, New York (1965)
80. Treves, D., *J. Appl. Phys.*, **36**, 1035 (1965).
81. Braun, P.J., *Philips Res. Repts*, **12**, 491 (1957).
82. Espinosa, G.P., *J. Chem. Phys.*, **37**, 2344 (1962).
83. Geller, S. and Gilleo, M.A., *J. Phys. Chem. Solids*, **3**, 30 (1957).
84. Geller, S., Remeika, J.P., Sherwood, R.C., Williams, H.J. and Espinosa, G.P., *Phys. Rev.*, **137A**, 1034 (1965).
85. Bertant, F. and Pauthenet, R., *Proc. IEE*, **B 104**, Suppl. 5, 261 (1957).

CHAPTER 6 Special Topics

1 Magneto-optics

A plane polarized (PP) light beam is one in which the $\mathbf{E}$ vector is confined to a fixed plane as in $\mathbf{E}^x(z,t) = \mathbf{i}A_x\cos(kz-\omega t)$ or $\mathbf{E}^y(z,t) = \mathbf{j}A_y\cos(kz-\omega t+\delta)$; or $\mathbf{E}(z,t) = \mathbf{E}^x+\mathbf{E}^y$ with $\delta = 0$ and $A_x = A_y = A$, the plane then containing $\mathbf{k}$ and OZ and lying at equal angles with OXZ and OYZ. If $\delta = -\pi/2 + 2n\pi$ and $\mathbf{E}^x = \mathbf{i}A\cos(kz-\omega t)$ then $\mathbf{E}^y = \mathbf{j}A\sin(kz-\omega t)$. $\mathbf{E}(0,0) = \mathbf{i}A$ but $\mathbf{E}(0,\tau/4) = -\mathbf{j}A$ and the plane rotates w.r.t. t at $z = 0$ or equally w.r.t z at fixed t, which is right circularly polarized (R) light: the rotation is clockwise viewed towards the source ($\tau = 2\pi/\omega$). If $\delta = \pi/2, 5\pi/2, \ldots$ the light is left circularly polarized (L). Conversely, linearly polarized light may be regarded as the sum of circular components. In general the sum of PP beams gives elliptical polarization with plane and circular polarization as special cases. Excited atoms emit PP waves of very short duration and with random phase relations, except in laser operation when the waves are coherent.

When PP light is passed through a quartz crystal, along a particular optic axis, the plane undergoes a rotation: quartz is said to be optically active. The sense depends on the direction of propagation and if the beam is reflected from the back of the crystal the rotations cancel.

The effect can be explained in terms of different refractive indices n_R and n_L or wave vectors k_R and k_L. A helical structure is required for optical activity, the charge oscillations caused by the L and R waves then differing. When a magnetic field is applied to an isotropic material the magnetic forces affect the oscillations caused by the L and R waves differently, as may be imagined in relation to the helical motion of charges moving in magnetic fields. The result is a magneto-optical activity or Faraday rotation α. If the field is reversed then the effects on L and R are interchanged and α, looking along the direction of propagation, is reversed. If a beam is reflected from the back of a crystal then α, looking along the *new* direction of propagation, is reversed. Thus the double passage doubles the rotation rather than annulling it, as stressed by Figure 6.1.

By considering the effect of the field on the classical oscillations (force constant g) of the charges e with mass m (N per unit volume), as by Möller [1], the Faraday rotation is found to be

$$\alpha = \frac{\pi}{n\lambda} \frac{\omega\omega_p^2\omega_c}{(\omega_0^2 - \omega^2) - \omega^2\omega_c^2} \rightarrow VBl \tag{1}$$

the second, with $\omega_c = eB/m$, to the extent that $\omega^2\omega_c^2$ can be ignored; ω_p is the plasma frequency, $\omega_p^2 = Ne^2/(\varepsilon_0 m)$ and $\omega_0 = \sqrt{g/m}$. V is the Verdet constant; Möller gives V (in 10^{-3} min Oe^{-1} cm^{-1}) for water with λ (μ m in brackets) as 12.6 (0.6), 7.0 (0.8), 4.4 (1.0), 2.9 (1.25). For ZnS, $V = 225 \times 10^{-3}$ min Oe^{-1} cm^{-1}. Faraday rotations in ordered materials are many orders higher than the above, depend on the ordering or magnetization rather than any fields directly and may be given as specific rotations per unit length $\parallel$ **M** (Table 6.1). They are crudely accounted for by using the exchange fields ($\sim 10^3$ T) in [1], but there is no simple dependence on $\mathbf{M}_s$ and rotations persist through the compensation points of

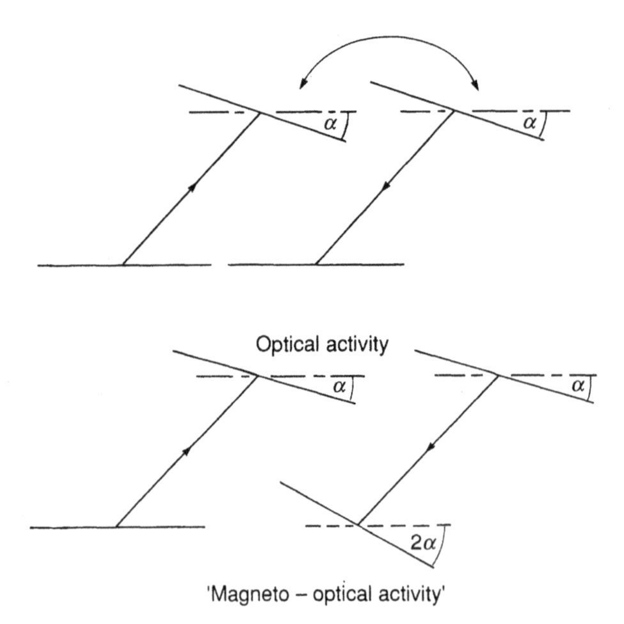

Figure 6.1 Contrasting optical activity and 'magneto-optical activity' or Faraday rotation, the curved arrow indicating reflection: a vital distinction for optical isolators

Table 6.1 Faraday rotations and absorption coefficients [63]. Due to the dependence on λ and T such a table can only give a general impression; e.g. for YIG, α is nearly constant from 100 to 400 K, but for TbIG at 1150 nm, $\alpha = 1500 \to 250$ as $T = 100 \to 450$ K; for DyIG, α changes from -40 to $+40$ at the compensation temperature of 220 K. Note the transmission 'window' for YIG

	T(K)	λ (nm)	α (deg cm^{-1})	a (cm^{-1})
Fe	300	546	3.5×10^5	7.6×10^5
Co	300	546	3.6×10^5	8.5×10^5
Ni	300	4000	7.2×10^5	2.1×10^5
MnBi	300	6300	1.4×10^5	3.4×10^6
YIG	300	12000	2.4×10^2	0.069
	300	6000	1.9×10^3	1.2×10^3
GdIG	300	5200	4×10^3	3×10^3
$CoFe_2O_4$	300	4000	3.7×10^4	1.7×10^5
EuO	60	12000	2×10^5	10^2
	5	6600	5×10^5	5×10^5
EuSe	4.2	6600	2.6×10^5	1.7×10^2

ferrimagnetics. α_F may be associated with the off-diagonal elements of the permittivity tensor, ε_{xy}, and at a microscopic level with the optical transitions, e.g. in ferrimagnetic oxides, as by Kahn *et al.* [2] or Matsumoto *et al.* [3].

Bi^{3+} substitutes for Y^{3+} in YIG, or R^{3+} in RIGs: $Y_{3-x}Bi_xFe_5O_{12}$, up to completion, with remarkable effects on α_F [4]. A 45° rotation may be effected in an epitaxially grown film only $\sim$ 10 μm thick. If a polarizer is placed before a '45° film' light reflected back will have undergone a 90° rotation and will not be passed by the polarizer. Takahashi *et al.* [5] obtained forward and backward attenuations of 4.2 and -35 dB. Such optical isolators are valuable in optical (fibre optic) communication systems, protecting laser diode sources from reflected light which would otherwise cause instabilities. The impact of such materials in this field generally should be substantial.

Though not strictly relevant to the present section, microwave isolators, in which a narrow rectangular waveguide defines the plane of polarization and the rotations are effected by a garnet rod in a circular-section waveguide, are analogous to the above. Although Faraday rotations are still involved the permeability tensor replaces the permittivity tensor at these frequencies, at which gyromagnetic rather than the optical effects are important.

Birefringence, by which is meant linear birefringence, involves differences in the velocities of waves with components $\parallel$ or $\perp$ to certain optic axes, or to the ordering axes in the magnetic case. The resulting phase difference clearly depends on the path length and may result in rotation while reproducing plane polarization, or the conversion of PP to circular polarization. Both the Faraday effect and birefringence may be utilized to observe domains, using suitably disposed polarizers and analysers, and some examples have already been given. Figure 6.2 shows a domain contrast Faraday micrograph for a (100) slice of $YFeO_3$, but this also exhibited birefringence, the contrast only being optimized by use of retarders and varying as the crystal was rotated about the beam axis, connected with the existence of the spin ordering axes $\perp M_s$ in this canted-spin material. When this crystal was relatively roughly polished faint domains in a surface layer remained when the bulk was saturated and

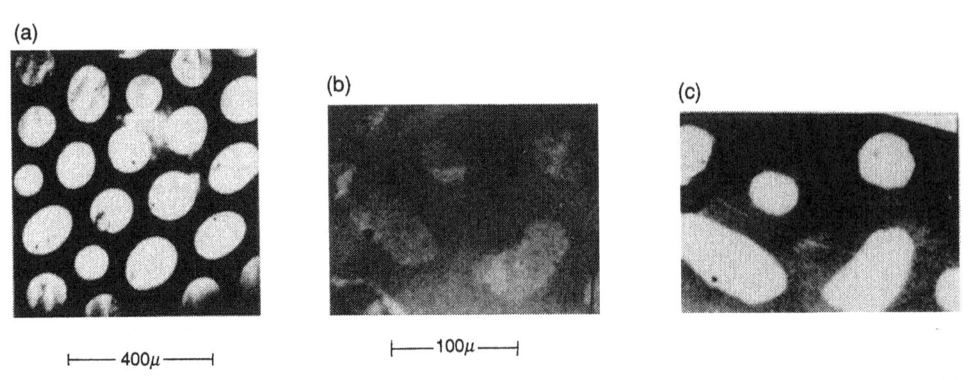

Figure 6.2 (a) Bubble domains in an ion-beam polished slice of $YbFeO_3$: Faraday contrast with compensation. The walls move freely. (b) Low-contrast 'ghost' domains in a nominally saturated crystal with poorly polished surfaces and (c) the way in which bulk domains nucleate at the positions of the ghosts, indicating a strong coupling between the normal material and anomalous layers

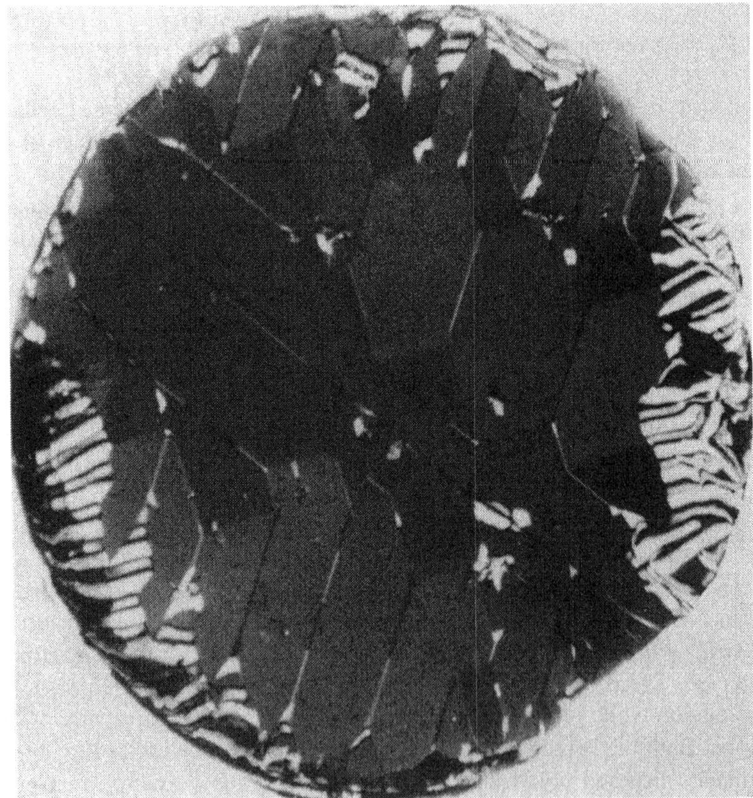

Figure 6.3 A single crystal (110) slice of YIG (two easy directions in plane) between crossed polarizers giving little Faraday contrast for the in-plane domains which are separated by 180°, 110° and 70° walls. The darker and lighter areas interchange on rotating the specimen and most of the domain contrast is due to birefringence. Alternate senses of rotation are shown for some of the walls, the contrast indicating that the polarizers were not perfectly crossed. Striking colour variations occur due to the dispersions (X75)

this exchange-coupled layer controlled the nucleation of new domains. Figure 6.3 shows a substantially strain-free crystal of YIG between nearly crossed polars with some Faraday contrast in the walls, which have positive and negative segments separated by Bloch lines, and birefringence contrast between the central region and the outer regions, which could be reversed by rotating the specimen.

1.1 Magneto-optical Effects on Reflection

Rotations, intensity changes and changes in ellipticity occur on reflecting light from a magnetized specimen. The polar Kerr effect is a relatively large rotation of PP light obtained when **M** is normal to the surface. It is maximum at normal incidence and reverses if **M** reverses. The longitudinal Kerr effect is a rotation observed when **M** is $\parallel$ the surface and the plane of incidence. If **M** is $\parallel$ the surface but $\perp$ the plane of incidence the transverse Kerr effect is observed, as a change in intensity dependent on the direction of **M** in the surface. A quantum theory giving Kerr effects linear in the magnetization was developed by Vonsovskii and Sokolov [6] and further reference should be made to Sokolov [7].

Refinements of laser systems favours high-density optical recording. A report by Roy Carey for the DTI in 1987 listed > 1000 publications on magneto-optical systems. A thin metallic magnetic film, e.g. ferrimagnetic GdCo with uniaxial anisotropy (axis $\perp$ the surface) due to a particular columnar microstructure, is laid down on a transparent substrate. Equipment has been available for some time for the production of films by sputtering or evaporation in UHV or controlled atmospheres, with precise compositions, multilayer or compositionally varied. A pulsed laser beam ($\sim$ 10 mW) is focused to a spot $\sim 1.0\ \mu$m on the film, heating it through 200 K. (It is even contrived that only some part of the thickness of multilayer films is substantially heated in some developments.) In the presence of reverse fields of $\sim 10^4\ \mathrm{A\,m^{-1}}$ a minute reversed region in the previously saturated film is formed, with assistance from the stray fields of the surrounding material in which M remains high, and with $H_c \sim 10^5\ \mathrm{A\,m^{-1}}$ this remains stable. A track of such domains, or lack of them, constitutes binary information written on to a spinning disc and may be detected (using a 1 mW laser) or observed as in Figure 6.4 by the Kerr effect. Reference to the recent literature will soon reveal the complexities of the processes occurring in the

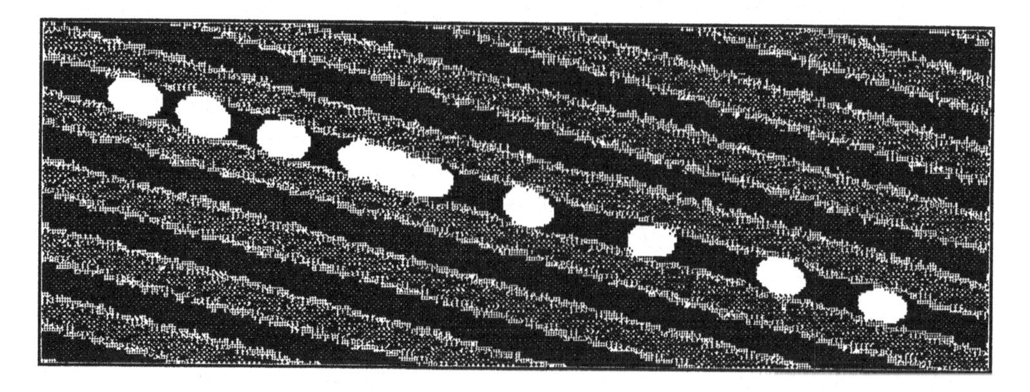

Figure 6.4 Reversed regions ($\sim 1.0\ \mu m$) of magnetic film on a grooved disc, as revealed by the Kerr effect and appropriate to magneto-optical recording [Courtesy Roy Carey, Coventry University]

exchange-coupled multilayer films studied. Apparently Saito *et al.* first demonstrated direct overwrite by light power modulation as such using TbFe/TbFeCo or TbFe/GdTbFe films [8]. Practical stores are available with advantages including the absence of the near-contact of the head and substrate involved in conventional high-density magnetic recording.

2 Geomagnetism, Palaeomagnetism and Archaeomagnetism

Many igneous and sedimentary rocks contain fine, high-H_c particles of, for example, magnetite, titanomagnetite $Fe_{3-x}Ti_xO_4$ and haemo-ilmenites $(1 - y)Fe_2O_3 \cdot yFeTiO_3$ (igneous) or γ-Fe_2O_3, α-Fe_2O_3, α-FeOOH and γ-FeOOH (in the latter). The magnetic field of the earth is that which would be given by a (geomagnetic) dipole at its centre, and this induces remanences on cooling through the Curie points (thermoremanent, TRM), since just below T_c the anisotropy is very low, on sedimentation (depositional, DRM), or on growth of the crystallites through the superparamagnetic blocking volumes (chemical, CRM). Values range from 10^{-2} kA m^{-1} (igneous) to 10^{-7} kA m^{-1} for sediments (see, for example, Creer *et al.* [9]). Measurements of $\mathbf{M}_r$ combine with fossil evidence etc. to give a past record of the geomagnetic field which is thus, in this respect, virtually a unique feature of the environment. The relevance of measurements of the current directions of $\mathbf{M}_r$, over large areas, to plate tectonics or continental drift is apparent.

Although direct measurements since the 1800s indicate a decrease of $\sim$ 5 per cent, the long-term average magnitude of the field seems to have been constant for about 3000 million years [10]. One of the most remarkable of all discoveries relating to the environment, however, is that measurements on successive slices cut from long cores from the sea bed or from solidified lava show that $\mathbf{M}_r$ remains roughly parallel to a common axis but reverses periodically w.r.t. the position along the core, i.e. the date of formation. Many ingenious studies, notably by Néel [11], succeeded in showing that a remanence could be induced antiparallel to a field, e.g. if one set of particles adopted a normal remanence direction but influenced, by stray fields or by exchange, a second set with lower T_c and eventually the higher magnetization, in such a way that the latter adopted a reverse remanence. Reverse remanence has even been achieved in model systems but it seems unlikely that the effect should of a sufficiently common occurrence to avoid the conclusion that the geomagnetic dipole has in fact reversed many times over the relevant period of several million years. There seem to be no indications as to whether a further reversal is imminent.

TRMs occur in hearths, kilns and pottery and if an independent time scale is afforded by carbon or K/Ar dating then measurements reinforce the geomagnetic record on a 'recent' time scale, or the last date of use of a site may be inferred if the record is accepted. In a related context, if a gradiometer is traversed across a site it is possible to detect where past soil disturbance has affected $\mathbf{M}_r$ locally and to map out historic foundations or ramparts without disturbance. Forensic applications are apparent.

3 Superconductivity: the Magnetization Model

Liquid He boils, at atmospheric pressure, at 4.2 K and N_2 at 77 K, and hot objects placed in such liquids are brought to T_B and held there until the liquid boils away. N_2 is much cheaper and more readily handled and has much higher latent heat, i.e. is more 'persistent'.

In 1911, with access to liquid He, Onnes found that the resistivity of mercury fell abruptly, apparently to zero, on cooling to a critical temperature $T_c \sim T_B^{He}$. Following this a great number of metals, alloys and even organics were found to superconduct with T_c values up to $\sim$ 20 K (for alloys of niobium for example). Currents set up in rings have persisted for decades. An early and obvious application was the construction of superconducting solenoids for which the fields are limited only according to the observation that critical current densities J_c and critical ambient fields B_c exist above which the superconductivity disappears, the two features interacting. The B_c values depend on the temperature and plots of $B_c(T)$ separate regions of normal and superconducting behaviour, as in Figure 6.5; materials with much higher critical fields are available and 5 T solenoids are commonplace, with 10 T solenoids being quite common. Advances in cryogenics have favoured widespread utilization, particularly for installations requiring infrequent cool-down, such as for routine n.m.r.

The observation that the presence of induction fields may destroy the superconductivity, together with certain magnetic measurements, suggests that B is excluded in the superconducting state, a phenomenon known as the Meissner effect [12]. (This is qualified later.) The most obvious way to account for this is to ascribe a relative permeability of $\mu_r = 0$ or equally $\chi = -1$ to the material in the superconducting state. However $M = \chi H = -H$ should not be taken to imply the existence of a magnetization comprising a dipole array. This is just the magnetization model, as discussed, for example, by Gregory [13] and in some ways it is the converse of the current-sheet model applicable to ferromagnetics. The statement that superconductors are perfect diamagnetics does not seem to be helpful: in the limit $S \to \infty$, $\chi \to \infty$ for a paramagnetic and the limiting diamagnetic might be supposed to be characterized by $\chi \to -\infty$.

It is convenient to introduce an observation that might be made on established superconductors but is greatly facilitated by the striking advances made, in 1986, by Bednorz and Muller [14], (see also Wu *et al.* [15]), introducing the high-temperature or 'N_2' materials, initially Y $Ba_2Cu_3O_{7-\delta}$, with T_c clearly above $T_B^{N_2}$. These are readily fabricated by simply mixing and sintering constituent oxides (or $BaCO_3$), through very porous unless particular care is taken, and lack the extreme sensitivity to paramagnetic impurities which applies to the metals. A block, say 2 × 2 × 1 cm, is placed in a dish into which N_2 is poured. When the boiling is quiescent a specimen magnet, say 5 × 5 × 2 mm, is held 1 cm above

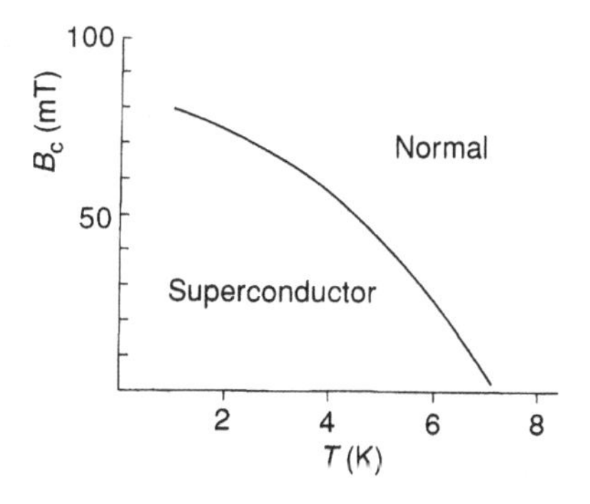

Figure 6.5 A critical curve, as for lead, indicating points at which the behaviour becomes normal on increasing the temperature or the field

the 'YBCO' and on release it remains in a state of stable levitation, returning to a central position after small lateral displacements. Accepting a long-standing conviction that stable levitation, without mechanical guidance or active feedback control, of a magnet (implicitly by another magnet or by a coil) is impossible, this is astonishing however often it is demonstrated. [Conventional levitation can be seen to be precluded by the solenoidal nature of the (static) fields involved.]

The levitation must involve magnetic fields, since clearly the forces are magnetic in nature, and such fields can only arise from dipoles or from current distributions generally. Regarding the specimen as a set of rings, the approach of the magnet might appear to induce currents in these rings which, by Lenz's law, should be such as to give fields and gradients appropriate for repulsion of the magnet, and perfect conductivity might perhaps give an adequate explanation of the observations, since then the currents would be persistent. This is not the case, for if the magnet is held in place when levitated, the temperature is raised above T_c so that the currents cease and T is brought down below T_c once more, the forces reappear without any motion of the magnet to induce the currents. Superconductivity is not just perfect conductivity but is a unique state involving perfect conductivity and other features including a unique magnetic response.

This response may be explained at a certain level by the (effective) magnetization model. The magnet rests with $\boldsymbol{\mu} \parallel$ the specimen surface. The image of a pole $\mathrm{d}p$ in a (semi-infinite) specimen with $\mu_r \to \infty$ is a pole $\mathrm{d}p' = -\mathrm{d}p$ at an equal distance from the interface, within the material. Only this ensures that $H_t = 0$ externally near the surface, as required. Integrating with $\mathrm{d}p = \mathbf{M}\cdot\mathbf{n}\,\mathrm{d}s$, the image of a magnet is a reflected magnet with reversed $\mathbf{M}$. If it is accepted that $\mathbf{B} = 0$ within a superconductor so that $B_n = 0$ externally adjacent to the surface, then the reader may see by a simple sketch that the external fields must be those that would be given by an image of a pole $\mathrm{d}p$ which is $\mathrm{d}p' = +\mathrm{d}p$ and that, on nominal integration, the image of a magnet is the true reflection of the magnet across the interface. It is then apparent why the forces are repulsive and why stability corresponds to $\boldsymbol{\mu} \parallel$ the surface. For a block of finite size numerical approximations are required. Figure 6.6 shows the close approach of the fields from a 'line dipole' reflection to the values to be expected for the perfect image, obtained by calculating the effective magnetization distribution in accordance with $\chi = -1$. Figure 6.7 shows the calculated forces on a levitated magnet according to a magnetization model. The measured height for a magnet of 0.37 g was 80 per cent of the prediction and while the force discrepancy was greater it could be largely accounted for by an extreme porosity.

It is to be stressed again that there is no magnetization here in the accepted sense: a distribution of current densities is created on the crystallite surfaces, which is equivalent to a magnetization distribution. The obvious question which arises is why it is not possible to reproduce the calculated distribution by real magnets, or by coils, and thus achieve stable, purely magnetic, levitation. However, such a system would be static and lack the appropriate response to displacements of the levitated magnet which arises in the superconducting case.

Any such account is both limited and approximate. It may be noted that a levitated magnet spins very freely about its axis so long as the cross-section is circular or square (the reader may show that in the latter case also the fields from the magnet remain static), but oscillations following lateral displacements are rapidly damped out. In poor specimens, identified as such by a low levitation height, there is no equilibrium magnet position and within limits the magnet remains in the position in which it is placed. The inability of the

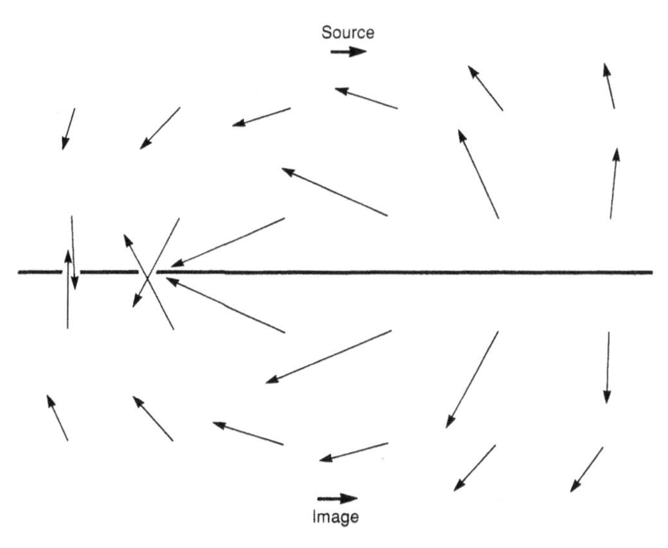

Figure 6.6 The arrows below the line show the field due to the source dipole in that region, not in the superconductor itself. Those above the line are the fields generated by the effective magnetization induced in the superconductor and are clearly equivalent to those from the image dipole indicated

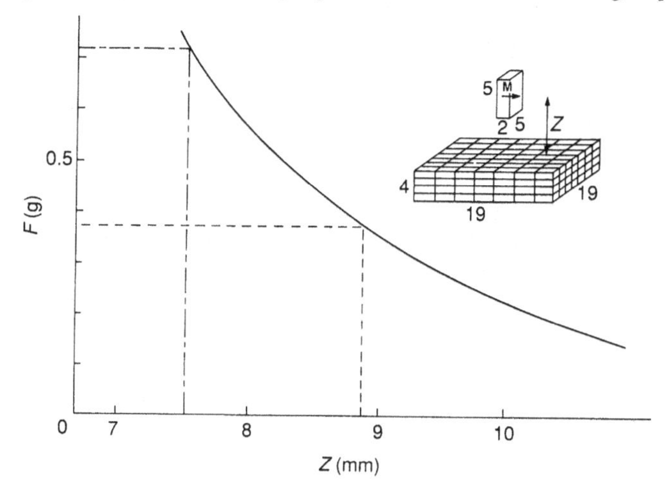

Figure 6.7 The levitation forces according to the magnetization model for the system inset, with lengths in mm. The predicted height was ~ 9 mm and the discrepancy with the observed 7.5 mm was ascribed to extreme porosity. Type I behaviour, in the relevant fields, was assumed

induction or flux $\phi(\mathbf{B} \times$ area $\perp \mathbf{B})$ to penetrate a superconductor leads to the possibility of 'compressing' or 'diluting' **B** by moving superconducting members. If a lead bag is initially folded flat, cooled below T_c and expanded, the initial internal flux remains and B is correspondingly reduced [16]. A second closed bag is inserted in the first and opened out and so forth, and when the initial B has a very small value it is abruptly reduced to zero (identically) by the operation. This occurs when the pre-existing flux corresponds to half a flux quantum or fluxon: $\phi_0 = h/(2e) = 2.07 \times 10^{-15}$ T m^2(2.07×10^{-7} G cm^2), i.e.

it is found that the flux through a closed superconducting ring is quantized. This follows from the relation (Chapter 2) $\mathbf{p} = m\mathbf{v} + q\mathbf{A}$, taking integrals around the ring, within the superconductor where $\int \mathbf{v} \cdot \mathrm{d}\mathbf{l} = 0$, transforming $\int \mathbf{A} \cdot \mathrm{d}\mathbf{l}$ to $\int \nabla \times \mathbf{A}\, \mathrm{d}\mathbf{s} = \int \mathbf{B} \cdot \mathrm{d}\mathbf{s} = \phi$ by Stokes and considering $\mathbf{p}$ to be quantized; $q = -2e$ here because the electrons form Cooper pairs, a blunt statement that points to the inadequacy of such a brief treatment. The tendency of ϕ through a ring to remain constant leads to flux trapping in porous specimens and this can even lead to the suspension of a magnet by a superconducting block, or vice versa, in certain circumstances, as well as the points noted above.

Since the abrupt change of $\mathbf{B}$ to zero on crossing an interface would require the existence of surface current sheets of zero thickness and thus $J \to \infty$, some sort of transition layer is to be expected. London and London [17] inferred that the current (Am^{-3}) should be related to the potential by $\mathbf{J} = -1/(\mu_0 \lambda^2)\, \mathbf{A}$ with λ a length. Taking $\mathbf{A} = \mathbf{j}A$ and varying with x only, $\mathbf{B} = \nabla \times \mathbf{A} = -\mathbf{k}\, \mathrm{d}A/\mathrm{d}x$. If $\mathbf{B}_\mathrm{a}$ in air is $\parallel OZ$ then B is reduced progressively in the material. $\nabla \times \mathbf{J} = -1/(\mu_0 \lambda^2)\, \mathbf{B}$ and also $\nabla \times \mathbf{B} = \mu_0 \mathbf{J}$ so that $\nabla \times (\nabla \times \mathbf{B}) = -(1/\lambda^2)\, \mathbf{B}$. On expanding the left-hand side of this and also applying ∇ to the scalar $\nabla \cdot \mathbf{B}$ and ∇^2 to the expansion of $\mathbf{B}$ it can be seen at a little length that $\nabla \times (\nabla \times \mathbf{B}) = \nabla(\nabla \cdot \mathbf{B}) - \nabla^2 \mathbf{B} = -\nabla^2 \mathbf{B}$ since $\nabla \cdot \mathbf{B} = 0$. Thus $\nabla^2 \mathbf{B} = 1/\lambda^2\, \mathbf{B}$ and in one dimension this gives $B = B_0 \mathrm{e}^{-x/\lambda}$ $B \to 0$ only at depths exceeding the London penetration depth which is, however, very small—$\sim$ 50 nm for metals.

The Cooper electron pairs are Bosons which can all be represented by a common wave function (with a common phase) within any one specimen. If two superconducting members are distinct but connected by a 'weak link', e.g. an oxide layer $\sim$ 1 nm thick or an extreme constriction in a film, the phases in the two are not the same but are related, as indicated by Josephson [18]. Moreover, the phases are exceptionally sensitive to the presence of magnetic fields with consequences as in the following section.

The introduction of the (relatively) high-T_c materials has stimulated interest since 1986 and focused attention on the excessive wastage of power in conventional power transmission and conversion which might be circumvented, the feasibility of levitated transport systems and potential uses in many existing and projected technologies. Choosing examples to rectify previous omissions, in magnetic filtration systems fields are applied to grids of fine stainless steel wires to give strong field gradients, due to the geometry and strong paramagnetism, so that magnetic particles are attracted to the wires and may be subsequently released by removing the fields. This applies to the separation and improvement of minerals, e.g. china clay, and organics may be removed via absorption on to colloids. Low-T_c materials have been used to produce the fields and replacement by high-T_c materials considered: $J_\mathrm{c} = 10^3\ \mathrm{A\,cm^{-2}}$ at $T = 1$–2 T is adequate here with $10^4\ \mathrm{A\,cm^{-2}}$ at 4 T required for commercial success in motors, transformers, generators and energy transmission and storage devices. (Non-magnetic metal particles can be separated by exploiting the eddy current dipoles induced by motion in a field.) Magnetohydrodynamics involves the thrust on a current-carrying liquid due to Lorenz forces and scattering processes and the new materials would favour realistic marine propulsion systems.

4 Magnetic Field Measurements

The current–voltage characteristics measured across a superconducting ring containing two Josephson junctions or weak links depend on the extent to which $\phi = n\phi_0$ is satisfied, the quantization condition being relaxed here. This permits field measurements by quantum counting with accuracy $\sim 10^{-11}$ T, by these superconducting quantum interference devices or SQUIDS. Measurements $\sim 10^{-15}$ T Hz$^{-1/2}$ may be achieved by feedback and locking methods and by detecting the flux through a larger superconducting ring, transformer-coupled to the SQUID. Since the flux specifically through the ring is involved, these are vector instruments. Gradiometers need not involve two SQUIDS but utilize interconnected pairs of pickup loops and second-derivative ($\partial^2 B/\partial z^2$) gradiometers, specially useful in unshielded environments, are described, for example, by Clarke [19]. Susceptometers of extreme sensitivity, detecting $\delta\mu \sim 10^{-11}$ A m^2 or 10^4 spins, have been contrived. High-T_c SQUIDS have sensitivities adequate for the detection of biomagnetic fields, as later, and with the development of 'plug-in' refrigeration these may become commonplace, in the teaching laboratory say, outmoding other techniques. Note the comparisons in Figure 6.8.

Hall effect (vector) magnetometers and electronic compasses utilize small crystals of suitable semiconductors carrying high-frequency currents. Small, low-power and low-cost devices consist of thin-film elements (e.g. Permalloy) in a bridge circuit utilizing magneto-conductivity. The angular variation of ρ w.r.t. the direction of **M** normally amounts to 1 per cent. Baibich *et al.* [20] found that the resistance of multilayer films [Fe($\sim$ 3 nm)/Cr($\sim$ 1 nm)] changed by $\sim$ 50 per cent when (large) fields were applied—the giant magneto-resistive effect (GMR). Cr is antiferromagetic and the Fe interlayer coupling is a.f.m. in nature but strong fields induce an overall f.m. state and it is the a.f.m. $\rightarrow$ f.m. transition that gives the GMR. In an unfilled band all energies and momenta may respond to an electric field and the conductivity is limited by scattering. In a field $\parallel OZ$, the electrons are divided into nearly equal groups, α and β; following White [21] these may be scattered differently by the polarized scattering centres. Cr atoms may diffuse into the Fe layers and it is known that Cr atoms in Fe scatter electrons with moments $\parallel$ the Fe moment some 7 times more effectively. The β electrons traverse all the Fe layers readily in the f.m. state and this is all that is needed for high conductivity, but for the a.f.m. state an unfavourable layer is encountered for both β and α electrons as the first or second crossed. The effect has potential for magnetoresistive read heads when the density exceeds 1 Gbyte inch^{-2}, on 2 inch discs, sensing the field rather than dH/dt being advantageously speed independent.

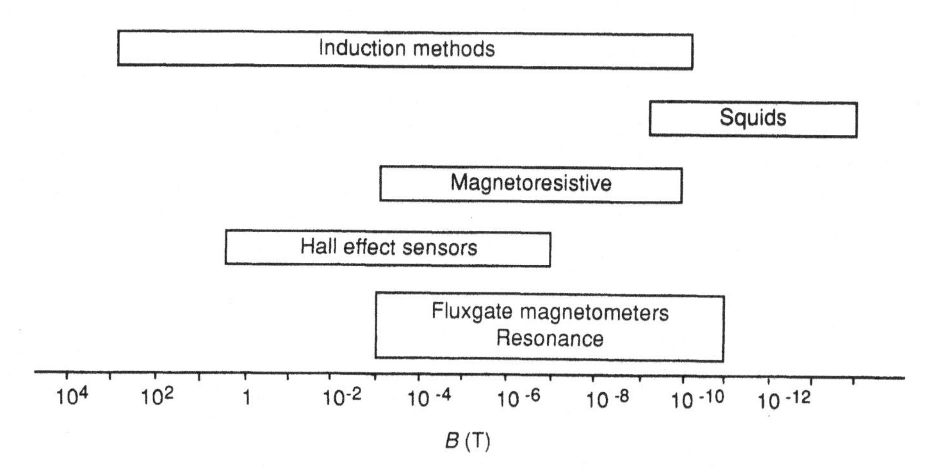

Figure 6.8 Indicating the scope of field measuring devices

With known γ's, e.s.r. and n.m.r. observations constitute field measurements, but the sensitivity would be considered very low for fields of ~ 0.1 mT, say the geomagnetic field. For n.m.r., sensitivity may be regained by using a large volume of water as the specimen. For e.s.r. an optimally concentrated spin system is afforded by, for example, the donor–acceptor complex Ac $(TCNQ)_2$, TCNQ, or tetracyanoquinodimethane, being the electron acceptor, as by Duret *et al.* [22], who obtained a resolution of 0.2 nT $Hz^{-1/2}$ with good temperature stability.

In fluxgate magnetometers || needles of Permalloy are driven in opposition to give zero signal in a common pickup coil, in $B = 0$. On drawing out the loops shifted relatively by a field along the axis it may be seen that a second harmonic signal arises, which may be annulled by a d.c. feedback in the coil. The needles may economically be replaced by a ring core about 1 cm in diameter and such a model was placed on the moon's surface by the astronauts to transmit readings of the ambient field; perhaps it still does so. Fluxgates with minimal power requirements utilize non-linearity of differently biased strips.

5 Molecular/Organic Magnetic Materials

Salts or charge-transfer complexes of electron donors D ($\rightarrow D^{\cdot +}$) and acceptors A ($\rightarrow A^{\cdot -}$) may exhibit high metallic or semiconductivity, or even superconductivity, and interesting magnetic properties associated with the unpaired electrons. Some are quasi-one-dimensional due to segregated stacking of the D and A molecules. $g = 2$ applies closely in most cases. A common magnetic response is that described in Chapter 1, near-Curie behaviour at high T with a peak in χ and a drop to very low values as T is reduced, indicative of Heisenberg coupling with a negative $\mathcal{J}$. χ depends on $\mathcal{J}/T$ and the peak indicates the value of $\mathcal{J}$, over several orders of magnitude. The materials have high diamagnetic contributions to χ, distinguished by the lack of temperature dependence, and may be net diamagnetics at high T.

In some materials $\mathcal{J} > 0$ apparently and χ is greatly enhanced. The behaviour of TCNQ–toluidene blue corresponds to a total spin $S \sim 50 \times \frac{1}{2}$, i.e. to superparamagnetism with a response adequate for the suspension of crystals from a magnet at room temperature [23]. Miller *et al.* [24] found magnetic order (with hysteresis and high exchange fields indicated by Mossbauer spectroscopy below 4.8 K) in materials comprising linear chains of alternating matalocenium donors and planar polycyano hydrocarbons : $[Fe^{III}(C_5Me_5)_2]^{\cdot +}[TCNE]^{\cdot -}$: Me = methyl, TCNE = tetracyanoethylene. Complexes of the donor with TCNQ were metamagnetic at low temperatures and antiferromagnetic in fields up to 130 kA m^{-1} with a sharp rises in M in higher fields. Sugano *et al.* [25] had also observed metamagnetic behaviour in organic polymers. Even pyrolytic carbon prepared by thermochemical (1000°C) deposition was shown by Tanaka *et al.* [26] to exhibit hysteresis at 4 K. $S = 20 \times \frac{1}{2}$ was indicated with a blocking temperature above 4 K.

6 Biology and Medicine

The only objective here can be to give a few examples to indicate the breadth of the topic, followed by separate sections on health hazards and on magnetic resonance imaging. In biology the topic that has probably most engaged general speculation is the possible existence of biological compasses as aids to the navigation of bees or pigeons, etc. It has been shown that certain bacteria are 'magnetotactic', i.e. sense (or rather respond to) the geomagnetic field [27]. These live in pools in caves, in total darkness. They have no interest in the declination but are concerned with the inclination, since this facilitates motion towards the surface or the bottom, a necessary feature of their life cycle. The mechanism of the magnetotaxis involves identifiable chains of minute magnetite particles. The effect is passive rather than cognitive and extrapolations would be rash.

Biomagnetism, or neuromagnetism in particular, concerns the study of the fields generated by biological functions involving electrical impulses, as of the (human) brain. Such fields can be measured, vectorially, by the use of SQUIDs and their study has been involved in the diagnosis of focal epilepsy for example.

Hyperthermia consists of raising the local temperature of tissues and some control of malignancy may be achieved. One method, as by Matsuki and Murakami [28], is the implantation of fine ferrite rods partly surrounded by metal rings. The rods concentrate an r.f. field and enhance induction heating of the metal rings. In particular 'soft heating' or automatic regulation of temperature is achieved by employing compositions, presumably MnZn ferrites with high Zn content, which have appropriate Curie points since above T_c the flux concentration disappears. It may be suggested that single-component needles of NiCu alloys might serve the same purpose since $T_c \sim$ body temperature could be achieved and the conductivity is high.

Further developments might just be possible. Albumin-coated magnetite particles (mean diameter of 2 μm) exhibit strong tissue selectivity in the human body in relation to hepatic cancers, e.g. on intravasal infusion [29] and in the gastrointestinal tract. Interest has been in exceptionally low concentrations in relation to contrast enhancement in MRI, as seen later. The author has shown that such colloids can be prepared with $T_c \sim$ room temperature and the question is whether concentrations could ever be adequate for any appreciable hyperthermia to be produced, perhaps by resonant absorption. The answer is probably no, but the potential benefits might still justify investigation, as suggested by the author some time ago. It is certainly possible to target the particles to leukaemia cells with concentrations adequate for the removal of such cells by magnetic filtration, a striking development in its own right [30].

6.1 Potential Hazards Due to Magnetic Fields

Speculation over the possible harmful effects of magnetic fields from power lines, industrial and domestic equipment, etc., has continued over many years. (Radiofrequency and microwaves introduce separate problems, leading to proposals to face whole buildings with ferrite sheets in the vicinity of radar at airports; see the National Radiological Protection Board reports NRPB 238, 239 and 240.) Epidemiology is notoriously difficult, but moves towards causal relationships have been made by McLauchlan and others [31, 32]. Even mT fields substantially affect certain reactions and small direct effects may be amplified by initiation of further reactions, although connections with physiological effects remain to be established. Free radical (f.r.) reactions proceed through short-lived f.r.'s which may be formed by bond breaking:

$$R_1 - \uparrow\downarrow - R_2 \rightarrow R_1 \uparrow + R_2 \downarrow \equiv {}^1RP \tag{2}$$

^{1}RP is a singlet radical pair. The reaction of the hydrocarbon pyrene with dicyanobenzene is photochemical and McLauchlan matched the rate (photomultiplier output) exactly with a slowly varying field of ~ 1 mT.

Most photochemical reactions involve the triplet state, the radicals being formed as a triplet pair with parallel spins: ^{3}RP. These are not reactive as such, antiparallel spins being required to form a stable bond, and one of the spins must reverse while the radicals separate and diffuse back again:

$$^3\mathrm{T} + \mathrm{M} \rightarrow {}^3[\mathrm{R_1R_2}] \rightarrow {}^1[\mathrm{R_1R_2}] \rightarrow \text{products} \tag{3}$$

^{3}T denotes a molecule in the triplet state which combines with a second molecule M to give the RP. As indicated, some singlet pairs return to the triplet state. The necessary triplet $\rightarrow$ singlet (T–S) conversion may occur because the spins on the two constituent radicals experience different local fields due to the fields from neighbouring nucleii (e.g. protons)—the hyperfine effect. The fields seen by the spins on different radicals may also differ in an applied field, presumably due to the different distortions of the orbitals as affecting the nuclear chemical shifts. At fields above 1.0 T the effective applied field differences predominate and 'provide an effective mechanism for T–S interchange' [31].

Low-field effects are to be treated separately. The triplet is Zeeman-split into T_{+1}, T_0 and T_{-1} according to the permitted M_s values [Figure 6.9(a)]. Accepting that, in the zero applied field, the hyperfine coupling can cause spin reversal, when the field exceeds a value at which the Zeeman energy is equal to the hyperfine coupling energy the primary interactions are those between the field and the T_{+1} and T_{-1} members of the triplet. The hyperfine interaction ceases to be effective in relation to these so that it only enables the conversion $T_0 \rightarrow S$. This decoupling takes out two-thirds of the T–S conversions and in view of the magnitude of the hyperfine effects can occur at as small a field as ~ 8.0 mT.

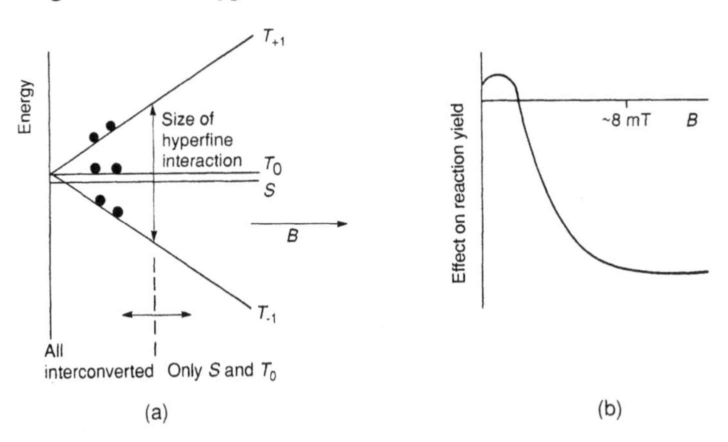

Figure 6.9 (a) Zeeman splitting for a radical pair, the dots indicating the populations produced during the reaction. Below the broken line the energy separations are less than the hyperfine coupling energy and all reaction products formed in the triplet state can be transformed into singlets and react. For fields above this line only the T_0–S separation is small enough for reaction to occur [31]. (b) The drop in yield due to decoupling the T_{+1} and T_{-1} levels from the S level in the higher, but still modest, fields. The small initial rise calls for further discussion [32]

At even lower fields the yield can, conversely, be increased. In the absence of a field the yield is limited by the conservation of the total (election + nuclear) spin, during conversion. For example, for a simple radical pair containing one H atom only and represented as [R_1 R_2 H], the (electronic) singlet state has a total spin $\frac{1}{2}$, due to the proton, and cannot interact with a triplet T_{+1} having total spin $1 + \frac{1}{2} = \frac{3}{2}$, leaving only S–$T_0$ available. However, this spin conservation rule is overcome by even very small fields and the yield may be increased a little as shown in Figure 6.9(b). The effect is general for radicals and corresponds to experimental observation as indicated in the figure. A fluctuating field can excite transitions between T_{+1} and T_{-1} and cause re-coupling so as to increase the yield. This is a resonance effect and is known as reaction-yield-detected magnetic resonance. The general interest of the topic is apparent even though the biological significance remains in doubt.

6.2 A Tentative Magnetotherapy

As a footnote, the following proposition might be considered. It is supposed to be established that the rates of f.r. and perhaps other reactions are substantially affected by magnetic fields, possibly with the further consideration of the frequencies, and that this does have a substantial effect on normal biological functions so as to constitute a health hazard. In this case the question that should be addressed is whether such fields can be applied so as to substantially affect the abnormal functioning of cells or tissues. If a certain phenomenon causes random disturbances of a system it does not follow that it can, conversely, be used to achieve substantial control over the system. However, the possibility exists that if certain, e.g. f.r., reactions were characteristic for abnormal cells only, then some control of the development of such cells might be achieved.

6.3 Magnetopharmaceuticals

MRI maps the (principally) proton concentration and also the spatially varying T_1 and T_2. Recovery of the nuclear magnetization between successive scans may be an essential feature and a short T_1 conventionally brightens the inversion recovery image, while T_2 weighted images may be produced by echo methods. Endogenous contrast may be adequate but otherwise the exogenous agents or magnetopharmaceuticals are employed, with an effect somewhat analogous to tissue staining in optical microscopy. The effect depends on the paramagnetic moments, the relaxation times of the electronic species and the extent to which the paramagnetic moments contact the nuclear moments. (Chelation in a hydrophobic pocket is disfavoured.)

With seven unpaired spins, Gd^{3+} ions have a special role, as when chelated in Gd–DPTA, or gadolinium diethylenetriamine pentacetic acid, an intravenous agent distributing primarily in the intravascular extracellular space and quickly diffusing into the extravascular interstitial space to enhance tissues with a high proportion of interstitial fluid, an ideal agent for brain abscesses and tumours [33]. Where the normal blood–brain barrier is broken down the agent passes into and contrast-enhances the abnormal tissue. The tissue-selective albumin-coated colloids [29] have been referred to: clearly the relaxation effects, in relation to those of the paramagnetics, call for consideration. Soluble t.m. complexes are administered orally and long-lived nitroxide f.r.'s have been extensively investigated [34]. O_2 with, unusually, $S = 1$, is a natural contrast agent, the moment being lost on metabolism with the formation of diamagnetic oxyhaemoglobin, and distinctions are made between the right and left ventricles of the heart after the inhalation of pure oxygen [35].

The continued very substantial development of these agents together with, for example, the EPI methods giving T_1, T_2 and flow weighting, described below, give great potential for functional studies in addition to routine diagnosis.

7 Magnetic Resonance Imaging (MRI)

On the grounds of general value and of the incredible richness of detail of the information afforded, and considering the underlying complexities and the difficulties overcome as well as the continued potential, it is suggested that MRI constitutes the crowning achievement in the field of magnetic studies. The present account may be supplemented by the book by Mansfield and Morris [36] and reviews by Ernst *et al.* [37] or Slichter [38] for example.

Reference will be made to fields substantially $\parallel OZ$, say, but having vector gradients that in certain cases have only one component. Taking, say, a cubic element of space with $\mathbf{B} = \mathbf{K}B$ normal to the lower surface, $\mathbf{B} = \mathbf{k}(B + \delta B)$ over the upper surface and $B_n = 0$ over the other faces it is clear that $\nabla \cdot \mathbf{B} \neq 0$. Interest is in small gradient fields, typically $b_z(\mathbf{r}) = b_z(0) + xG_x + yG_y + zG_z$ with $G_x = \partial b_z/\partial x, G_y = \partial b_z/\partial y, G_z = \partial b_z/\partial z; \mathbf{G} = \mathbf{i}G_x + \mathbf{j}G_y + \mathbf{k}G_z$. With $\mathbf{B}_0 = \mathbf{k}B_0$,

$$\begin{aligned} B &= [(B_0 + b_z)^2 + b_x^2 + b_y^2]^{1/2} \doteqdot (B_0^2 + 2B_0 b_z)^{1/2} \\ &= B_0(1 + 2b_z/B_0)^{1/2} \doteqdot B_0(1 + b_z/B_0) = B_0 + b_z \end{aligned} \quad (4)$$

To first order b_z adds to B_0 but b_x and b_y can be neglected (cf. the high field approximation). As pointed out by Mansfield (personal communication) $\mathbf{G}$ is in fact a tensor and the best approach to controlled linear field gradients is only achieved by minimizing extraneous components such as $\partial B_x/\partial y$, with great practical difficulty.

The realization that n.m.r. observations in linear field gradients could give structural information on a general spin system was reported independently by Mansfield and Grannell [39] and by Lauterbur [40] in 1973. Lauterbur's initial study applied to a space in which a certain region was occupied by water. Suppose for illustration that the gradient can be approximated by a series of increments so that the layers in Figure 6.10 experience fields B_1, B_2, B_3 and B_4. Absorption signals occur at $\omega_i = \gamma B_i$ and with an equal number of spins in each layer these will be of equal height (T_1, T_2 assumed constant). If the spin density $\rho(r)$ varies from sheet to sheet or if the widths of the sheets $\perp \mathbf{G}$ differ, the heights of the peaks differ and as the number of layers increases to represent the continuous field gradient a projection of the total spin density is obtained in terms of the signal height versus ω. Since $\omega \propto B$ and $B \propto z$ in this case, the frequency scale is readily converted to the coordinate z: the $S(\omega)$ plot becomes a plot of spin density in space or one-dimensional image, or rather a projection of the total spin density $\rho(z)$. Combining projections corresponding to differently directed gradients (numerically) a two-dimensional image is produced.

Considering later developments at a simple level, strong signals are observed from protons in water, fat or other soft components of living tissues. At a certain frequency and background field the application of a gradient G_z defines a plane over which the resonance condition applies. If G_z oscillates at a low frequency Ω the plane is replaced by a slice

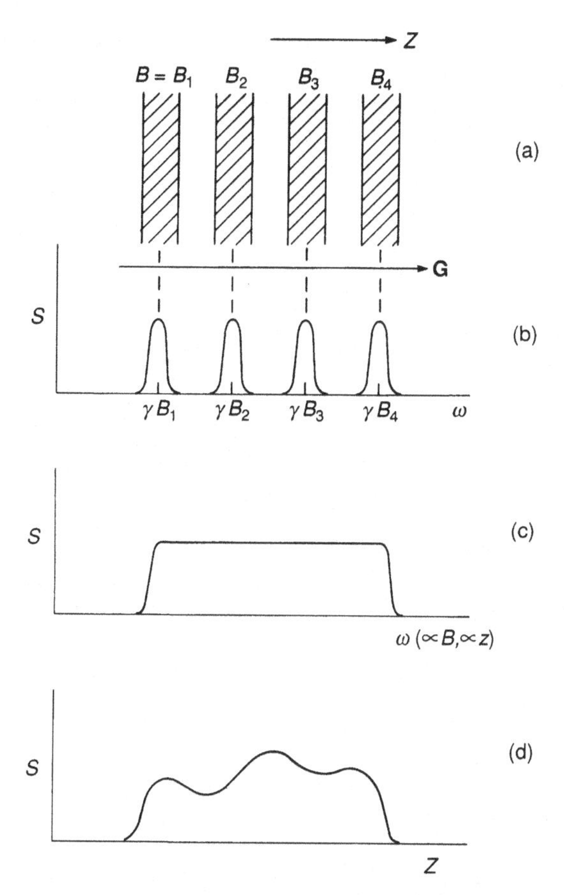

Figure 6.10 A simplified (one-dimensional) view of MRI. The four equivalent layers of material (a) with a certain proton density ρ are taken to be in different fields due to the presence of the gradient **G** and thus to give the four peaks in (b), merging to give (c) as the subdivision is indefinitely increased. With different ρ's (or relaxation mechanisms) for the layers, the peaks would be of different heights and, in the limit, a plot of $\rho(\omega)$ would be produced as in (d). Since $\omega \propto B$ and $B \propto z$, with a linear gradient, the plot becomes $S(z)$: a one-dimensional image of proton density. Differently directed **G**'s give different 'views' which re-combine, numerically, to the three-dimensional image

of thickness $\sim 2\Omega$ in terms of frequency or $2\Omega/\gamma$ in terms of length. Further gradient field components define 'active' elements of space in this slice for which the signal S, a function of proton density ρ and T_1, T_2, can be recorded. $S \approx \rho T_2/T_1$ with $T_2 \approx T_1$ for soft tissues but $T_1 \gg T_2$ for hard tissues, contrasting bones or teeth. With subdivision into $n \times n \times n$ elements image reconstruction requires the solution of n^3 equations and acquisition and processing are very slow. Lauterbur introduced the term zeugmatography for the reconstruction method, but this term is not now in common use.

Mansfield and Grannell commenced with the study of solids, choosing a model system of spaced sheets of camphor and minimizing dipolar and chemical shift interactions by multiple pulse f.i.d. methods. This work gave the basis of a general theory of imaging. Consider that slice selection has been achieved, as below: all the spins in a slice $\perp$ OZ have selectively responded to a 90° pulse and generate a free induction signal decaying according to T_2. Before this, a gradient with components $G_x(t)$, $G_y(t)$ is applied to give a range of fields in the chosen XYZ' plane at time t. The signal from the slice is first regarded as the sum over elements with different ρ_i in which the transverse magnetization precesses in the different local fields B_i with contributions to the signal $A_i e^{i\omega_i t} = A_i e^{i\gamma B_i t}$ with $A_i \propto \rho_i$. Generalizing to continuous variations, with $\rho(\mathbf{r})$ the spin density at $\mathbf{r}$, the response function or signal at time t is (ignoring relaxation)

$$S(t) = \int \rho(\mathbf{r}) e^{\left[i\gamma \int_0^t \mathbf{r}\cdot\mathbf{G}(t')\mathrm{d}t'\right]} \,\mathrm{d}\mathbf{r}$$

$$= \int \rho(\mathbf{r}) e^{i\mathbf{k}\cdot\mathbf{r}} \,\mathrm{d}\mathbf{r} \tag{5}$$

if

$$\mathbf{k} = \int_0^t \gamma \mathbf{G}(t')\mathrm{d}t' \tag{6}$$

Since $\mathbf{G}$ is a field gradient vector, $\mathbf{k}$ is a reciprocal space vector [$k \sim \gamma Gt \sim$ (frequency/length) $\times$ time $\sim$ 1/length]. The expression for $S(t)$, which can equally be denoted $S(k)$ since $\mathbf{k} = \mathbf{k}(t)$, has the form that describes the scattering of a plane wave with $\mathbf{k}$ the wave vector. Thus in the present case $\mathbf{k}$ can be considered the effective wave vector of the fictitious waves of wavelength

$$\lambda = \frac{2\pi}{|\mathbf{k}|} = \frac{2\pi}{k} \tag{7}$$

It is this wavelength, dependent on the field gradients applied, that determines the resolution and certainly not the wavelength associated with the r.f. radiation (in which case, as noted by Ernst *et al.* [37], MRI would just about detect the presence of an elephant). In a formal sense Mansfield thus gives a complete theory of MRI because, as can perhaps be seen more clearly by reducing the problem to one dimension and writing

$$S(k_x) = \int \rho(x) e^{ik_x x} \,\mathrm{d}x \tag{8}$$

this integral is of the Fourier form and thus $\rho(x)$ is the Fourier transform of $S(k_x)$. If $G_x(t) = G_x$ from 0 to t then $k_x = \gamma G_x t$ and in general by varying the times for which different gradient components are applied the whole of k-space can be covered. The formal treatment does not, of course, indicate the way in which the method is implemented in practice.

7.1 Slice-Selective Excitation and Imaging

For diagnosis the requirement is an image in a selected plane or slice through the body. Selective excitation of the spins in a defined slice can be achieved by a method put forward by Garroway *et al.* [41] and developed by Hoult [42] and others [43]. With $B_0 \parallel OZ$ and a selection gradient G_z, irradiation at a single frequency ω_0 would cause coherent rotations, in the ω_0 rotating frame, of the spins in a plane at $z = z'$ for which $\gamma(B_0 + z'G_z) = \omega_0$, giving a signal $\rho e^{i\omega_0 t}$ in an appropriate coil, with ρ the surface density of spins, with negligible intensity. The application of a pulse containing a continuous range of frequencies $\omega_0 \pm \delta\omega$ for an appropriate time can cause rotations through 90°, say, throughout a whole slice of thickness δz around z' with $\delta z \propto \delta\omega$. If all the frequencies had equal intensities the slice would be simply defined. Referring back to Section 2.8, Chapter 3 this can be achieved by employing a sinc-shaped r.f. pulse. All the spins in a slice with a near-rectangular profile can be excited, i.e. rotated through a chosen nutation angle in the rotating frame to give transverse magnetizations and thus free induction signals.

Mansfield and, later, others stressed the importance of basing observations on the f.i.d.'s or echoes. If the slice selected by applying G_z contained protons divided between p different sites, n_i with screening constants σ_i and effective fields B_i, then the FT of the f.i.d. would give p lines at $\omega_i = \gamma B_i$ with heights proportional to n_i, just as the CW spectrum is obtained by transformation in conventional n.m.r. If the protons were equivalent but the field changed regularly across the slice, along OX say, in p discrete steps, and strip i ($\parallel OY$) contained n_i protons, the FT would be analogous to the foregoing except that the different B_i would be those appropriate to the different strips rather than the chemical shift differences, and the signal heights would be proportional to the numbers of protons n_i in the strips $\parallel OY$. As the steps are taken to merge to give a continuous linear gradient G_x the transformed signal thus constitutes a projected image. (The chemical shift differences, e.g. between protons in water and in fats, are largely though not totally overwhelmed by the applied field gradients; see later.)

Conventional two-dimensional FT n.m.r. depends on the existence of two time scales such as the variable τ for spin echoes, giving a time scale t_1, and the detection time, giving t_2, the signal being denoted $S(t_1, t_2)$. If t_1 were fixed then $S(t_2)$ could be Fourier transformed in the usual way. This may again be transformed with a range of t_1 values or with t_1 as the variable to give the double $F(\omega_1, \omega_2) = \mathcal{F}[S(t_1, t_2)]$. In MRI the signal from an element $\mathrm{d}x \times \mathrm{d}y$ is

$$\mathrm{d}S = \rho(x, y)\mathrm{d}x\,\mathrm{d}y\ \mathrm{e}^{i\omega t} = \rho(x, y)\mathrm{d}x\,\mathrm{d}y\ \mathrm{e}^{i\gamma Bt} \tag{9}$$

This is taken to be from a volume element with $\mathrm{d}z$ implicit according to slice excitation. Time scales t_x and t_y are afforded by the times during which G_x and G_y are applied:

$$\begin{aligned} B &= B_0 + xG_x \qquad \text{during time } t_x,\ G_x(t) = G_x \\ B &= B_0 + yG_y \qquad \text{during time } t_y,\ G_y(t) = G_y \end{aligned} \tag{10}$$

Substituting for the value of B and integrating,

$$S(t_x, t_y) = \int\!\!\int \rho(x, y)\mathrm{e}^{i\gamma(xG_xt_x + yG_yt_y)}\,\mathrm{d}x\,\mathrm{d}y$$

Put $\omega_x = \gamma xG_x$, $\omega_y = \gamma yG_y$ so that

$$\mathrm{d}x = \frac{\mathrm{d}\omega_x}{\gamma G_x}, \qquad \mathrm{d}y = \frac{\mathrm{d}\omega_y}{\gamma G_y}$$

$$\rho(x, y) \equiv \rho(\omega_x, \omega_y) \tag{11}$$

and

$$S(t_x, t_y) = \frac{1}{\gamma^2 G_x G_y}\int\!\!\int \rho(\omega_x, \omega_y)\mathrm{e}^{i(\omega_xt_x + \omega_yt_y)}\,\mathrm{d}\omega_x\,\mathrm{d}\omega_y \tag{12}$$

The signal is proportional to the double FT of $\rho(\omega_x, \omega_y)$ and conversely

$$\rho(\omega_x, \omega_y) = \rho(x, y) \propto \mathcal{F}^{-1}[S(t_x, t_y)] \tag{13}$$

i.e. the image is given by the inverse FT of the signal. Two-dimensional FT imaging methods requiring N experiments for an $N \times N$ element image were developed by Kumar *et al.* [44] and Edelstein *et al.* [45] while Mansfield and co-workers developed methods whereby the slice image could be obtained in a single experiment, as below.

The f.i.d. must in general contain information relating to the two axes in the slice and Figure 6.11 (by Mansfield [46]) shows one appropriate timing sequence where (a) and (b) define the slice and (c) indicates a phase encoding gradient as illustrated by Figure 6.12. Figure 6.12(a) and (b) suggests selective excitation. In Figure 6.12(c), G_y is applied and if the spins or magnetizations at $y = 0$ are assumed stationary, in the frame corresponding to $B = B_0 + zG_z$, then after a certain time, with G_y on, the remainder have rotated as shown and phase information is associated with OY.

In Figure 6.11(e) the signal is recorded in the presence of G_x (after removing G_y which has served its purpose) and thus contains spatial x information also. The phase may be stepped by adjusting the magnitude of G_y or the time of application, a one-dimensional FT being carried out for each step. A second transform over the set then gives the image. The

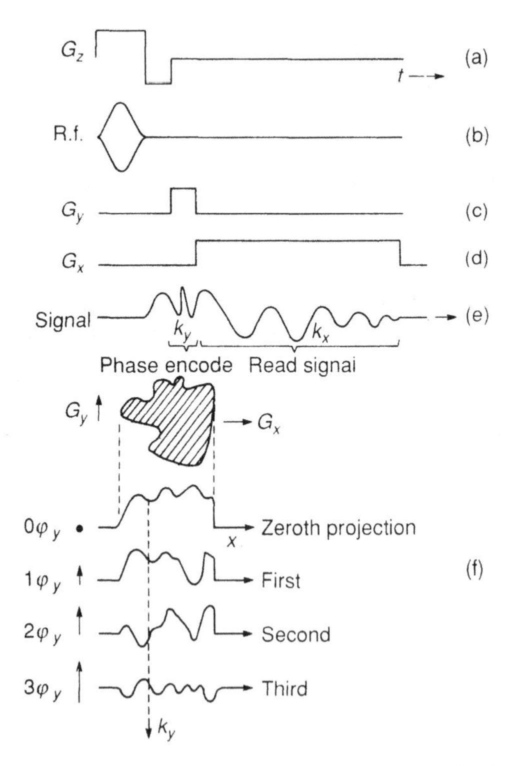

Figure 6.11 A timing diagram and illustration of two-dimensional Fourier transform imaging. (a) Slice selection gradient, (b) selective r.f. pulses, (c) phase-encoding gradient, (d) spatial encoding gradient, (e) f.i.d. signal and (f) an object with various degrees of initial preparation in the phase-encoding gradient [46]

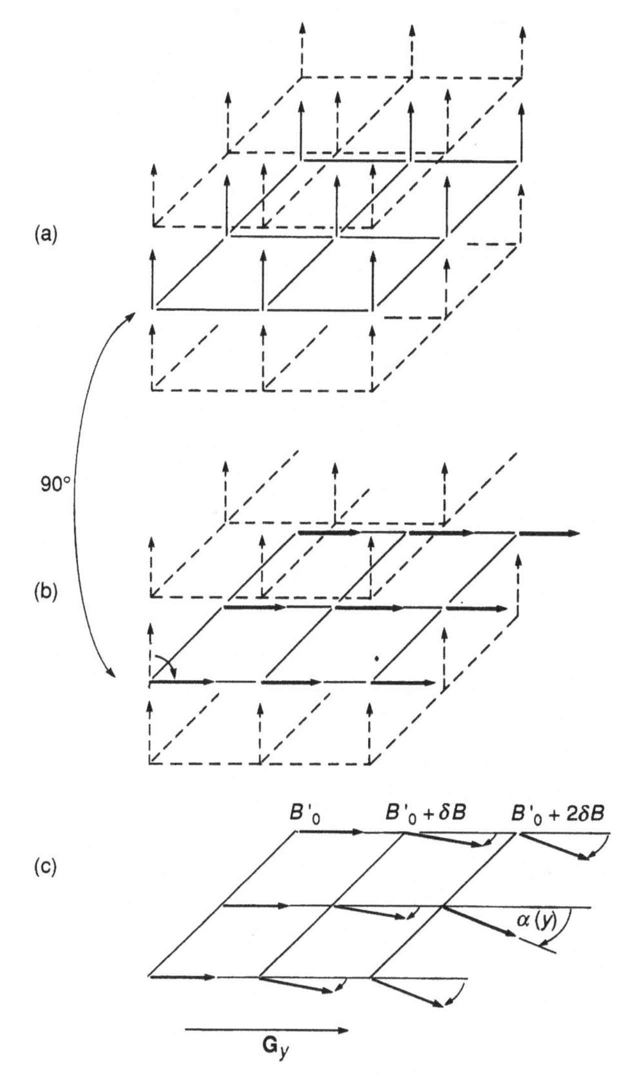

Figure 6.12 (a)→(b) illustrate slice selection, only one layer of 'spins' meeting the condition for coherent rotation with a narrowly defined range of frequencies (and B_0 varying vertically); (c) illustrates the acquisition of a spatially varying phase in an appropriate gradient, to be compared with the illustrations associating the f.i.d.'s with fanning out in random field gradients. (Viewing the arrows as individual spins constitutes a gross simplification). cp Figure 6.11, a, b, c

f.i.d. is sampled for 50–200 ms and, between excitations, time $T_R \sim T_1 \sim 1000$ ms must be allowed for recovery of the equilibrium M_0 in B. Compromise nutation angles, $< 90°$, reduce T_R.

An imaging method is described by the necessity to scan k space, equivalent to covering the plane corresponding to the slice. The two-dimensional FT method is represented by Figure 6.13(a). According to the time of application of G_y a certain point along the k_y axis is attained. The k_x axis is traversed while G_x is applied, followed by a transfer to a new

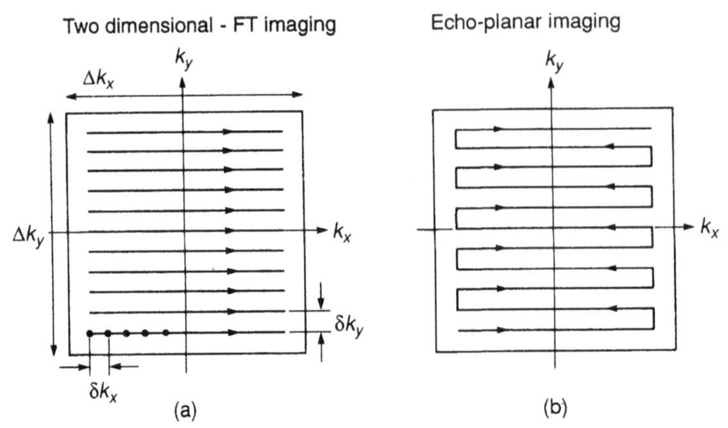

Figure 6.13 In conventional two-dimensional transform methods, (a) k psace is sampled in n consecutive acquisitions along lines parallel to the k_x axis, each line representing a digitally sampled spin or gradient echo. Nominal image resolution or pixel size P is determined by the highest spatial frequencies sampled, $P_x = 2\pi/\Delta k_x$, $P_y = 2\pi/\Delta k_y$. In echo planar imaging [46] the whole of k space is more efficiently sampled in a single continuous trajectory (b) within a fraction of a second. The k space data gives the image in real space by Fourier transformation

k_y value and so forth. By contrast, Figure 6.13(b) shows the continuous k-space trajectory appropriate to echo-planar imaging, or EPI, introduced by Mansfield as early as 1977 [47] (see also Haase *et al.* [48], Haacke and Tkach [49] and Waugh [50] for gradient-recalled echo techniques, GRE). The GREs, in 1985 and 1986, drastically reduced acquisition times from 1–2 hours to $\sim$ 15 s and improvements gave $\sim$ 1 s [51]. Meanwhile, development of EPI, giving images in a fraction of a second with good resolution and refinements to be discussed, was achieved in the early and mid-1980s by Mansfield's group and in the late 1980s by Rzedzian and Pykett [52]. Particular reference should be made to Stehling *et al.* [53] and to Guilfoyle and Mansfield [54] and the works cited. Severe difficulties were involved from the basic subtleties to problems associated with large and rapidly changing gradient fields, calling for active cancellation of eddy currents induced in the magnet structures.

7.2 Flow Encoding and Spin Preparation

Following a 90° pulse, all the magnetization vectors are taken to be $\parallel$ the (rotating) OY' and in G_x applied for time T they adopt a phase angle α, to OY', according to

$$\frac{\partial \alpha}{\partial t} = \omega = \gamma x G_x, \qquad \text{where } \alpha \equiv \alpha(x) = \int_0^T \gamma x G_x \, dt = \gamma x G_x T \tag{14}$$

assuming $\omega = \omega_0$, $B = B_0$ at $x = 0$ and that the spins are fixed in space. If a bipolar gradient is applied as

$$\begin{aligned} G &= G_x, \qquad 0 < t < T \\ G &= -G_x, \qquad T \leqslant t \leqslant 2T \end{aligned} \tag{15}$$

it is clear from equation (14) that $\alpha(x) = 0$: the phase created by the first half of the pulse is destroyed by the second half (Figure 6.14). However, if the spins are associated with flowing molecules, with a component of velocity v in the direction of the field gradient, and if the position is x at $t = 0$ and thus $x + vt$ at time t,

$$\begin{aligned}\alpha &= \int_0^T \gamma G_x(x + vt)\mathrm{d}t + \int_T^{2T} \gamma(-G_x)(x + vt)\mathrm{d}t \\ &= \gamma x G_x T - \gamma x G_x T + \tfrac{1}{2}\gamma G_x v T^2 - \tfrac{1}{2}\gamma G_x v(4T^2 - T^2) \\ &= -\gamma G_x v T^2 \qquad (16)\end{aligned}$$

as in Figure 6.14. The partial phase changes during the positive and negative halves of the bipolar pulse are not equal and opposite, and phase encoding may be achieved: only the static spins remain in phase and become re-focused in an echo. As described by Guilfoyle and Mansfield [54], EPI may then be used to map spin densities *and velocities* and produce quantitative flow or velocity images.

The above process comprises two parts or modules. In the first the bipolar pulse introduces the velocity-dependent phase angle and then an appropriate 90° pulse re-stores magnetizations along OZ which are not just equal to M_0 but are dependent on α and thus on v, so that subsequent image formation indicates the velocity dependence. Alternatively, a simple 90° pulse may be followed by an appropriate delay before re-storing and imaging and the resultant magnetization, and image, is then dependent on the local T_2, with similar procedures

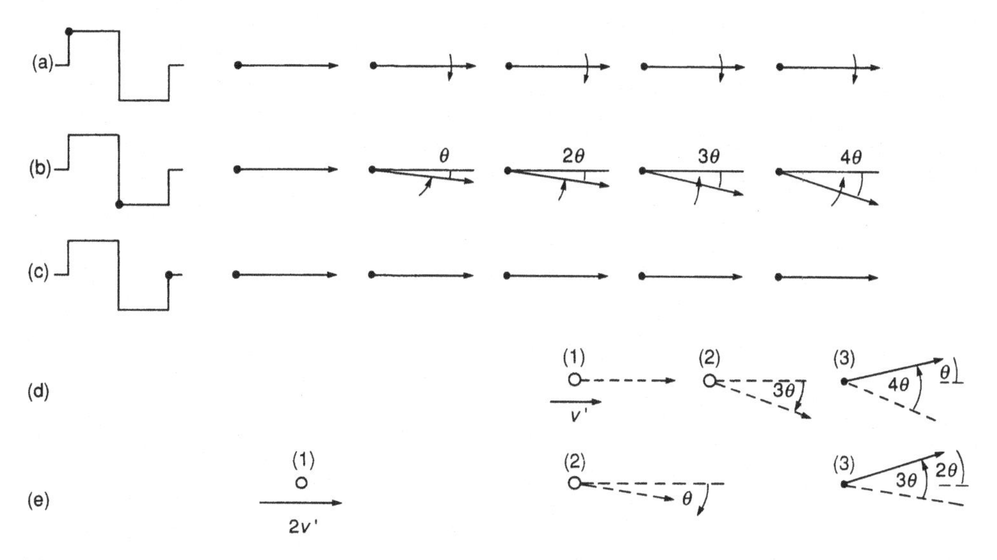

Figure 6.14 Flow encoding, where (a) shows a line of stationary spins at the instant of applying the gradient field as indicated by the dot on the gradient pulse. The progressive phase differences produced in (b) are annulled after the completion of the reverse part of the bipolar pulse (c). For the moving spins, the motion from position 1 to 2 is considered to take place in the positive part of the pulse and that from position 2 to 3 in the second half. In (d) the spin at position 3 has been exposed to different mean field magnitudes while moving from position 1 to 2 and from position 2 to 3 and thus acquires a net phase angle. (e) This indicates the way in which the phase angle acquired depends upon velocity

for T_1 weighting. These are spin preparation modules and with EPI their independence from the imaging module, and the speed, allows the choice of contrast mode, T_1, T_2, flow or velocity to be made interactively. Moreover, the chemical shift differences between water and the olefinic hydrocarbon chains of fat would effectively shift proton densities across the image and modules could be derived to selectively eradicate signals from either group, giving 'water images' or 'fat images' or consequently corrected total proton images by recombination.

7.3 Non-medical Applications

While the central interest is apparent, straightforward applications, as to diffusion for example, are important and will probably become more common, since traditional methods have substantial disadvantages. The diffusion of water into nylon, important to biomedical applications, has been studied [55]. Flow studies relating to water moving through rocks are of particular interest in the oil-recovery field. The high iron oxide contents and field disturbances shorten T_2^* and even though EPI gives a 128×128 element two-dimensional image in 100 ms the standard method, obtaining the echoes by gradient reversal, was inadequate. Thus special pulse sequences suggested by the work of Carr and Purcell were devised for studies of real-time fluid dynamics in porous media by Guilfoyle, Mansfield and Packer [56].

7.4 Medical Applications

Applications in the biomedical or specifically diagnostic field were foreseen by the originators. Recourse to opinions of practitioners confirms that statements such as that by Mansfield [57] that 'commercial (MRI) machines have transformed diagnostic imaging to the point where it is heralded by eminent radiologists as the greatest advance in diagnostic imaging since the discovery of X-rays ... it is the method of first choice in a whole range of pathologies ...' are not to be ascribed to an excess of proprietorial enthusiasm. Also, after years of development the field remains open and further inclusion of the study of ^{13}C and ^{31}P nuclei, etc., should elucidate metabolic processes as well as producing spectroscopic images that pinpoint metabolic disorders in tissues and organs.

In the usual case the spatially varying proton (water, fat) contents and relaxations, in conjunction with magnetopharmaceuticals, give, for example, contrast between soft and hard tissue and between normal and malignant tissue. The first clinical trial documenting intracranial pathology appears to have been carried out using a whole-body prototype resistive magnet instrument by Hawkes, Holland, Moore* and Worthington in Nottingham in 1980 [58]. Reference may also be made to early work by Andrew [59]. By 1987 Worthington [60] could report very favourably from the radiologist's point of view, using 0.15 T resistive systems with full examination times of some 20 minutes. With results as in Figure 6.15(a) and noting as further examples the identification of intervertebral disc protrusions and degeneration, which should replace invasive techniques requiring hospital admission [see Figure 6.15(b)], it was stated that 'MRI is superior to all other methods in assessing the

Figure 6.15 Images (a) of the normal brain and (b) of the spine, showing disc degeneration and protrusion, both by courtesy of B.S. Worthington, Queen's Medical Centre, Nottingham [61]

* Bill Moore's outstanding contributions ended with his untimely death, at 48, in 1984.

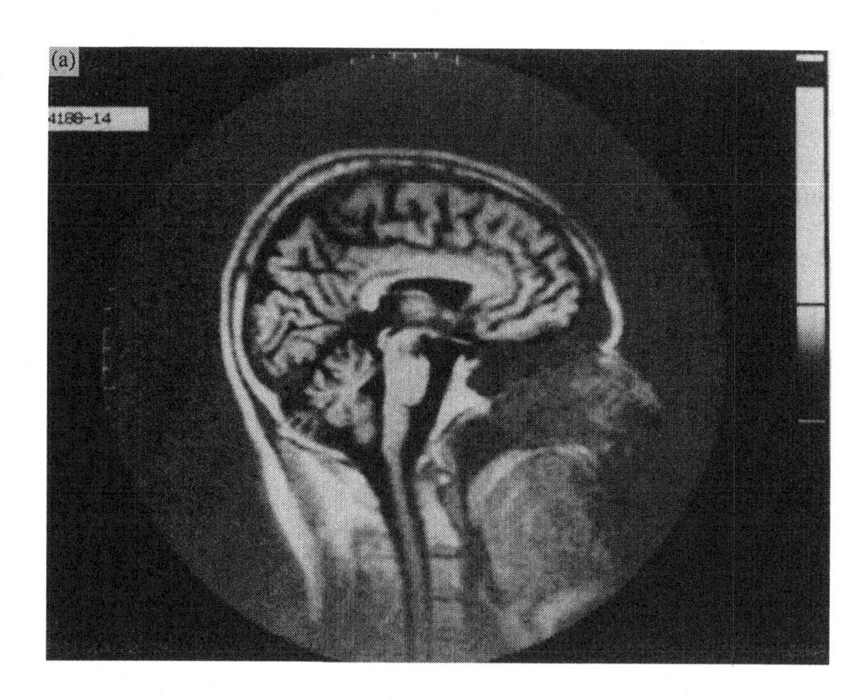
(a)

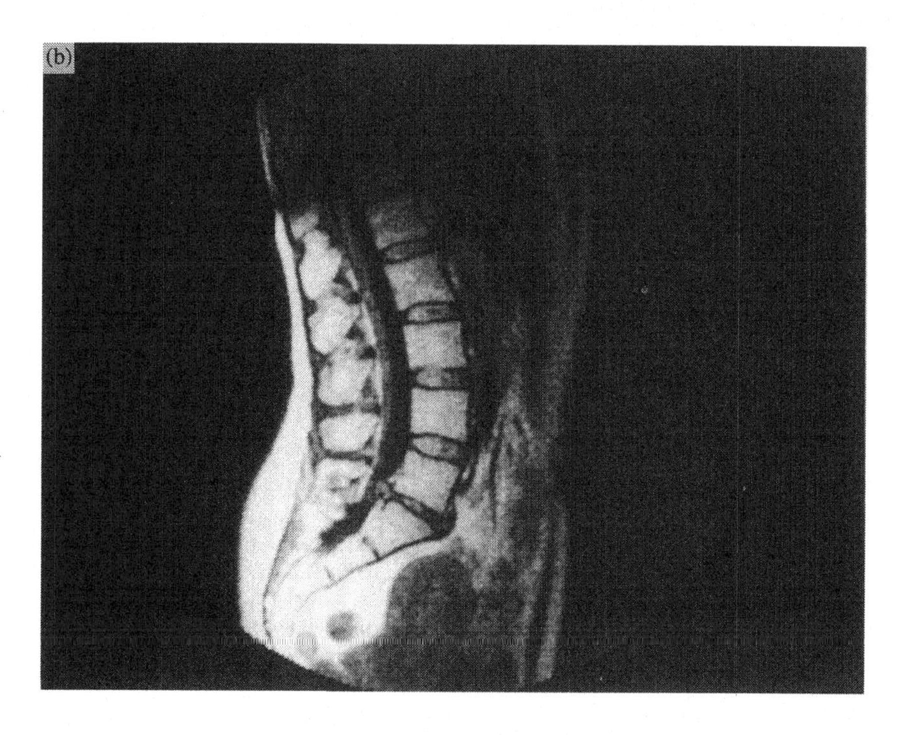
(b)

extension of bone tumours both inside and outside the parent bone' and 'MRI is the method of choice in the evaluation of the craniovertebral junction'. The relatively slow procedure at that time limited patient throughout and could introduce blurring due to involuntary body movements.

Stehling *et al.* [53] list areas in which the ultra-fast EPI has been found to be advantageous: multiplanar movie-loop images of the heart in real time; turbulent flow in heart disease and velocity maps in arteries; infusion rates in the liver, using contrast agents, for diagnosis; molecular diffusion, temperature mapping (via the diffusion constant) and perfusion in the brain. It is possible to depict the rapidly beating foetal heart in real time. The potential for monitoring invasive techniques is apparent, as are the difficulties and possible solutions.

To conclude with a later radiologist's assessment by Brady: 'The future of MR imaging is very bright. The ability to provide functional as well as biochemical information will greatly enhance basic research and clinical applications. It will not only become the cornerstone of radiology, but also will significantly change the focus of biomedical imaging.'

7.5 Concluding Notes

It must be stressed that many vital developments have not even been mentioned: references to fast methods other than EPI, for example, may be found in the works cited. Magnet design and field tailoring could well have received attention (see, for example, [62] and Section 4.5, Chapter 4). Any apparent bias is due to practical considerations. This is a topic of a special nature and its importance transcends questions of personal credit. Economic and social considerations cannot be ignored here and attention to installation cost might well parallel basic developments further.

References

1. Möller, K.D., *Optics*, University Science Books, Mill Valley, California (distributor Oxford University Press), 1988.
2. Kahn, F.J., Pershan, P.S. and Remeika, J.P., *Phys. Rev.*, **186**, 891 (1969).
3. Matsumoto, K., Sasaki, S., Haraga, K., Yamaguchi, K., Fujii, T. and Asahara, Y., *IEEE Trans. Magn.*, **28**, 2985 (1992).
4. Scott, G.B., Lacklison, D.E., Ralph, H.J. and Page, J.L., *Phys. Rev.*, **B12**, 2562 (1975); Shinagawa, K., *J. Magn. Soc. Japan*, **6**, 247 (1982); Simsa, Z., LeGall, H., Simsova, J., Kotacek, J. and LePaillier-Malecot, A., *IEEE Trans. Magn.*, **20**, 1001 (1984).
5. Takahashi, T., Toshima, H., Imaizumi, N., Okuda, T. and Miyazawa, Y., *IEEE Trans. Magn.*, **27**, 5396 (1991); see also Nakajima, K., Numajiri, Y. and Nomi, Y., *IEEE Trans. Magn.*, **27**, 5399 (1991).
6. Vonsovskii, S.V. and Sokolov, N. V., *J. Exper. Theor. Phys. USSR*, **19**, 703 (1949).
7. Sokolov, N. V., *Optical Properties of Metals*, Blackie and Son, London, 1967.
8. Saito, J., Akasaka, H., Birecki, H. and Perlov, C., *IEEE Trans. Magn.*, **28**, 2512 (1992); Saito, J., Sato, M., Matsumoto, H. and Akasaka, H., *Japan. J. Appl. Phys.*, **26**, Suppl. 26-4, 155 (1987).
9. Creer, K.M., Hedley, I.G. and O'Reilly, W., *Magnetic Oxides*, Ed. D.J. Craik, John Wiley and Sons, London and New York, p. 649, 1975.
10. Jacobs, J.A., *The Earth's Core and Geomagnetism*, Pergamon Press, Oxford, 1963; Doell, R.R. and Cox, A., *Advan. Geophys.*, **8**, 221 (1961); Bullard, E.C., Freedman, C., Gellman, H. and Nixon, J., *Proc. Roy. Soc.*, **A243**, 67 (1950); Dell, R.R., *J. Appl. Phys.*, **40**, 949 (1969).
11. Néel, L., *Compt. Rend.*, **254**, 598 (1962).
12. Meissner, W. and Ochsenfeld, R., *Naturwiss.*, **21**, 787 (1933).

13. Gregory, W.D., *The Science and Technology of Superconductors*, (Eds. W.D. Gregory, W.N. Matthews Jr and E.A. Edelsack, Plenum Press, New York and London, p. 25, 1973.
14. Bednorz, J.D. and Muller, K.A., *Z. Phys.*, **B64**, 189 (1986).
15. Wu, M.K., Ashburn, J.R., Torng, C.J., Hor, P.H., Meng, R.L., Gao, L., Huang, Z.J., Wang, Y.Q. and Chu, C.W. *Phys. Rev. Lett.*, **58**, 908 (1987).
16. Brown, R.E., *Rev. Sci. Inst.*, **39**, 547 (1968).
17. London, F. and London, H., *Proc. Roy. Soc.*, **A149**, 71 (1935).
18. Josephson, B.D., *Phys. Lett.*, **1**, 251 (1962).
19. Clarke, J., *IEEE Trans. Elect. Devices*, **ED-27**, 1896 (1980).
20. Baibich, M.N., Broto, J.M., Fert, A., Nguyen Van Dau, F., Petriff, F., Etienne, P., Creuzet, G., Frederick, A. and Chazelas, J., *Phys. Rev. Lett.*, **61**, 2472 (1988).
21. White, R.L., *IEEE Trans. Magn.*, **28**, 2482 (1992).
22. Duret, D., Beranger, M. and Mousavi, M., *IEEE Trans. Magn.*, **28**, 2187 (1992).
23. Batty, S.V., Clegg, D.W., Simmonds, D.J., Craik, D.J., Qureshi, S. and Willis, M.R., *Mol. Cryst. Liq. Cryst.*, **218**, 235 (1992).
24. Miller, J.S., Epstein, A.J. and Reiff, W.M., *Chem. Rev.*, **88**, 201 (1988); *Science*, **24**, 40 (1988).
25. Sugano, T., Kinoshita, M. and Shirotani, I., *Solid State Commun.*, **45**, 99 (1983).
26. Tanaka, K., Kobashi, M., Sunekata, H., Takata, A., Yamabe, T., Mizogami, S. and Kawahata, K., *J. Appl. Phys.* **71**, 836 (1992).
27. Frankel, R.B. and Blakemore, R.P., *J. Magn. Magn. Mater.*, **15–18**, 1562 (1980) and *Bioelectromangnetics*, **10**, 223 (1989); Blakemore, R.P., *Science*, **190**, 377 (1975) and *Annu. Rev. Microbiol.*, **36**, 217 (1982).
28. Matsuki, H. and Murakami, K., *IEEE Trans. Magn.*, **MAG-21**, 1927 (1985).
29. Olsson, M., Persson, B.R.B. and Salford, L.G., *Proceedings of Fourth Meeting of the Society of Magnetic Resonance in Medicine*, Berkely, California, p. 889, 1985; see also Mendonca-Dias, M.H., Bernado, M.L., and Muller, R.N., as above, p. 887.
30. Press release, British Cancer Research Council.
31. McLauchlan, K.A., *Physics World*, **5**, 41 (1992).
32. McLauchlan, K.A. and Steiner, U.E., *Molec. Phys.*, **241**, 73 (1991); Steiner, U.E. and Ulrich, T., *Chem. Rev.*, **51**, 89 (1989); Hoff, A.J., *Photochem. Photobiol.*, **727**, 43 (1991); Frolich, H. (Ed.), *Biological Coherence and Response to External Stimuli*, Springer-Verlag, Berlin, 1988.
33. Carr, D.H., Brown, J. and Leung, W-L., *J Comput. Assist. Tomog.*, **8**, 385 (1984); Carr, D.H. and Gadian, D.G., *Clin. Radiol.*, **36**, 561 (1985); Grossman, R.I., Wolf, G. and Biery, D., *J Comput. Assist. Tomogr.*, **8**, 204 (1984); Felix, R., Schorner, W. and Laniado, M., *Radiology*, **156**, 681 (1985).
34. Stone, T.J., Buckman, T. and Nordio, P.L., *Proc. Natl. Acad. Sci. USA*, **54**, 1010 (1965); Hoffman, A.K. and Henderson, A.T., *J. Am. Chem. Soc.*, **83**, 4671 (1961); Brasch, R.C., *Radiology*, **147**, 781 (1983); Brasch, R.C., London, D.A. and Wesbey, G.E., *Radiology*, **147**, 773 (1983); Brasch, R.C., Ehman, R.L. and Wesbey, G.E., *Radiology*, **149(P)**, 99 (1983); Koutcher, J.A., Burt, C.T. and Lauffer, R.B., *J. Nucl. Med.*, **25**, 506 (1984); Runge, V.M., Clanton, J.A. and Leukehart, C.M., *AJR*, **141**, 1209 (1983); Brasch, R.C., Nitecki, D.E. and Brand-Zawadzki, M.N., *AJR*, **141**, 1019 (1983); Bydder, G.M., Steiner, R.F. and Young, I.R., *AJR*, **139**, 215 (1983).
35. Alfidi, R.J., Haaga, J.R. and El Yousef, S.J., *Radiology*, **143**, 175 (1982).
36. Mansfield, P. and Morris, P.G., *NMR Imaging in Biomedicine*, Academic Press, New York, 1982.
37. Ernst, R.C., Bodenhausen, G. and Wokain, A., *Principles of Nuclear Magnetic Resonance in One and Two Dimensions*, Clarendon Press, Oxford, 1987.
38. Slichter, C.P., *Principles of Magnetic Resonance*, Harper and Row, New York, Evanston and London, 1989.
39. Mansfield, P. and Grannell, P.K., *J. Phys. C*, **6**, L422 (1973).
40. Lauterbur, P.C., *Nature*, **242**, 190 (1973).
41. Garroway, A.N., Grannell, P.K. and Mansfield, P., *J. Phys. C*, **7**, L457 (1974).
42. Hoult, D.I., *J. Magn. Res.*, **26**, 165 (1977) and **35**, 69 (1979).
43. Mansfield, P., Maudsley, A.A., Morris, P.G. and Pykett, I.L., *J. Magn. Res.*, **33**, 261 (1978); Aue, W.P., Mueller, S., Cross, T.A. and Seelig, J., *J. Magn. Res.*, **56**, 350 (1984); Ordidge, R.R., Connelly, A. and Lohman, J.A.B., *J. Magn. Res.*, **66**, 283 (1986).

44. Kumar, A., Welti, D. and Ernst, R.R., *J. Magn. Res.*, **18**, 69 (1975).
45. Edelstein, W.A., Hutchinson, J.M.S., Johnson, G. and Redworth, T., *Phys. Med. Biol.*, **25**, 751 (1980).
46. Mansfield, P., *J. Phys. E*, **21**, 18 (1988).
47. Mansfield, P., *J. Phys. C*, **10**, L55 (1977).
48. Haase, A., Frahm, J., Matthaei, D., Hanicke, W. and Merbolt, K.D., *J. Magn. Res.*, **67**, 258 (1986).
49. Haacke, E.M. and Tkach, J.A., *Am. J. Roentgenol.*, **155**, 952 (1990).
50. Waugh, J.S., *J. Mol. Spectros.* **35**, 298 (1970).
51. Atkinson, D.J. and Edelman, R.R., *Seventh Annual Meeting of Society of Magnetic Resonance in Medicine*, San Francisco, P. 137, 1988; Haase, A., *Magn. Res. Med.*, **13**, 77 (1990).
52. Rzedzion, R. and Pykett, I.L., *Am. J. Roentgenol.*, **149**, 245 (1987).
53. Stehling, M.K., Turner, R. and Mansfield, P., *Science*, **254**, 43 (1991).
54. Guilfoyle, D.N. and Mansfield, P., *J. Magn. Res.*, **97**, 342 (1992).
55. Mansfield, P., Bowtell, R.W., Blackbond, S.J. and Cawley, M., *Magn. Res. Imaging*, **9**, 763 (1991).
56. Guilfoyle, D.N., Mansfield, P. and Packer, K.J., *J. Magn. Res.*, **97**, 342 (1992).
57. Mansfield, P., *Medical Horizons*, **4**, 17 (1991).
58. Hawkes, R.C., Holland, G.N., Moore, W.S. and Worthington, B.S., *J. Compr. Assist Tomogr.*, **4**, 577 (1980).
59. Andrew, E.R., *Phil. Trans. Roy. Soc. B*, **289**, 471 (1980).
60. Worthington, B.S., *Magnetic Resonance Imaging*, Trent Regional Health Authority Seminar, p. 7, 1987.
61. Powell, M.C., Szpryt, P., Wilson, M., Symonds, E.M. and Worthington, B.S., *Lancet*, **ii**, 1366 (1986).
62. Miyamoto, T., Sakurai, H., Takabayashi, H. and Aoki, M., *IEEE Trans. Magn.*, **25**, 3907 (1989); Battocletti, J.H. and Myers, T.J., *IEEE Trans. Magn.*, **25**, 3910 (1989); Abele, M.G., Chandra, R., Rusinek, H., Leupold, H.A. and Potenziani, E., *IEEE Trans. Magn.*, **25**, 3904 (1989); Halbach, K., *Nucl. Instr. and Methods*, **169**, 1 (1980).
63. Johnson, B. and Tebble, R.S., *Proc. Phys. Soc.*, **87**, 935 (1966) Cooper, R.W., Crossley, W.A., Page, J.L. and Pearson, R.F., *J. Appl. Phys.*, **39**, 565 (1968) Tanake, S., Toshino, T. and Takahashi, T., Jap. *J. Appl. Phys.*, **5**, 994 (1966) Freiser, M.J., *IEEE Trans. Magn.*, MAG-4, 152 (1968) Mee, C.D. and Fan, G.I., *IEEE Trans. Magn.*, MAG-3, 72 (1967).

APPENDIX 1 Units: The SI and the C.G.S. (Gaussian) System

The dimensions and units of the magnetic quantities have been described throughout the text in the SI, but it may still be useful to collect and tabulate these. In particular, a great deal of the past and even the current literature (in books as well as periodicals) makes use of the c.g.s. system and a review of the relations between the two systems remains appropriate.

The c.g.s. system is based on the three units cm, gram and second in correspondence with the three dimensions L, M, T. All quantities may be alternatively expressed in either electrostatic units (e.s.u.) or in electromagnetic units (e.m.u.) cross-related by the factors c, c^{-1}, c^2 or c^{-2}. Since c has the dimensions of velocity such relations are made between unlike quantities. For the mixed c.g.s. or Gaussian system it is decided that all electrical quantities should be measured in e.s.u. and all magnetic quantities in e.m.u. Such a decision carries the implication that a distinction between electrical and magnetic behaviour can, in fact, be made and harks back to the days when this was thought to be reasonable. Thus, for example, if $e = e'$ e.s.u. is the proton charge then, in calculating magnetic moments, currents and charges are converted to e.m.u. and this leads to an expression for the Bohr magneton:

$$\beta^{\text{cgs}} = \frac{eh}{4\pi mc}, \qquad \text{cf. } \beta = \frac{eh}{4\pi m}$$

It would be equally acceptable to write $\beta^{\text{cgs}} = eh/(4\pi m)$ if it was separately specified that e was to be taken to be the charge in e.m.u., but this is not conventional. The [L M T] system is appropriate to mechanics and the dual nature of the c.g.s. system arose from the application to electricity and magnetism.

The vital feature of the MKSA or practical systems, developed by Maxwell and Giorgi and eventually formalized as the SI, is the introduction of a fourth dimension, current, and unit, the ampere (or, equally, charge or resistance), with the realization that no attempt should be made to express the ampere, dimensionally, in L, M, T. No duality is then necessary. It may be convenient to think of the SI as a m, kg, s, A, J system while realizing that the joule J is not, of course, independent ($1\ \mathrm{J} = 1\ \mathrm{kg\,m^2\,s^{-2}} \sim$ force times distance), i.e. to use J as if it were a base unit.

An equation need not involve the same units, so long as it is dimensionally consistent: the units may be balanced by the numbers involved. Thus $l = 1\ \mathrm{m} = 1000\ \mathrm{mm}$ is acceptable, or $l = l'\ \mathrm{m} = l''\ \mathrm{mm}$ with l' and l'' numbers. Accepting $1\ \mathrm{m} = 1000\ \mathrm{mm}$ and multiplying by l', $l'\ \mathrm{m} = 1000\ l'\ \mathrm{mm}$ so that $l'' = 1000\ l'$:

$$\frac{1\ \mathrm{m}}{1\ \mathrm{mm}} = 1000, \qquad \frac{l'}{l''} = \frac{1}{1000} \quad (l'\ \mathrm{m} = l''\ \mathrm{mm}) \tag{1}$$

and there are always the two equivalent ways to express relations between quantities measured in different units though having the same dimensions.

A key relation is that between the c.g.s. (e.m.u.) unit of current which became standardized as the abamp, designated $\mathcal{A}$ (cf. A for ampere) and the SI unit:

Current:
$$\begin{array}{cc} \text{(c.g.s.)} & \text{(SI)} \\ 1\mathcal{A} & = 10\ \mathrm{A} \end{array} \tag{2}$$

From the introduction of the practical systems, it has been accepted that this relation involves a number, whereas the relation between the e.s.u. and the ampere involves c.

There is, of course, no reason to associate the c.g.s. system with the pole concept and since each system is itself consistent, the starting point may, in view of equation (2), be taken as the dipole—directly from the definition:

$\boldsymbol{\mu}$:
$$\begin{array}{cc} \text{(c.g.s.)} & \text{(SI)} \\ 1\mathcal{A}\ \mathrm{cm^2} = (10\ \mathrm{A})(10^{-4}\ \mathrm{m^2}) & = 10^{-3}\ \mathrm{A\,m^2} \end{array} \tag{3}$$

For the intensity of magnetization M:

$\mathbf{M}$:
$$\begin{array}{cc} \text{(c.g.s)} & \text{(SI)} \\ 1\mathcal{A}\ \mathrm{cm^2\,cm^{-3}} = 1\mathcal{A}\ \mathrm{cm^{-1}} & = 10^3\ \mathrm{A\,m^{-1}} \end{array} \tag{4}$$

Recall the reciprocal statement, that if $M = M'\mathcal{A}\ \mathrm{cm^{-1}} = M''\ \mathrm{A\,m^{-1}}$ then $M' = 10^{-3}M''$.

The specific magnetizations, the dipole moment per unit mass, may be related remarkably simply as

$\boldsymbol{\sigma}$:
$$\begin{array}{cc} \text{(c.g.s.)} & \text{(SI)} \\ 1\mathcal{A}\ \mathrm{cm^2\,g^{-1}} = & 1\ \mathrm{A\,m^2\,kg^{-1}} \end{array} \tag{5}$$

For the magnetic field it is tempting to simply reproduce equation (4). However, it has been decided that a factor of $(4\pi)^{-1}$ should be introduced when calculating the field (SI) from a current distribution and thus from a dipole. Thus, for example, at 1 cm along the axis of a dipole $\mu = 1\mathcal{A}\ \mathrm{cm^2}$ the field is equivalently

$$H = \frac{2\mathcal{A}\ \mathrm{cm^2}}{1\ \mathrm{cm^3}} = \frac{2 \times 10^{-3}\ \mathrm{A\,m^2}}{4\pi \times 10^{-6}\ \mathrm{m^3}}$$

and so

$$\mathbf{H}: \qquad \underset{(1\ \text{Oe})}{\overset{\text{(c.g.s)}}{1\mathcal{A}\ \text{cm}^{-1}}} = \overset{\text{(SI)}}{\frac{10^3}{4\pi}\ \text{A}\,\text{m}^{-1}} \qquad (6)$$

Comparing equation (4), this cannot be a correct equation. It may be used as such but is really an equivalence applying only with an obvious qualification.

The susceptibility $\chi = M/H$ is correspondingly affected. In the absence of rationalization it would be assumed that the same numbers should give χ^{cgs} and χ. Suppose that $M'\mathcal{A}\ \text{cm}^{-1}$ or $M''\ \text{A}\,\text{m}^{-1}$ is induced by a field $H'\mathcal{A}\ \text{cm}^{-1}$ or $H''\ \text{A}\,\text{m}^{-1}$, the primes indicating numbers:

$$M' = 10^{-3}M'', \qquad H' = 4\pi \times 10^{-3}H''$$

$$M'' = 10^{3}M', \qquad H'' = (\tfrac{1}{4}\pi)10^{3}H'$$

so that

$$\chi^{\text{cgs}} = \frac{M'}{H'}, \qquad \chi = \frac{M''}{H''} = \frac{10^3 M'}{(1/4\pi)10^3 H'} = 4\pi\frac{M'}{H'} = 4\pi\chi^{cgs} \qquad (7)$$

Thus the number χ which is the susceptibility is 4π times the number χ^{cgs}, which is the suceptibility in the unrationalized c.g.s. system, and $\chi^{\text{cgs}} = (1/4\pi)\chi$. This is equivalent to saying that the 'unit' in which χ^{cgs} is measured is 4π times the unit in which χ is measured:

$$\chi: \qquad \overset{\text{(c.g.s.)}}{1\ \text{'unit of susceptibility'}} = \overset{\text{(SI)}}{4\pi \times \text{'unit of susceptibility'}} \qquad (8)$$

Since no units are, in fact, involved this is a formality giving consistency with the other relations.

The mass susceptibility is a quantity with units $\sim \sigma/H$ and so, more directly,

$$1(\mathcal{A}\ \text{cm}^2\,\text{g}^{-1})(\mathcal{A}\ \text{cm}^{-1})^{-1} \sim 1\ (\text{A}\,\text{m}^2\,\text{kg}^{-1})\left[\left(\frac{10^3}{4\pi}\right)\ \text{A}\,\text{m}^{-1}\right]^{-1}$$

or effectively

$$\chi_{\text{m}}: \qquad \overset{\text{(c.g.s.)}}{1\ \text{cm}^3\,\text{g}^{-1}} = \overset{\text{(SI)}}{4\pi \times 10^{-3}\ \text{m}^3\,\text{kg}^{-1}} \qquad (9)$$

again applying with qualification only, since it is clear that generally $1\ \text{cm}^3\,\text{g}^{-1} = 10^{-3}\ \text{m}^3\,\text{kg}^{-1}$.

For the molar susceptibilities χ_{M} and $\chi_{\text{M}}^{\text{cgs}}$:

$$1\mathcal{A}\ \text{cm}^2\,\text{mol}^{-1}(\mathcal{A}\text{cm}^{-1})^{-1} \sim 10^{-3}\ \text{A}\,\text{m}^2\,\text{mol}^{-1}\left[\left(\frac{10^3}{4\pi}\right)\ \text{A}\,\text{m}^{-1}\right]^{-1}$$

or effectively

$$\chi_{\text{M}}: 1\ \text{cm}^3\,\text{mol}^{-1} = 4\pi \times 10^{-6}\ \text{m}^3\,\text{mol}^{-1} \qquad (10)$$

Thus, for example, for one mole of spins ($S = \frac{1}{2}$, $g = 2$)

$$\chi^{\text{cgs}} = \frac{Lg^2\beta^2}{4kT} = \frac{0.375}{\text{T}}\ \text{cm}^3\,\text{mol}^{-1}$$

$$\chi = \frac{\mu_0 Lg^2\beta^2}{4kT} = \frac{4\pi \times 10^{-6} \times 0.375}{\text{T}}\ \text{m}^3\,\text{mol}^{-1}$$

making use of the values that are usually tabulated as $\beta = 9.274\ldots \times 10^{-24}\ \text{J}\,\text{T}^{-1}$ and $\beta = 9.274\ldots \times 10^{-21}$ erg Oe^{-1}. It is readily checked (see below) that

$$1\ \text{erg Oe}^{-1} = (10^{-7}\ \text{J})(10^4\ \text{T}^{-1}) = 10^{-3}\ \text{J}\,\text{T}^{-1} \tag{11}$$

consistent with equation (3). The dimensions of the e.m.u. current in the Gaussian system are $\text{M}^{1/2}\text{L}^{1/2}\text{T}^{-1}$. It follows that

$$(\text{c.g.s.})\ \mu \times H \sim \mathcal{A}^2\ \text{cm} \sim \text{M}\ \text{L}^2\ \text{T}^{-2} \sim \text{energy} \sim \text{erg}$$

hence the alternative units of erg Oe^{-1} for μ. In the SI, using $\mu_0 \sim \text{J}\,\text{A}^{-2}\text{m}^{-1}$ (or $\text{H}\,\text{m}^{-1}$) we have

$$\mu_0\mu H \sim \text{J}\,\text{A}^{-2}\text{m}^{-1}\text{A}\,\text{m}^2\text{A}\,\text{m}^{-1} = \text{J}$$

or with $\mu_0 H = B$ and thus $B \sim \text{J}\,\text{A}^{-1}\text{m}^{-2} \sim \text{T}$, $\mu B \sim \text{J}$ and hence $\mu \sim \text{J}\,\text{T}^{-1}$. In the c.g.s. system there is no need to introduce fields B and H having different units and dimensions, but since B and H are, in any case, to be distinguished carefully this can scarcely be considered advantageous. The relation

$$B = H + 4\pi M\ (\text{c.g.s.}), \qquad \text{cf. } B = \mu_0(H + M) \tag{12}$$

indicates that (c.g.s.) B, H and M have the same units and dimensions, though the units may be given different names: gauss for M and for B and Oe for H.

Because of the occurrence of (4π) in the relation (12),

$$\mu_\text{r} = \frac{B}{H} = 1 + 4\pi\chi^{\text{cgs}}, \qquad \text{cf. } \mu_\text{r} = \frac{1}{\mu_0}\frac{B}{H} = 1 + \chi$$

and from equation (8) the number giving the c.g.s. susceptibility is smaller by the factor $1/4\pi$ than that giving the SI susceptibility and so

$$\begin{array}{ccc} (\text{c.g.s.}) & & (\text{SI}) \\ \mu_\text{r} & = & \mu_\text{r} \end{array} \tag{13}$$

The units of B follow from its definition in terms of currents and forces, and are consistent with

$$B = \mu_0 H \sim (\text{J}\,\text{A}^{-2}\,\text{m}^{-1})(\text{A}\,\text{m}^{-1}) = \text{J}\,\text{A}^{-1}\,\text{m}^{-2} = \text{T (Tesla)} \tag{14}$$

noting that $BH \sim \text{J}\,\text{m}^{-3}$. Since, in the c.g.s. system $(HM) \sim \text{erg cm}^{-3}$, it is possible to write

$$H^{\text{cgs}} \sim \text{erg cm}^{-3}(\mathcal{A}\ \text{cm}^{-1})^{-1} = \text{erg}\ \mathcal{A}^{-1}\,\text{cm}^{-2}$$

in correspondence with equation (14). The direct conversion may then be made

$$\mathbf{B}: \quad \begin{array}{lcl} \text{(c.g.s.)} & & \text{(SI)} \\ 1\ \text{erg}\ \mathcal{A}^{-1}\ \text{cm}^{-2} & = & 10^{-7}\ \text{J}\ 10^{-1}\ \text{A}^{-1}\ 10^{4}\ \text{m}^{-2} \\ 1\ \text{gauss} & = & 10^{-4}\ \text{T} \\ (1\ \text{Oe} & = & 10^{-4}\ \text{T}) \end{array} \tag{15}$$

since the gauss and the oersted are equivalent in dimensions and units.

For the vector potential, $\mathbf{B} = \nabla \times \mathbf{A}$ indicates that $A \sim \text{J}\,\text{A}^{-1}\,\text{m}^{-1}$ whereas in the c.g.s. system $\mathbf{H} = \nabla \times \mathbf{A}$ indicates that $A \sim \text{erg}\ \mathcal{A}^{-1}\ \text{cm}^{-1}$ and the conversion is

$$\mathbf{A}: \quad \begin{array}{lcl} \text{(c.g.s.)} & & \text{(SI)} \\ 1\ \text{erg}\ \mathcal{A}^{-1}\ \text{cm}^{-1} & = & 10^{-6}\ \text{J}\,\text{A}^{-1}\,\text{m}^{-1} \\ 1\ \text{Oe cm} & = & 10^{-6}\ \text{T}\,\text{m} \end{array} \tag{16}$$

In describing a surface pole distribution as $\sigma = \mathbf{M} \cdot \mathbf{n}$, $\mathbf{n}$ is taken to be a dimensionless unit vector so that $\sigma \sim \text{A}\,\text{m}^{-1}$. For a volume pole density $\rho = -\nabla \cdot \mathbf{M}\ \text{A}\,\text{m}^{-2}$. Insofar as a nominal pole p is required, it follows that $p \sim \text{A}\,\text{m}$, consistent with the fictitious 'magnetostatic dipole' $\mu \sim p \times \text{distance} \sim \text{A}\,\text{m}^2$. The scalar magnetic potential corresponds to $\varphi \sim A$, from $\mathbf{H} = -\nabla\varphi$. Conversions are according to

$$\begin{array}{llcl} & \text{(c.g.s.)} & & \text{(SI)} \\ \sigma = \mathbf{M} \cdot \mathbf{n}: & 1\mathcal{A}\ \text{cm}^{-1} & = & 10^{3}\ \text{A}\,\text{m}^{-1} \\ \rho: & 1\mathcal{A}\ \text{cm}^{-2} & = & 10^{5}\ \text{A}\,\text{m}^{-2} \\ p: & 1\mathcal{A}\ \text{cm} & = & 10^{-1}\ \text{A}\,\text{m} \\ \varphi: & 1\mathcal{A} & = & \dfrac{10}{4\pi}\text{A} \end{array} \tag{17}$$

the last involving rationalization. Note $\mu_0\sigma\phi \sim \text{J}\,\text{m}^{-2}$.

Returning to the susceptibilities, Curie constants are given by $\chi = C/T$, $C = T\chi$ and it follows that, for example,

$$\mathbf{C}_\text{M}: \quad \begin{array}{lcl} \text{(c.g.s.)} & & \text{(SI)} \\ 1\ \text{cm}^3\ \text{mol}^{-1}\ \text{K} & = & 4\pi \times 10^{-6}\ \text{m}^3\ \text{mol}^{-1}\ \text{K} \end{array} \tag{18}$$

Anisotropy constants have the dimensions energy/volume and it is only necessary to note that

$$\mathbf{K}: \quad \begin{array}{lcl} \text{(c.g.s.)} & & \text{(SI)} \\ 1\ \text{erg cm}^{-3} & = & 10^{-7} \times 10^{6}\ \text{J}\,\text{m}^{-3} = 10^{-1}\ \text{J}\,\text{m}^{-3} \end{array} \tag{19}$$

Finally, magnetostriction coefficients $\delta l/l$ are dimensionless and since no rationalization can be involved no conversion arises.

APPENDIX 2 Vectors

1 Elementary Vector Analysis

With vector quantities, those having directional attributes as well as magnitudes, such as displacement, velocity and magnetic field, are associated mathematical vectors $\mathbf{r}$, $\mathbf{v}$, $\mathbf{H}$ and in turn points along a line or arrows on two-dimensional diagrams or in three-dimensional models. A vector is indicated by bold type, as $\mathbf{v}$, and its magnitude by $|\mathbf{v}|$ or just v. A position vector $\mathbf{r}_1$ may be taken to be the displacement of a point P_1 from a specified origin O and then the position of P_2 relative to P_1 [see Figure 1.1(a)] is given according to $\mathbf{r}_1 + \mathbf{r} = \mathbf{r}_2$ by $\mathbf{r} = \mathbf{r}_2 - \mathbf{r}_1$ while $\mathbf{r}' = \mathbf{r}_1 - \mathbf{r}_2 = -\mathbf{r}$ indicates the position of P_1 relative to P_2. $\mathbf{r}$ and $\mathbf{r}'$ clearly do not depend on the choice of O. The elementary rules corresponding to the diagrammatic representations will be assumed.

Vector equations have the obvious advantage of compactness. The simple inverse square law with a source m at $\mathbf{r}_1$ giving a field $\mathbf{h}$ at $\mathbf{r}_2$ is

$$\mathbf{h}(\mathbf{r}_2) = \frac{m(\mathbf{r}_2 - \mathbf{r}_1)}{|\mathbf{r}_2 - \mathbf{r}_1|^3} = \frac{m\mathbf{r}}{r^3} \tag{1}$$

but is otherwise only given as $h = m/r^2$ accompanied by a description of the direction.

A (three-dimensional) vector is also specified by listing its components as $\mathbf{r}_1 \sim (x_1, y_1, z_1) \sim (r_1, \theta_1, \phi_1)$, etc., i.e. by an ordered set of numbers, distinguished from a row matrix as such by the inclusion of commas. Careful distinction must be made between expressions such as $V(\mathbf{r})$ and $V(r)$; V depends on position generally or just on the one coordinate r, respectively. Conventionally $\mathbf{i}, \mathbf{j}, \mathbf{k}$ or $\mathbf{e}_x$, $\mathbf{e}_y$, $\mathbf{e}_z$ denote unit vectors drawn along OX, OY, OZ with $\mathbf{e}_r$, $\mathbf{e}_\theta$, $\mathbf{e}_\phi$ and $\mathbf{e}_\rho$, $\mathbf{e}_\phi$, $\mathbf{e}_z$ appropriate to spherical or cylindrical coordinate

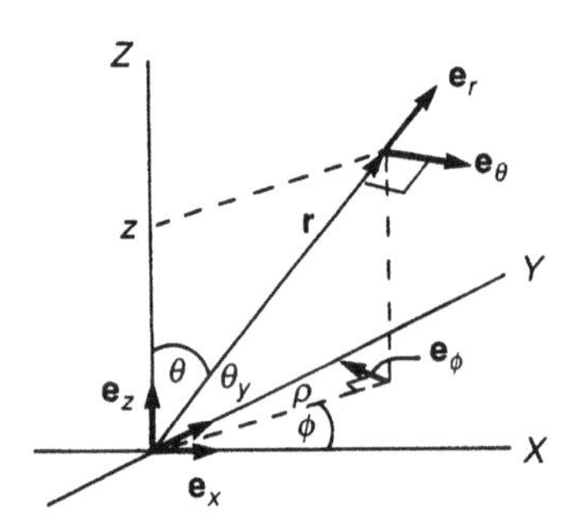

Figure A2.1 $\mathbf{e}_\theta$ is in the plane of $\mathbf{r}$ and OZ and $\mathbf{e}_\phi$ is in OXY. The cylindrical coordinates are ρ, ϕ, z with x often used in place of ρ when relations with x, y, z are not required

systems ($x = r\sin\theta\cos\phi$, $y = r\sin\theta\sin\phi$, $z = r\cos\theta$ and $x = \rho\cos\phi$, $y = \rho\sin\phi$, $z = z$ connecting the systems as illustrated by Figure A2.1).

Defining the scalar product by $\mathbf{v}\cdot\mathbf{u} = vu\cos\theta$ with θ the angle between $\mathbf{v}$ and $\mathbf{u}$, $\mathbf{i}\cdot\mathbf{i} = 1$, $\mathbf{i}\cdot\mathbf{j} = 0$, etc., so that $\mathbf{i}, \mathbf{j}, \mathbf{k}$ constitute an orthonormal set and on expanding $\mathbf{v}$ and $\mathbf{u}$ only three terms in the product are non-zero, giving

$$\mathbf{v}\cdot\mathbf{u} = v_x u_x + v_y u_y + v_z u_z \tag{2}$$

The vector product is given by $|\mathbf{v}\times\mathbf{u}| = vu\sin\theta$ with the direction given by the right-hand rule [see Figure 1.5(a) so that $\mathbf{u}\times\mathbf{v} = -\mathbf{v}\times\mathbf{u}$ (though $\mathbf{v}\cdot\mathbf{u} = \mathbf{u}\cdot\mathbf{v}$). Clearly $\mathbf{i}\times\mathbf{i} = 0$ and $\mathbf{i}\times\mathbf{j} = \mathbf{k}$, etc.

The most common vector operator is ∇ ('del'):

$$\nabla = \mathbf{i}\frac{\partial}{\partial x} + \mathbf{j}\frac{\partial}{\partial y} + \mathbf{k}\frac{\partial}{\partial z} \tag{3}$$

giving, as its scalar self-product, the Laplacian operator ('del squared')

$$\nabla\cdot\nabla = \frac{\partial^2}{\partial x^2} + \frac{\partial^2}{\partial y^2} + \frac{\partial^2}{\partial z^2} \equiv \nabla^2 \tag{4}$$

Clearly $\nabla\varphi \equiv \operatorname{grad}\varphi$ is a vector (field) if φ is a scalar (defined at all points and single-valued). ∇ operates on (i.e. forms, from the left, a scalar product with) a vector to give the (scalar) divergence, e.g.

$$\nabla\cdot\mathbf{v} \equiv \operatorname{div}\mathbf{v} = \frac{\partial v_x}{\partial x} + \frac{\partial v_y}{\partial y} + \frac{\partial v_z}{\partial z} \tag{5}$$

The alternative operation, involving the cross-product, is the curl

$$\nabla\times\mathbf{v} \equiv \operatorname{curl}\mathbf{v} = \left(\mathbf{i}\frac{\partial}{\partial x} + \mathbf{j}\frac{\partial}{\partial y} + k\frac{\partial}{\partial z}\right) \times (\mathbf{i}v_x + \mathbf{j}v_y + \mathbf{k}v_z) \tag{6}$$

and on expanding this and noting $\mathbf{i}\times\mathbf{j} = \mathbf{k}$, etc., it is seen to be given neatly by

$$\nabla\times\mathbf{v} = \begin{vmatrix} \mathbf{i} & \mathbf{j} & \mathbf{k} \\ \dfrac{\partial}{\partial x} & \dfrac{\partial}{\partial y} & \dfrac{\partial}{\partial z} \\ v_x & v_y & v_z \end{vmatrix} \tag{7}$$

Using spherical coordinates r, θ, ϕ:

$$\nabla\varphi = \mathbf{e}_r \frac{\partial\varphi}{\partial r} + \mathbf{e}_\theta \frac{1}{r}\frac{\partial\varphi}{\partial\theta} + \mathbf{e}_\phi \frac{1}{r\sin\theta}\frac{\partial\varphi}{\partial\phi} \tag{8}$$

$$\nabla^2\varphi = \frac{1}{r^2\sin\theta}\left(\frac{\partial}{\partial r}r^2\sin\theta\frac{\partial\varphi}{\partial r} + \frac{\partial}{\partial\theta}\sin\theta\frac{\partial\varphi}{\partial\theta} + \frac{\partial}{\partial\phi}\frac{1}{\sin\theta}\frac{\partial\varphi}{\partial\phi}\right) \tag{9}$$

$$\nabla\cdot\mathbf{A} = \frac{1}{r^2\sin\theta}\left(\frac{\partial}{\partial r}r^2\sin\theta A_r + \frac{\partial}{\partial\theta}r\sin\theta A_\theta + \frac{\partial}{\partial\phi}rA_\phi\right) \tag{10}$$

$$\nabla\times\mathbf{A} = \frac{1}{r^2\sin\theta}\begin{vmatrix} \mathbf{e}_r & r\mathbf{e}_\theta & r\sin\theta\mathbf{e}_\phi \\ \dfrac{\partial}{\partial r} & \dfrac{\partial}{\partial\theta} & \dfrac{\partial}{\partial\phi} \\ A_r & rA_\theta & r\sin\theta A_\phi \end{vmatrix} \tag{11}$$

and for cylindrical coordinates ρ, ϕ, z:

$$\nabla\phi = \mathbf{e}_\rho \frac{\partial\varphi}{\partial\rho} + \mathbf{e}_\phi \frac{1}{\rho}\frac{\partial\varphi}{\partial\phi} + \mathbf{e}_z \frac{\partial\varphi}{\partial z} \tag{12}$$

$$\nabla^2\varphi = \frac{1}{\rho}\frac{\partial}{\partial\rho}\left(\rho\frac{\partial\varphi}{\partial\rho}\right) + \frac{1}{\rho^2}\frac{\partial^2\varphi}{\partial\phi^2} + \frac{\partial^2\varphi}{\partial z^2} \tag{13}$$

$$\nabla\cdot\mathbf{A} = \frac{1}{\rho}\left[\frac{\partial}{\partial\rho}(\rho A_\rho) + \frac{\partial A_\phi}{\partial\phi} + \frac{\partial}{\partial z}(\rho A_z)\right] \tag{14}$$

$$\nabla\times\mathbf{A} = \frac{1}{\rho}\begin{vmatrix} \mathbf{e}_\rho & \rho\mathbf{e}_\phi & \mathbf{e}_z \\ \dfrac{\partial}{\partial\rho} & \dfrac{\partial}{\partial\phi} & \dfrac{\partial}{\partial z} \\ A_\rho & \rho A_\phi & A_z \end{vmatrix} \tag{15}$$

As an example, suppose that

$$\mathbf{A} \equiv (A_r, A_\theta, A_\phi) = \left(0, 0, \frac{m}{4\pi r}\tan\frac{\theta}{2}\right)$$

then from equation (11),

$$\nabla\times\mathbf{A} = \frac{1}{r^2\sin^2\theta}\left(\mathbf{e}_r\frac{\partial}{\partial\theta}r\sin\theta\frac{m}{4\pi r}\tan\frac{\theta}{2} - r\mathbf{e}_\theta\frac{\partial}{\partial r}r\sin\theta\frac{m}{4\pi r}\tan\frac{\theta}{2}\right) \tag{16}$$

The second term gives zero and using $\sin\theta = 2\sin(\theta/2)\cos(\theta/2)$ it is readily seen that

$$\nabla\times\mathbf{A} = \mathbf{e}_r\frac{m}{4\pi r^2} = \mathbf{H}$$

Thus $\mu_0\mathbf{A}$ can be considered the vector potential from an apparent 'source' m and to avoid fallacious sources this can be regarded as one end of a long solenoid of small diameter, so long as $\mathbf{A}$ is specified at points external to the solenoid.

If a vector $\mathbf{v} = \nabla\varphi$ then it is readily seen that

$$\nabla\times\mathbf{v} = \mathbf{i}\left(\frac{\partial^2\varphi}{\partial y\,\partial z} - \frac{\partial^2\varphi}{\partial z\,\partial y}\right) + \mathbf{j}(0) + \mathbf{k}(0) = 0 \tag{17}$$

all components being zero, i.e. $\nabla\times(\nabla\varphi) = 0$ always. If it can be shown that

$$\nabla\times\mathbf{H} = \text{curl}\,\mathbf{H} = 0 \qquad (\mathbf{H}\text{ is irrotational}) \tag{18}$$

then it may, conversely, be assumed that **H** may be related to a scalar potential by

$$\mathbf{H} = -\nabla\varphi = -\,\text{grad}\,\varphi \tag{19}$$

Such a field is also said to be conservative.

In the same way it can be seen that

$$\nabla\cdot(\nabla\times\mathbf{A}) = 0 \tag{20}$$

for any vector **A** and so it can be taken that if a field **B** is related to a (vector) potential by $\mathbf{B} = \nabla\times\mathbf{A}$ then

$$\nabla\cdot\mathbf{B} = \text{div}B = 0 \qquad (\mathbf{B}\text{ is solenoidal}) \tag{21}$$

Pictorially the field lines representing a solenoidal field (e.g. **B**) are continuous so that there are no associated sources or sinks whereas the lines of a conservative field (e.g. **H**) may originate and terminate at sources.

The operators ∇, $\nabla\cdot$ and $\nabla\times$ are clearly linear. The reader may confirm that the following formulae apply by expansions using rectangular coordinates; despite lack of generality it may hardly be imagined that a rule thus confirmed would not apply in other systems:

$$\nabla\cdot(\varphi\mathbf{A}) = (\nabla\varphi)\cdot\mathbf{A} + \varphi(\nabla\cdot\mathbf{A}) \tag{22}$$

$$\nabla\times(\varphi\mathbf{A}) = (\nabla\varphi)\times\mathbf{A} + \varphi(\nabla\times\mathbf{A}) \tag{23}$$

(as for the differential of a product: in $(\nabla\varphi)\cdot\mathbf{A}$ it is to be taken that ∇ operates on φ only.)

$$\nabla\cdot(\mathbf{A}\times\mathbf{B}) = \mathbf{B}\cdot(\nabla\times\mathbf{A}) - \mathbf{A}\cdot(\nabla\times\mathbf{B}) \tag{24}$$

$$\nabla\times(\mathbf{A}\times\mathbf{B}) = (\mathbf{B}\cdot\nabla)\mathbf{A} - \mathbf{B}(\nabla\cdot\mathbf{A}) - (\mathbf{A}\cdot\nabla)\mathbf{B} + \mathbf{A}(\nabla\cdot\mathbf{B}) \tag{25}$$

$$\nabla(\mathbf{A}\cdot\mathbf{B}) = (\mathbf{B}\cdot\nabla)\mathbf{A} + (\mathbf{A}\cdot\nabla)\mathbf{B} + \mathbf{B}\times(\nabla\times\mathbf{A}) + \mathbf{A}\times(\nabla\times\mathbf{B}) \tag{26}$$

$$\nabla\times(\nabla\times\mathbf{A}) = \nabla(\nabla\cdot\mathbf{A}) - \nabla^2\mathbf{A} \tag{27}$$

(the parentheses to be evaluated first). Some of the direct demonstrations become tedious and may be abbreviated.

2 Theorems of Gauss, Stokes and Green

Three integral theorems occur very commonly, i.e. the Gauss divergence theorem:

$$\int_V \nabla\cdot\mathbf{A}\,\mathrm{d}V = \int_S \mathbf{A}\cdot\mathbf{n}\,\mathrm{d}S = \int_S \mathbf{A}\cdot\mathrm{d}\mathbf{S} = \varphi_{\mathrm{A}} \tag{28}$$

The surface S is closed and bounds the volume over which the first integral is to be taken, **n** is the outward normal to S and φ_{A} is the flux of **A** over the surface.

Stokes theorem:

$$\int_C \mathbf{A}\cdot \mathrm{d}\mathbf{r} = \int_S (\nabla\times\mathbf{A})\cdot \mathbf{n}\,\mathrm{d}S = \int_S (\nabla\times\mathbf{A})\cdot \mathrm{d}\mathbf{S} \tag{29}$$

S is any open two-sided surface bounded by the closed non-intersecting curve C around which the first integral is to be taken. Green's theorem in the plane (in vector form):

$$\int_C \mathbf{A}\cdot \mathrm{d}\mathbf{r} = \int_R (\nabla\times\mathbf{A})\cdot \mathbf{k}\,\mathrm{d}R \tag{30}$$

where R is a closed region on the plane OXY bounded by the simple closed curve C.

As an illustration of Gauss's theorem, which is obviously applicable to conservative or 'source' fields, suppose that the deviation of $\mathbf{M}$ at the surface of a magnetized specimen produces a volume pole density ρ in a thin sheet extending from $z = -1$ to $z = 1$ and lying in OXY. The fields from an infinitesimal sheet with $\sigma = \rho\,\mathrm{d}z$ are $\pm\mathbf{k}\times\frac{1}{2}\rho\,\mathrm{d}z$ and thus at any point within the sheet $\mathbf{H} = \mathbf{k}\times\frac{1}{2}\rho[(1+z)-(1-z)] = \mathbf{k}\rho z$ by trivial integration and $\nabla\cdot\mathbf{H} = \rho$ as expected. Thus $\int \nabla\cdot\mathbf{H}\,\mathrm{d}v = \rho v$ for any volume v within the sheet. For example, for a right cylinder (section $\mathcal{A}$) with axis $\parallel OZ$ and with plane faces at $z = z'$ and at $z = z'+d$ the surface integral is $\rho\mathcal{A}[(z'+d)-z'] = \rho v$ and Gauss's theorem may be seen similarly to apply to any chosen volume. More realistically, but at a little greater length, a similar demonstration may be made for a long radially magnetized cylinder; alternatively, the theorem may be assumed to calculate the fields throughout, assuming $\nabla\cdot\mathbf{H} = \rho = -M/r$ and this is left as an exercise.

For a simple illustration of Stokes theorem, consider a long solenoid of very small section area $\mathcal{A}$ with surface current density I and axis $\parallel OZ$ to terminate at O. At points external to the solenoid the vector potential becomes $\mathbf{A} = \mathbf{e}_\phi(\mu_0/4\pi)(m/r)\tan\theta/2$ at distance r from the effective pole $m = \mathcal{A}I$, as $\mathcal{A}\to 0$. Take C to be a circle of radius a in OXY centred on O so that, with $\tan\theta/2 = 1$, $\int_C \mathbf{A}\cdot\mathrm{d}r = \mu_0 m/2$. Take S to be the hemisphere bounded by C and not intersecting the solenoid. Then $\nabla\times\mathbf{A} = \mathbf{B} = \mathbf{e}_r(\mu_0/4\pi)(m/a^a)$ and the surface integral is $m/2$ as required. C may be taken to lie at a certain distance above OXY and to form the periphery of a spherical cap. The demonstration is repeated as an exercise. (It might appear that the outward flux of $\mathbf{B}$ over a spherical surface enclosing the end of the solenoid is $\mu_0 m$, but the complete surface must intersect the narrow solenoid giving an inward flux of $\mathcal{A}\mu_0 H = \mu_0 m$ there, so there is no inconsistency with $\nabla\cdot\mathbf{B} = 0$.)

3 Vectors and Matrices

To recall the simpler rules applying to matrices, $\mathbf{A} = \mathbf{B}$ only if all $A_{ij} = B_{ij}$ and if $\mathbf{A}+\mathbf{B} = \mathbf{C}$ then $C_{ij} = A_{ij}+B_{ij}$; if $\mathbf{A} = kB$ all $A_{ij} = kB_{ij}$. No scalar value may be ascribed to a matrix, as opposed to the determinant of a matrix such as $\det\mathbf{A} = A_{11}A_{22}-A_{12}A_{21}$. Numbers may be associated with matrices, as eigenvalues, to the extent that $\mathbf{A}\mathbf{v}_i = a_i\mathbf{v}_i$, or $[A]\{v_i\} = a_i\{v_i\}$, where $\mathbf{A}$ is $n\times n$ and the $\mathbf{v}$, or $\{v\}$, are n-dimensional column matrices called eigenvectors. If $A^*_{ij} = A_{ji}$ (or $A_{ij} = A_{ji}$, if $\mathbf{A}$ is real), i.e. if $\mathbf{A}^\dagger = \mathbf{A}$ (or $A^\mathrm{T} = A$) then the a_i are real and if $a_{i'} \neq a_{i''}$ then $(v_{i'})\{v_{i''}\} = 0$, or $\mathbf{v}^\dagger_{i'}\mathbf{v}_{i''} = 0$, and the vectors are said to be orthogonal, or orthonormal if it is contrived that the self-products are unity. (Clearly these particular, scalar, products are numbers although products such as $\{v\}(v)$ are square matrices.) The examples $(1\ 0)^\mathrm{T}$ and $(0\ 1)^\mathrm{T}$, i.e. $\{1\}$ and $\{2\}$, constitute a linearly independent basis set in terms of which any column $(c_1c_2)^{\mathrm{T}}$ can be expressed as $c_1\{1\}+c_2\{2\}$ and so forth. It is apparent, on expansion, that $((a)\{b\})^* = (b)\{a\}$ and that $(a)\{a\}$ is real and if (a) is read, here, as $\{a\}^\dagger$.

A one-to-one correspondence can be established between Cartesian vectors and column matrices (vectors) by associating the columns $(c_1c_2c_3)^{\mathrm{T}}$ with the vectors (c_1, c_2, c_3) or $c_1\mathbf{e}_1 + c_2\mathbf{e}_2 + c_3\mathbf{e}_3$, due to the correspondence of the rules applying to identity, addition and multiplication by a scalar. To encompass scalar products, it is realized that the correspondence could equally have been made with the rows $(c_1c_2c_3)$ so that the scalar product $\mathbf{a}\cdot\mathbf{b}$ is associated with $(a_1a_2a_3)(b_1b_2b_3)^{\mathrm{T}}$. The rows and columns constitute distinct spaces since rows cannot be formed as linear combinations of columns, and vice versa, but the two spaces are associated in that for any $(c_1c_2)^{\mathrm{T}}$ in S_2, say, there is a transpose (c_1c_2) in S_2^{T}, the transpose space. The associations permit the formalization of transformations which include changes in direction as well as of magnitude (as by scalar multiplication). Denote as R_θ the operation of rotation through angle θ; $\mathbf{i}$ rotates to give the vector $\mathbf{i}\cos\theta - \mathbf{j}\sin\theta$ and $R_\theta\mathbf{j} = \mathbf{i}\sin\theta + \mathbf{j}\cos\theta$, so that writing $R_\theta\mathbf{a} = \mathbf{b}$ or

$$R_\theta(a_1\mathbf{i} + a_2\mathbf{j}) = b_1\mathbf{i} + b_2\mathbf{j}$$

and taking scalar products with $\mathbf{i}$ and with $\mathbf{j}$ and rearranging gives

$$\begin{pmatrix} \mathbf{i}\cdot R_\theta\mathbf{i} & \mathbf{i}\cdot R_\theta\mathbf{j} \\ \mathbf{j}\cdot R_\theta\mathbf{i} & \mathbf{j}\cdot R_\theta\mathbf{j} \end{pmatrix}\begin{pmatrix} a_1 \\ a_2 \end{pmatrix} = \begin{pmatrix} \cos\theta & \sin\theta \\ -\sin\theta & \cos\theta \end{pmatrix}\begin{pmatrix} a_1 \\ a_2 \end{pmatrix}$$

$$= \begin{pmatrix} a_1\cos\theta + a_2\sin\theta \\ -a_1\sin\theta + a_2\cos\theta \end{pmatrix} = \begin{pmatrix} b_1 \\ b_2 \end{pmatrix}$$

for the components of the rotated vector (in two dimensions).

4 Generalization

Having written $\mathbf{v} = \sum c_i\mathbf{e}_i$ with the $\mathbf{e}_i$ constituting a linearly independent, and preferably orthonormal, basis spanning the space, i.e. adequate in number for any vectors in the particular one-, two- or three-dimensional spaces to be so represented, and with the c_i real coefficients or components, it is natural to consider two generalizations: to an unrestricted dimension and to complex coefficients. Since the entities are still to be called vectors but are not associated with vector quantities, a particular notation is advisable and the choice is to write $|\ \rangle = \sum c_i|i\rangle$. With complex coefficients the adjoint replaces the transpose and the second space required for the formation of scalar products, and to be associated with row matrices or vectors becomes the adjoint space in which the vectors are written $\langle\ | = \sum c_i^*\langle i|$. Scalar products are written $\langle\ |\ \rangle$ and the basis is orthonormal if $\langle i'|i''\rangle = \langle i''|i'\rangle^* = \delta_{i'i''}$.

The associated column matrices, or column vectors, are those with elements $\langle i|\ \rangle = c_i$, which products simply pick out the appropriate coefficients, assuming orthonormality. Row matrices are associated with the adjoints according to $c_i^* = \langle\ |i\rangle$. For example, $n = 2$, $|\chi\rangle = c_1|1\rangle + c_2|2\rangle$, $\langle\chi| = c_1^*\langle 1| + c_2^*\langle 2|$, $\boldsymbol{\chi}$ or $\{\chi\} = (c_1c_2)^{\mathrm{T}}$, $\boldsymbol{\chi}^\dagger$ or $(\chi) = (c_1^*c_2^*)$ and the scalar self-product $\langle\chi|\chi\rangle = (c_1^*\langle 1| + c_2^*\langle 2|)(c_1|1\rangle + c_2|2\rangle) = c_1c_1^* + c_2c_2^*$, which is the matrix product $\boldsymbol{\chi}^\dagger\boldsymbol{\chi}$ or $(\chi)\{\chi\}$.

If a linear operator acts within a space so as to transform a vector $|\chi\rangle(= a_1|1\rangle + a_2|2\rangle$, say) to a different vector $|\varphi\rangle$ then $|\varphi\rangle$, in the same space, may be expressed as $|\varphi\rangle = b_1|1\rangle + b_2|2\rangle$

and due to linearity

$$A|\chi\rangle = a_1 A|1\rangle + a_2 A|2\rangle = b_1|1\rangle + b_2|2\rangle$$

Multiplying by $\langle 1|$ and then by $\langle 2|$ from the left gives

$$\begin{pmatrix} \langle 1|A|1\rangle & \langle 1|A|2\rangle \\ \langle 2|A|1\rangle & \langle 2|A|2\rangle \end{pmatrix} \begin{pmatrix} a_1 \\ a_2 \end{pmatrix} = \begin{pmatrix} b_1 \\ b_2 \end{pmatrix}, \qquad [A]\{\chi\} = \{\varphi\}$$

$[A]$ with $A_{ij} = \langle i|A|j\rangle$ is the matrix of the operator A represented in the chosen basis and the effect of A on $|\chi\rangle$ is found by noting the effect of [A] on $\{\chi\}$.

An adjoint vector $\langle\chi|$ may be taken to be that which is associated with the adjoint (χ) of $\{\chi\}$, i.e. with $\{\chi\}^\dagger$. Formally,

$$|\chi\rangle \sim \begin{pmatrix} c_1 \\ c_2 \\ \vdots \end{pmatrix} = \begin{pmatrix} \langle 1|\chi\rangle \\ \langle 2|\chi\rangle \\ \vdots \end{pmatrix} \qquad (c_1^* c_2^* \ldots) \sim \langle\chi|$$

The elements of (χ) are $\langle\chi|i\rangle$. Taking adjoints of $[A]\{\chi\} = \{\varphi\}$ gives $(\chi)[A]^\dagger = (\varphi)$, the matrix $[B] = [A]^\dagger$ having elements $B_{ij} = A_{ji}^* = \langle j|A|i\rangle^*$. Suppose that $A|i\rangle = |\psi\rangle$, some other vector, and that the adjoint equation can be written $\langle i|A^\dagger = \langle\psi|$ with $A^\dagger$ designating an operator on the adjoint space. It follows that the elements of $[A]^\dagger$ become $\langle i|A^\dagger|j\rangle$ and if $A^\dagger = A$ then the matrix $[A]$ is Hermitian (self-adjoint) and has real eigenvalues. Thus an operator on a general vector space is said to be Hermitian if $A^\dagger = A$, i.e. if the operator in the adjoint space is identical to that in the initially chosen space and its eigenvalues, according to $A|\chi_n\rangle = a_n|\chi_n\rangle$, are then real.

The rigid association of general vectors $|\ \rangle$ and operators A on these, with column vectors and matrix operators respectively, appears to be convenient but does not constitute the only approach. A vector space may be defined as a set of elements (vectors) combining according to certain rules whereby it is seen that Cartesian vectors, column or row matrices of a given n, and certain functions all constitute vector spaces. The relevant functions may be expanded as, for example, $g(x) = \sum c_i f_i(x)$, where the basis functions $f_i(x)$ may be chosen to be orthonormal according to $\int f_{i'}^*(x) f_{i''}(x)\mathrm{d}x = \delta_{i'i''}$ and column matrices associated with the functions according to $c_i = \int f_i^*(x) g(x)\mathrm{d}x$. Thus if $g = c_1 f_1 + c_2 f_2$ and $h = b_1 f_1 + b_2 f_2$ and also $Ag = h$ then the matrix of A in the f_i basis is seen [replacing 'multiply by $\langle i|$ from the left' as above by 'multiply by $f_i^*(x)$ and integrate'] to have elements $\int f_i^*(x) A f_j(x)\mathrm{d}x$ and the coefficients of h are found as before. An operator on the functions is Hermitian if its matrix is Hermitian, i.e. if $\int f_i^*(x) A f_j(x)\mathrm{d}x = (\int f_j^*(x) A f_i(x)\mathrm{d}x)^*$. In wave mechanics the generally discrete quantum states correspond to eigenfunctions distinguished by sets of quantum numbers such as n, l, m. The states may equally be associated with general vectors, state vectors, distinguished as $|n\ l\ m\rangle$ or as $|m_S\rangle$ for example. The reasons for this were hinted at in the section on the Stern–Gerlach experiments, but the best reason for accepting such a formulation may well be taken to be its success in predicting spectroscopic and other results. It is not reasonable to equate a vector to a function but associations may be made as $\psi_{nlm}(\mathbf{r}) \sim |n\ l\ m\rangle$, while matrix elements may be directly equated and scalar products of vectors equated to the appropriate integrals. The value of the vector formulation is particularly clear when the functions are not apparent, as in the case of spin.

Author Index

First-named authors, with reference numbers [ ], when names are not explicit in the text.

Subject Index

Note 'magnetic' is usually implicit: 'Field' ≡ 'Magnetic field', etc.

9 780471 954170